Protein Methylation

Editors

Woon Ki Paik
Professor of Biochemistry
Fels Institute for Cancer Research and Molecular Biology
Temple University School of Medicine
Philadelphia, Pennsylvania

Sangduk Kim
Associate Professor of Biochemistry
Fels Institute for Cancer Research and Molecular Biology
Temple University School of Medicine
Philadelphia, Pennsylvania

CRC Press, Inc.
Boca Raton, Florida

Library of Congress Cataloging-in-Publication Data

Protein methylation/editors, Woon Ki Paik, Sangduk Kim.
p. cm.
Includes bibliographies and index.
ISBN 0-8493-6818-9
1. Proteins--Methylation. I. Paik, Woon Ki.
[DNLM: 1. Methylation. 2. Protein Methyltransferases--metabolism.
QU 141 P967]
QP551.P6972 1990
574.19′25--dc20
DNLM/DLC 89-7139
for Library of Congress CIP

Direct all inquiries to CRC Press, Inc., 2000 Corporate Blvd., N.W., Boca Raton, Florida, 33431.

International Standard Book Number 0-8493-6818-9
Library of Congress Card Number 89-7139
Printed in the United States

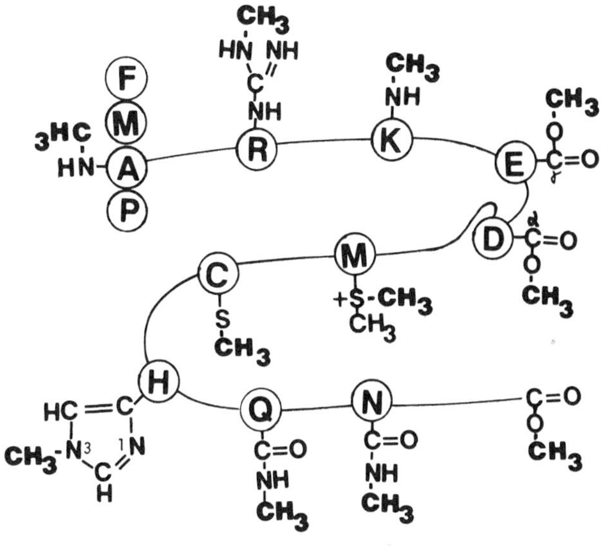

Posttranslational modifications of protein.

THE EDITORS

Woon Ki Paik, M.D., is Professor of Biochemistry at the Fels Institute for Cancer Research and Molecular Biology, Temple University School of Medicine, Philadelphia. He received his M. D. degree from Yonsei University School of Medicine, Seoul, Korea, in 1947 and M. Sc. degree from Dalhousie University, Halifax, Canada, in 1956. He was a postdoctoral fellow and visiting scientist at the National Cancer Institute, Bethesda, MD and the University of Wisconsin, Madison. He has published over 200 papers and has been the author of two books. His major research interest has been protein methylation for the past 25 years.

Dr. Sangduk Kim is an Associate Professor of Biochemistry at the Fels Institute for Cancer Research and Molecular Biology, Temple University School of Medicine, Philadelphia. She obtained her M. D. degree from Korea University College of Medicine, Seoul, Korea, in 1953 and Ph. D. degree from the University of Wisconsin, Madison, in 1960. She has published over 200 papers and has been the co-author of two books. Her major research interest has been protein methylation for the past 25 years, and jointly with Dr. Woon Ki Paik she discovered enzymatic methylation reaction in the early 1960s.

PREFACE

The natural occurrence of "methyllysine" was first observed in flagella protein of *Salmonella typhimurium* in 1959. This initial discovery 30 years ago was followed by rapid advances in protein methylation research which revealed that many highly specialized proteins are methylated, and this posttranslational modification reaction occurs ubiquitously in nature, ranging from eukaryotic to prokaryotic organisms. Lysine, arginine, histidine, proline, alanine, glutamine, and asparagine residues of certain proteins are postsynthetically *N*-methylated, glutamic and aspartic acid residues are *O*-methylesterified, and cysteine and methionine residues are *S*-methylated. In 1980, we published a monograph, *Protein Methylation* (John Wiley & Sons). Since this first publication, the scope as well as the understanding of the biological significance of protein methylation has been vastly expanded. The involvement of protein methylation reactions in bacterial chemotaxis, carnitine biosynthesis and DNA repair processes has been established, in addition to membrane protein repair and effect of methylated amino acids on cell growth. This expansion of our knowledge of protein methylation compelled us to plan a sequel to our previous monograph. Since we ourselves could not possibly cover all aspects of protein methylation research, we invited authorities in each respective field to contribute their reviews on specific topics. We emphasized that each topic should be critically reviewed and that some speculation would be welcomed. We are extremely thankful to them for their excellent contributions.

We arranged this volume as follows. First, we summarize current knowledge on the natural occurrence of methylated amino acids, and on the enzymology and biological significance of protein *N*-methyltransferases which are specific for proteins such as calmodulin, cytochrome *c*, myelin basic protein, nuclear/nucleolar acidic proteins, histones, heat shock proteins, and ribosomal proteins. We then cover protein *O*-carboxymethylation in both eukaryotes and prokaryotes. In addition, the involvement of protein *S*-methylation in DNA repair processes, in-depth studies on the biological transmethylation inhibitor, *S*-adenosyl-*L*-homocysteine, chemical synthesis of methylated amino acids, nonenzymatic methylation of protein, and effects of free methylated amino acids and peptides in various biological systems are also described.

We hope that this volume will serve as an up-to-date reference and guide to researchers, as well as graduate students interested in these postsynthetic modifications of macromolecules.

Woon Ki Paik
Sangduk Kim

CONTRIBUTORS

Dana W. Aswad
Associate Professor
Department of Molecular Biology and Biochemistry
Department of Psychobiology
University of California
Irvine, California

Jake Bello
Principal Research Scientist
Department of Biophysics
Roswell Park Institute
Buffalo, New York

N. Leo Benoiton
Professor
Department of Biochemistry
University of Ottawa
Ottawa, Ontario, Canada

Melvin L. Billingsley
Associate Professor
Department of Pharmacology
Center for Cell and Molecular Biology
Hershey Medical Center
Hershey, Pennsylvania

Harris R. Buttz
Department of Microbiology
University of North Dakota
Grand Forks, North Dakota

Latika P. Chanderkar
Fels Institute for Cancer Research and Molecular Biology
Temple University School of Medicine
Philadelphia, Pennsylvania

Steven Clarke
Associate Professor
Department of Chemistry and Biochemistry
U.C.L.A.
Los Angeles, California

Charles Colson
Laboratoire de Cytogenetique
Institut Carnoy
Universite Catholique de Louvain
Louvain-la-Neuve, Belgium

Frederick W. Dahlquist
Professor
Department of Chemistry
University of Oregon
Eugene, Oregon

Bruce Demple
Associate Professor
Department of Biochemistry and Molecular Biology
Harvard University
Cambridge, Massachusetts

Richard Desrosiers
Department of Medicine and Cell Biology
Laval University
Ste-Foy, Quebec, Canada

John Arden Duerre
Professor
Department of Microbiology and Immunology
University of North Dakota School of Medicine
Grand Forks, North Dakota

Blaise Frost
Fels Institute for Cancer Research and Molecular Biology
Temple University School of Medicine
Philadelphia, Pennsylvania

Claude Gagnon
Director
Urology Research Laboratory
Department of Urology
McGill University
Montreal, Quebec, Canada

Alice A. Heth
Department of Pediatrics
University of Wisconsin Medical School
Madison, Wisconsin

Subrata K. Ghosh
Fels Institute for Cancer Research and Molecular Biology
Temple University School of Medicine
Philadelphia, Pennsylvania

Brett A. Johnson
Postgraduate Researcher
Department of Psychobiology
University of California
Irvine, California

Sangduk Kim
Associate Professor
Fels Institute for Cancer Research and Molecular Biology
Department of Biochemistry
Temple University School of Medicine
Philadelphia, Pennsylvania

F. Lawrence
Chargee de Recherche
Institut de Chimie des Substances
C.N.R.S.
Gif-sur-Yvette, France

Jacques Lhoest
Senior Research Assistant
Laboratoire de Cytogenetique
Universite Catholique de Louvain
Louvain-la-Neuve, Belgium

Michael A. Lischwe
Biotechnology Group Leader
E. I. du Pont de Nemours and Company
Newark, Delaware

P. D. Lotlikar
Associate Professor
Fels Institute for Cancer Research and Molecular Biology
Temple University School of Medicine
Philadelphia, Pennsylvania

Tom L. Neal
Department of Physiological Chemistry
University of Wisconsin Medicial School
Madison, Wisconsin

David O. Nettleton
University of Illinois College of Medicine
Urbana, Illinois

George W. Ordal
Associate Professor
Department of Biochemistry
University of Illinois
Urbana, Illinois

Irene M. Ota
Department of Chemistry and Biochemistry
U.C.L.A.
Los Angeles, California

Woon Ki Paik
Professor
Fels Institute for Cancer Research and Molecular Biology
Department of Biochemistry
Temple University School of Medicine
Philadelphia, Pennsylvania

In Kook Park
Associate Professor
Department of Agrobiology
Dongguk University
Seoul, Korea

Malka Robert-Gero
Directeur de Recherche
Institut de Chimie des Substances
C.N.R.S.
Gif-sur-Yvette, France

Paul M. Rowe
Department of Physiological Chemistry
University of Wisconsin Medical School
Madison, Wisconsin

Frank L. Siegel
Professor
Departments of Pediatrics and Physiological Chemistry
University of Wisconsin
Madison, Wisconsin

Richard C. Stewart
Institute of Molecular Biology
University of Oregon
Eugene, Oregon

Jeffrey B. Stock
Assistant Professor
Department of Molecular Biology
Princeton University
Princeton, New Jersey

Béla Szende
Professor
1st Department of Pathological
Experimental Cancer Research
Semmelveis University Medical School
Budapest, Hungary

Robert M. Tanguay
Profesor
Department of Medicine
Laval University
Ste-Foy, Quebec, Canada

Ernö Tyihák
Associate Professor
Plant Protection Institute
Hungarian Academy of Sciences
Budapest, Hungary

Lajos Trézl
Associate Professor
Department of Organic Chemical
Technology
Technical University of Budapest
Budapest, Hungary

K. Veeraragavan
Research Associate
Department of Biochemical Engineering
Biotechnology Research Institute
Montreal, Quebec, Canada

Pamela Vincent
Departments of Physiological Chemistry
and Neurology
University of Wisconsin Medical School
Madison, Wisconsin

Lynda S. Wright
Department of Pediatrics
University of Wisconsin Medical School
Madison, Wisconsin

TABLE OF CONTENTS

Chapter 1

THE OCCURRENCE AND ANALYSIS OF METHYLATED AMINO ACIDS

In Kook Park and Woon Ki Paik

TABLE OF CONTENTS

I. INTRODUCTION

Over 150 posttranslationally modified amino acid derivatives in protein have thus far been reported,[1-3] and several comprehensive review articles on the analysis of these amino acids have recently been published, particularly in the *Methods of Enzymology*.[4-6] The objective of this review is to describe the current methods available for the analysis of posttranslationally methylated amino acid derivatives.[7-9] Additional information for the synthesis and analysis of methylated amino acids are available in articles written by Benoiton[10] and Huszar,[11] and Chapter 19 in this book.

II. OCCURRENCE OF VARIOUS METHYLATED AMINO ACID DERIVATIVES

Protein methylation, one of the posttranslational modifications, occurs in a wide variety of cell types ranging from prokaryotic to eukaryotic organisms. These modifications are either *N*-methylation or carboxylmethylation reactions. The former reaction usually involves *N*-methylation of lysine, arginine, histidine, alanine, proline, glutamine, phenylalanine, asparagine, and methionine, while the latter involves *O*-methylesterification of glutamic and aspartic acid. The sources of methylated amino acids in proteins including references are presented in Table 1.[12-93] In nature, methylated amino acids occur in highly specialized proteins such as histone, flagella proteins, myosin, actin, ribosomal proteins, hnRNA-bound protein, HMG-1 and HMG-2 protein, fungal and plant cytochrome *c*, myelin basic protein, opsin, EF-Tu, EF-lα, porcine heart citrate synthase, calmodulin, ferredoxin, α-amylase, heat shock protein, scleroderma antigen, nucleolar protein C23, IF-3l, pilin, *cheZ*, allophycocyanin, and phycobiliprotein.

III. ANALYSIS OF *N*-METHYLATED AMINO ACIDS

Except carboxyl methylesterified glutamic and aspartic acids, most of the *N*- and *S*-methylated amino acids are quite stable during treatment with 6 *N* HCl at 110°C, under which the hydrolysis of the protein is performed. This unique property of *N*- and *S*-methylated amino acids allows their identification relatively easily amenable to most of the common analytic methods.

Most of the analytical techniques currently available have been employed for the analysis of the methylated amino acids; paper chromatography and electrophoresis, thin-layer chromatography and electrophoresis, ion-exchange chromatography, gas chromatography, and high performance liquid chromatography (HPLC).

Since the methylated amino acids, in general, are present in minor quantities in protein, it is advisable to carry out an initial purification of protein hydrolyzates prior to the analysis. This process not only concentrates the methylated amino acids, but also eliminates possible overloading of the system with other amino acids.

In the following section, we will mainly focus our discussion only on the methylated lysines, arginines, and histidines since major research endeavors have been directed towards these amino acids.

A. PARTIAL PURIFICATION OF *N*-METHYLATED AMINO ACIDS FROM ACID HYDROLYZATES OF PROTEINS AND OTHER BIOLOGICAL SAMPLES

Prior to the analysis of the methylated amino acids in the acid-hydrolyzates of proteins or other biological materials, it is highly desirable to partly purify the amino acids, particularly when paper or thin-layer chromatography is to be employed. This partial purification of basic amino acids can be achieved by the use of an ion-exchange resin in H^+ form. In the following

TABLE 1
Natural Occurrence of Various Methylated Amino Acid Derivatives

Protein	Source	Authors and year of finding	Ref.
	ε-*N*-monomethyllysine		
Flagella protein	*Salmonella typhimurium*	Ambler and Rees (1959)	12
	Spirilum serpens	Glazer et al. (1969)	13
Histone	Thymus, wheat germ, kidney	Murray (1964)	14
	Ascites carcinoma	Comb et al. (1966)	15
	Pea	Fambrough and Bonner (1968)	16
Myosin	Skeletal muscle	Hardy and Perry (1969)	17
		Huszar and Elzinga (1969)	18
	Physarun polycephalum	Venkatesan et al. (1969)	19
Actin	Amoeba	Weihing and Korn (1970)	20
Ribosomal protein	*Blastocladiella emersonii*	Comb et al. (1966)	15
Opsin	Bovine retina	Reportet and Reed (1972)	21
Tooth matrix protein	Human	Kalasz et al. (1978)	22
Ferredoxin	*Sulfolobus acidocaldarius*	Minami et al. (1985)	23
Elongation factor Tu	*Salmonella typhimiurum* *Escherichia coli*	Ames and Niakido (1979)	24
Elongation factor lα	*Mucor* (fungus)	Hiatt et al. (1982)	25
Free amino acid	Brain (bovine)	Matsuoka et al. (1969)	26
	Plasma (human)	Perry et al. (1969)	27
	Plasma (sheep)	Weatherall and Haden (1969)	28
	Urine (human)	Asatoor (1969)	29
	ε-*N*-dimethyllysine		
Nuclear protein	Thymus (calf)	Kim and Paik (1965)	30
Histone	Thymus (calf)	Paik and Kim (1967)	31
Flagella protein	*Salmonella typhimurium*	Glazer et al. (1969)	13
Myosin	Soleus muscle (cat)	Kuehl and Adelstein (1970)	32
	Physarum polycephalum	Vankatesan et al. (1975)	19
Actin	Amoeba	Weihing and Korn (1970)	20
Myofibrilla protein	Skeletal muscle (rabbit)	Hardy et al. (1970)	33
Opsin	Bovine retina	Reporter and Reed (1972)	21
Ribosomal protein	HeLa cells	Chang et al. (1976)	34
Elongation factor lα	*Mucor* (fungus)	Hiatt et al. (1982)	25
Elongation factor Tu	*Salmonella typhimurium* *Escherichia coli*	Ames and Naikido (1979)	24
Free amino acid	Urine (human)	Kakimoto and Akazawa (1970)	35
	Plasma (human)	Kakimoto and Akazawa (1970)	35
Calmodulin	*Paramecium tetraurelia*	Schaefer et al. (1987)	36
	ε-*N*-trimethyllysine		
Histone	Thymus	Hempel et al. (1968)	37
Cytochrome *c*	Wheat germ *Neurospora crassa*	DeLange et al. (1969)	38
	Yeast	DeLange et al. (1970)	39
	Protozoa	Hill et al. (1971)	40
	Spinacea oleracea (spinach)	Brown and Boutler (1973)	41
Myosin	Skeletal muscle (rabbit)	Hardy et al. (1970)	33
	Physarum polycephalum	Venkatesan et al. (1975)	19
Actin	Amoeba	Weihing and Korn (1970)	20
Ribosomal protein	*Escherichia coli*	Alix and Hayes (1974)	42
		Chang et al. (1974)	43
	HeLa cells	Chang et al. (1976)	34

TABLE 1 (continued)
Natural Occurrence of Various Methylated Amino Acid Derivatives

Protein	Source	Authors and year of finding	Ref.
	ε-*N*-trimethyllysine		
Calmodulin	Testis (rat)	Jackson et al. (1977)	44
α-Amylase	Wheat	Motojima and Sakaguchi (1982)	45
Citrate synthase	Porcine heart	Bloxham et al. (1981)	46
Heat shock protein	Chicken fibroblast	Wang et al. (1982)	47
Elongation factor Tu		L'Italien and Laursen (1979)	48
		Ohba et al. (1979)	49
Elongation factor	*Mucor* (fungus)	Hiatt et al. (1982)	25
Free amino acid	*Laminariaceae* (kelp)	Takemoto et al. (1964)	50
	Seed of *Reseda luteola* L.	Larsen (1968)	51
	Urine (human)	Kakimtoto and Akazawa (1970)	35
	Plasma (human)	Tomita and Nakamura (1970)	35
	Placenta (human)	Tomita and Nakamura (1977)	52
	ε-*N*-trimethyl-δ-hydroxylysine		
Cell wall	Diatom	Nakajima and Volcani (1970)	53
	N^G-monomethylarginine		
Histone	Thymus (calf)	Paik and Kim (1967)	54
Acidic protein	Liver nuclei (rat)	Friedman et al. (1969)	55
Myelin basic protein	Myelin (human)	Badwin and Carnegie (1971)	56
Heat shock protein	Chicken fibroblast	Wang et al. (1982)	47
Histon H1[a]	*Euglena gracilis*	Tuck et al. (1985)	57
Free amino acid	Seed of *Vicia faba* L. (bean)	Kasai et al. (1976)	58
	Placenta (human)	Tomita and Nakamura (1977)	52
	N^G,N^G-dimethylarginine (asymmetric)		
Histone	Thymus (calf)	Paik and Kim (1967)	54
Myelin basic protein	Myelin (human)	Baldwin and Carnegie (1971)	56
Myosin	Leg muscle (rat and chicken)	Reporter and Corbin (1971)	59
HnRNP Protein	Liver nuclei (rat and chicken)	Boffa et al. (1977)	60
	Nucei (*Physarum polycephalum*)	Christensen et al. (1977)	61
HMG-2	Nuclei (rat)	Boffa et al.	62
C23 acidic phosphoprotein	Nucleolus (Novikoff hepatoma)	Lischwe et al. (1982)	63
Ribosomal protein	HeLa cells	Chang et al. (1976)	34
Tooth matrix protein	Human	Kalasz et al. (1978)	22
Scleroderma antigen	Nucleolus (Novikoff hepatoma)	Lischwe et al. (1986)	64
Free amino acid	Urine (human)	Kakimoto and Akazawa (1970)	35
	Serum (human)	Kakimoto and Akazawa (1970)	35
	Placenta (human)	Tomita and Nakamura (1977)	52
	Seed of *Vicia faba* L. (bean)	Kasai et al. (1976)	58
	N^G-N'^G-dimethlyarginine (symmetric)		
Histone	Thymus (calf)	Paik and Kim (1967)	54
Myelin basic protein	Myelin (human)	Baldwin and Carnegie (1971)	56

TABLE 1 (continued)
Natural Occurrence of Various Methylated Amino Acid Derivatives

Protein	Source	Authors and year of finding	Ref.
	N^{G}-N'^{G}-dimethlyarginine (symmetric)		
Tooth matrix protein	Human	Kalasz et al. (1978)	22
Free amino acid	Urine (human)	Kakimoto and Akazawa (1970)	35
	Serum (human)	Kakimoto and Akazawa (1970)	35
	Seed of *Vicia faba* L. (bean)	Kasai et al. (1976)	58
	δ-*N*-monomethylarginine		
	Root tuber (*Trichosanthes cucumeroides*)	Inukai et al. (1968)	65
	N^{G}-hydroxymethyl-L-arginine		
Free amino acid	Serum	Csiba et al. (1986)	66
	Urine	Csiba et al. (1986)	66
	δ-*N*-methylornithine		
Free amino acid	Brain (bovine)	Matsuoka et al. (1969)	26
	3-*N*-methylhistidine		
Myosin	Skeletal muscle (rabbit)	Johnson et al. (1967)	67
	Physarum polycephalum	Venkatesan et al. (1975)	19
Actin	Skeletal muscle (rabbit)	Johnson et al. (1967)	67
	Muscle (bovine)	Asatoor and Armstrong (1967)	68
Histone	Erythrocyte (avian)	Gershey et al. (1969)	69
Opsin	Retina (bovine)	Reporter and Reed (1972)	21
Free amino acid	Urine (human)	Tallan et al. (1954)	70
	δ-*N*-methylglutamine		
Ribosomal protein	*Escherichia coli*	Lhoest and Colson (1977)	71
	***N*-dimethylproline**		
Cytochrome *c*-557	*Chithidia oncopelti*	Pettigrew and Smith (1977)	72
Histone (H2B)	*Asteria rubens* (star fish)	Martinage et al. (1985)	73
	***N*-monomethylalanine**		
Ribosomal protein L11	*Escherichia coli*	Whittman-Liebold and Pannenbecker (1976)	74
	***N*-trimethylalanine**		
Ribosomal protein	*Escherichia coli*	Lederer et al. (1977)	75
Myosin light chain	Skeletal muscle (rabbit)	Henry et al. (1982)	76
Histone (H2B)	Tetrahymena (protozoan)	Nomoto et al. (1982)	77

TABLE 1 (continued)
Natural Occurrence of Various Methylated Amino Acid Derivatives

Protein	Source	Authors and year of finding	Ref.
	N-methylphenylalanine		
Pilin	*Moraxella nonliquefaciens*	Froholm and Sletten (1977)	78
	Neisseria gonorrhoeae	Hermodson et al. (1978)	79
	Pseudomonas aeruginosa K	Frost et al. (1978)	80
	Bacteroides nodosus (strain 198)	McKern et al. (1983)	81
	N-methylmethionine		
Ribosomal protein	*Escherichia coli*	Brasius and Chen (1976)	82
IF-31	*Escherichia coli*	Brauer and Wittman-Liebold (1977)	83
Che Z	*Escherichia coli*	Stock et al. (1987)	84
	Salmonella typhimurium	Stock et al. (1987)	84
	γ-N-methylasparagine		
Allophycocyanin β subunit	*Anabaena variabilis*	Klotz et al. (1986)	85
Phycobiliprotein	*Synechococcus* 6301	Klotz and Glazer (1987)	86
	S-methylcysteine[a]		
O[6]-Methylguanine methyltransferase	*Escherichia coli*	Olsson and Lindahl (1980)	87
Hemoglobin	Rat	Bailey et al. (1981)	88
	S-methylmethionine[a]		
Cytochrome *c*[a]	*Euglena gracilis*	Farooqui et al. (1985)	89
	Glutamic acid 5-methyl ester or (γ-glutamyl methyl ester)		
Chemotactic membrane protein	*Escherichia coli*	Kleene et al. (1977)	90
	Salmonella typhimurium	Van DerWerf and Koshland (1977)	91
	D-aspartic acid β-methyl ester		
Membrane protein	Human erythrocytes	Janson and Clarke (1980)	92
	L-isoaspartic acid α-carboxy methyl ester		
Adrenocorticotropin		Aswad (1984)	93

[a] (Presence is demonstrated in enzymatic incubation *in vitro*.)

Illustrative Procedure, the partial purification of basic methylated amino acids from brain and other organs by Kakimoto et al.[94] is described. One reason to choose this example is based on the fact that the authors also removed lysine and arginine by the treatment of individual specific enzymes after the isolation of basic amino acids and prior to the analysis. When ion-exchange chromatography was to be employed subsequently, the removal of lysine, arginine, or histidine by the enzymes might not be necessarily required.

Illustrative procedures[94] — One- to two hundred mg of the protein was hydrolyzed in 15 volumes of 6 N HCl at 110°C for 24 to 48 h in a sealed tube. The hydrolyzate was filtered, and the insoluble materials were washed with 30 ml of water. The combined filtrates were passed through a 1 × 4 cm column of Amberlite® IR-120 (100 to 200 mesh, H^+ form) and the resin was washed with 20 ml of water and then with 60 ml of 20% pyridine. This treatment elutes acidic, neutral, and imidazole amino acids. This fraction was used for the determination of 3-*N*-methylhistidine. The retained amino acids, lysine, arginine, and their methylated derivatives were eluted with 20 ml of 3 *N* NH_4OH. The dried eluate could be used for the analysis by paper, thin-layer, or ion-exchange chromatography. This sample could also be treated to remove lysine and arginine as follows; the sample was dissolved in 3 ml of 0.2 *M* phosphate buffer, pH 6.0, and 5 mg of lysine decarboxylase (specific activity, 15 μmol of lysine decarboxylated/20 min/mg enzyme protein), and 50 μg of pyridoxal phosphate in 0.25 ml of water was added. The mixture was incubated at 37°C for 2 h and the reaction was terminated by immersing the tube in boiling water for 3 min. After cooling, 10 mg of the arginine deiminase (EC 3.5.3.6) (specific activity, 5 μmol of arginine deiminated/20 min/mg enzyme protein) was added and incubated at 37°C for 2 h. Then 10 ml of 10% trichloroacetic acid was added, and the supernatant was passed through a 1 x 3 cm column of Amberlite® IR-120 (100 to 200 mesh, H^+ form). The resin was washed with 10 ml each of water and 1 *M* pyridine to remove citrulline formed from arginine. The basic amino acids were eluted with 20 ml of 3 *N* NH_4OH. Cadaverine formed from lysine stays on the resin under these conditions. The eluate was evaporated to dryness under reduced pressure.

B. PAPER CHROMATOGRAPHY

1. Paper Chromatography

Table 2 lists the R_f values of various methylated basic amino acids by one-dimensional ascending paper chromatography.[95-97] Combining solvent A for the first dimension and solvent B for the second, Kakimato and Akazawa[35] were able to resolve all of the aliphatic basic amino acids on two-dimensional paper chromatography.

2. Paper Electrophoresis

During the studies on the methylated arginines in myosin during muscle development, Reporter and Corbin[59] employed a paper electrophoresis method; the acid hydrolyzate of myosin was subjected to 4500 v for 40 min in a buffer containing pyridine/acetic acid/water (24:1:225), pH 6.4. Even though the method separates 1-*N*-methylhistidine, 3-*N*-methylhistidine, N^G,N^G-dimethylarginine, ε-*N*-monomethyllysine, and ε-*N*-trimethyllysine from their mother compounds, ε-*N*-dimethyllysine, and N^G,N'^G-dimethylarginine are not resolved by this method.

Chang and Chang[98] later employed a longer paper (110 cm) than that used by Reporter and Corbin,[59] and accomplished a better resolution of ε-*N*-dimethyllysine and N^G,N'^G-dimethylarginine. However, on closer examination of their results, one notices that the migration distance of three ε-*N*-methylated lysines are too close to be able to distinguish them with certainty. Thus, the paper electrophoresis method does not appear to be satisfactory for the identification of the methylated basic amino acids.

C. THIN-LAYER CHROMATOGRAPHY

There are two thin-layer chromatographic methods such as one- and two-dimensional chromatography to separate various methylated amino acid derivatives.

TABLE 2
R^f Values × 100 of Various Methylated Amino Acids on Paper Chromatogram

	Solvent[a]					
	$A_{(95)}$	$B_{(95)}$	$C_{(95)}$	$D_{(95)}$	$E_{(96)}$	$F_{(38)}$[b]
Lysine	46	46	14	19	43	18
ε-*N*-monomethyllysine	40	57	18	23	65	44
ε-*N*-dimethyllysine	58	61	16	23	75	76
ε-*N*-trimethyllysine	17	59	16	20	86	88
Arginine	20	52	19	26		30
N^G-monomethylarginine	30	64	24	29		
N^G,N^G-dimethylarginine	34	66	26	29		
N^G,N'^G-dimethylarginine	42	68	28	32		
α-*N*-monomethyllysine					66	
α-*N*-dimethyllysine					79	
α-*N*-trimethyllysine					89	

[a] Composition of the solvent: A — pyridine/acetone/3 *M* ammonia (10:6:5). B — isopropyl alcohol/formic acid/water (4:1:1). C — *n*-butanol/acetic acid/water (4:1:1:2). D — *n*-butanol/pyridine/acetic acid/water (4:1:1:2) E — *m*-Cresol/1% saturated NH_4OH. F — *m*-cresol/phenol/borate (165:190:45).

[b] Not listed, but Markiw[97] also carried out paper chromatography of the methylated amino acids with solvents A, B, C. The number in parentheses is the reference number.

[c] Some of the commercially available and/or laboratory synthesized standard methylated amino acids such as ε-*N*-trimethyllysine, N^G-monomethylarginine, N^G,N^G-di and N^G,N'^G-dimenthylarginine are obtained in the conjugated form of *p*-hydroxyazobenzene-*p*′-sulfonate (usually yellow) or flavianate. When these compounds are used as standard in an automatic amino acid analyzer, the conjugates need not be removed prior to use; however this is necessary for electrophoresis or paper chromatographic applications. Removal of the conjugates is easily achieved by passing the water solution through a Dowex-1 chloride resin. The conjugates will be retained on the resin, and free amino acids appear in the effluent.

During the course of an enzyme assay for the free L-lysine specific ε-*N*-methyltransferase, Rebouche and Broquist[99] analyzed the enzymatic products of a mixture of lysine, ε-*N*-mono, ε-*N*-di, and ε-*N*-trimethyllysine by one-dimensional thin-layer chromatography; the enzyme preparation was incubated with L-lysine substrate (L-lysine, ε-*N*-monomethyl-L-lysine or ε-*N*-dimethyl-L-lysine), bicarbonate buffer, and *S*-adenosyl-L-[methyl-^{3}H]methionine; the enzyme protein and unreacted *S*-adenosyl-L-[methyl-^{3}H]methionine were removed with charcoal after the reaction was over. An aliquot of the supernatant after the centrifugaton was then applied onto silica gel analytical thin-layer plates, and the chromatogram was developed to a height of 17 cm with methanol/concentrated ammonia (75:25, v/v). After visualization with ninhydrin, appropriate spots were scraped into scintillation vials containing scintillation fluid and the radioactivity was counted.

As listed in Table 3, the resolution of the compounds is excellent (this was confirmed in our laboratory), and the method is simple enough to handle a relatively large number of samples simultaneously. However, no information on methylated amino acids other than methylated lysines on the chromatographic system is available.

Tyihak et al.[100] developed a two-dimensional ion-exchange thin-layer chromatographic system to separate all of the other methylated amino acids including methylated lysine derivatives. The samples were applied on either Fixion 50-X8 (Na$^+$) containing Dowex 50-X8 type resin (Chinoin-Nagyteteny, Budapest, Hungary) or Ionex 25 SA (Na$^+$) chromatosheets (Macherey, Nagel and Co., Dueren, G.F.R.), and the chromatograms were developed in two of the three solvent systems (for the first and second dimension), whose compositions are listed in the footnote of Table 3. For example, the combination of solvent B for the first dimension and solvent C for the second dimension resolved all of the methylated amino acids form each other as well as from aromatic amino acids.

TABLE 3
R_f Values × 100 of Various Methylated Amino Acids on Thin-Layer Chromatogram

	Method of Tyihak et al.[100]a Solvent			Method of Rebouche and Broquist[99]
	A[b]	B	C	Methanol/ conc NH_4OH
Lysine	59	25	58	44.8
ε-*N*-monomethyllysine	38	19	39	36.8
ε-*N*-dimethyllysine	23	16	21	62.1
ε-*N*-trimethyllysine	14	12	13	15.5
Arginine	29	8	13	15.5
N^G-monomethylarginine	21	8	20	
N^G,N^G-dimethylarginine	14	9	13	
N^G,N'^G-dimethylarginine	15	8	16	
Histidine	47	16	48	
1-*N*-methylhistidine	37	15	36	
3-*N*-methylhistidine	26	12	24	

[a] Reference number.

[b] Composition per liter of developing solvent for thin-layer chromatography.

	A (pH 6.0)	B (pH 5.28)	C (pH 6.0)
Hydrated citric acid (g)	100.0	24.6	105.0
HCl (sp. gr. 1.19) (ml)	14.0	6.5	—
NaOH (g)	60.0	14.0	60.0
NaCl (g)	—	—	58.5
Na ion (*N*)	1.5	0.35	2.5

Recently, Csiba et al.[66] detected the presence of N^G-hydroxymethyl-arginines in human serum and urine using cation-exchange thin-layer chromatography. After human serum and urine were dialyzed against distilled water, the protein-free solution was dried. The sample dissolved in distilled waster was applied to a Sephadex G-15 column equilibrated with distilled water and eluted with distilled water. The pooled fractions from the column calibrated with the preparation of standard hydroxymethyl-L-arginines were dried under vacuum and dissolved in 1.0 ml of distilled water and 50 μl sample was spotted on Dowex 50-X8 cation-exchange resin sheet (20 × 20 cm, Chinoin, Budapest, Hungary). The duration of the chromatographic separation was 8 h and the sheet was developed by spraying ninhydrin solution. Hydroxymethyl derivatives of arginine have been identified in comparison with the standard mono-, di-, and trihydroxymethyl-L-arginines as well as after their decomposition by 3 *N* HCl.

An analytical method of methylated amino acids which combine both two-dimensional thin-layer chromatography and autoradiography has been introduced by Klagsbrun and Furano.[101] Mammalian cells or *Escherichia coli* were grown in the presence of L-[*methyl*-^{14}C]-methionine and subcelluar fractions were isolated after harvesting cells and the individual fractions were hydrolyzed in 6 *N* HCl at 110°C. The acid-hydrolyzates were subjected to two-dimensional chromatography with various methylated amino acids as carriers on cellulose thin-layer sheets. The solvent for the first detection was pyridine/acetone/H_2O/NH_4OH (15:9:6:1.5, v/v). After drying the plates and spraying with ninhydrin for visualization, autoradiography was carried out with Kodak RP/154 medical X-ray film. In order to determine the relative amount of radioactivity in various amino acids, the appropriate spots were either cut out or scraped off the plate, and the radioactivity was determined. Since the method is based on autoradiography, it is so sensitive that as little as 25 to 50 cpm of methylated amino acids can be detected and it allows one to screen many samples with relatively little effort. The resolution of methylated amino acids on the chromatogram, however, can be improved because N^G,N^G-dimethylarginine and N^G-monomethylarginine were not resolved. This is most likely due to the improper choice of solvent system during the chromatography since Tyihak et al.[100] were successful in achieving a clear resolution of these amino acids in another solvent system.

Macnicol[102] recently reported that the levels of *S*-methylmethionine in the plant tissue sample can be determined using the combined methods of thin-layer chromatography and dual-isotope technique. A 10% trichloroacetic acid homogenate of the tissue was spiked with [carboxyl-^{14}C]methylmethionine (2.2 nCi) and after centrifugation the clear supernatant was extracted with diethyl ether. The residual aqueous phase was evaporated to dryness, dissolved in 1 m*M* HCl and 10 m*M* HCl, and applied onto Cellex-P (H^+ form, Bio-Rad, Richmond, CA). *S*-Methylmethionine was eluted with 30 m*M* HCl after the initial washing with 1 m*M* HCl and 10 m*M* HCl, and eluates were dried. After the addition of 1 *N* NH_4OH to *S*-methylmethionine residue, the sample was heated in a boiling water bath for 1 h and upon dryness it was moistened with a small volume of 0.2 *M* $NaHCO_3$. To the solution was added 20 µl of 10 m*M* [^{3}H]dansyl chloride (13.5 mCi/mmol) in acetone, and the tube was incubated at 37°C for 3 h. A 5-µl aliquot of reaction mixture, together with carrier dansyl-homoserine, was spotted on a polyamide thin-layer and chromatographed in 0.5% formic acid in the first direction and 2 benzene:1 acetic acid in the second. The dansyl-homoserine spot was localized under UV light and eluted with 1% triethylamine in 95% ethanol. After the addition of scintillation fluid to the sample and the determination of the radioactivity, the ^{3}H/^{14}C ratio was converted to amount of *S*-methylmethionine using calibration lines obtained by substituting a series of dilutions of a standard solution containing *S*-methylmethionine for the homogenate.

Although paper and thin-layer chromatographic methods do not necessitate expensive equipment and many samples can be processed simultaneously within a shorter operating time, it is our conviction, nevertheless, that these methods lack the resolution obtainable with an automatic amino acid analyzer with an ion-exchange resin.

D. ION-EXCHANGE COLUMN CHROMATOGRAPHY

Precise estimation of the amounts of methylated amino acids in proteins often presents some analytical problems because of two factors. The first is that they elute from amino acid analyzers in positions very close to those of the parent amino acids. The second is that the methylated amino acids are usually present in small quantities relative to their parent amino acids. The analysis of various methylated amino acids by ion-exchange column chromatography can be divided into two categories, one employing an automatic amino acid analyzer and the other a manually operated system with the solvent gradient. The use of an automatic amino acid analyzer necessitates an expensive apparatus and may need a full-time technician. Although this method requires a long running time (more than 20 h) for a single sample, the resolution is quite superior by far. It is our recommendation from past experience that the paper or thin-layer chromatography should be used with great caution and that the identity of the unknown amino acid should eventually be confirmed by an automatic amino acid analyzer with an ion-exchange resin column.

1. The Use of an Automatic Amino Acid Analyzer

During the course of the study on the amino acid composition of calf thymus histone, Crampton et al.[103] observed an unidentified ninhydrin peak from the column, which followed closely behind lysine. This peak material was identified to be ε-*N*-methyllysine by Ambler and Rees[12] while they were analyzing the amino acid composition of flagella protein from *Salmonella typhimurium*. Complete separation of this peak from that of lysine was readily achieved when the amino acid analysis for basic amino acids was carried out on a larger column (40 to 50 cm) of resin with the usual 0.35 *M* sodium citrate buffer of pH 5.28 to 5.36.[14,18]

However, this elution system was later demonstrated to be inadequate to resolve three ε-*N*-methylated lysine derivatives, namely, ε-*N*-mono, ε-*N*-di, and ε-*N*-trimethyllysine.[31]

While investigating the origin of ε-*N*-methyllysine in histone in 1965 using paper chromatography with a *m*-cresol/phenol solvent system, we soon realized that the so-called ε-*N*-methyllysine was not a single identity but a mixture of ε-*N*-methylated lysines.[30] In order to confirm this observation, we modified the elution condition for the automatic amino acid analyzer.[31] Our original method consisted of eluting the amino acid analyzer column with 0.35 *M* sodium citrate buffer (in respect to sodium concentration) of pH 5.84 at 28°C at a flow rate of 30 ml/h. The buffer was made first by preparing 0.2 *M* sodium citrate of pH 5.28 followed by the addition of NaOH to raise the pH to 5.84, thus changing the pH as well as the ionic strength. Even if this method resolved three methylated lysines, it suffers from the incomplete separation of N^G-monomethylarginine from arginine.

Elzinga and Alonzo[104] employed Bio-Rad Aminex A-7 or Durrum DC-6A resin column (0.63 × 20 cm) to determine the amount of monomethyllysine, trimethyllysine, and N^{τ}-methylhistidine from the rabbit skeletal muscle myosin. The resolution of methylated amino acids was performed at 22°C and flow rates of 20 ml/h for buffer and 10 ml/h for ninhydrin. The buffer was 0.35 *M* sodium citrate, adjusted to pH 5.65, including 2 ml of octanoic acid and 25 ml of 25% Brij 35 per 20 l. The protein sample was hydrolyzate in 6 *N* HCl at 110°C for 22 h in a sealed, evacuated test tube. The hydrolyzate was dried under N_2 gas and dissolved in the buffer used to apply hydrolyzate to the amino acid analyzer. It has been suggested that the maintenance of the relatively low temperature during the column operation was the key to separation of methyl amino acids, and methyllysine and methylhistidine are best measured if about 500 to 1000 μg of hydrolyzed protein are applied.

Kakimato[105] determined the amount of N^G-monomethylarginine in the acid-hydrolyzate of protein after removing arginine by prior treatment with arginine deiminase. The elution condition was 0.51 *M* NaCl in 0.2 *M* sodium citrate buffer, pH 3.24, at a flow rate of 60 ml/h, and the temperature was raised from 35° to 58°C at 70 min running time. However, this method also suffers from an inability to resolve ε-*N*-monomethyllysine from lysine.

Deibler and Martenson[106] were successful in achieving a better resolution of N^G-monomethylarginine and arginine by lowering the pH of the elution buffer. We have repeated their method with a minor modification. A column (0.9 × 35 cm) of Bio-Rad A-5 resin with a particle size of 13 ± 2 μm was first eluted with 0.38 *M* sodium citrate buffer, pH 5.84 at 24°C at a flow rate of 45 ml/h, using a Perkin-Elmer KLA-3B amino acid analyzer. After 2 h or at tyrosine and phenylalanine positions, the buffer was changed to 0.35 *M* sodium citrate, pH 4.70. All the methylated lysines (ε-*N*-mono, ε-*N*-di, and ε-*N*-trimethyllysine) 3-*N*-methylhistidine, and methylated arginines (N^G,N^G-di-, N^G,N'^G-di-, and N^G-monomethylarginine) were well separated from each other as well as from their mother compounds, although 1-*N*-methylhistidine overlaps with ammonia.

γ-*N*-Methylasparagine in β-subunit of *Anabaena variabilis* allophycocyanin was determined using a Durrum amino acid analyzer with DC-5A resin plus *O*-phthalaldehyde detection. γ-*N*-methylasparagine containing decapeptide was isolated from *Anabaena variabilis* allophycocyanin β-subunit by the use of CNBr treatment. Amino acid analysis was carried out by hydrolyzing the samples *in vacuo* at 110 to 150°C for 22 h in 6 *N* HCl containing 0.1% (w/v) phenol. The measurement of γ-*N*-methylasparagine in a peptide is based on the yield of methylamine resulting from the acid hydrolysis. The presence of a peptide containing γ-*N*-methylasparagine was also confirmed by the fact that ^{1}H NMR spectrum of peptide exhibited a strong methyl singlet at 2.71 ppm characteristic of the methyl derivative of an amide nitrogen.[85]

Recently, Zarkadas et al.[107] developed an analytical single-column chromatographic method for the determination of all methylated basic amino acids and related compounds at picomole levels in protein and tissue hydrolysates. Amino acid analyses were carried out on a fully automated amino acid analyzer (Beckman Spinco Model 121 MB, Beckman Instruments, Inc.,

Palo Alto, CA) equipped with a Varian Vista 402 chromatographic data reduction system (Varian Instruments Group, Walnut Creek, CA). The complete separation of all methylated lysines and histidines, and related compounds was performed on an analytical 50 × 0.28 cm microcolumn of Dionex DC-4A cation-exchange resin (Dionex Chemical Co., Sunnyvale, CA) using a two-buffer system (pH 5.700 and 4.501). The major advantages of this single-column procedure over other methods constitute the high resolving power of the system, baseline separations of the eluting amino acids, and the accurate determination of unusual basic amino acids at picomole levels.

2. Manual Column Chromatography

While investigating the identities of various methylated amino acids in human urine, Kakimoto and Akazawa[35] analyzed the aliphatic amino acid on an Amberlite IR-120 column with a batchwise increase of ammonium hydroxide concentration from 0.2 to 4.0 *N*. The aliphatic amino acids were obtained by the method essentially similar to that under Section III A.

A more comprehensive isolation and identification method of various methylated amino acids from physiological fluids and protein hydrolyzates was carried out by Markiw.[97] Aliphatic amino acids partially purified from acid-hydrolyzates were applied on a 1.1 × 70 cm column of Dowex-50 W × 8 (NH_4^+), and eluted with 2000 ml of a linear gradient from 0.0 to 0.2 *M* NH_4OH in 10 ml fractions. After collecting about 150 fractions, the reservoir containing 0.2 *N* NH_4OH was substituted with 1.5 *N* NH_4OH, and elution was continued until the emergence of arginine. This method offers a means to prepare various methylated amino acid derivatives in comparatively large quantities and relatively pure qualities. However, it suffers from a poor reproducibility of the elution time as well as the elution sequence; for example, in contrast to the elution sequence reported by Markiw,[97] ε-*N*-trimethyllysine was first eluted followed by ε-*N*-dimethyllysine, ε-*N*-monomethyllysine, and lysine under the identical condition employed by us.[108] We did not pursue the problem further, however, this discrepancy suggests that every investigator should establish the elution condition of any particular amino acid on this column system.

E. GAS CHROMATOGRAPHY

Kalasz et al.[22] attempted the identification of various methylated amino acids by gas chromatography after preparing their corresponding ethyl ester *N*-trifluroacetyl derivatives. However, the resolution was not satisfactory, since the resolution of N^G,N^G-dimethylarginine (asymmetric) and N^G,N'^G-dimethylarginine (symmetric) was not achieved.

Recently, Rogoskin et al.[109] reported a gas chromatographic method for the determination of 1- and 3-*N*-methylhistidine in biological fluids after isolating methylhistidines using ion-exchange chromatography. The basic amino acids were eluted, except arginine, from the ion-exchange column (3 × 20 mm) containing resin KRS-8P (H^+, Soiuzchim-reactive, Riga, U.S.S.R.) with 4 *M* HCl. Methylene dichloroxide was added to the dry residue of amino acids and evaporated to dryness under vacuum. The esterification mixture (3 *M* acetyl chloride in isobutanol) was added to the amino acid residue and left to react for 30 min at 120°C. Once the amino acid samples were dried, they were derivatized to *N*-trifluoroacetyl-*O*-isobutyl esters by the treatment of tetrahydrofuran and trifluoroacetic acid at 120°C for 15 min. A LHM-8 MD gas chromatograph (U.S.S.R.) equipped with a flame ionization detector was used and the glass chromatographic column (1.5 m × 1.2 mm I.D.) was packed with 3% OV-17 on 80 to 100 mesh Gas Chrom Q (Applied Science Labs, State College, PA). The carrier gas was nitrogen at a flow rate of 20 ml/min and the flow rate of hydrogen and air to the detector were 30 and 300 ml/min, respectively. The injection port temperature was maintained at 250°C and the temperature of the column was programmed from 130 to 250°C at 20°C/min. The concentrations of 1- and 3-*N*-methylhistidine in serum and urine obtained using this method are reasonably comparable to those obtained using other methods such as HPLC.

F. HIGH PERFORMANCE LIQUID CHROMATOGRAPHY

Minkler et al.[110] determined the amount of ε-*N*-trimethyllysine in plasma and urine on HPLC using a postcolumn reaction (*o*-phthalic dicarboxaldehyde-2-mercaptoethanol) and fluorometric detection after the removal of interfering compounds from biological samples by sequential cation-exchange-anion-exchange chromatography. The sample was applied to a 3.5 × 0.5 cm column of Dowex-50 X8 (200 to 400 mesh, NH_4^+) cation-exchange resin. The column was washed with reagent-grade water, then placed directly above a 3.5 × 0.5 cm column of Dowex-1 X8 (200 to 400 mesh, OH^-) anion-exchange resin. The liquid chromatograph consisted of a Model 6000A pump, WISP-710B automatic sampler, and an RCM-100 radial compression module (Waters Assoc., Milford, MA). The chromatographic separation was accomplished using a 10 × 0.5 cm plastic cartridge containing Radial-Pak C_{18} of 10 μm particle diameter (Waters) and the chromatographic eluent containing amine mobile phase modifier was pumped through the HPLC system at a flow rate of 4.0 ml/min. The postcolumn derivatization reagent solution was pumped through solvent lines into the eluent stream at a flow rate of 2 ml/min and aliquots (5 to 80 μl) of isolated specimens were injected into the chromatograph. For the fluorescence measurement a fluorescence detector Model SF-970 (Kratos-Schoeffel Instruments, Ramsey, NJ). was operated at an excitation wavelength of 240 nm and with a 418 nm cut-off emission filter. With this method a reliable detection limit for ε-*N*-trimethylhistidine was 0.2 nmol/ml in 200 μl of human plasma. They also measured 3-*N*-methylhistidine in human urine on HPLC using postcolumn derivatization with *o*-phthalic dicarboxaldehyde-2-mercaptoethanol followed by fluorometric detection. Prior to the injection into HPLC system 3-*N*-methylhistidine was isolated from human urine using a 3 × 0.5 cm column of Dowex 50-X8 resin (200 to 400 mesh, pyridinium form).[111]

Lehman et al.[112] also reported an improved method for the measurement of ε-*N*-trimethyllysine in human plasma and urine using sequential cation-exchange-anion-exchange chromatography and isocratic HPLC. Urine hydrolysate or deproteinized plasma extract was applied to a column (0.7 × 1.6 cm) of cation-exchange resin (AG 50X-8, 200 to 400 mesh, BioRad, Richmond, CA; NH_4^+ form) which was subsequently washed with water. Once the cation-exchange column was placed over an anion-exchange column (AGIX-8, 200 to 400 mesh, BioRad; OH^- form), the two columns were eluted with 1 *N* NH_4OH. Amino acids in the column eluates were derivatized with *o*-phthalaldehyde and mercaptoethanol, and were separated by isocratic reversed-phase HPLC column (4.6 × 150 mm; Altex, Palo Alto, CA) in the presence of an ion-pairing reagent (sodium 1-heptane sulfonate). The eluant flow rate was 1.5 ml/min and fluorescent derivative of ε-*N*-trimethyllysine was quantified by fluorometry (λex = 360 nm; λem = 455 nm). In this method, the inclusion of the ion-paring reagent sodium 1-heptane sulfonate afforded a greatly improved separation of ε-*N*-trimethyllysine and arginine.

Recently Park et al.[113] reported an improved method for the measurement of N^G-monomethylaginine, N^G-dimethylarginine, and ε-*N*-trimethyllysine from human serum using a two-column HPLC system with fluorometric detection. Proteins were removed from human serum by TCA precipitation. The basic amino acids were essentially separated from acidic and neutral amino acids by ion-exchange chromatography on a column of Dowex 50 (H^+) (100 to 200 mesh; 0.9 cm × 2 cm) prior to HPLC analysis. The column was washed successively with 10 ml each of water and 1 *M* pyridine which resulted in the elution of acidic and neutral amino acids. The basic amino acids were eluted from the column with 10 ml of 3 *M* ammonium hydroxide, and were derivatized with phenylisothiocyanate to their corresponding phenylthiocarbamyl products.

The HPLC two column system which links μ-Bondapak CN(PR) (3.9 mm × 15 m) to μ-Bondapak C_{18} (3.9 mm × 30 cm) was employed to separate the phenylisothiocyanate adducts of methylated basic amino acids and appropriate internal standard amino acids. HPLC chromatography was monitored using a Model 420-AC fluorescence detector set at λ_{ex} = 338 mm and λ_{em} = 425 mm. The elution was carried out isocratically with 25 m*M* sodium acetate and 25 m*M* Na_2HPO_4 in methanol to tetrahydrofuran to water (34:1:65) at pH 7.4 with a flow rate

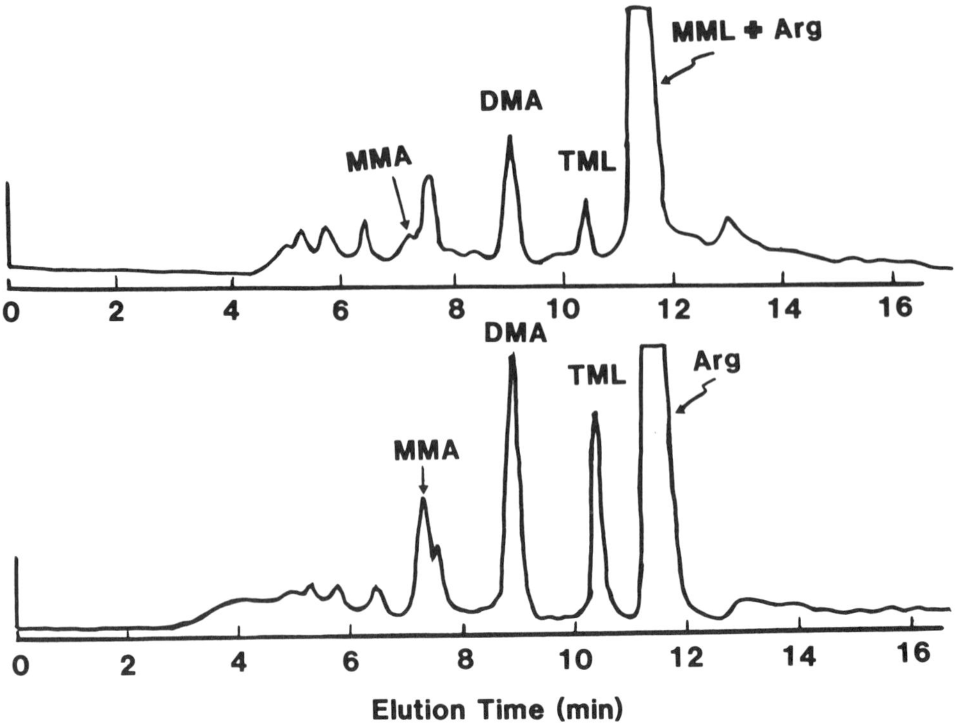

FIGURE 1. Separation profile of basic amino acids from human serum by a two-column HPLC system with fluorometric detection. Upper panel: normal serum. Lower panel: normal human serum fortified with standard amino acids at the concentration of 2.5 mmol each per ml serum prior to sample preparation. Detailed procedure for sample preparation and chromatographic method is described in the text.

of 0.9 ml/min. As illustrated in Figure 1, N^G-monomethylargenine and ε-*N*-trimethyllysine are well separated from each other as well separated from each other as well as from N^G-dimethylarginines. The use of this improved method allows a facile and rapid resolution of methylated basic amino acids and this technique can also be applicable to the analysis of those methylated amino acids in human sera obtained from patients with various diseased conditions.

A method for the determination of *S*-methylcysteine by reversed-phase HPLC is described by Baker et al.[114] Aqueous solution of *S*-methylcysteine was derivatized by the treatment of excess 1-fluoro-2,4-dinitrobenzene at 37°C for 70 min in the dark. The resulting *N*-(2,4-dinitrophenyl)-5-methylcysteine was analyzed by reversed-phase HPLC on a 5 μm Lichrosorb RP-18 column (12.5 cm × 4 mm I.D.) with a mobile phase consisting of 30 m*M* ammonium dihydrogen (pH 6.0) acetonitrile (4:1, v/v) at a flow rate of 1 ml/min. The column eluate was monitored at 420 n*M*. The concentration of *S*-methylcysteine was determined using the unknown peak height and reading the result directly off the standard calibration curve. This method has been applied to the detection of *S*-methylcysteine residues in albumin derived from rat hepatocyte cultures following exposure to radiolabeled dimethylnitrosamine.

IV. ANALYSIS OF CARBOXYL *O*-METHYLATED AMINO ACIDS

In contrast to *N*-methylated amino acid residue in protein, *O*-methylated residues, namely, carboxyl methylesters of glutamyl or aspartyl residues, are labile in both weak alkali and strong acid. Therefore, proteolytic digestion of the methylated protein is the only method available to hydrolyze peptide bonds while preserving such ester bonds. However, the recovery of the ester after proteolysis is not complete and is very much dependent on the kind of protein-methyl esters. This variability in the ester stability is most likely due to the side chain amino acid neighboring the ester linkage; free γ-glutamyl and β-aspartyl methyl esters are relatively more stable in alkaline than their protein-methyl esters.[115] Owing to this lability, when the protein is radiolabeled as carboxyl methylesters, the amount of alkali-hydrolyzable radiomethylmethanol can be measured to quantifying the amount of protein-methyl esters.[116]

A. GLUTAMYL γ-METHYL ESTER

Glutamyl γ-methyl ester is formed posttranslationally on specific glutamyl residues of chemotactic bacterial membrane proteins[90,91,117] by protein-carboxyl methyltransferase, a *cheR* gene product.[118] This carboxyl-methylation is known to be a biochemical signal by which bacterial flagella motility is controlled. In order to identify the presence of glutamyl methyl ester residues, methyl-accepting chemotactic protein (MCP) is radiomethyl-labeled, subjected to proteolysis, and the hydrolyzate is analyzed either on a standard automatic amino acid analyzer, paper chromatography, or on electrophoresis.[90,91] Carboxyl methylesterified MCP can be prepared from chemotactic bacteria either *in vivo* or *in vitro* according to the methods described.[91,119]

1. Enzymatic Digestion

In order to achieve the complete hydrolysis of peptide bonds, combinations of several proteases are used.[90,91,117] Typically, the radiomethyl-labeled MCP is suspended in 50 m*M* sodium acetate buffer, pH 5.2, containing 0.2 m*M* EDTA and 1 m*M* sodium acetate buffer, pH 5.2 containing 0.2 m*M* EDTA and 1 m*M* 2-mercaptoethanol. Papain (0.5 mg/ml) is added and the mixture is incubated at 37°C for 16 to 20 h. The solution is then acidified to pH 2 with 1 *M* HCl and freeze dried. The residue is resuspended in 5 m*M* Tris-HCl buffer, pH 8.0 to 8.5, containing 0.625 m*M* $MnCl_2$. Leucine aminopeptidase (100 μg/ml) and prolidase (200 μg/ml) are added, and this mixture is incubated at 37°C for 4 h. The sample is then acidified to pH 2.0 and freeze dried. The amount of radiolabel recovered throughout the entire digestion was reported to be 60[90] to 90%.[117]

2. Analysis

a. Ion-Exchange Chromatography

A standard automatic amino acid analyzer is suited for the analysis of radiomethyl-labeled carboxyl methylesters. A protein digest obtained from enzymatic proteolysis is mixed with standard glutamyl γ-methyl and aspartyl β-methyl esters (0.2 μmol each) and chromatographed on a 50 cm column using 0.2 *N* sodium citrate buffer, pH 5.23, at 55.5°C, and a flow rate of 66 ml/h. The fractions are collected and the aliquots are analyzed for nihydrin color reaction and radioactivity. Under these conditions, aspartyl β-methyl and glutamyl γ-methyl esters elute at 57 and 93 min, respectively.[91] Here, comigration of authentic ester and radioactive methyl ester should be observed. It should be cautioned, however, that when [*methyl*-^{3}H]-labeling is used, [^{3}H]methyl ester does not exactly coincide with authentic ester due to an isotope effect. Kleene et al.[90] reported that glutamyl γ-[*methyl*-^{3}H]ester elutes about 35 s earlier than the nonradioactive standard on their chromatogram.

b. Paper Chromatography and Electrophoresis

The protein digest is suspended in a minimal volume of water or 50% methanol (v/v) to minimize the salt content. Cloudy samples are filtered through a 0.45 μm Millipore filter. Descending paper chromatography has been carried out on Whatman 3 MM paper using two different solvent systems: (1) *n*-butanol/acetic acid/water (12:3:5, v/v), and (2) *n*-butanol/pyridine/acetic acid/water (20:10:5:2, v/v). A standard (30 to 50 μg) and sample are run in parallel and are located by spraying the chromatogram with ninhydrin in acetone. The radioactivity profile is determined by cutting the sample lane into 1-cm sections, eluting the paper into 0.5 ml of water for 30 min, and counting the radioactivity in a liquid scintillation spectrometer.

Electrophoresis is performed on Whatman 3 MM paper in a Savant flat-bed system with continuous cooling. Voltage is initially set at 100 V for 10 min, increased to 1000 V for 5 min, and then increased to 2000 V for 200 min. Electrode buffer is 1.5 *M* acetic acid/0.58 *M* formic acid, pH 1.9. The mobilities of authentic amino acids relative to the migration of glutamyl γ-methyl ester (^{R}Glu*O*Me) are listed in Table 4.[117]

B. ASPARTYL β-METHYL ESTER

Aspartyl β-methyl ester has been found as one of the products formed by mammalian protein carboxyl *O*-methyltransferase (EC 2.1.1.24). This ester is also extremely labile in weak alkali, and forms substoichiometric quantity in most cases. Interestingly, aspartyl β-methyl esters from human erythrocyte membrane proteins were shown to be in D-configuration,[120] and were implicated to be an intermediate for the repair of racemized D-aspartyl residues occurring in the aged protein. Because of the low extent of the ester formation in the membrane proteins, radiomethyl-labeled proteins have to be prepared. As in the case of bacterial MCP, both *in vivo* and *in vitro* methylesterification can be carried out.[121,122]

1. Enzymatic Digestion

Digestion of [*methyl*-^{3}H]erythrocyte membrane proteins was carried out with an excess amount of protease. Although the recovery of the ester after the digestion was low,[92] aspartyl β-methyl ester was the only product found (not glutamyl γ-methyl ester).

In a 400-μl plastic microfuge tube, 20 μl of membranes (0.1 mg protein, 3.2 pmol of [*methyl*-^{3}H] group) and 80 μl (0.16 mg) of carboxypeptidase γ (Sigma Chem. Co.) in citrate buffer (pH 5.0) are incubated at 37°C for 16 to 25 h. The reaction is terminated by the addition of 20 μl of 1 *M* HCl.

2. Analysis

Analysis of aspartyl β-methyl esters in the hydrolyzate can be performed by either automatic amino acid analyzer, paper chromatography, or electrophoresis in a similar manner as described for glutamyl γ-methyl ester.

O'Connor and Clarke[123] analyzed human erythrocyte protein methyl esters using two-dimensional gel electrophoresis. The procedure separates proteins under acidic conditions by isoelectric focusing in the first dimension and by sodium dodecyl sulfate electrophoresis at pH 2.4 in the second dimension. Isoelectric focusing gels were composed of 4.2% (w/v) acrylamide 0.145% (w/v) bisacrylamide, 9.2 *M* urea, 2% (w/v) NP40, and 8% (w/v) carrier ampholytes (Bio-Lytes 3/10, Bio-Rad). Electrophoresis was performed sequentially at 200 V for 30 min, 300 V for 30 min, 450 V for 16 to 18 h, and 800 V for 1 h. The washed isoelectric focusing gel was placed across the top of SDS-containing slab gel and covered with the running buffer, consisting of 50 m*M* sodium phosphate, 0.1% SDS, pH 2.4. The low pH is essential for preserving protein [^{3}H]methyl esters, but it limits the effective separating range of this system to proteins with isoelectric points between 4 and 8. With this system, they demonstrated that most, if not all, erythrocyte membrane and cytosolic proteins can act as substoichiometric methyl acceptors for an intracellular *S*-adenosylmethionine-dependent carboxyl methyltransferase and that protein carboxyl methylation may be the major methyl transfer reaction in erythrocytes.

To determine the optical configuration of isolated aspartyl β-methyl ester, the procedure of Manning and Moore[124] has been successfully applied, which involves the synthesis of dipeptide leucyl diastereomers.[120,124]

TABLE 4
Comparison of Relative Mobilities of Methylated Amino Acid Standards

	^{R}Glu*O*Me on paper chromatography — Solvent i	^{R}Glu*O*Me on paper chromatography — Solvent ii	^{R}Glu*O*Me on high voltage paper electrophoresis at pH 1.9 (origin at anode)
N^G-Methylarginine	0.62	0.74	2.08
Aspartate β-methyl ester	0.38	0.52	1.20
1-*N*-methylhistidine	0.39	0.53	2.27
3-*N*-methylhistidine	0.49	0.65	2.27
ε-*N*-methyllysine	0.81	0.88	2.46

[a] Composition of the solvent: i. *n*-Butanol/acetic acid/water (12:3:5, v/v).
ii. *n*-Butanol/pyridine/acetic acid/water (20:10:5:2, v/v).

C. ISOASPARTYL α-METHYL ESTER

The unusual aspartyl α-methyl ester is another recently identified product catalyzed by mammalian protein-carboxyl *O*-methyltransferase[93] and is also implicated as an intermediate in the repair process of age-damaged protein.[125] It has been reported that deamidation of asparagine-25 in adrenocorticotropin leads to the formation of an atypical isopeptide bond in which the resulting aspartyl residue is linked to the adjacent glycine-26 via its side chain β-carboxyl group. Thus, free α-carboxyl of aspartyl residue is not a methyl acceptor site for protein carboxyl *O*-methyltransferase, yielding isoaspartyl α-methyl ester (L-configuration). Indirect evidence, i.e., by identifying the weak alkali hydrolyzable methyl group, indicates that this methylation of α-carboxyl group of aspartyl isopeptide β-linkage is a stoichiometric reaction. Readers are referred to recent publications by Clarke[126] or Aswad and Johnson.[127]

D. REDUCTION OF PROTEIN-METHYL ESTER BY CALCIUM BOROHYDRIDE

As an alternate means to identify the labile protein-carboxyl methyl esters, protein-methyl esters can be reduced to corresponding stable alcohols with [^{3}H]labeled calcium borohydride.[128] Aspartyl β-methyl ester would yield homoserine whereas glutamyl γ-methyl ester yields α-amino-6-hydroxyvaleric acid. The resulting products are now stable to treatment with 6 *N* HCl.

Calcium borohydride is prepared by mixing 0.02 ml of 0.1 *M* NaB[^{3}H]$_4$ (8.8 Ci/mmol) in 0.05 *M* NaOH, 0.01 ml of 0.1 *M* $CaCl_2$ in dimethylsulfoxide (DMSO) and 6 μl of 1 *M* HCl. This mixture is then incubated with 0.4 ml of the methylesterified erythrocyte membrane proteins in DMSO for 6 h at room temperature under a stream of N_2 gas. The reaction is terminated by the addition of 0.1 ml of cold acetone. The membranes are washed four times by resuspending in a large volume of acetone, twice in 15% trichloroacetic acid, and once with ethanol.

The membrane protein is pelleted by centrifugation and hydrolyzed in 6 *N* HCl *in vacuo* for 36 h at 110°C. The hydrolyzate is dried *in vacuo*, taken up in 0.5 ml of water, and dried completely to remove any trace of exchangable tritium. This is now analyzed for the hydroxy amino acid in a standard amino acid analyzer. It has been shown that the radioactivity eluted only with homoserine, which had been derived from aspartyl β-methyl ester.[128]

REFERENCES

1. **Uy, R. and Wold, F.,** Posttranslational covalent modification of proteins, *Science,* 198, 890, 1977.
2. **Vickery, H. B.,** The history of the discovery of the amino acids. II. A review of amino acids described since 1931 as components of native proteins, in *Advances in Protein Chemistry,* Vol. 26, Afinsen, C.B., Edsall, J.T., and Richards, F.M., Eds., Academic Press, New York, 1972, 81.
3. **Wold, F.,** *In vivo* chemical modification of proteins (posttranslational modification), *Annu. Rev. Biochem.,* 50, 783, 1981.
4. **Horakova, M. and Deyl, Z.,** Chromatographic and electrophoretic behaviour of amino acids arising from post-translational reactions in proteins, *J. Chromatogr.,* 159, 227, 1978.
5. **Wold, F. and Moldave, K., Eds.,** *Methods in Enzymology,* Vol. 106, Academic Press, New York, 1984.
6. **Wold, F. and Moldave, K., Eds.,** *Methods in Enzymology,* Vol. 107, Academic Press, New York, 1984.
7. **Paik, W. K. and Kim, S.,** Protein methylation, *Science,* 174, 114, 1971.
8. **Paik, W. K. and Kim, S.,** Protein methylation: chemical, enzymological and biological significance, in *Advances in Enzymology,* Vol. 42, Meister, A., Ed., Academic Press, New York, 1975, 227.
9. **Paik, W.K. and Kim, S.,** *Protein Methylation,* John Wiley & Sons, New York, 1980.
10. **Benoiton, N. L.,** N^{ω}-Alkyl diamino acids: chemistry and properties, in *Chemistry and Biochemistry of Amino Acids, Peptides, and Proteins,* Weinstein, B., Ed., Marcel Dekker, New York, 1978, 163.
11. **Huszar, G.,** Methylated lysines and 3-methylhistidine in myosin: tissue and developmental differences, in *Methods in Enzymology,* Vol. 106, Wold, F. and Moldave, K., Eds., Academic Press, New York, 1984, 287.
12. **Ambler, R. P. and Rees, M.W.,** ε-*N*-Methyl-L-lysine in bacterial flagellar protein, *Nature,* 184, 56, 1959
13. **Glazer, A. N., DeLange, R. J., and Martinez, R. J.,** Identification of ε-*N*-methyllysine in *Spirillum serpens* flagella and of ε-*N*-dimethyllysine in *Salmonella typhimurium* flagella, *Biochim. Biophys. Acta,* 188, 164, 1969.
14. **Murray, K.,** The occurence of ε-*N*-methyllysine in histones, *Biochemistry,* 3, 10, 1964.
15. **Comb, D. G., Sarkar, N., and Pinzino, C.J.,** The methylation of lysine residues in protein, *J. Biol. Chem.,* 241, 1857, 1966.
16. **Fambrough, D.M. and Bonner, J.,** Sequence homology and role of cysteine in plant and animal arginine-rich histones, *J. Biol. Chem.,* 243, 4434, 1968.
17. **Hardy, M.F. and Perry, S.V.,** *In vitro* methylation of muscle proteins, *Nature,* 223, 300, 1969.
18. **Huszar, G. and Elzinga, M.,** ε-*N*-Methyllysine in myosin, *Nature,* 223, 834, 1969.
19. **Venkatesan, M., Nachmias, V., and McManus, I.R.,** *In vivo* methylation of contractile protein in *Physarum polycephalum, Fed. Proc.,* 34, 671, 1975.
20. **Weihing, R.R. and Korn, E.D.,** ε-*N*-Dimethyllysine in amoeba actin, *Nature,* 227, 1263, 1970.
21. **Reporter, M. and Reed, D.W.,** Methylation of bovine rhodopsin, *Nature New Biol.,* 239, 201, 1972.
22. **Kalasz, H., Kovacs, G.H., Nagy, J., Tyihak, E., and Barnes, W.T.,** Identification of *N*-methylated basic amino acids from human adult teeth, *J. Dent. Res.,* 57, 128, 1978.
23. **Minami, Y., Wakabayashi, S., Wada, K., Matsubara, H., Kerscher, L., and Oesterhelt, D.,** Amino acid sequence of a ferredoxin from thermo acidophilic archaebacterium, *Sulfolobus acidocaldarius.* Presence of an N^6-monomethyllysine and phyletic consideration of Archaebacteria, *J. Biochem. (Tokyo),* 97, 745, 1985.
24. **Ames, G.F. and Niakido, K.,** *In vivo* methylation of prokaryotic elongation factor Tu, *J. Biol. Chem.,* 254, 9947, 1979.
25. **Hiatt, W.R., Garcia, R., Merrick, W.C., and Sypherd, P.S.,** Methylation of elongation factor 1α from the fungus *Mucor, Proc. Natl. Acad. Sci. U.S.A.,* 79, 3433, 1982.
26. **Matsuoka, Y., Kumon, A., Nakajima, T., Kakimoto, Y., and Sano, I.,** Identification of ε-*N*-methyl-L-lysine and δ-*N*-methylornithine in bovine brain, *Biochim. Biophys. Acta,* 192, 136, 1969.
27. **Perry, T.L., Diamond, S., and Hansen, S.,** ε-*N*-Methyllysine: an additional amino acid in human plasma, *Nature,* 222, 668, 1969.
28. **Weatherall, I.L. and Haden, D.D.,** ε-*N*-Methyllysine in sheep plasma, *Biochim. Biophys. Acta,* 192, 553, 1969.
29. **Asatoor, A.M.,** The occurrence of ε-*N*-methyllysine in human urine, *Clin. Chim. Acta,* 26, 147, 1969.
30. **Kim, S. and Paik, W.K.,** Studies on the origin of ε-*N*-methyl-L-lysine in protein, *J. Biol. Chem.,* 240, 4629, 1965.
31. **Paik, W.K. and Kim, S.,** ε-*N*-Dimethyllysine in histones, *Biochem. Biophys. Res. Commun.,* 27, 479, 1967.
32. **Kuehl, W.M. and Adelstein, R.S.,** The absence of 3-methylhistidine in red cardiac and fetal myosins, *Biochem. Biophys. Res. Commun.,* 39, 956, 1970.
33. **Hardy, M.F., Harris, C.I., Perry, S.V., and Stone, D.,** Occurrence and formation of the N^{ε}-methyllysine in myosin and the myofibrillar proteins, *Biochem. J.,* 120, 653, 1970.
34. **Chang, F.N., Navivkas, I.J., Chang, C.N., and Dancis, B.M.,** Methylation of ribosomal proteins in HeLa cells, *Arch. Biochem. Biophys.,* 172, 627, 1967.
35. **Kakimoto, Y. and Akazawa, S.,** Isolation and identification of N^G,N^G, and $N^GN'^G$-dimethylarginine, N^{ε}-mono-, di-, and trimethyllysine, and glucosyl-galactosyl- and galactosyl-δ-hydroxylysine from human urine, *J. Biol. Chem.,* 245, 5751, 1970.

36. **Schaefer, W.H., Lukas, T.J., Blair, I.A., Schultz, J.E., and Watterson, D.M.,** Amino acid sequence of a novel calmodulin from *Paramecium tetraurelia* that contains dimethyllysine in the first domain, *J. Biol. Chem.*, 262, 1025, 1987.
37. **Hempel, K., Lange, H.W., and Birkofer, L.,** N^{ε}-Methylated lysine in histones from chicken erythrocytes, *Hoppe-Seyler's Z. Physiol. Chem.*, 349, 603, 1968.
38. **DeLange, R.J., Glazer, A.N., and Smith, E.L.,** Presence and location of an unusual amino acid, ε-*N*-trimethyllysine, in cytochrome *c* of wheat germ and *Neurospora, J. Biol. Chem.*, 244, 1385, 1969.
39. **DeLange, R.J., Glazer, A.N., and Smith, E.L.,** Identification and location of ε-*N*-trimethyllysine in yeast cytochrome *c, J. Biol. Chem.*, 245, 3325, 1970.
40. **Hill, G.C., Chan, S.K., and Smith, L.,** Purification and properties of cytochrome c_{555} from a protozoan, *Crithidia fasciculata, Biochim. Biophys. Acta,* 253, 78, 1971.
41. **Brown, R.H. and Boulter, D.,** The amino acid sequence of cytochrome *c* from *Allium porrum* L. (Leek), *Biochem. J.*, 131, 247, 1973.
42. **Allix, J.H. and Hayes, D.,** Properties of ribosomes and RNA synthesized by *Escherichia coli* grown in the presence of ethionine. III. Methylated proteins in 50S ribosomes of *E. coli* EA2, *J. Mol. Biol.*, 86, 139, 1974.
43. **Chang, F.N., Chang, C.N., and Paik, W.K.,** Methylation of ribosomal proteins in *Escherichia coli, J. Bacteriol.*, 120, 651, 1974.
44. **Jackson, R.L., Dedman, J.R., Schreiber, W.E., Bhatnagar, P.K., Knapp, R.D., and Means, A.R.,** Identification of ε-*N*-trimethyllysine in rat testis calcium-dependent regulatory protein of cyclic nucleotide phosphodiesterase, *Biochem. Biophy. Res. Commun.*, 77, 723, 1977.
45. **Motojima, K. and Sakaguchi, K.,** Part of the lysyl residues in wheat α-amylase is methylated as *N*-ε-trimethyllysine, *Plant Cell Physiol.*, 23, 709, 1982.
46. **Bloxham, D.P., Parmelee, D.C., Kumar, S., Wade, R.D., Ericsson, L.H., Neurath, H., Walsh, K.A., and Titani, K.,** Primary structure of porcine heart citrate synthase, *Proc. Natl. Acad. Sci. U.S.A.*, 78, 5381, 1981.
47. **Wang, C., Lazarides, E., O'Connor, C.M., and Clarke, S.,** Methylation of chicken fibroblast heat shock proteins at lysyl and arginyl residues, *J. Biol. Chem.*, 257, 8356, 1982.
48. **L'Italien, J.J. and Laursen, R.A.,** Location of the site of methylation of elongation factor Tu, *FEBS Lett.*, 107, 359, 1979.
49. **Ohba, M., Koiwai, O., Tanada, S., and Hayashi, H.,** *In vivo* methylation of elongation factor Tu of *Escherichia coli, J. Biochem. (Tokyo),* 86, 1233, 1979.
50. **Takemoto, T., Daigo, K., and Takagi, N.,** Studies on the hypotensine constituents of marine algae. I. A new basic amino acid "Laminine" and the other basic constituents isolated from *Laminaria angustata, Yakugaku Zassi,* 84, 1176, 1964.
51. **Larsen, P.O.,** N^6-Trimethyl-L-lysine betaine from seeds of *Reseda luteola* L., *Acta Chem. Scand.*, 22, 1369, 1968.
52. **Tomita, T. and Nakamura, K.,** Isolation and identification of N^G-monomethyl-,N^G,N^G-dimethylarginine and N^{ε}-trimethyllysine from human placenta, *Hoppe-Seyler's Z. Physiol. Chem.*, 358, 413, 1977.
53. **Nakajima, T. and Volcani, B.E.,** ε-*N*-Trimethyl-L-δ-hydroxylysine phosphate and its nonphosphorylated compund in diatom cell walls, *Biochem. Biophys. Res. Commun.*, 39, 28, 1970.
54. **Paik W.K. and Kim, S.,** Enzymatic methylation of protein fractions from calf thymus nuclei, *Biochem. Biophys. Res. Commun.*, 29, 14, 1967.
55. **Friedman, M., Shull, K.H., and Farber, E.,** Highly selective *in vivo* ethylation of rat liver nuclear protein by ethionine, *Biochem. Biophys. Res. Commun.*, 34, 857, 1969.
56. **Baldwin, G.S. and Carnegie, P.R.,** Specific enzymic methylation of an arginine in the experimental allergic encephalomyelitis protein from human myelin, *Science,* 171, 579, 1971.
57. **Tuck, M., Farooqui, J., and Paik, W.K.,** Two histone H1-specific protein-lysine *N*-methyltransferases from *Euglena gracilis, J. Biol. Chem.*, 260, 7114, 1985.
58. **Kasai, T., Sano, M., and Sakamura, S.,** N^G-Methylated arginines in broad bean seed, *Agric. Biol. Chem.*, 40, 2449, 1976.
59. **Reporter, M. and Corbin, J.L.,** N^G,N^G-Dimethylarginine in myosin during muscle development, *Biochem. Biophys. Res. Commun.*, 43, 644, 1971.
60. **Boffa, L.C., Karn, J., Vidali, G., and Allfrey, V.G.,** Distribution of N^G,N^G-dimethylarginine in nuclear protein fractions, *Biochem. Biophys. Res. Commun.*, 74, 969, 1977.
61. **Christensen, M.E., Beyer, A.L., Walker, B., and Lestourgeon, W.M.,** Identification of N^G,N^G-dimethylarginine in a nuclear protein from the lower eukaryote *Physarum polycephalum* homologous to major proteins of mammalian 40S ribonucleoprotein particles, *Biochem. Biophys. Res. Commun.*, 74, 621, 1977.
62. **Boffa, L.C., Sterner, R., Vidali, G., and Allfrey, V.G.,** Post-synthetic modification of nuclear proteins: high mobility group proteins are methylated, *Biochem. Biophys. Res. Commun.*, 89, 1322, 1979.
63. **Lischwe, M.A., Roberts, K.D., Yoeman, L.C., and Busch, H.,** Nuclear specific acidic phosphoprotein C23 is highly methylated, *J. Biol. Chem.*, 257, 14600, 1982.

64. **Lischwe, M.A., Ochs, R.L., Reddy, R., Cook, R.G., Yoeman, L.C., Tan, E.M., Reichlin, M., and Busch, H.,** Purification and partial characterization of a nucleolar scleroderma antigen (mw = 34,000; pI, 8.5) rich in N^G,N^G-dimethyl-arginine, *J. Biol. Chem.*, 260, 14304, 1985.
65. **Inukai, F., Suyama, Y., Inatome, H., Morita, Y., and Ozawa, M.,** Identification of δ-*N*-methylarginine abstr., *Annu. Meet. of the Agricultural Chemistry Soc. of Japan*, Nagoya, Japan, 1968, 183.
66. **Csiba, A., Trezl, L., and Rusznak, I.,** N^G-Hydroxymethyl-L-arginines: newly discovered serum and urine components — isolation and characterization on the basis of cation-exchange thin-layer chromatography, *Biochem. Med. Metab. Biol.*, 35, 271, 1986.
67. **Johnson, P., Harris, C.I., and Perry, S.V.,** 3-Methylhistidine in actin and other muscle proteins, *Biochem. J.*, 105, 361, 1967.
68. **Asatoor, A.M. and Armstrong, M.D.,** 3-Methylhistidine, a component of actin, *Biochem. Biophys. Res. Commun.*, 26, 168, 1967.
69. **Gershey, E.L., Haslett, G.W., Vidali, G., and Allfrey, V.G.,** Chemical studies of histone methylation, *J. Biol. Chem.*, 244, 4871, 1969.
70. **Tallan, H.H., Stein, W.H., and Moore, S.,** 3-Methylhistidine, a new amino acid from human urine, *J. Biol. Chem.*, 206, 825, 1954.
71. **Lhoest, J. and Colson, C.,** Genetics of ribosomal protein methylation in *Escherichia coli*, *Mol. Gen. Genet.*, 154, 175, 1977.
72. **Pettigrew, G.W. and Smith, G.M.,** Novel N-terminal protein blocking group identified as dimethylproline, *Nature*, 265, 661, 1977.
73. **Martinage, A., Briand, G., Van Dorsselaer, A., Turner, C.H., and Sautiere, P.,** Primary structure of histone H2B from gonads of the starfish *Asterias rubens:* identification of *N*-dimethylproline residue at the amino-terminal, *Eur. J. Biochem.*, 147, 351, 1985.
74. **Wittman-Liebold, B. and Pannenbecker, R.,** Primary structure of protein L33 from the large subunit of the *Escherichia coli* ribosome, *FEBS Lett.*, 68, 115, 1976.
75. **Lederer, F., Alix, J.H., and Hayes, D.,** *N*-Trimethylalanine, a novel blocking group found in *E. coli* ribosomal proteins L11, *Biochem. Biophys. Res. Commun.*, 77, 470, 1977.
76. **Henry, G.D., Dalgarno, D.C., Marcus, G., Scott, M., Levine, B.A., and Taryer, I.P.,** The occurrence of α-*N*-trimethlalanine as the *N*-terminal amino acid of some myosin light chains, *FEBS Lett.*, 144, 11, 1982.
77. **Nomoto, M., Kyogoku, Y., and Iwai, K.,** *N*-Trimethylalanine, a novel blocked *N*-terminal residue of Tetrahymena histone H2B, *J. Biochem.*, 92, 1675, 1982.
78. **Froholm, L.O. and Sletten, K.,** Purification and *N*-terminal sequence of a fimbrial protein from *Moraxella nonliquefaciens*, *FEBS Lett.*, 73, 29, 1977.
79. **Hermodson, M.A., Chen, K.C.S., and Buchanan, T.M.,** *Neisseria pili* proteins: amino-terminal amino acid sequences and identification of an unusual amino acid, *Biochemistry*, 17, 442, 1978.
80. **Frost, L.S., Carpenter, M., and Paranchych, W.,** *N*-Methylphenylalanine at the *N*-terminus of pilin isolated from *Pseudomonas aeruginosa K*, *Nature*, 271, 87, 1978.
81. **McKern, N.M., O'Donnell, I.J., Inglis, A.S., Steward, D.J., and Clark, B.L.,** Amino acid sequence of pilin from *Bacteroides nodosus* (strain 198), the causative organism of ovine foot rot, *FEBS Lett.*, 164, 149, 1983.
82. **Brasius, J. and Chen, R.,** The primary structure of protein L16 located at the peptidyltransferase center of *Escherichia coli* ribosome, *FEBS Lett.*, 68, 105, 1976.
83. **Brauer, D. and Wittman-Liebold, B.,** The primary structure of the initiation factor IF-3 from *Escherichia coli*, *FEBS Lett.*, 79, 269, 1977.
84. **Stock, A., Schaeffer, E., Koshland, D.E., Jr., and Stock, J.,** A second type of protein methylation reaction in bacterial chemotaxis, *J. Biol. Chem.*, 262, 8011, 1987.
85. **Klotz, A.V., Leary, J.A., and Glazer, A.N.,** Post-translational methylation of asparaginyl residues (identification of β-71 γ-*N*-methylasparagine in allophycocyanin), *J. Biol. Chem.*, 261, 15891, 1986.
86. **Klotz, A.V. and Glazer, A.N.,** γ-*N*-Methylasparagine in phycobiliproteins, *J. Biol. Chem.*, 262, 17350, 1987.
87. **Olsson, M. and Lindahl, T.,** Repair of alkylated DNA in *Escherichia coli*, *J. Biol. Chem.*, 255, 10569, 1980.
88. **Bailey, E., Connors, T.A., Farmer, P.B., Gorf, S.M., and Rickard, J.,** Methylation of cysteine in hemoglobin following exposure to methylating agents, *Cancer Res.*, 41, 2514, 1981.
89. **Farooqui, J.Z., Tuck, M.T., and Paik, W.K.,** Purification and characterization of enzymes of enzymes from *Euglena gracilis* that methylate methionine and arginine residues of cytochrome *c J. Biol. Chem.*, 260, 537, 1985.
90. **Kleene, S.J., Toews, M.L., and Adler, J.,** Isolation of glutamic acid methyl ester from an *Escherichia coli* membrane protein involved in chemotaxis, *J. Biol. Chem.*, 252, 3214, 1977.
91. **Van DerWerf, P. and Koshland, D.E., Jr.,** Identification of a γ-glutamyl methyl ester in bacterial membrane protein involved in chemotaxis, *J. Biol. Chem.*, 252, 2793, 1977.
92. **Janson, C.A. and Clarke, S.,** Identification of aspartic acid as a site of methylation in human erythrocyte membrane proteins, *J. Biol. Chem.*, 255, 11640, 1980.

93. **Aswad, D.W.,** Stoichiometric methylation of porcine adrenocorticotropin by protein carboxyl methyltransferase requires deamidation of asparagine 25, *J. Biol. Chem.*, 259, 10714, 1984.
94. **Kakimoto, Y., Matsuoka, Y., and Konishi, H.,** Methylated amino acid residues of proteins of brain and other organs, *J. Neurochem.*, 24, 893, 1975.
95. **Matsuoka, Y.,** *N*-Methylated lysine and N^G-methylated arginines, *Seikagaku* (in Japanese), 44, 252, 1972.
96. **Hempel, V.K. and Lange, H.W.,** Traktionierung und eigenschaften *N*-methylierter lysine, *Hoppe-Seyler's Z. Physiol. Chem.*, 350, 966, 1969.
97. **Markiw, R.T.,** Isolation of *N*-methylated basic amino acids from physiological fluids and protein hydrolyzates, *Biochem. Med.*, 13, 23, 1975.
98. **Chang, C.N. and Chang, F.N.,** Methylation of the ribosomal proteins in *Escherichia coli.* Nature and stoichiometry of the methylated amino acids in 50S ribosomal proteins, *Biochemistry*, 14, 468, 1975.
99. **Rebouche, C.J. and Broquist, H.P.,** Carnitine biosynthesis in *Neurospora crassa:* enzymatic conversion of lysine to ε-*N*-trimethyllysine, *J. Bacteriol.*, 126, 1207, 1976.
100. **Tyihak, E., Ferenczi, S., Hazai, I., Zoltan, S., and Patthy, A.,** Combined application of ion-exchange chromatography methods for the study of minor basic amino acids, *J. Chromatogr.*, 102, 257, 1974.
101. **Klagsbrun, M. and Furano, A.V.,** Methylated amino acids in the proteins of bacterial and mammalian cells, *Arch. Biochem. Biophys.*, 169, 529, 1975.
102. **Macnicol, P.K.,** Analysis of *S*-methylmethonine and *S*-adenosylmethionine in plant tissue by a dansylation, dual-isotope method, *Anal. Biochem.*, 158, 93, 1986.
103. **Crampton, C.F., Stein, W.H., and Moore, S.,** Comparative studies on chromatographically purified histone, *J. Biol. Chem.*, 225, 363, 1957.
104. **Elzinga, M. and Alonzo, N.,** Analysis for methylated amino acids in proteins, in *Methods in Enzymology*, Vol 91, Hirs, C.H.W. and Timasheff, S.N., Eds., Academic Press, New York, 1983, 8.
105. **Kakimoto, Y.,** Methylation of arginine and lysine residues of cerebral proteins, *Biochim. Biophys. Acta*, 243, 31, 1971.
106. **Deibler, G.E. and Martenson, R.E.,** Determination of methylated basic amino acids with amino acid analyzer, *J. Biol. Chem.*, 248, 2387, 1973.
107. **Zarkadas, C.G., Rochemont, J.A., Zarkadas, G.C., Karatzas, C.N., and Khalili, A.D.,** Determination of methylated basic, 5-hydroxylysine, elastin crosslinks, other amino acids, and the amino sugars in proteins and tissues, *Anal. Biochem.*, 160, 251, 1987.
108. **Paik, W.K., DiMaria, P., Pearson, E., and Kim, S.,** Preparation of radioactive ε-*N*-methylated lysine, not involving elaborate organic synthesis, *Anal. Biochem.*, 90, 262, 1978.
109. **Rogoskin, V.A., Krylov, A.I., and Khlebnikova, N.S.,** Gas chromatographic determination of 1- and 3-methylhistidine in biological fluids, *J. Chromatogr.*, 423, 33, 1987.
110. **Minkler, P.E., Erdos, E.A., Ingalls, S.T., Griffin, R.L., and Hoppel, C.L.,** Improved high-performance liquid chromatographic method for the determination of 6-*N,N,N*-trimethyllysine in plasma and urine: biomedical application of chromatographic figures of merit and amine mobile phase modifiers, *J. Chromatogr.*, 380, 285, 1986.
111. **Minkler, P.E., Ingalls, S.T., Griffin, R.L., and Hoppel, C.L.,** Rapid high-performance liquid chromatography of 3-methylhistidine in human urine, *J. Chromatogr.*, 413, 33, 1987.
112. **Lehman, L.J., Olson, A.L., and Rebouche, C.J.,** Measurement of ε-*N*-trimethyl-lysine in human blood plasma and urine, *Anal. Biochem.*, 162, 137, 1987.
113. **Park, K.S., Lee, H.W., Hong, S.-Y., Shin, S., Kim, S., and Paik, W.K.,** Analysis of methylated amino acids in human serum by HPLC, *J. Chromatogr.*, 440, 225, 1988.
114. **Baker, I., Shuker, D.E.G., and Tannenbaum, S.R.,** A method for the determination of *S*-methylcysteine by high-performance liquid chromatography: application to the study of carcinogen methylating agents, *J. Chromatogr.*, 329, 202, 1985.
115. **Kim, S. and Paik, W.K.,** Labile protein-methylester: comparison between chemically and enzymatically synthesized, *Experientia*, 32, 982, 1976.
116. **Kim, S. and Paik, W.K.,** New assay method for protein methylase II activity, *Anal. Biochem.*, 42, 255, 1971.
117. **Ahlgren, J.A. and Ordal, G.W.,** Methyl esterification of glutamic acid residues of methyl-accepting chemotaxis proteins in *bacillus subtilis*, *Biochem. J.*, 213, 759, 1983.
118. **Springer, W.R. and Koshland, D.E., Jr.,** Identification of a protein methyltransferase as the *che* R gene product in the bacterial sensing system, *Proc. Natl. Acad. Sci. U.S.A.*, 74, 533, 1977.
119. **Kort, E.N., Goy, M.F., Larsen, S.H., and Adler, J.,** Methylation of a membrane protein involved in bacterial chemotaxis, *Proc. Natl. Acad. Sci. U.S.A.*, 72, 3939, 1975.
120. **McFadden, P.N. and Clarke, S.,** Methylation of D-aspartyl residues in erythrocytes: possible step in the repair of aged membrane proteins, *Proc. Natl. Acad. Sci. U.S.A.*, 79, 2460, 1982.
121. **Galleti, P., Paik, W.K., and Kim, S.,** Methyl acceptors for protein methylase II from human erythrocyte membrane, *Eur. J. Biochem.*, 97, 221, 1979.

122. **Frietag, C. and Clarke, S.,** Reversible methylation of cytoskeletal and membrane proteins in intact human erythrocytes, *J. Biol. Chem.*, 256, 6102, 1981.
123. **O'Connor, C.M. and Clarke, S.,** Analysis of erythrocyte protein methyl esters by two-dimensional gel electrophoresis under acidic separating conditons, *Anal. Biochem.*, 148, 79, 1985.
124. **Manning, J.M. and Moore, S.,** Determination of D- and L-amino acids by ion exchange chromatography as L-D and L-L dipeptides, *J. Biol. Chem.*, 243, 5591, 1968.
125. **O'Connor, C.M., Aswad, D.W., and Clarke, S.,** Mammalian brain and erythrocyte carboxyl methyltransferase are similar enzymes that recognize both D-aspartyl and L-isoaspartyl residues in structurally altered protein substrates, *Proc. Natl. Acad. Sci. U.S.A.*, 81, 7757, 1984.
126. **Clarke, S.,** Protein carboxyl methyltransferases: two distinct classes of enzymes, *Annu. Rev. Biochem.*, 54, 479, 1985.
127. **Aswad, D.W. and Johnson, B.A.,** The unusual substrate specificity of eukaryotic protein carboxyl methyltransferases, *Trends Biochem. Sci.*, 12, 155, 1987.
128. **Kwon, D.S., Jun, G.-J., and Kim, S.,** Calcium borohydride method for identification of a site of methylation in human erythrocyte membrane proteins, *J. Korean Res. Inst. Better Living*, 34, 29, 1984.

Chapter 2

REEVALUATION OF THE ENZYMOLOGY OF PROTEIN METHYLATION

Woon Ki Paik and Sangduk Kim

TABLE OF CONTENTS

I. INTRODUCTION

In this chapter, we will focus our attention on two subjects concerning protein methylation: evidence of multiplicity of methyltransferase activities and whether methylation of some protein as a *co*translational process.

II. MULTIPLICITY OF PROTEIN-SPECIFIC METHYLTRANSFERASES

A. BACKGROUND CONSIDERATION

Since the first report by Ambler and Rees[1] on the presence of protein-bound methyllysine residues in flagellar protein of *Salmonella typhimurium*, nearly 3 decades have passed. During the intervening years, the list of naturally occurring methylated amino acid residues found in proteins has steadily expanded as the analytical methods have improved and as numerous investigators with keen insight have joined this research endeavor. As illustrated in Table 1 of Chapter 1, the side chains of lysine, arginine, histidine, glutamine, and asparagine as well as the α-amino groups of terminal alanine, proline, phenylalanine and methionine residues of certain proteins are *N*-methylated, the free carboxyl groups of glutamyl and aspartyl residues are *O*-methylated, and the side chains of methionine and cysteine residues are *S*-methylated. As this table also indicates, methylated amino acids in proteins occur quite commonly. Despite this variety, early theory maintained that a single enzyme was responsible for methylation of a particular type of amino acid, regardless of protein species.

When described first in 1970, the protein methylase I (*S*-adenosyl-L-methionine:protein-arginine *N*-methyltransferase; EC 2.1.1.23) was assumed to methylate guanidino groups of arginine residues of all the proteins which had previously been shown to contain this modified residue on their side chains.[2] Similarly, subsequent identifications of protein methylase II (EC 2.1.1.24) and III (EC 2.1.1.43) were naturally assumed to be responsible for methylating the free carboxyl groups of dicarboxylic amino acids and the ε-amino groups of lysine residues, respectively, regardless of the protein species. However, it has been increasingly evident in recent years that protein methylase III, for example, represents a class of enzymes which are not only specific for the specific lysine residues to be methylated, but also specific for the substrate protein species. To date, three subtypes of protein methylase III and two subtypes of protein methylase I have well been characterized. This has the very important biochemical implication that each protein listed in Table 1 of Chapter 1 could potentially have a specific methyltransferase.

The second subject to be discussed in this chapter is a question of whether the protein methylation reaction should be considered as a *post*- or *co*translational modification. Originally, protein methylation reaction was classified as one of more than a dozen posttranslational side chain modification reactions of protein.[3] However, recent evidence led us to strongly suggest that some of the protein methylation reactions are *co*translational. This consideration would have important consequences in studies on the biological significance of any particular methylation reaction.

It should be mentioned here that general characteristics and biochemical significance of protein methylase I have been described in Chapter 5, and those of protein methylase II in Chapters 10 and 11. In addition, for more detailed information, readers are referred to our recent review on the enzymology of protein methylation.[4]

B. INDIVIDUAL ENZYMES WHICH METHYLATE SPECIFIC PROTEIN SPECIES

Table 1 lists general reaction types of several well-characterized protein-specific methyltransferases. The most conspicuous example of multiple subtypes among these enzymes

TABLE 1
Known Protein-Specific Methyltransferases

	Trivial name	Systematic nomenclature (EC number)	Methylation product: Amino acid and product	Methylation product: Stability
N-Methylation	Protein methylase I	AdoMet:protein-arginine *N*-methyltransferase (EC 2.1.1.23)	Arginyl: N^G-monomethyl, N^G,N^G-dimethyl, N^G,N'^G-dimethyl	Stable in 6 *N* HCl but unstable in 0.2 *N* NaOH
	Protein methylase III	AdoMet:protein-lysine *N*-methyltransferase (EC 2.1.1.43)	Lysyl: ε-monomethyl, ε-dimethyl, ε-trimethyl	Stable in both 6 *N* HCl and 0.2 *N* NaOH
	Protein methylase IV	AdoMet:protein-histidine *N*-methyltransferase	Histidyl: 3-methyl	Stable in both 6 *N* HCl and 0.2 *N* NaOH
O-Methylation	Protein methylase II	AdoMet:protein-glutamate *O*-methyltransferase (EC 2.1.1.24)	Glutamyl:methyl ester	Unstable in both 6 *N* HCl and 0.2 *N* NaOH
	O-methyltransferase	AdoMet:protein-D-aspartate (EC 2.1.1.77)	D-Aspartyl:methyl ester $t_{1/2}$ = 30 min at pH	7.1 at 37°C Unstable in both 6 *N* HCl and 0.2 *N* NaOH
S-Methylation		AdoMet:protein-methioine *S*-methyltransferase	Methionyl: *S*-methyl	Fairly stable in acid, unstable in alkaline
		O^6-Methylguanine-DNA methyltransferase (EC 2.1.1.63)	Cysteinyl: *S*-methyl	Stable in 6 *N* HCl

Note: AdoMet; abbreviation of *S*-adenosyl-L-methionine.

is protein methylase III. This enzyme (*S*-adenosyl-L-methionine:protein-lysine *N*-methyltransferase; EC 2.1.1.43) catalyzes the transfer of methyl group from *S*-adenosyl-L-methionine (AdoMet) to the ε-amino group of lysine residues in protein, and several forms of this enzyme exist as illustrated in Tables 2 and 3. Besides differences in regard to pH optima and molecular weight, etc. between the preparations isolated from various sources, the most obvious important difference is the substrate protein specificity. In addition to the specificity concerning the protein species such as histone, calmodulin and cytochrome *c*, however, the specificity extends further to the subclasses of the same protein species. For example, two distinct enzymes have been resolved, both of which are specific towards arginine-rich histones,[6,7] one of them specifically methylates histone H3 while the other methylates histone H4.

Furthermore, we have earlier purified a protein methylase III from *Neurospora crassa* approximately 3500-fold. This enzyme methylated *in vitro* the lysine residue at position 72 of horse heart cytochrome *c* only while all other lysine residues were unmethylated, thereby mimicking methylation occurring *in vivo* in *Neurospora*. Since lysine-72 and lysine-86 residues of wheat germ cytochrome *c* are methylated *in vivo*, this suggests the highly likely possibility of the existence of an additional enzyme in wheat germ which recognizes the lysine at residue 86 specifically.

Since ε-*N*-methylated lysines exist in numerous and widely diverse protein species, it is not unrealistic to expect to find many more protein methylase IIIs which are specific for any particular individual protein. Presently, three subtypes of protein methylase III have been well

TABLE 2
Properties of Protein Methylase III from Higher Organisms[4]

Properties	Sources of enzymes: Calf brain	Calf thymus	Rat brain	Chicken embryo	Wheat germ
Subcellular location	Chromatin	Chromatin	Cytosol	Nuclei	Cytosol
In vitro protein substrate	Histones	Histone H4	Calmodulin	Histones	Cytochrome *c*
Residues modified	—[a]	Lys-20	Lys-115	—	Lys-72
Purification achieved (-fold)	10	100	—	5	135
Molecular weight	>200,000	—	—	~150,000	—
pH optimum	8.2—8.7	7.5—9.0	8.1	8.4	9.0
K_m for *S*-adenosyl-L-methionine ($\times 10^{-5}$ *M*)	12	3	—	3.06	48
K_m for protein substrate ($\times 10^{-6}$*M*)	5.7	—	—	—	—
Specific activity of the purified enzyme[b]	68	—	—	0.38	508
Requirement for maxiumum enzyme activity	—	Mg^{++}	Mn^{++}	Mg^{++}, dithiothreitol	SH reagents
Stability	6 months at -40°C, 90% destroyed in 7 d at 4°C	Extremely unstable	—	50% destroyed at 0°C in 1 min	Quite stable in 2-mercaptoethanol
Ratio of ε-mono: di:trimethyllysine in products	64:33:3	—	Mostly tri-, some di- and mono-	—	—

[a] Not available.
[b] Picomoles of *S*-adenosyl-L-[*methyl*-^{14}C]methionine used/min/mg enzyme protein.

characterized. *S*-Adenosyl-L-methionine:histone-lysine *N*-methyltransferase (EC 2.1.1.43), *S*-adenosyl-L-methionine:cytochrome *c*-lysine *N*-methyltransferase (EC 2.1.1.59), and *S*-adenosyl-L-methionine:calmodulin-lysine *N*-methyltransferase (EC 2.1.1.60) have been described in detail in Chapters 7, 4, and 3, respectively.

The protein methylase I catalyzes the transfer of methyl group from *S*-adenosyl-L-methionine to guanidino group of arginine residues, and was originally assumed to be responsible for methylating all the proteins which were known to have N^G-methylarginine residues on their side chain. However, the enzyme in bovine brain has recently been resolved into two distinct protein peaks, one of which methylates only histones *in vitro*, and the other which methylates myelin basic protein (MBP) preferentially with a significantly lower K_m value for MBP than for histone (K_m values of 2×10^{-7} *M* vs. 1×10^{-4} *M*).[9] Antibodies raised against each enzyme preparation did not cross-react with the other antigen. A detailed description on this very recent and exciting development has been presented in Chapter 5. Extracts of *Euglena gracilis* also contains a protein methylase I which methylates cytochrome *c* mainly with some slight activity towards myoglobin[10] (and no activity on histone or MBP).

The third example of the presence of protein methyltransferase subtypes is concerned with protein methylase II. The enzyme (*S*-adenosyl-L-methionine:protein-carboxyl *O*-methyltransferase; EC 2.1.1.24) was initially considered to methylate both glutamyl and aspartyl residues. However, studies by many investigators revealed that the mammalian protein methylase II methylates aspartyl residues (EC 2.1.1.24), while the bacterial enzymes are

TABLE 3
Properties of Protein Methylase III from Microorganisms[4]

Properties	*Physalum polycephalum*	*Euglena*[a5] gracilis (V-A)	*Euglena gracilis* (V-B)	*Neurospora crassa*	*Saccharomyces cerevisiae*	*Crithidia oncopelti*	*Escherichia coli*
	Sources of enzymes						
Subcellular location	Cytosol	Cytosol	Cytosol	Cytosol	Cytosol	—	Ribosome
In vitro protein substrate	Histones	Histone H1	Histone H1	Cytochrome *c*	Cytochrome *c*	Cytochrome *c*-557 (*Crithidia*)	L11 ribosomal protein of 50S subunit
Residues modified	—	—	—	Lys-72	Lys-72	Lys-2	—
Purification achieved (-fold)	40	48	214	3,500	63	—	380
Molecular weight	—	55,000	34,000	120,000	97,000	—	31,000
pH optimum	8.0	9.0	9.0	9.0	9.0	9.0	7.8—8.2
K_m for *S*-adenosyl-L-homocysteine ($\times 10^{-6}$ *M*)	7.3	27	34.5	19	40	—	3.2
K_m for protein substrate ($\times 10^{-5}$ *M*)	—	0.031	0.044	170	133	—	—
K_i for *S*-adenosyl-L-homocysteine ($\times 10^{-6}$ *M*)	70% inhibition at 0.12	14.8	1.6	2.0	2.7	—	—
Specific activity of the purified enzyme[c]	257	225	1,012	29,500	117	—	2
Requirement for maxiumum enzyme activity	SH reagents	—	—	None	None	—	Mg^{++}, dithiothreitol
Stability	Destroyed at 60°C for 30 min	—	—	Stable for several months in 50% ammonium sulfate	Relatively stable in 50% ammonium sulfate	Stable for 10 d	—
Ratio of ε-mono: di: trimethyl-lysine in products	66:17:17	37:30:19	38:39:23	12:38:50	9:17:74	Mostly tri-	Mainly tri-, small amounts of mono-

[a] There are two distinct enzymes in *Euglena gracilis* which methylate histone H1 at different sites.[5]
[b] Not available.
[c] Picomoles of *S*-adenosyl-L-[*methyl*-^{14}C]methionine used/min/mg enzyme protein.

responsible for methylation of glutamyl residues of membrane protein (EC 2.1.1.77). A more detailed description on these enzyme activities has been presented in Chapter 10.

Although enzymatic participation in the formation of 3-*N*-methylhistidine in muscle proteins has been speculated since 1969,[11] the enzyme *S*-adenosyl-L-methionine:protein-histidine *N*-methyltransferase (protein methylase IV) has only recently been purified.[12] The 22-fold purified enzyme from rabbit skeletal muscle methylates histidine residues of cardiac and skeletal muscle actins only, and is completely inactive towards myosin. The obvious implication is that there may be an other enzyme for methylating myosin specifically.

C. ENZYMES WHICH METHYLATE THE SAME PROTEIN SPECIES AT DIFFERENT SITES

One of the more perplexing observations has been our recent finding that there is more than a single enzyme to methylate the same protein species *in vitro* at different residues. For example, there exists two protein methylases III in Euglena gracilis,[5] both of which are specific towards very lysine-rich histone H1. As indicated in Table 3, their general characteristics are quite different from each other and further studies revealed that they may methylate lysine residues located at two entirely different sites of the histone H1.[13]

In addition to the above, we recently identified and partially purified two enzymes from *Euglena gracilis,*[10] both of which methylate cytochrome *c* primarily. However, one of the enzymes methylates the arginine residue at position 38 of horse heart cytochrome *c,* whereas the other methylates the methionine residue at position 65 of this hemoprotein. Therefore, there are three known enzymes which methylate cytochrome *c in vitro* acting at lysine 72, arginine-38, and methionine-65.

D. ENZYMES WHICH METHYLATE THE SAME RESIDUE MORE THAN ONCE

In the case of multiple methylation of the same residues such as ε-NH_2 trimethylation of lysine residue or dimethylation of either lysine or arginine residues, a question arises whether a single (or more than one) enzyme is able to complete the methylation process. Earlier, Burdon and Garven[14] prepared a soluble fraction from chromatin of Krebs 2 ascites carcinoma cells that methylates exogenously added histone. Upon acid hydrolysis of the reaction product, ε-*N*-trimethyllysine was the only methylated amino acid detectable. Furthermore, ε-trimethyllysine has also been shown to be the only methylated lysine in cytochrome *c* isolated from wheat germ, yeast, protozoa and fungi,[15-17] and in the myosin of rabbit skeletal muscle.[18] However, in other systems a different situation is seen. The ε-NH_2 group of the lysine-9 and lysine-27 of pea embryo histone H3 or opsin of bovine retina was found to be methylated only as either ε-*N*-mono- or ε-*N*-dimethyllysine, but not as ε-*N*-trimethyllysine.[19,20] Here, the existence of such an uneven distribution of the various methylated lysine residues suggests that more than one enzyme is involved in the ε-NH_2 methylation of lysine.

In contrast to the above contention, the ratio of ε-*N*-mono, di-, and trimethyllysine in the reaction products by cytochrome *c*-specific protein methylase III during 3500-fold purification from *Neurospora crassa* remained constant.[21] This result strongly supports the notion that a single enzyme is responsible for all three methylation steps. If more than one enzyme were necessary for trimethylation of the lysine residue, one would expect loss or enrichment of these enzymes during purification in relation to each other to cause a shift in the ratio of the methylated lysine products. A similar observation indicating N^G-mono-, N^G,N^G-di- (asymmetric), and N^G,N'^G-di- (symmetric)-methylarginines are produced by a single protein methylase I has also been obtained.[22]

III. CONSIDERATION OF PROTEIN METHYLATION AS *CO*TRANSLATIONAL MODIFICATION REACTION OF PROTEIN

During the early phase of investigation,[23] protein methylation reaction was considered to occur subsequent to polypeptide formation and was one of the *post*translational reactions of protein. However, increasing evidence suggests that some of the protein methylation reactions are *co*translational.

Reporter was the first to demonstrate that enzymatic methylation of histidine residues in primary muscle cell cultures occurred at the level of polysomes.[24] When [*methyl*-^{14}C] or [*methyl*-^{3}H]methionine-derived labeling was studied during the early period after the introduction of the precursors in the primary culture of rat leg muscle cells, the methylation of polysome-bound, presumably nascent, proteins was unaffected by addition of cycloheximide to the culture medium. The antibiotic, however, inhibited incorporation of methionine, and as a consequence increased the ratios of the incorporated methylated, to methionine residues and the ratio of ribosome-bound to free radioactivity. The methylated, polysome-bound proteins were decreased when puromycin was added to the culture medium. These results led the author to propose that selective methylation of nascent proteins could begin at the level of polysomes.

A similar conclusion that protein methylation occurs on nascent polypeptide chain was also obtained by Chen and Liss.[25] In the *in vitro* protein synthesizing system (wheat germ) with chicken oviduct mRNA, the methylation of the translation product by bovine pituitary protein methylase II (EC 2.1.1.24) was inhibited by the presence of puromycin, indicating the requirement for *de novo* protein synthesis. Ultracentrifugal analysis showed that carboxymethylated proteins were associated with ribosomal absorption peaks.

Recently, we have investigated this problem with our own experimental system. The heme-free apocytochrome *c* served as a much better substrate for *S*-adenosyl-L-methionine:cytochrome *c*-lysine *N*-methyltransferase (EC 2.1.1.59) than holocytochrome *c* [low K_m (0.019 m*M* vs. 1.35 m*M*) and high V_{max} (0.86 vs. 0.58 nmol/min/mg enzyme protein)].[26] However, when bound to mitochondria in low ionic conditions, both the apo- and holo- forms of cytochrome *c* lost their substrate capability. This capability was restored when the proteins were released from the mitochondria by KCl treatment. Cycloheximide also inhibited both protein backbone synthesis and methylation. These results suggested that cytochrome *c* is methylated before heme-attachement and binding to the mitochondria at a stage concomitant with or very shortly after peptide backbone synthesis. This suggestion has been further supported by pulse-chase studies showing methylation to be tightly coupled with protein synthesis.[27]

In extended studies on cytochrome *c* methylation, we observed that only 0.2% of "native" horse heart cytochrome *c* could maximally be methylated *in vitro* by *N. crassa* cytochrome *c*-lysine *N*-methyltransferase.[21] When the "native" cytochrome *c* was denatured with ethanol treatment, the extent of maximum methylation increased to 3.6%. In contrast, however, at least 17% of newly *in vitro* synthesized apocytochrome *c* could be trimethylated by the simultaneous presence of cytochrome *c*-lysine *N*-methyltransferase in the *in vitro* translation system of apocytochrome *c* mRNA together with *S*-adenosyl-L-[*methyl*-^{3}H]methionine.[28] Considering the extremely minute amount of newly synthesized apocytochrome *c* as a substrate for the methylating enzyme, this extent of methylation is quite remarkable. These results not only demonstrate that the extent of methylation of protein is greatly dependent on the tertiary structure of the substrate protein, but also strongly suggest that enzymatic methylation of apocytochrome *c in vivo* occurs concomitant with its synthesis. Once the polypeptide chain synthesis is completed and the synthesized protein takes up a final tertiary structure, the extent of methylation of the protein appears to be greatly diminished.

One obvious exception to the proposed *co*translational nature of protein methylation is the enzymatic methylation of histone in higher organisms. It has been well established that histones are synthesized on cytoplasmic polysomes and subsequently transported into the nuclei,[29] and that mammalian histone-specific protein methylase III (EC 2.1.1.43) is chromatin bound.[6,7,30] This difference in the site of histone biosynthesis and the subcelluar location of the methylating enzyme constituted strong background for the contention that protein methylation is a *post*translational modification reaction of protein.

IV. CONCLUSION

Looking back 3 decades since the first report on the presence of methylated amino acid residues in flagellar protein, there have been a few important "milestones" in the path of protein methylation research such as the discoveries of ε-*N*-methyllysines in histones, N^G-methylarginine residue in MBP, and membrane protein-carboxymethylation during bacterial chemotaxis.

In the early 1960s, histones had been postulated to be a modulator for genetic expression, and the finding of the methylation of histone molecules in 1964 generated a great deal of excitement, since addition or removal of the methyl groups from histones could offer a "fine" tuning mechanism to switch "on and off" of the genetic function. However, soon thereafter, this hypothesis was fatally dampened by the observation that the histone methyl groups were metabolically quite stable.

Despite this early disappointment, slow but steady progress in protein methylation research has been achieved. As illustrated in this series of monographs, much information has been gained through many diverse experimental model systems. During this progress, we realized that some of the earlier ideas have had to be altered, and, therefore, we chose two topics worthy of new consideration. We strongly envisage that most of the proteins, if not all, should have their own methyltransferase and confidently predict a tremendous expansion of the enzymology in protein methylation. Considering the vast array of methylated proteins, the horizon is wide open.

REFERENCES

1. **Ambler, R.P. and Rees, M.W.,** ε-*N*-Methyl-lysine in bacterial flagellar protein, *Nature (London)*, 184, 56, 1959.
2. **Paik, W.K. and Kim, S.,** Protein methylase I, *J. Biol. Chem.*, 243, 2108, 1968.
3. **Paik, W.K. and Kim, S.,** Protein methylation, in *Biochemistry: A Series of Monographs*, Vol. 1, John Wiley & Sons, New York, 1980, 5.
4. **Paik, W.K. and Kim, S.,** Protein methylation, in *The Enzymology of Post-Translational Modification of Proteins*, Vol. 2, Freedman, R.B. and Hawkins, H.C., Eds., Academic Press, London, 1985, 187.
5. **Tuck, M.T., Farooqui, J.Z., and Paik, W.K.,** Two histone H1-specific protein-lysine *N*-methyltransferase from *Euglena gracilis*. Purification and characterization, *J. Biol. Chem.*, 260, 7114, 1985.
6. **Sarnow, P., Rasched, I., and Knippers, R.,** A histone H4-specific methyltransferase. Properties, specificity and effects on nucleosomal histones, *Biochim. Biophys. Acta*, 655, 349, 1981.
7. **Duerre, J.A. and Onisk, D.V.,** Specificity of the histone-lysine methyltransferases from rat brain chromatin, *Fed. Proc.*, 41 (Abstr. 6962), 1461, 1982.
8. **Valentine, J. and Pettigrew, G.W.,** A cytochrome *c* methyltransferase from *Crithidia oncopelti*, *Biochem. J.*, 201, 329, 1982.
9. **Ghosh, S.K., Paik, W.K., and Kim, S.,** Purification and molecular identification of two protein methylase I from calf brain: myelin basic protein- and histone-specific enzyme, *J. Biol. Chem.*, 263, 19024, 1988.
10. **Farooqui, J.Z., Tuck, M., and Paik, W.K.,** Purification and characterization of enzymes from *Euglena gracilis* that methylate methionine and arginine residues of cytochrome *c*, *J. Biol. Chem.*, 260, 537, 1985.
11. **Reporter, M.,** 3-Methylhistidine metabolism in proteins from cultured mammalian muscle cells, *Biochemistry*, 8, 3489, 1969.
12. **Vijayasarathy, C. and Narasinga Rao, B.S.,** Partial purification and characterization of *S*-adenosylmethionine:protein-histidine *N*-methyltransferase from rabbit skeletal muscle, *Biochim. Biophys. Acta*, 923, 156, 1987.
13. **Frost, B.F., Park, K.S., Desi, S., Kim, S., and Paik, W.K.,** Site-specificity of histone H1 methylation by two H1-specific protein-lysine *N*-methyltransferases from *Euglena gracilis*, *Int. J. Biochem.*, in press.
14. **Burdon, R.H. and Garven, E.V.,** Enzymic modification of chromsomal macromolecules. II. The formation of histone ε-*N*-trimethyl-L-lysine by a soluble chromatin methylase, *Biochim. Biophys. Acta*, 232, 371, 1971.
15. **DeLange, R.J., Glazer, A.N., and Smith, E.L.,** Presence and location of an unusual amino acid, ε-*N*-trimethyllysine in cytochrome *c* of wheat germ and *Neurospora crassa*, *J. Biol. Chem.*, 244, 1385, 1969.

16. **DeLange, R.J., Glazer, A.N., and Smith, E.L.,** Identification and location of an unusual amino acid, ε-*N*-trimethyllysine in yeast cytochrome *c*, *J. Biol. Chem.*, 245, 3325, 1970.
17. **Hill, G.C., Chan, S.K., and Smith, L.,** Purification and properties of cytochrome c_{555} from a protozoan, *Crithidia fasciculata*, *Biochim. Biophys. Acta*, 253, 78, 1971.
18. **Huszar, G.,** Amino acid sequence around the two ε-*N*-trimethyllysine residues in rabbit skeletal muscle myosin, *J. Biol. Chem.*, 247, 4057, 1972.
19. **Reporter, M. and Rees, D.W.,** Methylation of bovine rhodopsin, *Nature (London)*, 239, 201, 1972.
20. **Patthy, L., Smith, E.L., and Johnson, J.,** Histone III. V. The amino acid sequence of pea embryo histone III, *J. Biol. Chem.*, 248, 6834, 1973.
21. **Durban, E., Nochumson, S., Kim, S., Paik, W.K., and Chan, S.-K.,** Cytochrome *c*-specific protein-lysine methyltranferase from *Neurospora crassa*, *J. Biol. Chem.*, 253, 1427, 1978.
22. **Lee, H.W., Kim, S., and Paik, W.K.,** *S*-Adenosylmethionine:protein-arginine methyltransferase. Purification and mechanism of the enzyme, *Biochemistry*, 16, 78, 1977.
23. **Paik, W.K. and Kim, S.,** Protein methylation, in *Biochemistry: A Series of Monographs*, Vol. 1, John Wiley & Sons, New York, 1980, 92.
24. **Reporter, M.,** Protein synthesis in cultured muscle cells: methylation of nascent proteins, *Arch. Biochem. Biophys.*, 158, 577, 1973.
25. **Chen, J.-K. and Liss, M.,** Evidence of the carboxymethylation of nascent peptide chains on ribosomes, *Biochem. Biophys. Res. Commun.*, 84, 261, 1978.
26. **DiMaria, P., Polastro, E., DeLange, R.J., Kim, S., and Paik, W.K.,** Studies on cytochrome *c* methylation in yeast, *J. Biol. Chem.*, 254, 4645, 1979.
27. **Farooqui, J., Kim, S., and Paik, W.K.,** *In vivo* studies on yeast cytochrome *c* methylation in relation to protein synthesis, *J. Biol. Chem.*, 255, 4468, 1980.
28. **Park, K.S., Frost, B., Tuck, M., Ho, L.L., Kim, S., and Paik, W.K.,** Enzymatic methylation of *in vitro* synthesized apocytochrome *c* enhances its transport into mitochondria, *J. Biol. Chem.*, 262, 14702, 1987.
29. **Borun, T.W., Schariff, M., and Robbins, E.,** Rapidly labeled polyribosome-associated RNA having properties of histone messenger, *Proc. Natl. Acad. Sci. U.S.A.*, 58, 1977, 1967.
30. **Paik, W.K. and Kim, S.,** Solubilization and partial purification of protein methylase III from calf thymus nuclei, *J. Biol. Chem.*, 245, 6010, 1970.

Chapter 3

CALMODULIN AND PROTEIN METHYLATION

Frank L. Siegel, Pamela L. Vincent, Tom L. Neal, Lynda S. Wright, Alice A. Heth, and Paul M. Rowe

TABLE OF CONTENTS

I. CALMODULIN STRUCTURE AND FUNCTIONS

A. BIOLOGICAL ACTIVITIES OF CALMODULIN

A primary challenge facing workers in the areas of protein *N*-methylation and protein carboxylmethylation is one of determining the functional significance of these posttranslational modifications. Calmodulin is a protein which serves as an excellent substrate for both *N*-methylation and carboxylmethylation and is ideally suited for structure-function studies by virtue of its multiple biological activities. Calmodulin would thus appear to be a good candidate for attempts to learn more about what methylation really means in terms of protein function. Calmodulin is a highly conserved calcium-binding protein with the unique ability to activate a wide range of enzymes and physiological processes;[1] it is found in all eukaryotic organisms, although the amount of this protein varies greatly from tissue to tissue and from species to species. In the rat, for example, brain, testis and adrenal medulla have the highest calmodulin levels; kidney and spleen are intermediate, liver, heart, and skeletal muscle have low levels.[2] The richest known source of calmodulin is electroplax from the electric eel *Electrophorus electricus*;[3] in this tissue, 30% of the mRNA has been reported to be calmodulin message.[4] The high levels of calmodulin in electric tissue, brain and adrenal medulla may reflect key roles of calmodulin in neural function; evidence has been obtained for calmodulin stimulation of neurotransmitter biosynthesis[5] and release[6] and also for stimulation of neurotransmitter activated postsynaptic events, such as cyclic AMP formation.[7]

Most known actions of calmodulin involve the stimulation of phosphotransferase and phosphohydrolase enzymes; these are listed in Table 1. In most instances, calmodulin is thought to bind calcium and then calcium-activated calmodulin binds to target enzymes, producing enzyme activation as a result of conformational changes. An exception to this mechanism is phosphorylase kinase, which has calmodulin as one of four subunits comprising the enzyme structure. In addition to underwriting the activation of these calcium-dependent enzymes, calmodulin appears to be required in a variety of physiological processes where the molecular mechanism involved in the action of calmodulin is not clearly understood. These calmodulin-activated events are listed in Table 2.

B. CHEMICAL PROPERTIES AND STRUCTURE OF CALMODULIN

The discovery of multiple biological activities of calmodulin spurred an interest in structure-function relationships in this protein as well as an interest in mechanisms which may regulate its activities. Research in this area has been aided by the relative ease of isolating large amounts of calmodulin from animal tissues and by the availability of several inhibitors of calmodulin activity, including trifluoperazine, calmidazolium, compound 48/80, and the naphthalenesulfonamides W-7 and W-9.

Calmodulin is a protein of 16,800 Da, with an isoelectric point of 4.0, reflecting its high content of glutamate and aspartate. Calmodulin does not contain tryptophan or cysteine and has a high phenylalanine to tyrosine ratio, which is responsible for its characteristic "buried phenylalanine"-type UV spectrum. The amino acid sequence of calmodulin was determined by Vanaman and Watterson,[28] the gene cloned and sequenced by Means,[29] and the crystal structure recently determined by Cook and co-workers.[30] The structure is characterized by long central helix and paired, dumbbell-like calcium-binding domains at either end of the molecule. One calcium is bound in each of the four calcium-binding domains and the K_D of binding in each domain is about 1 mm. Most species of calmodulin contain a single residue of trimethyllysine in position 115, but both plant and animal species without methylated lysine have been identified, as seen in Table 3. An exception to the rule that the only *N*-methylated residue of calmodulin is a single trimethyllysine at residue 115 was found in *Paramecium,* which has, in addition to trimethyllysine-115, dimethyllysine at residue 13.[49]

TABLE 1
Calmodulin Activated Enzymes

Enzyme	Ref.
Cyclic nucleotide phosphodiesterase	8
Adenylate cyclase	7
Guanylate cyclase	9, 10
Ca^{2+}, Mg^{2+}-ATPase	11,12
Calmodulin-dependent protein kinases	13
Phosphoprotein phosphatase (calcineurin)	14
Phosphofructokinase	15
Phospholipase A_2	16
Myosin light chain kinase	17
Phosphorylase kinase	18
NAD kinase	19
Calmodulin-stimulated protein methyltranferases	20
Chitin synthetase	21

TABLE 2
Calmodulin Activated Physiological Processes

Process	Ref.
Release of hormones and neurotransmitters from secretory vesicles	22
Cell cycle progression from G_1 to S phase (DNA synthesis)	23
DNA repair	24
Regulation of gene transcription (prolactin)	25
Microtubule disaggregation	26
Ciliary motility (*Paramecia*)	27

C. REGULATION OF CALMODULIN ACTIVITY

Since the intracellular calcium concentration in unstimulated animal cells is less than 0.1 μ*M*, calmodulin probably has no bound calcium in resting cells. Upon stimulation by a variey of agents, depending upon cell type, intracellular calcium concentrations may reach 10 μ*M* and calcium will bind to calmodulin, changing its solution structure to a more helical, activated conformation, facilitating the binding of calmodulin to target enzymes and other calmodulin-binding proteins. Thus, the availability of calcium is a major factor regulating the biological activity of calmodulin. Our laboratory has been testing the hypothesis that posttranslational modification of calmodulin is another mechanism for governing the activity of this important protein. Table 4 indicates the known posttranslational modifications affecting calmodulin structure. This review will consider both the *N*-methylation and carboxylmethylation of calmodulin and their possible functional significance.

II. *N*-METHYLATION OF CALMODULIN

The presence of trimethyllysine in calmodulin but not in troponin C, a calcium-binding protein in muscle which shares considerable amino acid sequence homology with calmodulin, has given rise to speculation that *N*-methylation may contribute to the biological activities of calmodulin. This suggestion was first made by Klee and Vanaman,[54] who wrote:

TABLE 3
Methylated Lysines in Calmodulin

Species	Residues TML[a]	Ref.
Dictyostelium discoideum	0	31
Neurospora crasa	0	32
Saccharomyces cerevisiae	0.5	33
Mushroom	0	34
Spinach	1	35
Citrus sinensis	1	36
Peanut seeds	1	37
Zucchini hypocotyls	1	38
Barley	0	39
Renilla reniformis	1	40
Sea anemone	1	41
Ascarus suum	1	42
Scallop	1	43
Octopus (whole animal)	0.1	44
Octopus (optic tube)	0.6	45
Crayfish	0	46
Bovine brain and testis	1	47
Bovine skeletal muscle	0.7	48

[a] TML, trimethyllysine.

The methylation of calmodulin at lysine 115 may be a prerequisite for one or another of calmodulin's functions. The availabiltiy of calmodulin synthesized *in vivo* that is not methylated, as well as that of a methylase in brain extracts that catalyzes the methylation of this residue in calmodulin, will permit elucidation of the role of this specific and highly conserved modification of the calmodulin molecule.

Although calmodulin posseses limited troponin C-like functional properties, most of the Ca^{2+}-dependent activities of these two proteins are unique. The differences in their structures must specify the ability to troponin C to make proper functional contacts with the other troponin subunits, while they must provide calmodulin with the specific domains needed for the reversible Ca^{2+}-dependent interactions with a host of enzymes and proteins.

The enzyme which is responsible for *N*-methylation of calmodulin is a specific methylase III class enzyme according to the Paik and Kim nomenclature. In this review, we will refer to this enzyme, *S*-adenosylmethionine:calmodulin (lysine) *N*-methyltransferase (EC 2.1.1.60), as calmodulin *N*-methyltransferase. The substrate specificity of calmodulin *N*-methyltransferase is reflected by the lack of amino acid sequence homology between the region of the *N*-methylation site of calmodulin and other *N*-methylated proteins, as shown in Figure 1.

Several questions prompted our investigation of calmodulin *N*-methylation:

1. Does *N*-methylation effect the ability of calmodulin to activate calcium-dependent enzymes, and are these effects consistent from enzyme to enzyme?
2. Is the *N*-methylation of calmodulin enzymatically reversible?
3. Is calmodulin *N*-methyltransferase activity subject to physiological regulation, and if so, do tissue levels of calmodulin or other factors control the activity of this enzyme?
4. In transformed cells and other rapidly proliferating tissues where calmodulin levels are elevated, does calmodulin *N*-methyltransferase activity increase in parallel with calmodulin?
5. Do any mammalian tissues contain significant amounts of nonmethylated calmodulin?

To address these questions, the following eight-part strategy was adopted:

TABLE 4
Posttranslational Modifications of Calmodulin

Modification	Ref.
N-acetylation	28
N-methylation	50
Carboxylmethylation	51
Proteolytic processing by pituitary carboxypeptidase B	52
Phosphorylation	53

Cytochrome C (yeast)	Y	L	E	N	P	TML	K	Y	I	P	G
EL-1 (shrimp)											
Site 1	C	G	S	I	D	TML	R	T	I	E	K
Site 2	D	I	A	L	W	TML	E	E	T	A	K
Myosin	N	I	N	P	Y	TML	R	L	D	I	Y
ATP Carrier Protein	A	E	K	Q	Y	TML	G	I	I	D	C
Calmodulin	T	N	L	G	E	TML	L	T	D	E	E

FIGURE 1. Amino acid sequences surrounding *N*-methylation sites of calmodulin and four other proteins containing trimethyllysine (TML) residues.

1. Develop a sensitive and specific assay for calmodulin *N*-methyltransferase.
2. Using this assay, determine the tissue with highest enzyme activity.
3. Purify calmodulin *N*-methyltransferase from this tissue.
4. Using purified mammalian calmodulin *N*-methyltransferase, methylate des(methyl)calmodulin.
5. Determine the amino acid residue(s) methylated *in vitro* by purified calmodulin *N*-methyltransferase as well as the stoichiometry of this methylation.
6. Compare the biological activities of methylated, sham-methylated, and des(methyl)calmodulins with the same amino acid sequences.
7. Determine the substrate specificity of calmodulin *N*-methyltransferase.
8. If this enzyme can be shown to have strict substrate specificity for calmodulin, use calmodulin *N*-methyltransferase as a probe for des(methyl)calmodulin substrate in mammalian tissues.

A. DEMONSTRATION OF CALMODULIN *N*-METHYLTRANSFERASE ACTIVITY IN BRAIN

The *in vitro* *N*-methylation of calmodulin was first shown by Sitaramayya et al.[50] In this study, rat brain cytosol was incubated with [*methyl*-^{3}H]-*S*-adenosyl-L-methionine and the methylated peptides were resolved on SDS-PAGE and visualized with fluorography. The pH of this electrophoresis system was 8.6; at this pH, eukaryotic carboxylmethyl esters spontaneously hydrolyze and the surviving methylated peptides are presumed to represent *N*-methylation.[55] At an incubation pH of 7.2, peptides of M_r 17,000, 21,000, 29,000, and 34,000 were visible on fluorograms, with the M_r 17,000 peptide methylation much stronger than that of the other peptides. If EGTA was included in the incubation buffer, methylation of the M_r 17,000 peptide did not take place, indicating a calcium dependence of this reaction. If EGTA was added after incubation but before electrophoresis, the M_r 17,000 band showed the shift to an apparent higher molecular weight that characterizes calmodulin.

The *N*-methylation and carboxylmethylation of added calmodulin substrate by a partially purified brain cytosolic methyltransferase fraction was also measured by TCA precipitation of methylated proteins from incubation mixtures. Methylation blanks contained all components except calmodulin and duplicate methylation mixtures were incubated with sodium borate (pH 11) to hydrolyze carboxylmethyl groups, allowing the selective determination of stable *N*-methyl groups. Carboxylmethylation was calculated as the difference between total methylation and stable *N*-methylation. This experiment demonstrated that calmodulin can serve as a substrate both for protein carboxylmethyltransferase (PCMT) and for an *N*-methyltransferase. The pH optimum for the *N*-methylation of calmodulin was found to be 8.0, while carboxylmethylation of calmodulin increased with decreasing pH, plateauing at pH of about 6. Hydrolysis of the putative [*methyl*-^{3}H]-*N*-methylated calmodulin followed by amino acid analysis and fraction collection of [methyl-^{3}H]-amino acids indicated that the trimethyllysine was the only methylated residue. N-methylation of calmodulin decreased with increasing amounts of added calmodulin substrate, indicating that the substrate, beef testis calmodulin, being greater than 95% *N*-methylated at lysine 115, inhibits the reaction by product inhibition. The calmodulin *N*-methylation which was observed in this study is presumed to be a small amount of newly synthesized and as yet unmethylated calmodulin. Having identified *N*-methylating activity, the next step was to develop an assay for this enzyme.

B. DEVELOPMENT OF AN ASSAY FOR CALMODULIN *N*-METHYLTRANSFERASE

1. Principle

The *in vitro* methylation of calmodulin can be assayed by three different methods. In each case, des(methyl)calmodulin is incubated with a source of methyltransferase and [*methyl*-^{3}H]-*S*-adenosyl-L-methionine in a suitable buffer at pH 8. The three methods differ by the means employed to determine the incorporation of the radiolabelled methyl group into calmodulin. In each of the three assays, blanks contain all components except des(methyl)calmodulin.

In the first method, the proteins in the incubation mixture are resolved by SDS-polyacrylamide gel electrophoresis and the gel is stained with Coomassie Blue. The calmodulin band, identified by its electrophoretic mobility, is cut out of the gel with a razor blade, digested in 0.75 ml of 30% hydrogen peroxide at 80°C for 3 h in a closed glass scintillator vial, then scintillation fluid is added and the radioactivity of methylated calmodulin is determined. The corresponding value from an incubation mixture lacking des(methyl)calmodulin is subtracted to correct for endogenous substrate and the results are expressed as pmol^{-1} mg protein^{-1}. This method is not readily adaptable to the assay of large numbers of samples and suffers from the inherent uncertainties in cutting stained bands from gels and incomplete extraction of radioactivity from gel bands in the digestion process.

The second method is based upon trichloroacetic acid (TCA) precipitation of [*methyl*-^{3}H]-calmodulin from incubation mixtures.[56] TCA precipitates are dissolved in borate buffer at pH 11 and incubated at this pH to hydrolyze [^{3}H]-carboxylmethyl esters of calmodulin. Calmodulin is then reprecipitated with TCA and dissolved in formic acid for liquid scintillation determination of radioactivity. Incubation mixtures contained, in a final volume of 100 μl, 0.1 *M* glycylglycine, pH 8.0, 0.1 *M* NaCl, 2 m*M* $MgCl_2$, 10 μ*M* [*methyl*-^{3}H]-*S*-adenosyl-L-methionine (1 μCi) and 1 μg des(methyl)calmodulin. Assay tubes are incubated for 10 min at

37°C and the reactions are stopped by placing tubes in an ice bath and adding 0.5 ml of a 1 mg/ml solution of bovine serum albumin and 3 ml of 7.5% trichloroacetic acid. Tubes are agitated on a vortex mixer and centrifuged at 3000 × *g* for 15 min to pellet precipitated proteins. The supernatants are discarded and the protein pellets are dissolved in 200 µl of saturated sodium borate, pH 11, with shaking for 5 min in a 37°C water bath. Proteins are reprecipitated by the addition of 3 ml of 7.5% trichloroacetic acid and pelleted by centrifugation. The solubilization in sodium borate and reprecipitation steps are repeated two additional times and the final protein pellets are dissolved in 250 µl 88% formic acid, 750 µl of water is added and the solutions are transferred to liquid scintillation vials for the determination of radioactivity. One unit of calmodulin *N*-methyltransferase activity is the amount of enzyme that transfers 1 pmol of methyl groups per minute and specific activity is expressed as units per milligram of protein.

The third method employs calcium-dependent hydrophobic interaction chromatography to isolate [*methyl*-^{3}H]-calmodulin, followed by heating to destroy carboxylmethyl esters.[57] This procedure, like the acid precipitation method, can be used for large numbers of samples, but has several advantages when compared to the other two methods. Blanks are considerably lower in the column method and the precision is significantly greater. The specificity of the procedure also ensures that only calmodulin methylation is determined. In addition, the column assay requires fewer manipulations of samples; 100 samples may be assayed in almost the same time as 20. The capacity of the acid precipitation assay, in contrast, is limited to the capacity of the centrifuge employed.

2. Substrate Purification

Preliminary experiments indicated that beef testis or brain calmodulin had poor methyl acceptor activity at low substrate concentrations and inhibited the reaction as the calmodulin concentration was increased. Des(methyl)calmodulin from the slime mold *Dictyostelium discoideum*[31] is an excellent substrate, but we have found it far more convenient to isolate the nonmethylated calmodulin from fresh mushrooms.[34]

Fresh mushrooms (1200 g) obtained from local markets were homogenized in three volumes of ice-cold 5% trichloroacetic acid (TCA) in a Waring blender at low speed, with three 30 s bursts. The homogenate was stirred for 30 min in a cold room, centrifuged at 20,000 × *g* for 30 min and the supernatant discarded. The TCA pellet was taken up in 1500 ml Buffer A (50 *M*) Tris-HCl, pH 7.4, containing 1 m*M* phenylmethylsulfonyl fluoride (PMSF) and the pH adjusted to 7.4 with 1 *N* NaOH. This suspension was stirred for 15 min in the cold and centrifuged at 20,000 × *g* for 30 min. The pellet was suspended in a minimum volume of Buffer A and the extraction and centrifugation steps were repeated. The second pellet was discarded and the two supernatant fractions are pooled and stirred for 10 min with 500 ml DEAE-cellulose which had been washed previously with Buffer B (50 m*M* Tris-HCl, pH 7.4, 0.1 m*M* EGTA, 1 m*M* PMSF). The cellulose slurry was collected on a Buchner funnel by vacuum filtration and washed with 2 liters of Buffer B, then with 1 liter of 50 m*M* Tris-HCl, pH 7.4, 150 m*M* NaCl and 1 m*M* PMSF. Des(methyl)calmodulin was eluted with two 300-ml washes of 50 m*M* Tris-HCl, pH 7.4, 0.1 m*M* EGTA, 0.5 *M* NaCl, and 1 m*M* PMSF. The pooled eluates were adjusted to 2 m*M* Ca^{2+} by the addition of 1 *M* $CaCl_2$ and then filtered through a 0.45 µm Millipore filter. This des(methyl) calmodulin-containing solution was pumped onto an 18 ml column of phenyl-Sepharose which had previously been equilibrated with 50 m*M* Tris-HCl, pH 7.4, 0.5 m*M* $CaCl_2$, 0.5 *M* NaCl, and 0.1 m*M* PMSF. The column effluent was monitored at 280 nm and the column was washed with equilibration buffer until a stable baseline was reached. Des(methyl)calmodulin was then eluted with 50 m*M* Tris-HCl, pH 7.4, 2 m*M* EGTA, and the des(methyl)calmodulin containing fractions were pooled and dialyzed overnight against 0.5 m*M* $CaCl_2$, centrifuged 15 min at 100,000 × *g* and concentrated on an Amicon Diaflow apparatus with a YM-50 filter. Aliquots of this solution were lyophillized and stored at –80°C.

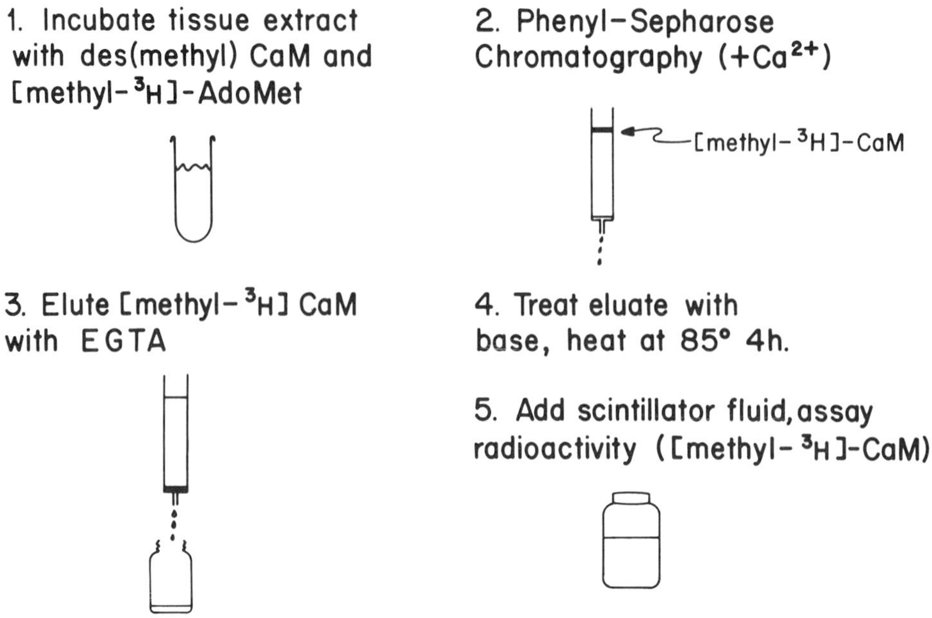

1. Incubate tissue extract with des(methyl) CaM and [methyl-³H]-AdoMet
2. Phenyl-Sepharose Chromatography (+Ca²⁺)
[methyl-³H]-CaM
3. Elute [methyl-³H] CaM with EGTA
4. Treat eluate with base, heat at 85° 4h.
5. Add scintillator fluid, assay radioactivity ([methyl-³H]-CaM)

FIGURE 2. Radiometric assay for calmodulin *N*-methyltransferase.

3. Assay

The chromatographic assay for calmodulin *N*-methyltransferase is illustrated in Figure 2 and detailed below.

a. Reagents

- Homogenization buffer: 25 m*M* sodium-Hepes, pH 7.4, 0.25 *M* sucrose.
- Termination buffer: 50 m*M* Tris-HCl, pH 8.0, 0.3 *M* NaCl, 0.5 m*M* $CaCl_2$, 20 µg/ml bovine calmodulin and 0.02% sodium azide.
- Binding buffer: 50 m*M* Tris-HCl, pH 8.0, 0.3 *M* NaCl, 0.1 m*M* $CaCl_2$, and 0.02% sodium azide.
- Elution buffer: 0.1 *M* NH_4HCO_3, pH 8.0, 2 m*M* EDTA, and 0.02% sodium azide.
- Incubation buffer: 0.1 *M* glyclyglycine, pH 8.0, 0.1 *M* NaCl, 2 m*M* $MgCl_2$, 10 µ*M* [*methyl*-^{3}H]-*S*-adenosyl-L-methionine (0.5 µCi/nmol), tissue cytosolic protein and 8 µg mushroom des(methyl)calmodulin. The final assay volume is 100 µl.

b. Procedure

Rat tissues were homogenized in 4 volumes of cold homogenizing buffer, using a Polytron homogenizer. Cytosolic fractions, prepared by centrifuging the homogenates for 60 min at 100,000 × *g*, were stored at –80°C; calmodulin *N*-methyltransferase is stable for 1 month at this temperature. Assay tubes were incubated for 10 min at 37°C, termination buffer was added to each tube and the tube contents transferred to polyethylene columns (8 × 200 mm; Kontes), each containing 0.75 ml of phenyl-Sepharose which had been previously washed two times with 4 ml of binding buffer. The columns were washed sequentially with 5 ml of binding buffer, 3 ml of binding buffer, two times with 3 ml of 0.1 m*M* $CaCl_2$ containing 0.02% sodium azide, and finally two times with 1.5 ml elution buffer. The last two washes were collected together in vials containing 0.02% sodium dodecyl sulfate and the uncapped vials were heated for 4 h at 85°C to hydrolyze protein carboxylmethyl esters and to volatilize the resulting [^{3}H]-methanol. The

TABLE 5
Comparison of CLNMT Activities in Rat Tissue Cytosolic Fractions as Measured by Three Methods[59]

[^{3}H]methyl incorporation (pmol min^{-1}mg protein^{-1})

	Blank	With added *Dictyostelium* CaM	CLNMT activity
Brain			
Acid precipitation	0.66 ± 0.065	3.48 ± 0.12	2.82 ± 0.14
Phenyl-Sepharose columns	0.14 ± 0.03	3.34 ± 0.08	3.21 ± 0.09
Gel electrophoresis	0.03	2.87	2.84
Liver			
Acid precipitation	0.63 ± 0.03	0.83 ± 0.05	0.20 ± 0.06
Phenyl-Sepharose columns	0.08 ± 0.02	0.30 ± 0.01	0.22 ± 0.03
Gel electrophoresis	0.025	0.219	0.193
Skeletal muscle			
Acid precipitation	0.153 ± 0.015	0.190 ± 0.003	0.037 ± 0.016
Phenyl-Sepharose columns	0.049 ± 0.004	0.089 ± 0.003	0.040 ± 0.006
Gel electrophoresis	0.023	0.061	0.038

vials were allowed to cool, scintillation fluid was added, and radioactivity determined. Blanks contained all components of the assay except the des(methyl)calmodulin substrate. Data are expressed as picomoles methyl incorporated per minute per milligram of protein. Phenyl-Sepharose columns were regenerated by washing with 12 volumes of 8 *M* urea and 12 volumes of 0.02% sodium azide and stored in the sodium azide solution.

The phenyl-Sepharose assay is linear with time for 20 min and with enzyme concentration over a range of 25 to 200 μg of brain cytosolic protein and up to 800 μg of liver cytosolic protein. Using plexiglass frames drilled to accept the Kontes plastic columns, 100 assays can be completed within 90 min. Table 5 summarizes the results obtained with this assay compared with those from gel electrophoresis and TCA precipitation assays, both of which are more cumbersome and less precise.

C. PURIFICATION OF CALMODULIN *N*-METHYLTRANSFERASE

Rowe et al.[58] have reported the partial purification of calmodulin *N*-methyltransferase (EC 2.1.1.60) from testis, the rat tissue with highest enzyme activity. Rat testis (57 g) was homogenized in 5 volumes of 25 m*M* Hepes-NaOH, pH 7.4, 0.25 *M* sucrose, using a Polytron homogenizer and centrifuged for 1 h at 100,000 × *g*. The resulting supernatant fraction was brought to 35% saturation with solid ammonium sulfate, stirred an additional 45 min and centrifuged for 30 min at 15,000 × *g*. This supernatant was brought to 70% saturation with ammonium sulfate and centrifuged to pellet the enzyme. The pellet was suspended in 30 ml of homogenizing buffer without sucrose and dialyzed overnight against three changes of this buffer. The dialyzed preparation was brought to 0.01% Triton X-100, an additional 1 m*M* dithiothreitol was added, and the solution was clarified by centrifugation for 30 min at 15,000 × *g*. This clarified solution was applied to a column of DEAE-cellulose (2.6 × 16 cm; Whatman DE-53) which had previously been equilibrated with 25 m*M* Hepes-NaOH, pH 7.4, 2 m*M* dithiothreitol, and 0.01% Triton X-100. The absorbance of the column effluent was monitored at 280 nm and the column throughput which contained the calmodulin *N*-methyltransferase was

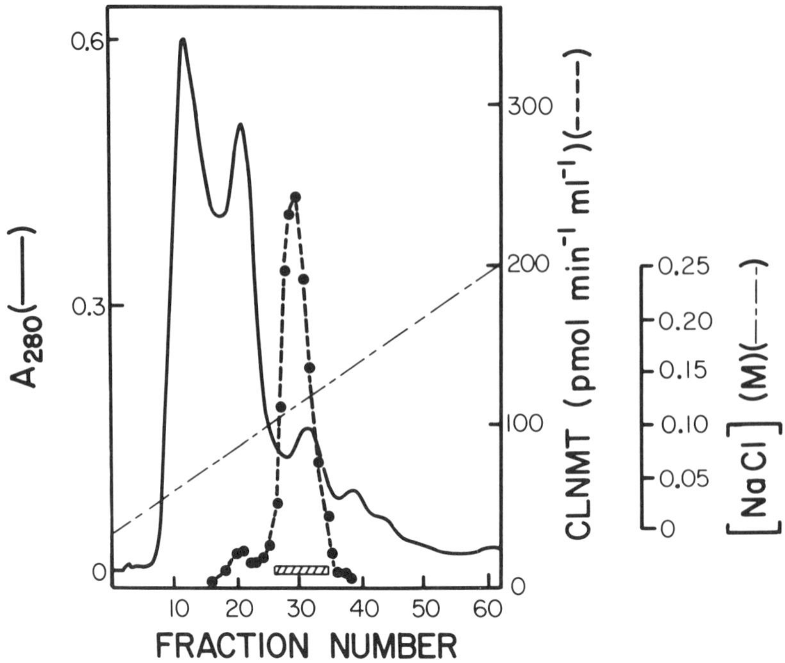

FIGURE 3. CM-Sepharose chromatography of calmodulin *N*-methyltransferase. The wash-through protein from DEAE-celluose chromatography was applied to a column of CM-sepharose CL-6B and eluted with a NaCl gradient. (From Rowe, P. M., Wright, L. S., and Siegel, F., *J. Biol. Chem.*, 261, 7060, 1986. With permission.)

collected. This active fraction was applied directly to a column of CM-Sepharose CL-6B (1.5 × 13 cm) which had been equilibrated with 25 m*M* Hepes-NaOH, pH 7.4, 2 m*M* dithiothreitol, and 0.01% Triton X-100. The absorbance of the column effluent was monitored at 280 nm and the column was washed with the equilibration buffer until the absorbance returned to baseline. Calmodulin *N*-methyltransferase was eluted with a linear gradient to 0.25 *M* NaCl. The elution profile (Figure 3) indicates one major peak of activity as well as a minor component which was not studied further. Fractions containing the major peak of enzyme activity were pooled, concentrated by ultrafiltration (Amicon PM-10 membrane), and applied to a column of Sephadex G-100 (1.5 × 90 cm) previously equilibrated with 25 m*M* Hepes-NAOH, pH 7.4, 0.15 *M* NaCl, 2 m*M* dithiothreitol, and 0.01% Triton X-100. The column was eluted at a flow rate of 5 ml/h, active fractions were pooled, made 10 m*M* in dithiothreitol, concentrated by ultrafiltration and stored at –80°C.

The key to this procedure is the inclusion of Triton X-100 and dithiothreitol in all buffers to stabilize the enzyme. In early experiments, purification of only 20-fold was associated with a loss of 95% of the enzyme activity in the course of the procedure. Among a number of compounds tested for their ability to stabilize this enzyme, only bovine serum albumin, Triton X-100, Lubrol PX, and polyethylene glycol were effective; these agents all cause slight enzyme activation, Of these agents, Triton X-100 proved to be most effective, at an optimal concentration of 0.01%; this corresponds to a concentration of 0.15 m*M*, or about half of the critical micelle concentration. In crude cytosolic fractions of rat testis, this concentration of Triton X-100 afforded a 30% stimulation of enzyme activity and also stabilized the enzyme. Sulfhydryl reagents also have marked effects on enzyme activity. A 50-fold purified enzyme was found to be stimulated 50% by 5 m*M* dithiothreitol in the presence of Triton X-100; 2-mercaptoethanol is less effective. As a result of these findings, we include 2 m*M* dithiothreitol in all chromatography buffers and store the purified enzyme in 10 m*M* dithiothreitol. Enzyme assay buffers contain a final concentration of 0.01% Triton X-100 and 5 m*M* dithiothreitol.

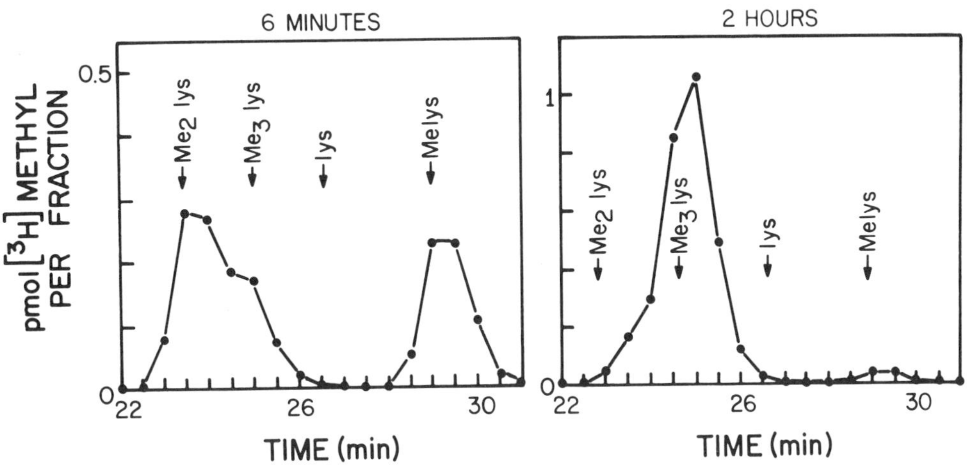

FIGURE 4. Methylated amino acids in calmodulin. *Dictyostelium* calmodulin was incubated with calmodulin *N*-methyltransferase and [*methyl*-^{3}H]-*S*-adenosyl-L-methionine; reactions were stopped in either 6 min or 2 h and calmodulin was reisolated by phenyl-Sepharose chromatography. Calmodulin was then hydrolyzed in 5.6 *N* HCl for amino acid analysis with fractions collected for the determination of radioactivity. Only the lysine region of the chromatogram is shown; no radioactivity eluted elsewhere. (From Rowe, P. M., Wright, L. S., and Siegel, F., *J. Biol. Chem.*, 261, 7060, 1986. With permission.)

Morino and co-workers have added an additional calmodulin-Sepharose affinity chromatography step to this procedure and have purified calmodulin *N*-methyltransferase 7800-fold to apparent homogeneity.[59] They reported K_m values of 2.2×10^{-8} *M* for calmodulin and 0.8×10^{-6} *M* for *S*-adenosyl-L-methionine and a molecular weight of 57,000 Da for the enzyme, as estimated by gel filtration. This molecular weight estimate agrees closely with the 55,000 Da reported by Rowe et al.

D. METHYLATION OF DES(METHYL)CALMODULIN WITH CALMODULIN *N*-METHYLTRANSFERASE

Des(methyl)calmodulin, isolated from *Dictyostelium discoideum* was methylated to completion by incubation with *S*-adenosyl-L-methionine and 470-fold purified beef testis calmodulin *N*-methyltransferase for 4 h at 37°C in a glycylglycine buffer at pH 8.0. The buffer contained dithiothreitol, Triton X-100, and $MgCl_2$, to afford maximal stabilization of the enzyme. In experiments designed to indicate the methylation site on calmodulin, [*methyl*-^{3}H] *S*-adenosyl-L-methionine was used as the tracer. Methylated calmodulin was reisolated from incubation mixtures by calcium-dependent chromatography on phenyl-Sepharose, with EGTA elution. The radiochemical purity of [*methyl*-^{3}H] calmodulin was demonstrated by reverse-phase HPLC using acetonitrile gradient elution from a C3 300 Å pore size column. In studies of the time course of methylation, methylated calmodulins formed after 6 min and after 2 h incubation were hydrolyzed and their content of trimethyllysines was determined. At the short incubation time, almost equal amounts of mono-, di-, and trimethyllysines were formed, whereas after 2 h, 1 mol of trimethyllysine per calmodulin was the only methylated amino acid (Figure 4). This result is significantly different than was found by Durban et al.[60] in their studies of the *N*-methylation of cytochrome *c*. In the latter case, methylation stopped after the addition of only a small fraction of the substrate, suggesting that the methyltransferase was inactivated during the incubation.[65]

Calmodulin methylated *in vitro* with [*methyl*-^{3}H] *S*-adenosyl-L-methionine was digested with trypsin at an enzyme substrate ratio of 1:100 overnight at 37°C and the resulting tryptic peptides were separated by reverse-phase HPLC. Fractions were taken for the determination of

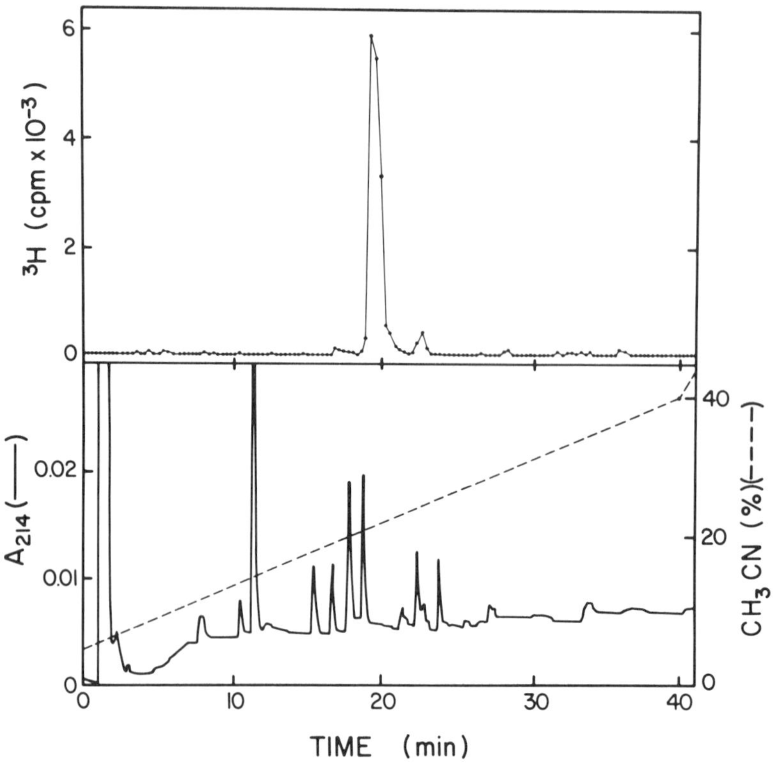

FIGURE 5. HPLC of [*methyl*-^{3}H] tryptic peptides of calmodulin. *Dictyostelium* calmodulin was methylated *in vitro* with calmodulin *N*-methyltransferase and *S*-adenosyl-L-methionine, reisolated by reverse-phase HPLC and digested with trypsin (1:100) overnight at 37°C. Tryptic peptides were resolved by reverse-phase HPLC using a linear gradient of acetonitrile. The upper trace (cpm/fraction) has not been corrected for lag time between the UV detector and fraction collector relative to the lower trace. (From Rowe, P. M., Wright, L. S., and Siegel, F., *J. Biol. Chem.*, 261, 7060, 1986. With permission.)

radioactivity, and the resulting chromatograms indicate only a single methylated peptide, as shown in Figure 5. Amino acid analysis of this radioactive peptide indicated that it corresponded to residues 107 to 126 of calmodulin. This peptide contains a single trimethyllysine residue and no other lysines, indicating that lysine 115 was the methylation site.

E. EFFECT OF METHYLATION ON THE BIOLOGICAL ACTIVITIES OF CALMODULIN

The first suggestion that methylation might affect calmodulin function came from Chau and co-workers,[61] who reported that des(methyl)calmodulin is a substrate for ubiquitination and subsequent ATP-dependent proteolysis, whereas *N*-methylated calmodulin is not degraded in this pathway. This interesting finding suggests that the biological half-life of methylated calmodulin may be greater than that of des(methyl)calmodulin, and that *N*-methylation provides a mechanism for conserving calmodulin.

Another suggestion as to effects of *N*-methylation on the activity of calmodulin came from studies which indicated that nonmethylated plant calmodulins have a greater ability to activate plant NAD kinase than do methylated (animal) calmodulins.[19] It was not clear from this study if the differences between the methylated and nonmethylated calmodulins were due solely to *N*-methylation, or if they were attributable to differences in amino acid sequence between des(methyl)calmodulin and calmodulin, which has been isolated from different species. This

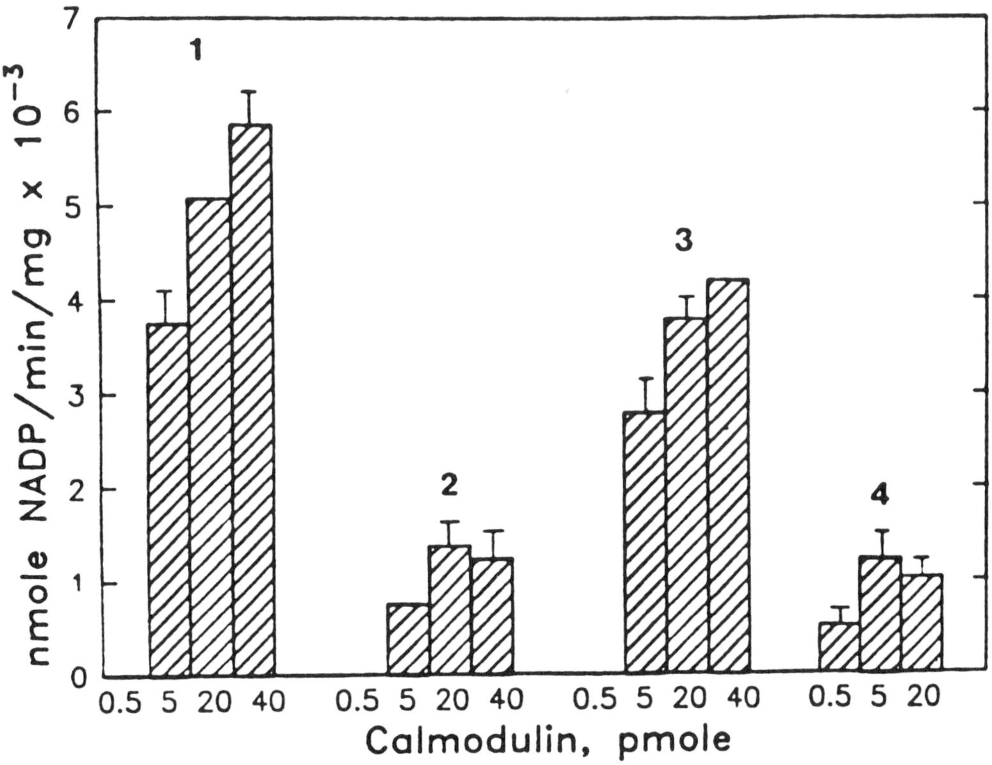

FIGURE 6. NAD-kinase activation by VU-1, methylated VU-1, and VU-1 R115 calmodulins. The VU-1 R115 calmodulin is the product of a mutant gene which codes for a calmodulin in which lysine 115 is replaced by arginine. 1. VU-1 (nonmethylated) calmodulin; 2. VU-1 calmodulin methylated *in vitro;* 3. VU-1 R115 calmodulin, and 4. spinach (methylated) calmodulin. (From Roberts, D. M., Rowe, P. M., Siegel, F. L., Lukas, T. J., and Watterson, D. M., *J. Biol. Chem.*, 261, 1491, 1986. With permission.)

question was resolved when Rowe et al. compared the ability of methylated, sham-methylated, and nonmethylated calmodulins to activate cyclic nucleotide phosphodiesterase[58] and NAD kinase.[62] In these experiments, des(methyl)calmodulins isolated from *Dictyostelium* and *Chlamydomonas* or made by expression of the VU-1 synthetic calmodulin gene VU-1 in *E. coli* were methylated by incubation with calmodulin *N*-methyltransferase and *S*-adenosyl-L-methionine; sham incubation mixtures lacked *S*-adenosyl-L-methionine. These experiments showed that methylated and nonmethylated calmodulins have an equal ability to activate calmodulin-stimulated cyclic nucleotide phosphodiesterase prepared from beef brain, but that methylation of calmodulin impairs the ability of this protein to activate pea seedling NAD kinase (Figure 6). Molla had previously shown that des(methyl)calmodulin from octopus activates myosin light chain kinase, indicating that the trimethyllysine is not required, but no quantitative comparisons of methylated and nonmethylated calmodulins were made by these workers.[44]

Carafoli has shown that the carboxyl terminal large tryptic fragment of calmodulin activates erythrocyte Ca^{2+}, Mg^{2+}-ATPase (albeit less efficiently than does intact calmodulin), whereas this calmodulin peptide does not activate cyclic nucleotide phosphodiesterase.[63] This finding indicates that different regions of calmodulin participate in the activation of different enzymes and suggests that posttranslational modifications of specific regions of calmodulin may selectively affect the abiltiy of calmodulin to activate specific enzymes and cellular processes. The differential effects of *N*-methylation on activation of cyclic nucleotide phosphodiesterase and NAD kinase provide one more example of this concept of specialized regions of the calmodulin structure.

F. DEMONSTRATION OF DES(METHYL)CALMODULIN IN MAMMALIAN TISSUES

1. Enzyme Specificity

The purification of calmodulin *N*-methyltransferase made it possible to use this enzyme as a probe to ask the question "is the substrate for this enzyme present in any mammalian tissues?" If the enzyme can be shown to methylate only calmodulin, then the inference can be made that any substrate detected is des(methyl)calmodulin. The prevailing view is that calmodulin is either not *N*-methylated or it contains a single trimethyllysine (*Paramecium* being the exception to this generalization). Perry has reported that skeletal muscle calmodulin contains 0.7 mol of trimethyllysine,[48] while Seamon and Moore found that octopus optic lobe calmodulin contains 0.6 mol of trimethyllysine,[45] and calmodulin from whole octopus contains 0.1 mol each of the three methylated lysine species.[44] These earlier results suggest the possible presence of undermethylated calmodulin or a mixture of trimethyl and des(methyl)calmodulins.

Two approaches to the question of substrate specificity of calmodulin *N*-methyltransferase were made. In the first, the ability of calmodulin *N*-methyltransferase to methylate purified substrates was determined. There was no activity toward protamine, cytochrome *c*, or histone, which are known to be methylated by other *N*-methyltransferases, or toward bovine serum albumin, even at concentrations of up to 1 mg/ml. Troponin C had low methyl acceptor activity; less than 20% that of calmodulin. [64] Additional data came from Morino, who reported that purified calmodulin *N*-methyltransferase had either very low or undetectable activity toward myosin light chains, bovine brain S-100 protein or calf thymus histones f_1, f_{2a}, f_{2b}, and f_3.[59]

The substrate specificity of calmodulin *N*-methyltransferase was also determined by adding partially purified enzyme to cytosolic extracts of rat tissues and incubating at pH 8 with [*methyl*-^{3}H]-*S*-adenosyl-L-methionine under conditions at which *N*-methylation is optimal. Four incubations are run for each tissue; one tube contained neither added enzyme nor added calmodulin, a second tube contained added enzyme, a third tube contained added calmodulin and the fourth tube contained both calmodulin and calmodulin *N*-methyltransferase. Following incubation, methylated proteins were resolved by SDS-polyacrylamide gel electrophoresis and visualized with fluorography. The results of surveying five rat tissues are shown in Figure 7. It can be seen that addition of calmodulin *N*-methyltransferase increases only the methylation of endogenous calmodulin and has no effect on the methylation of any other protein, confirming the strict substrate specificity of this *N*-methyltransferase. Demonstration of this specificity made it possible to use calmodulin *N*-methyltransferase as a probe for des(methyl)calmodulin in tissue extracts.

2. Substrate Assay

Calmodulin *N*-methyltransferase was added to cytosolic extracts of rat tissues using optimal methylating conditions, which included [*methyl*-^{3}H]-*S*-adenosyl-L-methionine, and methylated calmodulins were isolated by calcium-dependent chromatography on phenyl-Sepharose. SDS-polyacrylamide gel electrophoresis of EGTA eluates from phenyl-Sepharose columns confirmed the identity of the methylation product as calmodulin. Calmodulin *N*-methyltransferase substrate content of tissues was assayed as the ability of proteins in a boiled tissue extract containing 1 μg of calmodulin to accept [^{3}H]-methyl groups when incubated with this enzyme. Boiled tissue extracts were incubated with 0.1 *M* glycylglycine-NaOH, pH 8, 120 m*M* NaCl, 2 m*M* $MgCl_2$, 5 m*M* dithiothreitol, 1.6 m*M* EGTA, 0.4% Lubrol PX, 2 m*M* $CaCl_2$, 5 μCi [*methyl*-^{3}H]-*S*-adenosyl-L-methionine (5 Ci/mmol) with or without 0.5 pmol min^{-1} of purified calmodulin *N*-methyltransferase (purified through the CM-Sepharose step) for 5 min at 37°C in a total volume of 110 μl. Reactions were stopped by the addition of a terminator buffer containing 50 *mM* Tris-HCl, pH 8, 0.3 *M* NaCl, 0.5 m*M* $CaCl_2$, 20 μg/ml bovine testis calmodulin, 0.02% sodium azide, and 1 m*M* *S*-adenosylhomocysteine. Blanks contained tissue extracts but not the methyltransferase. Substrate levels were expressed as a percentage of the value obtained with 1 μg of des(methyl)calmodulin.

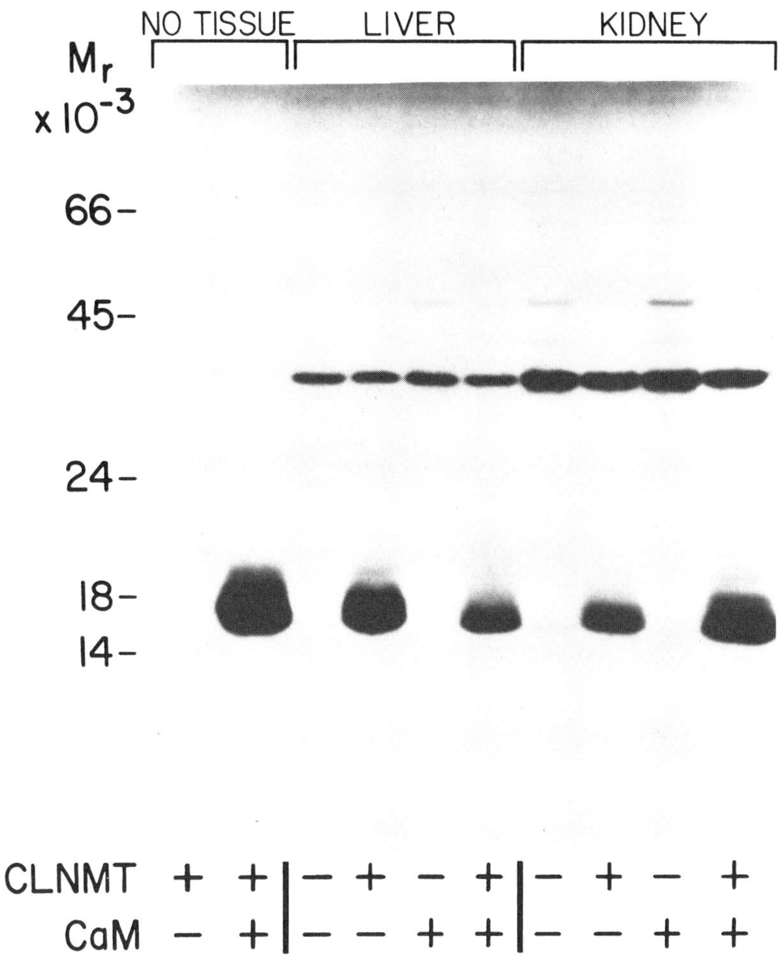

FIGURE 7. Electrophoretic assay for substrate specificity of calmodulin *N*-methyltransferase (CLNMT). Tissue cytosolic fractions were incubated with or without calmodulin *N*-methyltransferase and bovine testis calmodulin prior to electrophoresis and fluorography. (From Rowe, P. M., Wright, L. S., and Siegel, F., *J. Biol. Chem.*, 261, 7060, 1986. With permission.)

No detectable substrate methyl accepting activity was found in cytosolic extracts of rat testis, brain, spleen, or kidney, all tissues with high levels of both calmodulin and calmodulin *N*-methyltransferase. In tissues with low levels of calmodulin *N*-methyltransferase, significant levels of substrate, presumed to be des(methyl)calmodulin, were found. Liver, heart, and skeletal muscle were in this latter group (Table 6). Since *N*-methylation of calmodulin has been shown to affect at least one of its many biological activities, the finding of variable amounts of nonmethylated calmodulins in animal tissues is of interest as a possible control point in calmodulin function. This finding would assume a greater potential importance if calmodulin *N*-methyltransferase activity is also subject to physiological regulation.

G. REGULATION OF CALMODULIN *N*-METHYLTRANSFERASE

Preliminary experiments indicate that calmodulin *N*-methyltransferase activity is elevated in rapidly proliferating tissues (Wright and Siegel, manuscript in preparation). In fetal and neonatal brain and liver, enzyme activities are three times higher than in the corresponding adult tissues. Enzyme activities were also found to be higher in regenerating liver and in rapidly

TABLE 6
Tissue Distribution of *S*-Adenosylmethionine-Calmodulin(lysine) *N*-methyltransferase (CLNMT), CNLMT Substrate, and Calmodulin[58]

	CLMNT activity		Total CaM	CLNMT/CaM	CLNMT substrate
	pmol min^{-1}mg protein^{-1}	pmol^{-1} min^{-1} g^{-1}, wet weight	μg g^{-1}, wet weight	pmol min^{-1} μg	% relative to *Dictyostelium stelium* CaM
Testis	4.75 ± 0.21	137 ± 6.0	420	0.33	<3%
Brain	4.54 ± 0.26	107 ± 6.1	360	0.30	<3%
Spleen	2.30 ± 0.06	152 ± 4.0	67	2.3	<3%
Kidney	1.86 ± 0.19	107 ± 11.0	100	1.1	<3%
Liver	0.22 ± 0.02	17 ± 1.5	84	0.20	12%
Heart	0.15 ± 0.05	6.8 ± 2.3	41	0.17	21%
Skeletal muscle	0.034 ± 0.002	1.7 ± 0.1	38	0.045	68%

Note: Samples for CLNMT and calmodulin were prepared from pooled tissues from 4 male rats (300—330 g). CLNMT values represent average of triplicates ± SD. Samples for CLNMT substrate determinations were prepared from pooled tissues from four male rats (280—320 g). Accuracy of calmodulin determinations was checked by adding purified calmodulin to tissue extracts; recovery of added calmodulin was 102 ± 33% (mean ± SD of ten determinations).

growing hepatomas, compared to host liver. One explanation for these findings is a possible greater turnover of calmodulin in rapidly dividing cells, with a greater need for the methylation of newly synthesized calmodulin.

III. CARBOXYLMETHYLATION OF CALMODULIN

A. INTRODUCTION

Carboxylmethylation of calmodulin was first shown by Gagnon et al.;[51] these workers incubated bovine brain calmodulin with a partially purified protein carboxylmethyltransferase and *S*-adenosyl-L-[*methyl*-^{3}H]-methionine at pH 6.0 and measured the incorporation of methyl groups into calmodulin. The methyl acceptor properties of calmodulin were compared with those of a number of other proteins and calmodulin was shown to be the most active methyl acceptor tested. It was also shown that calmodulin is carboxylmethylated cultured Walker ascites carcinoma cells. Incubation of calmodulin with *S*-adenosyl-L-methionine and protein carboxylmethyl transferase reduced its ability to activate cyclic nucleotide phosphodiesterase by about 50%.[51] The carboxylmethylation of calmodulin differs from its *N*-methylation in three major respects. First, protein carboxylmethyltransferase has very low substrate specificity and there appears to be only one enzyme which methylates all substrates. Aswad resolved two protein carboxylmethyltransferase activities on DEAE-cellulose, but no differences in substrate specificity between these activities were found. By contrast, there appears to be a family of protein *N*-methyltransferases, each with a high degree of substrate specificity. Second, all eukaryotic carboxylmethyl groups are quite unstable, especially at pH values greater than 6, while *N*-methyl groups are chemically stable. Third, whereas *N*-methylation is stoichiometric, all reported eukaryotic protein carboxylmethylations are substoichiometric. This may be due to the nature of the methylated amino acids; glutamate in prokaryotic organisms and aspartate in eukaryotes.

B. CARBOXYLMETHYLATION OF CALMODULIN IN CULTURED CELLS

In a recent study in the author's laboratory, cultured pituitary GH_3 cells were incubated with [*methyl*-^{3}H] methionine and carboxylmethylated proteins were resolved by sequential FPLC gel filtration, FPLC Mono Q anion exchange chromatography, and reverse-phase HPLC.[65] These experiments were designed to determine if calmodulin is methylated *in vivo* and if calmodulin is a major substrate for protein carboxylmethylation. Two HPLC analyses were run, with elution buffer containing EGTA and with buffer containing calcium. The retention of calmodulin on C3 reverse phase columns was shown to be reduced by the presence of calcium, which caused calmodulin to elute 5 min earlier than elution with EGTA-containing elution buffer. This "EGTA shift" facilitated the identification of calmodulin as one of two major methyl acceptor proteins.

1. Cell Culture

Pituitary GH_3 cells were maintained at 37°C in Ham's F10 culture medium supplemented with 15% horse serum and 2.5% fetal bovine serum in a humidified atmosphere of 5% CO_2, 95% air. For labeling experiments, cells were plated at a density of 2×10^5 cells/35 mm dish or 5×10^5 cells/100 mm dish and grown as monolayer cultures. Culture medium (2 ml for 35 mm dishes, 5 ml for 100 mm dishes) was changed every 3 to 4 d. Experiments utilized replicate dishes inoculated with an equal number of cells from a single donor culture. Cells were grown for 7 to 8 d after plating before experiments were performed. The cells were in the early stationary phase of growth; each 35 mm dish contained between 0.2 and 0.5 mg protein and each 100 mm dish between 0.8 and 1.0 mg protein.

2. [*Methyl*-^{3}H]-L-Methionine Labeling of GH_3 Cells

Replicate dishes of cells were plated and grown as described above. Prior to labeling, cells were rinsed two times with prewarmed (37°C) Hepes buffered balanced salts (HBBS) containing 118 m*M* NaCl, 4.6 m*M* KCl, 0.5 m*M* $CaCl_2$, 1.0 m*M* $MgCl_2$, 10 m*M* glucose, 5 m*M* Hepes, pH adjusted to 7.4 with 1 *M* NaOH. All incubations were carried out in the HBBS solution at 37°C using 1.0 to 1.5 ml HBBS/35 mm dish or 3 ml HBBS/100 mm dish.

Aliquots of [*methyl*-^{3}H]-L-methionine, which is supplied as a 70% ethanol solution, were put into test tubes and evaporated to dryness in a Speed-Vac (Savant Instruments) at room temperature, then dissolved in prewarmed HBBS to a final concentration of 50 μCi/ml; this solution was added to culture dishes. The isotope-containing medium was removed and replaced by fresh, warm HBBS after 1 to 1.5 h for all protein fractionation experiments. Cell viability was always greater than 93%, as determined by trypan blue exclusion. [^{3}H]-methyl incorporation was terminated by aspirating the incubation medium, rinsing the dishes in ice cold HBBS and rapid freezing on a block of dry ice. Frozen cells were thawed in appropriate buffers as described for individual experiments; if they were not to be processed immediately, they were stored at –80°C until just prior to use. Cells were scraped off the dishes with a rubber policeman after being lysed by refreezing and thawing in buffer. The cell suspension was homogenized with 5 manual passes of a Teflon-glass homogenizer, and centrifuged at $100{,}000 \times g$ for 60 min. The $100{,}000 \times g$ supernatant was either immediately used for chromatography, assayed for label incorporation or quick-frozen and stored at –80°C.

3. *In Vitro* Protein Carboxylmethylation in GH_3 Cytosol

GH_3 cells were grown using standard conditions for maintenance of the culture. Cells from one T-25 tissue culture flask were washed three times with ice cold HBBS, a high-speed supernatant was prepared by freezing and thawing in buffer (10 ml of 10 m*M* Tris, 0.32 *M* sucrose, pH 7.4), followed by homogenization and centrifugation as described above. *In vitro* methylation assays were done in 383 μl volumes which contained 0.13 *M* bisTris, pH 6.5, 2 μCi [*methyl*-^{3}H]-*S*-adenosyl-L-methionine (15 Ci/mmol) and 89 μg supernatant protein.

Incubations, at 37°C for 30 min, were terminated by rapid freezing on dry ice. Samples were stored frozen at –80°C until fractionated.

4. FPLC-Superose Chromatography

Samples were prepared as described for the Mono Q fractionation of GH_3 methyl acceptor proteins labeled in culture (see Section III.B.5). Samples (500 µl) containing 0.6 to 0.7 mg protein were chromatographed on a Superose 12 HR 10/30 FPLC gel filtration column (Pharmacia) which had been preequilibrated with 50 m*M* L-histidine, 5 *M* urea, pH 5.3. This sample size was greater than the recommended 200 µl maximum sample size; some resolution was sacrificed for the advantage of rapid separation of labile carboxylmethyl esters in GH_3 supernatant components. The column was eluted at a flow rate of 0.1 ml/min for a total volume of 25 ml; 0.3 ml fractions were collected. Fractions containing [^{3}H]-labeled protein methyl esters were detected by the extraction method, using 0.1 ml aliquots of the column fraction; the remaining 0.2 ml fractions were quickly frozen and stored at –80°C. The total time of each Superose 12 gel filtration run was 250 min.

5. FPLC Mono Q Chromatography

Dishes (100 mm) of GH_3 cells were grown and labeled as described. Cells were homogenized in 750 µl of 50 m*M* L-histidine, 5 *M* urea, pH 5.4, at 4°C and centrifuged 60 min at 100,000 × *g* to prepare a cytosolic fraction. Samples were kept cold (4°C) and at pH 5.5 or below to minimize spontaneous decarboxylmethylation, which increases significantly above pH 5.5 (a 25% decrease in decarboxylmethylation was found when the pH was lowered from 5.5 to 5.0). At pH values below 5.5, the decreased solubility of calmodulin becomes a problem, although the addition of 5 *M* urea helps to keep calmodulin in solution at this pH. An aliquot (400 µl, approximately 0.5 mg protein) of this supernatant was brought to a volume of 500 µl with Buffer A (20 m*M* L-histidine, 6 *M* urea, pH 5.5). The sample was fractionated on a Mono Q HR 5/5 anion exchange column at 4°C. The column was pre-equilibrated with buffer A and proteins were eluted with a linear gradient of sodium chloride from 0 to 0.8 *M* in buffer A at a flow rate of 1.0 ml/min. One-milliliter fractions were collected and assayed for protein carboxylmethylesters by a modification of the microdistillation method of Stock and Koshland,[66] or by extraction into organic solvents.

For distillation, aliquots of 0.5 ml were transferred to minivials and 200 µl of saturated sodium borate, pH 11, was added. The distillation vials were tightly capped with serum bottle stoppers. For samples which were to be fractionated by HPLC, a 0.25-ml aliquot was assayed by the organic extraction method and the remaining 0.75 ml was used for HPLC. Each stopper held a Pyrex glass tube (2 mm) and a tuberculin syringe filled with 1 ml of methanol. Attached to the glass tubing was a 9-cm length of tygon tubing leading to a second minivial, the collection vial, which contained 2.5 ml of ice-cold Triton-toluene PPO. The collection vials were kept in an ice bath. Distillation vials were incubated in an 80°C waterbath for 30 min, a time sufficient to allow complete hydrolysis of protein methyl esters. Methanol (0.5 ml) was then injected into each distillation vial and was allowed to distill; after 30 min the second 0.5 ml of methanol was added and distillation continued for another 30 min. Distilled radioactivity was measured by liquid scintillation counting after the addition of another 2.5 ml of scintillation cocktail. The efficiency of this method, as determined by recovery of [^{14}C]-methanol, is greater than 90%, compared to an efficiency of only about 60% using the organic extraction assay. Fractionation of methyl acceptor protein labeled *in vitro* was done in essentially the same way except that the sample consisted of the 383 µl reaction mixture which was brought to 6 *M* urea and a volume of 1 ml by the addition of 600 µl of 10 *M* urea and 17 µl of water. Some samples were fractionated on Mono Q after an inital FPLC gel filtration on Superose 12, in which case the samples were first adjusted to the Mono Q starting buffer conditions. The total time of Mono Q chromatography was 42 min.

6. HPLC

Fractions corresponding to radioactive peaks from the ion exchange or gel filtration chromatography steps were pooled. Since the samples were in elution buffers containing either 5 or 6 *M* urea, they were diluted by the addition of water to a urea concentration of 3 *M* or less. The pooled peaks were then fractionated at room temperature by reverse-phase high pressure liquid chromatography on a Beckman ultrapore C3/300 Å column (0.4 × 7.5 cm) on a Waters HPLC system. Two different HPLC elution buffers were used, one containing EGTA and the other containing calcium. The EGTA buffer consisted of 95% 10 m*M* potassium phosphate, 2 m*M* EGTA, pH 6.1, containing 5% acetonitrile as the starting (A) buffer and 25% 10 m*M* potassium phosphate, 2 m*M* EGTA, pH 6.1 containing 75% acetonitrile as the B buffer. The calcium buffer was identical except that it contained 0.2 m*M* calcium phosphate in place of EGTA;[58] calmodulin shows a characteristic shift in elution volume depending on the presence or absence of calcium or EGTA in the buffer. The column was eluted by a linear gradient of 5 to 37.5% acetonitrile at a flow rate of 1.0 ml/min; 1.0 ml fractions were collected and assayed for radioactivity by either distillation or extraction of [^{3}H]-methanol. The total time for each HPLC run was 50 min.

7. Identification of Carboxylmethylated Proteins

Mono Q anion exchange chromatography of extracts of cells which had been incubated with [*methyl*-^{3}H]-methionine indicated significant differences in elution profiles of carboxylmethylated proteins derived from *in vivo* as compared to *in vitro* methylation. In each of these two profiles, however, the major labeled peak coeluted with calmodulin. Gel filtration of extracts of cells labeled *in vivo* also showed that the major carboxylmethylated fraction coelutes with calmodulin. Subsequent fractionation of the major labeled fraction from Superose 12 gel filtration on Mono Q yielded two carboxylmethylated species; calmodulin radioactivity was about half that of the major peak. Each of these fractions gave single peaks on reverse-phase HPLC (Figure 8); the identity of the smaller of the two peaks as calmodulin was confirmed by its EGTA shift on HPLC. These results indicate that despite the broad *in vitro* specificity of protein carboxylmethyl transferase, *in vivo* methylation is rather selective, with a few major methyl acceptor proteins and that native calmodulin is methylated *in vivo*.

Several important questions remain unanswered. Is the *in vivo* carboxylmethylation of calmodulin substoichiometric? What residues in calmodulin are carboxylmethylated *in vivo?* Both of these questions are difficult to address due to the inherent instability of eukaryotic protein carboxylmethyl esters. Carboxylmethyl groups of calmodulin hydrolyze rapidly at neutral pH, but with biphasic kinetics (Vincent, unpublished observations); the half-life of the first phase of this curve is 11 min. Thus, studies of methylated calmodulins measure only the surviving carboxylmethyl groups, and no reliable estimate of the methylation state of calmodulin *in situ* is yet available. Treatment of calmodulin with base was shown to make this protein a substrate for stoichiometric carboxylmethylation by protein carboxylmethyltransferase.[67] Base treatment deamidates asparagine residues and produces isoaspartate at asparagine-glycine couples in the protein structure, prompting the suggestions by Aswad[68] and Clarke[69] that calmodulin is only methylated following deamidation and that isoaspartate residues are the methylation sites. Support for this hypothesis comes from the finding that treatment of calmodulin with bis-(I,I-trifluoroacetoxy)iodobenzene, BTI), a reagent which converts asparagine or glutamine amides to the corresponding primary amines without affecting the carboxyl groups of aspartic or glutamic acids reduces the number of base-induced methylation sites by 71%.[67] According to this attractive hypothesis, the carboxylmethylation of calmodulin (and other eukaryotic proteins) is substoichiometric, because isoaspartate residues are substoichiometric. The function of carboxylmethylation is postulated to be either repair of "damaged" proteins or the targeting of these proteins to lysosomes for proteolysis. It is also uncertain whether calmodulin is carboxylmethylated in its native state or whether prior

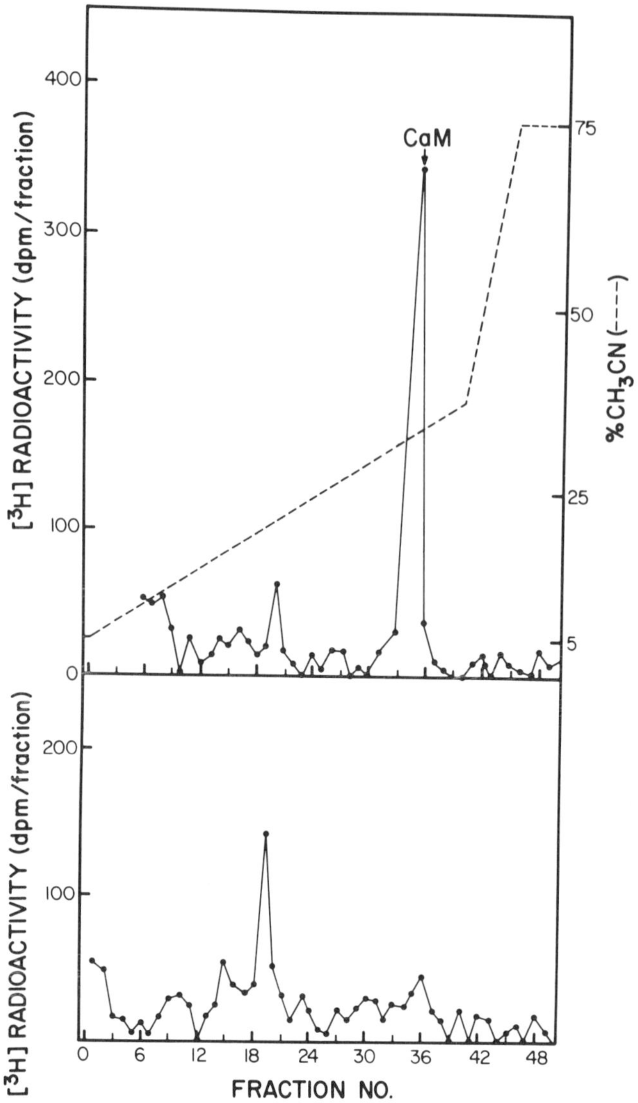

FIGURE 8. Reverse-phase HPLC of methylated proteins from Mono Q chromatography peak fraction. The upper panel is the elution profile of the Mono Q fraction which eluted with authentic calmodulin and the lower panel is the elution profile of the major peak of methylated proteins from the Mono Q chromatography. Elution was in an EGTA-containing buffer and a linear gradient of acetonitrile. Protein carboxylmethyl esters were assayed by microdistillation. (From Vincent, P. L. and Siegel, F. L., *J. Neurochem.*, 49, 1613, 1987. With permission.)

deamidation to produce isoaspartate residues is a necessary prerequisite to carboxylmethylation.

IV. CALMODULIN AS AN ACTIVATOR OF PROTEIN METHYLATION

In the course of investigating the *N*-methylation of calmodulin we have found that calmodulin stimulates the *in vitro* methylation of several proteins in cytosolic extracts of liver, kidney, lung,

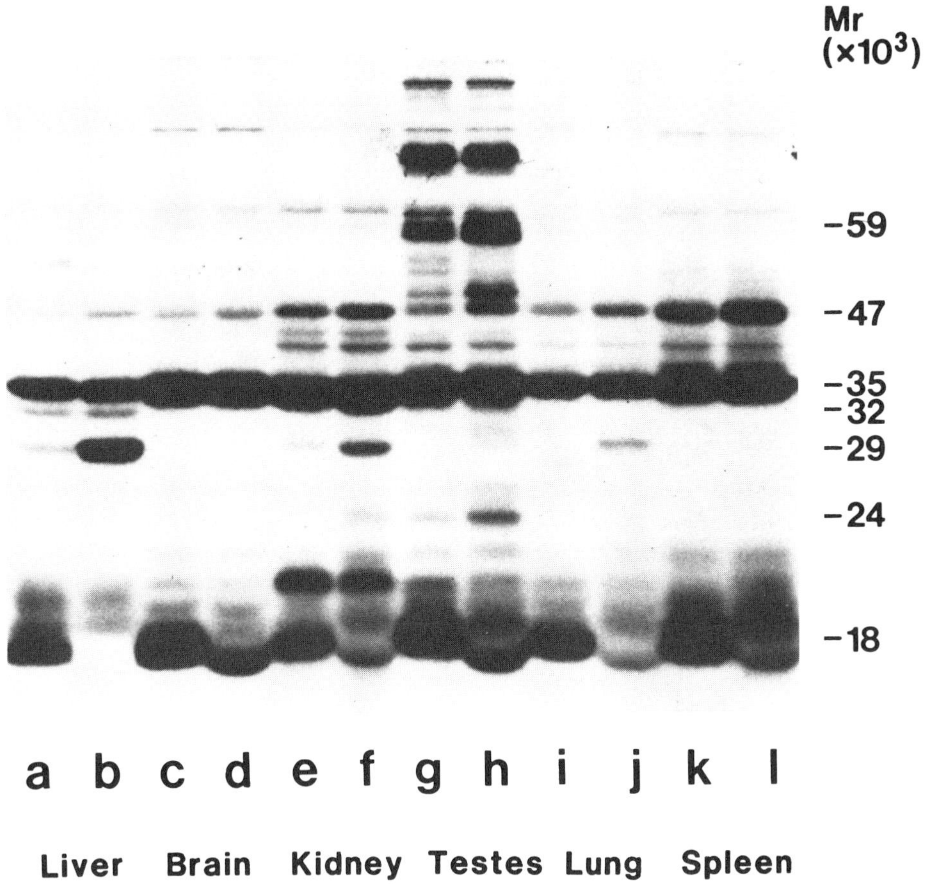

FIGURE 9. Fluorogram of SDS-polyacrylamide gel electrophoresis of protein methylation in cytosolic extracts of rat tissues. Cytosolic protein (100 μg) was incubated with [*methyl*-^{3}H]-*S*-adenosyl-L-methionine and methylated proteins were resolved by electrophoresis. Samples in alternate lanes were incubated in the presence of 20 μg of added calmodulin. (From Siegel, F. L. and Wright, L. S., *Arch. Biochem. Biophys.*, 237, 347, 1985. With permission.)

testis, and prostate (Figure 9; Reference 20). No calmodulin stimulation of protein methylation was detected in cytosol from brain or spleen. This experiment utilized endogenous methyltransferases and endogenous methyl acceptor protein substrates. The *in vitro* methylation incubation mixtures contained cytosolic protein (100 μg), 0.1 *M* Tris-HCl, pH 8.0, 1 m*M* dithiothreitol, 2 m*M* EDTA, 2.5 m*M* $MnCl_2$, and 5μCi [*methyl*-^{3}H]-*S*-adenosyl-L-methionine (80 Ci/mmol) in a final volume of 50 μl. Calmodulin was added to some tubes prior to incubation at 37°C for 60 min and some incubation mixtures were treated with sodium borate, pH 11 for 15 min to hydrolyze carboxylmethyl esters and thus assure that only *N*-methylated proteins were seen on electrophoresis. Methylation was terminated by the addition of 50 μl of double-strength electrophoresis sample buffer and methylated proteins were separated by SDS-polyacrylamide gel electrophoresis and visualized by fluorography.

Prior dialysis of liver cytosol stimulated the methylation of proteins of M_r 29,000, M_r 32,000, and M_r 47,000 even without added calmodulin. Adding calmodulin to dialyzed cytosol produced and additive stimulation of the methylation of only the M_r 29,000 protein. These findings indicate the presence of a low molecular weight (<2000 Da) inhibitor of calmodulin-stimulated protein methylation. We have focused on the calmodulin-stimulated methylation of the M_r 29,000 peptide in liver; this peptide is the major substrate for calmodulin-stimulated protein methylation.

Preliminary experiments with developing rat liver[20] and transplanted hepatoma lines indicate that calmodulin-stimulated protein methylation does not take place in rapidly proliferating liver. Under conditions where rapid cell division is correlated with marked increases in calmodulin *N*-methyltransferase activity, methylation of the M_r 29,000 methyl acceptor protein is reduced to the threshold of detection.

The M_r 29,000 substrate was purified to apparent homogeneity, as evidenced by the appearance of a single band on SDS-polyacrylamide gel electrophoresis, and the methyltransferase which methylates this protein has been partially purified. Gel overlay experiments indicate that this M_r 29,000 protein is also a calmodulin-binding protein. Preliminary amino acid sequence data from the M_r 29,000 methyl acceptor protein showed heterogeneity at several amino acid residues, suggesting that this M_r 29,000 protein is a mixture of isoforms. We have recently resolved the M_r 29,000 protein into three isoforms, using FPLC chromatofocusing followed by reverse-phase HPLC. Partial amino acid sequences have been obtained on all three forms; one of the isoforms shows 100% homology of the first 22 N-terminal residues with the Yb_1 subunit of glutathione-*S*-transferase and the other two isoforms show total homology with the first 20 residues of the Yb_2 subunit of this enzyme (Neal et al., manuscript in preparation). Glutathione *S*-transferase subunits have several structural characteristics which have recently been proposed as characterizing calmodulin-binding proteins,[72] including: (1) a strong preponderance of lysine or arginine residues in clusters, (2) a preponderance of hydrophobic residues, especially in the first (amino terminal) half of the calmodulin-binding domain, (3) a tryptophan in the first half of the domain, and (4) the presence in the second half of the domain of one or more serines or threonines.

Glutathione *S*-transferase is a major liver enzyme, constituting about 10% of total liver protein; this enzyme catalyzes the detoxification of drugs and other xenobiotics by catalyzing their conjugation with glutathione[70] and it also has steroid reductase activity.[71] There are at least five subunits of gluathione *S*-transferase in liver, which can form at least 13 homo- and heterodimers, the active isoforms of this enzyme. Present work is directed toward determining the effects of methylation and calmodulin on glutathione *S*-transferase activity. The methylation of the M_r 29,000 glutathione *S*-transferase subunits is strongly inhibited by glutathione, providing evidence that glutathione is the dialyzable inhibitor of calmodulin-stimulated protein methylation. Glutathione *S*-transferase thus has binding sites for glutathione, substrate, calmodulin, and methyltransferase. A model consistent with the data presently available is shown in Figure 10.

V. SUMMARY AND CONCLUSIONS

Calmodulin is a substrate for both *N*-methylation and carboxylmethylation, making this protein ideally suited for studies of the effects of methylation on protein structure and function. While *N*-methylation is not required for the enzyme-activating functions of calmodulin that have been studied at this writing, methylation does selectively inhibit at least one of these activities, providing evidence that different regions of calmodulin may interact with different target enzymes. It will be of interest to survey all of the known calmodulin-stimulated enzymes to determine the effects of methylation to modulate the ability of calmodulin to activate these enzymes. It will also be of interest to learn if *N*-methylation affects the binding of calcium to calmodulin or the inhibition of calmodulin activity by classical inhibitors of calmodulin function.

The finding that calmodulin stimulates the methylation of at least one protein, glutathione *S*-transferase, has interesting implications. First, we can no longer think that the biological activities of calmodulin are restricted to the activation of phosphotransferases and phosphohydrolases, as was previously thought. This raises questions concerning the possible modulation of other methylation reactions by calmodulin; particularly methylation of proteins

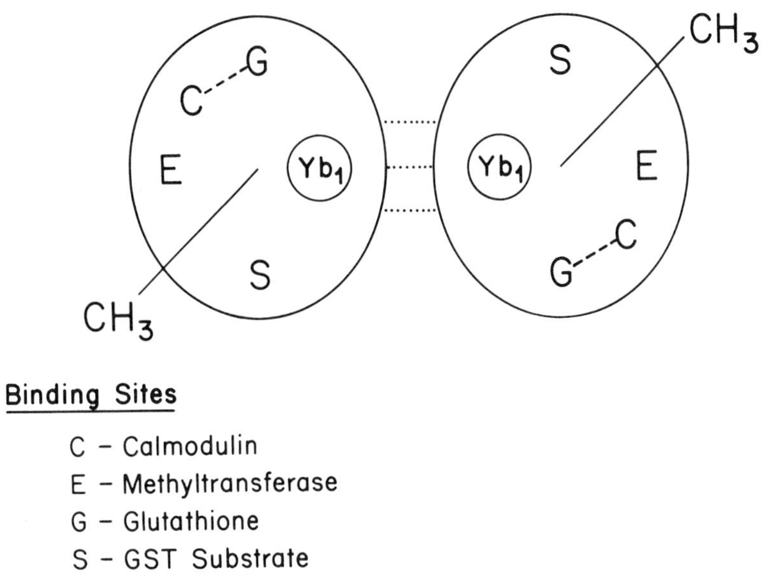

FIGURE 10. Binding sites on glutathione *S*-transferase—a model.

or nucleic acids. Are the reported effects of calmodulin on transcription, for example, due in whole or in part to effects on gene methylation? We also wonder if calmodulin will be found to be involved in other group transfer reactions, such as myristylation or acetylation. Since glutathione inhibits glutathione *S*-transferase methylation *in vitro* and since the glutathione concentration in liver is in the millimolar range, the significance of this methylation and its *in vivo* occurrence are open to speculation.

ACKNOWLEDGMENTS

Research in the authors' laboratory was supported by Public Health Service Grants NS 11652, NS24969, GM 38497, and HD 03552. Major contributions to the early stages of this research were made by Drs. Ari Sitaramayya and Timothy J. Murtaugh and expert technical assistance was provided by Kenneth Mishark, Michael Hayward, Mark Shahidi, and Rochelle Navis.

REFERENCES

1. **Cheung, W. Y.,** Calmodulin plays a key role in cellular regulation, *Science,* 207, 213, 1980.
2. **Smoake, J. A., Song, S. Y., and Cheung, W. Y.,** Cyclic 3′,5′-nucleotide phosphodiesterase. Distribution and developmental changes of the enzyme and its protein activator in mammalian tissues and cells, *Biochim. Biophys. Acta,* 341, 402, 1974.
3. **Childers, S. R. and Siegel, F. L.,** Isolation of and purification of a calcium-binding protein of *Electrophorus electricus, Biochim. Biophys. Acta,* 405, 99, 1975.
4. **Lagace, L., Chandra, T., Woo, S. L. C., and Means, A. R.,** Identification of multiple species of calmodulin messenger RNA using a full length complementary DNA, *J. Biol. Chem.,* 258, 1684, 1983.
5. **Vuillet, P. R.,** Direct activation of tyrosine hydroxylase by calmodulin, *Proc. West. Pharmacol. Soc.,* 28, 27, 1985.
6. **DeLorenzo, R. J.,** Calmodulin in neurotransmitter release and synaptic function, *Fed. Proc.,* 41, 2265, 1982.

7. **Brostrom, C. O., Huang, Y. C., Breckenridge, B. M., and Wolff, D. J.,** Identification of a calcium-binding protein as a calcium-dependent regulator of brain adenylate cyclase, *Proc. Natl. Acad. Sci. U.S.A.*, 72, 64, 1975.
8. **Lin, Y. M., Liu, Y. P., and Cheung, W. Y.,** Cyclic 3′, 5′-phosphodiesterase. Ca^{2+}-dependent formation of bovine brain enzyme-activator complex, *FEBS Lett.*, 49, 356, 1975.
9. **Kakiuchi, S., Sobue, K., Yamazaki, R., Nagao, S., Umeki, S., Nozawa, Y., Yazawa, M., and Yagi, K.,** Ca^{2+}-dependent modulator proteins from *Tetrahymena pyriformis*, sea anemone, and scallop and guanylate cyclase activation, *J. Biol. Chem.*, 256, 19, 1981.
10. **Klumpp, S., Kleefeld, G., and Schultz, J. E.,** Calcium/calmodulin-regulated guanylate cyclase of the excitable ciliary membrane from *Paramecium*. Dissociation of calmodulin by La^{3+}:calmodulin specificity and properties of the reconstituted guanylate cyclase, *J. Biol. Chem.*, 258, 12455, 1983.
11. **Jarrett, H. W. and Penniston, J. T.,** Partial purification of the Ca^{2+}-Mg^{2+} ATPase activator from human erythrocytes: its similarity to the activator of 3′, 5′-cyclic nucleotide phosphodiesterase, *Biochem. Biophys. Res. Commun.*, 77, 1210, 1977.
12. **Gopinath, R. M. and Vincenzi, F. F.,** Phosphodieterase protein activator mimics red blood cell cytoplasmic activator of Ca^{2+}-Mg^{2+} ATPase, *Biochem. Biophys. Res. Commun.*, 77, 1203, 1977.
13. **Browning, M., Huganir, R., and Greengard, P.,** Protein phosphorylation and neuronal function, *J. Neurochem.*, 45, 11, 1985.
14. **Klee, C. B., Krinks, M. H., Manalan, A. S., Cohen, P., and Stewart, A. A.,** Isolation and characterization of bovine brain calcineurin: a calmodulin-stimulated protein phosphatase, in *Methods in Enzymology*, Vol. 102, Means, A. R. and O'Malley, B. W., Eds., Academic Press, New York, 1983, 227.
15. **Mayr, G. W.,** Interaction of calmodulin with phosphofructokinase:binding studies and evaluation of enzymatic and physicochemical changes, in *Methods in Enymology*, Vol. 139, Means, A. R. and Conn, P. M., Eds., Academic Press, New York, 1987, 745.
16. **Wong, P. Y. K. and Cheung, W. Y.,** Calmodulin stimulates human platelet phospholipase A_2, *Biochem. Biophys. Res. Commun.*, 90, 473, 1979.
17. **Dabrowska, R., Sherry, J. M. F., Aromatorio, D. K., and Hartshorne, D. J.,** Modulator protein as a component of the myosin light chain kinase from chicken gizzard, *Biochemistry*, 17, 253, 1978.
18. **Cohen, P., Burchell, P., Foulkes, J. G., and Cohen, P. W. T.,** Identification of the Ca^{2+}-dependent modulator protein as the fourth subunit of phosphorylase kinase, *FEBS Lett.*, 92, 287, 1978.
19. **Roberts, D. M., Burgess, W. H., and Watterson, D. M.,** Comparison of the NAD kinase and myosin light chain kinase activator properties of vertebrate, higher plant and animal calmodulins, *Plant Physiol.*, 75, 796, 1984.
20. **Siegel, F. L. and Wright, L. S.,** Calmodulin-stimulated protein methylation in rat liver cytosol, *Arch. Biochem. Biophys.*, 237, 347, 1985.
21. **Martinez-Cadena, G. and Ruiz-Herrera, J.,** Activation of chitin synthetase from *Phycomyces blakesleeanus* by calcium and calmodulin, *Arch. Microbiol.*, 148, 280, 1987.
22. **Sletholt, K., Haug, E., Gordeladze, J., Sand, O., and Gautvik, K. M.,** Effects of calmodulin antagonists on hormone release and cyclic AMP levels in GH_3 pituitary cells, *Acta Physiol. Scand.*, 130, 333, 1987.
23. **Veigl, M. L., Sedwick, W. D., and Vanaman, T. C.,** Calmodulin and Ca^{2+} in normal and transformed cells, *Fed. Proc.*, 41, 2283, 1982.
24. **Charp, P. A.,** DNA repair in human cells: methods for the determination of calmodulin involvement, in *Methods in Enzymology*, Vol. 139, Means, A. R. and Conn, P. M., Eds., Academic Press, New York, 1987, 715.
25. **White, B. A. and Bancrift, C.,** Ca^{2+}/calmodulin regulation of prolactin gene expression, in *Methods in Enzymology*, Vol. 139, Means, A. R. and Conn, P. M., Eds., Academic Press, New York, 1987, 655.
26. **Lee, Y. C. and Wolff, J.,** Calmodulin and cold-labile microtubules, in *Methods in Enzymology*, Vol. 139, Means, A. R. and Conn, P. M., Eds., Academic Press, New York, 1987, 834.
27. **Burgess-Cassler, A., Hinrichsen, R. D., Maley, M. E., and Kung, C.,** Biochemical characterization of a genetically altered calmodulin in *Paramecium, Biochim. Biophys. Acta*, 913, 321, 1987.
28. **Watterson, D. M., Sharief, F., and Vanaman, T. C.,** The complete amino acid sequence of the Ca^{2+}-dependent modulator protein (calmodulin) of bovine brain, *J. Biol. Chem.*, 255, 962, 1980.
29. **Epstein, P. D., Simmen, R. C. M., Tanaka, T., and Means, A. R.,** Isolation and structural analysis of the chromosomal gene for chicken calmodulin, in *Methods in Enzymology*, Vol. 139, Means, A. R. and Conn., P. M., Eds., Academic Press, New York, 1987, 217.
30. **Babu, Y. S., Sack, J. S., Greenough, T. J., Bugg, C. E., Means, A. R., and Cook, W. J.,** Three-dimensional structure of calmodulin, *Nature*, 315, 37, 1985.
31. **Bazari, W. L. and Clarke, M.,** Characterization of calmodulin from *Dictyostelium discoideum*, *J. Biol. Chem.*, 256, 3598, 1981.
32. **Cox, J. A., Ferraz, C., Demaille, J. G., Perez, R. O., Van Tuinen, D., and Marme, D.,** Calmodulin from *Neurospora crasa*. General properties and conformational changes, *J. Biol. Chem.*, 257, 10694, 1982.
33. **Ohya, Y., Uno, I., Ishikawa, T., and Anraku, Y.,** Purification and biochemical properties of calmodulin from *Saccharomyces cerevisiae*, *Eur. J. Biochem.*, 168, 13, 1987.
34. **Cormier, M.,** personal communication.

35. **Van Eldik, L. J., Grossman, A. R., Iverson, D. B., and Watterson, D. M.,** Isolation and characterization of calmodulin from spinach leaves and *in vitro* translation mixtures, *Proc. Natl. Acad. Sci. U. S. A.*, 77, 1912, 1980.
36. **Dubery, I. A. and Schabort, J. C.,** Calmodulin from *Citrus sinensis:* purification and characterization, *Phytochemistry*, 26, 37, 1987.
37. **Anderson, J. M., Charbonneau, H., Jones, H. P., McCann, R. O., and Cormier, M. J.,** Characterization of the plant nicotinamide adenine dinucleotide kinase activator protein and its identification as calmodulin, *Biochemistry*, 19, 3113, 1980.
38. **Marme, D. and Dieter, P.** Role of Ca^{2+} and calmodulin in plants, in *Calcium and Cell Function*, Vol. IV, Cheung, W.Y., Ed., Academic Press, New York, 1983, 263.
39. **Grand, R. J. A., Nairn, A. C., and Perry, S. V.,** The preparation of calmodulins from barley (*Hordeum sp.*) and basidiomycyte fungi, *Biochem. J.*, 185, 755, 1980.
40. **Jones, H. P., Matthews, J. C., and Cormier, M. J.,** Isolation and characterization of Ca^{2+}-dependent modulator protein from the marine invertebrate *Renilla reniformis, Biochemistry*, 18, 55, 1979.
41. **Takagi, T., Nemoto, T., Konishi, K., Yazawa, M., and Yagi, K.,** The amino acid sequence of the calmodulin obtained from sea anemone (*Metridium senile*) muscle, *Biochem. Biophys. Res. Commun.*, 96, 377, 1980.
42. **Masaracchia, R. A., Hassell, T. C., and Donahue, M. J.,** Structural analysis of the calcium-binding protein from *Ascarus suum* obliquely striated muscle, *J. Parisitol.*, 72, 299, 1986.
43. **Yazawa, M., Sakuma, M., and Yagi, K.,** Calmodulins from muscles of marine invertebrates, scallop and sea anemone. Comparison with calmodulins from rabbit skeletal muscle and pig brain, *J. Biochem.*, 87, 1313, 1980.
44. **Molla, A., Kilhofer, M. C., Ferraz, C., Audermard, E., Walsh, M. P., and Demaille, J. G.,** Octopus calmodulin. The trimethyllysine residue is not required for myosin light chain kinase activation, *J. Biol. Chem.*, 256, 18, 1981.
45. **Seamon, K. B. and Moore, B. W.,** Octopus calmodulin. Structural comparison with bovine brain calmodulin, *J. Biol. Chem.*, 255, 11644, 1980.
46. **Hergenhahn, H. G., Gunter, K., and Sedlmeier, D.,** Ca^{2+}-binding proteins in crayfish abdominal muscle. Evidence for a calmodulin lacking trimethyllysine, *Biochim. Biophys. Acta*, 787, 196, 1984.
47. **Jackson, R. L., Dedman, J. R., Schreiber, W. E., Bhatnagar, P. K., Knapp, R. D., and Means, A. R.,** Identification of ε-N-trimethyllysine in a rat testis calcium-dependent regulatory protein of cyclic nucleotide phosphodiesterase, *Biochem. Biophys. Res. Commun.*, 77, 723, 1977.
48. **Nairn, A. C., Grand, R. J. A., and Perry, S. V.,** The amino acid sequence of rabbit skeletal muscle calmodulin, *FEBS Lett.*, 167, 215, 1984.
49. **Schaefer, W. H., Lukas, T. J., Blair, I. A., Schultz, J. E., and Watterson, D. M.,** Amino acid sequence of a novel calmodulin from *Paramecium tetraurelia* that contains dimethyllysine in the first domain, *J. Biol. Chem.*, 262, 1025, 1987.
50. **Sitaramayya, A., Wright, L. S., and Siegel, F. L.,** Enzymatic methylation of calmodulin in rat brain cytosol, *J. Biol. Chem.*, 255, 8894, 1980.
51. **Gagnon, C., Kelly, S., Manganiello, V., Vaughn, M., Odya, C., Strittmatter, W., Hoffman, A., and Hirata, F.,** *Nature*, 291, 515, 1981.
52. **Murtaugh, T. J., Wright, L. S., and Siegel, F. L.,** Posttranslational modification of calmodulin in rat brain and pituitary, *J. Neurochem.*, 47, 53, 1986.
53. **Fukami, Y., Nakamura, A., and Takeharu, K.,** Phosphorylation of tyrosine residues of calmodulin in Rous sarcoma virus-transformed cells, *Proc. Natl. Acad. Sci. U. S. A.*, 83, 4190, 1986.
54. **Klee, C. B. and Vanaman, T. C.,** Calmodulin, in *Advances in Protein Chemistry*, Vol. 35, Academic Press, New York, 1982, 213.
55. **Kim, S. and Paik, W. Y.,** Labile protein-methyl ester: comparison between chemically and enzymatically synthesized, *Experientia*, 32, 982, 1976.
56. **Murtaugh, T. J., Rowe, P. M., Vincent, P. L., Wright, L. S., and Siegel, F. L.,** in *Methods in Enzymology*, Vol. 102, Means, A. R. and O'Malley, B. W., Eds., Academic Press, 1983, 158.
57. **Rowe, P. M., Murtaugh, T. J., Bazari, W. L., Clarke, M., and Siegel, F. L.,** Radiometric assay of *S*-adenosyl-L-methionine:calmodulin(lysine) *N*-methyltransferase by calcium-dependent hydrophobic interaction chromatography, *Anal. Biochem.*, 133, 394, 1983.
58. **Rowe, P. M., Wright, L. S., and Siegel, F. L.,** Calmodulin *N*-methyltransferase. Partial purification and characterization, *J. Biol. Chem.*, 261, 7060, 1986.
59. **Morino, H., Kawamoto, T., Miyake, M., and Kakimoto, Y.,** Purification and properties of calmodulin-lysine *N*-methyltransferase from rat brain cytosol, *J. Neurochem.*, 48, 1201, 1987.
60. **Durban, E., Nochumson, S., Kim, S., Paik, W. K., and Chan, S. K.,** Cytochrome c-specific protein-lysine methyltransferase from *Neurospora crassa*. Purification, characterization and substrate requirements, *J. Biol. Chem.*, 253, 1427, 1978.
61. **Gregori, L., Marriott, D., West, C. M., and Chau, V.,** Specific recognition of calmodulin from *Dictyostelium discoideum* by the ATP, ubiquitin-dependent degradative pathway, *J. Biol. Chem.*, 260, 5232, 1985.

62. **Roberts, D. M., Rowe, P. M., Siegel, F. L., Lukas, T. J., and Watterson, D. M.,** Trimethyllysine and protein function. Effect of methylation and mutagenesis of lysine 115 of calmodulin on NAD kinase activation, *J. Biol. Chem.,* 261, 1491, 1986.
63. **Guerini, D., Krebs, J., and Carafoli, E.,** Stimulation of the purified Ca^{2+}-ATPase by tryptic fragments of calmodulin, *J. Biol. Chem.,* 259, 15172, 1984.
64. **Dedman, J. R., Potter, J. D., and Means, A. R.,** Biological cross reactivity of rat testis phosphodiesterase activator protein and rabbit skeletal muscle troponin-C, *J. Biol. Chem.,* 252, 2437, 1977.
65. **Vincent, P. L. and Siegel, F. L.,** Carboxylmethylation of calmodulin in cultured pituitary cells, *J. Neurochem.,* 49, 1613, 1987.
66. **Stock, J. B. and Koshland, D. E.,** Changing reactivity of receptor carboxyl groups during bacterial sensing, *J. Biol. Chem.,* 256, 10826, 1981.
67. **Johnson, B. A., Freitag, N. E., and Aswad, D.,** Protein carboxylmethyltransferase selectively modifies an atypical form of calmodulin, *J. Biol. Chem.,* 260, 10913, 1985.
68. **Aswad, D. H.,** The unusual substrate specificity of eucaryotic carboxylmethyl transferases, *TIBS,* 12, 155, 1987.
69. **Clarke, S.,** Protein carboxylmethyltransferases: two distinct classes of enzymes, in *Annu. Rev. Biochem.,* Richardson, C. C., Boyer, P. D., Dawid, I. B. and Meister, A., Eds., Annual Reviews, Palo Alto, CA, 1985, 479.
70. **Jakoby, W. B.,** The glutathione *S*-transferases: a group of multifunctional detoxification proteins, *Advances in Enzymology,* Vol. 46, Meister, A., Ed., 1978, 383.
71. **Benson, A. M., Talalay, P., Keen, J. H., and Jakoby, W. B.,** Relationship between the soluble glutathione-dependent delta-5-3-ketosteroid isomerase and the glutathione S-transferases of the liver, *Proc. Natl. Acad. Sci. U. S. A.,* 74, 158, 1977.
72. **James, P., Maeda, M., Fischer, R., Verma, A. K., Krebs, J., Penniston, J. T., and Carafoli, E.,** Identification and primary structure of a calmodulin-binding domain of the Ca^{2+} pump of human erythrocytes, *J. Biol. Chem.,* 263, 2905, 1988.

Chapter 4

CYTOCHROME *c* METHYLATION

Blaise Frost and Woon Ki Paik

TABLE OF CONTENTS

I. POSTTRANSLATIONAL MODIFICATION OF PROTEINS

Posttranslational modification represents a possible rapid and reversible method of manipulating the activity of a protein. This action can sidestep the usual control of the more time-consuming process of transcription and translation which is desirable since the energy cost of each modification is significantly less than protein biosynthesis *de novo*.[1] Research in this area has led to the identification of several types of modification, including phosphorylation, glycosylation, acetylation, ADP-ribosylation, and methylation. These alterations can act to produce change in the overall charge, hydrophobicity, molecular weight, solubility and structure of the protein, and can ultimately alter the function as well, as mentioned above. For example, phosphorylation of some enzymes causes an increase in activity,[2] and glycosylation inhibits degradation or varies the antigenicity of certain proteins.[3] While the putative purpose of certain modifications has been determined, the exact cause and biological effect of many others is, as yet, unknown. Posttranslational modification occurs in a wide variety of proteins and has been observed in a number of organisms.[4]

Methylation is one such posttranslational modification which occurs throughout nature, and though its exact biological effect is not known in most cases, the ubiquity of this modification suggests it has a role in protein-protein and cellular interactions.[4] The reaction involves the substitution of a methyl group from *S*-adenosyl-L-methionine (AdoMet) for a hydrogen on the side chain of an amino acid in a protein. Methylated proteins are found in many eukaryotic and prokaryotic organisms, and perform a vast array of functions in these organisms (see Chapter 2). The site of methylation varies with amino acid;[4] it occurs on an oxygen of the side chain carboxyl group in aspartate and glutamate, on a nitrogen of the side chains of histidine, arginine, lysine, asparagine, and glutamine, on the alpha-nitrogen of proline, alanine, and methionine, and on the sulfur of methionine. A more detailed account of the specific proteins and reactions involved can be found throughout the text of this book.

II. AN INTRODUCTION TO CYTOCHROME *c*

Although the list of proteins which are methylated is quite lengthy, the focus of this treatise is on the methylation of cytochrome *c*. Cytochrome *c* is a relatively small protein with a molecular weight around 12,000, and consists of approximately 100 to 110 amino acids. It is found in the mitochondria of all eukaryotic organisms where its role as an electron carrier in the electron transport chain is highly conserved. The tertiary structure, isoelectric point (pI), and other properties of cytochrome *c* from different sources are also very similar (for a review, see References 5 and 6).

Cytochrome *c* is somewhat unique with respect to other mitochondrial proteins: though it is transcribed and translated from a nuclear gene, it is synthesized as a complete apoprotein with no "prepiece", and released into the cytosol prior to being imported into the mitochondria.[7] Furthermore, its import and processing to holocytochrome *c* are independent of a membrane potential, similar to other imported mitochondrial proteins which need only pass through the outer mitochondrial membrane.[8] Cytochrome *c*, as the electron carrier between cytochrome *c* oxidase complex and cytochrome *c* reductase, is loosely associated with the outside of the inner mitochondria membrane (intermembrane space) and appears to pass only through the outer membrane for import.[9]

Import is apparently dependent upon at least two proteins: an apocytochrome *c*-binding protein ("receptor") associated with the other membrane[10,11] and the heme-attaching protein, cytochrome *c* heme lyase, associated with the intermembrane space.[12-14] The two activities act to draw membrane-associated apocytochrome *c* into the intermembrane space and subsequently (or simultaneously) to attach the heme moiety covalently as holocytochrome *c* is formed.[10-14] It has been suggested that these two proteins are either the same or very closely associated in the outer membrane/intermembrane space region of the mitochondria.

TABLE 1
A Comparison of Some Methylases Which Act on Cytochrome *c*

Amino acid (source)	Arginine (*Euglena*)	Methionine (*Euglena*)	Lysine (*Saccharomyces*)
K_m for			
Cytochrome *c*	8.3×10^{-5}	1.7×10^{-5}	1.3×10^{-5}
AdoMet	4.0×10^{-5}	1.7×10^{-5}	4.0×10^{-5}
K_i for			
AdoHcy	1.2×10^{-5}	8.1×10^{-6}	2.7×10^{-6}
Molecular weight estimation (Da)	36,000	28,000	97,000
pH optimum	7.0	7.0	9.0
Location	Cytosol	Cytosol	Cytosol
pI change for horse heart cytochrome *c*, from 10.06 to	9.33	9.23	9.49

Although the function, processing, and other characteristics of cytochrome *c* from different sources are very much alike, there is significant deviation in the primary sequence of the protein from distantly related organisms. For example, the primary sequence of human cytochrome *c* differs from that of *Neurospora crassa* and *Tetrahymena pyriformis* by 40 and 54%, respectively.[6] In spite of these overall sequence differences, certain regions of the protein have been conserved very well throughout evolution. Clearly, the best example of this conservation is found in the undecapeptide region of residues 70 to 80, where sequence homology is virtually universal in all species compared.[6] This region corresponds to an area which appears to be quite necessary for import into the mitochondria and is also involved with binding of the heme iron in the holoprotein.[5,15] It is of interest in the study of protein methylation since DeLange et al.[16] first observed that one lysine residue (Lys-72) in this conserved region of *Neurospora crassa, Saccharomyces cerevisiae,* and wheat germ had been replaced by the methylated residue, trimethyllysine. Since the cytochrome *c* of higher eukaryotes contains no methylated lysine, the methylation of cytochrome *c* in some organisms warranted further investigation.

III. CYTOCHROME c METHYLATION

A. ENZYMOLOGY OF CYTOCHROME c METHYLATION

The initial observation of methylated lysine-72 in the cytochrome *c* of the lower eukaryotes led Paik and co-workers to the discovery of several enzymes which methylate this and other amino acids.[17-20] These enzymes, termed protein methyltransferases or methylases, have been purified to various extents and characterized. As has been observed in the methylation of some other proteins, the enzymes appear to be very specific for cytochrome *c* and for the particular amino acid. As Table 1 elucidates, three different enzymes have been observed which methylate arginine, methionine, or lysine of cytochrome *c,* specifically.

1. Arginine

An enzyme which methylates an argininyl residue of cytochrome *c* was discovered in *Euglena gracilis* extracts by Farooqui et al.[20] and was purified 50-fold upon DEAE-cellulose and Sephadex G-200 chromatography. The activity was associated predominantly (80%) with the cytosol and was highly specific for cytochrome *c*. Among the proteins which were not substrates for the methylase, several, including some histones, are known to contain methylated arginine,[4] while this amino acid has not been identified to date in cytochrome *c*. The enzyme was inhibited by the known methylase inhibitors, *S*-adenosylhomocysteine, (AdoHcy), a reaction product, and sinefungin, an AdoHcy analog. The methylation reaction lowered the pI of cytochrome *c* from 10.06 to 9.33, when horse heart protein was used as a substrate.

Using horse heart cytochrome *c*, it was determined that the methylase acted most likely upon the argininyl residue at position 38. This conclusion was drawn from mapping of peptides of chymotryptic digests of cytochrome *c* methylated with radioactive AdoMet. The spot possessing radioactivity corresponded roughly to the peptide which contained arginine-38.[20] This is, as yet, not very conclusive evidence and, even more important is the fact that *E. gracilis* cytochrome *c* methylated at *any* argininyl residue has yet to be reported. Still, the identification of a methylase which apparently prefers an arginine residue of cytochrome *c* suggests that methylation of this protein at this site may play a role in its metabolism.

2. Methionine

In the same study which elucidated the cytochrome *c*:arginine methylase, an additional cytochrome *c* methylase which acts specifically at a methioninyl residue was also identified.[20] The methylase was purified 100-fold, as above, and was also associated primarily with the cytosolic fraction. Like the arginine methylase, AdoHcy was inhibitory, as was sinefungin, and this methylase had no significant activity when other proteinaceous substrates were employed. The methylation of horse heart cytochrome *c* by this methylase lowered the pI of the protein from 10.06 to 9.23.

Evidence for the site of methylation of this methylase was also somewhat circumstantial: two-dimensional paper chromatography of chymotryptic digests of methylated horse heart cytochrome *c* yielded two spots which contained radioactivity. These spots correlated to those of the methionine-65-containing peptides and not to that of the methionine-80 containing peptide. Naturally, this suggests that methionine-65 is the substrate, although exact sequence data would be more confirmatory.

Like methylated arginine, methylated methionine has not been observed in cytochrome *c* from any source. Indeed, *no* proteins have been shown to contain methylated methionine. With this in mind, any projected significance for this methylase and its product would be highly speculative. Until either of these methylated amino acids is observed in *Euglena* cytochrome *c*, the further characterization of either enzyme responsible for their formation may not be too productive.

3. Lysine

The discovery of trimethyllysine as a residue of cytochrome *c* in some fungi[16] led Paik and co-workers to search for the enzymatic activity which catalyzed the formation of this residue. Through their efforts, a cytochrome *c*: lysine protein methylase was identified in *Saccharomyces cerevisiae*[17] and *Neurospora crassa*,[18] which they further purified and characterized. Table 2 shows a comparison of the purification schemes and some significant properties of this enzyme, *S*-adenosyl-L-methionine:cytochrome *c*:lysine *N*-methyltransferase (EC 2.1.1.59), from these two organisms. The enzyme is also known as protein methylase III or PMIII (see Reference 1).

Both enzymes have similar pH optima, are associated with the cytosolic fraction, and are inhibited by AdoHcy with similar K_i values. The K_m values for AdoMet for each are also quite similar, as are their isoelectric points (Table 2). Both enzymes exhibit exceptional specificity for cytochrome *c* and do not act very efficiently on other proteins (Table 3), including histones, which are among the best-known methylated proteins in mammalian tissues.[4] The apparent difference in activity on the various cytochromes *c* is not understood at present; it may be an

TABLE 2
Purification and Characterization of PM III from *Saccharomyces cerevisiae* and *Neurospora crassa*

Source	*Saccharomyces*		*Neurospora*	
(Purification step)	Purification	Yield	Purification	Yield
Crude homogenate	1.0	100	1.0	100
105,000 × *g* supernatant	2.9	106	1.3	58
Ammonium sulfate precipitation	7.2	101	28.0	46
Calcium phosphate gel	17.4	58	39.0	20
DEAE-cellulose chromatography	63.1	19	280.0	8
Ratio of products	MML:DML:TML[a]		MML:DML:TML	
Homogenate	12.2: 19.4: 68.3:		10.1: 36.4: 53.5:	
Ammonium sulfate	7.2: 14.6: 78.0:		14.2: 41.5: 44.3:	
DEAE-cellulose	8.9: 15.8: 74.5:		14.7: 40.4: 44.9:	
pI	ND[b]		4.8	
pH optimum	9.0		9.0	
Molecular weight estimation (Da)	97,000		120,000	
k_m for AdoMet		4.0×10^{-5}		1.9×10^{-5}
k_i for AdoHcy		2.7×10^{-6}		2.0×10^{-6}
Site of action		Lysine-72		Lysine-72

[a] MML, monomethyllysine; DML, dimethyllysine; TML, trimethyllysine.
[b] ND, not determined.

artifact caused by purification processes, or it may be due either to the slight sequence variations from species to species, or to subtle changes in secondary or tertiary conformation.

Since lysine can be mono- and dimethylated as well as trimethylated, it was of interest to determine whether the same enzyme forms all three residues. As can be seen in Table 2, the enzymes appear to be capable of forming mono-, di-, and trimethyllysine, and the similar ratio of formation of these lysine derivatives, regardless of enzyme purity, suggests that the same enzyme forms all three residues. From this enzyme purification data and the natural occurrence of trimethyllysine alone in cytochrome *c*,[1] it has been speculated that the methylations take place sequentially, with trimethyllysine being the *in vivo* product.

The specificity of the enzymes for residue 72 was determined by several methods. First, the methylases had little or no activity on naturally methylated cytochrome *c* from fungal sources, which is already methylated at lysine-72 (Table 3). Second, paper chromatography of a chymotryptic digest of horse heart cytochrome *c* which was methylated with radiolabeled AdoMet by PMIII yielded only one radioactive spot: that corresponding to the 68 to 74 residue fragment.[17] Since this fragment contains seven amino acids (two of which are lysine), Edman degradation of the labeled fragment was performed to determine which was (were) methylated. The results clearly indicate that only one lysine, that at residue 72, was methylated.[17,18] It was interesting that the methylases were only recognizing lysine-72 and ignoring a neighboring lysine at residue 73 completely. Furthermore, the sequence of x-Lys-Lys-y appears a total of five times in yeast cytochrome *c* and four times in horse heart cytochrome *c*,[6] yet only lysine-72 is methylated.

In vitro methylation of various cytochrome *c* fragments, formed by cyanogen bromide, by these PMIIIs yielded confusing evidence on the specificity.[17,18] The enzymes acted on a peptide

TABLE 3
Specificity of Protein Methylase III with Respect to Different Protein Substrates

Substrate	*Saccharomyces* PMIII (percent activity)[a]	*Neurospora* PMIII (percent activity)[b]
Cytochrome *c* from		
Horse heart	100	100
Cow	109	112
Rabbit	142	117
Chicken	152	66
Pigeon	82	120
Yeast	16	4
Histone		
Lysine-rich	0	0
Arginine-rich	0	0
Polylysine	0	0
Protamine	0	0
Pancreatic ribonuclease	0	0
Bovine serum albumin	0	0
Trypsin inhibitor	0	0

[a] 100% activity corresponds to 102 pmol of methyl group incorporated per minute per milligram protein methylase III preparation.
[b] 100% activity corresponds to 1.3 nmol of methyl group incorporated per minute per milligram protein methylase III preparation.

consisting of residues 1 to 65, which does *not* contain lysine-72, and not on a peptide consisting of residues 66 to 104, which does! It was further noted that these methylases have a much higher affinity for the apocytochrome, compared to the holoprotein, although the V_{max} of the *Neurospora* enzyme is lower with apocytochrome *c*. These results suggest certain conformational requirements must be met before methylation can occur, even if the precise sequence which is methylated *in vivo* is present.[4]

Although *Neurospora,* yeast, and wheat germ all have trimethyllysine at residue 72 of cytochrome *c,* wheat germ cytochrome *c* possesses another trimethylated lysine at residue 86. Thus, considering the high specificity of the purified fungal protein methylase III, Paik and coworkers[21] extracted wheat germ to determine whether this organism had separate PMIIIs for each residue. Using horse heart or yeast cytochrome *c* as substrate, they attempted to isolate the possible two PMIIIs, following the fungal purification procedures. The rationale was that horse heart cytochrome *c* had no methyl groups and hence could be methylated at both sites, while yeast cytochrome *c,* already methylated at residue 72, could only be methylated at residue 86. Thus, as shown in Table 4, the ratio of methylase activity on horse heart cytochrome to yeast cytochrome was an appropriate meter of the putative two methylases. Although purification of both enzymes was ultimately unsuccessful, the evidence suggested that two distinct methylases, each specific for only one residue, do exist. This observation warrants the further investigation of wheat germ for lysine-86 specific PMIII.

B. RELATIONSHIP BETWEEN CYTOCHROME *c* LEVELS AND PROTEIN METHYLASE III ACTIVITY

From a purely scientific standpoint, the arginine- and methionine-cytochrome *c* methylases discovered are quite interesting in terms of enzymology, labeling techniques, and possibly, blocking or inactivation studies. However, since these methylated amino acids have not been observed as naturally occurring residues in any cytochrome *c,* the biological relevance of the

TABLE 4
Separation of Wheat Germ Protein Methylases III Specific for Lysine-72 or Lysine-86

Purification step	Activity on horse heart cytochrome c[a] (lys-72 and lys-86 methylation)	Activity on yeast cytochrome c (lys-86 methylation only)	Ratio of activities (horse heart to yeast)
Crude homogenate	1.62	1.15	1.41
Ammonium sulfate precipitation	1.59	1.71	0.93
Calcium phosphate gel	7.94	0.90	10.80
DEAE-cellulose chromatography	132.80	0.0	—

[a] Activity is in nanomole methyl group incorporated per minute per milligram enzyme prep.

protein methylases responsible for their formation remains to be seen. The significance of these methylases will, therefore, have to be discussed in some future review as information becomes available. For the remainder of this review, the focus will be entirely upon the methylation of lysine-72, not lysine-86, of cytochrome *c* since this residue is naturally occurring in several organisms and since methylases for this residue have been purified and characterized.

Initially, it was suggested that PMIII activity may be controlled in conjunction with cytochrome *c* metabolism. During anaerobic growth, cytochrome *c* activity is greatly reduced, so it follows that PMIII might also be reduced. Paik and co-workers,[17] and, separately, Sherman and co-workers[22] observed a decrease in PMIII activity exhibited by yeast cells grown anaerobically. It is important to note, however, that while cytochrome *c* levels were reduced drastically (approximately 90 to 95%), PMIII activity was only decreased approximately 50% (Table 5), according to both reports.

Both groups extended these studies to cytochrome *c*-deficient mutants, in the hope of observing some direct link in regulation between decreased cytochrome *c* and lowered PMIII activity. Unfortunately, no solid correlation between levels of the two proteins was established; yeast strains containing mutations in genes associated with cytochrome metabolism are deficient, to various extents, in cytochrome *c* protein or activity, yet their PMIII activity does not correlate with cytochrome *c* deficiency and is actually higher in some cases. This may be fortuitous since these mutations are presumably due to some defect other than PMIII regulation.

The combination of these results implies that respiration metabolism may exercise *some* control over PMIII levels, though this control is not nearly as complete as that over cytochrome *c* metabolism. Anaerobic conditions almost entirely repress cytochrome *c* synthesis while only partially reducing PMIII activity.

C. EFFECT OF MITOCHONDRIA ON CYTOCHROME *c* METHYLATION

As mentioned in the introduction, apocytochrome *c* is synthesized in the cytoplasm, imported into the mitochondria, and processed to holocytochrome *c*. Studies on PMIII indicated that the enzyme is also cytoplasmic; it seemed natural that methylation takes place in the cytoplasm, before import of the protein into the mitochondria or formation of holoprotein. To examine this hypothesis, DiMaria et al.[17] subjected holocytochrome *c* and apocytochrome *c* (made by heme removal from that of holocytochrome *c* according to Reference 23) to 0.2 *M* KCl or mitochondria or both, before methylation. The rationale was quite simple: if PMIII acted in the mitochondria, methylation in each test condition should be similar. However, if PMIII methylated cytochrome *c* prior to uptake by the mitochondria, the pretreatment of substrate with mitochondria alone

TABLE 5
Relationship Between Cytochrome *c* Levels and Protein Methylase III Activity in Yeast Grown Under Aerobic and Anaerobic Conditions

Media	Conditions	Cytochrome *c* (percent present)	PM III Activity (percent present)	Ref.
2% glucose	Aerobic	100[a]	100[c]	22
2% glucose	Aerobic	100[b]	100[d]	17
3% glucose	Anaerobic	8	43	22
10% glucose	Anaerobic	2	54	17

[a] Corresponds to 0.275 mg cytochrome *c*/g dry cells.
[b] Corresponds to 0.453 mg cytochrome *c*/g wet cells.
[c] Corresponds to 0.8 pmol methyl group incorporated per minute per milligram PMIII preparation.
[d] Corresponds to 5.2 pmol methyl group incorporated per minute per milligram PMIII preparation.

would lower the amount of methylation observed. Salt (KCl) was used to extract cytochrome *c* from the mitochondria, according to the method of Jacobs and Sanadi.[24]

The results expressed in Table 6 definitely support the latter scenario; after binding to the mitochondria (and presumably after import into the same), both apo- and holocytochrome *c* were not "available" for methylase action. This complete inhibition of activity was reversible upon treatment with 0.2 *M* KCl. The KCl treatment in the presence or absence of mitochondria had a slight inhibitory effect on PMIII activity when holocytochrome *c* was examined. However, the effect was similar whether mitochondria were present or not, so the inhibition was not considered detrimental to the experiment and may be within the range of experimental error. Additionally, this data, like that involving the methylation of the cyanogen bromide fragments,[17,18] suggests that apocytochrome *c* is the preferred, although not absolute, substrate of PMIII.

D. PHYSICAL AND CHEMICAL EFFECT OF METHYLATION

The previous data provides a very strong case for the existence of a cytoplasmic PMIII which acts specifically and solely on lysine-72 of apocytochrome *c* and, to a somewhat lesser extent, holocytochrome *c*. The question is: what effect, if any, does this methylation have on the molecule itself?

Initial observations by Mitchell and co-workers using *Neurospora* revealed that the cytochrome *c* population of this mold was a mixture of two chromatographically separable proteins, CI and CII.[25] The two cytochromes *c* are identical in terms of sedimentation coefficient, and ultraviolet and visible absorption spectra. They also noted that CI and CII were temporally related; pulse chase experiments showed CII was made and disappeared as CI appeared.[25] The sequences of the two proteins were identical except for residue-72, which was lysine in CII and trimethyllysine in CI.

Polastro and co-workers extended the similarities between the two proteins to several other parameters, including circular dichroic spectra and denaturation by exposure to heat, acid, and guanidinium hydrochloride.[26,27] Though there were some slight differences, data from the two proteins were mostly indistinguishable indicating that the helical content, overall structure, and stability were identical or nearly so.

The first major difference between methylated and unmethylated cytochrome *c* was reported by Paik and co-workers when they observed a decrease in the pI of methylated horse heart cytochrome *c* as compared to unmethylated form 10.03 to 9.49 (Figure 1).[28] This revelation was particularly surprising since in theory, trimethylation should cause an increase in pI by changing the lysyl amine to an ammonium ion. Also, according to pI determination of the individual amino acids, trimethyllysine is more basic than lysine.[29] This indicated that methylation had some appreciable conformational effect not detected by the previous physical methods.

TABLE 6
Effect of Mitochondria on Protein Methylase III Activity

Assay mixture	Apocytochrome *c* (pmol CH_3 incorporated)	%	Holocytochrome *c* (pmol CH_3 incorporated)	%
Control	86.40	100	50.90	100
Control plus 0.2 *M* KCl	86.86	101	43.60	86
Control plus mitochondria	6.12	7	1.71	3
Control plus mitochondria and 0.2 *M* KCl	81.07	94	37.70	74

Furthermore, the decrease in pI could be eradicated by treating the proteins with increasing amounts of urea during isoelectricfocusing (Figures 1 and 2),[30] suggesting that methylation is affecting the hydrogen bonding of the apoprotein. The addition of three methyl groups to the lysyl residue does stabilize the positive charge, however, it also eliminates the hydrogen-bonding capabilities of the residue. The unmethylated lysine can theoretically hydrogen bond with neighboring asparagine-70 while trimethyllysine cannot, which may free this asparagine to hydrogen bond elsewhere.[30] Upon addition of urea, the "secondary" hydrogen bond is broken thus returning the cytochrome *c* to a more native, i.e., unmethylated, state.

This decrease in pI was also observed in yeast apocytochrome *c* which was translated *in vitro* and then methylated.[31] A comparison of the pI values for purified methylated and unmethylated holocytochrome *c* and *in vitro*-translated methylated apocytochrome *c* can be seen in Table 7. The decrease in pI due to methylation occurs in either holo- or apocytochrome *c*, although the difference between the two holoproteins is not as pronounced. This suggests that apocytochrome *c* is the *in vivo* form most affected by methylation which indirectly supports apocytochrome *c* being the *in vivo* substrate of the methylase.

E. BIOLOGICAL EFFECT OF CYTOCHROME *c* METHYLATION

1. Resistance to Proteolytic Degradation

Determining the effect of methylation on cytochrome *c in vivo* is of paramount importance and would give great significance to these studies. As yet, no mutant lacking PMIII activity alone has been isolated which could mean that either: (1) all methylase mutations are extremely lethal, or (2) lack of methylase does not confer a detectable phenotype. Until a suitable mutant is isolated, the effect of methylation has to be studied by measuring the differences in metabolism of methylated and unmethylated protein *in vivo* and *in vitro*.

When Polastro et al.[26] reported that the stability of methylated cytochrome *c* was not appreciably different from the unmethylated form, they also observed that both forms were significantly less resistant than horse heart cytochrome *c* to denaturation. This relative instability led them to theorize that methylation of yeast cytochrome *c* may confer resistance to proteolytic attack and degradation. Along this line of reasoning, Poncz and Dearborn used oligopeptides and myoglobin, both containing methylated or unmethylated lysine, to show that trypsin is less active on the methylated sequences.[32] This was not totally unexpected since lysine residues act as a recognition site for this protease, however, some proteins, including histone,[33] pancreatic ribonuclease,[33] flagella protein from *Salmonella typhimurium*,[34] and elongation factor EF-Tu from *Escherichia coli*,[35] were not protected from proteolytic degradation when methylated either *in vivo* or *in vitro*. Of course, these studies dealt with overall proteolysis of the protein and may not have observed resistance to protease action at one particular site. Furthermore, different proteases have different recognition sites, and unlike trypsin, these may not include or may not be affected by lysine.

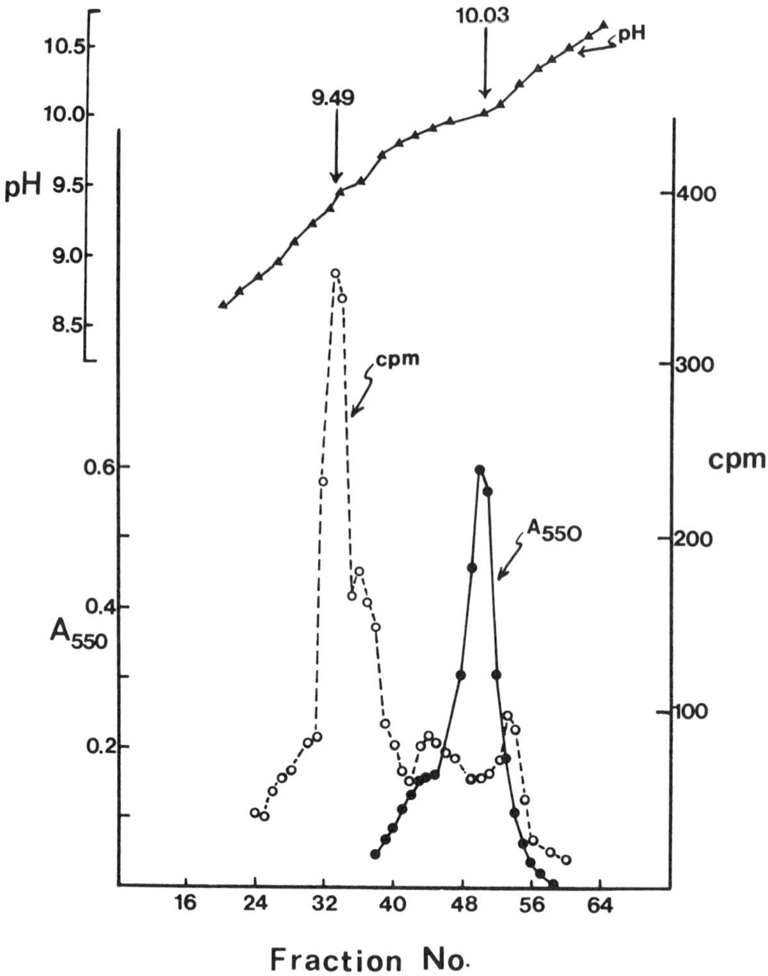

FIGURE 1. Isoelectric focusing of [^{14}C]methylated and unmethylated horse heart cytochrome *c*. Horse heart cytochrome *c* was electrofocused in Ampholines of pH 9 to 11 after treatment with (or without, control) *N. crassa* PMIII and [^{14}C]AdoMet.[28] The open circles represent radioactivity, and thus, methylated cytochrome *c*, while the closed circles represent the absorbance at 550 nm of unmethylated protein. The triangles represent pH.

Polastro and co-workers reported that the half-digestion times of methylated and unmethylated cytochrome *c* from yeast were essentially the same when trypsin and yeast proteases A and B were used,[36] as can be seen in Table 8. However, it should be noted that they used holocytochrome *c* in this study. When apocytochrome *c* (translated *in vitro*) was employed, the results suggested a similar effect with trypsin and chymotrypsin, but methylated cytochrome *c* was more resistant to proteolysis by yeast extract (Figure 3). The crude yeast extract used in this experiment is presumably a more natural protease preparation, and apocytochrome *c*, methylated or not, is probably exposed to a similar mixture prior to uptake by the mitochondria.

This latter result supported *in vivo* data of Farooqui et al.[34] who noted that unmethylated cytochrome *c* disappeared more rapidly than methylated protein (Figure 4) under anaerobic conditions. Unfortunately, this may be due to preferential degradation during the 4 to 7 d separation process (see Reference 37 for detailed method) or may be due to decreased import of unmethylated cytochrome *c* into the mitochondria. In either situation, however, methylation appears to exhibit a positive effect on cytochrome *c* stability, though it may be indirectly.

TABLE 7
The Effect of Methylation on the pI Values of Cytochrome *c* from Different Sources

Cytochrome *c*	Methylated	Unmethylated
Horse heart[a]	9.49	10.06
Yeast[b]	9.68	9.72
Yeast[c]	8.70	9.60

[a] Horse heart holocytochrome *c* was purified, then methylated *in vitro*.
[b] Yeast holocytochrome *c*, methylated and unmethylated, was purified from cell.
[c] Yeast apocytochrome *c* was translated *in vitro*, then methylated.

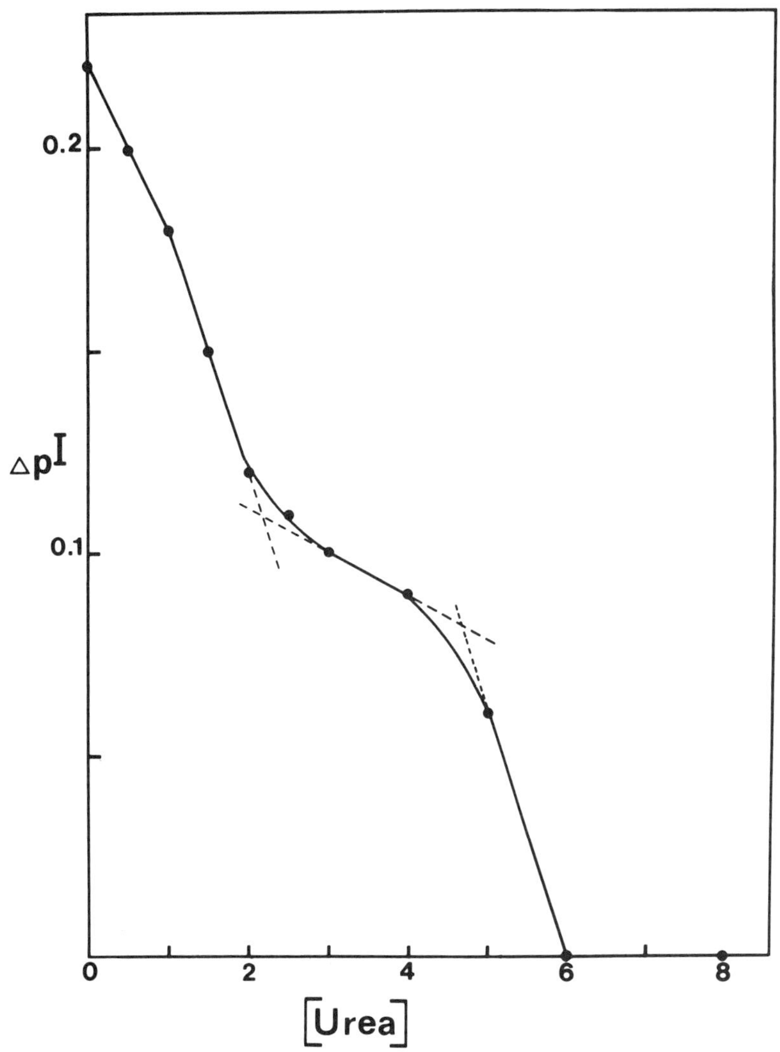

FIGURE 2. The effect of urea on the difference in pI between methylated and unmethylated cytochrome *c*. The pI values of methylated and unmethylated horse heart cytochrome *c* were determined by isoelectric focusing in the presence of various concentrations of urea.[30]

TABLE 8
The Effect of Methylation on the Susceptibility of Cytochrome *c* to Proteolytic Degradation

	Half-digestion time[a] in minutes	
Protease	Methylated cytochrome *c*	Unmethylated cytochrome *c*
Bovine trypsin	297	300
Yeast protease A	113	115
Yeast protease B	>420	>420

[a] The half-digestion time is the time necessary to digest 50% of the protein.

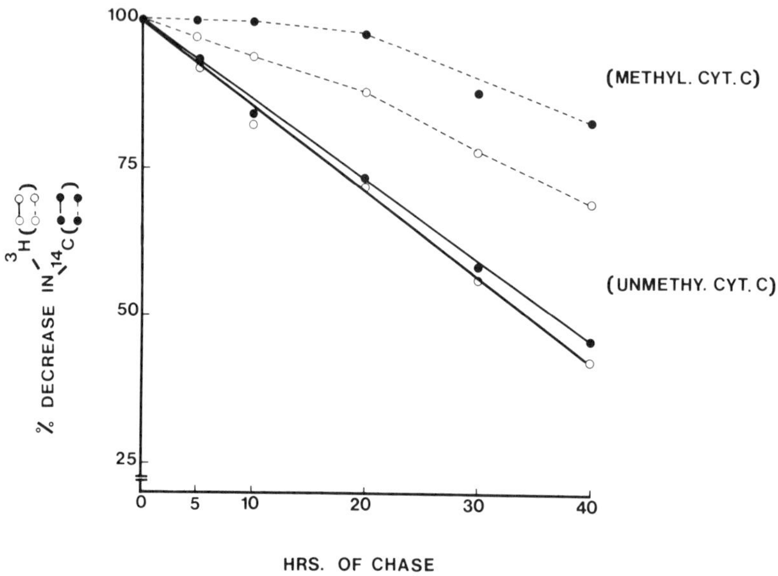

FIGURE 3. *In vivo* stability of methylated and unmethylated cytochrome *c*. Yeast cells were grown aerobically in the presence of [*methyl*-^{3}H]-methionine and [^{14}C]methionine for 24 h, washed, and then grown anaerobically in unlabeled media containing cycloleucine to reduce the amount of incorporation of label into cytochrome *c*. Cytochrome *c* was then extracted from aliquots removed at the given times, and methylated and unmethylated cytochrome *c* were separated as described.[37] The closed circles represent [^{14}C] radioactivity (protein synthesis), and the open circles represent [^{3}H] radioactivity (protein synthesis plus methylation). Solid lines are unmethylated cytochrome *c*, and broken lines depict methylated protein.

Figure 4 also shows a loss of methyl group, without loss of methylated protein, over a 2-d period. This suggests that demethylation followed by remethylation may be occurring although the significance of this event remains a mystery. An identical occurrence was observed by Paik and co-workers[38] when they examined the disappearance of methylated cytochrome *c* alone. They suggested that an as yet unidentified demethylase may be responsible.

2. Increased Import into Mitochondria

As an alternative to protease protection, Mitchell and co-workers suggested that methylation facilitates the binding of cytochrome *c* to mitochondria.[25] They first described the build-up of unmethylated, and to a lesser extent methylated, cytochrome *c* in a *poky* mutant of *Neurospora* and could only extract methylated cytochrome *c* from the mitochondria. Polastro and colleagues

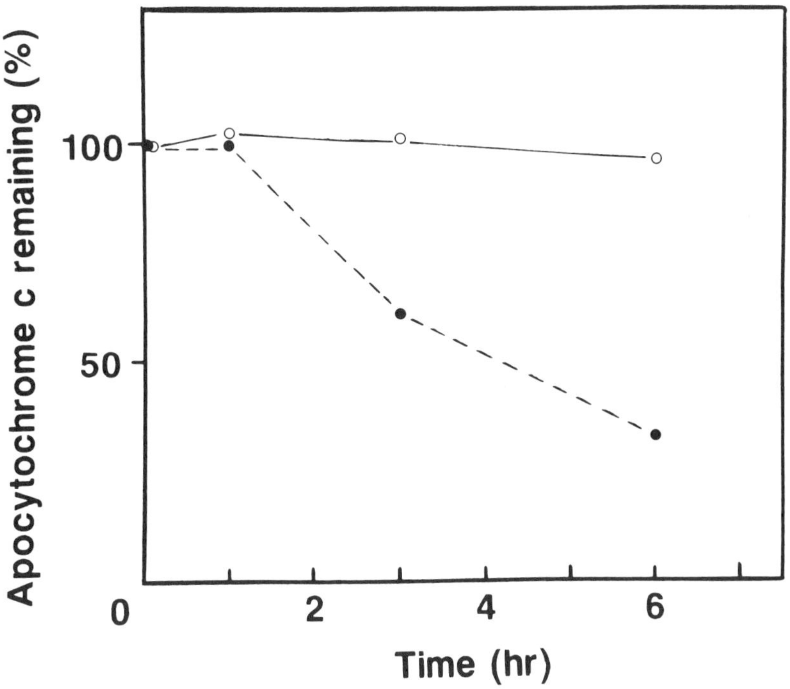

FIGURE 4. The effect of methylation on apocytochrome *c* stability *in vitro*. Methylated or unmethylated *in vitro*-translated apocytochrome *c* was treated with crude yeast extract. Aliquots were removed at the times indicated and isoelectricfocused. Percent apocytochrome *c* remaining was determined by measuring the radioactivity of the apocytochrome *c* band and comparing this value for each time with the zero time value.

took these observations one step further by measuring the binding of yeast cytochrome *c* either methylated or unmethylated, to yeast or horse heart mitochondria.[39,40] They observed that methylated cytochrome *c* had a significantly higher affinity than unmethylated protein for yeast mitochondria, while the two proteins had much more similar affinities for horse heart mitochondria. Their results, shown in Table 9, suggested the putative cytochrome *c* receptor in yeast mitochondria recognized methylated cytochrome *c* preferentially. However, this study employed holocytochrome *c* which may not be imported into mitochondria very well, if at all.[14]

In order to dispel this possible discrepancy, Paik and co-workers employed an *in vitro* translation system to synthesize labeled apocytochrome *c* which could then be methylated as necessary.[41] Thus, the effect of methylation of apocytochrome *c* on binding to mitochondria could be examined *in vitro*. This system circumvented the problems of chemical preparation of apocytochrome *c* from holocytochrome *c* which may be causing irreparable damage to the protein, and of low methylase activity on chemically prepared substrates. Prior to the development of this system, methylation of holocytochrome *c* was very poor with a yield of less than 1% protein methylated; the *in vitro* translation and concomitant methylation of apocytochrome *c* yields over 15% protein methylated, typically.[41] Their observations also support the contention that methylation is a posttranslational (or perhaps cotranslational) modification.

Using this procedure, Paik and colleagues observed a two- to fourfold increase in uptake of methylated apocytochrome *c* by yeast mitochondria, over unmethylated apoprotein (Table 10). This increase was not observed in rat liver mitochondria which, along with the observation of Polastro et al.,[39,40] suggests that the cytochrome *c* receptor in yeast mitochondria recognizes methylated apocytochrome *c* preferentially. Of course, it is possible that the increase in observed uptake of methylated protein is due to resistance to proteolytic degradation, as mentioned earlier. However, the degradation data (Figure 3) showed no difference in sensitivity to protease action

TABLE 9
The Effect of Methylation on the Interaction of Purified Yeast Cytochrome *c* with Mitochondria

Mitochondria	Methylated cytochrome *c*[a] (μ*M*)	Unmethylated cytochrome *c*[a] (μ*M*)
Yeast	0.05[b]	0.10[b]
	0.04[c]	0.12[c]
	0.05[d]	0.11[d]
Horse heart	0.08[b]	0.11[b]
	0.09[c]	0.11[c]
	0.08[d]	0.10[d]

[a] The numbers presented are the dissociation constants for the interaction of cytochrome *c* with the mitochondria.
[b] These figures were determined by measuring the oxygen consumption rate (see Reference 36 for complete details).
[c] These figures were determined by a fluorescence quenching method (see Reference 36 for complete details).
[d] These figures were determined by measuring the direct binding of the methylated or unmethylated protein (see Reference 36 for complete details).

during short (i.e., less than 1 h) incubations, which were used in the import study. Furthermore, this study employed a partially purified yeast mitochondrial fraction, which has very little, if any, cytosolic proteases. While methylation may confer long-term resistance to proteolytic digestion upon apocytochrome *c*, this may be a secondary phenomenon and may not be physiologically relevant. The real effect of methylation appears to be aiding the import of cytochrome *c* into the mitochondria.

IV. DISCUSSION

Since DeLange et al. observed that cytochrome *c* in some lower eukaryotes contains trimethyllysine at residue-72,[16] Paik and colleagues searched for and identified enzymes which catalyzed the formation of this modified amino acid in preformed cytochrome *c*.[17-20] The specificities of this enzyme, *S*-adenosyl-L-methionine:cytochrome *c*-lysine *N*-methyltransferase are quite striking and include preference for (1) a certain amino acid (lysine *only*) in a particular sequence, (2) tertiary structure, and (3) overall protein conformation (cytochrome *c* only and apoprotein in particular).

The purification and characterization of two other apparent cytochrome *c* methylases, one specific for methionine and the other for arginine, have also been reported. However, since these methylated amino acids have not been identified in any cytochrome *c*, it is difficult to project any biological significance for these enzymes. While this fact may be due to degradation of the modified amino acids during sequencing or purification, it is also possible that these enzymes target cytochrome *c* for degradation. It is also possible that cytochrome *c* is not the major methyl acceptor *in vivo* of these methylases and that some alternate methyl group acceptor will be identified.

The use of purified lysine methylase (PMIII) has been inextricably involved in the quest for a biological significance attributable to methylation. Of all the physical parameters examined to date, only the isoelectric point was appreciably altered by *in vitro* methylation at lysine-72, and this effect could be eliminated by treatment with increasing concentrations of urea. The lower pI caused by methylation is directly contradictory to the expected increase predicted by data on the methylation of free lysine. Since urea abolition of this effect is biphasic, it is conceivable that methylation of lysine-72 causes a shift in hydrogen bonding of the unmethyl-

TABLE 10
The Effect of Methylation of Apocytochrome *c* on its Import into the Mitochondria from Yeast and Rat Liver

Mitochondria	Unmethylated apocytochrome *c* imported[a]	%[b]	Methylated apocytochrome *c* imported[a]	%[b]
Yeast[c]	0.106	100	0.243	230
Yeast[d]	0.022	100	0.081	375
Rat liver[d]	0.011	100	0.015	133

[a] pmole *in vitro* translated cytochrome *c* imported per milligram mitochondria protein.
[b] Percent of unmethylated apocytochrome *c* imported.
[c] Yeast mitochondria pretreated with KCl for cytochrome *c* depletion.
[d] No KCl pretreatment.

ated apoprotein. This possibility is even more attractive when it is considered that trimethylation at residue 72 disrupts any possible hydrogen bonding of this lysine.

Methylation also promotes the stability of apocytochrome *c* when the protein is exposed to yeast cytosolic protease activities. This resistance apparently only affects the long-term stability since short (1 h) incubations yielded little difference in susceptibility to protease activity between methylated or unmethylated apocytochrome *c*. This difference was not extended to holocytochrome *c* when trypsin or yeast proteases A and B were employed which suggests that methylation only affects apocytochrome *c* and that once holocytochrome *c* is formed, methylation does not play a major role in its metabolism. It should be emphasized that this protease resistance effect may not be physiologically relevant due to the length of time required to observe a difference. In other words, apocytochrome *c* may be imported into the mitochondria under normal circumstances before there is a need for protease resistance.

Methylated apocytochrome *c* is imported into the mitochondria preferentially over unmethylated protein,[41] and it follows that the receptor, or the possibly associated heme lyase,[7] might also be recognizing some change induced by methylation. Studies dealing with import into the mitochondria have suggested that the region near lysine-72 (residues 65 to 80) is of the utmost importance,[14] and it is clearly possible that methylation of this residue is involved. It is very tempting to hypothesize that as methylation lowers the pI and alters the hydrogen bonding of apocytochrome *c*, these changes serve to facilitate import by the receptor/heme lyase.

The exact structure of methylated apocytochrome *c* with respect to unmethylated apocytochrome *c* has not been elucidated. It is possible that methylation lends stability to a more "open" or unfolded structure which can be imported into the mitochondria more readily. This theory is attractive since it has been proposed that posttranslational import of the precursors of other mitochondrial proteins requires some unfolding of the proteins to be imported.[42] This unfolding is apparently at the expense of high-energy phosphate, either ATP or GTP. In yeast and other lower eukaryotes, apocytochrome *c* "unfolding" may be accomplished by methylation of lysine-72, though apocytochrome *c* does not require ATP for import in any known organism.[7,8,42]

Methylation also indirectly requires energy in the form of high-energy phosphate, but once methylated, apocytochrome *c* could conceivably remain in the "proper" import conformation indefinitely until import could take place. Thus, apocytochrome *c* could be formed and maintained in an import-competent form for short periods (with respect to natural proteolysis) during anaerobic growth, allowing the organism to shift back to aerobic growth more quickly. It is possible that methylated apocytochrome *c* may also be more resistant to protease action for longer periods which would allow a pool of transport-competent apocytochrome *c* molecules to be preserved more easily.

According to this scheme, the methylation process would not be absolutely required for growth, and inhibition of methylation or loss of methylase activity would not severely cripple the organism. Rather, methylation of cytochrome *c* would only serve to enhance a process that would occur albeit more slowly. Indeed, under normal circumstances, the effect of methylation might not be so noticeable, since unmethylated apocytochrome *c* is also imported into the mitochondria, just not as quickly. On the other hand, methylation may be useful in times of stress, perhaps during reoxygenation or during mitochondrial synthesis, and may even be essential.

A somewhat similar situation may exist for the role of receptor methylation in chemotaxis, the movement of bacteria up a stimulus gradient. It requires a precise balance in the amount of random movement (tumbling) and movement toward the stimulus (running), and observations had indicated that methylation, followed by demethylation, of certain receptors was necessary to maintain this proper balance.[43] Mutations in either the methylating or demethylating enzyme genes caused gross aberrations in the chemotactile response, consisting of tumbling or running continuously. However, double mutants had proper chemotactic response although the response was not as competent as in the wild-type cells.[44] After analysis of these results, it has been suggested that methylation of the receptors serves as a fine-tuning mechanism for the response of the cell to a particular environment.[45] (A much more detailed review can be found in Chapter 16.) Thus, methylation in this case is not required, especially when no stimulus gradient is present, but it provides the organism with a better means of adapting to its environment, and provides an advantage over individuals with a less sensitive adaptation mechanism.

Though the situation is far from completely analogous, methylation of cytochrome *c* may also provide an advantage by facilitating its import into mitochondria, however, it may not be an essential modification. Using oligonucleotide-directed mutagenesis, Sherman and co-workers developed a yeast cytochrome *c* mutant which had its lysine-72 residue replaced with an arginine residue.[46] Presumably, this substitution should not alter the protein to any great extent since the charge and size of lysine and arginine are similar, but this residue cannot be methylated in the same way. Growth of the arginine-substituted mutant under aerobic conditions was very similar to their "wild-type" parental strain which had normal cytochrome *c*. (It should be noted that the "wild-type" parental strain is not a real wild-type yeast; it contains several mutations and has been manipulated in such a way as to facilitate the generation of the mutant.) Several molecular parameters of the arginine-substituted protein compared very well with normal cytochrome *c*, with the possible exception of its binding to cytochrome b_2 which was decreased slightly. (Lysine-72 has been implicated in the interaction of cytochrome *c* with cytochrome b_2.[47])

While the replacement of lysine by arginine in cytochrome *c* may slightly affect the molecule, overall, the results suggest that lysine-72, and its methylation, is not absolutely required. It is possible that the arginine residue acts in a manner analogous to trimethyllysine, in terms of pI, hydrogen bonding, or steric effects, or that arginine does not behave like lysine and sufficiently disrupts the normal external and internal interactions of the unmethylated apoprotein (especially in this mutant), though this is not too likely. Their observations dispute the need for methylation of cytochrome *c*, even though this modification clearly enhances the metabolism of cytochrome *c*.

Though it has not been proven categorically, all evidence supports the hypothesis that apocytochrome *c* is the primary target for methylation *in vivo*. This is not to say that holocytochrome *c* cannot be methylated or that methylation has no effect on the holoprotein. Indeed, there is some evidence that lysine-72 is involved in cytochrome *c* interaction, and future studies may delineate the role of methylation in this interaction. However, at present, since the methylase is a cytosolic protein, and since methylation affects the pI and stability of apocytochrome *c* much more than holocytochrome *c*, conjecture that the effect of this modification is to allow the differentiation between the two forms of apoprotein is not unreasonable.

The occurrence of methylated cytochrome *c* and highly specific enzymes which catalyze its formation in several lower eukaryotes, and the distinct absence of the same in mammals and insects, poses an intriguing question of biological significance in the lower organisms. Indeed, if methylation served no purpose or gave no advantage, its prevalence in these slightly disparate organisms would seem unlikely, even wasteful. On the other hand, if some great advantage was conferred by this modification, it would be reasonable to assume that (some) higher organisms might also possess this trait. It seems the solution to this paradox must be very subtle, such that the answer probably lies in a "relative" advantage which is employed in certain environments. It is proposed that methylation of apocytochrome *c* at lysine-72 causes a shift in the hydrogen bonding which allows the molecule to maintain a more readily imported conformation. This conformation is not required since unmethylated apocytochrome *c* is also imported, however it allows the molecule to be imported more easily. The modification might be advantageous during the shift from anaerobic to aerobic growth or during mitochondria synthesis.

REFERENCES

1. **Paik, W.K., Polastro, E., and Kim, S.,** Cytochrome *c* methylation: enzymology and biologic significance, in *Current Topics in Cellular Regulation,* Vol. 16, Horecker, B.L. and Stadtman, E.R., Eds., Academic Press, New York, 1980, 87.
2. **Rosen, O.M. and Krebs, E.G.,** in *Protein Phosphorylation,* Vol. 8, Cold Spring Harbor Laboratories, Cold Spring Harbor, NY, 1981.
3. **Hughes, R.C.,** in *Glycoproteins,* Chapman and Hall, New York, 1983, chap. 4.
4. **Paik, W.K. and Kim, S.,** in *Protein Methylation,* John Wiley & Sons, New York, 1980.
5. **Smith, E.L.,** in *The Enzymes,* Vol. 1, Boyer, P.D., Ed., Academic Press, New York, 1970, 267.
6. **Schwartz, R.M. and Dayhoff, M.O.,** Cytochromes, in *Atlas of Protein Structure and Sequence,* Vol. 5, National Biomedical Research Foundation, Washington, D.C., 1978, Chap. 4.
7. **Henning, B. and Neupert, W.,** Biogenesis of cytochrome *c* in *Neurospora crassa,* in *Methods in Enzymology,* Vol. 97, Academic Press, New York, 1983, 261.
8. **Hay, R., Bohni, P., and Gasser, S.,** How mitochondria import proteins, *Biochim. Biophys. Acta,* 779, 65, 1984.
9. **Neupert, W. and Schatz, G.,** How proteins are transported into mitochondria, *Trends in Biochem. Sci.,* 6, 1, 1979.
10. **Gonzalez-Busch, C., Miralles, V.J., Hernandez-Yago, J. and Grisolia, S.,** Apocytochrome *c* competes with pre-ornithine carbamoyltransferase for transport in mitochondria, *Biochem. Biophys. Res. Commun.,* 146, 1318, 1987.
11. **Hennig, B. and Neupert W.,** Assembly of cytochrome *c*. Apocytochrome *c* is bound to specific sites on mitochondria before its conversion to holocytochrome *c, Eur. J. Biochem.,* 121, 203, 1981.
12. **Nicholson, D.W., Kohler, H., and Neuport, W.,** Import of cytochrome *c* into mitochondria, *Eur. J. Biochem.,* 64, 147, 1987.
13. **Visco, C., Taniuchi, H., and Berlett, B.S.,** On the specificity of cytochrome *c* synthetase in recognition of the amino acid sequence of apocytochrome *c, J. Biol. Chem.,* 260, 6133, 1985.
14. **Taniuchi, H., Basile, G., Taniuchi, M., and Veloso, D.,** Evidence for formation of two thioether bonds to link heme to apocytochrome *c* by partially purified cytochrome *c* synthetase, *J. Biol. Chem.,* 258, 10963, 1983.
15. **Matsuura, S., Arpin, M., Hannum, C., Margoliash, E., Sabatini, D.D., and Morimoto, T.,** *In vitro* synthesis and posttranslational uptake of cytochrome *c* in isolated mitochondria: role of a specific addressing signal in the apocytochrome, *Proc. Natl. Acad. Sci. U.S.A.,* 78, 4368, 1981.
16. **DeLange, R.J., Glazer, A.N., and Smith, E.L.,** Presence and location of an unusual amino acid, ε-*N*-trimethyllysine in cytochrome *c* of wheat germ and *Neurospora crassa, J. Biol. Chem.,* 244, 1385, 1969.
17. **DiMaria, P., Polastro, E., DeLange, R.J., Kim S., and Paik, W.K.,** Studies on cytochrome *c* methylation in yeast, *J. Biol. Chem.,* 254, 4645, 1979.
18. **Durban, E., Nochumson, S., Kim, S., Paik, W.K., and Chan, S.-K.,** Cytochrome *c*-specific protein-lysine methyltranferase from *Neurospora crassa, J. Biol. Chem.,* 253, 1427, 1978.
19. **Nochumson, S., Durban, E., Kim, S., and Paik, W.K.,** Cytochrome *c*-specific protein methylase III from *Neurospora crassa, J. Biol. Chem.,* 165, 11, 1977.

20. **Farooqui, J.Z., Tuck, M., and Paik, W.K.,** Purification and characterization of enzymes from *Euglena gracilis* that methylate methionine and arginine residues of cytochrome *c*, *J. Biol Chem.*, 260, 537, 1985.
21. **Shin, S. and Paik, W.K.,** manuscript in preparation.
22. **Liao, H.H. and Sherman, F.,** Yeast cytochrome *c*--specific protein-lysine methyltransferase: coordinate regulation with cytochrome *c* and activities in *cyc* mutants, *J. Bacteriol.*, 138, 853, 1979.
23. **Margoliash, E. and Lustgarten, J.,** Interconversion of horse heart cytochrome *c* monomers and polymers, *J. Biol. Chem.*, 237, 3397, 1962.
24. **Jacobs, E.E. and Sanadi, D.R.,** The reversible removal of cytochrome *c* from mitochondria, *J. Biol. Chem.*, 235, 531, 1960.
25. **Scott, W.A. and Mitchell, H.K.,** Secondary modification of cytochrome *c* from *Neurospora crassa*, *Biochemistry*, 8, 4282, 1969.
26. **Polastro, E., Looze, Y., and Leonis, J.,** Study of biological significance of cytochrome *c* methylation, *Biochim. Biophys. Acta*, 446, 310, 1976.
27. **Looze, Y., Polastro, E., Gielens, C. and Leonis, J.,** Isocytochrome *c* species from Baker's yeast, *Biochem. J.*, 157, 773, 1976.
28. **Kim, C.-S., Kueppers, F., DiMaria, P., Farooqui, J.Z., Kim, S. and Paik, W.K.,** Enzymatic trimethylation of residue-72 lysine in cytochrome *c*, *Biochim. Biophys. Acta*, 622, 144, 1980.
29. **Paik, W.K., Farooqui, J.Z., Roy, T., and Kim, S.,** Determination of pI values of variously methylated amino acids by isoelectric focusing, *J. Chromatogr.*, 256, 331, 1983.
30. **Paik, W.K., Farooqui, J.Z., Gupta, A., Smith, H.T., and Millett, F.,** Enzymatic trimethylation of lysine-72 in cytochrome *c*, *Eur. J. Biochem.*, 135, 259, 1983.
31. **Park, K.S., Frost, B., Shin, S., Park, I.K., Kim, S., and Paik, W.K.,** Effect of enzymatic methylation of yeast iso-1-cytochrome *c* on its isoelectric point, *Arch. Biochem. Biophys.*, 267, 195, 1988.
32. **Poncz, L. and Dearborn, D.G.,** The resistance to tryptic hydrolysis of peptide bonds adjacent to *N,N*-dimethyllysine residues, *J. Biol. Chem.*, 258, 1844, 1983.
33. **Paik, W.K. and Kim, S.,** Effect of methylation on susceptibility of protein to proteolytic enzymes, *Biochemistry*, 11, 2589, 1972.
34. **Martinez, R.J., Shaper, J.H., Lundh, N.P., Bernard, P.D., and Glaser, A.N.,** Effect of proteolytic enzymes on flagella, *J. Bacteriol.*, 109, 1239, 1972.
35. **Van Noort, J.M., Kraal, B., Sinjorgo, K.M.C., Persoon, N.L.M., Johanns, E.S.D., and Bosch, L.,** Methylation *in vivo* of elongation factor EF-Tu at lysine-56 decreases the rate of tRNA-dependent GTP hydrolysis, *Eur. J. Biochem.*, 160, 557, 1986.
36. **Polastro, E., Schnek, A.G., Leonis, J., Kim, S., and Paik, W.K.,** Cytochrome *c* methylation, *Int. J. Biochem.*, 9, 795, 1978.
37. **Farooqui, J.Z., DiMaria, P., Kim, S., and Paik, W.K.,** Effect of methylation on the stability of cytochrome *c* of *Saccharomyces cerevisiae in vivo*, *J. Biol. Chem.*, 256, 5041, 1981.
38. **Farooqui, J.Z., Kim, S., and Paik, W.K.,** *In vivo* studies on yeast cytochrome *c* methylation in relation to protein synthesis, *J. Biol. Chem.*, 255, 4468, 1980.
39. **Polastro, E., Looze, Y., and Leonis, J.,** Biological significance of methylation of cytochrome from ascomycetes and yeast, *Phytochemistry*, 16, 39, 1977.
40. **Polastro, E., Deconinck, M.M., DeVogel, M.R., Mailier, E.L., Looze, Y.R., Schnek, A.G., and Leonis, J.,** Evidence that trimethylation of iso-1-cytochrome *c* from *Saccharomyces cerevisiae* affects interaction with mitochondria, *FEBS Lett.*, 86, 17, 1978.
41. **Park, K.S., Frost, B., Tuck, M., Ho, L.L., Kim, S., and Paik, W.K.,** Enzymatic methylation of *in vivo* synthesized apocytochrome *c* enhances its transport into mitochondria, *J. Biol. Chem.*, 262, 14702, 1987.
42. **Hurt, E.C.,** Unravelling the role of ATP in post-translational protein translocation, *Trends in Biochem. Sci.*, 12, 369, 1987.
43. **Stock, J. and Koshland, D.E. Jr.,** in *Microbial Development*, Shapiro, L. and Losick, R., Eds., Cold Spring Harbor Laboratories, Cold Spring Harbor, NY, 1984, 117.
44. **Stock, J., Borszuk, A., Chiou, F., and Burchenal, J.E.B.,** Compensatory mutations in receptor function: a re-evaluation of the role of methylation in bacterial chemotaxis, *Proc. Natl. Acad. Sci. U.S.A.*, 82, 8364, 1985.
45. **Stock, J. and Stock, A.,** What is the role of receptor methylation in bacterial chemotaxis?, *Trends Biochem. Sci.*, 12, 371, 1987.
46. **Holzschu, D., Principio, L., Conklin, K.T., Hickey, D.R., Short, J., Rao, R., McLendon, G., and Sherman, F.,** Replacement of the invariant lysine-77 by arginine in yeast iso-1-cytochrome *c* results in enhanced and normal activities *in vitro* and *in vivo*, *J. Biol. Chem.*, 262, 7125, 1987.
47. **Guiard, B. and Lederer, F.,** The "cytochrome b_5 fold": structure of a novel protein superfamily, *J. Mol. Biol.*, 135, 639, 1979.

Chapter 5

PROTEIN-ARGININE METHYLATION: MYELIN BASIC PROTEIN AS A MODEL

Sangduk Kim, Latika P. Chanderkar, and Subrata K. Ghosh

TABLE OF CONTENTS

I. INTRODUCTION

Since the discovery of methyllysine in flagella protein[1] and of its origin in nuclear proteins via posttranslational methylation reaction,[2] several amino acid side chains, such as arginine, histidine, and aspartic/glutamic acid have also been shown to be methylated at the polypeptide level.[3-6] The formation of these methylated amino acids is catalyzed by methyl-group transfer reaction utilizing *S*-adenosyl-L-methionine (AdoMet) as the methyl donor. It is increasingly evident from the past 2 decades of investigations in several laboratories that the significance of protein methylation varies from eukaryote to prokaryote and from structural/membrane proteins to functional proteins such as carnitine[7] and calmodulin.[8] The diversity and wide occurrence of methylated amino acids in nature consequently led to uncover the multiplicity of the methyltransferases specific for each methyl acceptor protein. It is now clear that each protein methyltransferase with a given amino acid residue specificity can be further subclassified based on the specific methyl acceptor protein.

Protein methylase I[9] (PMI; *S*-adenosylmethionine:protein-arginine *N*-methyltransferase; EC 2.1.1.23) is one such enzyme for which evidence suggested more than one subtype existed, however, an unequivocal molecular evidence for the presence of subclasses of the enzyme (myelin basic protein-specific and histone-specific) has been obtained only in recent years. In 1971, the specific occurrence of methylarginine in Res-107 of myelin basic protein (MBP), which is catalyzed by PMI, was reported independently by Baldwin and Carnegie,[10] and Brostoff and Eylar.[11] Since this initial discovery, several laboratories have studied the MBP-arginine methylation in conjunction with myelination,[12-18] as MBP is the major protein constituent of the myelin membrane. The modification of the side chain arginine is expected to play some role in the structure-function relationship of this membrane protein. Furthermore, the membrane system is suitable for the study of the posttranslational modification, since membrane assembly is predominantly a postsynthetic phenomenon. Several specific questions can be asked — What is the effect of the methyl-modification of MBP on its primary, secondary, and tertiary structure? What is the effect of methylation on the subsequent interaction of MBP with other membrane components (phospholipid, proteolipid proteins, and proteins other than MBP) and/or between MBP? Is there any temporal relationship between the methylating activity and myelination process?

In this article, we have reviewed two major areas of recent development in the study of protein-arginine methylation: (1) the molecular distinction between subclasses of PMI (namely, MBP-specific and histone-specific) from mammalian brain and other eukaryotic tissue; and (2) the biochemical significance of MBP-specific methylation as it relates to myelination in the central nervous system.

II. ENZYMOLOGY OF PROTEIN-ARGININE METHYLATION

A. REACTION OF PROTEIN METHYLASE I

Enzymatic methylation on the guanidinium group of arginyl residue is catalyzed by PMI utilizing AdoMet as the methyl donor. The enzyme was initially discovered in 1968 in calf thymus during the course of the studies on enzymatic methylation of lysyl residues.[9] The alkali-labile nature of the enzymatic products, N^G-methylarginine, resulted in the release of methylurea and methylamine upon 2 *M* NaOH hydrolysis (in contrast to the alkali-stable methyllysine), and served as a clue to the identification of protein-arginine methyltransferase reaction. The enzymatic reaction is known to yield three methylated arginine derivatives:[9,19] N^G-monomethylarginine [MMeArg], N^G,N'^G-dimethylarginine [Di(sym)MeArg] and N^G,N^G-dimethylarginine [Di(asym)MeArg], as shown in Figure 1. The demethylated AdoMet, *S*-adenosyl-L-homocysteine (AdoHcy) which is a potent product inhibitor for all known AdoMet-dependent transmethylation reaction,[20-22] also inhibits PMI with a K_i value of approximately $10^{-6}M$.[23]

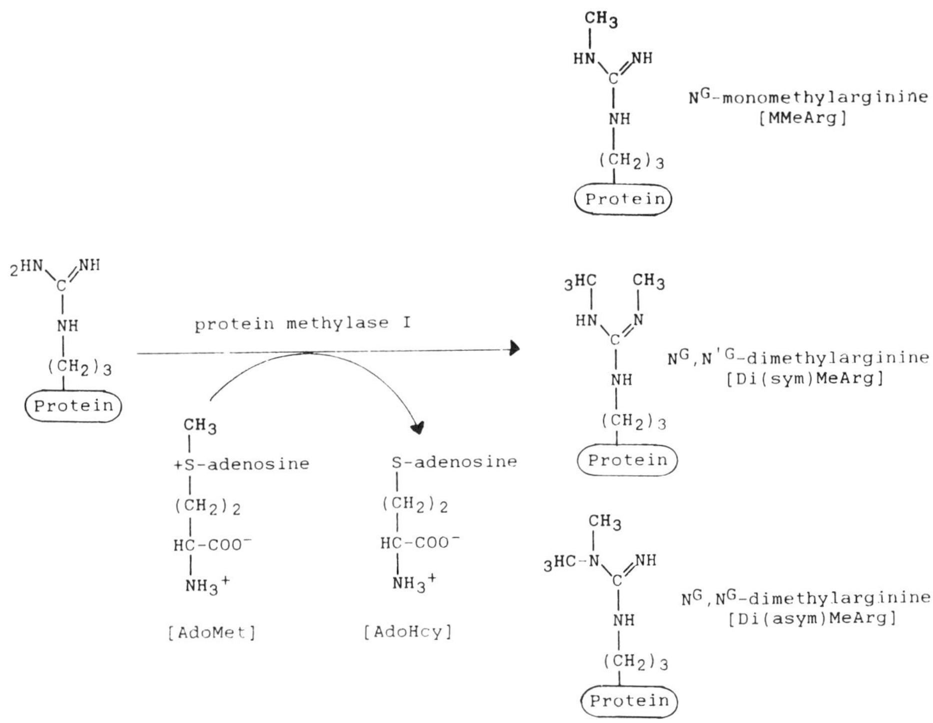

FIGURE 1. Reaction of protein methylase I.

The natural occurrence of these methylated amino acids in proteins is limited to a number of highly specialized proteins, such as MBP[10,11] and nuclear[24-26] and contractile proteins[27] (see Chapter 1), which are primarily structural or membrane proteins. The methylation reaction is highly specific for the amino acid sequence around the methylation site and the kind of dimethylarginine formed. Among 18 arginine residues in MBP, only arginine at residue-107, in the sequence of Lys-Gly-Arg-Gly-Leu, is present as a mixture of MMeArg and Di(sym)MeArg.[10,11] Additionally, several nuclear/nucleolar proteins and muscle proteins contain MMeArg and Di(asym)MeArg, but Di(sym)MeArg is not detectable in these proteins. The presence of several Di(asym)MeArg residues in nonhistone nuclear proteins, such as nucleolin, 34 kDa nucleolar protein, HnRNP (A-1 protein) and high mobility group 1 and 2 is of interest since these proteins contain clusters of glycine and Di(asym)MeArg interspersed with phenylalanine (refer to Chapter 6 for further details).

B. PROTEIN METHYLASE I FROM MAMMALIAN ORGANS

PMI activity is widely present in mammalian organs, having higher activity in the order of the testis, brain, thymus, spleen, kidney, and liver.[3-5] Although cytosol has been shown to be the major site for the enzyme in most mammalian tissues, recent study indicates that particulate/membrane fraction of the brain also contains a significant amount of the enzyme. The measurement of the enzyme activity in crude tissue extract has been complicated, since two other protein *N*-methyltransferase (protein-lysine and protein-histidine *N*-methyltransferases) as well as protein-carboxyl *O*-methyltransferase also yield the $C1_3CCOOH$-precipitable products. However, as a rule, the different optimal pH of each reaction and the alkali-lability of the products have been a useful means to distinguish between the different methylation reactions, prior to final confirmation of the methylated product by amino acid analysis. In most cases, pH optimum of mammalian PMI is about 7.2, whereas that of protein lysine methyltransferase is 9.0, and of protein carboxylmethyltransferase is 6.0.

TABLE 1
Purification of MBP-Specific Protein Methylase I

Purification steps	Total protein (mg)	Enzyme activity[a] S.A.[b]	Enzyme activity[a] Total activity	Purification (-fold)
Supernatant at 78,500 × *g*	2660	1.24	3300	1
$(NH_4)_2SO_4$ precipitates (0–40%)	840	2.1	1790	1.69
Acetone precipitates	630	1.23	775	0.99
DE-52 batchwise treatment	40	13.4	536	10.8
DE-52 column chromatography	5.6	60.4	338	48.7
1st Sephadex G-200 chromatography	1.04	80.9	84.0	65.2
Hydroxyapatite chromatography	0.178	198	35.2	160
2nd Sephadex G-200 chromatography	0.070	501	35.0	404

[a] Enzyme activity was measured using MBP as the methyl acceptor protein.
[b] S.A. is defined as picomole of methyl/group transferred/minute/milligram protein under the assay condition.

Earlier studies on the mammalian PMI showed both histone and MBP to be good substrates for the enzyme. However, it has become increasingly apparent that the enzyme responsible for methylating MBP is distinctly different from the histone-methylating enzyme. It has earlier been observed that the ratio of the enzyme activities determined using MBP and histone as substrates varied significantly during purification of PMI from calf brain.[23] Nevertheless, purification of MBP-PMI from brain encountered great difficulty due to the fact that the level of the histone-PMI activity in the brain is more than seven times higher than that of MBP-PMI, causing both methylases to be easily copurified. In 1985, Park et al.[28] demonstrated that a partially purified histone-specific PMI from calf brain was not only incapable of methylating MBP, but was also inhibited by MBP.

Recently, we have successfully purified both PMIs from calf brain cytosol to near homogeneity utilizing differential ammonium sulfate precipitation and chromatographies on DE-52, hydroxyapatite and Sephadex G-200 (Tables 1 and 2; Figure 2). It is remarkable to observe that although both enzymes methylate arginine residues in protein substrates, there are more differences than similarities in their respective properties. The most notable differences are respective molecular weight, affinity for protein substrate, and immunological recognition. The MBP-specific PMI (500 kDa) preferentially methylates MBP ($K_m = 2 \times 10^{-7}$ *M*) but also histone to a much lesser extent ($K_m = 1 \times 10^{-4}$ *M*), while the histone-specific enzyme (275 kDa) methylates histone only. Both enzymes exhibit two major subunit bands on SDS-polyacrylamide gel electrophoresis; 100 and 75 kDa for the former, and 110 and 72 kDa for the latter (Figure 3). Sensitivity toward various chemicals and higher temperatures are also different, as summarized in Table 3. Western immunoblot analysis of the purified PMIs following nondenaturing PAGE indicated that the corresponding enzyme band is only immunoreactive against its own respective antibodies (Figure 4), indicating no cross-reactivity between the two methylases.

C. PROTEIN METHYLASE I FROM CULTURED MAMMALIAN CELLS

Several mammalian cells in culture have been shown to contain a significant amount of PMI activity. Casellas and Jeanteur[29] partially purified PMI from Krebs II ascites cell nuclei by lysing the cells in 0.4 *M* NaCl followed by phosphocellulose, DE-52, and hydroxyapatite chromatography. The substrate specificity indicated that only histones were efficient substrates. The reaction was quite sensitive to ionic strength, 50% of the activity being inhibited at 0.1 *M* NaCl. The molecular weight of the enzyme was approximately 500 kDa determined by gel

TABLE 2
Purification of Histone-Specific Protein Methylase I

Purification steps	Total protein (mg)	Enzyme activity[a] S.A.[b]	Enzyme activity[a] Total activity	Purification (-fold)
Supernatant at 78,000 × *g*	1760	2.52	4440	1
$(NH_4)_2SO_4$ precipitates (40–70%)	480	3.57	1710	1.42
DE-52 chromatography	45	20.5	923	8.13
Sephadex G-200 chromatography	9.8	31.5	309	12.5
Hydroxyapatite chromatography	1.6	75.0	120	29.8

[a] Enzyme activity was measured using histone as the methyl acceptor protein.

[b] S.A. is defined as picomole of methyl/group transferred/minute/milligram protein under the assay condition.

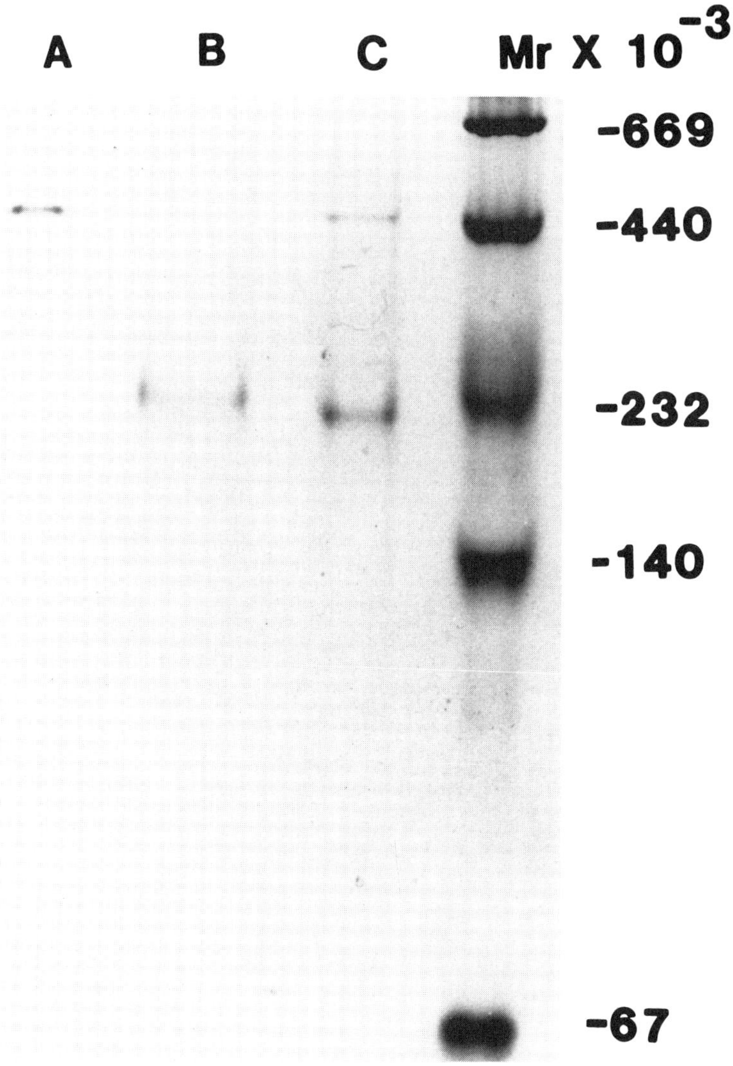

FIGURE 2. Nondenaturing polyacrylamide gel electrophoresis of purified MBP-specific and histone-specific protein methylase I. Lanes indicate: A,MBP-specific; B, histone-specific; and C, a mixture of MBP- and histone-specific enzymes.

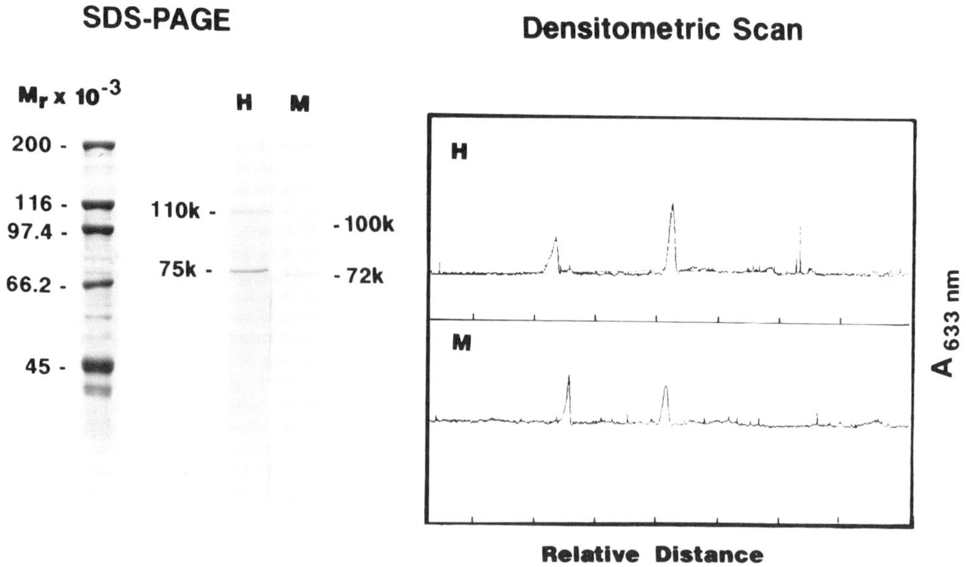

FIGURE 3. SDS-polyacrylamide gel electrophoresis of purified protein methylase I. Symbols indicate: H, histone-specific; and M, MBP-specific protein methylase I. Densitometric scan of the Coomassie blue stained gel indicates that the ratio is 40:60 for 110 kDa; 75 kDa (histone-enzyme subunits) and 49:51 for 100 kDa:72 kDa (MBP-enzyme subunits).

filtration, indicating a possible multimeric structure or aggregation, however, all attempts to dissociate the complex into smaller enzymatically active molecules have failed. The methylated products were identified as MMeArg and Di(asym)MeArg.

In order to probe PMI binding site for AdoMet, the inhibition kinetics of ascites cell PMI were carried out using various analogs of AdoHcy,[30] and it was found that *S*-isobutyladenosine was the most potent inhibitor.[30] Similar studies were also carried out by Lederer and co-workers[31-33] with PMI isolated from chick embryo fibroblast. The most significant observation in this study was that there existed a correlation between inhibition of PMI and that of the virus-induced chick embryo fibroblast transformation by AdoHcy analogs i.e., all good inhibitors of PMI were also good inhibitors of the fibroblast transformation. Characteristically, *S*-isobutyladenosine and sinefungin inhibited transformation, while AdoHcy had no effect on cell transformation.[33] Readers should refer to Chapter 18 for a comprehensive review on this subject.

D. PROTEIN METHYLASE I FROM OTHER EUKARYOTES

PMI has also been identified and partially purified from wheat germ[34] and *Euglena gracilis*.[35] The former enzyme is specific for histone and the latter for cytochrome *c*. Both methylases do not methylate MBP, which is quite logical since these organisms naturally do not contain MBP. A unique feature of wheat germ PMI is that intracellular adenosine acts as an endogenous inhibitor of the enzyme. Thus, the recovery of the enzyme activity increased to 160% of the whole homogenate activity after 90-fold purification. The enzyme was shown to require a low molecular weight, dialyzable and heat-labile cofactor present in cell extract, but its identity is not yet known. The molecular weight of the enzyme was estimated to be 28 kDa and the optimum pH of the reaction is 9.0, unlike that of mammalian PMI which is pH 7.2. The *in vitro* methylation site of the histone H4 substrate by the enzyme was identified as the Res-35 arginine yielding only N^G-monomethylarginine.[34] The AdoMet analogs such as AdoHcy, 9145C, sinefungin and *S*-inosyl(2-hydroxy-4-methyl-thio)-butyrate on the PMI from wheat germ showed the K_i value of about 10^{-6} *M*. The PMI that specifically methylates horse heart cytochrome *c* at the Res-38 arginine has also been identified and purified from *Euglena gracilis* cytosol.[36] A more detailed account on this enzyme is presented in Chapter 4.

TABLE 3
Comparative Properties of MBP-Specific and Histone-Specific Protein Methylase I from Calf Brain

Characteristics	MBP specific	Histone specific
M_r (by Sephadex G-200)	500 kDa	275 kDa
Subunit (by SDS-PAGE)	100 kDa, 72 kDa	110 kDa, 75 kDa
pI	5.09	5.68
K_m values		
MBP	2.3×10^{-7} *M*	—
Histone	1.0×10^{-4} *M*	1.1×10^{-5} *M*
AdoMet	4.4×10^{-6} *M*	8.0×10^{-6} *M*
K_i values		
MBP	—	3.42×10^{-5} *M*[a]
S-adenosyl-L-homocysteine	1.8×10^{-6} *M*	2.3 × 10-6 *M*
Sinefungin	7.0×10^{-6} *M*	6.6×10^{-6} *M*
Dialysis	Easily inactivated	Not inactivated
50% inactivation		
p-chloromercuribenzoate	0.46 m*M*	0.15 m*M*
Guanidine-HCl	3.1 m*M*	0.3 m*M*
At 50°C for 5 min	99% activity remained	60% activity remained

[a] Data is from Reference 28.

III. BIOLOGY OF MYELIN BASIC PROTEIN METHYLATION AND MYELINATION

A. BACKGROUND

1. Myelin Basic Protein as Myelin Component

Myelin is a membrane characteristic of the nervous tissue laid down in segments along the selected nerve fibers; it functions as an insulator to increase the velocity of the stimuli being transmitted between a nerve cell body and its target.[37] Large numbers of neurological diseases result in the degeneration of myelin. Myelin is present in both the central and peripheral nervous system, although their origins are different; the former originates from oligodendrocytes, and the latter from Schwann cells.

The myelination process can be broadly divided into two stages of development.[38,39] In the case of the mouse, the first stage is between 8 to 15 d of postnatal period, while the second stage is between 16 to 30 d of age. In the first stage, oligodendrocytes proliferate, enlarge, and plasma membrane is formed which then starts to wrap loosely around the nerve axon to form a multilamellar structure. In the second stage, an active deposition of different proteins takes place for the compaction of the lamellar structure.

The cytoplasmic surfaces of these bilayers interact to form a major dense line of myelin where MBP is appositioned, while the extra cytoplasmic surfaces interact to form the intraperiod line where proteolipid proteins are located. Thus, structurally, myelin is recognized as a lipid bimolecular leaflet, with proteins sandwiched between the two layers, which is wrapped in a spiral fashion around a segment of axon. A model, based on available data, illustrating the molecular organization of myelin macromolecules, indicates that proteolipid protein is a transbilayer component in monomeric/polymeric form, whereas MBP is an extrinsic protein having one or more domains in contact with hydrophobic interior of the bilayer.[40] Thus, the basic protein is exclusively located at the cytoplasmic layer, probably as a dimer, a head-to-tail (i.e., antiparallel) orientation.

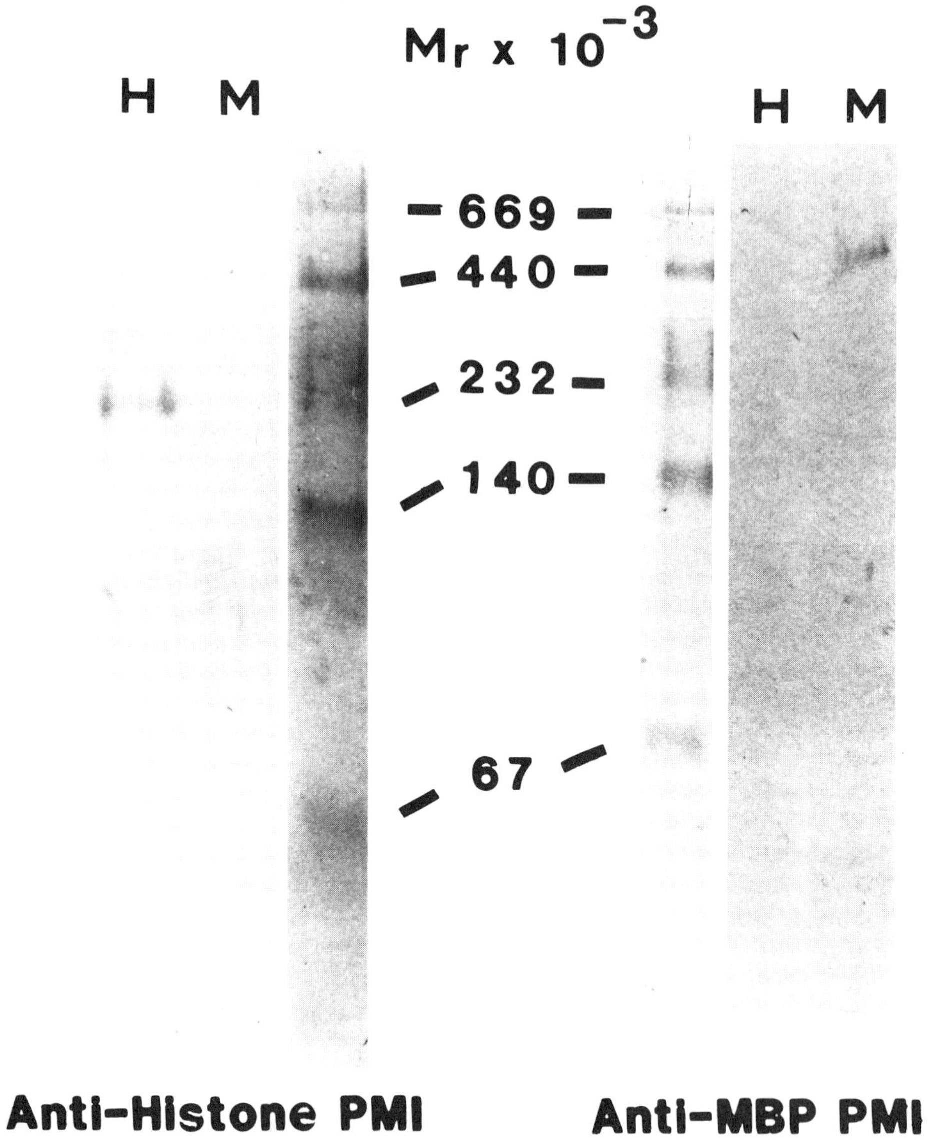

FIGURE 4. Western immunoblot analysis of protein methylase I subsequent to nondenaturing polyacrylamide gel electrophoresis. Lane H indicates histone-specific, and M, MBP-specific enzyme.

2. Characteristics of Myelin Basic Protein Structure

Myelin isolated from human and bovine nervous tissue is composed of approximately 80% lipid and 20% protein. The protein fraction is composed of 30% MBP, 50% proteolipid, and 20% high molecular weight Wolfgram protein.[41] The MBP is a highly conserved protein and occurs as a single polypeptide chain with a molecular weight of approximately 18,000 (170 amino acids) in many species, including bovine and human.[42,43] In rodents, however, several different molecular species of MBP are present, and these are encoded by distinct mRNA formed by alternative RNA splicing pathways acting on a common primary transcript.[44,45]

There are several unusual properties of MBP: it is a highly basic protein (pI = 10.5), and has a high proline content with lack of tertiary structure.[46] MBP is known to undergo rapid proteolysis[47,48] particularly when it is dissociated from the membrane structure.[49] During the course of MBP primary sequence analysis, Baldwin and Carnegie[10] and Brostoff and Eylar[11] found an unusual amino acid, N^G-methylarginine, at Res-107. The amino acid was identified as

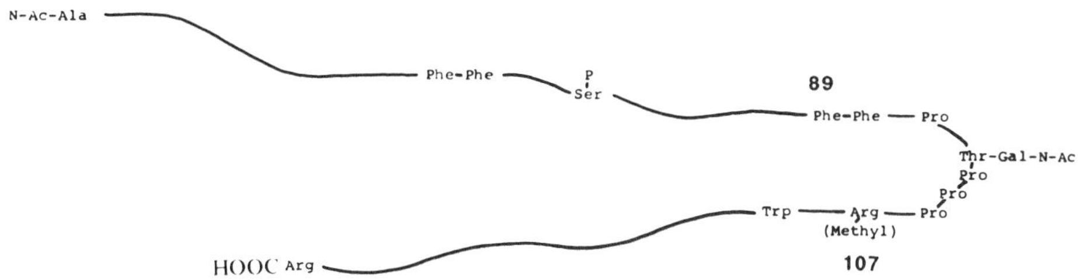

FIGURE 5. Schematic representation of MBP in the hair-pin configuration.

a mixture of two methylarginine derivatives, MMeArg and Di(sym)MeArg. Partial methylation of the methylatable arginine residue in MBP has been reported and the ratio of methylated to unmethylated arginine differed from species to species.[50,51] The rat has a lower proportion of Di(sym)MeArg than the human. It has been suggested that the extent of methylation must be genetically regulated within the species.[51]

MBP contains a single tryptophan residue near the midpoint of the sequence[41-43] which has been implicated as a focus for the immunological properties of MBP.[52] When the protein is injected the animal develops neurological symptoms, known as "experimental allergic encephalomyelitis", a model often used for the study of multiple sclerosis in the human.[41,46,53] Another unique feature of the protein is the presence of a relatively rare triproline sequence (Res 99 to 101) in the region close to the tryptophan residue. It has been speculated that the triproline sequence bends the protein into a double chain configuration and that the methylation of arginine at Res-107 could provide stabilization of the "hair-pin" conformation[11,46] (Figure 5), either by interaction with lipids or in conjugation with an adjacent phenylalanine side chain, helping to stabilize the insertion of MBP into myelin sheath.

B. INVOLVEMENT OF METHYLATION IN MYELINATION

An involvement of biological methylation in the integrity and maintenance of myelin has been suggested earlier through the animal experiments which produced subacute combined degeneration (SCD). SCD is found in man with untreated vitamin B_{12} deficiency, and is characterized by the degeneration of the myelin sheath. When mice[54] or monkeys[55,56] were exposed to an atmospheric environment containing 15% nitrous oxide, the animal became ataxic and neurological disorder developed over a period of 2 to 3 weeks until the animals were moribund. Microscopic examination of the nitrous oxide-treated spinal cord and peripheral nervous system showed the classical changes of SCD.[56] Nitrous oxide oxidizes cobalamin (vitamin B_{12}), thus blocking the formation of methylcobalamin which is an essential intermediate for the biosynthesis of methionine in animals (the overall biological transmethylation pathway is depicted in Figure 6). When the experimental diet was supplemented with methionine, the animals were free of any detectable clinical changes.[56]

Jacobson and co-workers further showed that a neurological syndrome similar to vitamin B_{12} deficiency can be produced in mice by injecting cycloleucine.[57] Cycloleucine is an analog of methionine which inhibits the biosynthesis of AdoMet, a biological methyl donor for a variety of compounds including proteins and other macromolecules.[58] Indeed, Crang and Jacobson[18,57] demonstrated that cycloleucine inhibited the activity of *S*-adenosylmethionine synthetase (EC 2.5.1.6) as well as methylation of MBP in the brain. These results thus argue against the previously suspected methylation of phospholipid hypothesis[55,60] which involves the synthesis of methylmalonyl-CoA:[59] Cyclolcucine does not block the CoA biosynthesis. Utilizing cycloleucine to produce myelopathy in mice, Small et al.[61] suggested that inhibition of MBP methylation might be responsible for this condition, since cycloleucine strongly depressed the formation of N^G-methylarginine *in vivo,* while the incorporation of methionine directly into myelin protein was not affected, thus suggesting that MBP-methylation and not the synthesis of the protein was inhibited.

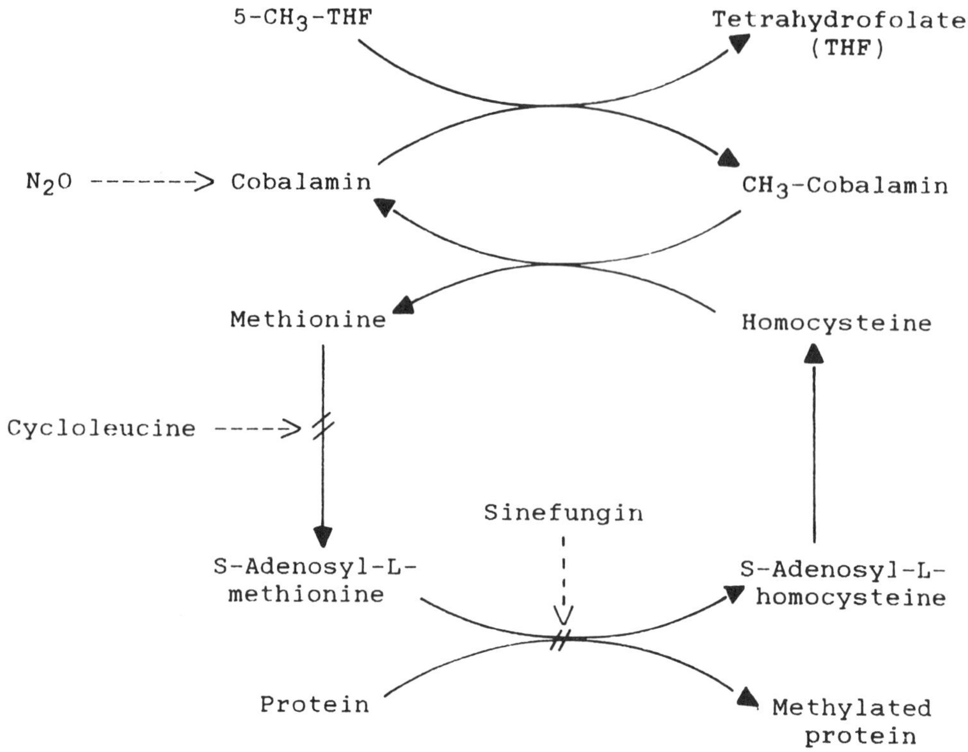

FIGURE 6. Biological transmethylation pathway.

C. *IN VIVO* BIOSYNTHESIS AND METHYLATION OF MYELIN BASIC PROTEIN

The developmental expression of the mouse MBP was studied by Campagnoni and coworkers using intracerebral injection of L-[^{3}H]lysine;[39] these studies showed that the biosynthesis of MBP parallels with the myelination process with maximal synthesis at 18 d of the mouse brain, with the increased rate of the 14-kDa MBP synthesis in relation to that of the 18.5-kDa species during the developmental period. Unlike bovine and human MBP (M_r =18.0 kDa), the MBP from rodents displays polymorphism, as shown by several investigators,[62-64] with apparent molecular weights of 34, 30, 29, 23.6, 21, 18.5, 17, and 14.5K, which are all immunoreactive against MBP antibody. We have also demonstrated that differences in the rate of accumulation of each MBP species correlated with brain development:[64] the higher molecular weight species were predominant in younger brain, while the smaller MBPs were the major species in older brain (Figure 7). Although the genetic basis of this developmental pattern of MBP polymorphism is not known, it is speculated that each species may play a significant role during the myelination process. The differential developmental pattern of MBPs may be due to their different rates of incorporation into myelin, their turnover rates, or their functional relationship within the membrane.

Employing chicken as an experimental animal, Small and Carnegie[65] studied the *in vivo* methylation of MBP. The incorporation of methyl groups of injected [*methyl*-^{3}H]methionine into methylarginines in myelin was found to occur readily in 2-d-old chickens at a ratio of Di(sym)MeArg to MMeArg to Arg of 1.3:0.9:1.0, and these arginine derivatives were confined only in MBP. Di(asym)MeArg was not detected in MBP, but could only be derived from other brain proteins, confirming exclusive occurrence of Di(sym)MeArg in MBP. DesJardins and Morell[66] further studied the metabolic relationship between MBP backbone and the methyl groups of methylarginine using [*methyl*-^{3}H]-methionine as a precursor. Turnover rates of the

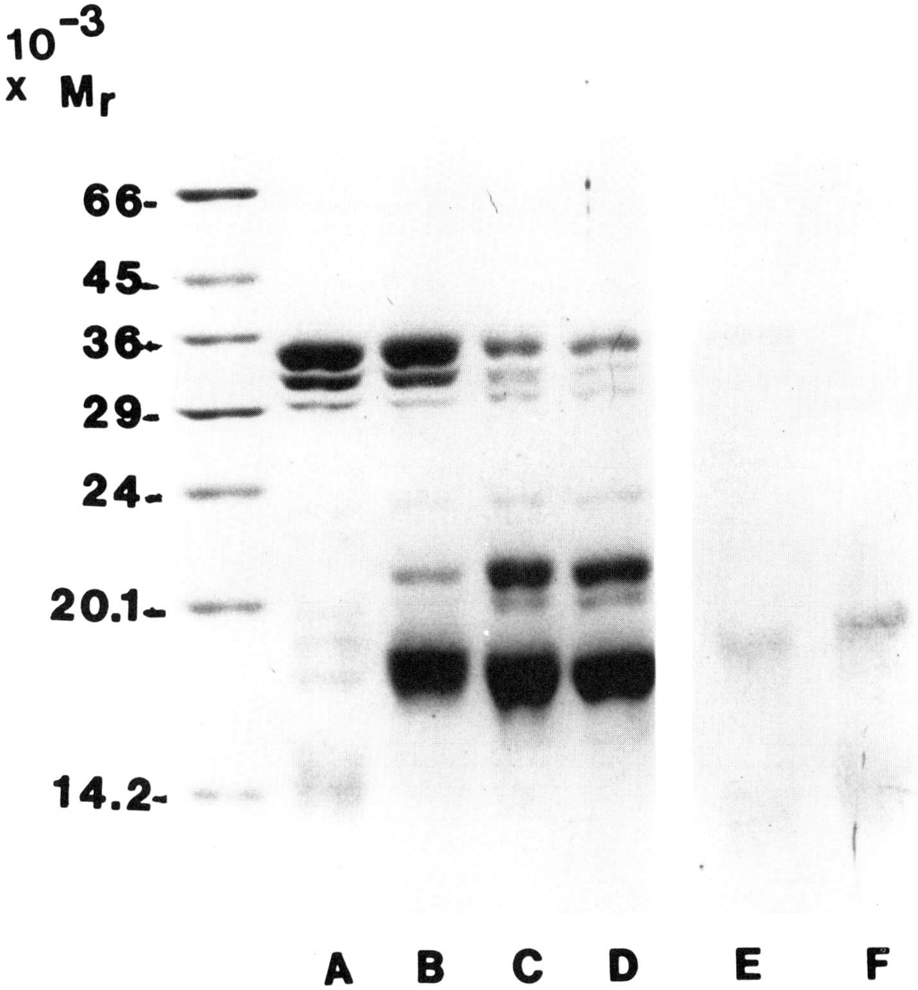

FIGURE 7. Electrophoretic evaluation of different species of MBP during mouse brain development. Lanes A, B, C, and D represent MBP isolated from 13-, 17-, 49-, and 59-d-old mouse brains, respectively. Lanes E and F represent the immunoblot of lanes C and D, respectively.

incorporated [^{3}H]-labels into MBP and the methylarginine were shown to have similar half-lives, indicating the parallel synthesis and methylation of MBP.

Recently, Chanderkar et al.[64] carried out a systematic study on the MBP biosynthesis and methylation during mouse brain development. An initial study was carried by injecting L-[*methyl*-^{3}H]methionine to mice brains at different ages from 12 to 60 d. MBP was purified from the brain "acid extract" by CM-52 chromatography utilizing NaCl gradient as described by Chou et al.[67] As shown in Figure 8, after "breakthrough" fractions MBP can be subfractionated into three major components which elute between 0.08 and 0.132 *M* NaCl concentrations; component 1 was mainly small MBP (M_r = 14 kDa), whereas 2 and 3 are a mixture of small and large MBPs. The radioactivity incorporation into MBP fractions were maximum at younger age compared to the mature brain (Table 4).

In order to distinguish the extent of methyl-group incorporation and protein backbone synthesis of MBP, a double-labeling technique was used; L-[^{35}S]-methionine for backbone synthesis and L-[*methyl*-^{3}H]methionine for both backbone synthesis and methylation via Ado[*methyl*-^{3}H]Met. The ratio of ^{3}H/^{35}S incorporated into MBP thus serves as a "methylation index"; the higher the ratio, the greater the extent of methylation. As shown in Table 5, the ^{3}H/

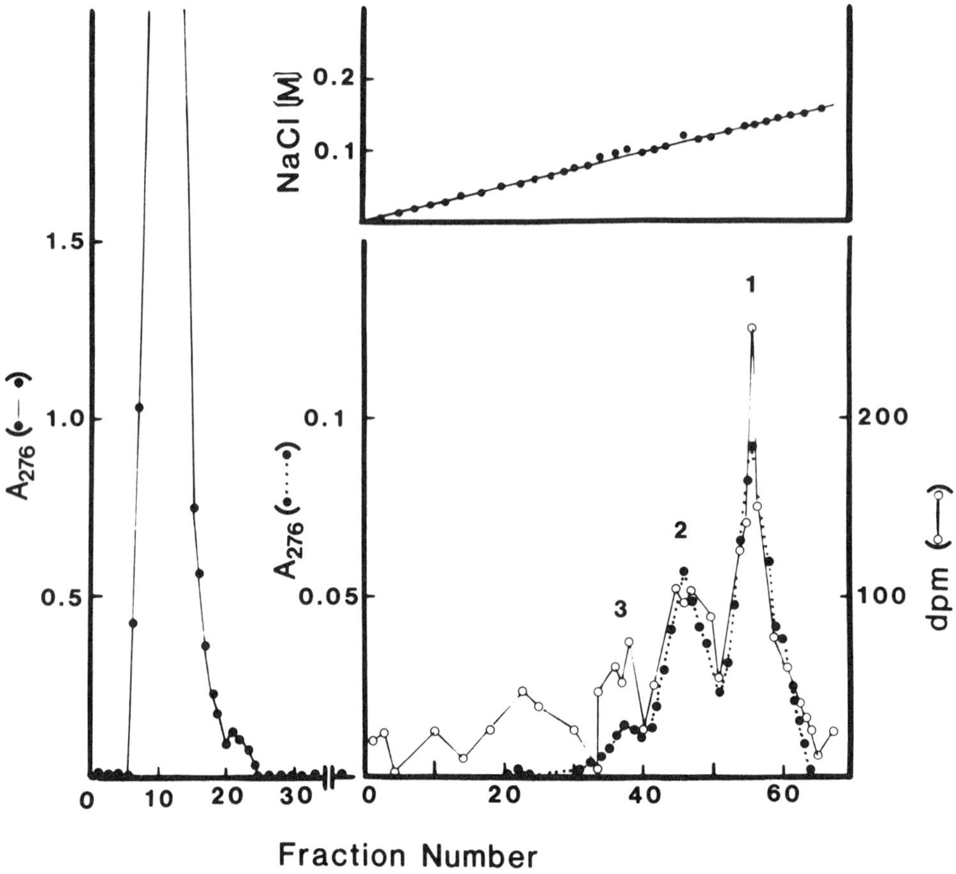

FIGURE 8. Elution profile of the mouse brain "acid extract" on DE-52 chromatography. MBP was eluted with a linear gradient of 0.1 to 0.3 *M* NaCl as described. 1, 2, and 3 indicate MBP components.

^{35}S in the "acid extract" and "breakthrough" fractions remained about the same as the originally injected ratio of 3.6, whereas values in the MBP fractions were much higher: the ratio was the highest at 17 d of age when myelination and protein synthesis were at maximum, and decreased thereafter, remaining at about the ratio of 6 to 8 in mature animals.This high ratio indicates a higher methylation rate corresponding to the active MBP-backbone synthesis during brain development and, in turn, myelination. These results strongly demonstrate that MBP-arginine methylation is indeed a myelination-associated process.

D. PROTEIN METHYLASE I ACTIVITY DURING BRAIN DEVELOPMENT

Several investigators studied the level of PMI during brain development,[12,13,68] the period when active myelination takes place. However, earlier studies had been flawed to measure MBP-specific PMI activity because only histone was used as methyl acceptor substrate. Nevertheless, the high enzyme activity in the neonatal rat brain decreased progressively as a function of age correlating well with the decrease in the capacity of cell proliferation, but not with myelination.[12,13,68] In 1973, Miyake and Kakimoto[16] suggested that cytoplasmic fraction of rat brain contains two different PMIs based on the evidence that differential ammonium sulfate precipitation yielded different levels of enzyme activities when MBP or histone was used as methyl acceptor. Subsequently, Miyake,[17] and Crang and Jacobson[18] employing the developing rat brain and the mouse spinal cord, respectively, have demonstrated the parallel increase of the MBP-PMI during active myelination, whereas histone-specific PMI activity decreased during

TABLE 4
Incorporation of L-[*Methyl*-^{3}H]Methionine into Mice Brain MBP

Age of animal (d)	Component[a]	Amount of MBP (mg/ten mouse brains)	[^{3}H]-incorporation (dpm/mg)
13	1	0.162	23682
	2	0.167	12768
	3	0.173	20590
20	1	0.486	13900
	2	0.381	11600
	3	0.091	9400
33	1	1.13	901
	2	0.743	916
	3	0.161	510
45	1	2.010	401
	2	1.230	341
	3	0.144	070
60	1	1.640	923
	2	1.070	786
	3	0.243	813

[a] The numbers, 1, 2, and 3 indicate the MBP components shown in Figure 8. Amount of MBP was estimated by using the absorption coefficient of 5.89 (1% solution) at 276 nm.

this period. Recently, Chanderkar et al.[64] have also confirmed the temporal correlation between MBP-PMI and myelination in mouse brain: the enzyme activity increased during brain development and reached its peak level at the age of 17 d when myelination and MBP-synthesis is maximal. Employing primary embryonic brain cells which is often used for the study of differentiation of oligodendroglia, Amur et al.[69] further demonstrated an effect of thyroid hormone as well as temporal correlation between expression of MBP-PMI and myelin marker enzymes[70] such as 2′, 3′-cyclic nucleotide 3′-phosphohydrolase and 5′-nucleotidase during the culture period (see Section III. F).

E. PROTEIN METHYLASE I ACTIVITY IN DYSMYELINATING BRAIN

There are a number of dysmyelinating mutant mice which possess abnormalities in the structure, composition, and/or metabolism of myelin. These animal models are useful tools in studying biochemical lesions which are associated with molecular defects of myelination, since the pathology is limited to the myelinated fiber tracts. These include jimpy (jp) which is a sex-linked recessive, and quaking (qk), myelin deficient (mld) and shiverer (shi), which are autosomal recessive mutations. These mutants are characterized by reductions in the levels of myelin in their brains, although the reasons for its deficit are different in each case.[71]

The jimpy mouse is the most severely affected dysmyelinating mutant, and is characterized by failure to incorporate MBP into myelin sheath, since the deficit of MBP in myelin is greater than the total reduction of the basic protein in whole brain.[72,73] In 1984, using both histone and MBP as the methyl acceptor substrates, Kim et al.[74] have studied PMI activity in the jimpy brain as a function of age. The histone-specific PMI activity in the jimpy brain exhibited an age-dependent decrease as normal brain. However, the MBP-specific PMI activity in 15-, 18-, and 21-d-old hemizygous jimpy mice (jp/y) brains decreased by 20, 50, and 75%, respectively, from those of normal littermates (Figure 9). The heterozygous jimpy mice (jp/+) which are phenotypically normal showed unaltered normal enzyme levels.

Another hypomyelinating mutant mouse, shiverer, also displayed an altered pattern of PMI activity,[75] however, quite differently from that of jimpy mouse. The activity in homozygous shiverer (shi/shi) mutant mouse brain is significantly higher at the onset of myelination than in

TABLE 5
Incorporation of L-[*Methyl*-^{3}H]Methionine and L-[^{35}S]Methionine into Protein Fractions during Mouse Brain Development

Age of animal (d)	$^3H/^{35}S$ Acid extract	Breakthrough	MBP Fraction
13	3.8	3.4	9.1
17	3.7	3.7	9.4
21	3.8	3.5	5.7
49	3.8	3.3	7.5
59	3.5	3.3	6.9

Note: The ratio of $^3H/^{35}S$ (methylation index) was calculated by counting the fractions in a liquid scintillation counter programmed for dual-label counting.

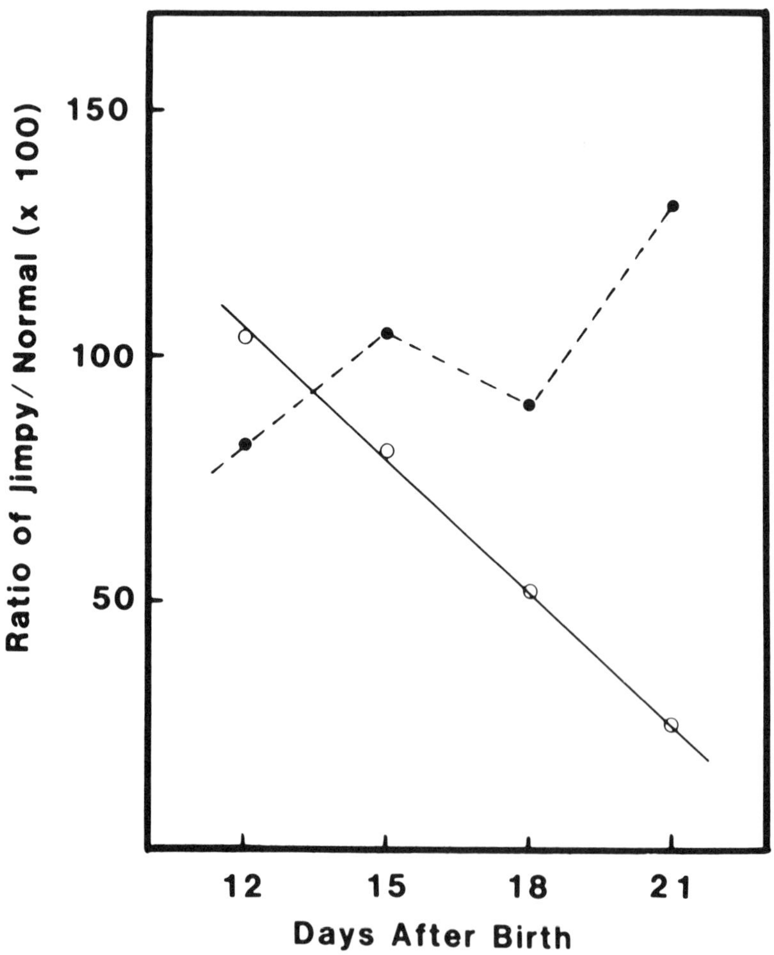

FIGURE 9. Alteration in protein methylase I activity between jimpy (jimpy/y) and normal mice. •—• represents the ratio of enzyme activities using histone as the protein substrate, and ○—○ for MBP as the substrate. Jimpy/normal represents the ratio of protein methylase I activity of jimpy (hemizygous) to that of normal.

the normal littermate brain but decreases rapidly during the period of myelination, compared to normal control brain. Since there is no difference neither in the weight nor protein concentration of the normal and shiverer mutant brains, the decrease in enzyme activity is not a mere reflection of changes in the protein concentration in the mutant brain. Although the reason for this anomalously high MBP-PMI activity in the younger shiverer brain is not known, it may be related to the finding that the number of oligodendrocytes in shiverer brain was found to be increased (Knobler, unpublished observation).

Among the mutant mice, myelin deficient quaking is known to have a normal life span[71] and is characterized by the reduction of 14-kDa MBP as compared to the 18.5-kDa species in the brain. In homozygous quaking mutant (qk/qk) brain, both histone-specific and MBP-specific PMI activities were shown to be normal.[74]

F. HORMONAL EFFECT ON MYELINATION AND PROTEIN METHYLASE I

The functional maturation of the mammalian central nervous system is dependent on the presence of the thyroid hormone. It has been shown that both neonatally induced hypo- and hyperthyroidism cause marked reduction in the number of oligodendroglia,[76,77] as well as impairment of myelination in rat brains.[78] In 1981, Walters and Morell[79] studied the developmental pattern of myelin proteins in both hypo- and hyperthyroid status by intracranial injection of [^{3}H]glycine. The results indicated that hypothyroidism retards the developmental program for myelinogenesis, whereas in hyperthyroid state myelin synthesis is initiated earlier but is also terminated earlier. Recently, employing myelinogenic cultures of cells dissociated from embryonic mouse brain, the effect of thyroid hormone on the expression of MBP-PMI was studied. Amur et al.[69] observed that the MBP-PMI activity in the cell is highly dependent on the presence of the thyroid hormone, but no such dependency was seen for histone-PMI expression. These investigators have also investigated several adrenergic effectors and neurotransmitters as a potential regulator for MBP- and histone-specific PMI.[80] Both enzymes were stimulated by β-adrenergic agonists (propranolol), via an increase in the cAMP/adenyl cyclase system. However, the agonist did not block the stimulation of MBP-enzyme triggered by thyroid hormone, indicating that the effect of the hormone is not mediated via a β-adrenergic-dependent system. Therefore, the methylation of MBP seems to be regulated by the thyroid hormone and/or by neurotransmitter, whereas the methylation of histone is only regulated by an adrenergic system.

G. EFFECT OF MBP-METHYLATION ON MYELIN STRUCTURE

A direct approach to elucidate a possible involvement of MBP-arginine methylation in the formation of compact myelin has been carried out at an ultrastructural level: the myelin-like membranes isolated from sinefungin-treated and -untreated myelogenic cells in culture were compared using a transmission electron microscope.[81] Sinefungin is an AdoHcy analog and a well-known inhibitor for PMI (see Chapter 18). In the myelogenic cells, MBP-PMI activity was shown to be 50% inhibited at 25 μM of the inhibitor, although the ratio of MBP to the total protein concentration was not significantly altered. Convincingly, the sinefungin-treated membrane exhibited a ring-like structure that was devoid of multilamellar periodicity and compactness characteristic of normal myelin which resembled the vacuolated myelin observed in vitamin B_{12}-deficient membrane (Figure 10). In support of these findings, Young et al.[82] have recently studied an effect of MBP-methylation in the dimer formation of unilamellar vesicle: the fully methylated bovine MBP was shown to be more efficient in inducing the dimerization of the vesicle than by the carp MBP which is known to be unmethylated.

IV. CONCLUDING REMARKS

Among the enzymatic post-/cotranslational methylation of arginine in proteins. the modification of MBP-arginine in the central nervous system is one of the best-studied biological

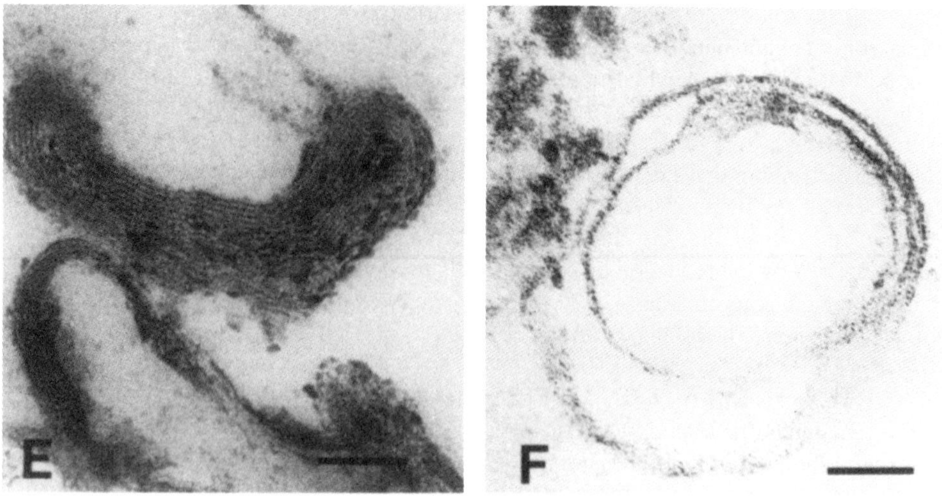

FIGURE 10. Electron micrographs of myelin-like membranes obtained from brain cell cultures. E is from normal control culture, and F is from sinefungin (30 μ*M*)-treated cultures. Bar = 100 nm.

systems thus far. There are several evidences which strongly point out a possible involvement of MBP-arginine methylation in the formation of compact myelin sheath. (1) Historically, the methyl-deficiency in animals is known to induce myelopathy[83] and the recent studies indicate that the methylation of MBP, not that of phospholipid, is linked to the biochemical lesion associated with the methyl-deficiency. (2) The specific location of methylarginine in MBP (residue-107) together with the chemical nature of hydrophobicity enhanced by the methylation has been implicated to help in stabilizing the proper myelin structure. (3) In several experimental systems, the MBP-specific PMI activity, but not histone-specific PMI, is shown to be temporarily correlated with myelination during brain development. Finally, (4) cells treated with transmethylation inhibitor (sinefungin) yielded membrane structure which lacks the compactness, similar to vitamin B_{12}-deficient myelin with concurrent reduction in the MBP-specific PMI.

In addition to the above, a possible alteration in the susceptibility of the protein toward intracellular proteases by arginine-methylation should also be considered in relation to membrane structure-function. While the maintenance of compact myelin sheath requires an intact maintenance of its membrane components, MBP is known to be easily metabolized by the MBP-associated neutral and acidic proteases. Since the methylated arginine residue in MBP is partially resistant to trypsin,[11] it is tempting to speculate that the rate of MBP degradation may be influenced by the level of MBP-arginine methylated, inducing the stability of myelin.

Recent advances in enzymology on PMI have clarified the earlier confusion concerning the identity of MBP-specific and histone-specific PMI. Although both enzymes methylate arginine residues, they are completely different proteins in terms of molecular, catalytic, and immunological properties. In light of these findings, it is now possible to study the molecular mechanism by which PMI activities are regulated, since MBP-specific PMI is low at postnatal and increases during myelination, while histone-specific enzyme is high in early postnatal and decreases progressively thereafter. In this connection, a recent interesting finding should be pointed out in which MBP was shown to be an inhibitor for histone-specific PMI *in vitro*.[28] What role, if any, MBP plays on the *in vivo* expression of two coexisting PMI activities during myelination is not known.

REFERENCES

1. **Ambler, R.P. and Rees, M.W.**, ε-*N*-Methyl-L-lysine in bacterial flagellar protein, *Nature*, 184, 56, 1959.
2. **Kim, S. and Paik, W.K.**, Studies on the origin of ε-*N*-methyl-L-lysine in protein, *J. Biol. Chem.*, 240, 4629, 1965.
3. **Paik W.K. and Kim, S.**, Protein methylation, *Science*, 174, 114, 1971.
4. **Kim, S. and Paik, W.K.**, Protein methylation: perspectives in central nervous system, in *Biochemical and Pharmacological Roles of Adenosylmethionine and the Central Nervous System*, Zappia, Z., Usdin, E., and Salvatore, F., Eds., Pergamon Press, Oxford, 1978, 37.
5. **Paik, W.K. and Kim, S.**, Protein methylation, in *Biochemistry: A Series of Monographs*, Vol. 1, John Wiley & Sons, New York, 1980, 112.
6. **Paik, W.K. and Kim, S.**, Methylation and demethylation of protein, in *The Enzymology of Post-Translational Modification of Proteins*, Vol. 2, Freedman, R.B. and Hawkins, H.C., Eds., Academic Press, London, 1985, 187.
7. **Paik, W.K., Nochumson, S., and Kim, S.**, Carnitine biosynthesis *via* protein methylation, *TIBS*, 2, 159, 1977
8. **Sitaramayya, A., Wright, L.S., and Siegel, F.L.**, Enzymatic methylation of calmodulin in rat brain cytosol, *J. Biol. Chem.*, 255, 8894, 1980.
9. **Paik, W.K. and Kim, S.**, Protein methylase I, *J. Biol. Chem.*, 243, 2108, 1968.
10. **Baldwin, G.S. and Carnegie, P.R.**, Specific enzymic methylation of an arginine in the experimental allergic encephalomyelitis protein from human myelin, *Science*, 171, 579, 1971.
11. **Brostoff, S. and Eylar, E.H.**, Localization of the methylated arginine in the A1 protein from myelin, *Proc., Natl. Acad. Sci. U.S.A.*, 68, 765, 1971.
12. **Paik, W.K. and Kim, S.**, Protein methylases during the development of rat brain, *Biochim. Biophys. Acta*, 313, 181, 1973.
13. **Paik, W.K. and Kim, S.**, Protein methylation during the development of rat brain, *Biochem. Biophys. Res. Commun.*, 46, 993, 1972.
14. **Jones, G.M. and Carnegie, P.R.**, Methylation of myelin basic protein by enzymes from rat brain, *J. Neurochem.*, 23, 1231, 1974.
15. **Small, D.H. and Carnegie, P.R.**, *In vivo* methylation of an arginine in chicken myelin basic protein, *J. Neurochem.*, 38, 184, 1982.
16. **Miyake, M. and Kakimoto, Y.**, Protein methylation by cerebral tissue, *J. Neurochem.*, 20, 859, 1973.
17. **Miyake, M.**, Methylases of myelin basic protein and histone in rat brain, *J. Neurochem.*, 24, 909, 1975.
18. **Crang, A.J. and Jacobson, W.**, The relationship of myelin basic protein (arginine) methyltransferase to myelination in mouse spinal cord, *J. Neurochem.*, 39, 244, 1982.
19. **Nakajima, T., Matsuoka, Y., and Kakimoto, Y.**, Isolation and identification of N^G-monomethyl, N^G,N^G-dimethyl- and N,N'^G-dimethylarginine from the hydrolysate of proteins of bovine brain, *Biochim. Biophys. Acta*, 230, 212, 1971.
20. **Zappia, V., Zydek-Cwick, C.R., and Schlenk, F.**, The specificity of *S*-adenosylmethionine derivatives in methyl transfer reactions, *J. Biol. Chem.*, 244, 4499, 1960.
21. **Deguchi, T. and Barchas, J.**, Inhibition of transmethylations of biogenic amines by *S*-adenosylhomocysteine, *J. Biol. Chem.*, 246, 3175, 1971.
22. **Borchardt, R.T. and Pugh, C.S.G.**, Analogues of *S*-adenosyl-L-homocysteine as inhibitors of viral mRNA methyltransferases, in *Transmethylation, Developments in Neuroscience*, Vol. 5, Usdin, E., Borchardt, R.T., and Creveling, C.R., Eds., Elsvier/North-Holland, New York, 197, 1979.
23. **Lee, H.W., Kim, S., and Paik, W.K.**, *S*-adenosylmethionine:protein-arginine methyltransferase, purification and mechanism of the enzyme, *Biochemistry*, 16, 78, 1977.
24. **Karn, J., Vidali, G., Boffa, L., and Allfrey, V.G.**, Characterization of the non-histone nuclear proteins associated with rapidly labeled heterogeneous nuclear RNA, *J. Biol, Chem.*, 252, 7307, 1977.
25. **Lischwe, M.A., Cook, R.G., Ahn, Y.S., Yeoman, L.C., and Busch, H.**, Clustering of glycine and N^G,N^G-dimethylarginine in nucleolar protein C23, *Biochemistry*, 24, 6025, 1985.
26. **Lischwe, M.A., Ochs, R.L., Reddy, R., Cook, R.G., Yeoman, L.C., and Busch, H.**, Purification and partial characterization of a nucleolar scleroderma antigen (M_r = 34,000; pI, 8.5) rich in N^G,N^G-dimethylarginine, *J. Biol. Chem.*, 260, 14304, 1985.
27. **Reporter, M. and Corbin, J.L.**, N^G, N^G-Dimethylarginine in myosin during muscle development, *Biochem. Biophys. Res. Commun.*, 43, 644, 1971.
28. **Park, G.-H., Chanderkar, L.P., Paik, W.K., and Kim, S.**, Myelin basic protein inhibits histone-specific protein methylase I, *Biochim. Biophys. Acta*, 874, 30, 1986
29. **Casellas, P. and Jeanteur, P.**, Protein methylation in animal cells. I. Purification and properties of *S*-adenosyl-L-methionine;protein (arginine) *N*-methyltransferase form Krebs II ascites cells, *Biochim. Biophys. Acta*, 519, 243, 1978.

30. **Casellas, P. and Jeanteur, P.,** Protein methylation in animal cells. II. Inhibition of *S*-adenosyl-L-methionine:protein (arginine) *N*-methyltransferase by analogs of *S*-adenosyl-L-homocysteine, *Biochim. Biophys. Acta,* 519, 255, 1978.
31. **Vedel, M., Lawrence, F., Robert-Gero, M., and Lederer, E.,** The anti-fungal antibiotic sinefungin as a very active inhibitor of methyltransferases and of the transformation of chick fibroblasts by Rous sarcoma virus, *Biochem. Biophys. Res. Commun.,* 85, 371, 1978.
32. **Enouf, J., Lawrence, F., Tempete, C., Robert-Gero, M., and Lederer, E.,** Relationship between inhibition of protein methylase I and inhibition of Rous sarcoma virus-induced cell transformation, *Cancer Res.,* 39. 4497, 1979.
33. **Robert-Gero, M., Lawrence, F., Farrugia, G., Berneman, A., Blanchard, P., Vigier, P., and Lederer, E.,** Inhibition of virus-induced cell transformation by synthetic analogues of *S*-adenosyl homocysteine, *Biochem. Biophys. Res. Commun.,* 65, 1242, 1975.
34. **Gupta, A., Jensen, D., Kim, S., and Paik, W.K.,** Histone-specific protein-arginine methyltransferase from wheat germ, *J. Biol. Chem.,* 257, 9677, 1982.
35. **Farooqui, J.Z., Tuck, M., and Paik, W.K.,** Purification and characterization of enzymes from *Euglena gracilis* that methylate methionine and arginine residues of cytochrome *c, J. Biol. Chem.,* 260, 537, 1985.
36. **Disa, S.G., Gupta, A., Kim, S., and Paik, W.K.,** Site specificity of histone H4 methylation by wheat germ protein-arginine *N*-methyltransferase, *Biochemistry,* 25, 2443, 1986.
37. **Raine, C.S.,** Morphology of myelin and myelination, in *Myelin,* 2nd ed., Morell, P., Ed., Plenum Press, New York, 1984, 1.
38. **Norton, W.T. and Podeslo S.E.,** Myelination in rat brain: changes in myelin composition during brain maturation, *J. Neurochem.,* 21, 759, 1973.
39. **Campagnoni, C.W., Carey, G.D., and Campagnoni, A.T.,** Synthesis of myelin basic proteins in the developing mouse brain, *Arch. Biochem. Biophys.,* 190, 118, 1978.
40. **Braun, P.E.,** Molecular organization of myelin, in *Myelin,* 2nd ed., Morell, P., Ed., Plenum Press, New York, 1984, 97.
41. **Eylar, E.H.,** The structure and immunological properties of myelin, *Ann. N.Y. Acad. Sci.,* 195, 481, 1972.
42. **Carnegie, P.R.,** Amino acid sequence of the encephalitogenic basic protein from human myelin, *Biochem. J.,* 123, 57, 1971.
43. **Eylar, E.H.,** Amino acid sequence of the basic protein of the myelin membrane, *Proc. Natl. Acad. Sci. U.S.A.,* 67, 1425, 1970.
44. **Takahashi, N., Roach, A., Teplow, D.B., Prusiner, S.B., and Hood, L.,** Cloning and characterization of the MBP gene from mouse: one gene can encode both 14 kd and 18.5 kd MBPs by alternate use of axons, *Cell,* 42, 139, 1985.
45. **Campagnoni, A.T.,** Molecular biology of myelin proteins from the central nervous system, *J. Neurochem.,* 51, 1, 1988.
46. **Carnegie, P.R.,** Properties, structure and possible neuroreceptor role of the encephalitogenic protein of human brain, *Nature,* 229, 25, 1971.
47. **Kornguth, M.L., Tomasi, M.L., Keyes, D.L., and Kornguth, S.E.,** The proteolytic degradation of a purified pit brain protein, *Biochim. Biophys. Acta,* 229, 167, 1971.
48. **Sato, S., Quarles, R.H., and Brady, R.O.,** Susceptibility of the myelin-associated glycoprotein and basic protein to a neutral protease in highly purified myelin from human and rat brain, *J. Neurochem.,* 39, 97, 1982.
49. **Glynn, P., Chantry, A., Groome, N., and Cuzner, M.L.,** Basic protein dissociating from myelin membranes at physiological ionic strength and pH is cleaved into three major fragments, *J. Neurochem.,* 48, 752, 1987.
50. **Deibler, G.E. and Martenson, R.E.,** Determination of methylated basic amino acids with the amino acid analyzer, *J. Biol. Chem.,* 248, 2387, 1973.
51. **Dunkley, P.R. and Carnegie, P.R.,** Amino acid sequence of the smaller basic protein from rat brain myelin, *Biochem. J.,* 141, 243, 1974.
52. **Brostoff, S.W.,** Immunological responses to myelin and myelin components, in *Myelin,* 2nd ed., Morell, P., Ed., Plenum Press, New York, 1984, 405.
53. **Suckling, A.J., Kirby, J.A., Wilson, N.R., and Rumsby, M.G.,** The generation of the relapse in chronic relapsing experimental allergic encephalomyelitis and multiple sclerosis – parallels and differences, in *Experimental Allergic Encephalomyelitis: A Useful Model for Multiple Sclerosis,* Alvord, J. E.C., Kies, M.W., and Suckling, A.J., Eds., Alan R. Liss, New York, 1984, 7.
54. **Jacobson,W., Gandy, G., and Disman, R.L.,** Experimental subacute combined degeneration of the cord in mice, *J. Pathol.,* 109, 243, 1973.
55. **Dinn, J.J., Weir, D.G., McCann, S., Reed, B., Wilson, P., and Scott, J.M.,** Methyl group deficiency in nerve tissue: a hypothesis to explain the lesion of subacute combined degeneration, *Irish J. Med. Sci.,* 149, 1, 1980.
56. **Scott, J.M., Wilson, P., Din, J.J., and Weir, D.G.,** Pathogenesis of subacute combined degeneration: a result of methyl group deficiency, *Lancet,* 2, 334, 1981.

57. **Crang, A.J. and Jacobson, W.,** The methylation *in vitro* of myelin basic protein by arginine methylase from mouse spinal cord, *Biochem. Soc. Trans.*, 8, 611, 1980.
58. **Lombardini, J.B., Coulter, J.B., and Talalay, P.,** Analogues of methionine as substrates and inhibitors of the methionine adenosyl-transferase reaction: Deductions concerning the conformation of methionine, *Mol. Pharmacol.*, 6, 481, 1970.
59. **Frenkel, E.P., Kitchens, R.L., and Johnson, J.M.,** The effect of vitamin B_{12} deprivation on the enzyme of fatty acid synthesis, *J Biol. Chem.*, 248, 7540, 1973.
60. **Siddons, R.C., Spence, J.A., and Dayan, A.D.,** Experimental vitamin B_{12} deficiency in the baboon, *Adv. Neurol.*, 10, 239, 1975.
61. **Small, D.H., Carnegie, P.R., and Anderson, R. McD.,** Cycloleucine-induced vacuolation of myelin is associated with inhibition of protein methylation, *Neurosci. Lett.*, 21, 287, 1981.
62. **Yu, Y.-T. and Campagnoni, A.T.,** *In vitro* synthesis of the four mouse myelin basic proteins: evidence for the lack of a metabolic relationship, *J. Neurochem.*, 39, 1559, 1982.
63. **Carson, J.H., Nielson, M.L., and Barbarese, E.,** Developmental regulation of myelin basic protein expression in mouse brain, *Dev. Biol.*, 96, 485, 1983.
64. **Chanderkar, L.P., Paik, W.K., and Kim, S.,** Studies on myelin basic protein methylation during mouse brain development, *Biochem. J.*, 240, 471, 1986.
65. **Small, D.H. and Carnegie, P.R.,** *In vivo* methylation of an arginine in chicken myelin basic protein, *J. Neurochem.*, 38, 184, 1982.
66. **DesJardins, K.C. and Morell, P.,** Phosphate groups modifying myelin basic proteins are metabolically labile: methyl groups are stable, *J. Cell Biol.*, 97, 438, 1983.
67. **Chou, C.H.J., Shapira, R., and Friz, R.B.,** Encephalitogenic activity of the small form of mouse myelin basic protein in the SjL/J mouse, *J. Immunol.*, 130, 2183, 1983.
68. **Jones, G.M. and Carnegie, P.R.,** Methylation of myelin basic protein by enzymes from rat brain, *J. Neurochem.*, 23, 1231, 1984.
69. **Amur, S.C., Shanker, G., and Pieringer, R.A.,** Regulation of myelin basic protein (arginine) methyltransferase by thyroid hormone in myelinogenic cultures of cells dissociated from embryonic mouse brain, *J. Neurochem.*, 43, 494, 1984.
70. **Bhat, N.R., Rao, G.S., and Pieringer, R.A.,** Investigations of myelination *in vitro*, regulation of sulfolipid synthesis by thyroid hormone in cultures of dissociated brain cells from embryonic mice, *J. Biol. Chem.*, 256, 1167, 1981.
71. **Hogan, E.L.,** Animal models of genetic disorders of myelin, in *Myelin*, Morell, P., Ed., Plenum Press, New York, 1977, 489.
72. **Campagnoni, A.T., Campagnoni, C.W., Huang, A.L., and Sampugna, J.,** Developmental changes in the basic proteins of normal and jimpy mouse brain, *Biochem. Biophys. Res. Commun.*, 46, 700, 1972.
73. **Mattjieu, J.-M., Windmer, S., and Herschkowitz, N.,** Jimpy, an anomaly of myelin maturation: biochemical study of myelination phases, *Brain Res.*, 55, 403, 1972.
74. **Kim, S., Tuck, M., Kim, M., Campagnoni, A.T., and Paik, W.K.,** Studies on myelin basic protein-protein methylase I in various dysmyelinating mutant mice, *Biochem. Biophys, Res. Commun.*, 123, 468, 1984.
75. **Kim, S., Tuck, M., Ho, L.-L., Campagnoni, A.T., Barbarese, E., Knobler, R.L., Lublin, F.D., Chanderkar, L.P., and Paik, W.K.,** Myelin basic protein-specific protein methylase I activity in shiverer mutant mouse brain, *J. Neurosci. Res.*, 16, 357, 1986.
76. **Bass, N.H. and Young, E.,** Effects of hypothyroidism on the differentiation of neurons and glia in developing rat cerebrum, *J. Neurol. Sci.*, 18, 155, 1973.
77. **Pelton, E.W. and Bass, N.H.,** Adverse effects of excess thyroid hormone on the maturation of rat cerebrum, *Arch. Neurol.*, 29, 145, 1973.
78. **Dalal, K.B., Valcana, T., Timiras, P.S., and Einstein, E.R.,** Regulatory role of T_4 on myelinogenesis in the developing rat, *Neurobiology*, 1, 211, 1971.
79. **Waters, S.N. and Morell, P.,** Effect of altered thyroid states on myelinogenesis, *J. Neurochem.*, 36, 1792, 1981.
80. **Amur, S.G., Shanker, G., and Pieringer, R.A.,** β-Adrenergic stimulation of protein (arginine) methyltransferase activity in cultured cerebral cells from embryonic mice, *J. Neurosci. Res.*, 16, 377, 1986.
81. **Amur, S.G., Shanker, G., Cochran, J.M., Ved, H.S., and Pieringer, R.A.,** Correlation between inhibition of myelin basic protein (arginine) methyltransferase by sinefungin and lack of compact myelin formation in cultures of cerebral cells from embryonic mice, *J. Neurosci. Res.*, 16, 367, 1986.
82. **Young, P.R., Vacante, D.A., and Waickus, C.M.,** Mechanism of the interaction between myelin basic protein and the myelin membrane: the role of arginine methylation, *Biochem. Biophys. Res. Commun.*, 145, 1112, 1987.
83. **Russell, J.S.R., Batten, F.E., and Collier, J.,** Subacute combined degeneration of the spinal cord, *Brain*, 23, 39, 1900.

Chapter 6

AMINO ACID SEQUENCE OF ARGININE METHYLATION SITES

Michael A. Lischwe

TABLE OF CONTENTS

I. INTRODUCTION

Arginine methylation is catalyzed by a group of enzymes classified as protein methylase I [*S*-adenosyl-L-methionine:protein (arginine) *N*-methyltransferase; EC 2.1.1.23].[1] These enzymes can modify arginine residues to N^G-mono-, N^G,N'^G-di-, and N^G,N^G-di-methylarginines.[1] The enzymes are primarily cytosolic and use *S*-adenosylmethionine as the methyl donor.[1-4] The methylation of the arginine residues apparently occurs while the protein is being synthesized and appears to be an irreversible cotranslational modification.[1,5]

It has been known for some time that myelin basic protein contains a methylated arginine. A methyltransferase modifies arginine 107 in the myelin basic protein to N^G-monomethylarginine or N^G,N'^G-dimethylarginine.[2,3,6] The high-mobility groups 1 and 2,[7] histones,[1] heterogeneous nuclear RNA (hnRNA) binding proteins,[8-10] single-stranded nucleic acid binding protein,[10-18] a 34-kDa nucleolar proteins which is apparently associated with U3 RNA,[19,20] and protein C23[21,22] (also called nucleolin)[23,24] are nuclear nucleic acid binding proteins which contain methylarginines. The 34-kDa nucleolar protein, which contains 4.1 mol% N^G,N^G- dimethylarginine (DMA), is the most highly arginine methylated protein found, thus far, in mammalian cells.[19,20]

A conserved domain which contains clusters of glycine and N^G,N^G-dimethylarginine interspersed with phenylalanine residues have been found in the 34-kDa nucleolar protein,[19] and in nucleolin.[22-24] Nucleolin contains a domain near its COOH-terminal which has ten dimethylarginine residues.[22-24] The NH_2-terminal of the 34-kDa nucleolar protein contains 6 N^G,N^G-dimethylarginine and 16 glycine residues which are interspersed by phenylalanine residues.[19] The presence of a glycine, DMA-rich domain interspersed by phenylalanine residues in two distinct proteins suggests that this domain has been evolutionarily conserved.[19,22] This sequence array may represent a conserved domain characteristic of a certain class of nuclear proteins.[19,22]

Three other nucleic acid binding proteins are known to contain glycine, arginine-rich regions interspersed with phenylalanine residues. The hnRNP core protein A1 contains a glycine-rich domain interspersed by phenylalanine residues which contains a N^G,N^G-dimethylarginine residue and several other arginines which may be methylated.[14,16,17,25] This protein contains approximately three residues of N^G,N^G-dimethylarginine per mole.[8,9,14,16,17,25] A single site has been sequenced at position 194.[14] A yeast single-stranded nucleic acid binding protein[26] and a putative RNA binding protein from *Artemia salina*[27] were found to have a glycine, arginine-rich domain interspersed with phenylalanines. It is not known if these arginine residues are methylated. The sequence was obtained from the cDNA.[26,27]

The function of the domains which contain clusters of glycine and N^G,N^G-dimethylarginine/arginine interspersed with phenylalanines is currently not known. Several groups have proposed a nucleic acid binding role.[19,21-28] It has been further speculated that the difference in actual sequence of the three amino acids may give rise to different protein-nucleic acid structures[26] and/or determine specificity for certain nucleic acid molecules.[22] This domain may also be involved in cooperative protein-protein interactions as well as direct interactions with nucleic acid.[28]

At least a single glycine has been found adjacent to all methylated arginine residues sequenced to date. This suggests that a single adjacent glycine to the arginine residue is required for methylation.[19,22] Access of certain arginine methylases to arginine residues may be sterically possible because of the lack of a side chain on the adjacent glycine residue(s).[19,22]

The modified arginine residues in nucleolin (protein C23), UP1 and in myelin basic protein were found to be resistant to trypsin.[2,22,29] One role of this protein modification may be protection from intracellular proteases.[22]

II. ENZYMOLOGY

A. ARGININE METHYLATION SITES

A summary of the arginine methylation sites which have been sequenced to date is given in Table 1. The dimethylarginines in nucleolin, the 34-kDa nucleolar protein and in the hnRNP core protein A1, are N^G,N^G-dimethylarginine.[8,14,19,22] No monomethylarginine or arginine were detected at any of these sites in these three proteins.[14,19,22] Although a very small amount of N^G-monomethylarginine was found when the amino acid composition of nucleolin was determined,[21] no N^G-monomethylarginine was found in the 34-kDa nucleolar protein.[19] These results are consistent with earlier reports which showed that the predominant modified arginine residue in nuclear proteins was N^G,N^G-dimethylarginine.[8,30] Position 107 in myelin basic protein contains arginine, N^G-monomethylarginine, and N^G,N'^G-dimethylarginine.[2,6] The arginine residues in the CHO cell nucleolin peptide shown are presumed to be methylated. This is based on the similarity of this domain to that found in Novikoff hepatoma nucleolin.[22-24] The underlined amino acids in the CHO cell nucleolin are additional residues not found in Novikoff hepatoma nucleolin.[22,24]

It is likely that additional arginine residues in the sequence shown from the hnRNP core protein A1 are methylated. The protein is known to contain at least three residues of N^G,N^G-dimethylarginine per molecule.[8,25]

A glycine, arginine-rich domain interspersed with phenylalanines has been found in a yeast single stranded nucleic acid binding protein, SSB1.[26] The sequence from residues 125 to 160 is shown in Table 2. It is not known if the arginine residues are methylated.

Two glycine, arginine-rich zones were found in a putative RNA binding protein, GRP33, from *Artemia salina*.[27] The sequence of these domains which do contain phenylalanine and/or tyrosine residues are shown in Table 2. This protein is closely related to HD40, the major protein component of *Artemia* heterogeneous nuclear ribonucleoprotein particles.[10,12,27] It is known that the HD40 protein contains 1.4 mol% of N^G,N^G-dimethylarginine.[10,12] It is likely that some or all of the arginine residues in this putative RNA binding protein that are underlined in Table 2 are methylated.[27]

The 100-kDa adenovirus late protein contains the sequence Gly-Arg-Gly-Gly-Ile-Leu-Gly-Gln-Ser-Gly-Arg-Gly-Gly-Phe-Gly-Arg-Gly-Gly-Gly-Asp from residues 726 to residue 745.[31] An Epstein-Barr virus nuclear antigen and the pen repeat sequence, p19, contain glycine, arginine clusters.[32,33] Several keratins have been found to contain domains rich in glycine interspersed by phenylalanine residues at there COOH-terminal and/or NH_2-terminal.[34-37] There are some arginine residues present in some of these domains although they are not interspersed like those in nucleolin, the 34-kDa nucleolar protein, GRP33, SSB1, and hnRNP protein A1. It is not known if these arginines are methylated.

B. PROTEIN METHYLASE I RECOGNITION SITES

All 18 arginine methylation sites reported to date have at least one adjacent glycine residue. Twelve of these sites are flanked by glycine residue. Seventeen of the 18 residues are found in clusters of glycine residues. It appears that at least a single adjacent glycine to the arginine residue is required for methylation.[19,22] The lack of a side group on the glycine residue(s) may facilitate access of the methylase to the arginine residues. The glycine residue(s) would not sterically hinder the protein methylase.[19,22]

The one methylated arginine site that is not found in glycine clusters interspersed with phenylalanines is position 107 of myelin basic protein.[2] This residue is flanked by glycines. It is only partially modified to N^G-monomethylarginine and N^G,N'^G-dimethylarginine.[2,6] This partial modification may be do to a lack of accessibility of the methylase to this residue. Lack of easy access may also account for the conversion of arginine to N^G-monomethylarginine and N^G,N'^G-dimethylarginine instead of complete conversion to N^G,N^G-dimethylarginine as is seen

TABLE 1
Arginine Methylation Sites

Novikoff Hepatoma Nucleolin (Protein C23)[22]

Gly-Glu-Gly-Gly-Phe-Gly-Gly-DMA-Gly-Gly-Gly-DMA-Gly-Gly-Phe-Gly-
Gly-DMA-Gly-Gly-Gly-DMA-Gly-Gly-DMA-Gly-Gly-Phe-Gly-Gly-DMA-Gly-
DMA-Gly-Gly-Phe-Gly-Gly-DMA-Gly-Gly-Phe-DMA-Gly-Gly-DMA-Gly-Gly-
Gly-Gly-Asp-Phe-Lys

CHO Cell Nucleolin (Protein C23)[23,24]

648
Gly-Glu-Gly-Gly-Phe-Gly-Gly-DMA-Gly-Gly-Gly-DMA-Gly-Gly-Phe-Gly-

664
Gly-DMA-Gly-Gly-Gly-DMA-Gly-Gly-Gly-DMA-Gly-Gly-Phe-Gly-Gly-DMA-

680
Gly-DMA-Gly-Gly-Phe-Gly-Gly-DMA-Gly-Gly-Phe-DMA-Gly-Gly-DMA-Gly-

696
Gly-Gly-Gly-Gly-Gly-Gly-Asp-Phe-Lys

Nucleolar 34 KDa protein[19]

1
Met-Lys-Pro-Gly-Phe-Ser-Pro-DMA-Gly-Gly-Gly-Phe-Gly-Gly-DMA-Gly-

17
Gly-Phe-Gly-Asp-DMA-Gly-Gly-DMA-Gly-Gly-Gly-DMA-Gly-Gly-DMA

Heterogeneous Nuclear Ribonucleoprotein Complex Protein AI[16,17]

191
Ser-Ser-Gln-DMA-Gly-Arg-Ser-Gly-Ser-Gly-Asn-Phe-Gly-Gly-Gly-Arg-

207
Gly-Gly-Gly-Phe-Gly-Gly-Asn-Asp-Asn-Phe-Gly-Arg-Gly-Gly-Asn-Phe-

223
Ser-Gly-Arg-Gly-Gly-Phe-Gly-Gly-Ser-Arg-Gly-Gly-Gly-Gly-Tyr

Human Myelin Basic Protein[2,6]

DMA*
105 MMA
Lys-Gly-Arg-Gly-Leu-Ser-Leu-Ser-Arg-Phe

DMA*, N^G,N'^G-dimethylarginine; DMA, N^G,N^G-dimethylarginine

TABLE 2
Putative Arginine Methylation Sites

Yeast single stranded nucleic acid binding protein (SSB1)[26]

125
Arg - Gly - Gly - Phe - Arg - Gly - Arg - Gly - Gly - Phe - Arg - Gly - Arg - Gly - Gly -

140
Phe - Arg - Gly - Gly - Phe - Arg - Gly - Gly - Tyr - Arg - Gly - Gly - Phe - Arg -Gly -

155
Arg-Gly-Asn-Phe-Arg-Gly

Putative RNA binding protein from *Artemia,* GRP33[27]

209
Gly-Arg-Gly-Arg-Gly-Arg-Gly-Arg-Gly-Gly-Phe-

272
Gly - Arg - Gly - Ala - Gly - Ala - Gly - Ala - Arg - Gly - Ala - Arg - Gly - Gly - Leu -

287
Asp - Gln - Ser - Arg - Gly - Gly - Gly - Lys - Phe - Pro - Ser - Ala - Arg - Gly - Gly -

302
Arg-Gly-Arg-Ala-Ala-Pro-Tyr

at the other sequenced sites. Other possibilities such as a different methylase may account for the partial and unique modification observed in myelin basic protein.[1,2,6]

The phenylalanine residues which intersperse the domains which contain the modified arginines may also be essential for enzyme recognition.

III. SPECIFIC PROTEINS THAT CONTAIN METHYLATED ARGININES

A. NUCLEOLIN (PROTEIN C23)

Nucleolin (M_r 110,000, pI = 5.5) is the major protein found in the nucleolus of exponentially growing eukaryotic cells.[21-24,38-42] It has been immunochemically localized to the nucleolus,[39,40,42] found to be concentrated in the fibrillar region of the nucleolus,[39,42,43] and present on the nucleolus organizer region of metaphase chromosomes.[39] It has also been proposed that nucleolin is, in part, responsible for silver uptake at the nucleolus organizer region.[39,40,42,44]

Nucleolin is a multifunctional protein that is thought to play a direct role in pre-rRNA transcription and ribosome assembly.[21-24,38-50] This protein has been found associated with chromatin and with preribosomal particles,[44] suggesting a role in the regulation of chromatin structure and in pre-rRNA processing.[21-24,39] *In vitro* nucleolin binds single- and double- stranded DNA with a higher affinity for the nontranscribed rDNA spacers than for the transcribed portions of the rDNA.[45,47] Nucleolin also was found to bind RNA *in vitro*.[46,48] It has been postulated that phosphorylation and protease maturation of nucleolin regulates its involvement in chromatin conformation and ribosome assembly.[24,41,49,50]

Nucleolin is a highly phosphorylated protein which contains acidic tryptic phosphopeptides.[24,51] Nucleolin is also a highly methylated protein.[21,22] It contains 1.3 mol% of N^G,N^G-dimethylarginine and a trace of N^G-monomethylarginine. Nucleolin is the only large, acidic protein known to contain large amounts of methylated arginines. All other proteins, identified to date, which contain modified arginines are small and basic.[19,21] Figure 1 shows a chromatography of the basic amino acids in a hydrolysate of nucleolin.[21] The elution position of N'^G-monomethylarginine, N^G,N'^G-dimethylarginine and N^G,N^G-dimethylarginine is distinct as shown in panels A, B, and C. The undotted line in panel D shows a hydrolysate of unlabeled nucleolin. The N^G,N^G-dimethylarginine peak is nearly equivalent in height to that of histidine. The dotted line represents a hydrolysate of radiolabeled nucleolin. Nucleolin was labeled by growing cells in the presents of L-[*methyl*-^{14}C] methionine. Both N^G,N^G-dimethylarginine and N^G-monomethylarginine were observed.[21]

To elucidate the sequence of the arginine methylation sites, purified nucleolin was treated with trypsin.[22] The resulting peptides were resolved by reverse phase column chromatography with a C18 column. One predominant peptide was found to contain the N^G,N^G-dimethylarginine. In addition, this peptide was rich in glycine and phenylalanine residues. The entire amino acid sequence of this peptide was determined by Dr. Richard G. Cook on a gas phase Applied Biosystems Model 470A protein sequencer.[22] HPLC analysis of the PTH-amino acid standards including PTH-N^G,N^G-dimethylarginine are shown in panel A of Figure 2. PTH-N^G-monomethylarginine eluted after PTH-arginine. PTH-N'^G,N'^G-dimethylarginine could not be separated from PTH-N^G,N^G-dimethylarginine with the gradient used. This was not a concern with nucleolin since it does not contain N^G,N'^G-dimethylarginine. The PTH-amino acids obtained at steps 7 through 9 in this peptide are shown in Figure 2. Quantitative amino acid sequence results for the entire tryptic peptide are illustrated in Figure 3. The peptide contains 34 glycine, 10 N^G,N^G-dimethylarginine and 6 phenylalanine residues and has a cluster of glycines and N^G,N^G-dimethylarginines interspersed with phenylalanine residues.[22] While this work was in progress a similar domain was found in a 34-kDa U3 RNA-associated protein which will be discussed later.[19] This domain in nucleolin was found to be located at the COOH-terminal.[23,24]

The complete primary structure of CHO cell nucleolin was determined in 1987 and supports a multifunctional role for this protein[24] (Figure 4). Nucleolin, a 713 residue protein, has at least three other distinct sequence domains.[24] The repetitive sequence, Hy-Thr-Pro-Hy-Lys-Lys-Hy-Hy, in which Hy is a nonpolar residue, is found six times near the NH_2-terminal.[24] This sequence array is followed by three acidic stretches containing 25, 25, and 33 glutamic and aspartic acid residues.[24] The next striking feature in the primary structure of nucleolin is the four "RNP consensus sequences" that are present between residues 344 and 617.[24,45] Near the COOH-terminal of nucleolin is the glycine, N^G,N^G-dimethylarginine-rich zone interspersed by phenylalanine residues.[22-24] Each of these distinct domains likely has a unique role in ribosome biogenesis and nucleolar organization.[21-24,47-50]

The repetitive sequence located near the NH_2-terminal and the following acidic clusters of glutamic and aspartic acid have been postulated to be the portion of nucleolin which binds chromatin.[24] There are significant analogies between this zone and the HMGs, small nuclear

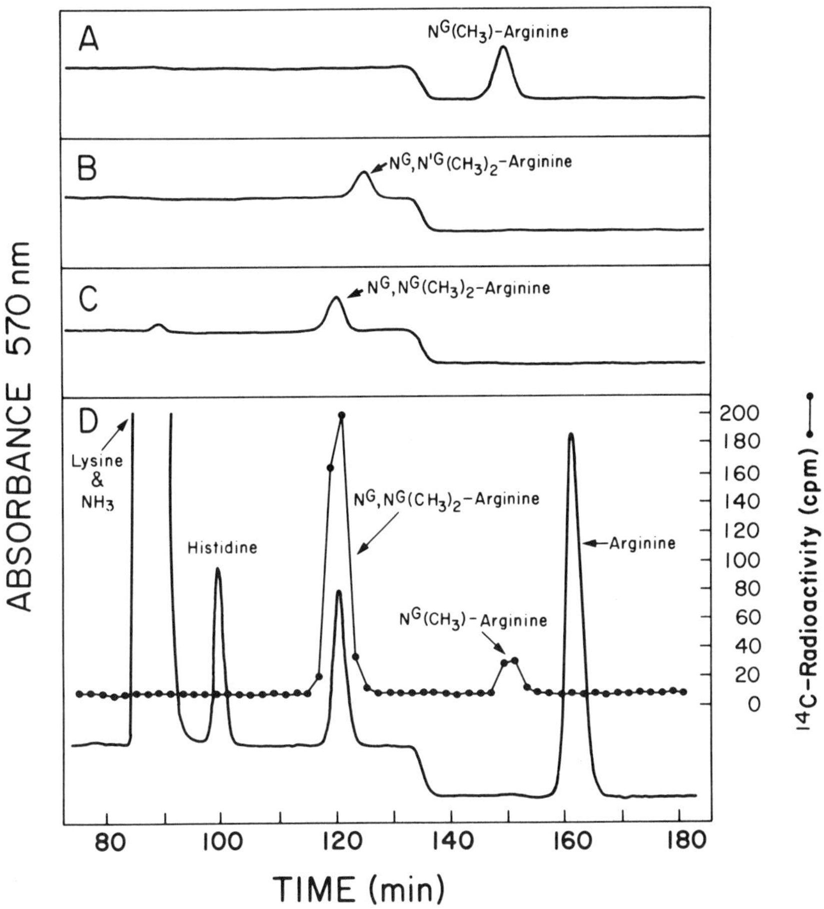

FIGURE 1. N^G,N^G-dimethylarginine in nucleolin (protein C23). Novikoff hepatoma tissue culture cells (N_1S_1-73) were grown in Dulbecco's minimum essential medium with 5% fetal calf serum and antibiotics. To label the methylated arginine residues, the cells were incubated for 1 h in methionine-free medium and then for 4 h in medium containing L-[*methyl*-^{14}C] methionine. The cells were harvested and combined with Novikoff hepatoma ascites cells. Purified nucleolin[39] was hydrolyzed in 6 *N* HCl for 22 h at 110°C *in vacuo*. The hydrolysate was analyzed on a Beckman Model 121 MB amino acid analyzer with a program designed to extend the basic region.[21] (A) N^G-monomethylarginine; (B) N^G,N'^G-dimethylarginine; (C) N^G,N^G-dimethylarginine; (D) hydrolysate of protein C23. Approximately 2000 cpm of the hydrolysate of ^{14}C-labeled protein C23 were loaded on the column; 0.5 ml fractions were collected.[21]

proteins which are thought to replace histone Hl.[24,49,52,53] One of the phosphorylation sites in nucleolin in this domain is identical to the sequence of an HMG 14 phosphorylation site.[24,54] In addition, acidic stretches are found at the COOH-terminal of the HMGs.[52,53] CNBr peptides from nucleolin representing this region were found to exhibit DNA and histone binding properties.[24]

Near the center of nucleolin are four "RNP consensus sequences".[24,48] The "RNP consensus sequence" is a 11 to 13 residue domain that has been found in a poly (A)-binding protein,[55,56] the U1 RNA-associated protein,[57] and the hnRNP proteins Al,[16,17] C1,[58] C2,[58] and A2.[58] The "RNP consensus sequence" is believed to be involved in RNA binding.[55-58] It has been postulated that these domains in nucleolin have a similar function.[48] Peptides from nucleolin which contain this zone have been shown to bind RNA.[48]

The function of the COOH-terminal domain in nucleolin which is rich in glycine and arginine residues interspersed with phenylalanines is currently not clear. It has been proposed to also be

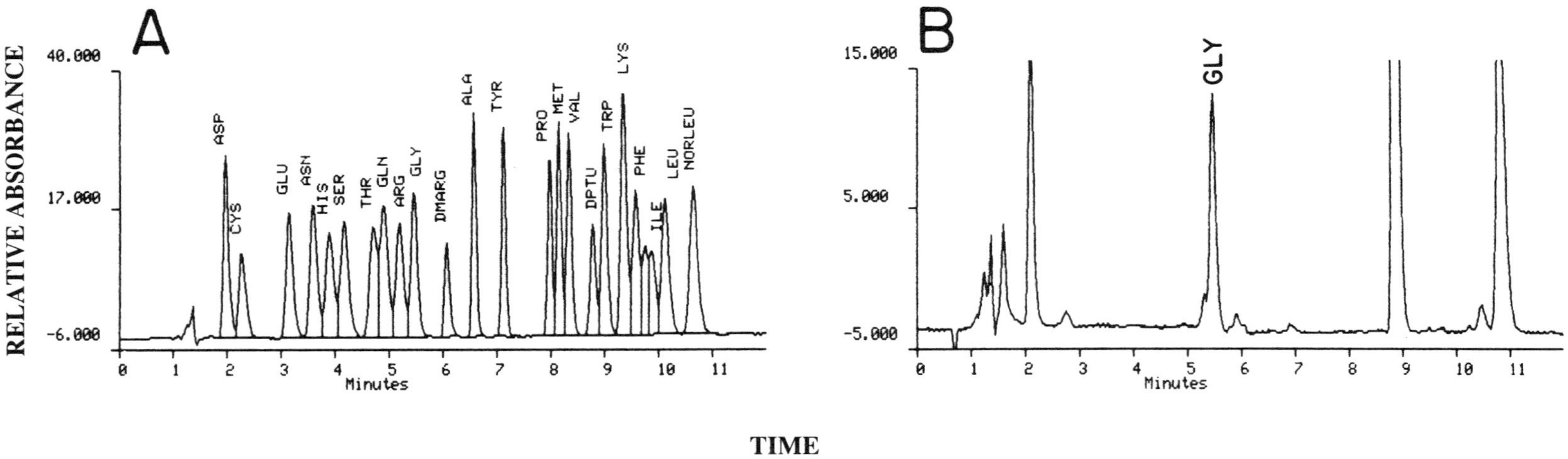

FIGURE 2. HPLC analysis of the PTH-amino acids. Purified nucleolin from Novikoff hepatoma cells[39] was digested with trypsin.[22] The tryptic peptides were resolved by reverse phase column chromatography. One predominant peptide was found to contain the N^G,N^G-dimethylarginine as determined by amino acid analysis. This peptide was sequenced on a gas phase Applied Biosystems Model 470A protein sequencer as described.[22] The PTH-amino acids were identified and quantitated on a Waters HPLC system.[22] The elution positions for the standards (250 pmol) including PTH-N^G,N^G-dimethylarginine are shown in panel A. The chromatographs for steps 7 to 9 of the tryptic peptide from nucleolin are shown in panels B through D. DMARG is N^G,N^G-dimethylarginine.[22]

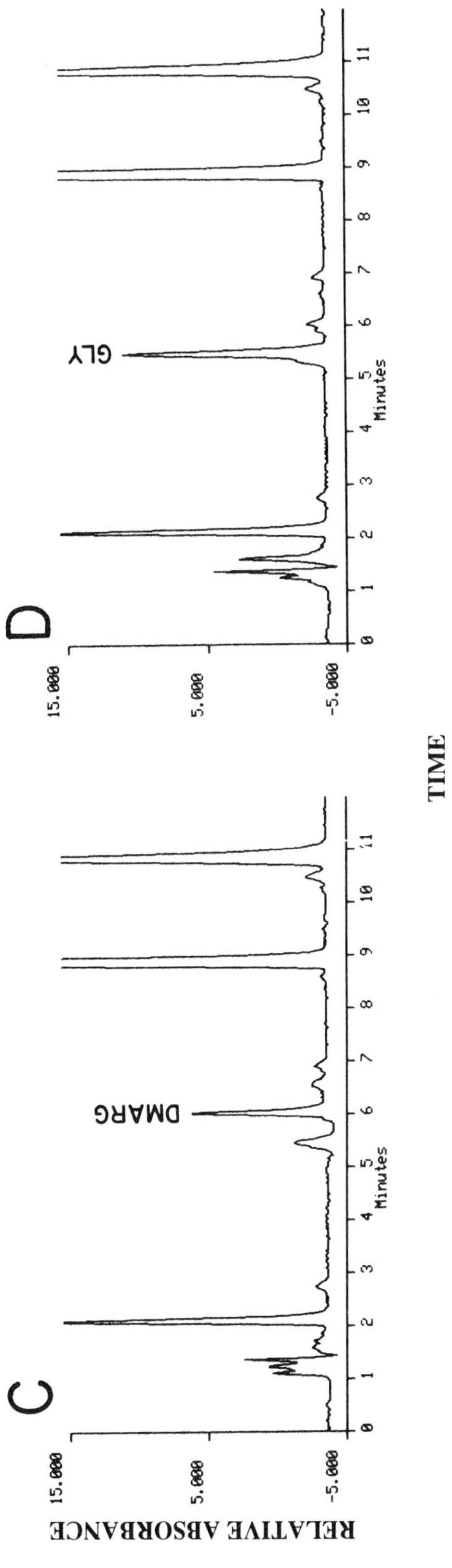

FIGURE 2 (continued)

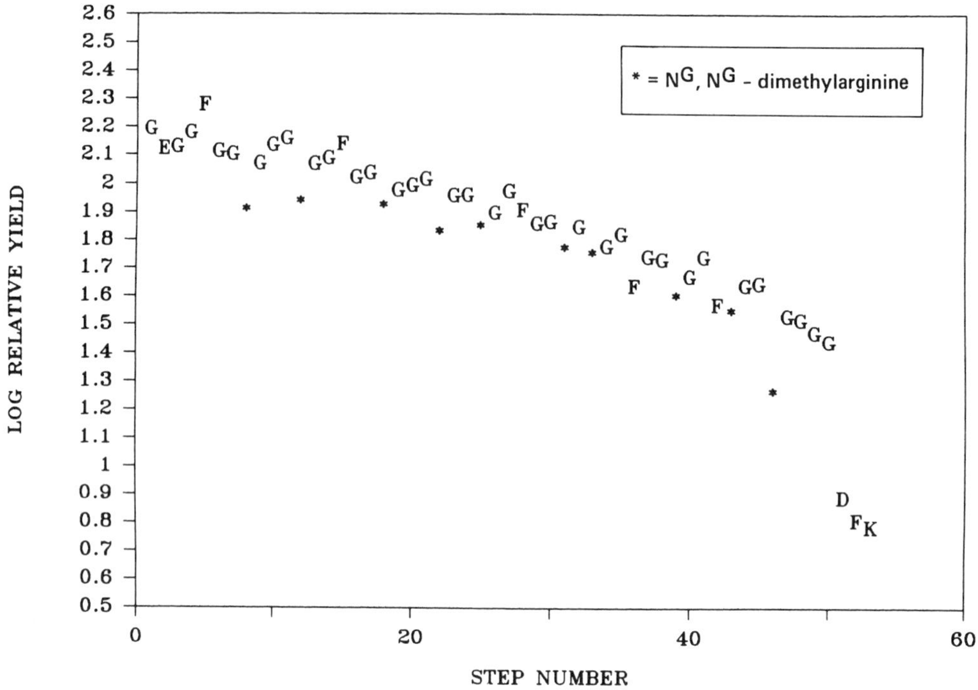

FIGURE 3. Quantitative amino acid sequence results of a tryptic peptide from nucleolin. The yield of the PTH-amino acids was normalized on the bases of the recovery of the internal standard (norleucine). The logarithm of the yield, in picomoles, of the PTH-amino acids is plotted vs. the step number. The asterisks denote N^G,N^G-dimethylarginine.[22]

a nucleic acid binding zone.[22,24] Proposed functional roles for this conserved domain will be given later in this chapter.

B. THE 34-kDa U3 RNA-ASSOCIATED NUCLEOLAR PROTEIN

Antinuclear autoantibodies are found in many patients with systemic rheumatic diseases.[59,60] The antinuclear autoantibodies are directed against DNA, RNA, histones, nuclear ribonucleoprotein particles, or nucleolar antigens.[59-64] Antinuclear and/or antinucleolar antibodies have been demonstrated in sera from 97% of the patients with scleroderma.[62] Antinucleolar antibodies have been found in 47% of the patients with scleroderma.[62] The sera of some patients with scleroderma were found to precipitate particles containing nucleolar U3 RNA, 7-2 RNA, or 8-2 RNA.[64] U3 RNA is the nucleolar specific snRNA (small nuclear RNA), which is thought to have a role in preribosomal RNA processing.[65-70]

Four patients sera which precipitated U3 RNA-containing particles from whole cell sonicates have been characterized.[19] The sera from these patients did not precipitate purified U3 RNA which suggests that the immunoreactivity requires a protein moiety. All four sera reacted with a protein of 34,000 molecular weight.[19] One patients serum also reacted with a 70,000 molecular weight protein. Affinity purified human and monoclonal antibodies to the 34-kDa protein also precipitate U3 RNA-containing particles.[19,70,71] These observations suggest that the 34-kDa protein is directly associated with U3 RNA, although indirect association is still a possibility. Parker and Steitz have used antibodies directed against the 34-kDa protein to reveal a conserved sequence available for base pairing with pre-rRNA in the U3 ribonucleoprotein particle.[70]

The 34-kDa protein was immunochemically localized to the nucleolus.[19,71,72] By immunoelectron microscopy, the 34-kDa protein was localized in HeLa nucleoli primarily in the fibrillar regions, with little, if any, localization in the granular region.[19,72] It has accordingly been named "fibrillarin".[72]

The 34-kDa nucleolar protein has been purified to homogeneity from Novikoff hepatoma cells.[19] It was found to have a pI of approximately 8.5.[19] The amino acid composition of the 34-kDa protein, illustrated in Table 3, showed a high content of glycine (22.8 mol%) and N^G,N^G-dimethylarginine (4.1 mol%).[19] No N^G-monomethylarginine or N^G,N'^G-dimethylarginine were detected in this protein.[19] This protein is the most highly arginine methylated nuclear protein, thus far, detected in higher eukaryotes.[19] Approximately 45% of the arginine residues in the 34-kDa protein are methylated.

The NH_2-terminal 31 residues of this protein have been sequenced.[19] The sequence is Met-Lys-Pro-Gly-Phe-Ser-Pro-DMA-Gly-Gly-Gly-Phe-Gly-Gly-DMA-Gly-Gly-Phe-Gly-Asp-DMA-Gly-Gly-DMA-Gly-Gly-Gly-DMA-Gly-Gly-DMA. In the first 31 residues, there are 16 glycine, 6 DMA, and 3 phenylalanine residues. The glycine, DMA clusters interspersed with phenylalanine residues is evident.[19] This domain likely extends further in the molecule since this protein contains approximately ten DMA residues based on its amino acid composition. All ten DMA residues are likely cluster as they are in nucleolin.

The 34-kDa nucleolar protein is apparently highly conserved.[19,20] The composition of this protein is very similar to a *Physarum polycephalum* protein previously identified (Table 3).[20,30] Monoclonal antibodies produced against the *Physarum* 34-kDa protein (called B-36) cross-reacted with the purified rat 34-kDa protein.[20] The *Physarum* protein is also a nucleolar protein.[20] A NH_2-terminal domain of clusters of glycine and N^G,N^G-dimethylarginine interspersed with phenylalanines has been found in the *Physarum* protein.[74] Although the sequence is not identical to that found at the NH_2-terminal end of the rat protein, it is very similar.[74] These observations illustrate that the 34-kDa nucleolar protein, the protein methylase I responsible for methylating this protein and the domain of clusters of glycine and N^G,N^G-dimethylarginine interspersed by phenylalanines are highly conserved.[19,20]

The function of the glycine, DMA-rich domain interspersed with phenylalanines in the 34-kDa nucleolar protein is not clear. It has been speculated that this domain is a nucleic acid binding site, more specifically that this domain binds U3 RNA or pre-rRNA.[19] The function of this zone will be discussed in more detail later in this chapter.

C. HETEROGENEOUS NUCLEAR RIBONUCLEOPROTEIN COMPLEX PROTEIN A1

Single-stranded nucleic acid binding proteins have been a topic of extensive investigation in several laboratories.[11-18] These proteins were found to bind both single-stranded RNA and DNA.[11-17] The proteins have been classified as helix-destabilizing proteins (HDPs),[13,16] nucleic acid unwinding proteins (UPs),[11,15] and nucleic acid single-stranded binding proteins (SSBs).[11,26,75] Recently by cloning, it was determined that one of these HDPs was identical to the heterogeneous nuclear ribonucleoprotein complex protein A1.[14,16,17]

Heterogeneous nuclear RNA becomes associated with at least six proteins following transcription.[8-10] The complexes formed, hnRNPs, can be recovered from purified nuclei as substructures with a relatively homogeneous sedimentation coefficient of 30 to 40S.[8-10,76] Based on migration characteristics in SDS gels these proteins have been classified as doublets A1/A2, B1/B2, and C1/C2.[8] Further complexity was observed with two-dimension gels, showing multiple species.[77-79] The A, B, and C hnRNP proteins are believed to play a role in packaging, transport, processing, and stabilization of hnRNA.[8,9,73,77-80] The majority of these proteins associated with hnRNA are rich in glycine and contain N^G,N^G-dimethylarginine.[8,9,73,77-81]

The nucleotide sequence and deduced amino acid sequence of rodent protein A1 is shown in Figure 5.[16] Residues 2 to 196 of this protein are identical to the reported sequence of UPI.[14,16,17] Protein A1 contains two "RNP consensus sequences".[16,17,55] These regions of the molecule are believed to be involved in RNA binding.[48,55-58] A region, boxed off, near the middle of the molecule contains glycine, arginine clusters interspersed with phenylalanines. It is known that the arginine at position 194 is modified to N^G,N^G-dimethylarginine.[14] It is likely that other arginine residues in this region are also methylated since it is known that the protein contains

1	Val	Lys	Leu	Ala	Lys	Ala	Gly	Lys	Thr	His	Gly	Glu	Ala	Lys	Lys	Met	Ala	Pro	Pro	Pro	Lys	Glu	Val	Glu	Glu	Asp	Ser	Glu	Asp	Glu
										10										20										30
					GAA	GAA	GAT	GAC	AGC	AGT	GGA	GAA	GAG	GTT	GTC	ATC	CCT	CAG	AAA	AAA	GGA	AAG	AAG	GCT	ACA	GCA	ACT	CCA	GCA	AAG
2	Glu	Met	Ser	Glu	Glu	Glu	Asp	Asp	Ser	Ser	Gly	Glu	Glu	Val	Val	Ile	Pro	Gln	Lys	Lys	Gly	Lys	Lys	Ala	Thr	Ala	Thr	Pro	Ala	Lys
										40										50										60
	AAG	GTG	GTT	GTT	TCA	CAA	ACA	AAA	AAA	GTT	GCA	GTT	CCC	ACA	CCA	GCT	AAG	AAA	GCT	GCT	GTT	ACT	CCA	GGT	AAA	AAG	GCA	GCA	GCC	ACA
3	Lys	Val	Val	Val	Ser	Gln	Thr	Lys	Lys	Val	Ala	Val	Pro	Thr	Pro	Ala	Lys	Lys	Ala	Ala	Val	Thr	Pro	Gly	Lys	Lys	Ala	Ala	Ala	Thr
										70										80										90
	CCA	GCC	AAG	AAG	GCA	GTT	ACA	CCA	GCC	AAA	GCA	GTT	GCA	ACA	CCT	GGA	AAG	AAG	GGA	GCC	ACA	CAA	GCA	AAA	GCA	TTG	GTA	GCA	ACC	CCT
4	Pro	Ala	Lys	Lys	Ala	Val	Thr	Pro	Ala	Lys	Ala	Val	Ala	Thr	Pro	Gly	Lys	Lys	Gly	Ala	Thr	Gln	Ala	Lys	Ala	Leu	Val	Ala	Thr	Pro
										100										110										120
	GGT	AAA	AAG	GGA	GCT	GTC	ACT	CCA	GCC	AAG	GGG	GCT	AAG	AAT	GGT	AAG	AAT	GCT	AAG	AAG	GAA	GAC	AGT	GAT	GAA	GAT	GAA	GAT	GAC	GAC
5	Gly	Lys	Lys	Gly	Ala	Val	Thr	Pro	Ala	Lys	Gly	Ala	Lys	Asn	Gly	Lys	Asn	Ala	Lys	Lys	Glu	Asp	Ser	Asp	Glu	Asp	Glu	Asp	Asp	Asp
										130										140										150
	GAT	GAT	GAA	GAC	GAC	AGT	GAT	GAG	GAT	GAA	GAG	GAT	GAG	GAA	GAG	GAT	GAA	TTT	GAG	CCA	CCA	GTA	GTG	AAA	GGA	AAA	CAA	GGA	AAG	GTA
6	Asp	Asp	Glu	Asp	Asp	Ser	Asp	Glu	Asp	Glu	Glu	Asp	Glu	Glu	Glu	Asp	Glu	Phe	Glu	Pro	Pro	Val	Val	Lys	Gly	Lys	Gln	Gly	Lys	Val
										160										170										180
	GCT	GCT	GCT	GCT	CCT	GCC	TCA	GAG	GAT	GAA	GAC	GAG	GAA	GAG	GAT	GAA	GAG	GAG	GAG	GAG	GAG	GAT	GAA	GAG	GAG	GAG	GAT	GAC	TCT	GAA
7	Ala	Ala	Ala	Ala	Pro	Ala	Ser	Glu	Asp	Glu	Asp	Glu	Glu	Glu	Asp	Glu	Glu	Glu	Glu	Glu	Glu	Asp	Glu	Glu	Glu	Glu	Asp	Asp	Ser	Glu
										190										200										210
	GAA	GAA	GAA	GCT	ATG	GAG	ATC	ACA	CCT	GCC	AAA	GGA	AAG	AAA	GCT	CCT	GCA	AAA	GTT	GTT	CCT	GTG	AAA	GCC	AAG	AAT	GTG	GCT	GAG	GAG
8	Glu	Glu	Glu	Ala	Met	Glu	Ile	Thr	Pro	Ala	Lys	Gly	Lys	Lys	Ala	Pro	Ala	Lys	Val	Val	Pro	Val	Lys	Ala	Lys	Asn	Val	Ala	Glu	Glu
										220										230										240
	GAT	GAT	GAT	GAT	GAA	GAG	GAG	GAT	GAA	GAT	GAG	GAG	GAG	GAT	GAA	GAA	GAG	GAA	GAA	GAT	GAG	GAG	GAG	GAG	GAA	GAG	GAA	GAA	GAG	GAA
9	Asp	Asp	Asp	Asp	Glu	Glu	Glu	Asp	Glu	Asp	Glu	Glu	Glu	Asp	Glu	Glu	Glu	Glu	Glu	Asp	Glu	Glu	Glu	Glu	Glu	Glu	Glu	Glu	Glu	Glu
										250										260										270
	GAG	CCT	GTT	AAA	CCA	GCA	CCT	GGA	AAA	CGG	AAG	AAG	GAA	ATG	ACC	AAA	CAG	AAA	GAA	GTT	CCT	GAA	GCA	AAG	AAA	CAG	AAA	GTG	GAA	GGT
10	Glu	Pro	Val	Lys	Pro	Ala	Pro	Gly	Lys	Arg	Lys	Lys	Glu	Met	Thr	Lys	Gln	Lys	Glu	Val	Pro	Glu	Ala	Lys	Lys	Gln	Lys	Val	Glu	Gly
										280										290										300
	TCA	GAA	TCA	ACT	ACA	CCT	TTC	AAT	CTG	TTC	ATT	GGA	AAC	CTT	AAC	CCA	AAC	AAG	TCT	GTT	GCT	GAA	TTA	AAA	GTT	GCC	ATC	AGT	GAA	CCT
11	Ser	Glu	Ser	Thr	Thr	Pro	Phe	Asn	Leu	Phe	Ile	Gly	Asn	Leu	Asn	Pro	Asn	Lys	Ser	Val	Ala	Glu	Leu	Lys	Val	Ala	Ile	Ser	Glu	Pro
										310										320										330
	TTT	GCT	AAA	AAT	GAT	CTT	GCT	GTT	GTG	GAT	GTC	AGA	ACT	GGT	ACA	AAT	AGG	AAA	TTT	GGT	TAT	GTG	GAC	TTT	GAG	TCA	GCT	GAA	GAC	CTA
12	Phe	Ala	Lys	Asn	Asp	Leu	Ala	Val	Val	Asp	Val	Arg	Thr	Gly	Thr	Asn	Arg	Lys	Phe	Gly	Tyr	Val	Asp	Phe	Glu	Ser	Ala	Glu	Asp	Leu
										340										350										360
	GAA	AAG	GCC	TTG	GAG	CTC	ACC	GGT	TTG	AAA	GTG	TTT	GGC	AAT	GAA	ATT	AAA	CTA	GAA	AAA	CCA	AAA	GGA	AGA	GAT	AGT	AAA	AAA	GTT	CGA
13	Glu	Lys	Ala	Leu	Glu	Leu	Thr	Gly	Leu	Lys	Val	Phe	Gly	Asn	Glu	Ile	Lys	Leu	Glu	Lys	Pro	Lys	Gly	Arg	Asp	Ser	Lys	Lys	Val	Arg
										370										380										390
	GCT	GCA	AGG	ACA	CTT	TTA	GCA	AAA	AAT	CTT	TCT	TTC	AAC	ATC	ACT	GAG	GAT	GAA	TTA	AAA	GAA	GTG	TTT	GAA	GAT	GCC	TTG	GAG	ATC	AGA
14	Ala	Ala	Arg	Thr	Leu	Leu	Ala	Lys	Asn	Leu	Ser	Phe	Asn	Ile	Thr	Glu	Asp	Glu	Leu	Lys	Glu	Val	Phe	Glu	Asp	Ala	Leu	Glu	Ile	Arg
										400										410										420

```
   TTG GTT AGC CAG GAT GGG AAG AGT AAA GGG ATT GCT TAT ATT GAA TTT AAG TCT GAA GCT GAT GCA GAG AAA AAC TTG GAA GAA AAG CAA
15 Leu Val Ser Gln Asp Gly Lys Ser Lys Gly Ile Ala Tyr Ile Glu Phe Lys Ser Glu Ala Asp Ala Glu Lys Asn Leu Glu Glu Lys Gln
                                      430                                     440                                     450

   GGA GCA GAA ATT GAT GGG CGT TCT GTT TCA CTT TAC TAT ACT GGG GAG AAG GGA CAA AGG CAA GAA AGA ACT GGA AAG AAT AGC ACT TGG
16 Gly Ala Glu Ile Asp Gly Arg Ser Val Ser Leu Tyr Tyr Thr Gly Glu Lys Gly Gln Arg Gln Glu Arg Thr Gly Lys Asn Ser Thr Trp
                                      460                                     470                                     480

   AGT GGT GAA TCA AAA ACT TTG GTT TTA AGT AAC CTT TCC TAC AGT GCA ACA GAA GAA ACT CTT CAG GAA GTG TTT GAG AAA GCA ACT TTT
17 Ser Gly Glu Ser Lys Thr Leu Val Leu Ser Asn Leu Ser Tyr Ser Ala Thr Glu Glu Thr Leu Gln Glu Val Phe Glu Lys Ala Thr Phe
                                      490                                     500                                     510

   ATC AAA GTG CCC CAG AAC CAA CAG GGC AAA TCT AAA GGG TAT GCG TTT ATA GAA TTT GCT TCA TTT GAA GAT GCT AAA GAA GCT TTA AAT
18 Ile Lys Val Pro Gln Asn Gln Gln Gly Lys Ser Lys Gly Tyr Ala Phe Ile Glu Phe Ala Ser Phe Glu Asp Ala Lys Glu Ala Leu Asn
                                      520                                     530                                     540

   TCC TGT AAC AAA ATG GAA ATT GAG GGC AGA ACA ATC AGG TTG GAG TTA CAA GGA CCC AGG GGG TCA CCT AAT GCG AGA AGT CAG CCA TCC
19 Ser Cys Asn Lys Met Glu Ile Glu Gly Arg Thr Ile Arg Leu Glu Leu Gln Gly Pro Arg Gly Ser Pro Asn Ala Arg Ser Gln Pro Ser
                                      550                                     560                                     570

   AAA ACA CTG TTT GTC AAG GGT CTG TCT GAG GAT ACC ACT GAA GAG ACC TTA AAA GAA TCA TTT GAG GGC TCT GTT CGT GCA AGA ATA GTC
20 Lys Thr Leu Phe Val Lys Gly Leu Ser Glu Asp Thr Thr Glu Glu Thr Leu Lys Glu Ser Phe Glu Gly Ser Val Arg Ala Arg Ile Val
                                      580                                     590                                     600

   ACT GAT CGG GAA ACT GGT TCC TCT AAA GGG TTT GGT TTT GTA GAC TTC AAC AGT GAG GAA GAT GCC AAA GCT GCC AAG GAG GCC ATG GAA
21 Thr Asp Arg Glu Thr Gly Ser Ser Lys Gly Phe Gly Phe Val Asp Phe Asn Ser Glu Glu Asp Ala Lys Ala Ala Lys Glu Ala Met Glu
                                      610                                     620                                     630

   GAT GGA GAG ATT GAT GGG AAT AAA GTT ACT TTG GAC TGG GCC AAA CCT AAG GGT GAA GGT GGC TTT GGT GGC CGA GGT GGA GGC AGA GGA
22 Asp Gly Glu Ile Asp Gly Asn Lys Val Thr Leu Asp Trp Ala Lys Pro Lys Gly Glu Gly Gly Phe Gly Gly Arg Gly Gly Gly Arg Gly
                                      640                                     650                                     660

   GGT TTT GGG GGC CGA GGT GGT GGA AGA GGA GGA GGC CGA GGT GGA TTT GGC GGC AGA GGC CGG GGC GGC TTT GGA GGT AGA GGA GGC TTC
23 Gly Phe Gly Gly Arg Gly Gly Gly Arg Gly Gly Gly Arg Gly Gly Phe Gly Gly Arg Gly Arg Gly Gly Phe Gly Gly Arg Gly Gly Phe
                                      670                                     680                                     690

   CGA GGT GGC CGA GGA GGA GGA GGA GGA GGA GGA GAC TTC AAG CCA CAA GGA AAG AAG ACG AAG TTT GAA TAA TTCCTTCCCAATCTTTCCCACTT
24 Arg Gly Gly Arg Gly Gly Gly Gly Gly Gly Gly Asp Phe Lys Pro Gln Gly Lys Lys Thr Lys Phe Glu xxx
                                      700                                     710
```

FIGURE 4. DNA sequence of nucleolin cDNA clones isolated from a Chinese hamster ovary cell library.[24] The encoded amino acid sequence is shown below the DNA sequence (residues 35 to 713). The NH_2-terminal sequence of nucleolin was determined by automatic protein sequencing and by sequencing two exons corresponding to the NH_2-terminal proximal protein of nucleolin. The boxed region shows the domain rich in glycine and N^G,N^G-dimethylarginine residues interspersed by phenylalanine residues. The arginine residues are assumed to be methylated based on the similarity of this region to Novikoff hepatoma nucleolin.[22,23] Nucleolin contains four "RNP consensus sequences", several serine phosphorylation sites, acidic domains, and possible *N* glycosylation sites.[24] (Permission to use this figure was generously provided by Dr. Francois Amalric.)

TABLE 3
Amino Acid Composition of the Rat 34 kDa Nucleolar Protein

Amino acid	34 kDa protein	*Physarum* 34 kDa Protein[73]
Aspartic acid	8.0	6.6
Threonine	2.6	3.1
Serine	5.3	4.4
Glutamic acid	8.0	8.0
Proline	4.1	4.7
Glycine	22.8	22.6
Alanine	7.4	10.1
Valine	6.8	6.9
Methionine	1.2	2.0
Isoleucine	4.8	3.6
Leucine	4.8	4.4
Tyrosine	1.3	2.4
Phenylalanine	4.4	4.6
Lysine	6.5	6.0
Histidine	3.3	1.8
Arginine	4.6	4.7
N^G,N^G-dimethylarginine	4.1	4.1

Note: The 34 kDa protein was hydrolyzed in 5.7*N* HCl for 22 h. The mol% values for Thr and Ser were corrected for destruction during hydrolysis.

at least three N^G,N^G-dimethylarginines per molecule[8,25] and this region is similar to other arginine methylation sites.[19,22] This zone contains a considerably higher content of two other amino acids, serine and asparagine, than the domains found in nucleolin and the 34-kDa nucleolar protein.

In vitro nucleic acid binding characteristics of UP1, protein A1, recombinant protein A1, and synthetic peptide analogous to the COOH-terminal of protein A1 have been studied.[14-17,28] It was determined that the UP1 protein, the NH_2-terminal 195 residues of protein A1, bound single-stranded nucleic acid in a noncooperative manner.[28] This is the region of the molecule that contains the "RNP consensus sequences". The entire A1 protein bound nucleic acid with a higher association constant and showed cooperativity.[28] It was concluded that the COOH-terminal domain strongly influences nucleic acid binding by A1.[28] An oligopeptide of approximately 12,000 molecular weight that was polymerized from the peptide Gly-Asn-Phe-Gly-Gly-Gly-Arg-Gly-Gly-Asn-Phe-Gly-Gly-Ser-Arg-Gly was found to bind tightly to nucleic acid.[28] This peptide represents two of the repeats found in the COOH-terminal end of protein A1. It was thus postulated that the COOH-terminal domain contributes to A1 binding through both cooperative protein-protein interactions and direct interaction with nucleic acid.[28] The COOH-terminal domain in protein A1 may contain two regions, one for protein-protein interaction and the other for nucleic acid interactions. It is not known what role the glycine, arginine/DMA clusters interspersed with phenylalanine residues plays in the function of protein A1. The recombinant protein A1 used in one of these studies did not contain any N^G,N^G-dimethylarginine suggesting that this residue is not essential for *in vitro* nucleic acid binding.[25,28]

D. PUTATIVE RNA BINDING PROTEIN FROM *ARTEMIA SALINA*

In the brine shrimp *Artemia salina* there is a major hnRNP protein component, HD40.[10,12,27,81,82] This protein is believed to be a helix destabilizing protein that unwinds most of the secondary structure of RNA, and compacts the unwound RNA into a structure that has a "beads-on-a-string" appearance.[10,12,81,82] This protein has been purified to homogeneity and binds strongly to single-stranded nucleic acid.[12] It is biochemically similar to the hnRNP core proteins from higher eukaryotes, having a high glycine content, the modified amino acid N^G,N^G-dimethylarginine and a blocked amino-terminal.[10,12,81,82] In the 30S hnRNP from *Artemia* three or four antigenically related proteins of molecular weight 30,000 to 40,000 to the HD40 were found.[27] The cDNA containing the coding sequence for one of these antigenically related proteins has been isolated and sequenced.[27] The protein has a molecular weight of 32,992 and has been designated GRP33. The cDNA and protein sequence is shown in Figure 6. The COOH-terminal (123 residues) is glycine rich as is the COOH-terminal of the hnRNP protein A1.[27] Two regions were found in this COOH-terminal zone to be rich in glycine and arginine residues interspersed with primarily phenylalanine residues (Figure 6, boxed). It is not known if these arginine residues are methylated. The antigenically related HD40 protein from *Artemia* does contain 1.4 mol% N^G,N^G-dimethylarginine.[10,12] This GRP33 protein also likely contains this modified amino acid.[27] It was postulated that this COOH-terminal domain was on the exterior of the molecule and is likely involved in nucleic acid binding.[27] It should be noted that this putative RNA binding protein does not contain an "RNP consensus sequence".[27]

E. *SACCHAROMYCES CEREVISIAE* SSB1

Saccharomyces cerevisiae contain several single-stranded nucleic acid binding proteins.[83,84] A 45-kDa SSB, SSB1, has been isolated on the basis of its preferential binding for ssDNA.[84] It binds RNA and stimulates yeast DNA polymerase I on ssDNA templates.[83,84] Strains of *Saccharomyces cerevisiae* which contained an SSB1 gene disruption were found to grow normally. This gene was not required for sporulation, spore germination, or recombination.[26,83,84] It was postulated that SSB1 played a role in nuclear RNA metabolism.[26] This protein was immunochemically localized to the nucleus, predominantly located in the nucleolus.[26]

The amino acid sequence of SSB1 has been obtained from the nucleotide sequence and is shown in Figure 7.[26] It was found to have a domain rich in glycine and arginine interspersed with phenylalanine residues from position 125 to 160. It is not known if these arginine residues are methylated. In this region there are 17 glycine, 6 phenylalanine, and 10 arginine residues. The arginine and phenylalanine residues are dispersed throughout this domain. At position 148 a tyrosine residue is found where one may expect a phenylalanine residue. This does not disrupt the fairly even distribution of charged and hydrophobic amino acids. This region is more like the domains found in the 34-kDa nucleolar protein and nucleolin than the domain in protein A1. This is likely significant since SSB1 is predominantly a nucleolar protein.[26] Unlike the domains in the 34-kDa nucleolar protein and nucleolin this region in SSB1 is located at the center of the molecule. In addition it was postulated to have a helical secondary structure[26] as compared to the extended conformation proposed for the other two nucleolar proteins. This region in SSB1 has the tandem repeat Arg-Gly-Gly-Phe-Arg-Gly which is found once in nucleolin.

It has been postulated that this domain in SSB1 is a nucleic acid binding region.[26] It was further speculated that the actual sequence differences of the three common amino acids, glycine, arginine, and phenylalanine, may give rise to different protein-nucleic acid structures and hence specific function.[26] The SSB1 protein has a "RNP consensus sequence" which is believed to be a RNA binding region.[26,55-58] In addition, other sequence homologies between SSB1 and hnRNP proteins A1 and A2 were found.[26]

IV. FUNCTION

A. PROPOSED FUNCTION OF THE CONSERVED DOMAIN RICH IN GLYCINE AND ARGININE/DMA RESIDUES INTERSPERSED WITH PHENYLALANINE RESIDUES

A zone rich in glycine and N^G,N^G-dimethylarginine/arginine interspersed with phenylalanine residues has been found in nucleolin,[22-24] the 34-kDa nucleolar protein from Novikoff hepatoma cells[19] and *Physarum polycephalum*,[74] in the heterogeneous nuclear ribonucleoprotein particle

```
                     10          20          30          40          50          60
   cgctgaacgctctcatcatcctaccgtc ATG TCT AAG TCA GAG TCC CCC AAG GAA CCG GAA CAG
                                Met Ser Lys Ser Glu Ser Pro Lys Glu Pro Glu Gln  12
    70          80          90          100         110         120
CTG CGG AAG CTC TTC ATT GGA GGG CTG AGC TTC GAA ACA ACC GAC GAG AGT CTG AGG AGC
Leu Arg Lys Leu Phe Ile Gly Gly Leu Ser Phe Glu Thr Thr Asp Glu Ser Leu Arg Ser  32
    130         140         150         160         170         180
CAT TTT GAG CAA TGG GGA ACA CTC ACC GAC TGT GTG GTA ATG AGA GAT CCA AAC ACC AAA
His Phe Glu Gln Trp Gly Thr Leu Thr Asp Cys Val Val Met Arg Asp Pro Asn Thr Lys  52
    190         200         210         220         230         240
AGA TCC AGA GGC TTT GGG TTT GTC ACA TAT GCC ACT GTG GAG GAA GTG GAT GCT GCC ATG
Arg Ser Arg Gly Phe Gly Phe Val Thr Tyr Ala Thr Val Glu Glu Val Asp Ala Ala Met  72
    250         260         270         280         290         300
AAT GCA AGA CCA CAC AAA GTG GAT GGA AGA GTT GTG GAA CCT AAG AGA GCT GTG TCA AGA
Asn Ala Arg Pro His Lys Val Asp Gly Arg Val Val Glu Pro Lys Arg Ala Val Ser Arg  92
    310         320         330         340         350         360
GAA GAT TCT CAG AGA CCA GGT GCC CAC TTA ACT GTG AAG AAG ATC TTT GTT GGC GGT ATT
Glu Asp Ser Gln Arg Pro Gly Ala His Leu Thr Val Lys Lys Ile Phe Val Gly Gly Ile 112
    370         380         390         400         410         420
AAA GAA GAC ACT GAA GAA CAT CAC CTA CGA GAT TAT TTT GAG CAG TAT GGG AAA ATT GAA
Lys Glu Asp Thr Glu Glu His His Leu Arg Asp Tyr Phe Glu Gln Tyr Gly Lys Ile Glu 132
    430         440         450         460         470         480
GTG ATT GAA ATT ATG ACT GAC AGA GGC AGT GGA AAA AAG AGG GGA TTT GCG TTT GTC ACC
Val Ile Glu Ile Met Thr Asp Arg Gly Ser Gly Lys Lys Arg Gly Phe Ala Phe Val Thr 152
    490         500         510         520         530         540
TTT GAT GAC CAT GAC TCT GTG GAT AAG ATT GTT ATT CAG AAA TAC CAT ACT GTG AAT GGC
Phe Asp Asp His Asp Ser Val Asp Lys Ile Val Ile Gln Lys Tyr His Thr Val Asn Gly 172
    550         560         570         580         590         600
CAC AAC TGT GAA GTA AGA AAG GCT CTG TCG AAA CAA GAG ATC GCT AGT GCT TCA TCC AGC
His Asn Cys Glu Val Arg Lys Ala Leu Ser Lys Gln Glu Met Ala Ser Ala Ser Ser Ser 192
    610         620         630         640         650         660
CAG AGA GGT CGA AGT GGT TCC GGA AAC TTT GGT GGT GGT CGT GGA GGT GGT TTC GGT GGC
Gln Arg Gly Arg★Ser Gly Ser Gly Asn Phe Gly Gly Gly Arg Gly Gly Gly Phe Gly Gly 212
```

```
    670                 680                 690                 700                 710                 720
AAT GAC AAT TTT GGT CGA GGA GGG AAC TTC AGT GGT CGT GGT GGC TTT GGT GGC AGC CGT
Asn Asp Asn Phe Gly Arg Gly Gly Asn Phe Ser Gly Arg Gly Gly Phe Gly Gly Ser Arg 232
    730                 740                 750                 760                 770                 780
GGT GGT GGT GGA TAT GGT GGC AGT GGG GAT GGC TAT AAT GGA TTT GGC AAT GAT GGA AGC
Gly Gly Gly Gly Tyr Gly Gly Ser Gly Asp Gly Tyr Asn Gly Phe Gly Asn Asp Gly Ser 252
    790                 800                 810                 820                 830                 840
AAT TTT GGA GGT GGT GGA AGC TAC AAT GAT TTT GGC AAT TAC AAC AAC CAG TCA TCA AAT
Asn Phe Gly Gly Gly Gly Ser Tyr Asn Asp Phe Gly Asn Tyr Asn Asn Gln Ser Ser Asn 272
    850                 860                 870                 880                 890                 900
TTT GGA CCG ATG AAA GGA GGA AAC TTT GGA GGC AGG AGC TCT GGC CCT TAT GGT GGT GGA
Phe Gly Pro Met Lys Gly Gly Asn Phe Gly Gly Arg Ser Ser Gly Pro Tyr Gly Gly Gly 292
    910                 920                 930                 940                 950                 960
GGC CAG TAC TTT GCT AAA CCA CGA AAC CAA GGT GGC TAT GGA GGT TCC AGC AGC AGC AGT
Gly Gln Tyr Phe Ala Lys Pro Arg Asn Gln Gly Gly Tyr Gly Gly Ser Ser Ser Ser Ser 312
    970                 980                 990           1000           1010           1020          1030
AGC TAT GGC AGT GGC AGG AGG TTC TAA ttacagccaggaaacaaagcttagcaggagagccagagaagtg
Ser Tyr Gly Ser Gly Arg Arg Phe ***
```

FIGURE 5. The nucleotide sequence and deduced amino acid sequence of the 320 amino acid 34,315-D rodent hnRNP complex protein A1.[10] The cDNA was purified from a newborn rat brain library using a synthetic oligonucleotide probe corresponding to the N-terminal region of the calf helix-destabilizing protein, UP1.[16] It was later determined that UP1 is a proteolytic degradation product derived from the hnRNP core protein A1.[14,17] Residues 2 to 196 of the deduced protein are identical to the reported sequence of calf thymus UP1, the end of which is marked (*). The glycine, arginine/N^G,N^G-dimethylarginine-rich domain interspersed with phenylalanines is boxed. The arginine at positions 194 is methylated. Other arginine residues in this domain may be methylated.[8,25] Protein A1 contains two "RNP consensus sequences".[14,16]

10 20 30 40 50 60 70 80
cgctacaagtgtggttgtatttgataattaaggtataaaaaa ATG GCT GCC AAA CCC GAG CAA GAG CCT GTG TAT GTC CGA
Met Ala Ala Lys Pro Glu Gln Glu Pro Val Tyr Val Arg
90 100 110 120 130 140 150
GAT TTG GTG AAA GAT TAT GAT GAT GCT CGT CAA ATG CTA ACT CAA GCA GGA GTA TCT GAA GCA GTA CTT GGA ACA
Asp Leu Val Lys Asp Tyr Asp Asp Ala Arg Gln Met Leu Thr Gln Ala Gly Val Ser Glu Ala Val Leu Gly Thr
160 170 180 190 200 210 220 230
ATA GAT GCA GAA ATC AAG CAC ATA AAA ACT GGA AGT CGG CCG AAA ACC GTG CCA AAT ACA GAT GGA TCT GGA TTT
Ile Asp Ala Glu Ile Lys His Ile Lys Thr Gly Ser Arg Pro Lys Thr Val Pro Asn Thr Asp Gly Ser Gly Phe
240 250 260 270 280 290 300
ATG GAT CTT TAC AAT GAC ACC AAA GTT AAA CTT GTT TCA AGA TGT TGC TTG CCT GTT GAT CAA TTC CCC AAG TAC
Met Asp Leu Tyr Asn Asp Thr Lys Val Lys Leu Val Ser Arg Cys Cys Leu Pro Val Asp Gln Phe Pro Lys Tyr
310 320 330 340 350 360 370 380
AAC TTC CTT GGT AAA CTT CTT GGA CCT GGT GGA AGC ACC ATG AAA CAA CTT CAA GAT GAA ACG ATG ACT AAG ATT
Asn Phe Leu Gly Lys Leu Leu Gly Pro Gly Gly Ser Thr Met Lys Gln Leu Gln Asp Glu Thr Met Thr Lys Ile
390 400 410 420 430 440 450
TCA ATC CTT GGA AGA GGC TCA ATG AGA GAT AGG AAC AAG GAA GAA GAA TTG AGG AAT TCA GGA GAC GTC AAA TAT
Ser Ile Leu Gly Arg Gly Ser Met Arg Asp Arg Asn Lys Glu Glu Glu Leu Arg Asn Ser Gly Asp Val Lys Tyr
460 470 480 490 500 510 520 530
GCC CAC TTG AAC GAG CAG CTC CAC ATT GAG ATC ATT TCT ATT GCT AGT CCT GCT GAG GCT CAT GCC CGT ATG GCC
Ala His Leu Asn Glu Gln Leu His Ile Glu Ile Ile Ser Ile Ala Ser Pro Ala Glu Ala His Ala Arg Met Ala
540 550 560 570 580 590 600
TAT GCT CTC ACT GAA ATC AAA AAG TAT ATC ACC CCA GAA GAG GAT CCA AAC TAC ATG ATG ATG GCC GGT CAT GGC
Tyr Ala Leu Thr Glu Ile Lys Lys Tyr Ile Thr Pro Glu Glu Asp Pro Asn Tyr Met Met Met Ala Gly His Gly
610 620 630 640 650 660 670 680
GCT GGT CCA ATG ATG GGC ATG GGA GGT ATG ATG GGA GGT CCA GGG CCA ATG GGA CCA CAA GGC CGT GGT CGT GGA
Ala Gly Pro Met Met Gly Met Gly Gly Met Met Gly Gly Pro Gly Pro Met Gly Pro Gln Gly Arg Gly Arg Gly
690 700 710 720 730 740 750
AGA GGA CGA GGT GGT TTC AGT GGA CCT GAT AGG ACA TTT GAC TTA TTA GAA AAG GCA AGA ATG AAC ACA AGC GAA
Arg Gly Arg Gly Gly Phe Ser Gly Pro Asp Arg Thr Phe Asp Leu Leu Glu Lys Ala Arg Met Asn Thr Ser Glu
760 770 780 790 800 810 820 830
ACC ATG GAC CCT GGC TAT GGT TTC GAC GAG TCC TAT GGT GGA ATG GGA GGA GGA TAT GAA ATG CCA TAC AAC GGC
Thr Met Asp Pro Gly Tyr Gly Phe Asp Glu Ser Tyr Gly Gly Met Gly Gly Gly Tyr Glu Met Pro Tyr Asn Gly
840 850 860 870 880 890 900
AAT GCA GGA TGG ACA GCA TCT CCT GGC CGC GGG GCT GGC GCC GGT GCC CGT GGT GCA CGA GGT GGA CTT GAC CAG
Asn Ala Gly Trp Thr Ala Ser Pro Gly Arg Gly Ala Gly Ala Gly Ala Arg Gly Ala Arg Gly Gly Leu Asp Gln
910 920 930 940 950 960 970 980
TCA AGA GGA GGT GGA AAA TTT CCC TCC GCA CGC GGA GGA AGA GGC CGC GCA GCA CCC TAC TGA gttgccctatggcaa
Ser Arg Gly Gly Gly Lys Phe Pro Ser Ala Arg Gly Gly Arg Gly Arg Ala Ala Pro Tyr ***

FIGURE 6. Full-length cDNA sequence and deduced amino acid sequence of a putative RNA binding protein from *Artemia salina*.[27] The clone was purified from a cDNA library constructed in λ gt11 from *Artemia* total poly(A)$^+$ RNA. The protein is 308 amino acids long and has a molecular weight of 32,992. It has been termed GRP33 (glycine-rich protein, M_r 33,000).[27] The two glycine, arginine-rich regions are boxed. It has been speculated that some, if not all, of these arginines are methylated since this protein is immunologically related to the dimethylarginine containing HD40 protein and since the glycine, arginine rich domains interspersed with phenylalanine is known to contain N^G,N^G-dimethylarginine in proteins A1, nucleolin and the 34-kDa nucleolar protein.[8,19,22,27]

```
-50         -40         -30         -20         -10                                      15
GAAGAAGTTT  CCCCCAAAAG  AAAGAAGAAA  ACCCTCAAAC  GAAGAAAAAT  ATG TCT GCT GAA ATT GAA GAA GCT ACT
                                                            Met Ser Ala Glu Ile Glu Glu Ala Thr

 30              45              60              75              90
AAT GCC GTA AAC AAC TTG AGC ATC AAC GAC TCC GAA CAG CAA CCA AGG GCT CCT ACT CAT AAG ACA GTA ATT
Asn Ala Val Asn Asn Leu Ser Ile Asn Asp Ser Glu Gln Gln Pro Arg Ala Pro Thr His Lys Thr Val Ile

    105             120             135             150             165
GAC CCC GAG GAC ACA ATC TTT ATT GGT AAT GTT GCT CAC GAA TGT ACC GAA GAC GAC TTG AAG CAA TTG TTT
Asp Pro Glu Asp Thr Ile Phe Ile Gly Asn Val Ala His Glu Cys Thr Glu Asp Asp Leu Lys Gln Leu Phe

        180             195             210             225             240
GTG GAG GAA TTC GGG GAT GAA GTC AGC GTA GAG ATC CCA ATT AAG GAA CAC ACC GAC GGT CAC ATT CCA GCT
Val Glu Glu Phe Gly Asp Glu Val Ser Val Glu Ile Pro Ile Lys Glu His Thr Asp Gly His Ile Pro Ala

            255             270             285             300             315
AGT AAA CAC GCT CTA GTC AAG TTC CCA ACC AAG ATT GAT TTT GAT AAT ATC AAG GAG AAT TAT GAC ACG AAA
Ser Lys His Ala Leu Val Lys Phe Pro Thr Lys Ile Asp Phe Asp Asn Ile Lys Glu Asn Tyr Asp Thr Lys

                330             345             360             375
GTC GTT AAG GAC AGA GAA ATT CAT ATT AAG AGA GCT AGA ACT CCA GGC CAA ATG CAA AGA GGA GGT TTC AGA
Val Val Lys Asp Arg Glu Ile His Ile Lys Arg Ala Arg Thr Pro Gly Gln Met Gln Arg Gly Gly Phe Arg

390             405             420             435             450
GGC AGA GGC GGT TTC AGA GGC AGA GGA GGT TTT AGA GGA GGT TTC AGA GGC GGC TAC AGA GGA GGT TTC AGA
Gly Arg Gly Gly Phe Arg Gly Arg Gly Gly Phe Arg Gly Gly Phe Arg Gly Gly Tyr Arg Gly Gly Phe Arg

    465             480             495             510             525
GGC AGA GGG AAC TTC AGA GGT AGA GCG GCG CCA GAG GTG GTT TCA ATG GAC AAA AAA GGG AAA GAT TCC ATT
Gly Arg Gly Asn Phe Arg Gly Arg Ala Ala Pro Glu Val Val Ser Met Asp Lys Lys Gly Lys Asp Ser Ile

        540             555             570             585             600
AGA CCA ATG GAA AGA TCA AAG GAT ACC TTA TAT ATT AAT AAC GTC CCA TTC AAA GCT ACC AAA GAG GAG GTC
Arg Pro Met Glu Arg Ser Lys Asp Thr Leu Tyr Ile Asn Asn Val Pro Phe Lys Ala Thr Lys Glu Glu Val

            615             630             645             660             675
GCT GAA TTT TTC GGT ACT GAC GCC GAC TCC ATC TCT TTG CCA ATG AGA AAA ATG AGA GAC CAA CAC ACT GGT
Ala Glu Phe Phe Gly Thr Asp Ala Asp Ser Ile Ser Leu Pro Met Arg Lys Met Arg Asp Gln His Thr Gly

                690             705             720             735
AGG ATC TTC ACA TCC GAT TCT GCT AAT AGA GGT ATG GCA TTT GTC ACT TTC AGT GGT GAA AAC GTT GAT ATT
Arg Ile Phe Thr Ser Asp Ser Ala Asn Arg Gly Met Ala Phe Val Thr Phe Ser Gly Glu Asn Val Asp Ile

750             765             780             795             810
GAA GCT AAA GCT GAA GAA TTT AAA GGC AAG GTT TTC GGT GAC AGG GAG TTA ACT GTA GAT GTT GCT GTT ATT
Glu Ala Lys Ala Glu Glu Phe Lys Gly Lys Val Phe Gly Asp Arg Glu Leu Thr Val Asp Val Ala Val Ile

    825             840             855             870             885
AGA CCA GAA AAT GAT GAA GAA GAA ATT GAG CAA GAA ACT GGT TCT GAA GAA AAG CAA GAA TAA TTACTTCT
Arg Pro Glu Asn Asp Glu Glu Glu Ile Glu Gln Glu Thr Gly Ser Glu Glu Lys Gln Glu
```

FIGURE 7. Full-length cDNA sequence and deduced amino acid sequence of a single-stranded nucleic acid binding protein, SSB1, from *Saccharomyces cerevisiae*.[26] The amino acid sequence contains 293 amino acid residues and has a molecular weight of 32,853.[26] The boxed region shows the domain rich in glycine and arginine residues interspersed with phenylalanine residues. It is not known if these arginine residues are methylated. The SSB1 protein contains an "RNP consensus sequence", an acidic stretch near the COOH-terminal, and a region near the NH_2-terminal which has sequence homology to the hnRNP protein A1.[26]

protein A1,[16,17] in an *Artemia* putative RNA binding protein[27] and in a single stranded nucleic acid binding protein, SSB1, from yeast.[26] This domain is also likely present in other hnRNP proteins which are rich in glycine and contain N^G,N^G-dimethylarginine as well as other proteins.[19,22] The function of this domain is currently not clear. All of the above proteins are believed to bind nucleic acid *in vitro* and are associated with nucleic acid, primarily RNA, *in vivo*.[8-30,38-51,70-86] It has accordingly been proposed that this domain is involved in nucleic acid binding.[16,17,19,22-28] The high concentration of aromatic and positively charged amino acid observed in this domain has been found in several other ssDNA binding proteins.[86-89] It was postulated that the positively charged amino acids interact with the negatively charged phosphodiester backbone of ssDNA while the aromatic amino acids intercalate with the nucleotide bases.[86] The regular spacing of N^G,N^G-dimethylarginine/arginine and phenylalanines found in these domains leans support for a nucleic acid binding role. The presences of glycine between the aromatic and charged residues in this zone would allow freedom of motion and glycine does not contain a side group to interfere with the interaction of this domain with nucleic acid.

The zone enriched in glycine and arginine/DMA interspersed with phenylalanines is found at the COOH-terminal of nucleolin,[23,24] near the COOH-terminal of protein A1,[16,17] at the COOH-terminal of the putative RNA binding protein from *Artemia*[27] and at the NH_2-terminal of the 34-kDa nucleolar protein.[19,74] It has been proposed that this domain is in an extended conformation which would allow structural flexibility in interactions with nucleic acid. Inconsistent with an end-terminal location of this domain is that found in the yeast SSB1 protein.[26] The glycine, arginine clusters interspersed with phenylalanines is located near the center of the molecule. An alpha helical structure was proposed which had the appropriate dimensions for interaction with the A-form ssDNA.[26] It was accordingly postulated that this domain in SSB1 is involved in binding to single-stranded nucleic acid.

The *in vitro* nucleic acid binding properties of hnRNP protein A1 have been extensively studied.[11-17,25,28,85] The NH_2-terminal region of this molecule, which corresponds to UP1 and does not contain the glycine, arginine/DMA clusters interspersed with phenylalanines was found to bind nucleic acid in a noncooperative fashion.[28] The entire A1 protein which contains the glycine, arginine domain from residues 191 to 236 and a following glycine-rich region interspersed primarily with aromatic amino acids throughout the rest of the molecule, was found to bind nucleic acid with a much higher association constant than UP1.[28,85] In addition protein A1 binding was much less salt sensitive and showed cooperatively.[28] These results were further supported when an oligopeptide made up of two repeats from the COOH-terminal of A1, Gly-Asn-Phe-Gly-Gly-Gly-Arg-Gly-Gly-Asn-Phe-Gly-Gly-Ser-Arg-Gly, was found to have a high affinity for nucleic acid.[28] The above results are consistent with a nucleic acid binding role for the domain rich in glycine and arginine/DMA interspersed with phenylalanine residues. Involvement in cooperative protein-protein interactions cannot be ruled out. The COOH-terminal of the A1 protein does have some sequence similarity to domains found at the COOH- and NH_2-terminal of several keratins.[16,35-39] These regions at the ends of the keratin molecules are believed to be involved in protein-protein interactions.[34-37,90]

Complicating the role of this conserved domain in nucleic acid binding is the presence of a "RNP consensus sequence(s)" in the A1 protein, nucleolin, the yeast SSB1 and other proteins which do not contain glycine arginine/DMA clusters interspersed with phenylalanine but are known to bind nucleic acid.[24,26,46,55-58,85] The "RNP consensus sequence" in addition to other essential phenylalanine residues in an approximately 90 amino acid domain is believed to be involved in RNA binding.[55,85] In genetic studies, it was determined that the complete repeat of approximately 90 amino acids is apparently not required in yeast polyadenylate-binding protein for cell viability.[91] Protein A1, nucleolin and the yeast SSB1 thus have a domain believed to bind RNA, the "RNP consensus sequence" in combination with adjacent residues. This does not rule out the glycine arginine/DMA clusters interspersed with phenylalanines as a nucleic acid binding domain. The putative RNA binding protein from *Artemia,* GRP33, does not contain an "RNP consensus sequence".[27]

The most likely role for this domain is interaction with nucleic acid, presumably single-stranded RNA. It has been speculated that the glycine, arginine/DMA cluster interspersed with phenylalanines may modulate specificity for the protein to bind RNA over ssDNA.[27] It has also been postulated that the actual sequence of the three amino acids may give rise to different protein-nucleic acid structures[26] and/or determine specificity for certain nucleic acid molecules.[19] The precise function of this conserved domain will have to await further characterization.

B. PROTEASE RESISTANCE

The peptide bond adjacent to the N^G-monomethylarginine and N^G,N'^G-dimethylarginine residue in myelin basic protein were reported to be resistant to trypsin.[2,6] The peptide bond(s) adjacent to the N^G,N^G-dimethylarginine(s) in nucleolin and in the A1 protein were also found to be trypsin resistant.[22,29] A 53 amino acid peptide which contains the ten dimethylarginine sites was generated from nucleolin following trypsin digestion.[22] One possible role for the methylation of arginine residues is to provide protection from intracellular proteases.

V. SUMMARY

A new domain rich in glycine and arginine/DMA interspersed with phenylalanine residues has been found in several nuclear nucleic acid binding proteins.[16,17,19-24,26,27] The presence of this domain in *Physarum* B-36[74] and yeast SSB1[26] illustrates that this domain is evolutionarily conserved at least to lower eukaryotes. The presence of N^G,N^G-dimethylarginine in this domain in the *Physarum* protein also illustrates that the enzyme responsible for methylating the arginine residues is also conserved.[19,74] The most likely role for this zone in these proteins is nucleic acid binding, although a role in protein-protein interactions can not be ruled out.[16,17,19-28,74]

All of the arginine residues that are methylated have at least one adjacent glycine residue. Access of certain arginine methylases to arginine residues may be sterically possible because of the lack of a side chain on the adjacent glycine residues(s).[19,22] The interspersed phenylalanines may also be involved in recognition of the target arginines.

The peptide bonds adjacent to the methylated arginine residues have been found to be resistant to trypsin.[2,6,22,29] One role of this modification may be to protect the protein from intercellular proteases.

ACKNOWLEDGMENTS

I would like to thank K. R. Williams, M. E. Christensen, and J. O. Thomas for their helpful discussions and for making available unpublished results. The excellent amino acid sequencing work of R. G. Cook is acknowledged. I am grateful to L. C. Yeoman, Y. S. Ahn, R. Reddy, and R. L. Ochs for their help in the identification and characterization of nucleolin and the 34-kDa nucleolar protein. Human autoantibodies were generously provided by E. M. Tan, M. Reichlin, and G. Reimer. I am especially grateful to H. Busch for his advice while we were working on arginine methylation. Preparation of this chapter was supported by E. I. DuPont de Nemours & Company, Inc.

REFERENCES

1. **Paik, W. K. and Kim, S.,** *Protein Methylation,* John Wiley & Sons, New York, 1980.
2. **Baldwin, G. S. and Carnegie, P. R.,** Specific enzymic methylation of an arginine in the experimental allergic encephalomyelitis protein from human myelin, *Science,* 171, 579, 1971.
3. **Crang, A. J. and Jacobson, W.,** The relationship of myelin basic protein (arginine) methyltransferase to myelination in mouse spinal cord, *J. Neurochem,* 39, 244, 1982.
4. **Farooqui, J. Z., Tuck, M., and Paik, W. K.,** Purification and characterization of enzymes from *Euglena gracilis* that methylate methionine and arginine residues of cytochrome c, *J. Biol. Chem.,* 260, 537, 1985.
5. **DesJardins, K. C. and Morell, P.,** Phosphate groups modifiying myelin basic proteins are metabolically labile; methyl groups are stable, *J. Cell Biol.,* 97, 438, 1983.
6. **Carnegie, P. R.,** Amino acid sequence of the encephalitogenic basic protein from human myelin, *Biochem. J.,* 123, 57, 1971.
7. **Boffa, L. C., Sterner, R., Vidali, G., and Allfrey, V. G.,** Post-synthetic modification of nuclear proteins high mobility group proteins are methylated, *Biochem. Biophys. Res. Commun.,* 89, 1322, 1979.
8. **Beyer, A. L., Christensen, M. E., Walker, B. W., and LeStrourgeon, W. M.,** Identification and characterization of the packaging proteins of core 40S hnRNP particles, *Cell,* 11, 127, 1977.
9. **Karn, J., Boffa, L. C., Vidali, G., and Allfrey, V. G.,** Characterization of the nonhistone nuclear proteins associated with rapidly labeled heterogeneous nuclear RNA, *J. Biol. Chem.,* 252, 7307, 1978.
10. **Thomas, J. O., Glowacka, S. K., and Szer, W.,** Structure of complexes between a major protein of heterogeneous nuclear ribonucleoprotein particles and polyribonucleotides, *J. Mol. Biol.,* 171, 439, 1983.
11. **Herrick, G. and Alberts, B.,** Purification and physical characterization of nucleic acid helix-unwind proteins form calf thymus, *J. Biol. Chem.,* 251, 2124, 1976.
12. **Marvil, D. K., Nowak, L., and Szer, W.,** A single-stranded nucleic acid-binding protein from *Artemia salina, J. Biol. Chem.,* 255, 6466, 1980.
13. **Planck, S. R. and Wilson, S. H.,** Studies on the structure of mouse helix-destabilizing protein-1, *J. Biol. Chem.,* 255, 11547, 1980.
14. **Williams, K. R., Stone, K. L., LoPresti, M. B., Merrill, B. M., and Planck, S. R.,** Amino acid sequence of the UPI calf thymus helix-destabilizing protein and its homology to an analogous protein from mouse myeloma, *Proc. Natl. Acad. Sci. U.S.A.,* 82, 5666, 1985.
15. **Merrill, B. M., LoPresti, M. B., Stone, K. L., and Williams, K. R.,** High pressure liquid chromatography purification of UP1 and UP2, two related single-stranded nucleic acid-binding proteins from calf thymus, *J. Biol. Chem.,* 261, 878, 1986.
16. **Cobianchi, F., SenGupta, D. N., Zmudzka, B. Z., and Wilson, S. H.,** Structure of rodent helix-destabilizing protein revealed by cDNA cloning, *J. Biol. Chem.,* 261, 3536, 1986.
17. **Riva, S., Morandi, C., Tsoulfas, P., Pandolfo, M., Biamonti, G., Merrill, B., Williams, K. R., Multhaup, G., Beyreuther, K., Werr, H., Henrich, B., and Schafer, K. P.,** Mammalian single-stranded DNA binding protein UP1 is derived from the hnRNP core protein A1, *EMBO J.,* 5, 2267, 1986.
18. **Lahiri, D. K. and Thomas, J. O.,** A cDNA clone of the hnRNP C proteins and its homology with the single-stranded DNA binding protein UP2, *NAR,* 14, 4077, 1986.
19. **Lischwe, M. A., Ochs, R. L., Reddy, R., Cook, R. G., Yeoman, L. C., Tan, E. M., Reichlin, M., and Busch, H.,** Purification and partial characterization of a nucleolar scleroderma antigen (M_r = 34,000; pI, 8.5) rich in N^G,N^G-dimethylarginine, *J. Biol Chem.,* 260, 14304, 1985.
20. **Christensen, M. E., Moloo, J., Swischuk, J. L., and Schelling, M. E.,** Characterization of the nucleolar protein, B-36, using monoclonal antibodies, *Exp. Cell Res.,* 166, 77, 1986.
21. **Lischwe, M. A., Roberts, K. D., Yeoman, L. C., and Busch, H.,** Nucleolar specific acidic phosphoprotein C23 is highly methylated, *J. Biol. Chem.,* 257, 14600, 1982.
22. **Lischwe, M. A., Cook, R. G., Ahn, Y. S., Yeoman, L. C., and Busch, H.,** Clustering of glycine and N^G,N^G-dimethylarginine in nucleolar protein C23, *Biochemistry,* 24, 6025, 1985.
23. **Lapeyre, B., Amalric, F., Ghaffari, S. H., Venkataram Rao, S. V., Dumbar, T. S., and Olson, M. O. J.,** Protein and cDNA sequence of a glycine-rich, dimethylarginine-containing region located near the carboxyl-terminal end of nucleolin (C23 and 100 KDa), *J. Biol. Chem.,* 261, 9167, 1986.
24. **Lapeyre, B., Bourbon, H., and Amalric, F.,** Nucleolin, the major nucleolar protein of growing eukaryotic cells: an unusual protein structure revealed by the nucleotide sequence, *Proc. Natl. Acad. Sci. U.S.A.,* 84, 1472, 1987.
25. **Williams, K. R.,** personal communication, 1988.
26. **Jong, A. Y.-S., Clark, M. W., Gilbert, M., Oehm, A., and Campbell, J. L.,** *Saccharomyces cerevisiae* SSB1 Protein and its relationship to nucleolar RNA-binding proteins, *Mol. Cell. Biol.,* 7, 2947, 1987.
27. **Cruz-Alvarez, M. and Pellicer, A.,** Cloning of a full-length complementary DNA for an *Artemia salina* glycine-rich protein, *J. Biol. Chem.,* 262, 13377, 1987.

28. **Cobianchi, F., Karpel, R. L., Williams, K. R., Notario, V., and Wilson, S. H.,** Mammalian heterogeneous nuclear ribonucleoprotein complex protein A1, *J. Biol. Chem.,* 263, 1063, 1988.
29. **Merrill, B. M., LoPresti, M. B., Stone, K. L., and Williams, K. R.,** Amino acid sequence of UP1, an hnRNP-derived single-stranded nucleic acid binding protein from calf thymus, *Int. J. Peptide Protein Res.,* 29, 21, 1987.
30. **Christensen, M. E., Beyer, A. L., Walker, B., and LeStrourgeon, W. M.,** Identification of N^G,N^G-dimethylarginine in a nuclear protein from the lower eukaryote *Physarum polycephalum* homologous to the major proteins of mammalian 40S ribonucleoprotein particles, *Biochem. Biophys. Res. Commun.,* 74, 621, 1977.
31. **Galibert, F., Herisse, J., and Courtois, G.,** Nucleotide sequence of the *Eco* RI-F fragment of adenovirus 2 genome, *Gene,* 6, 1, 1979.
32. **Hennesy, K. and Kieff, E.,** One of two Epstein-Barr virus nuclear antigens contains a glycine-alanine copolymer domain, *Proc. Natl. Acad. Sci. U.S.A.,* 80, 5665, 1983.
33. **Haynes, S. R., Rebbert, M. L., Mozer, B. A., Forquignon, F., and Dawid, I. B.,** pen repeat sequences are GGN clusters and encode a glycine-rich domain in a Drosophila cDNA homologous to the rat helix destabilizing protein, *Proc. Natl. Acad. Sci. U.S.A.,* 84, 1819, 1987.
34. **Lersch, R. and Fuchs, E.,** Sequence and expression of a type II keratin, K5, in human epidermal cells, *Mol. Cell. Biol.,* 8, 486, 1988.
35. **Rieger, M., Jorcano, J. L., and Franke, W. W.,** Complete sequence of a bovine type I cytokeratin gene: conserved and variable intron positions in genes of polypeptides of the same cytokeratin subfamily, *EMBO J.,* 4, 2261, 1985.
36. **Jorcano, J. L., Franz, J. K., and Franke, W. W.,** Amino acid sequence diversity between bovine epidermal cytokeratin polypeptides of the basic (type II) subfamily as determined from cDNA clones, *Differentiation,* 28, 155, 1984.
37. **Steinert, P. M., Parry, D. A. D., Idler, W. W., Johnson, L. D., Steven, A. C., and Roop, D. R.,** Amino acid sequences of mouse and human epidermal type II keratins of M_r 67,000 provide a systematic basis for the structural and functional diversity of the end domains of keratin intermediate filament subunits, *J. Biol. Chem.,* 260, 7142, 1985.
38. **Orrick, L. R., Olson, M. O. J., and Busch, H.,** Comparison of nucleolar proteins of normal rat liver and Novikoff hepatoma ascites cells by two-dimensional polyacrylamide gel electrophoresis, *Proc. Natl. Acad. Sci. U.S.A.,* 70, 1316, 1973.
39. **Lischwe, M. A., Richards, R. L., Busch, R. K., and Busch, H.,** Localization of phosphoprotein C23 to nucleolar structures and to the nucleolus organizer regions, *Exp. Cell Res.,* 136, 101, 1981.
40. **Lischwe, M. A., Smetana, K., Olson, M. O. J., and Busch, H.,** Proteins C23 and B23 are the major nucleolar silver staining proteins, *Life Sci.,* 25, 701, 1979.
41. **Bugler, B., Caizergues-Ferrer, M., Bouche, G., Bourbon, H., and Amalric, F.,** Detection and localization of a class of proteins immunologically related to a 100 KDa nucleolar protein, *Eur. J. Biochem.,* 128, 475, 1982.
42. **Ochs, R., Lischwe, M. A., O'Leary, P., and Busch, H.,** Localization of nucleolar phosphoprotein B23 and C23 during mitosis, *Exp. Cell Res.,* 152, 260, 1983.
43. **Spector, D. L., Ochs, R. L., and Busch, H.,** Silver staining, immunofluorescence, and immunoelectron microscopic localization of nucleolar phosphoproteins B23 and C23, *Chromosoma,* 90, 139, 1984.
44. **Ochs, R. L. and Busch, H.,** Further evidence that phosphoprotein C23 (110 KD/pI 5.1) is the nucleolar silver staining protein, *Exp. Cell Res.,* 152, 260, 1984.
45. **Olson, M. O. J., Rivers, Z. M., Thompson, B. A., Kao, W.-Y., and Chase, S. T.,** Interaction of nucleolar phosphoprotein C23 with cloned segments of rat ribosomal deoxyribonucleic acid, *Biochemistry,* 22, 3345, 1983.
46. **Herrera, A. H. and Olson, M. O. J.,** Association of protein C23 with rapidly labeled nucleolar RNA, *Biochemistry,* 25, 6258, 1986.
47. **Sapp, M., Knippers, R., and Richter, A.,** DNA binding properties of a 110 KDa nucleolar protein, *NAR,* 14, 6803, 1986.
48. **Bugler, B., Bourbon, H., Lapeyre, B., Wallace, M. O., Chang, J.-H., Amalric, F., and Olson, M. O. J.,** RNA Binding fragments from nucleolin contain the ribonucleoprotein consensus sequence, *J. Biol. Chem.,* 262, 10922, 1987.
49. **Jordan, G.,** At the heart of the nucleolus, *Nature,* 329, 489, 1987.
50. **Sommerville, J.,** Nucleolar structure and ribosome biogenesis, *Trends Biol. Sci.,* 11, 438, 1986.
51. **Mamrack, M. D., Olson, M. O. J., and Busch, H.,** Amino acid sequence and sites of phosphorylation in a highly acidic region of nucleolar nonhistone protein C23, *Biochemistry,* 18, 3381, 1979.
52. **Walker, J. M., Goodwin, G. H., and Johns, E. W.,** The primary structure of the nucleosome-associated chromosomal protein HMG14, *FEBS Lett.,* 100, 394, 1979.

53. **Levy, W. B., Wong, N. C. W., and Dixon, G. H.,** Selective association of the trout specific H6 protein with chromatin regions susceptible to DNase I and DNase II, Possible location of HMG-T in the spacer region between core nucleosomes, *Proc. Natl. Acad. Sci. U.S.A.,* 74, 2810, 1977.
54. **Walton, G. M., Spiess, J., and Gill, G. N.,** Phosphorylation of high mobility group protein 14 by casein kinase II, *J. Biol. Chem.,* 260, 4745, 1985.
55. **Adam, S. A., Nakagawa, T., Swanson, M. S., and Dreyfuss, G.,** mRNA polyadenylated-binding protein: gene isolation and sequencing and identification of a ribonucleoprotein consensus sequence, *Mol. Cell. Biol.,* 6, 2932, 1986.
56. **Sachs, A. B., Bond, W. M., and Kornberg, R. D.,** A single gene from yeast for both nuclear and cytoplasmic polyadenylate-binding proteins: domain structure and expression, *Cell,* 45, 827, 1986.
57. **Theissen, H., Etzerodt, M., Reuter, R., Schneider, C., Lottspeich, F., Argos, P., Luhrmann, R., and Philipson, L.,** Cloning of the human cDNA for U1 RNA-associated 70K protein, *EMBO J.,* 5, 3209, 1986.
58. **Swanson, M. S., Nakagawa, T. Y., LeVan, K., and Dreyfuss, G.,** Primary structure of human nuclear ribonucleoprotein particle C proteins: conservation of sequence and domain structures in the heterogeneous nuclear RNA, mRNA and pre-rRNA-binding proteins, *Mol. Cell. Biol.,* 7, 1731, 1987.
59. **Tan, E. M.,** Autoantibodies to nuclear antigens (ANA): their immunobiology and medicine, *Adv. Immunol.,* 33, 167, 1982.
60. **Kohler, P. F. and Vaughan, J.,** The autoimmune diseases, *JAMA,* 248, 2646, 1982.
61. **Lerner, M. R. and Steitz, J.,** Antibodies to small nuclear RNAs complexed with proteins and produced by patients with systemic lupus erythematosus, *Proc. Natl. Acad. Sci. U.S.A.,* 76, 5496, 1979.
62. **Bernstein, R. W., Steigerwald, J. C., and Tan, E. M.,** Association of antinuclear and antinucleolar antibodies in progressive systemic sclerosis, *Clin. Exp. Immunol.,* 48, 43, 1982.
63. **Stetler, D. A., Rose, K. M., Wenger, M. E., Berlin, C. M., and Jacob, S. T.,** Antibodies to distinct polypeptides of RNA polymerase I in sera from patients with rheumatic autoimmune disease, *Proc. Natl. Acad. Sci. U.S.A.,* 79, 7499, 1982.
64. **Reddy, R., Tan, E. M., Henning, D., Nohga, K., and Busch, H.,** Detection of a nucleolar 7-2 ribonucleoprotein and a cytoplasmic 8-2 ribonucleoprotein with autoantibodies from patients with scleroderma, *J. Biol. Chem.,* 258, 1383, 1983.
65. **Prestayko, A. W., Tonato, M., and Busch, H.,** Low molecular weight RNA associated with 28S nucleolar RNA, *J. Mol. Biol.,* 47, 505, 1970.
66. **Calvet, J. P. and Pederson, T.,** Base-pairing interactions between small nuclear RNAs and nuclear RNA precursors as revealed by psoralen cross-linking *in vivo, Cell,* 26, 363, 1981.
67. **Epstein, P., Reddy, R., and Busch, H.,** Multiple states of U3 RNA in Novikoff hepatoma nucleoli, *Biochemistry,* 23, 5421, 1984.
68. **Busch, H., Reddy, R., Rothblum, L., and Choi, Y. C.,** snRNAs, snRNPs and RNA processing, *Annu. Rev. Biochem.,* 51, 617, 1982.
69. **Hughes, J. M. X., Konings, D. A. M., and Cesareni, G.,** The yeast homologue of U3 snRNA, *EMBO J.,* 6, 2145, 1987.
70. **Parker, K. A. and Steitz, J. A.,** Structural analysis of the human U3 ribonucleoprotein particle reveal a conserved sequence available for base pairing with pre-rRNA, *Mol. Cell. Biol.,* 7, 2899, 1987.
71. **Reimer, G., Pollard, K. M., Penning, C. A., Ochs, R. L., Lischwe, M. A., Busch, H., and Tan, E. M.,** Monoclonal autoantibody from a New Zealand Black-X-New Zealand white F1 mouse and some human scleroderma sera target a M_r 34,000 nucleolar protein of the U3 RNP particle, *Arthritis Rheum.,* 30, 793, 1987.
72. **Ochs, R. L., Lischwe, M. A., Spohn, W. H., and Busch, H.,** Fibrillarin: a new protein of the nucleolus identified by autoimmune sera, *Biol. Cell,* 54, 123, 1985.
73. **LeStrourgeon, W. M., Beyer, A. E., Christensen, M. E., Walker, B. W., Pourpore, S. M., and Daniels, L. P.,** The packaging proteins of core hnRNP particles and the maintenance of proliferative cell states, *Cold Spring Harbor Symp. Quant. Biol.,* 42, 885, 1978.
74. **Christensen, M. E.,** personal communication, 1988.
75. **Kowalczykowski, S. T., Bear, D. G., and vonHippel, P. H.,** *The Enzymes,* Vol. 14, Academic Press, New York, 1981, 373.
76. **Pederson, T.,** Proteins associated with heterogeneous nuclear RNA in eukaryotic cells, *J. Mol. Biol.,* 83, 163, 1974.
77. **Wilk, H.-E., Werr, H., Friedrich, D., Kiltz, H. H., and Schafer, K. P.,** The core proteins of 35S hnRNP complexes, *Eur. J. Biochem.,* 146, 7, 1985.
78. **Lothstein, L., Arenstorf, H. P., Chung, S.-Y., Walker, B. W., Wooley, J. C., and LeStrourgeon, W. M.,** General organization of protein in HeLa 40S nuclear ribonucleoprotein particles, *J. Cell Biol.,* 100, 1570, 1985.
79. **Fuchs, J.-P., Judes, C., and Jacob, M.,** Characterization of glycine-rich proteins from the ribonucleoproteins containing heterogeneous nuclear ribonucleic acid, *Biochemistry,* 19, 1087, 1980.
80. **LeStrourgeon, W. M., Lothstein, M. L., Walker, B. W., and Beyer, A. L.,** *The Cell Nucleus,* Vol. 9, Academic Press, New York, 1981, 47.

81. **Nowak, L., Marvil, D. K., Thomas, J. O., Boublik, M., and Szer, W.,** A single-stranded nucleic acid-binding protein from *Artemia salina, J. Biol. Chem.,* 255, 6473, 1980.
82. **Thomas, J. O., Raziuddin, Sobota, A., Boublik, M., and Szer, W.,** A RNA helix-destabilizing protein is a major component of *Artemia salina* nuclear ribonucleoproteins, *Proc. Natl. Acad. Sci. U.S.A.,* 78, 2888, 1981.
83. **Jong, A. Y. and Campbell, J. L.,** Isolation of the gene encoding yeast single-stranded nucleic acid binding protein 1, *Proc. Natl. Acad. Sci. U.S.A.,* 83, 877, 1986.
84. **Jong, A. Y., Aebersold, R., and Campbell, J. L.,** Multiple species of single-stranded nucleic acid-binding proteins in *Saccharomyces cerevisiae, J. Biol. Chem.,* 260, 16367, 1985.
85. **Merrill, B. M., Stone, K. L., Cobianchi, F., Wilson, S. H., and Williams, K. R.,** Phenylalanines that are conserved among RNA-binding proteins form part of a nucleic acid-binding pocket in the A1 hnRNP protein, *J. Biol. Chem.,* in press.
86. **Chase, J. W. and Williams, K. R.,** Single-stranded DNA binding proteins required for DNA replication, *Annu. Rev. Biochem.,* 55, 103, 1980.
87. **O'Conner, T. P. and Coleman, J. E.,** Proton nuclear magnetic resonance (500 MHz) of mono-, di-, tri- and tetradeoxynucleotide complexes of gene 5 protein, *Biochemistry,* 22, 3375, 1983.
88. **Prigodich, R. V., Cases-Finet, J., Williams, K. R., Konigsberg, W., and Coleman, J. E.,** ^{1}H NMR (500 MHz) of gene 32 protein-oligonucleotide complexes, *Biochemistry,* 23, 522, 1985.
89. **Williams, K. R., Spicer, E. K., LoPresti, M. B., Guggenheimer, R. A., and Chase, J. W.,** Limited proteolysis studies on the *Escherichia coli* single-stranded DNA binding protein, *J. Biol. Chem.,* 258, 3346, 1983.
90. **Steinert, P. M., Rice, R. H., Roop, D. R., Trus, B. L., and Steven, A. C.,** Complete amino acid sequence of a mouse epidermal keratin subunit and implications for the structure of intermediate filaments, *Nature,* 302, 794, 1983.
91. **Sachs, A. B., Davis, R. W., and Kornberg, R. D.,** A single domain of yeast polyadenylate-binding protein is necessary and sufficient for RNA-binding and cell viability, *Mol. Cell. Biol.,* in press.

Chapter 7

HISTONE METHYLATION AND GENE REGULATION

John A. Duerre and Harris R. Buttz

TABLE OF CONTENTS

I. INTRODUCTION: CHROMATIN STRUCTURE

Our knowledge of eukaryotic chromatin has increased markedly over the past few years. The model of chromatin as "beads on a string" first proposed by Olins and Olins[1] has been greatly expanded. The basic chromatin fiber (10 nm diameter) is a linear repeating array of nucleosomes or nucleosome filaments.[2] Nucleosomes are disc-shaped particles consisting of about 200 base pairs (bp) of DNA and five histones. The core contains two molecules each of histone H3, H4, H2A, and H2B around which two turns of DNA (83 bp each) are wound in a left-hand superhelix of pitch 28 Å.[1-6] The two turns are sealed by a molecule of histone H1, which lies outside the core and migrates between sites on different chromatin fragments.[3] It binds at the point where the DNA strand continues beyond 146 bp (1 $^{3}/_{4}$ turn) of the core particle protecting an extra 10 bp from nuclease attack at each end. Beyond bp 166 is linker DNA.[2-6]

The length of the linker DNA which connects one particle to another varies from near zero to 100 bp.[3,6] The four core histones are complexed into an $(H2A\text{-}H2B\text{-}H3\text{-}H4)_2$ octomer.[6] At low ionic strength, or at nonneutral pH, the histone octomer dissociates into H2A-H2B dimer and a $(H3\text{-}H4)_2$ tetramer.[7] When these subunits reassociate, they form a (H2A-H2B) $(H3\text{-}H4)_2$ hexamer intermediate to which a second H2A-H2B dimer binds forming the octomer.[8] These and other studies suggest that the octomer is a dynamic tripartite entity.[9]

Various models have been proposed for the alignment of the core histones along the major and minor groves of DNA.[3-6,9] One of the common features of all these models is that the N-terminal residues of the core histones are flexible and can bind electrostatically to DNA (Figure 1). All the N-terminal sequences in all four histones are accessible to proteases, while only H2A and to a lesser extent H3 undergo C-terminal cleavage.[10] Histone H1 and the N-terminal sequences in H3 and H4 appear to be the most accessible, thus these regions are exposed in a different configuration than the remainder of the chain. The trypsin digestible portions of the core histones do not appear to be vital to the overall stability of the core particle. There is growing evidence that these N-terminal domains are not located within their own core and may function in higher order structure.[10] Reconstituted oligosomes from which the N-terminal tails have been removed with trypsin reconstitute correctly with histone H1, but remain essentially open structures which do not fold into solenoids and show increased template activity.[11]

The N-terminal domains, particularly of the $(H3\text{-}H4)_2$ tetramer appear to be the main agent in generating the DNA supercoil.[5,10] However, limited proteolytic digestion of chromatin fails to yield a homogeneous material, e.g., only H3 or H4. Hence, it is premature to establish the importance of specific histones in stabilizing the higher-order structure. However, histone H3 and H4 sequences are extremely well conserved and probably interact in a specific manner with another conserved feature. Histone H4 is perhaps the most highly conserved of all the histones. The 102 residues remain constant from such sources as calf,[12,13] pea,[12] and trout.[14] The sequence from such divergent sources as sea urchin and calf thymus vary only in the substitution of a cysteine for threonine at position 73.[15] Histone H4 from calf thymus has an amino acid sequence which differs only in the substitution of a lysine for an arginine and a valine for an isoleucine from the corresponding histone in pea seedling.[16]

Calf thymus histones H2A, H2B, and H3 show a great deal of structural similarity to the histones from a variety of species including fish,[17] *Neurosporax,*[8] mollusks,[19] plants,[20] and birds.[21-23] However, genetic polymorphism within the species has been noted for these three histones. These histones have been resolved into variants, which exhibit tissue-specificity variation, by polyacrylamide gel electrophoresis in the presence of nonionic detergents.[24] Histone H2A has been resolved into two variants H2A.1 and H2A.2. Their sequences are identical, except H2A.1 contains serine at position 16 and methionine at position 51.[24,25] Histone H2B has been resolved into three variants. Their amino acid sequences differ only at positions 75 and 76. Histone H2B.1 contains the sequence Gly-Glu, H2B.2 the sequence Ser-Glu and H2B.3 the sequence Gly-Gln. Histone H3 also consists of three variants. The amino acid

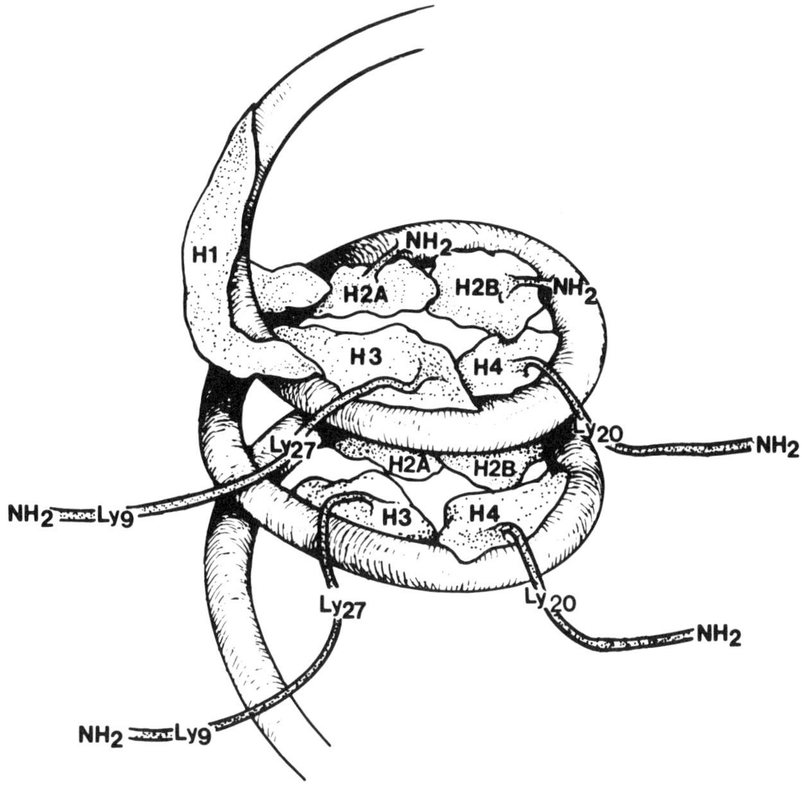

FIGURE 1. Possible arrangement of histones within the nucleosome.

sequences of H3.1 and H3.2 are identical except H3.2 contains serine at position 96 instead of cysteine.[24,26-28] Histone H3.3 from calf and sheep contains the same serine-cysteine substitution at position 96 and also contains the residues Ile-Gly at positions 88-89 instead of the Val-Met sequence which is present in both variants 1 and 2.[24] Except for H2B.2, which has been found only in mice,[24] the histone variants have been found in every tissue of all mammals examined, although in different relative amounts.[29] The simultaneous presence of all these variants in tissue culture lines derived from inbred mice indicates that the genes of different histones are nonallelic.[24] All of the amino acid substitutions appear to be conservative, except the glycine-methionine substitution in histones H3.3. This substitution has been found to occur rarely.[30]

Histone H1 is the most divergent histone and appears to be almost species specific in its sequence.[31] From one to eight subfractions of histone H1 have been observed in various species. Within a species the quantitative amount varies from organ to organ.[31] Most of the amino acid substitutions are not conservative and frequently involve the interchange of lysine, alanine, proline, and serine. This histone lies outside the nucleosome core and migrates between sites on different chromatin fragments.[3] It binds at the point where the DNA strand continues beyond 146 bp ($1^3/_4$ turn) of the core particle protecting an extra 10 bp from nuclease attack at each end, beyond bp 166 is linker DNA.[2-6]

One of the most interesting features of histone structure is the presence of substantial amounts of modifiers in the form of phosphorylated serine, histidine, lysine and threonine, acetylated serine and lysine, methylated lysine, and the presence of ADP-ribosyl groups derived from NAD. All major *in vivo* methylation, acetylation, and phosphorylation sites in H2A, H2B, H3, and H4 are located in the protease sensitive N-terminal sequence.[10] These modifications could provide a mechanism whereby chromatin structure is altered for gene expression or for replication and mitosis.

II. DISTRIBUTION OF N^{ε}-METHYL GROUPS IN HISTONES

An N^{ε}-methyl derivative of lysine was observed first by Murry.[32] Later it was shown by others[12,13,16,33,34] that the mono-, di-, and trimethyllysine derivatives were present (for a complete historical review see Reference 35). Of the five major histones, only the arginine-rich histones H3 and H4 are methylated in chordates. Histone H3 from trout testis,[14,36] calf thymus,[37] and carp testis[38] contain major sites at lysine 9 and 27, while trout has an additional minor site at lysine 4.[36] In trout testis, unlike most other tissues, the histones are replaced by protamine during spermatogenesis. Furthermore, histone H6, a minor trout-specific histone,[39] contains a sequence for residues 3 to 5, -Arg-Lys-Ser-, which is the same as the one methylated in histone H3 at lysine 9 and 27. However, histone H6 yields only trace amounts of methyllysine.[36] Histone H3 from chicken erythrocytes also contains major methylation sites at lysine 9 and 27 and a minor site at lysine 36.[40,41]

Histone H3 from calf thymus,[12,37] fish,[36,38] and birds[40,41] contain N^{ε}-mono-, N^{ε}-di-, and N^{ε}-trimethyllysine at both positions 9 and 27, while the trimethyllysine derivative is absent from histone H4. The methylation sites are highly conserved.[12,16,27,35-38] However, methylation of a particular lysyl residue need not occur. Sequence analysis of peptides from thymic histone revealed the presence of unmethylated lysine at position 20.[12] Even more striking is the complete absence of methylated lysine in histone H4 and trimethyllysine in H3 from peas.[20,42,43]

Some of the early investigators had reported the presence of methylhistidine[44,45] and methylarginine in histones.[44,46] However, sequence analysis and pulse-labeled studies with L-[*methyl*-^{3}H]methionine have failed to reveal any detectable methylated arginine or histidine in such diverse sources as plants,[16,20] fish,[36,38-39] mammals,[13,16,37,47-50] birds,[40-41] or insects.[51] However, there has been a recent report of the presence of methylarginine in histone H3 from *Drosophila* after heat or chemical shock.[52]

III. TURNOVER OF N^{ε}-METHYL GROUPS OF LYSINE RESIDUES OF HISTONES

After the simultaneous administration of [4,5-^{3}H]lysine and L[*methyl*-^{3}H]methionine to 12-d-old rat pups the turnover of the [^{14}C]methyl groups in brain histones did not differ significantly from that of the [^{3}H]lysyl residues.[48] Nor did histones H3 or H4 turn over in the absence of cell division. Similar studies carried out with rat liver, an organ with a relatively high rate of cellular replacement, revealed that all histones turnover throughout development and adulthood (Table 1). However, there was no significant difference between the turnover of the DNA strands and the core histones H3 and H4 (Duerre, unpublished data). Nor was there any difference in turnover rates of the lysine and N^{ε}-methyllysine residues. Byvoet[53] came to similar conclusions employing Novikoff hepatomas cells. Attempts to detect turnover of methyl groups of histones *in vivo* with Ehrlich-ascites tumor cells,[54] Chinese hamster ovary cells,[55] and developing trout testis[36] have also yielded similar results.

In contrast the ε-*N*-methyl group in histones from cat kidney appears to turnover independently of the polypeptide chain.[56] This reaction has been found to be catalyzed by the ε-alkyllysinase which also catalyzes the dimethylation of ε-*N*-methyllysine in kidney.[57]

IV. KINETICS OF HISTONE METHYLATION *IN VIVO*

The methylation of histones appears to be a highly specific event. Unlike acetylation which is an early event,[58] methylation is a relatively late event in cellular division occurring for some time after DNA synthesis has stopped.[48,58,59] After adult rat livers had been perfused with puromycin for 1 h, the incorporation of amino acids into chromosomal proteins was essentially nil (Table 2). Under such conditions the methylation of histones continues at a rate of about 40% of that observed in the controls. This might suggest that the methyl groups turn over

TABLE 1
Biological Half-Lives of Proteins from Rat Liver

Fraction	T½ [^{3}H]lysine	T½ [*methyl*-^{14}C]Met
Cytoplasmic proteins		
Soluble	8.6 ± 0.9	8.3 ± 0.7
Microsomal	7.5 ± 1.2	7.5 ± 0.8
Mitochondrial	9.8 ± 0.8	10.1 ± 0.9
Nuclear proteins		
Nucleoplasm	7.7 ± 1.2	8.0 ± 0.7
NHCP	0.9 ± 0.1	0.7 ± 0.2
Histones (total)	36.0 ± 2.5	60.4 ± 7.2
H1	9.7 ± 1.2	—
H2A-H2B	44.5 ± 4.5	47.5 ± 6.0
H3	78.2 ± 6.2	86.5 ± 6.5
H4	80.3 ± 5.3	82.4 ± 7.1

TABLE 2
Effect of Puromycin on Synthesis and Methylation of Proteins in Isolated Rat Liver

	Control		Puromycin		
	nmol/mg		nmol/mg		% inhibition
Fraction	^{3}H	^{14}C	^{3}H	^{14}C	of ^{14}C
Cytoplasmic	0.30	0.34	0.022	0.015	97
Microsomal	0.76	0.99	0.011	0.010	99
Mitochondrial	0.36	0.42	0.022	0.016	98
Nuclear					
Nucleoplasm	0.29	0.29	0.013	0.023	92
Nonhistone chromosomal proteins	0.54	0.57	0.172	0.035	94
Histones	0.38	0.33	0.105	0.023	94

Note: Normal adult rat livers were perfused with or without puromycin (125 mg/ml) for 30 min. All essential amino acids were added at twice normal plasma levels. L-Methionine (6.4 mmol) contained 15.6 mCi/mmol [*methyl*-^{3}H]methionine and 3.2 mCi/mmol L-[carboxyl-^{14}C]methionine. After 2 h, the protein fractions were prepared and specific radioactivity determined. Results are the averages of two separate determinations.

independently of the polypeptide chain; however, if the livers were perfused with unlabeled methionine and puromycin for 3 h prior to the addition of L-[*methyl*-^{3}H]methionine, methylation of histone H3 and H4 was essentially nil (data not presented).

Thomas et al.[60] in studies of the kinetics of methylation of histones in Ehrlich ascites tumor cells, found that methylation of H4 proceeded more slowly than the methylation of H3. Monomethylation of histone H3 begins in the S phase just after histone biosynthesis. Dimethylation of histone H4 appeared to occur mainly in G2 phase. By the end of one generation all the methylation sites in H3 and H4 are fully methylated at a ratio of about 1:3:1 for mono-, di-, and trimethyllysine in histone H3. Histone H4 contained essentially all dimethyllysine. In the rat, methylation of the arginine-rich histones does not vary from organ to organ, but varies significantly with age (Tables 3 and 4). In histone H3 from neonatal rat liver, the mol ratio of mono- to di- to trimethyllysine is 0.65:0.90:0.30.[49] In histone H4 the mol ratio of mono- to

TABLE 3
Moles of N^{ε}-Methyllysine Residue in Histone H3 from Adult Rat[49]

Tissues	Monomethyllysine	Dimethyllysine	Trimethyllysine
Cerebrum	0.52	1.01	0.35
Cerebellum	0.63	1.02	0.26
Thymus	0.60	0.91	0.31
Kidney	0.56	0.96	0.32
Liver	0.55	1.00	0.35

TABLE 4
Moles of N^{ε}-Methyllysine Residues in Histone H4 from Different Tissues

Tissue	N^{ε}-monomethyllysine	N^{ε}-dimethyllysine	Ref.
Novikoff hepatoma	0.63	0.37	63
Human leukemia lymphoblast	0.65	0.35	63
Bovine lymphosarcoma	0.30	0.60	63
Developing trout	0.43	0.27	36
Liver, neonatal rat	0.1	0.80	49
Liver, adult rat	trace	0.86	49
Cerebrum, rat	trace	0.91	49
Cerebellum, rat	trace	0.93	49
Kidney, rat	trace	0.93	49
Thymus, rat	trace	0.84	49
Erythrocytes, chicken	trace	1.0	41
Thymus, calf	0.15	0.6	38

dimethyllysine is 0.1:0.9. This ratio shifts toward the more highly methylated forms with age (Tables 3 and 4).

After partial hepatectomy, histone methylation in regenerating rat liver occurs at a time when rates of histone and DNA synthesis have already begun to decline.[58] Under these conditions the monomethylation predominated early, shifting to the higher methylated forms with time. Thus it appears that the methyl groups are added sequentially. This conclusion has been verified by *in vitro* experiments.[61] Methylation of the lysyl residues in rat brain nuclei proceeds stepwise, progressing from mono- to di- to trimethyllysine in histone H3 and from mono- to dimethyllysine in histone H4. Hence, the lower methylated forms serve as precursors for the higher methylated derivatives. This would account for the predominance of undermethylated forms in rapidly proliferating tissues, as well as various tumors. In the heteroploid malignant HeLa S3 cell the ratio of di- to monomethyllysine in histone fractions was about half the ratio of these amino acids found in the corresponding histones of normal cells.[63] Similar ratios have been reported for histone H4 from human leukemic lymphoblasts and Novikoff hepatomas (Table 4).

V. *IN VITRO* STUDIES ON HISTONE METHYLATION IN INTACT NUCLEI

One of the major difficulties in studies on the *in vitro* methylation of histones is obtaining a suitable substrate. Histones H3 and H4 do not turnover in adult tissue, nor do the methyl groups turnover independently of the polypeptide chain.[17,48] Consequently, histones from such tissues should be fully methylated. This has been substantiated by both *in vitro* and *in vivo* studies. When brain nuclei from adult animals are incubated with radiolabeled *S*-adenosyl-L-methionine, histones H3 and H4 failed to incorporate significant amounts of labeled methyl groups.[47] Furthermore, when adult rats are given radiolabeled lysine and methyl-labeled methionine only trace quantities of radioactivity were incorporated into brain histones (Duerre, unpublished data). In contrast, significant amounts of ^{3}H-methyl groups from labeled *S*-adenosyl-L-methionine were incorporated into histones when nuclei were prepared from the brains of 10-d-old rats.[33,61] Even in these nuclei, the number of unmethylated lysyl residues in histones H3 and H4 appear to be limited. From saturation data it appeared that 0.024% of the total methylation sites on histone H3 and 0.013% of the sites on histone H4 were unmethylated at the time the nuclei were isolated.[61] Apparently a small fraction of the cells were in a division state in which the newly synthesized histones had condensed with DNA, but were not yet fully methylated. Methylation of these sites was found to proceed stepwise, progressing to a stable ratio of 0.93:1.0:0.17 for N^{ε}-mono-, N^{ε}-di-, and N^{ε}-trimethyllysine in histone H3 and 0.19:1.0 for N^{ε}-mono- and N^{ε}-dimethyllysine in histone H4.[61] These ratios are comparable to those obtained *in vivo;* however, there does appear to be excess monomethyllysine formed *in vitro.*

In intact nuclei the K_m values of the histone lysine methyltransferase for *S*-adenosyl-L-methionine were 11.5 and 12.5 μM with histones H3 and H4 as methyl acceptors, respectively.[61] The V_{max} values were 11.1 and 5.3 pmol [^{3}H] methyl incorporated/min/mg histone H3 and H4, respectively. Since histone H3 contains 2 mol N^{ε}-methyllysine/mol and histone H4 1.0 mol N^{ε}-methyllysine/mol, no difference in the overall rates of methylation can be deduced from the kinetic data.

Nuclei isolated from sea urchin embryoes also readily incorporate radioactive methyl groups from *S*-adenosyl-L-[*methyl*-^{3}H]-methionine into histones.[64,65] Maximum activity occurred in the mesenchymal blastula, gradually decreased throughout the gastrula stage, and was virtually nil by the pluteus stage. The *in vitro* methyl acceptor activity of histone H3 was threefold greater than histone H4, while the major, perhaps only product, was ε-*N*-methyllysine. However, these authors did not make a detailed analysis of the ε-*N*-methyllysine derivative.

VI. HISTONE LYSINE METHYLTRANSFERASE

An enzyme which methylated the ε-amino group of lysine residues in histones was first reported in nuclei of Ehrlich ascites carcinoma cells,[66] rat liver,[67] and calf thymus.[33] Paik and Kim[68] designated this enzyme protein methylase III. The enzyme has since been designated *S*-adenosylmethionine:protein lysine methyltransferase (EC 2.1.1.25). This enzyme has been found in a wide variety of tissues and species, including calf,[68,69] rat,[70-72] mouse,[73] *Leishmania,*[74] and tumor cells.[66,75] Enzyme activity has been found to be elevated whenever cellular proliferation occurs, i.e., in fetal brain,[47,48] fetal kidney,[72] regenerating liver,[58,59] and fast-growing tumor cells.[66]

The enzyme is firmly bound to chromatin.[61,67,76,77] Excess washing of the chromatin with dilute buffers fails to elute the enzyme while it has been removed with high salt[69,76] and detergents.[68] Unfortunately, NaCl also elutes histones, particularly histone H3, which forms insoluble complexes with the enzyme.[70] The enzyme prepared from acetone powders of calf thymus nuclei by Means 0.4% sodium deoxycholate is also extremely unstable.[68]

A. SPECIFICITY OF HISTONE LYSINE METHYLTRANSFERASE

The enzymes solubilized via detergents,[68] NaCl,[76] or repeated extraction of chromatin with distilled water,[70] catalyze the methylation of lysyl and arginyl residues in all soluble histone subfractions, especially the arginine-rich histones. Apparently, soluble histones serve as substrates for the various methyltransferases. In contrast, only the lysyl residues of arginine-rich histones serve as methyl acceptors when chromatin is used as a substrate.[70]

One of the most efficient means of solubilizing the histone methyltransferase, is by limited digestion (10 to 15%) of chromosomal DNA from rapidly proliferating tissue with micrococcal nuclease.[77,78] After limited digestion of chromatin from newborn rat brains, an enzyme specific for histone H3 has been purified to near homogeneity by gel filtration, ammonium sulfate fractionation and DEAE-cellulose chromatography.[77] This enzyme remained associated with a short DNA fragment throughout purification. Dissociation of the enzyme from the DNA fragment with high salt or DNAse 1 digestion resulted in complete loss of enzyme activity. However, when this enzyme remained associated with DNA it was quite stable indicating that DNA is required for stability and/or activation. After DEAE-cellulose chromatography the histone H3 methyltransferase is extremely unstable losing all activity within 24 h. If sheared calf thymus or *E. coli* DNA is added, insoluble complexes formed and come out of solution. Attempts to recover the enzyme by salt elution or nuclease digestion of these complexes were unsuccessful.

However, the histone H3 lysine methyltransferase from a DEAE-cellulose column remained quite stable when it was added back to filtrates, which had been obtained by concentrating the nuclease digested chromatin on a YM-10 membrane. It appears that DNA fragments of less than 50 bp (less than 10,000 molecular weight) are required to prevent hydrophobic interactions.

The histone-H3-lysine methyltransferase was found to methylate lysyl residues of chromosomal bound or soluble histone H3, while H3 associated with nucleosomes was not methylated.[77] The molar ratio of mono- to di- to trimethyllysine in the soluble system was 1.0:2.1:1.0, while the ratio with chromosomal bound histone H3 was 1.9:1.0:0.8. The pH optimum of this enzyme was 8.5 with little variation from 8.2—8.7. The K_m value of this enzyme for *S*-adenosylmethionine was $11.2 \pm 0.7\ \mu M$ with soluble histone H3 as methyl acceptor. *S*-adenosylhomocysteine, one of the products of the reaction, was a competitive inhibitor with respect to *S*-adenosylmethionine ($K_i = 5.3\ \mu M$). The kinetic constants obtained with the enzyme from nuclease digests were quite similar to those that have been obtained with water extracted enzyme,[70] and intact nuclei.[61,64]

After passage of the nuclease digests through a Sepharose 6B-100 column, the histone H4 lysine methyltransferase remains active; however, further attempts to fractionate this enzyme resulted in complete loss of activity.[77] This enzyme has a very high affinity for histone H4 and forms insoluble complexes which aggregate.

B. ARGININE HISTONE METHYLTRANSFERASE

Sarrnow et al.[69] partially purified a histone-H4 specific methyltransferase from nuclei from calf thymus and lymphocytes which catalyzed the methylation of amino acids in the sequence Lys-Val-Leu-Arg in the N-terminal peptide (amino acids 20-23 of histone H4). The primary amino acid acceptor appeared to be arginine, although methylated lysine also was detectable. An arginine histone methyltransferase also has been isolated from micrococcal digests of rat brain chromatin.[77] This enzyme catalyzes the methylation of arginine residues in soluble histone H4, but not nucleosomal or chromosomal bound histones.[69,77] The enzyme is considerably more stable than the histone lysine methyltransferases and does not bind to sheared calf thymus or *E. coli* DNA.[77] Gupta et al.[79] have reported the presence of a histone-specific arginine methyltransferase in wheat germ, while Lee et al.[80] reported the presence of this enzyme in calf brain.

However, numerous investigations[12-14,37,38,47-50] have been unable to detect methylarginine in histones from a variety of plant or animal sources. The enzyme may methylate only a few select histone H4 molecules on chromatin during replication or repair, or the enzyme might play some role in transcription. Methylation of arginyl residues in histone H3 has been observed in *Drosophila* after chemical or heat shock.[52] The methylation of histone H4 via arginine methyltransferase could also be an artifact. A soluble arginine methyltransferase catalyzes the methylation of various proteins, particularly in the microsome.[35] Consequently the role, if any, that this enzyme plays in the methylation of histone H4 remains open.

C. REGULATION OF METHYLTRANSFERASE REACTIONS

S-adenosylhomocysteine (AdoHcy), one of the products of all methyltransferase reactions, has been proposed by several investigators, including the author, to be a bioregulatory compound.[81] With histone lysine methyltransferase, AdoHcy acts as a competitive inhibitor with respect to AdoMet. The inhibition constants (K_i) for AdoHcy are 5.5 and 5.9 μM with histone H3 and H4 as methyl acceptors, respectively.[61,70,77] The enzyme apparently has a higher affinity for AdoHcy than AdoMet (K_m = 11.5 to 12.4 μM).

In the liver, the concentration of AdoHcy has been reported to be 10 to 15 μM while the AdoMet concentration is 80 to 110 μM.[82] The concentration of AdoMet in the brain is 50 to 60 μM during development and decreases to 25 to 30 μM in the adult, while the concentration of AdoHcy never exceeds 1 μM.[82] The ratio of AdoMet to AdoHcy is so large in these organs that it is doubtful that the latter compound would have any effect on transmethylase reactions.

However, this does not preclude the fact that various drugs, nutritional deficiencies, or certain metabolic defects which alter the level of AdoMet could result in altered methylation patterns. The carcinogen, ethionine, inhibits the *in vivo* methylation of lysine in histones by 50% in regenerating livers.[83] Ethionine forms the respective adenosyl derivative in the presence of methionine activation enzyme and ATP.[84] *S*-adenosyl-L-ethionine has been found to compete with AdoMet resulting in the transfer of ethyl groups to histones, but at a very low efficiency.[85]

Methylation of histones is also depressed in livers from animals which have been fed a zinc-deficient diet for prolonged periods of time.[86] The reduced rate of methylation of histones cannot be attributed to the availability of a methyl donor, since there was ample AdoMet available in the liver from the zinc-deprived animal. This effect is probably related to the availability of a methyl acceptor. Histone biosynthesis was approximately twofold less in the livers from zinc-deficient animals than in controls.[86]

Under normal conditions, histone gene transcription is switched on in mid G-2 phase prior to DNA synthesis.[87] During S phase of the cell cycle, histone gene expression and DNA replication are tightly coupled. The cellular level of the histone methyltransferase would appear to parallel histone biosynthesis. Synthesis of these enzymes remains constant in late S phase reaching maximum levels by the end of G-2 phase.[75] Hence, both histone and histone methyltransferase mRNA synthesis appear to be controlled by cell cycle regulators.

VII. FUNCTION OF HISTONE METHYLATION

All the methylation sites in the core histones are located in the N-terminal sequences of histone H3 and H4. The protease cutting points in histones H3 and H4 from duck erythrocytes are adjacent to very basic N^{ε}-methyl-lysine residues at positions 27 and 19, respectively.[10] The cutting points in erythrocyte histones H3 and H4 are quite close to those in cycad pollen, particularly with regard to the sites of methylation.[50] These authors proposed that the methylation of lysine residue 20 in H4 and residue 27 in H3, which increase the basicity of the lysine, anchor the histone chains as they exit from the inside of the DNA fold. However, additional methylation sites are present both above (at lysine 36) and below (at lysine 4 and 9) lysine 27, in histone H3.[36] Furthermore, all these sites contain varying degrees of methylated lysine in the form of N^{ε}-mono-, N^{ε}-di-, and N^{ε}-trimethyllysine, as well as unmethylated lysine. Histone H4 has only one site at lysine 20, which contains N^{ε}-mono- and N^{ε}-dimethyllysine, as well as unmethylated lysine. Methylation of histone H4 is not an essential feature of chromatin structure, since it is unmethylated in plants.[42,43] Hence, the role that methylation plays in chromatin structure is far from resolved.

Methylation of lysine residues in histone may play a role in stabilizing the superhelix (solenoid). With increasing ionic strength, rat liver chromatin folds up progressively from a filament of nucleosomes through some higher order helical structure, having a fairly constant pitch, with an increasing number of nucleosomes per turn until at 60 mM Na^+ (or 0.3 mM Mg^{2+}) a thick fiber of 250 Å diameter is formed. This corresponds to a well organized superhelix (solenoid) of pitch 110 Å.[3-6] When nuclei from developing rat brain were incubated with *S*-adenosyl-L-[*methyl*-^{3}H]methionine in the presence of 0.5 to 1.0 mM Mg^{2+} all the methylation

TABLE 5
Effects of Polyamines, NaCl and $MgCl_2$ on the Extent of Methylation of Histone H4

Additions	μmol [^{3}H-*methyl*]methyllysine/mol histone H4		
	Monomethyllysine	Dimethyllysine	Ratio di- to mono-
1 m*M* $MgCl_2$	196	219	1.12
10 m*M* $MgCl_2$	37	53	1.43
1.5 m*M* Spermidine	64	139	2.17
1.0 m*M* $CaCl_2$	68	145	2.13
0.1 *M* NaCl	685	686	1.00

Note: Rat brain nuclei were incubated with 15 μ*M* *S*-adenosyl-L-[*methyl*-^{3}H]methionine (1.0 Ci/nmol), 2.0 m*M* Tris-HCl buffer, (final pH of 7.0), 0.32 *M* sucrose and the above additions for 60 min at 37°.[78]

sites appear to be saturated within 1 h (Table 5). However, 5 to 10 m*M* Mg^{2+} inhibits the methylation of histone H4 some fivefold, while the methylation of histone H3 was unaffected. This effect is mimicked with either spermidine or Ca^{2+}. When Na^+ (0.1 to 0.15 *M*) was added, both the rate and extent of methylation of histone H4 increased markedly. Again the methylation of histone H3 was not altered significantly. At the concentrations listed above these cations have no effect on the methylation of soluble histones.[78] However, high concentrations of Na^+ (>0.2 *M*) inhibit the methylation of both bound and free histones. Cations and polyamines are known to bind to DNA resulting in structural alterations. Apparently the N-terminal region of histone H4 is located in a region within the solenoid that is relatively inaccessible to histone-H4-lysine methyltransferase. At 20°C a decrease in winding angle of about 0.1° per base occurs when the concentration of Na^+ changes from 0.2 to 0.05 *M*.[88] Relaxing the superhelical structure not only results in an increase in the rate and extent of methylation, but also results in a shift toward the more highly methylated form, dimethyllysine.[78] The flexible N-terminal region of histone H3 apparently is located in a region which is unaffected by the conformation of the nucleosomes in the solenoid, perhaps they extend outside this structure. This suggests independent functions for the N-terminal regions of histones H3 and H4. This region of H4 may be involved in superhelical formation while H3 may be involved in some higher order of folding. If methylation of lysine residues plays a role in stabilizing the supercoil, then demethylation could occur prior to relaxation, which precedes transcription or replication. Failure to detect methyl group turnover in nonproliferating cells, such as brain,[47] would in no way negate this possibility. The majority of operons which are actively transcribing would be turned on at the time the cell differentiates and should continue to translate throughout the life span of that cell.

VIII. SUMMARY AND PROSPECTUS

Of the five major histones found associated with chromatin, only the arginine-rich histones are methylated. All of these methylation sites are located in N-terminal sequences which appear to lie outside the nucleosome. Histone H3 contains two major sites at lysine 9 and 27, while histone H4 contains only one major site at lysine 20. Both lysine 9 and 27 on histone H3 contain N^ε-mono-, N^ε-di-, and N^ε-trimethyllysine, while the trimethyllysine derivative is absent from histone H4. The methylation sites appear to be highly conserved; however, differences in the extent of methylation of these sites has been observed. This is particularly striking in the pea where histone H4 is unmethylated and trimethyllysine is absent from histone H3.

The methylation of histones appears to be a highly specific event, occurring after the histones are bound to chromatin. Kinetics of methylation have shown that methylation proceeds stepwise, with the lower methylated forms serving as substrates for the more highly methylated forms. The enzymes which catalyze these reactions are located in the nucleus and are firmly bound to chromatin. The chromosomal bound enzymes do not methylate free or nonspecifically associated histones, while histones H3 and H4 within newly synthesized nucleosomes are methylated. Dissociation of the enzymes from DNA with salts or detergents results in complete loss of activity. However, a histone H3 lysine methyltransferase has been solubilized by limited digestion of chromatin with micrococcal nuclease. This enzyme catalyzes the methylation of lysine residues in both soluble and chromosomal bound histone H3. In both systems the molar ratio of mono- to di- to trimethyllysine was similar to that observed *in vivo*. To date it is not known what factor or factors determine the extent to which the various lysine residues in histone H3 are methylated. It is possible that the amino acid substitutions in the histone H3 variants (H3.1, H3.2, and H3.3) could determine the extent to which the lysyl residues are methylated. The presence of these histone variants and the varying degrees to which the lysyl residues are methylated would allow for a considerable amount of heterogeneity.

Another interesting feature of histone methylation is the presence of unmethylated lysine residues at positions 9 and 27 in histones H3, and at position 20 in histone H4 from resting cells. If methylation plays a role in assembly and/or condensation of chromatin, then the unmethylated and/or undermethylated molecules could lie in regions which are flagged for transcriptional events. Thus histone methylation could play an active role in gene regulation. Clearly histone methylation is of a nontrivial nature, if only an extremely limited number of highly conservative amino acid substitutions have been observed throughout evolution. The introduction of a methyl group must have a marked effect on some aspect of histone interaction.

REFERENCES

1. **Olins, A. L. and Olins, D. E.,** Spheroid chromatin units (v bodies), *Science*, 183, 330, 1974.
2. **Felsenfeld, G. and McGhee, J. D.,** Structure of the 30 nm chromatin fiber, *Cell*, 44, 375, 1986.
3. **Thomas, J. O.,** The higher order structure of chromatin and histone H1, *J. Cell Sci. Suppl.*, 1, 1, 1984.
4. **Klug, A. and Butler, P. J.,** The structure of nucleosomes and chromatin, *Horiz. Biochem. Biophys.*, 7, 1, 1983.
5. **McGhee, J. D. and Felsenfeld, G.,** Nucleosome structure, *Annu. Rev. Biochem.*, 49, 1115, 1980.
6. **Kornberg, R. D.,** Structure of chromatin, *Annu. Rev. Biochem.*, 46, 931, 1977.
7. **Eickbush, T. H. and Moudrianakis, E. N.,** The histone core complex: an octamer assembled by two sets of protein-protein interactions, *Biochemistry*, 17, 4955, 1978.
8. **Benedict, R. C., Moudrianakis, E. N., and Ackers, G. K.,** Interactions of the nucleosomal core histones: a calorimetric study of octamer assembly, *Biochemistry*, 23, 1214, 1984.
9. **Burlingame, R. W., Love, W. E., Wang, B. C., Hamlin, R., Nguyen, H. X., and Moudrianakis, E. N.,** Crystallographic structure of the octameric histone core of the nucleosome at a resolution of 3.3 Å, *Science*, 228, 546, 1985.
10. **Bohm, L. and Crane-Robinson, C.,** Proteases as structural probes for chromatin: the domain structure of histones, *Biosci. Rep.*, 4, 365, 1984.
11. **Allan, J., Harborne, N., Rau, D. C., and Gould, H.,** Participation of core histone "tails" in the stabilization of the chromatin solenoid, *J. Cell Biol.*, 93, 285, 1982.
12. **DeLange, R. J., Fambrough, D. M., Smith, E. L., and Bonner, J.,** Calf and pea histone IV. II. The complete amino acid sequence of calf thymus histone IV; presence of ε-*N*-acetyllysine, *J. Biol. Chem.*, 244, 319, 1969.
13. **Ogawa, Y., Quagliarotti, G., Jordan, J., Taylor, C. W., Starbuck, W. C., and Busch, H.,** Structural analysis of the glycine-rich, arginine-rich histone. III. Sequence of the amino-terminal half of the molecule containing the modified lysine residues and the total sequence, *J. Biol. Chem.*, 244, 4387, 1969.
14. **Dixon, G. H., Candido, E. P. M., Honda, B. M., Louie, A. J., Macleod, A. R., and Sung, M. T.,** The biological roles of post-synthetic modifications of basic nuclear proteins, *Ciba Found. Symp.*, 28, 229, 1975.
15. **Strickland, M., Strickland, W. N., Brandt, W. F., and Von Holt, C.,** Sequence of the cysteine-containing portion of the histone F2al from the sea urchin *Parechinus angulosus*, *FEBS Lett.*, 40, 346, 1974.

16. **DeLange, R. J., Fambrough, D. M., Smith, E. L., and Bonner, J.,** Calf and pea histone IV: comparison with the homologous calf thymus histone, *J. Biol. Chem.,* 244, 5669, 1969.
17. **Palau, J. and Butler, J. A. V.,** Trout liver histones, *Biochem. J.,* 100, 779, 1966.
18. **Hsiang, M. W. and Cole, R. D.,** The isolation of histone from *Neurospora crassa, J. Biol. Chem.,* 248, 2007, 1973.
19. **Palau, J., Ruiz-Carrillo, A., and Subirana, J. A.,** Histones from sperm of the sea urchin *Arbacia lixula, Eur. J. Biochem.,* 7, 209, 1969.
20. **Fambrough, D. M. and Bonner, J.,** Limited molecular heterogeneity of plant histones, *Biochim. Biophys. Acta,* 175, 113, 1969.
21. **Hnilica, L. S. and Bess, L. G.,** Fractionation of calf thymus histone fractions 2a and 3 on Sephadex, *Anal. Biochem.,* 8, 521, 1964.
22. **Vidali, G. and Neelin, J. M,** A comprehensive fractionation procedure for avian erythrocyte histones, *Eur. J. Biochem.,* 5, 330, 1968.
23. **Lindsay, D. T.,** Histones from developing tissues of the chicken: heterogeneity, *Science,* 144, 420, 1964.
24. **Franklin, S. G. and Zweidler, A.,** Non-allelic variants of histones 2a, 2b and 3 in mammals, *Nature,* 266, 273, 1977.
25. **Laine, B., Sautiere, P., and Biserte, G.,** Primary structure and microheterogeneities of rat chloroleukemia histone H2a (histone ALK, II_{b1} or F_{2a2}), *Biochemistry,* 15, 1640, 1976.
26. **Brandt, W. F., Strickland, W. N., and Von Holt, C.,** The primary structure of histone F3 from shark erythrocytes, *FEBS Lett.,* 40, 349, 1974.
27. **Patthy, L. and Smith, E. L.,** Histone III. VI. Two forms of calf thymus histone III, *J. Biol. Chem.,* 250, 1919, 1975.
28. **Marzluff, W. F., Jr., Sanders, L. A., Miller, D. M., and McCarty, K. S.,** Two chemically and metabolically distinct forms of calf thymus histone F3, *J. Biol. Chem.,* 247, 2026, 1972.
29. **Zwiedler, A.,** Resolution of histones by polyacrylamide gel electrophoresis in presence of nonionic detergents, *Methods Cell Biol.,* 17, 223, 1978.
30. **Dayhoff, M. D.,** *Atlas of Protein Sequence and Structure, 4,* National Biomedical Research Foundation, Maryland, 1969.
31. **Elgin, S. C. R. and Weintraub, H.,** Chromosomal proteins and chromatin structure, *Annu. Rev. Biochem.,* 44, 725, 1975.
32. **Murry, K.,** The occurrence of N^{ε}-methyl lysine in histones, *Biochemistry,* 3, 10, 1964.
33. **Kim, S. and Paik, W. K.,** Studies on the origin of ε-*N*-methyl-L-lysine in protein, *J. Biol. Chem.,* 240, 4629, 1965.
34. **Paik, W. K. and Kim, S.,** ε-*N*-Dimethyllysine in histones, *Biochem. Biophys. Res. Commun.,* 27, 479, 1967.
35. **Paik, W. K. and Kim, S.,** *Protein Methylation,* Vol. 1, Meister, A., Ed., John Wiley & Sons, New York, 1980.
36. **Honda, B. M., Dixon, G. H., and Candido, E. P. M.,** Sites of *in vivo* histone methylation in developing trout testis, *J. Biol. Chem.,* 250, 8681, 1975.
37. **DeLange, R. J., Hooper, J. A., and Smith, E. L.,** Histone III. III. Sequence studies on the cyanogen bromide peptides, complete amino acid sequence of calf thymus histone III, *J. Biol. Chem.,* 248, 3261, 1973.
38. **Hooper, J. A., Smith, E. L., Sommer, K. R., and Chalkley, R.,** Histone III. IV. Amino acid sequence of histone III of the testes of the carp, *Letiobus bubalus, J. Biol. Chem.,* 248, 3275, 1973.
39. **Wigle, D. T. and Dixon, G. H.,** A new histone from trout testis, *J. Biol. Chem.,* 246, 5636, 1971.
40. **Brandt, W. F. and Von Holt, C.,** The complete amino acid sequence of histone F3 from chicken erythrocytes, *Eur. J. Biochem.,* 46, 419, 1974.
41. **Hempel, V. K. and Lange, H. W.,** N^{ε}-methylierte lysine in histonen aus Huhner-erythrozyten, *Hoppe-Seyler's Z. Physiol. Chem.,* 349, 603, 1968.
42. **Fambrough, D. M., Fujimura, F., and Bonner, J.,** Quantitative distribution of histone components of the pea plant, *Biochemistry,* 7, 575, 1968.
43. **Patterson, B. D. and Davies, D. D.,** Specificity of the enzymatic methylation of pea histone, *Biochem. Biophys. Res. Commun.,* 34, 791, 1969.
44. **Byvoet, P.,** Uptake of label into methylated amino acids from rat tissue histones after *in vivo* administration of [$Me^{14}C$]methionine, *Biochem. Biophys. Acta,* 238, 375a, 1971.
45. **Gershey, E. L., Haslett, G. W., Vidali, G., and Allfrey, V. G.,** Chemical studies of histone methylation: evidence for the occurrence of 3-methylhistidine in avian erythrocyte histone fractions, *J. Biol. Chem.,* 244, 4871, 1969.
46. **Paik, W. K. and Kim, S.,** ω-*N*-Methylarginine in histones, *Biochem. Biophys. Res. Commun.,* 40, 224, 1970.
47. **Lee, C. T. and Duerre, J. A.,** Changes in histone methylase activity of rat brain and liver with ageing, *Nature,* 251, 240, 1974.
48. **Duerre, J. A. and Lee, C. T.,** *In vivo* methylation and turnover of rat brain histones, *J. Neurochem.,* 23, 541, 1974.

49. **Duerre, J. A. and Chakrabarty, S.,** Methylated basic amino acid composition of histones from various organs from the rat, *J. Biol. Chem.,* 250, 8457, 1975.
50. **Bohm, L., Briand, G., Sautiere, P., and Crane-Robinson, C.,** Proteolytic digestion studies of chromatin core-histone structure: identification of the limit peptides of histones H3 and H4, *Eur. J. Biochem.,* 119, 67, 1981.
51. **Alfageme, C. R., Zweidler, A., Mahowald, A., and Cohen, L. H.,** Histones of *Drosophila* embryos: electrophoretic isolation and structural studies, *J. Biol. Chem.,* 249, 3729, 1974.
52. **Desrosiers, R. and Tanguay, R. M.,** Methylation of *Drosophila* histones at proline, lysine and arginine residues during heat shock, in press.
53. **Byvoet, P.,** *In vivo* turnover and distribution of radio-*N*-methyl in arginine-rich histones from rat tissues, *Arch. Biochem. Biophys.,* 152, 887, 1972.
54. **Thomas, G., Lange, H. W., and Hempel, K.,** Relative stabilitat lysin-gebundener methylgruppen bei den argininreichen histonen und ihren unterfraktionen von Ehrlich-Ascites-Tumorzellen *in vitro, Hoppe-Seyler's Z. Physiol. Chem.,* 353, 1423, 1972.
55. **Byvoet, P., Shepherd, G. R., Hardin, J. M., and Noland, B. J.,** The distribution and turnover of labeled methyl groups in histone fractions of cultured mammalian cells, *Arch. Biochem. Biophys.,* 148, 558, 1972.
56. **Hempel, K., Thomas, G., Roos, G., Stocker, W., and Lange, H.,** N^{ε}-Methyl groups on the lysine residues in histones turn over independently of the polypeptide backbone, *Hoppe-Seyler's Z. Physiol. Chem.,* 360, 869, 1979.
57. **Paik, W. K. and DiMaria, P.,** Enzymatic methylation and demethylation of protein-bound lysine residues, *Methods Enzymol.,* 106, 274, 1984.
58. **Tidwell, T., Allfrey, V. G., and Mirsky, A. E.,** The methylation of histones during regeneration of the liver, *J. Biol. Chem.,* 243, 707, 1968.
59. **Lee, H. W. and Paik, W. K.,** Histone methylation during hepatic regeneration in rat, *Biochim. Biophys. Acta,* 277, 107, 1972.
60. **Thomas, G., Lange, H. W., and Hempel, K.,** Kinetics of histone methylation *in vivo* and its relation to the cell cycle in Ehrlich ascites tumor cells, *Eur. J. Biochem.,* 51, 609, 1975.
61. **Duerre, J. A., Wallwork, J. C., Quick, D. P., and Ford, K. M.,** *In vitro* studies on the methylation of histones in rat brain nuclei, *J. Biol. Chem.,* 252, 5981, 1977.
62. **Borun, T. W., Pearson, D., and Paik, W. K.,** Studies of histone methylation during the HeLa S-3 cell cycle, *J. Biol. Chem.,* 247, 4288, 1972.
63. **Desai, L. S. and Foley, G. E.,** Homologies in amino acid composition and structure of histone F2al isolated from human leukemic cells, *Biochem. J.,* 119, 165, 1970.
64. **Branno, M. and Tosi, L.,** Methylation of nuclear proteins during early embryogenesis in sea urchin, *Boll. Soc. It. Biol. Sper.,* 15, 1778, 1980.
65. **Branno, M., De Franciscis, V., and Tosi, L.,** *In vitro* methylation of histones in sea urchin nuclei during early embryogenesis, *Biochim. Biophys. Acta,* 741, 136, 1983.
66. **Comb, D. G., Sarkar, N., and Pinzino, C. J.,** The methylation of lysine residues in protein, *J. Biol. Chem.,* 241, 1857, 1966.
67. **Sekeris, C. E., Sekeri, K. E., and Gallwitz, D.,** The methylation of the histones of rat liver nuclei in vitro, *Hoppe-Seyler's Z. Physiol. Chem.,* 348, 1660, 1967.
68. **Paik, W. K. and Kim, S.,** Solubilization and partial purification of protein methylase III from calf thymus nuclei, *J. Biol. Chem.,* 245, 6010, 1970.
69. **Sarnow, P., Rasched, I., and Knippers, R.,** A histone H4-specific methyltransferase: properties, specificity and effects on nucleosomal histones, *Biochim. Biophys. Acta,* 655, 349, 1981.
70. **Wallwork, J. C., Quick, D. P., and Duerre, J. A.,** Properties of soluble rat brain histone lysine methyltransferase, *J. Biol. Chem.,* 252, 5977, 1977.
71. **Kaye, A. M. and Sheratzky, D.,** Methylation of protein (histone) *in vitro:* enzymic activity from the soluble fraction of rat organs, *Biochim. Biophys. Acta,* 190, 527, 1969.
72. **Paik, W. K., Benditt, M., and Kim, S.,** Protein methylase III of rat kidney during the early development, *FEBS Lett.,* 35, 236, 1973.
73. **Lee, N. M. and Loh, H. H.,** Phosphorylation and methylation of chromatin proteins from mouse brain nuclei, *J. Neurochem.,* 29, 547, 1977.
74. **Paolantonacci, P., Lawrence, F., Lederer, F., and Robert-Gero, M.,** Protein methylation and protein methylases in *Leishmania donovani* and *Leishmania tropica* promastigotes, *Mol. Biochem. Parasitol.,* 21, 47, 1986.
75. **Lee, H. W., Paik, W. K., and Borun, T. W.,** The periodic synthesis of *S*-adenosylmethionine: protein methyltransferases during the HeLa S-3 cell cycle, *J. Biol. Chem.,* 248, 4194, 1973.
76. **Burdon, R. H. and Garven, E. V.,** Enzymic modification of chromosomal macromolecules. II. The formation of histone ε-*N*-trimethyl-L-lysine by a soluble chromatin methylase, *Biochim. Biophys. Acta,* 232, 371, 1971.
77. **Duerre, J. A. and Onisk, D. V.,** Specificity of the histone lysine methyltransferases from rat brain chromatin, *Biochim. Biophys. Acta,* 843, 58, 1985.

78. **Duerre, J. A., Quick, D. P., Traynor, M. D., and Onisk, D. V.,** Effect of polyamines and cations on the *in vitro* methylation of histones, *Biochim. Biophys. Acta,* 719, 18, 1982.
79. **Gupta, A., Jensen, D., Kim, S., and Paik, W. K.,** Histone-specific protein-arginine methyltransferase from wheat germ, *J. Biochem.,* 257, 9677, 1982.
80. **Lee, H. W., Kim, S., and Paik, W. K.,** *S*-Adenosylmethionine: protein-arginine methyltransferase. Purification and mechanism of the enzyme, *Biochemistry,* 16, 78, 1977.
81. **Duerre, J. A. and Walker, R. D.,** Metabolism of *S*-adenosyl homocysteine, in *The Biochemistry of S-Adenosylmethionine,* Salvatory, F., Borek, E., Zappia, V., Williams-Ashman, H. G., and Schlenk, F., Eds., Columbia University Press, New York, 1977, 430.
82. **Hoffman, D. R., Cornatzer, W. E., and Duerre, J. A.,** Relationship between tissue levels of *S*-adenosylmethionine, *S*-adenosylhomocysteine and transmethylation reactions, *Can. J. Biol. Chem.,* 57, 56, 1979.
83. **Cox, R. and Tuck, M. T.,** Alteration of methylation patterns in rat liver histones following administration of ethionine, a liver carcinogen, *Cancer Res.,* 41, 1253, 1981.
84. **Stekol, J. A.,** Formation and metabolism of *S*-adenosyl derivatives of *S*-alkylhomocysteines in the rat and mouse, in *Transmethylation and Methionine Biosynthesis,* Shapiro, S. K. and Schlenk, F., Eds., The University of Chicago Press, Chicago, 1965, 231.
85. **Baxter, C. S. and Byvoet, P.,** Effects of carcinogens and other agents on histone methylation in rat liver nuclei by endogenous histone lysine methyltransferase, *Cancer Res.,* 34, 1424, 1974.
86. **Wallwork, J. C. and Duerre, J. A.,** Effect of zinc deficiency on methionine metabolism, methylation reactions and protein synthesis in isolated perfused rat liver, *J. Nutr.,* 115, 252, 1985.
87. **Carrino, J. J., Keung, V., Braun, R., and Laffler, T. G.,** Distinct replication-independent and -dependent phases of histone gene expression during the *Physarum* cell cycle, *Mol. Cell. Biol.,* 7, 1933, 1987.
88. **Record, M. T., Jr., Mazur, S. J., Melancon, P., Roe, J. H., Shaner, S. L., and Unger, L.,** Double helical DNA: conformations, physical properties and interactions with ligands, *Annu. Rev. Biochem.,* 50, 977, 1981.

Chapter 8

POSTTRANSLATIONAL METHYLATION OF HISTONES AND HEAT SHOCK PROTEINS IN RESPONSE TO HEAT AND CHEMICAL STRESSES

Robert M. Tanguay and Richard Desrosiers

TABLE OF CONTENTS

I. INTRODUCTION

Exposure of prokaryotic and eukaryotic cells to supraoptimal temperatures causes several modifications in various cellular activities. This response referred to as the heat shock response is characterized by a rapid increase in the synthesis of a restricted number of proteins, the heat shock proteins (HSP) and a concomitant decrease in the synthesis of normal cellular proteins. A similar response can also be elicited by many chemical agents such as oxidizing agents, metal ions, sulfydryl reagents, ethanol, amino acid analogs, etc. (reviewed in References 1 through 5) The induction of HSP by a large variety of physical, chemical, and environmental insults has been observed in bacteria, fungi, plants, insects, amphibians, and vertebrates suggesting that it is a universal response to cellular stress. While the precise function(s) of the various HSP is still obscure, they have been suggested to be involved in cellular repair and protection mechanisms such as thermotolerance[6] and in various cellular activities during recovery from stress.[7]

The heat shock response has proved to be a particularly suitable system for studying mechanisms of gene regulation operating during genetic information transfer. Thus, in *Drosophila* where these mechanisms have been extensively studied, the expression of heat shock genes is regulated both at the transcriptional and translational levels. At the nuclear level, the rapid transcriptional activation of heat-shock genes (reviewed in Reference 8) is accompanied by regulation in the transcription, processing, and transport of normal cellular genes.[9,10] In the cytoplasm, heat shock mRNAs are preferentially translated while the normal cellular mRNAs are sequestered in an inactive form during heat shock.[11,12] Although the mechanisms of transcriptional activation of the heat shock genes are beginning to be understood,[8] little is known about the molecular events responsible for the transcriptional repression of normal genes during heat shock.

Posttranslational modification of proteins has been suggested to be an important component in the regulation of numerous physiological processes, including gene function.[13] In this chapter, we review the current data on the methylation of the heat shock proteins and on its modulation by stress. The changes in methylation of core histones which are observed in thermally or chemically stressed cells are also summarized and their possible significance discussed in relation to chromatin structure and function.

II. METHYLATION OF HSP

Heat shock proteins can undergo several posttranslational modifications such as phosphorylation, ADP-ribosylation, glycosylation, and methylation. The functional significance of these modifications is unknown in most cases but the changes have been correlated with various physiological events. The methylation of heat shock proteins has only been investigated in a few cell types. Wang et al.[14] originally reported the methylation of certain HSP in both chicken embryonic cells and baby hamster kidney cells (BHK-21). In chick embryo fibroblasts, both HSP 83 and 68 (later referred to as HSP 70 isovariants A and B) are methylated under normal growth conditions. Exposure of these cells to sodium arsenite, an inducer of the heat shock response, resulted in an increased methylation of HSP 70A and B while the methylation level of HSP 83 is unaffected.[14] The increase in the methylation of the HSP 70 variants results from the new synthesis of these HSP in induced cells rather than from an unspecific activation of preexisting methyltransferase. The weak induction of HSP 83 by arsenite in these cells could explain the apparent absence of a significant increase in the methylation level of this HSP. Whether other stresses and in particular heat shock can induce similar changes in the methylation of HSP in these cells is presently unknown. In tomato cells, HSP 17, 70, 74, and 80 have also been reported to be methylated in heat-shocked cells[15] but changes in their methylation levels comparatively to those observed under normal conditions or in response to other stresses have not been measured.

The nature of the methylated residues and the stoichiometry of the methylation of HSP 83 and

HSP 70A and B have also been investigated in chick embryo fibroblasts. Analysis of HSP 83 from arsenite-treated cells shows that methylation occurs on lysine residues mainly as ε-*N*-trimethyllysine.[16] Methylated lysyl and arginyl residues are found in the two HSP 70 variants.[16,17] As in HSP 83, the major form of methylated lysine is ε-*N*-trimethyllysine but mono- and dimethyllysine forms are also found in HSP 70. The ratio of the different methylated lysine forms in the variant HSP 70B is similar in control and arsenite-treated cells but a fivefold decrease in the level of N^G-monomethylarginine is observed after an arsenite treatment.[17] In the case of HSP 70A, arsenite induces a decrease in the relative amount of ε-*N*-trimethyllysine and an increase of the mono- and dimethylated forms. Thus, the changes induced by arsenite are specific for each HSP.

While the internal localization of the methylated amino acids in these HSP is unknown, each appears to contain multiple methylation sites. In the case of HSP 70, methylation is probably substoichiometric.[17] Furthermore, methylation occurs at an early stage of the translation process and appears irreversible as suggested by a measured half-life of 65 h in these culture cells.[16] The functional significance of cellular protein methylation remains elusive in most cases.[18,19] The specificity of the methylation reactions affecting the various HSP during an arsenite treatment warrants further investigations to test whether these changes are related to the dynamic intracellular localization of HSP,[20] or to other unknown function(s) of these ubiquitous proteins.

III. CHANGES IN THE METHYLATION OF HISTONES IN STRESSED CELLS

Posttranslational modifications of histones within the nucleosome have been suggested to modify chromatin structure and function.[13] Many of these modifications have been correlated with changes in transcriptional or replicational activities. The function of histone methylation in chromatin structure and function is largely unknown although the enzymology and the site specificity of methylation is well documented in many biological systems (reviewed in References 21 and in Chapter 7 of this volume). Briefly, two types of *N*-methylation have been identified in histones: the first is at the N-terminal end of histone H2B in *Tetrahymena*,[22] *Asterias rubens*,[23] and in *Drosophila*.[24] The other type is the side chain methylation of basic amino acids, mainly lysine, in H3 and H4. Methylation of H3 and H4 is a late cellular event during the cell cycle[25] and is generally believed to be an irreversible modification.[26] The exact function of these methylating reactions in chromatin is unknown.

Heat shock induces major changes in chromatin function. Thus, in *Drosophila* cells where the response has been most extensively studied, there are striking modifications in the transcription of various classes of genes. The exact mechanisms by which heat shock genes are rapidly activated while many genes active prior to the heat shock (such as those coding for actin and many other cellular proteins) are rapidly repressed, are still unclear. Heat-induced variations in the phosphorylation[27] and acetylation[28,29] levels of histones have been reported but the most striking posttranslational modification of histones which is modulated by stress is methylation. In *Drosophila* cells cultured at normal growth temperatures, H3 and H4 are the major methylated histones. H2B is also lightly methylated.[28-30] Upon heat shock, one observes an arrest in the methylation of H3 and a small decrease in the level of methylation of H4[29,30] (Figure 1). Arsenite also induces a light decrease in the methylation level of H3 but only after a prolonged treatment (4 h).[28] Under these conditions, there is a 3-fold decrease in the methylation level of H3 as compared to a 20-fold decrease during heat shock.[29] Both heat shock and arsenite induce an increase in the level of methylation of H2B.[28-30] The heat-induced changes in the methylation levels of H2B and H3 are very rapid being detected as early as 5 min after the initiation of the heat treatment.[31] They are also dependent on the intensity of the heat stress and the changes in the methylation of H3 and H2B show different temperature response curves.[30] The increase in H2B methylation starts at 33°C while the decrease in the methylation of H3 is only observed from 35°C with a maximum at 37°C.[30] An interesting question is whether the fall in the level of

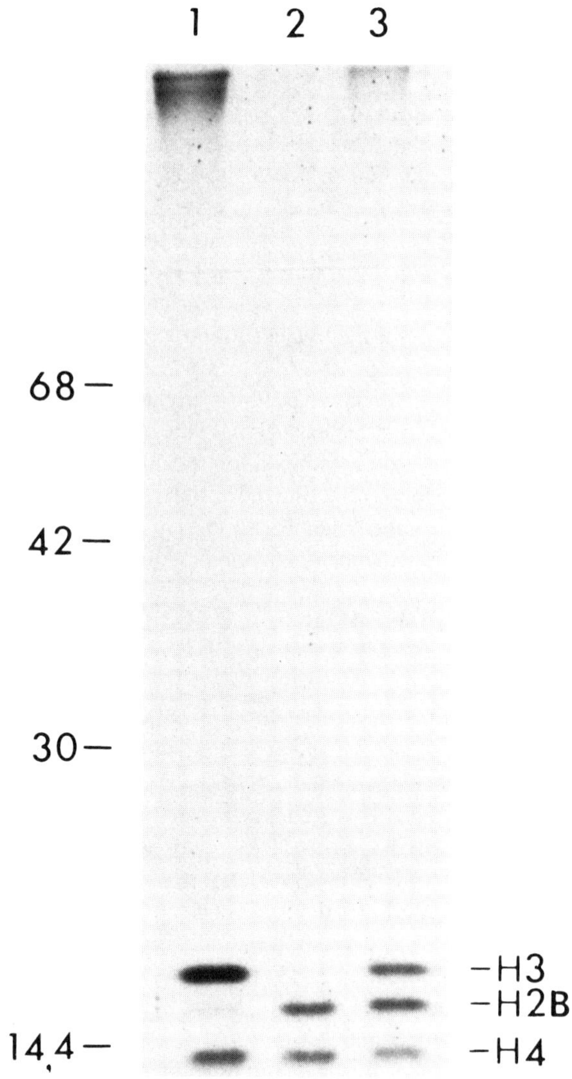

FIGURE 1. Analysis of methylated nuclear proteins on SDS gel. *Drosophila* cells were preincubated for 30 min at 23°C (lane 1), at 37°C (lane 2) or in presence of 50 μ*M* arsenite at 23°C (lane 3). Cycloheximide and chloramphenicol were added to inhibit protein synthesis and cells labeled for 3 h with L-[*methyl*-^{3}H] methionine. Nuclear proteins were resolved on SDS gel and visualized by fluorography.

methylation of H3 is caused by an arrest in methylation or by an accelerated demethylation of H3 by a heat-induced demethylase such as that previously found in mammalian cells.[32] Cells labeled with the methyl donor at 23°C and cold chased at 23°C or under heat shock conditions (37°C) show no differences in their level of H3 methylation suggesting that the reduction of H3 methylation results from an arrest in methylation rather than from an heat-induced demethylation.[29]

In the case of an arsenite treatment, the changes in the level of methylation of H3 and H2B are slower and are only seen after 2 h of treatment.[31] Interestingly the kinetics of the changes of H2B methylation parallel those of the inhibition of RNA synthesis[31] (and see discussion in Section VI). As shown in Figure 2, the amplitude of the changes of the methylation patterns of H2B and H3 is also dose dependent between 25 and 200 μ*M* sodium arsenite. The absence of

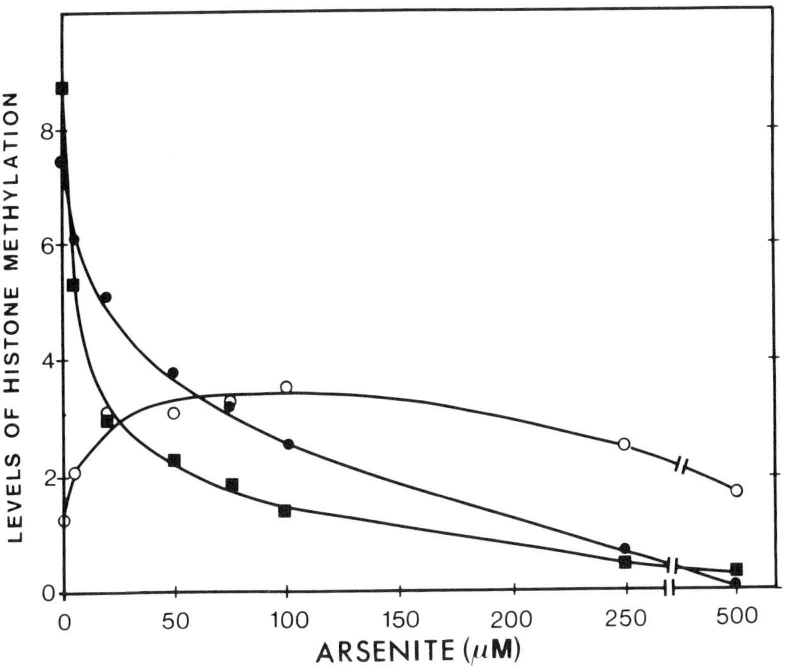

FIGURE 2. Effects of arsenite on the levels of histone methylation in *Drosophilia* cells. They were incubated for 1 h at 23°C with various concentrations of arsenite. Cells were then labeled for 2 h with L-[*methyl*-^{3}H] methionine in the presence of protein synthesis inhibitors. Nuclear proteins were separated on SDS gel and methylated proteins detected by fluorography with preflashed film. The radioactive bands corresponding to H2B (○), H3 (■), and H4 (●) were quantified by scanning the fluorogram with a densitometer.

methylation of core histone observed at 500 μ*M* arsenite is probably due to cellular death under these severe conditions.

In *Drosophila* cells, the synthesis of H2B is induced by heat shock (but not by chemical stresses) while the expression of the other nucleosomal histones is inhibited.[33-35] Addition of protein synthesis inhibitors prior to an heat shock to prevent the synthesis of newly heat-induced H2B, does not affect the level of methylation of H2B indicating that the heat-induced increase in methylation occurs on preexisting nucleosomal H2B.[30] Moreover it suggests that the methyltransferase responsible for the methylation of H2B is present prior to the heat shock. The increased H2B methyltransferase activity may be determined through activation by stress or alternatively by an increased accessibility of nucleosomal H2B as a result of chromatin reorganization.

As shown in Figure 3, all the forms of H2B resolved on acetic acid-urea-Triton X100/acetic acid-urea-CTAB gels are similarly methylated in response to heat shock or arsenite treatments. Since these forms are most likely generated by charge modifications such as those produced by acetylation and phosphorylation, it appears that the changes in methylation induced by stress occur concomitantly with other modifications.[29]

So far, the changes in the methylation pattern of core histones induced by heat shock have only been reported in *Drosophila* cells.[28-31] However, similar heat-induced changes in the methylation patterns of H3 and H4 are observed in another insect, *Chironomus*, and in both avian and mammalian cell lines (Figure 4). This common response from insect to mammalian cells suggests that the methylation of H3 and H4 may have an important function in chromatin function or structure. As discussed in Section V, the increased methylation of H2B may be a property of lower eukaryotes.

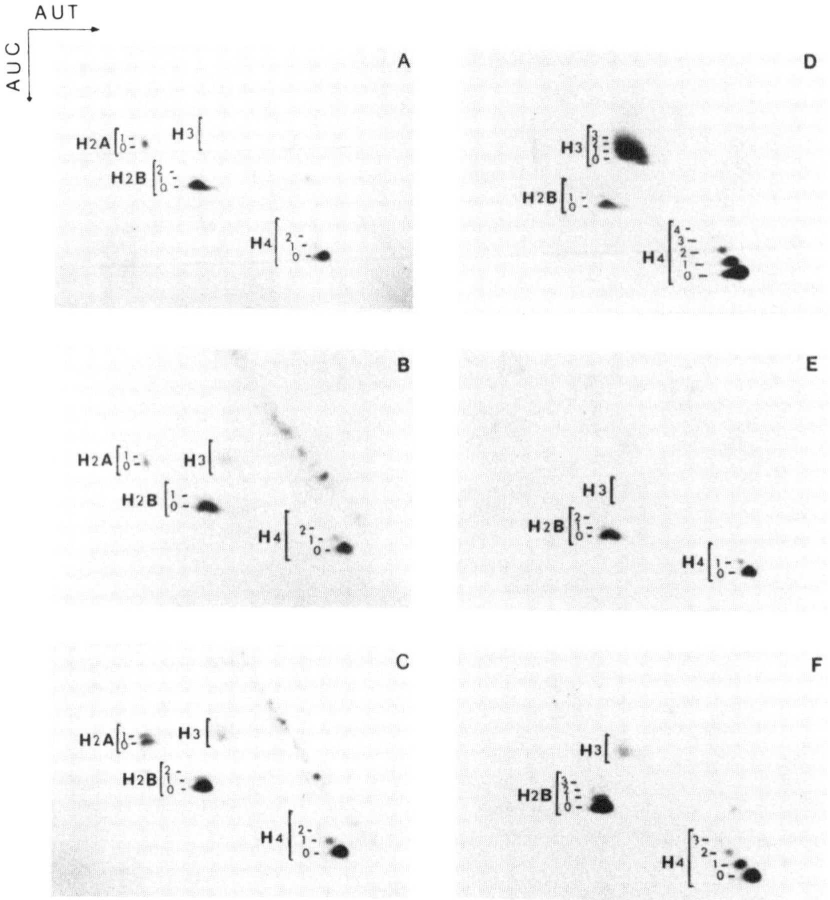

FIGURE 3. AUT-AUC two-dimensional gel electrophoresis of the various methylated forms of histones. Cells preincubated for 1 h at 23°C, 37°C, or in the presence of 50 μM arsenite at 23°C were treated and labeled as described in the legend of FIGURE 1. A to C: Coomassie blue stained gels of nuclear proteins from cells in normal growth conditions at 23°C (A) heat-shocked at 37°C (B) or arsenite-treated (C). D to F are the corresponding fluorograms. (From Desrosiers, R. and Tanguay, R. M., *Biochem. Cell Biol.*, 64, 750, 1986. With permission.)

IV. METHYLATION OF LYSINE, ARGININE, AND PROLINE RESIDUES DURING STRESS

The important and rapid decrease in the level of H3 methylation and the increase in that of H2B induced by heat shock and arsenite treatment in *Drosophila* incited us to identify the amino acid residues which show changes in methylation under these conditions.[24] These results are summarized in Table 1. Histones H3 and H4 are methylated on lysine residues with mono-, di-, and trimethyl forms in control (23°C) as well as in heat-shocked and arsenite-treated cells. After heat shock, there is an important decrease in the methylation of lysine forms in H3 and the appearance of newly methylated arginine residues tentatively identified as N^G-N^G-dimethylarginine (symmetrical and asymmetrical).[24] This residue is not observed in H3 of arsenite-treated cells where only methylated lysine forms are found.

In H2B, the methylated residue has been identified as α-*N*-methylproline thus located at the amino terminal end of H2B.[24] N-terminal methylated residues have been reported in H2B of *Tetrahymena*[22] (*N*-trimethylalanine) and the starfish *Asterias*[23] (dimethylproline). Other H2B in sea urchin,[36] plants,[37,38] and in another starfish, *Marthasterias*[39] also have blocked N-terminal amino acids but the nature of the blocking group has not been identified.

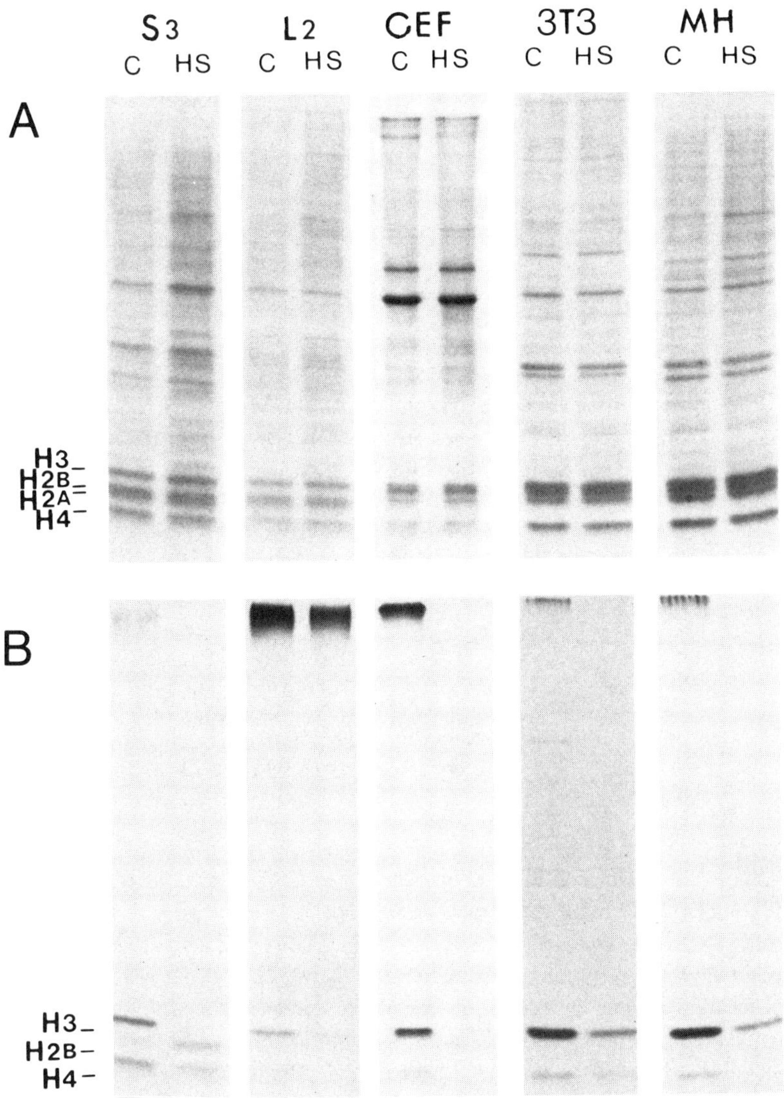

FIGURE 4. Patterns of histone methylation in various cell types during heat shock. Insect cells were preincubated for 30 min in control conditions (C) at 23°C or heat-shocked (HS) at 37°C. The *Drosophila* cells (S3) were labeled at 23 or 37°C according to their conditions while the *Chironomus* cells (L2) were labeled at 23°C with L-[*methyl*-^{3}H] methionine for 90 min. Mouse (3T3) and rat (MH) cells were kept at 37°C or heat-shocked at 43°C for 30 min and returned to their normal growth temperature for 1 h before labeling for 2 h with L-[*methyl*-^{3}H] methionine.The chicken fibroblast (CEF) cells were kept at 37°C or heat-shocked at 45°C for 2 h followed by labeling for 2 h with L-[*methyl*-^{3}H] methionine at 37°C. Cycloheximide and chloramphenicol were present to inhibit protein synthesis. (A) Coomassie blue stained gels of nuclear proteins from various cell types and (B) corresponding fluorograms of the methylated nuclear proteins in these cells. (From Desrosiers, R. and Tanguay, R. M., unpublished).

V. METHYLATION OF H2B: ENZYMOLOGY AND POSSIBLE FUNCTION

A. IN SEARCH OF THE H2B TERMINAL METHYLTRANSFERASE

Methylation of H2B is a rare event as this histone usually possess a free N-terminal proline

TABLE 1
Methylated Amino Acids in *Drosophila* Histones in Control (23°C) Cells during Heat Shock (37°C) and after Arsenite Treatment

Histones	Control	Heat shock	Arsenite
H2B	Proline (+)	Proline (++)	Proline (++)
H3	Lysine (+++)	Lysine (+), Arginine (+)	Lysine (++)
H4	Lysine (++)	Lysine (+)	Lysine (+)

TABLE 2
Sequences of Eukaryotic Proteins Methylated at their N-Termini and of Certain Histones H2B which are Methylated or Potential Substrates for N-Methylation

Proteins	Organisms	N-terminal sequences	Ref.
Cytochrome *c*-557	*Crithidia*	$(Me)_2$ Pro-Pro-Lys-Ala-Arg	41
Myosin light chains 1 and 2	Rabbit Chicken	$(Me)_3$ Ala-Pro-Lys-Lys-Asn	42,43
H2B	*Tetrahymena*	$(Me)_3$ Ala-Pro-Lys-Lys-Ala	22
H2B	*Asterias*	$(Me)_2$ Pro-Pro-Lys-Pro-Ser	23
H2B	*Drosophila*	(Me) Pro-Pro-Lys-Thr-Ser	24
H2B	*Marthasterias*	(X,Pro_2,Lys)-Ser	39
H2B	*C. elegans*	Ala-Pro-Pro-Lys-Pro-Ser	46
H2B	*Patella*	Pro-Pro-Lys-Val-Ser	47

residue.[40] Other cellular proteins such as cytochrome *c*-557 in *Crithidia*[41] and the myosin light chains (LC1 and LC2)[42,43] of all vertebrate striated muscles examined, also have methylated N-terminal ends. The N-terminal sequence of LC1 and LC2 (Me_3 Ala-Pro-Lys-Lys) is recognized by a monoclonal antibody which cross-reacts with histone H2B of *Tetrahymena* which has an identical N-terminal sequence suggesting common features. Table 2 shows the N-terminal end sequence of some eukaryotic proteins whose N-terminal amino acid is methylated. All these sequences begin with the three amino acids Pro-Pro-Lys or Ala-Pro-Lys. On the basis of this common feature, Stock et al.[44] postulated the existence of a specific methyltransferase (designated as the PK methyltransferase) which would be responsible for all N-terminal methylation in eukaryotes. Up to now, there is only a single report which suggests the existence of this hypothetical methyltransferase. In *Crithidia,* different fractions of a crude enzyme preparation methylate the N-terminal proline residue and lysine-3 of cytochrome *c*-557 with different ratios suggesting the existence of two specific methyltransferases.[45] Unfortunately no data is presently available about the methyltransferase involved in N-terminal methylation of histone H2B. Preliminary results on the localization of methyltransferases in *Drosophila* cells suggest that the enzyme responsible for the methylation of H2B is nuclear (Labarre, Desrosiers, and Tanguay, unpublished). Such a localization is similar to that of the lysine methyltransferases of H3 and H4.[19,21]

As shown in Table 2, other H2B are potentially good candidates for N-terminal methylation on the basis of their sequences. The N-terminal residue of histone H2B from sperm of the starfish *Marthasterias* is blocked by an unidentified group which is not acetyl or formyl.[39] Two other potential H2B candidates are those from the nematode *C. elegans*[46] and the mollusk *Patella granatina*.[47] Interestingly, all these H2B as well as those which have been shown to be methylated at the N-terminal are from lower eukaryotes. Moreover, there is a strong homology of the N-terminal portion of H2B from the mollusk *Patella,* the starfish *Asterias,* the nematode

C. elegans, and the dipteran insect *Drosophila*.[48] Such a homology is somewhat surprising as the N-terminal tails of H2B are known to have undergone considerable variations during evolution. Thus, N-terminal methylation of H2B may be a specific property of invertebrates, a suggestion compatible with our preliminary observations on the absence of methylation of H2B in rat, mouse, and chicken (Figure 4).

H2B shares a common characteristic with other N-terminal methylated proteins such as cytochrome *c*-557 and myosin LC1 and LC2. Each of these methylated proteins has a mobile extended end detached from a more central globular structure. *Crithidia*'s cytochrome *c*-557 is ten residues longer than vertebrates cytochromes *c* (which are not methylated).[45] Similarly the methylated myosin LC1 and LC2 are approximately 40 residues longer than LC3 (not methylated).[42] Histone H2B also has tails which seem to interact with DNA in the solenoid structure of chromatin. The conformation of the terminal methylated hexapeptide of histone H2B from gonads of *Asterias* has been recently shown to be highly flexible at any pH.[49]

B. H2B METHYLATION AND PROTEOLYSIS

The function of the increase in the terminal methylation of proline in H2B in response to heat and arsenite stress is presently unknown. We recently suggested that the increased methylation of H2B may be involved in its protection from increased proteolysis in stressed cells.[24] Although still speculative, such a model is compatible with many observations. First there are several reports showing the importance of blocked α- or ε-amino groups on the susceptibility of proteins to various proteolytic pathways.[50-53] Cytochrome *c*- 557 which has an N-terminal dimethylproline residue is particularly resistant to various peptidases.[41] Methylation of internal lysine in calmodulin has been proposed to protect it from ATP-ubiquitin dependent proteolysis.[50] *In vitro* reduction methylation and carbamoylation of the α-NH2 group of lysozyme and globin inhibits their conjugation with ubiquitin and prevents proteolytic breakdown by this degradation pathway.[51] Moreover, many proteins with naturally modified NH2-terminal (such as by acetylation) are not degraded by the ubiquitin-dependent protease.[51] Recently the importance of the nature of the N-terminal amino acid for protein stability has been suggested using chimeric protein constructions fused to ubiquitin at the N-terminal position.[54] Interestingly, proline is the only amino acid which does not fit into this end-terminal rule since it cannot be deubiquinated leading to its rapid degradation by the ubiquitin-mediated proteolytic pathway.[54]

In mammalian cells, heat shock has also been reported to cause a transient increase in proteolytic activity[55] (presumably to degrade abnormal proteins accumulating in these cells[56]). A similar increase in proteolysis has been observed in heat-shocked *Drosophila* S3 cells (Desrosiers and Tanguay, unpublished). Blocking of the α-NH2 by methylation could protect it from degradation by the ubiquitin-dependent proteolytic system. Interestingly, Carlson et al. recently reported that less microinjected labeled ubiquitin bound to histones under heat shock conditions than in control cells,[57] a finding also consistent with a putative role of methylation in histone H2B protection. However, as discussed below, it cannot be excluded that this modification is also involved in some aspects of chromatin structure and activity.

VI. HISTONE METHYLATION AND GENE ACTIVITY

The rapid changes in the methylation patterns of histones in heat-shocked cells could be related either to the rapid activation of the heat shock genes or to the repression of the cellular genes active prior to heat shock or to other chromatin-associated processes. To differentiate between these possible sites, we used different conditions of activation of heat shock genes and studied the effects of various drugs known to act on different chromatin functions.

A. METHYLATING CHANGES ARE CORRELATED WITH GENE REPRESSION

As shown in Figure 5, treatment of *Drosophila* cells at suboptimal heat shock temperatures

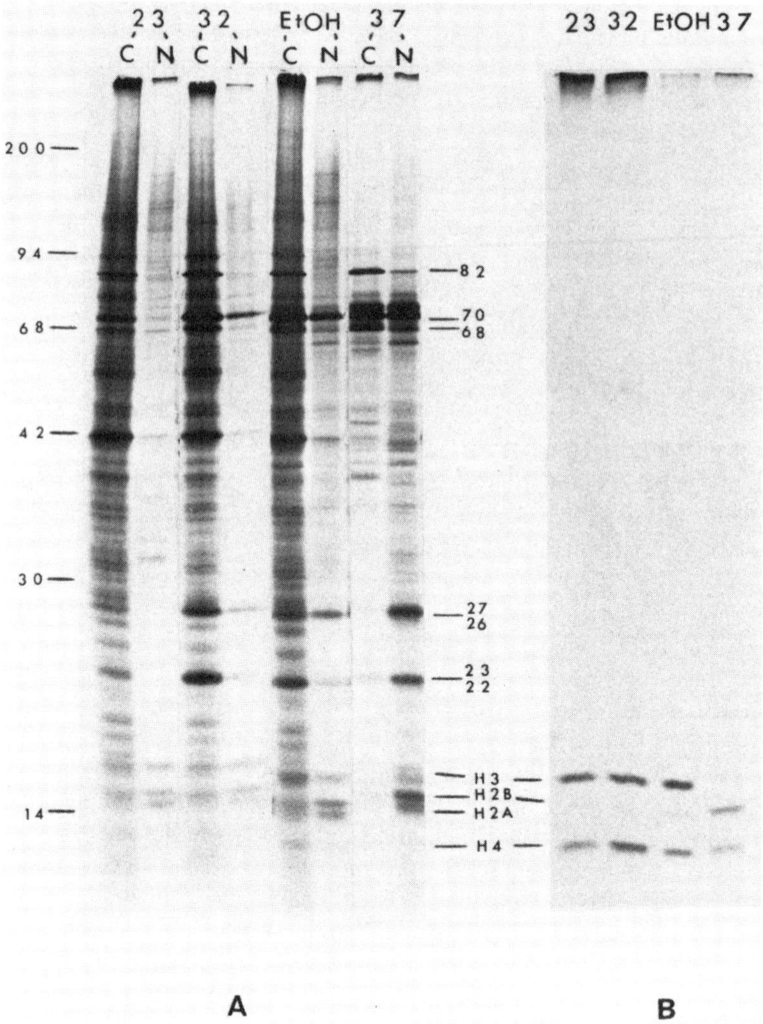

FIGURE 5. Patterns of protein synthesis and histone methylation in *Drosophila* cells heat-shocked at different temperatures or ethanol-treated. Cells were preincubated for 30 min at 23, 32, 37°C or in presence of 8% ethanol (ETOH) at 23°C and labeled for 2 h with either L-[^{35}S] methionine or L-[*methyl*-^{3}H] methionine. In methylation experiments, the inhibitors of protein synthesis were included. (A) autoradiogram of protein synthesis from cytoplasm (C) and nuclei (N) and (B) fluorogram of methylated nuclear proteins. (From Desrosiers, R. and Tanguay, R. M., *Biochem. Biophys. Res. Commun.*, 133, 823, 1985. With permission.)

(below 33°C) or with ethanol, results in the activation of the heat shock genes without the concomitant arrest in normal gene expression observed under more severe conditions.[58] No changes in the methylation patterns of histones are observed under these mild conditions suggesting that the changes are related to inactivation of certain genes rather than to the transcriptional activation of heat shock genes.[58] Arsenite also induces a time- and dose-dependent inhibition of RNA synthesis which shows a close correlation with the changes in histone methylation patterns.[31] Moreover, when cells are allowed to recover following a heat shock, the kinetics of recovery of the normal pattern of protein synthesis parallels that of the return to a normal level and pattern of histone methylation.[58] Thus, the changes in histone methylation during heat shock appear to be associated with the process of gene inactivation.

TABLE 3
Methylation of H2B and H3 in *Drosophila* Cells Treated with Drugs Affecting Transcription and/or Chromatin Structure

Treatments	Site of action	Inhibition of transcription	Methylation H2B	Methylation H3
23°C	—	0	-	+
37°C	Unknown	70	+	-
Arsenite	Unknown	60	+	+
AMD	Intercalation	99	+	-
DRB	Polymerase II	75	-	+
5-FUR	rRNA synthesis and processing	85	-	+
CPT	Topoisomerase I	80	-	+
VM-26	Topoisomerase II	70	+	-
NVB	Topoisomerase II, histone-histone interactions, histone-DNA interactions	80	+	-

Note: Cells incubated for 15 min with different drugs at various concentrations were labeled for 2 h with [5,6-^{3}H] uridine or L-[*methyl*-^{3}H] methionine to analyze the level of RNA synthesis and the patterns of histone methylation respectively. In the methylation experiments, protein synthesis inhibitors were added. RNA synthesis was estimated by measurement of the acid-insoluble radioactivity and the histone methylation patterns by electrophoresis of nuclear proteins and detection of methylated histones by fluorography. The abbreviations used are AMD, actinomycin D; DRB, 5,6-dichloro-1-ß-D-ribofuranosylbenzimidazole; 5-FUR, 5-fluorouridine; CPT, camptothecin; VM-26, epipodophyllotoxin teniposide and NVB, novobiocin.

B. EFFECTS OF DRUGS AFFECTING CHROMATIN FUNCTION

Inhibition of transcription by actinomycin D induces changes in histone methylation patterns similar although less intense than those observed in heat-shocked cells.[58] The heat shock mimicking effect of actinomycin D suggests a possible relationship between transcriptional inhibition and the observed changes in the methylation patterns of histones. Such a relationship was further investigated using various inhibitors acting at different levels of the transcription process. Table 3 summarizes these results.

DRB (5,6-dichloro-1-β-D-ribofuranosylbenzimidazole), an inhibitor of transcription initiation[59] presumably acting at the level of polymerase II[60] has no effect on the methylation patterns of histones.[31] 5-Fluorouridine (5-FUR), an inhibitor of ribosomal RNA synthesis and processing in these cells,[61] similarly does not affect the pattern of methylation. These results indicate that inhibition of transcription *per se* is not sufficient to trigger the changes in the methylation of histones. Alternatively these changes may be associated with certain aspects of chromatin restructuration in response to heat or arsenite treatments. DNA conformation is an important component of transcriptional and replicational processes. This conformation is modulated by various enzymes (topoisomerases, helicases)[62,63] as well as by DNA-protein and protein-protein interactions. Although the role of DNA topology in the control of gene expression is not clearly established, there are reports on the interaction of certain topoisomerases with transcriptionnally active chromatin.[64-66] Since actinomycin D has also been reported to interfere with DNA topoisomerase II activity,[63] we tested a number of topoisomerases inhibitors to see if they induce changes in histone methylation similar to those induced by actinomycin D.

Camptothecin (CPT), an inhibitor of topoisomerase I does not affect the histone methylation patterns although it inhibits transcription (Table 3). However, the epipodophyllotoxin teniposide VM-26, a specific inhibitor of topoisomerase II, induced changes in histone methylation similar to those observed during heat shock (Desrosiers and Tanguay, submitted). This suggest that the changes observed in stressed cells may be linked to topoisomerase II activity under these conditions Novobiocin (NVB), an unspecific inhibitor of topoisomerase II, induced at high doses (25 to 100 μg/ml) changes in the pattern of histone methylation similar although less intense than those induced by stress. Although the mechanism of inhibition of topoisomerase II by novobiocin[63] differs from that of VM-26, their similar effects on methylation suggest that this enzyme may be the primary target involved. Interestingly, novobiocin also interferes with the formation and maintenance of histone octamers by disruption of histone-histone interactions[67] at a dose (25 to 50 μg/ml) in the range of that which induces changes in the pattern of histone methylation in *Drosophila* cells. These data suggest that changes in chromatin structure induced either by topoisomerase II and/or by modification of protein-protein interaction in the nucleosome are involved in inducing methylation changes. Further experiments will be necessary to elucidate if actinomycin D, VM-26 and novobiocin act on methylation by a common or different mechanisms from those involved during heat or arsenite stress in these cells.

C. HISTONE METHYLATION AND CHROMATIN CONDENSATION

Heat shock and arsenite treatments induce changes in the ratios of the various methylated forms of lysine in H4. In heat shocked cells, there is a decrease of the mono- and trimethyllysine forms while in arsenite treated cells, monomethyllysine is predominant. In the case of H3, the ratio of the three forms of methyllysine remains unchanged under all conditions. Following heat shock, there is a dramatic decrease in the methylation level of lysine with the appearance of new methylation sites on arginine residues. Most posttranslational modifications of histones H3 and H4 including methylation occur on the N-terminal tails of these histones. The high degree of conservation of the N-terminal sequences of H3 and H4 and their susceptibility to numerous posttranslational modifications which appear correlated with various cellular processes suggest an important role of these regions in chromatin structure and function. These tails have been proposed to be involved in the organization of the higher order structure of chromatin such as the solenoid.[68] Heat shock causes a general decrease in the level of acetylation of histones[28-29] as well as changes in the phosphorylation level of H1, H3, and H4.[27,69,70] The amount of ubiquinated H2A and H2B[27,55,57] also decreases during heat shock, a phenomenon also observed during chromosome condensation at mitosis.[71] Many of these changes have been correlated with chromatin condensation mainly during mitosis (reviewed in Reference 13). The heat-induced changes in the levels and specificity of histone methylation coupled with the decreased acetylation and ubiquination are consistent with a function of these changes in chromatin reorganization and/or compaction possibly accompanying the repression of most normal genes under these conditions. Such a function is also compatible with the observation that chromatin is more difficult to shear after heat shock.[72]

VII. SUMMARY

Cells exposed to supraoptimal temperatures exhibit marked changes in gene activity. These changes are accompanied by an important decrease in the methylation level of histone H3 and are observed in various cells from insect to mammals. Moreover, in *Drosophila* cells, where this phenomenon was originally observed, heat shock and arsenite induce a marked increase in the methylation of the N-terminal amino acid residue of histone H2B, identified as *N*-methylproline. These changes are rapid and their magnitude dependent on the intensity of the stress. In addition, heat shock induces a shift in the site of methylation of H3 from lysine to arginine.

The biological functions of histone methylation are presently unknown. In *Drosophila* cells, the stress-induced changes in the methylation of histones as well as in other posttranslational modifications appear to be correlated with chromatin reorganization accompanying the transcriptional repression of many genes under these conditions. Moreover, the hypermethylation of H2B in stressed cells may be involved in its protection from increased proteolysis. Further experiments are underway to delineate the exact functions of these changes in relation to chromatin structure and function during heat or chemical stress.

ACKNOWLEDGMENTS

Research in the author's laboratory was supported by the Medical Research Council of Canada (Program grant PG-35). R. D. is a recipient of a studentship from the FCAR of Quebec. We are indebted to Dr. P. Rogers and S. Côté for reviewing the manuscript and to Mary Ann Godbout for her patient editorial assistance.

REFERENCES

1. **Ashburner, M. and Bonner, J. J.,** The induction of gene activity in *Drosophila* by heat shock, *Cell,* 17, 241, 1979.
2. **Tanguay, R. M.,** Genetic regulation during heat shock and function of heat-shock proteins: a review, *Can. J. Biochem. Cell Biol.,* 61, 387, 1983.
3. **Atkinson, B. G. and Walden, D. B., Eds.,** *Changes in Eukaryotic Gene Expression in Response to Environmental Stress,* Academic Press, Orlando, 1985.
4. **Craig, E.,** The heat shock response, *CRC Crit. Rev. Biochem.,* 18, 239, 1985.
5. **Lindquist, S.,** The heat shock response, *Annu. Rev. Biochem.,* 55, 1151, 1986.
6. **Landry, J., Lamarche, S., and Chrétien, P.,** Heat shock proteins: a lead to the understanding of cell thermotolerance in *Thermotolerance and Thermophily,* Henle, K. J., Ed., CRC Press, Boca Raton, FL, 1987, 145.
7. **Carper, S. W., Duffy, J. J., and Gerner, E. W.,** Heat shock proteins in thermotolerance and other cellular processes, *Cancer Res.,* 47, 5249, 1987.
8. **Tanguay, R. M.,** Transcriptional activation of heat shock genes in eukaryotes, *Biochem. Cell Biol.,* 66, 587, 1988.
9. **Findley, R. C. and Pederson, T.,** Regulated transcription of the genes for actin and heat-shock proteins in cultured *Drosophila* cells, *J. Cell Biol.,* 88, 323, 1981.
10. **Yost, H. J. and Lindquist, S.,** RNA splicing is interrupted by heat shock and is rescued by heat shock protein synthesis, *Cell,* 45, 185, 1986.
11. **Storti, R. V., Scott, M. P., Rich, A., and Pardue, M. L.,** Translational control of protein synthesis in response to heat shock in *D. melanogaster* cells, *Cell,* 22, 825, 1980.
12. **DiDomenico, B. J., Bugaisky, G. E., and Lindquist, S.,** The heat shock response is self-regulated at both the transcriptional and post-transcriptional levels, *Cell,* 31, 593, 1982.
13. **Wu, R. S., Panusz, H. T., Hatch, C. L., and Bonner, W. M.,** Histones and their modifications, *CRC Crit. Rev. Biochem.,* 20, 201, 1986.
14. **Wang, C., Gomer, R. H., and Lazarides, E.,** Heat shock proteins are methylated in avian and mammalian cells, *Proc. Natl. Acad. Sci. U.S.A.,* 78, 3531, 1981.
15. **Nover, L. and Scharf, K. D.,** Synthesis, modification and structural binding of heat-shock proteins in tomato cell cultures, *Eur. J. Biochem.,* 139, 303, 1984.
16. **Wang, C., Lazarides, E., O'Connor, C. M., and Clarke, S.,** Methylation of chicken fibroblast heat shock proteins at lysyl and arginyl residues, *J. Biol. Chem.,* 257, 8356, 1982.
17. **Wang, C. and Lazarides, E.,** Arsenite-induced changes in methylation of the 70,000 dalton heat shock proteins in chicken embryo fibroblasts, *Biochem. Biophys. Res. Commun.,* 119, 735, 1984.
18. **Paik, W. K. and Kim, S.,** *Protein Methylation,* Meister, A., Ed., John Wiley & Sons, New York, 1980.
19. **Paik, W. K. and Kim, S.,** Protein methylation, in *The Enzymology of Post-Translational Modification of Proteins,* Vol. 2, Freedman R. B. and Hawkins, H. C., Eds., Academic Press, London, 1985, 187.
20. **Tanguay, R. M.,** Intracellular localization and possible functions of heat shock proteins, in *Changes in Eukaryotic Gene Expression in Response to Environmental Stress,* Atkinson, B. G. and Walden, D. B., Eds., Academic Press, Orlando, 1985, 91.
21. **Duerre, J. A. and Onisk, D. J.,** Specificity of the histone lysine methyltransferase from rat brain chromatin, *Biochim. Biophys. Acta,* 843, 58, 1985.
22. **Nomoto, M., Kyogoku, Y., and Iwai, K.,** N-trimethylalanine, a novel blocked N-terminal residue of *Tetrahymena* histone H2B, *J. Biochem. Tokyo,* 92, 1675, 1982.
23. **Martinage, A., Briand, G., Van Dorsselaer, A., Turner, C. H., and Sautière, P.,** Primary structure of histone H2B from gonads of the starfish *Asterias rubens, Eur. J. Biochem.,* 147, 351, 1985.
24. **Desrosiers, R. and Tanguay, R. M.,** Methylation of *Drosophila* histones at proline, lysine and arginine residues during heat shock, *J. Biol. Chem.,* 263, 4686, 1988.

25. **Shepherd, G. R., Hardin, J. M., and Noland, B. J.,** Methylation of lysine residues of histone fractions in synchronized mammalian cells, *Arch. Biochem. Biophys.,* 143, 1, 1971
26. **Byvoet, P., Shepherd, G. R., Hardin, J. M., and Noland, B. J.,** The distribution and turnover of labeled methyl groups in histone fractions of cultured mammalian cells, *Arch. Biochem. Biophys.,* 148, 558, 1972.
27. **Glover, C. V. C.,** Heat shock effects on protein phosphorylation in *Drosophila,* in *Heat Shock: From Bacteria to Man,* Schlesinger, M. J., Ashburner, M., and Tissières, A., Eds., Cold Spring Harbor Laboratory, Cold Spring Harbor, NY, 1982, 227.
28. **Arrigo, A.-P.,** Acetylation and methylation patterns of core histones are modified after heat or arsenite treatment of *Drosophila* tissue culture cells, *Nucleic Acids Res.,* 11, 1389, 1983.
29. **Desrosiers, R. and Tanguay, R. M.,** Further characterization of the posttranslational modifications of core histones in response to heat and arsenite stress in *Drosophila, Biochem. Cell Biol.,* 64, 750, 1986.
30. **Camato, R. and Tanguay, R. M.,** Changes in the methylation pattern of core histones during heat-shock in *Drosophila* cells, *EMBO J.,* 1, 1529, 1982.
31. **Tanguay, R. M. and Desrosiers, R.,** Histone methylation and modulation of gene expression in response to heat shock and chemical stress in *Drosophila,* in *Advances in the Post-Translational Modification of Proteins and Ageing,* Zappia, V., Ed., Plenum Press, New York, 1988, 353.
32. **Paik, W. K. and Kim, S.,** ε-alkyllysinase new assay method, purification, and biological significance, *Arch. Biochem. Biophys.,* 165, 369, 1974.
33. **Tanguay, R. M., Camato, R., Lettre, F., and Vincent, M.,** Expression of histone genes during heat shock and in arsenite-treated *Drosophila* Kc cells, *Can. J. Biochem. Cell Biol.,* 61, 414, 1983.
34. **Farrell-Towt, J. and Sanders, M. M.,** Noncoordinate histone synthesis in heat-shocked *Drosophila* cells is regulated at multiple levels, *Mol. Cell. Biol.,* 4, 2676, 1984.
35. **Spadoro, J. P., Capertino, D. W., and Strausbaugh, L. D.,** Differential expression of histone sequences in *Drosophila* following heat shock, *Dev. Genet.,* 7, 133, 1986.
36. **Brandt, W. F., Strickland, W. N., Strickland, M., Carlisle, L., Woods, D., and Von Holt, C.,** A histone programme during the life cycle of sea urchin, *Eur. J. Biochem.,* 94, 1, 1979.
37. **Hayashi, H., Iwai, K., Johnson, J. D., and Bonner, J.,** Pea histones H2A and H2B, *J. Biochem.,* 82, 503, 1977.
38. **Von Holt, C., Strickland, W. N., Brandt, W. F., and Strickland, M. S.,** More histone structures, *FEBS Lett.,* 100, 201, 1979.
39. **Strickland, M. S., Strickland, W. N., and Von Holt, C.,** The histone H2B from the sperm cell of the starfish *Marthasterias glacialis, Eur. J. Biochem.,* 106, 541, 1980.
40. **Wells, D. E.,** Compilation analysis of histones and histone genes, *Nucleic Acids Res.,* 14, r119, 1986.
41. **Pettigrew, G. W. and Smith, G. M.,** Novel N-terminal protein blocking group identified as dimethylproline, *Nature,* 265, 661, 1977.
42. **Henry, G. D., Trayer, I. P., Brewer, S., and Levine, B. A.,** The widespread distribution of α-N-trimethylalanine as the N-terminal amino acid of light chains from vertebrate striated muscle myosins, *Eur. J. Biochem.,* 148, 75, 1985.
43. **Tokunaga, M., Suzuki, M., Saeki, K., and Wakabayashi, T.,** Position of the amino terminus of myosin light chain 1 and light chain 2 determined by electron microscopy with monoclonal antibody, *J. Mol. Biol.,* 194, 245, 1987.
44. **Stock, A., Clarke, S., Clarke, C., and Stock, J.,** N-terminal methylation of proteins: structure, function and specificity, *FEBS Lett.,* 220, 8, 1987.
45. **Valentine, J. and Pettigrew, G. W.,** A cytochrome c methyltransferase from *Crithidia oncopelti, Biochem. J.,* 201, 329, 1982.
46. **Vanfleteren, J. R., Van Bun, S. M., and Delcambe, L. L.,** Multiple forms of histone H2B from the nematode *Caenorhabditis elegans, Biochem. J.,* 235, 769, 1986.
47. **Van Helden, P. D., Strickland, W. N., Brandt, W. F., and Von Holt, C.,** The complete amino acid sequence of histone H2B from the mollusc *Patella granatina, Eur. J. Biochem.,* 93, 71, 1979.
48. **Elgin, S. C. R., Schilling, J., and Hood, L. E.,** Sequence of histone 2B of *Drosophila melanogaster, Biochemistry,* 18, 5679, 1979.
49. **Helbecque, N., Bernier, J. L., Hénichart, J. P., Martinage, A., and Sautière, P.,** Synthesis and conformation of the amino-terminal hexapeptide of histone H2B from gonads of the starfish *Asterias rubens, Int. J. Peptide Protein Res.,* 30, 689, 1987.
50. **Gregori, L., Marriott, D., West, C. M., and Chau, V.,** Specific recognition of calmodulin from *Dictyostelium discoideum* by the ATP, ubiquitin-dependent degradative pathway, *J. Biol. Chem.,* 260, 5232, 1985.
51. **Hershko, A., Heller, H., Eytan, E., Kaklij, G., and Rose, I. A.,** Role of the α-amino group of protein in ubiquitin-mediated protein breakdown, *Proc. Natl. Acad. Sci. U.S.A.,* 81, 7021, 1984.
52. **Breslow, E., Daniel, R., Ohba, R., and Tate, S.,** Inhibition of ubiquitin-dependent proteolysis by non-ubiquitinatable proteins, *J. Biol. Chem.,* 261, 6530, 1986.
53. **Murakami, K. and Etlinger, J. D.,** Degradation of proteins with blocked amino groups by cytoplasmic proteases, *Biochem. Biophys. Res. Commun.,* 146, 1249, 1987.

54. **Bachmair, A., Finley, D., and Varshavsky, A.,** *In vivo* half-life of a protein is a function of its amino-terminal residue, *Science,* 234, 179, 1986.
55. **Parag, H. A., Raboy, B., and Kulka, R. G.,** Effect of heat shock on protein degradation in mammalian cells: involvement of the ubiquitin system, *EMBO J.,* 6, 55, 1987.
56. **Ananthan, J., Goldberg, A. L., and Voellmy, R.,** Abnormal proteins serve as eukaryotic stress signal and trigger the activation of heat shock genes, *Science,* 232, 522, 1986.
57. **Carlson, N., Rogers, S., and Rechsteiner, M.,** Microinjection of ubiquitin: changes in protein degradation in HeLa cells subjected to heat shock, *J. Cell Biol.,* 104, 547, 1987.
58. **Desrosiers, R. and Tanguay, R. M.,** The modifications in the methylation patterns of H2B and H3 after heat shock can be correlated with the inactivation of normal gene expression, *Biochem. Biophys. Res. Commun.,* 133, 823, 1985.
59. **Mukherjee, R. and Molloy, G. R.,** 5,6-dichloro-1-ß-D–ribofuranosylbenzimidazole inhibits transcription of the ß-hemoglobin gene *in vivo* at initiation, *J. Biol. Chem.,* 262, 13697, 1987.
60. **Zandomeni, R. and Weinmann, R.,** Inhibitory effect of 5,6-dichloro-1-ß-D–ribofuranosylbenzimidazole on a protein kinase, *J. Biol. Chem.,* 259, 14804, 1984.
61. **Lengyel, J. and Penman, S.,** hnRNA size and processing as related to different DNA content in two dipterians: *Drosophila* and *Aedes, Cell,* 5, 281, 1975.
62. **Vosberg, H. P.,** DNA topoisomerases: enzymes that control DNA conformation, *Curr. Topics. Microbiol. Immunol.,* 114, 19, 1985.
63. **Wang, J. C.,** Recent studies of DNA topoisomerases, *Biochim. Biophys. Acta,* 909, 1, 1987.
64. **Gilmour, D. S., Pflugfelder, G., Wang, J. C., and Lis, J. T.,** Topoisomerase I interacts with transcribed regions in *Drosophila* cells, *Cell,* 44, 401, 1986.
65. **Gilmour, D. S. and Elgin, S. C. R.,** Localization of specific topoisomerase I interactions within the transcribed region of active heat shock genes by using the inhibitor camptothecin, *Mol. Cell. Biol.,* 7, 141, 1987.
66. **Stewart, A. F. and Schütz, G.,** Camptothecin-induced *in vivo* topoisomerase I cleavages in the transcriptionally active tyrosine aminotransferase gene, *Cell,* 50, 1109, 1987.
67. **Sealy, L., Cotten, M., and Chalkey, R.,** Novobiocin inhibits passive chromatin assembly *in vitro, EMBO J.,* 5, 3305, 1986.
68. **Allan, J., Harborne, N., Rau, D. C., and Gould, H.,** Participation of core histone "tails" in the stabilization of the chromatin solenoid, *J. Cell Biol..,* 93, 285, 1982.
69. **Glover, C. V. C., Vavra, K. J., Guttman, S. D., and Gorovsky, M. A.,** Heat shock and deciliation induce phosphorylation of H1 in *T. pyriformis, Cell* , 23, 73, 1981.
70. **Pekkala, D., Heath, B., and Silver, J. C.,** Changes in chromatin and the phosphorylation of nuclear proteins during heat shock of *Achlya ambisexualis, Mol. Cell. Biol.,*4, 1198, 1984.
71. **Mueller, R. D., Yasuda, H., Hatch, C. L., Bonner, W. M., and Bradbury, E. M.,** Identification of ubiquinated histones H2A and 2B in *Physarum polycephalum.* Disappearance of these proteins at metaphase and reappearance at anaphase, *J. Biol. Chem.* , 260, 5147, 1985.
72. **Arrigo, A. P., Fakan, S., and Tissières, A.,** Localization of the heat shock-induced proteins in *Drosophila melanogaster* tissue culture cells, *Dev. Biol.,* 78, 86, 1980.

Chapter 9

RIBOSOMAL PROTEIN METHYLATION

Jacques Lhoest and Charles Colson

TABLE OF CONTENTS

I. INTRODUCTION

The ribosome is an intricate particle consisting of a complex association of RNA molecules and proteins. This particle interacts with many other molecules during protein synthesis. Moreover, the nascent ribosomal components are subjected to a series of modifications. In this context, the study of methylation of ribosomal proteins might appear to be of incidental interest; attributing a function to individual r-proteins remains a matter of discussion. However, in a few instances, it has been shown that methylation of ribosomal protein contributed to the building of mature ribosomes. The main objective of this review is to make clear what is known and unknown about ribosomal protein methylation. First, a summary of the structure and the function of ribosome as well as its multiple postsynthetic modifications will be outlined. This chapter is intended only to help understanding the frame in which ribosomal protein methylation takes place. More information on ribosome can be obtained from other publications.[1]

Since many direct or indirect methods have been used to unveil methylated ribosomal proteins, a section will be devoted to summarize these methods. Next, a "catalog" will be presented, reviewing the discovery and characterization of methylated ribosomal proteins and their methylated amino acid content. A short survey on what is known or unknown on the ribosomal protein methylating enzymes is presented afterwards. Finally, the basic question is evoked; why are ribosomal proteins methylated?

II. STRUCTURE, FUNCTION, AND MODIFICATIONS OF THE RIBOSOME

A. COMPOSITION

The translational process is carried out by ribosomes along with other factors. In exponentially growing bacteria, nearly all the ribosomes are actively engaged in protein synthesis and they account for 40% of the total dry cell mass.

The ribosome is a complicated assembly of RNA molecules and proteins, all organized into a large and a small subunit. The proteins in the two subunits differ, as do the molecules of ribosomal RNA (rRNA). The length of the rRNA, the amount of proteins in each subunit and consequently the sizes of the subunits vary between prokaryotic and eukaryotic cells.

In prokaryotes, about two thirds of the mass of each ribosomal subunit are RNA and only one third is proteins. Ribosomes from eukaryotes are bigger and more complicated than those from prokaryotes, where the total protein mass amounts to roughly 50% of the particle.

The subunits and the rRNA are specified in terms of Svedberg units (S). The prokaryotic ribosomes which sediment at a velocity of 70S dissociate into two subunits when partially starved of magnesium ions. The 50S subunit contains two rRNA molecules, a 23S and a 5S RNA and about 33 proteins. The 30S subunit contains a 16S RNA and about 20 proteins. All the components are present in one copy except two proteins, L7 and L12, which are present in four copies.[2] The cytoplasmic eukaryotic 80S ribosome is formed by the 60S and 40S subunits. The large subunit contains one species of high molecular weight RNA, 28S RNA, and two types of lower molecular weight RNA, 5S and 5.8S RNAs. The small subunit contains an 18S RNA. The eukaryotic ribosome is composed of more than 70 proteins, about 40 in the large and 30 in the small subunit. Most of the proteins are also present in one copy. Chloroplastic ribosomes are closely related to the prokaryotic ribosomes while the mitochondrial ribosomes show a high degree of heterogeneity, both in size and composition of RNA and protein.

The function and the structure of the ribosomal individual components are better understood in prokaryotes than in eukaryotes because they are more extensively studied and are less complex. Moreover, genetical approach has allowed to obtain numerous mutants affected in a single component.

Some proteins associate temporarily with ribosomes during protein synthesis. These proteins are called factors and can be separated from constitutive ribosomal proteins (r-proteins) by salt treatment.

B. ASSEMBLY OF RIBOSOMAL SUBUNITS

1. *In Vitro* Assembly

Both subunits of *Escherichia coli* ribosome have been entirely reconstituted *in vitro* from the various macromolecular components. Traub and Nomura[3] isolated the 16S RNA and 21 proteins of the small subunit. Self-assembly was performed by mixing all the components. An assembly map was established which shows that the reconstitution is a highly cooperative process which involves an ordered sequential addition of single proteins or groups of proteins to the 16S rRNA.[4] All the information needed for the assembly is contained in the structure of its components. However, a heating step is necessary to activate a conformational rearrangement of an intermediate precursor particle for further addition of proteins in order to obtain a mature ribosomal subunit.

The assembly of the 50S subunit is more complex. The first 50S subunit *in vitro* assembly has been performed at high temperature in *Bacillus stearothermophilus*,[5] and later in *E. coli*.[6] The *in vitro* assembly of the 50S subunit of *E. coli* also requires high temperature incubation steps leading to conformational rearrangements of intermediate particles.[6] Partial reconstitution of cytoplasmic eukaryotic ribosomes has also been achieved.[7]

2. *In Vivo* Assembly

The *in vivo* assembly also proceeds by the formation of precursors with sedimentation coefficients similar to those observed *in vitro*.[8] The requirement of heating activation steps *in vitro* suggested to search cold-sensitive mutants. Some of these mutants appeared to be impaired in ribosome assembly and accumulated precursors.[9-13] The mutations were located either in genes coding for a ribosomal protein,[10,12] or in genes not coding for a structural component of the ribosome.[11] This led to the idea that "assembly factors" were implicated in the ribosome assembly but neither their nature nor their role was elucidated.

Genes responsible for the synthesis of the rRNA are clustered in operons, with the exception of the 5S RNA in eukaryotic cells. Transcription gives rise to a 30S precursor RNA in prokaryotic cells and a 45S precursor RNA in the eukaryotic cells. These precursors undergo cleavages and chemical modifications. These posttranslational processes lead into the mature rRNA. The initial transcript is immediately associated with a group of proteins to constitute ribonucleoprotein particles. In eukaryotes, the ribosomal proteins are synthesized on cytoplasmic ribosomes and the majority of them are bound to the pre-rRNA molecules after their transportation into the nucleus within the nucleolus. The final maturation of the precursor particles takes place in the cytoplasm.

C. FUNCTION OF THE RIBOSOME

Protein synthesis is carried out in a stepwise fashion. The codons of the mRNA are read from the 5′ end towards the 3′ end by the ribosomes which allow synthesis of the polypeptide, starting from the N-terminal amino acid and ending with the C-terminal.

Three stages can be distinguished during translation. During initiation, the small ribosomal subunit recognizes the AUG codon of the mRNA and interacts with a tRNA and proteins called initiation factors. All these components form the initiation complex. Then the large subunit associates with this complex to form the active ribosome. The large subunit possesses two sites, the peptidyl site and aminoacyl site. The first tRNA binds to the peptidyl site. During elongation, the tRNA carrying the amino acid specific to the next codon positions itself at the aminoacyl site. A peptide bond is formed (peptidyl transferase reaction). Then, during translocation the whole ribosome moves along the mRNA and the peptidyl-tRNA translocates from the A site to the P site. A new tRNA specific to the next codon arrives into the ribosome. Approximately 15 amino acids are incorporated into the growing polypeptide every second. Several ribosomes may translate the mRNA simultaneously and form polysomes. When the ribosome reaches a termination codon (UAA, UAG, or UGA), it releases the polypeptide and dissociates into its two subunits. This phase constitutes the termination.

Each of these steps requires proteins called factors as well as ATP or GTP hydrolysis.

D. RIBOSOMAL RNA METHYLATION

During and after the maturation of rRNA, posttranscriptional modifications occur, mainly pseudouridylation and methylation.

E. coli 16S and 23S RNA contain only a small number of 2′-*O*-methylated ribose residues and most of the methyl groups occur as methylated bases (9 and 10, respectively).[14] In eukaryotes, the number of methyl groups is higher and reach about 100.[15] Most of them are added to 2′-*O*-ribose of different nucleotides before the maturation of the precursor RNA. A few bases are methylated after the first steps of maturation.[16-18] Experiments using methionine starvation have indicated that at least in HeLa cell, methylation is essential for ribosome maturation. Ribose methylation appears to be important for an efficient completion of the subsequent maturation steps of the 45S RNA to the mature cytoplasmic products 18S and 28S.[18] It would in particular play a role in protecting the rRNA precursors against enzymatic degradation. Inhibition of methylation leads to extensive degradation.[19] Base methylation appears to occur later during the maturation and partially in the cytoplasm.[17,18,20]

The methylated residues are found in very evolutionarily conserved regions of the 45S precursor RNA. Extensive homologies between the methylated nucleotide sequences have been reported in the rRNA from HeLa cells, hamster cells, mouse, chick embryo fibroblasts, and yeast. Methyl groups are added only in the sequences corresponding to mature 28S, 18S, and 5.8S RNAs whereas spacer sequences are not methylated. The sequence $Gm6_2Am6_2ACCUG$ at the 3′ end of the rRNA of the small subunit is extremely conserved and has been identified in all prokaryotic and eukaryotic organisms except in organelle ribosomes of a few eukaryotes.[21-23]

In prokaryotes, methylation does not seem to play an important role in the maturation process. When *E. coli* was incubated under conditions which block methylation, normal molecules of 16S, 23S, and 5S RNA were obtained, but they were undermethylated.[24] Therefore, in *E. coli*, methylation is not necessary for the maturation of the pre-rRNA to produce normal mature RNA.

Björk and Isaksson[25] isolated a mutant lacking a single m1G residue normally present in the 23S RNA which is not impaired in either the assembly or in the function of the ribosome. There are several cases where changes in the resistance or sensitivity status to an antibiotic were attributed to a small specific change in the pattern of rRNA methylation. Loss by mutation of a methylase activity can lead to the resistance to an antibiotic. For example, Helser et al.[26] found that in *E. coli,* resistance to kasugamycin was due to undermethylation in the $m6_2Am6_2ACCUG$ sequence of the 16S RNA resulting from the mutational inactivation of a specific methylase. The 30S ribosome assembly is quite normal but the initiation step of protein synthesis is a little impaired.[27] In *B. stearothermophilus,*[28] the same mutational change was also found to confer resistance to kasugamycin. On the other hand, hypermethylation of rRNA can also result in antibiotic resistance. For instance, Thompson et al.[29] have well documented that antibiotic producing *Streptomyces* immunize their own ribosome target by an rRNA methylation absent in sensitive strains. Pentose methylation in 23S RNA confers autoimmunity of *Streptomyces azureus* to thiostrepton[30] whereas dimethylation at a single adenine in the same rRNA is responsible for erythromycin resistance in *S. erythraeus.*[31] A gene cloned from *Micromonospora purpurea,* the producer of gentamicin, was shown to code for a 16S RNA methylase. When heterologous ribosomes of *Streptomyces* or *E. coli* were hypermethylated with this enzyme, their protein synthetic capacity became resistant *in vitro* against gentamicin and kanamycin.[32]

E. RIBOSOMAL PROTEINS (r-PROTEINS)

1. Analysis and Nomenclature

Nonribosomal proteins, mainly factors, remain bound to the ribosomes during their isolation. They can be removed by washing with a high concentration of salt. r-Proteins are often extracted from the ribosomes by acetic acid[33] and separated by two-dimensional polyacrylamide gel electrophoresis. Most of the r-proteins are very basic and are separated according to their charge in the first dimension and according to their molecular weight in the second dimension. This system was first elaborated by Kaltschmidt and Wittmann in *E. coli.*[34] The first dimension is realized at pH 8.6 in the presence of urea and the second at pH 4.5 in the presence of sodium dodecyl sulfate. Most r-proteins have a molecular weight between 10,000 and 35,000. Each protein is attributed a number preceded by "S" or "L" indicating that it belongs to the small or to the large subunit, and according to its position in the gel. The protein pattern is read from the anodic end towards the cathodic end in both dimensions. So, a higher number is bestowed to proteins with a lower molecular weight and which are more basic. For instance, S21 refers to a protein which belongs to the small subunit and which has a molecular weight lower than S10. Other nomenclatures and analysis methods were used before but the system of Kaltschmidt and Wittmann prevailed. r-Proteins of other prokaryotes as well as eukaryotes have also been separated by similar electrophoresis systems.[35,36] The same rules of nomenclature are used. Sometimes the initial letter of the organism precedes the name of the protein. For instance, BL11 and YL11 designate proteins of the large subunit from *Bacillus* and yeast, respectively. In the case of some organisms, i.e., yeast, proteins were analyzed by distinct two-dimensional gel electrophoresis systems leading to different nomenclatures,[37-39] and correlations between them have been shown.[39] For example, YL23 according to Otaka and Osawa[39] corresponds to L15 of Bollen et al.[38]

2. Homologies Between r-Proteins

Electrophoretic patterns of r-proteins display variability according to evolution distance between species. In the case of a few r-proteins, strong homologies exist between phylogenetically distant organisms. The detection of such homologies is based on electrophoretic mobility, immunological cross-reactions, structural relationship, and functional equivalence.

Only four proteins of the 50S subunit migrated identically in eight Enterobacteriaceae,[40] while minor differences were found between r-proteins from vertebrates.[36] Antibodies raised against L7-L12 from *E. coli* cross-react with an acidic r-protein of rat liver and Krebs II ascites cells.[41] Immunological evidence for structural homology among *Drosophila* S14, rat S12, yeast S25, *Bacillus* S6, and *E. coli* S6 has also been reported.[42] In yeast, five cytoplasmic r-proteins share immunological determinants with mitochondrial r-proteins.[43] The *E. coli* S11 protein has a sequence which is highly conserved among organisms including yeast and Chinese hamster.[44] *E. coli* L11 and *B. megaterium* L11 are not only serologically related, but they are both required for the tight binding of thiostrepton.[45] The interchangeability of the acidic proteins of the large subunit of bacteria and several eukaryotic species has been well established.[36]

3. Ribosomal Factors

Protein synthesis requires factors which temporarily bind to specific compounds of the ribosome. These factors are proteins which are called initiation factors (IF), elongation factors (EF), and termination or release factors (RF) according to the phase of protein synthesis in which they take part. The factor EF-Tu in prokaryotic organisms is found in abundance and fulfills numerous activities. Besides its implication in protein synthesis it has also been reported as a component of Qβ RNA polymerase,[46] as a regulator factor for the transcription of rRNA genes[47] and associated with the bacterial membrane.[48] This factor is very well conserved with evolution. The elongation factor EF-1α from the brine shrimp *Artemia salina,* which also binds aminoacyl-tRNA to the ribosome, contains amino acid sequences homologous to distinct regions in EF-Tu from *E. coli.*[49]

4. Posttranslational r-Protein Modifications

Several proteins are subjected to various posttranslational modifications, which include acetylation, phosphorylation, and methylation.

E. coli L7 differs from L12 by acetylation of the N-terminal serine[50] while S5[51] and S18[52] contain an acetylated N-terminal alanine. Mutants lacking an enzyme acetylating specifically either L12,[53] S5,[54] or S18[55] have been isolated and none of them showed a noticeable growth impairment. Acetylated r-proteins have also been found in eukaryotes.[36,41] Experiments in which inhibition of acetylation was obtained by sodium fluoride suggested that acetylation could play a role in the formation of the initiation complex.[56]

In *E. coli,* a peculiar modification has been described in protein S6, which consists of an additional four glutamic acid residues to the C-terminus of the protein.[57] The growth of a mutant lacking this modification was not altered.[58]

Phosphorylation occurs in eukaryotes, especially protein S6 which is heavily phosphorylated in ribosomes of many organisms.[36] The degree of phosphorylation depends on the physiological conditions. Phosphorylation of S6 has been reported to be stimulated by cyclic-AMP[59] and puromycin.[60] Viral infections[61] and heat shock[62] can also influence phosphorylation. A mutant of *Saccharomyces cerevisiae* without the phosphorylated form of r-protein S10, equivalent to protein S6 in other organisms, grows reasonably well.[63] The actual effect of phosphorylation on the function of r-proteins is not yet known in spite of the fact that phosphorylation activity is highly regulated during the growth cycle

III. METHODS USED TO DETECT r-PROTEIN METHYLATION

Demonstration of ribosomal protein methylation has been achieved by many different experimental approaches. These can be as crude as to show evidence that the ribosome or its subunits contain an unknown number of methyl groups in undefined proteins and amino acids. On the other extreme, the analysis can lead to the precise identification of which modified residue is present in a particular protein. None of the methods used can be considered as specifically designed for *ribosomal* protein methylation. However, since many different methods have been used, we though it worthwhile to summarize them briefly in this review.

A. DETECTION BY PROTEIN SEQUENCING AND AMINO ACID ANALYSIS

The most definitive method to identify a methylated protein and its qualitative and quantitative content of the methylated amino acid is sequencing of the purified protein. This has been achieved for all ribosomal proteins known to be methylated in *E. coli*. However, it was only in the case of monomethyllysine in L7-L12,[50] that this method was the first to demonstrate that the protein was methylated. For some other proteins, Edman-type degradation revealed that methylation resulted from the presence of N-terminal monomethylated alanine or methionine.[64] Finally, particular sequencing strategies were designed to locate methylated amino acid residues in proteins of whom the methylated nature was already established or suggested. This was the case for L11, blocked at its N-terminus by a residue of *N*-trimethylalanine[65,66] and for L3 where a single N^5-methylglutamine was attributed to residue 150 by means of seeking overlapping peptides yielding both glutamic acid and methylamine after acid hydrolysis.[67] Methylated lysine was located at residue 56 in EF-Tu by the same technique.[68] Conversely, sequencing under appropriate conditions ruled out that some proteins were methylated although other methods had suggested that they could be.

B. RADIOACTIVE LABELING *IN VIVO*

To detect the presence of methylated nucleic acids, growing cells can be exposed to methionine radioactively labeled in the methyl group. After purification of the macromolecule, all the radioactivity found reflects the transfer of methyl groups into it. The same is not true for proteins, since a large part of the label is incorporated as methionine during protein synthesis. Several strategies have been used to circumvent this problem.

1. Double Labeling

The method consists of labeling the cells with two isotopes, one of which is in the methyl group of methionine, the other is not. After labeling, the ribosomes are purified and the r-proteins are separated by two-dimensional gel electrophoresis. Finally, the ratio of the two isotopes is determined in individual proteins by scintillation counting. A higher than average ratio of the methyl-labeling isotope is taken as an indication that the protein is methylated. This method has been used to detect methylation in ribosomal proteins of bacteia,[69] yeast,[70-72] and higher eukaryotes.[73-75]

Although sensitive and accurate in principle when the protein has a known methionine content, this method sometimes proved to yield false-positive results and underestimated stoichiometric ratios.[76]

2. Single Labeling

a. Single Labeling with Methionine

If no special technique is used to dissociate protein synthesis from methylation, labeling with [*methyl*-^{3}H]- or [*methyl*-^{14}C]-methionine *in vivo* can be used to demonstrate protein methylation only if an insignificant portion of the radioactivity is found in other amino acids than methionine upon amino acid analysis by paper electrophoresis or chromatography,[77-78] or by using an automated amino acid analyzer coupled with a radioactivity counter.[80]

Alix and Hayes[81] used sublethal concentrations of ethionine, an analog of methionine, to block methyltransferase activities while allowing a residual protein synthesis in *E. coli* cultures. In a second step the cells were washed free of ethionine and *in vivo* methylation of ribosomal protein using labeled methionine was performed in the presence of chloramphenicol to block protein synthesis. A similar strategy was used to demonstrate methylation of the EF-Tu factor. In this case, however, the first step consisted of only methionine starvation of a Met-RelA$^-$ mutant of *E. coli*.[82]

b. Single Labeling with S-Adenosyl-L-Methionine

Active uptake of *S*-adenosyl-L-methionine from the medium and its accumulation into the cell vacuole has been demonstrated and studied in detail in *Saccharomyces cerevisiae*.[83] When cultures of this yeast were grown in the presence of labeled *S*-adenosyl-L-methionine, it was found that virtually all the radioactivity in acid hydrolysates of ribosomal proteins was in methylated amino acids and not in methionine. Thus, this *in vivo* labeling method was used to identify yeast methylated proteins and amino acids[84] as an alternative and most definite method than the double labeling used by others.[70-72]

C. RADIOACTIVE LABELING *IN VITRO*

For *in vitro* methylation, a system must be set up in which are present the methyl group donor, a methylable recipient substrate, and a suitable enzymatic source. In the case of ribosomal proteins these conditions have been fulfilled on many but not all occasions. *In vitro O*-methylation, where reversibility implies that the substrate is not saturated *in vivo,* has been demonstrated.[85] *N*-Methylated proteins reported to be methylated at a lower than stoichiometric level or to vary in methyl-group content depending on growth phase or growth conditions have not been methylated *in vitro* (e.g., L7-L12 in *E. coli*). On the other hand, some but not all proteins demonstrated or presumed to be saturated *in vivo* were methylated *in vitro* after extraction from cells followed by treatments intended to reduce the level of methylation *in vivo*. Thus, treatments described in Section B.2.a, followed immediately by ribosome extraction and protein methylation were successful to detect *in vitro* methylation of several r-proteins. Direct starvation of the methyl-group donor was possible in yeast using a mutant strain lacking *S*-adenosyl-L-methionine synthetase.[84,86] Comparison of methylated proteins and amino acids found *in vivo* and *in vitro* were in good agreement.[73,81,84,87]

The ideal conditions for *in vitro* methylation require an entirely nonmethylated protein. These conditions have been fulfilled in only a few cases. Colson and Smith[88] and Lhoest and Colson[89] have described mutants completely lacking in methylation of *E coli* L11 and L3, respectively. They yielded an abundant source of ribosomes where either of these two proteins could be *specifically* methylated *in vitro* to an almost stoichiometric ratio (9CH_3 for L11, 1 CH_3 for L3). No mutant lacking in another ribosomal protein methylating enzyme has been reported. This surely is due to the very difficult and hazardous task required to undertake in seeking such mutants.[88]

More recently, the DNA from a specialized transduction phage ($\lambda rif^{d}18$) was used *in vitro* to synthesize and methylate small quantities of *E. coli* protein L11 and factor EF-Tu in the presence of labeled *S*-adenosyl-L-methionine. In these experiments, the methylated proteins were identified by immunoprecipitation.[90-91]

IV. OCCURRENCE OF METHYLATED r-PROTEINS AND AMINO ACIDS

The methylated r-proteins and amino acids in the various organisms studied are presented in Table 1.

A. *ESCHERICHIA COLI*

Many r-proteins of *E. coli* have been reported to be "significantly" methylated. However, we will describe here only those in which methylation has been convincingly demonstrated and documented.

1. Methyl Esterification of Proteins S3 and S9

Kim et al.[85] detected that proteins in the 30S subunit — but not the 50S — could be methyl-esterified *in vitro*. Analysis by one-dimensional gel electrophoresis indicated that two proteins, presumably S3 and S9, were specifically esterified. Due to the labile nature of the ester bond, they could not confirm the identity of these two proteins by two-dimensional gel electrophoresis.

TABLE 1
Occurrence of Methylated Ribosomal Proteins and Amino Acids

Organism	Ribosomal protein or subunit	Amino acid[a]	Method of[b] detection	Authors and year
Escherichia coli	S3, S9	Methyl-esters	C1	Kim et al. (1977)
	L7-L12	ε-*N*-monomethyllysine	A	Terhorst et al. (1973)
			B1	Chang (1981)
	L3	N D	B1, C2	Alix and Hayes (1974)
		N^5-methylglutamine	C4	Lhoest and Colson (1977)
			A	Muranova et al. (1978)
	L33	*N*-monomethylalanine and	B1	Chang and Budzilowicz (1977)
		N-monomethylmethionine	A	Chen et al. (1977)
	L16, S11	*N*-monomethylalanine	A	Chen et al. (1977)
	L11	ε-*N*-trimethyllysine and	B1, C2	Alix and Hayes (1974)
		neutral	B2	Chang et al. (1974)
			C4	Colson and Smith (1977)
			A	Dognin and Wittmann-Liebold (1977)
			C5	Jerez and Weissbach (1980)
		N-trimethylalanine	B1, C2	Lederer et al. (1977)
			A	Dognin and Wittmann-Liebold (1980)
	IF-3ℓ	*N*-monomethylmethionine	A	Brauer and Wittmann-Liebold (1977)
	EF-Tu	ε-*N*-monomethyllysine and	B1	Ferro-Luzzi Ames and Miakido (1979)
		ε-*N*-dimethyllysine	C2, C5	Toledo and Jerez (1985)
Bacillus megaterium	BM-L11	N D	B2	Cannon and Cundliffe (1979)
Bacillus subtilis and	L11	ε-*N*-trimethyllysine	B2	Mardones et al. (1980)
Bacillus stearothermophilus				Amaro and Jerez (1984)
Saccharomyces cerevisiae	L42 (YL42)	N D	B2	Cannon et al. (1977)
	L3	N D	B2	Cannon et al. (1977)
			B3	Lhoest et al. (1984)
	L15 (YL23)	N D	B2	Cannon et al. (1977)
		ε-*N*-dimethyllysine and	B3, C3	Lhoest et al. (1984)
		ε-*N*-trimethyllysine		

TABLE 1 (continued)
Occurrence of Methylated Ribosomal Proteins and Amino Acids

Organism	Ribosomal protein or subunit	Amino acid[a]	Method of[b] detection	Authors and year
	YL32	ε-*N*-monomethyllysine and ε-*N*-dimethyllysine	B3, C3	Lhoest et al. (1984)
	YS1, YS5, YS28	N D	B3	Lhoest et al. (1984)
	EF-1	N D		Fonzi et al. (1985)
Saccharomyces carlsbergiensis	L15(YL23), L41(YL27), S31(YS23), S2(YS28)	N D	B2	Kriuswijk et al. (1978)
Blastociadiella emersonii	70S	Methylated lysine	C1	Comb et al. (1966)
Mucor racemosus	EF-1	ε-*N*-monomethyllysine, ε-*N*-dimethyllysine, and ε-*N*-trimethyllysine	B1	Fonzi et al. (1982) Fonzi et al. (1985)
Tetrahymena pyriformis	L13, L15, L19, L20, S36, S36a	N D	B2	Cyrne et al. (1981)
Euglena gracilis	90S	ε-*N*-trimethyllysine and 3-methylhistidine	A	Reporter (1972)
Artemia salina	EF-1	ε-*N*-trimethyllysine	A	Amons et al. (1983)
Mouse cells	EF-1	N D	B1	Coppard et al. (1983)
Mouse fibroblasts	60S	ε-*N*-monomethyllysine, ε-*N*-dimethyllysine, ε-*N*-trimethyllysine, and N^G,N'^G-dimethylarginine	B1	Klagsbrun and Furano (1975)
	40S	N^G,N^G-dimethylarginine and N^G,N'^G-dimethylarginine		
HeLa cells	40S, 60S	N^G,N^G-dimethylarginine, ε-*N*-trimethyllysine, and ε-*N*-dimethyllysine	B1	Chang et al. (1976)
	Protein spots 3,20,51,51, 52,55,56, protein spot 28,	N^G,N^G-dimethylarginine ε-*N*-trimethyllysine	B1, B2	Chang et al. (1978)

	protein spots 1,38,	Methylated arginine	B1, B2	Goldenberg and Eliceiri (1977)
	protein spots			
	30,62,63, and	Methylated lysine		
	protein spots 18,42	N D		
	40S = 1	N D	B2	Scolnik and Eliceiri (1979)
	60S = 42	N D	B2	
	40S = 18,19,38	N D	B2, C1	
	60S = 30	N D	B2, C1	
Spinacia oleracea	Chloroplast			
	r-protein, L2	*N*-monomethylalanine	A	Kamp et al. (1987)

[a] N D = not determined.

[b] A: Protein sequencing. B: *In vivo* labeling. B1: single labeling; B2: double labeling; B3: labeling with *S*-adenosyl-L-methionine. C: *In vitro* labeling. C1: without desaturation of the substrate; C2: desaturation with ethionine or methionine starvation; C4: mutant lacking the methylase activity; C5: during *in vitro* protein synthesis (for explanations, see text).

The modified residues were not identified. To our knowledge this is the only report of *O*-methylation of ribosomal proteins. Moreover, *O*-methylation was found in the small subunit whereas *N*-methylation in *E. coli* and other prokaryotes concern mainly proteins of the large subunit.

2. ε-*N*-Monomethyllysine in Protein L7-L12

L7 and L12 are the same polypeptide. They differ only by the presence of an α-*N*-acetyl group in L7[50] which has a slower electrophoretic mobility than L12 in the second dimension of the Kaltschmidt and Wittmann electrophoretic system.[34] L7-L12 is exceptional in that it is the most acidic protein in the 50S and is present in more than one copy per particle.[2] ε-*N*-monomethyllysine was found by Terhorst et al.[50] during the amino acid analysis of peptides derived from L7-L12. This methylation was attributed to residue 82 and was not stoichiometric (75% of methylated lysine). Later work demonstrated a temperature-dependent proportion of the lysine methylation in this protein,[92] and some implication of these results is discussed in Sections V and VI.

3. α-*N*-Methylation at the N-Terminus of L16, L33, and S11

N-Monomethylalanine and *N*-monomethylmethionine were first found in protein L33.[93,94] The methylated methionine amounted to no more than 25% of the stoichiometric ratio. This suggested that a methylating enzyme competed with those that release formyl and/or formylmethionine from the nascent L33 polypeptide and that this enzyme did not discriminate between a methionine or an alanine at the N-terminal position.[94] When analyzing the product of first step Edman degradation, Chen et al.[64] pointed out that among the bulk of r-proteins, the N-terminus of L16, L33, and S11 was *N*-monomethylalanine and that a portion of L33 had an *N*-monomethylmethionine. For a short while, S11 was thought to have an awkward N-terminus consisting of an N-terminal lysine with an ε-*N* linkage to a monomethylalanine.[95] However, it proved to be an artifact caused by unusual Edman reaction on α-*N*-monomethyl amino acids.[96] Further sequencing work confirmed that the N-terminal residue of S11 is *N*-monomethylalanine.[97] Amino acid analysis indicates clearly that the *in vivo* methylation of the N-terminus of these proteins is stoichiometric.

4. N^5-Methylglutamine in L3

Results of double labeling experiments suggested that protein L3 of *E. coli* could be methylated.[69,81] However, which kind of methylated amino acid was present in that protein was not sorted out in these investigations. Lhoest and Colson[89] isolated a mutant of *E. coli* lacking a methyltransferase specifically methylating L3. Hence, they found that L3 contains a single stoichiometric residue of a methylated amino acid which had not previously been found in protein. The strong basic nature and the volatility of its methylated acid hydrolysis product (methylamine) led them to identify it as N^5-monomethylglutamine. This single methylated glutamine was attributed to residue 150 of protein L3.[67] No other methylation of protein L3 has been found. The significance of this glutamine methylation is discussed in Section VI.

5. *N*-Trimethylalanine and ε-*N*-Trimethyllysine in L11

Among the ribosomal proteins of *E. coli*, L11 is the most heavily methylated. It is firmly established by sequence data that it contains nine methyl groups; three at the N-terminal alanine, three at lysine 3, and three at lysine 39.[98] Before these protein sequencing results were published, several laboratories reported a more or less accurate estimation on the nature and stoichiometry of methylated amino acids in L11. The presence of ε-*N*-trimethyllysine and that of a neutral unidentified methylated amino acid were first detected by Alix and Hayes.[81] This neutral amino acid was identified as *N*-trimethylalanine, a residue blocking the N-terminus of protein L11.[65] A mutant lacking methylation of L11 was isolated by Colson and Smith.[88] *In vitro* methylation

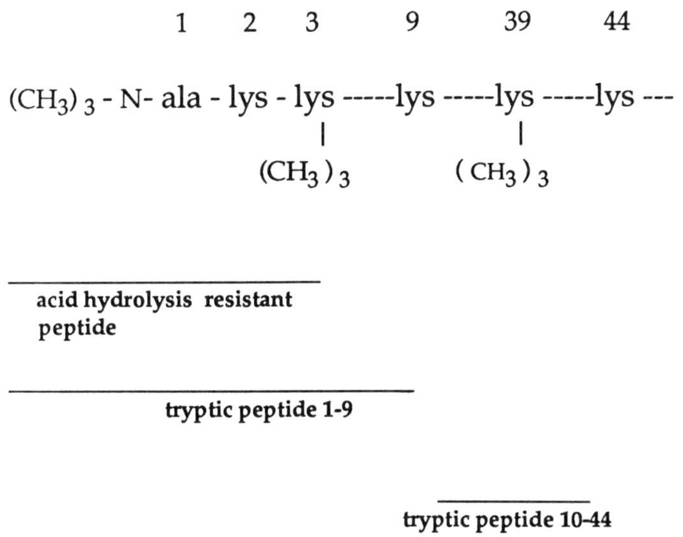

FIGURE 1. Terminal portion of the primary structure of *E. coli* ribosomal protein L11. The tryptic map shows that peptide bonds 2-3, 3-4, and 39-40 are not cleaved by trypsin.

and analysis of the acid hydrolysate of 50S proteins from this mutant suggested the presence of ε-*N*-monomethyllysine in addition to trimethylalanine and ε-*N*-trimethyllysine in L11.[88] Further work demonstrated that this compound was actually the N-terminal dipeptide trimethyl Ala-Lys of protein L11 (Figure 1), that was only partially cleaved under standard conditions of acid hydrolysis.[80] It is agreed that L11 which is the most heavily methylated r-protein of *E. coli*, has the methylation pattern shown in Figure 1.

6. Methylation of Initiation Factor IF-3ℓ and Elongation Factor EF-Tu

IF-3ℓ of *E. coli* has an N-terminal monomethylmethionine,[99] as found in L33.[93,94] Again, sequencing studies indicated that not all of the molecule contained this methylated amino acid. However, in IF-3ℓ, N-terminal truncated protein starting with either Lys 2 or Val 7 were not methylated.

EF-Tu, a protein known to be associated with ribosome and involved in elongation, was shown to be methylated at its lysine 56[68] by *in vivo*[82] and *in vitro*[91] labeling and by sequencing. Both ε-*N*-monomethyllysine and ε-*N*-dimethyllysine were observed. Whether the ratio of these two methylated forms of the same lysine residue vary according to the cellular location of this abundant protein has not yet been investigated.

B. OTHER PROKARYOTES

The possible occurrence of a similar pattern of methylated proteins in bacteria related to *E. coli* has not been investigated. However, it has indirectly been shown that several members of the group Enterobacteriaceae have a specific methylation activity towards L11.[100] Investigations in the genus *Bacillus* were performed by several groups using the double-labeling technique. A general conclusion is that, again, *N*-methylation principally affects proteins in the large subunit and that a single protein possibly related to *E. coli* L11 is the most heavily methylated.[101-103] In *Bacillus megaterium,* a mutant resistant to thiostrepton was found to lack this L11 related protein.[101]

C. YEAST

Since several nomenclatures are presently in use, correlations are given in brackets according to that of Otaka and Osawa.[39] Detection of the methylated r-proteins was first realized by Cannon et al.[70] by using the double-labeling technique. One protein of the 60S subunit was found to be heavily methylated L15 (YL23), and to a lesser extent proteins L42 (YL42) and L3 (YL1). The methylation level observed in the 40S subunit was very low.[71] Using the same technique, Kruiswijk et al.[72] reported the presence of two methylated proteins in both subunits, L15 (YL23), L41 (YL27), S31 (YS23), and S2 (YS28) of *S. carlsbergiensis*. Direct *in vivo* labeling by incorporation of *S*-adenosyl-L-[*methyl*-^{3}H]-methionine was performed by Lhoest et al.[84] and they estimated about ten methyl groups in the large subunit, distributed mainly among YL23, YL32, and YL1. The small subunit was shown to contain only a few methyl groups (two to four) shared by YS28 and YS23. These three laboratories agree that one protein, YL23, is methylated. It is interesting to point out that structural analogy and functional equivalence of protein YL23 and L11 from *E. coli* have been demonstrated.[104]

In vitro methylation was achieved by obtaining a submethylated substrate. A yeast strain which lacks the two isoenzymes of *S*-adenosyl-L-methionine synthetase and therefore auxotrophic for exogenous *S*-adenosyl-L-methionine,[86] was starved from *S*-adenosyl-L-methionine. This provided submethylated substrate during residual growth. The enzyme source was a crude extract filtered on Sephadex G-25 in order to remove a low molecular weight compound which appeared to be a nonspecific inhibitor of methylation of guanosine (5) triphospho (5) guanosine, protein L11 from *E. coli*, and horse heart cytochrome *c*.[84,105] The principal proteins methylated *in vitro*, i.e., YL23 and YL32, were the same as those found *in vivo*. Moreover, the methylated amino acids of YL23 and YL32 were the same and in the same ratio as *in vivo*.[84] YL23 contained ε-*N*-dimethyllysine and ε-*N*-trimethyllysine while YL32 contained ε-*N*-monomethyllysine and ε-*N*-dimethyllysine. It is also striking that L11 from *E. coli* which is related to YL23 also contains ε-*N*-trimethyllysine.[81,88,98]

D. OTHER LOWER EUKARYOTIC ORGANISMS

Comb et al.[106] were the first to discover a methylated r-protein in 1966. They found methylated lysine in the ribosomes of a water fungus, *Blastocladiella emersonii*. Other fungi beside yeast were also examined for protein methylation. The most heavily methylated protein of all eukaryotic organisms reported until now is the elongation factor 1 (EF-1α) from the dimorphic fungus *Mucor racemosus*.[107,108] It is posttranslationally modified with the formation of ε-*N*-mono-, di-, and trimethyllysine at as many as 16 sites. This modification occurs particularly in mycelia but not in the sporangiospores. Fonzi et al.[108] have found that EF-1α from *Neurospora crassa* or *Saccharomyces cerevisiae* is also methylated. EF-1α is the functional equivalent of prokaryotic EF-Tu which also contains methylated lysine residues.[68,91]

Methylated proteins in the cytoplasmic ribosomes of the protozoan *Tetrahymena pyriformis* have been detected by the double-labeling technique.[75] Two basic proteins (L15 and L20) and two acidic proteins (L13 and L19) of the large subunit as well as two basic proteins (S36 and S36a) of the small subunit contain methylated amino acids. Reporter[109] discovered by amino acid analysis of total r-proteins that *Euglena gracilis* contained large amounts of ε-*N*-trimethyllysine and small amounts of 3-methylhistidine.

It appears from these studies that lower eukaryotes including yeast show mainly methylation of lysine residues. On the other hand, the most heavily methylated ribosomal subunit is the large one, and a single protein is the main target (like L11 in *E. coli*). In these aspects, lower eukaryotes behave like prokaryotes.

E. HIGHER EUKARYOTES

r-Proteins from higher eukaryotic organisms differ from those of prokaryotes and lower eukaryotes by the presence of N^G,N^G-dimethylarginine as the major methylated amino acid. Methylated lysine residues generally are much less abundant. So is the case in the 40S subunit of a mouse fibroblast line (3T3) labeled with L-[*methyl*-^{14}C]-methionine, where small amounts of N^G,N'^G-dimethylarginine (symmetric) was also found. The three types of methylated lysine residues as well as N^G,N^G-dimethylarginine appeared in the 60S subunit.[77] ε-*N*-dimethyllysine has been reported in cultured muscle cell ribosomes.[110] r-Proteins are also methylated in rat liver cells.[111] Only HeLa cell ribosomes have been extensively investigated. Chang et al.[78] and Goldenberg and Eliceiri[112] presented evidence from double-labeling studies that both ribosomal subunits were methylated and contained predominantly N^G,N^G-dimethylarginine (asymmetric). In addition, a smaller amount of methylated lysine was found, mainly ε-*N*-trimethyllysine.[78] At least seven proteins appeared to be clearly methylated, i.e., spots 3, 20, 28, 50, 51, 52, and 55 according to the nomenclature of Chang et al.[74] and spots 1, 18, 19, 38 (small subunit), 30, and 42 (large subunit) according to the nomenclature used by Scolnik and Eliceiri.[73] When a cell-free system was incubated with [*methyl*-^{3}H]-methionine, ^{3}H-incorporation was detected in four r-proteins (18, 19, 30, 38) which are among the most methylated *in vivo*.[73]

As in lower eukaryotes, the elongation factor 1α (EF-1α) in mouse cells[79] and in *Artemia salina* is methylated.[113] In the latter case, amino acid sequences of polypeptide fragments showed the presence of three ε-*N*-trimethyllysine residues. This confirms the evolutionaly conservation of methylated lysine residues in this elongation factor (EF-1α or EF-Tu) in all organisms.

The chloroplast ribosomes display the prokaryotic ribosome motif in many of their structural and functional properties, although many of their components are encoded by the nuclear DNA. An *N*-methylalanine is the N-terminal residue of an r-protein of spinach chloroplast,[114] shown to be immunologically homologous to *E. coli* L2. *N*-Methylalanine is a typical prokaryotic methylated amino acid. However, *E. coli* L2 is not a methylated protein.

V. ENZYMES FOR r-PROTEIN METHYLATION

Ribosomal protein methylase has not yet been extensively purified and characterized. When methylation was observed *in vitro,* the assay was merely carried out to demonstrate that the proteins and amino acids methylated were the same as in *in vivo*. Since completely unmethylated proteins are usually not available for these assays, quantitative estimations are hindered by the lack of an accurate knowledge of the level of *in vivo* undermethylation. Still, some results were obtained to characterize r-protein methylases in *E. coli* and yeast.

A. METHYLASES OF *E. COLI* r-PROTEINS L11 AND L3

An enzyme which is probably specific for L11 methylation was partially purified and characterized.[115] In a mass screening, two mutants of *E. coli* lacking methylation of L11 and L3, respectively, were isolated.[88,89] Ribosomes from these mutants were shown to accept methyl groups specifically from different fractions of an *E. coli* wild-type extract chromatographed on DEAE-cellulose.[116] Moreover, the genes coding for these enzymes were mapped at minute 50 for L3 (gene *prm*B) and at minute 71 for L11 (gene *prm*A).[117] Absence of methylation at the single glutamine residue 150 of L3 could be easily understood as a deficiency of a single enzyme coded by gene *prm*B. Mutant *prm*A was proved to fully lack methylation at the three methylated residues of L11 (see Figure 1), as the result of a single mutation.[80,88] It is rather difficult to conclude that a single enzyme could perform triple methylation at lysine 3 and 39 as well as triple methylation at the N-terminal alanine on the same protein. However, as shown below, independent genetical observations suggest that it should be a single enzyme. First, mutation *prm*A was mapped at a single chromosomal locus.[117] Second, a "leaky" *prm*A mutant was isolated which displayed the same level of leakiness for methylation of both *N*-trimethylalanine and ε-*N*-trimethyllysine in L11. Third, two independent *prm*A mutants were isolated in a quite unexpected way, both fully lacking *N*-trimethylalanine and ε-*N*-trimethyllysine in L11.[100] A conclusive proof that a single enzyme methylates L11 will be obtained only when the gene is cloned and characterized. Finally, it should be noted that the enzymes which specifically methylate L11 and L3 are quite stable, at least at a purification level of about 200-fold.

B. OTHER r-PROTEIN METHYLASES OF *E. COLI*

The only other methylations of r-proteins of *E. coli* detected *in vitro* are the *O*-methylation of two 30S proteins[85] and that of factor EF-Tu.[91] In the case of *O*-methylation, the enzyme has a K_m value for *S*-adenosyl-L-methionine of 1.96×10^{-6} *M* and *S*-adenosyl-L-homocysteine was found to be a competitive inhibitor with a K_i value of 1.75×10^{-6} *M*. At least one more methylase must be present in *E. coli* to account for methylation at the N-terminal residue of L16, L33, S11, and IF-3, and for Lys 82 in L7—L12. A strain was constructed in our laboratory with the phenotype $Met^-RelA^-PrmA^-PrmB^-$. Many attempts to methylate ribosomal proteins of this strain *in vitro* were unsuccessful. Trivial explanations to this failure, i.e., lack of efficient methionine starvation *in vivo* or inadequate conditions of reaction *in vitro* are unlikely. A possible explanation is that methylation of these proteins occurs on nascent chains and cannot be mimicked *in vitro* so that fully synthesized polypeptides may not provide a primary or secondary structure of the nascent chain required for enzyme activity. This may particularly be true with L7—L12 where *in vivo* methylation is not stoichiometric and is temperature dependent.[92] Whether a nascent chain is required for methylation of these proteins could be tested *in vitro* using a cloned gene and a coupled system of protein synthesis and methylation. This has been performed recently of L11.[90] Another possibility is that *S*-adenosyl-L-methionine, used as methyl donor *in vitro,* is not the genuine methyl group donor for these proteins. This possibility has been suggested for alanine methylation,[65] by the fact that in *Bacillus subtilis* a tRNA methylase using N^5-methyltetrahydrofolate as methyl donor has been reported.[11]

C. r-PROTEIN METHYLASES IN YEAST

Among yeast r-proteins reported to be methylated, two belonging to the large subunit were demonstrated to be methylable *in vitro,* YL23 and YL32 in ribosomes from Sam^- yeast cultures starved from *S*-adenosyl-L-methionine.[84] Using this assay, two distinct enzymes were partially purified and characterized. Both were found to methylate lysine residue, one specifically produced ε-*N*-dimethyllysine in YL32 and the other ε-*N*-trimethyllysine in YL23. The enzyme for YL32 seemed to methylate lysine sequentially to ε-*N*-monomethyllysine and ε-*N*-dimethyllysine. Similar results were obtained by Durban et al.[119] in their study of the trimethylation of the lysine 72 of *Neurospora crassa* cytochrome *c*. Further studies on these enzymes were hindered because of their increased unstability upon purification.[120,121] It was observed that methylation of yeast nucleic acids[105] as well as proteins[84] *in vitro* was inhibited by a compound of moderately low molecular weight present in yeast crude extract. It could be removed by G-25 Sephadex filtration. This compound was not particularly specific for yeast methyltransferase reactions, since its action on *E. coli* L11 methylation was also observed.[84] Although its peptide nature has not yet been demonstrated, it is tempting to correlate its structure to that of the peptide inhibitor found in rabbit and rat liver.[122,123] In yeast and perhaps in other higher eukaryotes, an *in vivo* function of this molecule could be to participate in the transfer of *S*-adenosyl-L-methionine molecules across cellular compartments and to stock it away from the pool available for transmethylation reactions.[83]

D. r-PROTEIN METHYLASES IN HIGHER EUKARYOTES

In vitro methylation of r-proteins has been demonstrated in HeLa cells.[73] A good correlation was found in respect of the proteins most methylated *in vitro* and *in vivo*. However, when *S*-adenosyl-L-methionine was used in the *in vitro* assay instead of methionine, two of the three major methylated proteins were not labeled. In addition, the methylation *in vivo* of the third protein but not the two others was inhibited by cycloleucine. These results suggest that methylation of some r-proteins in eukaryotes might involve a methyl group donor other than *S*-adenosyl-L-methionine.

VI. THE ROLE OF r-PROTEIN METHYLATION

A. ROLE OF L11 METHYLATION

As protein L11 was found to be heavily methylated, one was tempted to attribute a significant role to this modification to the putative functions that this protein might play in the ribosome. These methylations were found to confer *in vitro* resistance of peptide bonds adjacent to methyl groups to trypsin digestion or even acid hydrolysis, as shown in Figure 1. Once thought to be essential for peptidyl transferase activity, L11 was found to be dispensable in ribosome. It was demonstrated that subparticles of *E. coli* lacking L11 had peptidyl transferase activity.[124,125] In addition, viable mutants lacking *E. coli* protein L11 or its equivalent in *Bacillus megaterium* have been isolated.[126,127] The mutant *prm*A1, fully lacking methylation of *E. coli* L11 was isolated with the purpose to identify any phenotype trait resulting from absence of methylation. It was isolated under conditions allowing a conditionally lethal phenotype, i.e., methylation activity at 25°C but not at 37°C. However, *prm*A1 was not temperature sensitive even though methylation of L11 is fully lacking.[88] Moreover, no phenotypic trait different from an isogenic wild-type strain was ever observed. This was reviewed briefly elsewhere.[110,116] No defect could have been in the stringent control of RNA synthesis, since L11 was demonstrated to be one of the partners in this intricate process in both *E. coli* and *B. megaterium*.[128,129] In particular, a mutant with a relaxed phenotype (RelC) was found to be affected in the primary structure of L11.[128] We have found that strains with *prm*A1 mutation does not display some of the typical phenotypes of a relaxed mutant.[100] In addition, it was shown that mutant *prm*A1 was not affected in the synthesis of ppGpp, the signal molecule which triggers the stringent response.[130] The target of the antibiotic thiostrepton is protein L11. Ribosome lacking this protein had thiostrepton resistant protein synthesis either *in vivo* in *B. megaterium*,[126] or *in vitro* in *E. coli*. A reasonable hypothesis was that *in vitro* protein synthesis by *prm*A1 ribosomes might be resistant to thiostrepton, however, it was not the case.

A subtle and somewhat irrational function of L11 methylation was found in a "crooked" way. Dabbs[131] isolated a large number of *E. coli* "revertants" from drug dependence to independence. Some of them displayed an alteration in the electrophoretic mobility of L11 while not being affected in the structural gene of this protein. It was shown that two of the "revertants" lacked L11 methylation and were actually independent mutants in gene *prm*A.[116] Thus, omission of methylation of L11 can reconstitute a functional ribosome to mutant strains which became previously dependent for protein synthesis in the presence of an antibiotic (streptomycin or kasugamycin). Although interesting, this role of absence of L11 methylation is not likely to play a significant function in a natural environment.

B. METHYLATION OF L3 ACTS AS A RIBOSOME ASSEMBLY FACTOR IN *E. COLI*

N^5-Methylglutamine, a methylated amino acid not found in any other protein, has been shown to be present in protein L3 and located at position 150.[67,89] This methylation constitutes the only modification of this protein. L3 is a protein essential for a correct assembly of the ribosome and is known to interact with rRNA early in the assembly.[132] L3 has also been suggested to be involved in the elongation step of protein synthesis, especially in the formation of the peptidyl transferase site.[133] A mutant of *E. coli*, *prm*B2,[89] lacks the methylase activity which modifies specifically the glutamine residue at position 150 of protein L3. The mutation *prm*B was mapped at minute 50, very close to the gene *aro*C.[117] No gene coding for a ribosomal component has been detected in this region of the map. The *prm*B2 mutant displays a cold-sensitive phenotype: it grows at a significantly lower rate than wild-type at low temperature.[134] At 22°C, the mutant cells synthesize abnormal ribosomal particles. These are very unstable and are quickly degraded. However, once assembled, ribosomes of this mutant are stable and fully active in protein synthesis either *in vivo* or *in vitro*. This strongly indicates that methylation of L3 is involved in ribosome assembly but has no detectable effect on ribosome stability or in various aspects of protein synthesis. Thus, the methylase of L3 would qualify as an enzymatic assembly factor as postulated by the isolation of cold-sensitive mutants.[11] Methylation of L3 should play a role very early in the ribosome assembly since abnormal particles of both subunits accumulate in the *prm*B 2 mutant. This is not surprising for the following reasons. The rRNA operons are transcribed in the order 16S-23S-5S and association of proteins with rRNA begins already during transcription.[135-137] Early steps in assembly are common to the 30S and the 50S subunits.[8,138] A mutant whose protein L22 is impaired accumulated particles similar to those formed by the *prm*B 2 mutant.[12] L3 and L22 are among the first proteins which interact with the 23S RNA *in vivo*.[132] Moreover, both proteins contribute to alter the conformation of an intermediate particle *in vitro*.[139] Branlant et al.[140] suggested that protein L3 interacts with the 23S RNA near nucleotide 2000 but has a secondary binding site close to the 5′ extremity where L22 binds specifically. It seems that L3 interacts with L22.[140]

In vitro methylation of L3 is inefficient when free proteins or mature intact 50S subunits are used as substrates. In contrast, addition of RNA to free proteins stimulates the incorporation of methyl groups.[134] This supports the idea that methylation of L3 does not occur *in vivo* on free proteins nor on completed ribosomal subunits, but on a rRNA.L3 complex accessible to the methylase early during the ribosome assembly.

C. METHYLATION OF OTHER r-PROTEINS

Little progress, if any, has been made in the study of the role of methylation in the case of most of the r-proteins. However, some results have been obtained about the regulation of this modification and about the time of its appearance during the ribosome assembly.

1. Methylation of L7 and L12 is Temperature Dependent in *E. coli*

Chang[92] observed that methylation of L7 and L12 in *E. coli* was dependent upon the cell growth temperature. Lowering the temperature during cell growth led to a higher extent of methylation. Independent of the growth phase, cells grown at 25°C contained approximately 0.75 molecule of ε-*N*-monomethyllysine in both proteins while very little methylation was found in either protein when the cells were grown at 37°C.[92,141] After a temperature shift-down, proteins L7 and L12 of the only newly synthesized ribosomes were methylated. On the other hand, the possibility that L7-L12 are methylated at the free stage might explain why they are better methylated at low temperature, since they are incorporated into the ribosome at a lower rate.[142] These results support the assumption that the methylation might be required either to protect these proteins from degradation by proteolytic enzymes or to render the ribosomes more efficient at lower temperatures.

2. Methylation in HeLa Cells

In HeLa cells, the methylation of the 40S subunit occurs heavily in the late G1 phase whereas methylation in the 60S subunit is most pronounced in the early S phase.[74] The levels of methylation differ among several proteins in relation to suppression of protein synthesis or ribosome formation.[112] Some of the modifications occur during ribosome processing while others take place in mature ribosomes. On the other hand, the r-proteins which appeared to be methylated in nascent ribosomes contained methyllysine while methylarginine was found in those methylated in mature ribosomes. These results could only suggest that some types of methylation are not essential for the ribosome assembly and would rather play a role in the protein synthesis or its regulation.

3. Methylation of Elongation Factors EF-Tu and EF-1α

The significance of the methylation of EF-Tu is yet unexplained. Still, the posttranslational modification of this elongation factor appears universal and well conserved during evolution.

The methylation of the homologous factor EF-1α in *Mucor racemosus* has been shown to be highly regulated.[107,108] *Mucor racemosus* is a dimorphic fungus which grows as a budding yeast in CO_2 and as a mycelium in air. EF-1α is less methylated in the yeast form than in the mycelium form while spores contain EF-1α that is essentially unmethylated. On the other hand, the EF-1α methylation activity rises six times during spore germination and in parallel with the extent of methylation. However, the EF-1α mRNA level or the relative amount of EF-1α compared to total protein do not change significantly during this period. Therefore, Fonzi et al.[108] made a tentative conclusion that methylation could play a role in increasing EF-1α activity. The methylation of EF-1α in mouse 3T3B is also enhanced when cells are transformed by SV40.[79] Whether this change is concomitant with viral transformation itself or is related to the increased rate of cell division associated with the viral transformation remains still unclear. The elongation factor EF-1α in eukaryotes, homologous to the factor EF-Tu in prokaryotes, also interacts with a variety of substrates including GTP, aminoacyl-tRNAs, and ribosomes and it has a high affinity for nucleic acids. The various functions in which it is implicated make it difficult to find out precisely at which level methylation might play a role.

VII. CONCLUSION

The reader might have noted that most publications cited herein are rather old. There was a burst of papers in the 1970s describing methylated proteins and amino acids in the ribosome of *Escherichia coli* and other organisms. However, little has been published to pursue to explain why these proteins are methylated. In some instances (e.g., L7-L12 in *E. coli* and EF-1α in *Mucor*), correlations between the extent of methylation and growth conditions or growth stage have been made. More studies of their type could be done, especially with higher eukaryotic systems where cells of various lineages from the same organism are available, or with cells subjected to various stress, e.g., viral infection or heat shock. Comparing the behavior, e.g., in protein synthesis of ribosomes from cells deprived from the methyl group donor with those from normal cells cannot lead to unequivocal interpretations. Undermethylated RNA and protein components present in the same ribosome constitute a hodge podge from which a single ingredient cannot be smelled.

Sorting out a single protein lacking methylation can be done most effectively by isolating a mutant deficient of the corresponding methylase. This approach has been rewarding with mutant *prm*B2; the absence of a single methyl group at residue 150 of L3 could perturb the ribosome assembly of *E. coli*. However, how this absence of a single methyl group on the glutamine 150 affects ribosome assembly has not been worked out in molecular terms, e.g., in terms of protein-RNA interaction.

Mutant *prm*A1 was isolated to investigate whether *E. coli* protein L11 methylation had a significant function. None was yet found. It is quite possible that the significance of this protein methylation eludes our comprehension because, for example, the conditions required for an effect of the absence of L11 methylation to be detected are not met in usual laboratory conditions.

Ribosomal protein methylation is widespread among living cells and organisms and probably does not exercise a single function. Even if these functions remain uncovered in most cases, one cannot refrain from postulating that they must exist since the life processes are supposedly energetically very efficient, and retained these energy-spending mechanisms across evolution.

ACKNOWLEDGMENTS

Work in our laboratory was supported by grants n°2.455.76 and n°2.4537.80 from the Belgian Fonds de la Recherche Scientifique (FNRS).

REFERENCES

1. **Hardesty, B. and Kramer, G.,** *Structure, Function and Genetics of Ribosomes,* Springer-Verlag, Heidelberg, 1986.
2. **Hardy, S. J. S.,** The stoichiometry of the ribosomal proteins of *Escherichia coli, Mol. Gen. Genet.,* 140, 253, 1975.
3. **Traub, P. and Nomura, M.,** Structure and function of *E. coli* ribosomes. V. Reconstitution of functionally active 30S ribosomal particles from RNA and proteins, *Proc. Natl. Acad. Sci. U.S.A.,* 59, 777, 1968.
4. **Held, W. A., Ballou, B., Mizushima, S., and Nomura, M.,** Assembly mapping of ribosomal proteins from *E. coli:* further studies, *J. Biol. Chem.,* 249, 3103, 1974.
5. **Nomura, M. and Erdmann, V. A.,** Reconstitution of 50S ribosomal subunits from dissociated molecular components, *Nature (London),* 228, 744, 1970.
6. **Nierhaus, K. H. and Dohme, F.,** Total reconstitution of functionally active 50S ribosomal subunits from *Escherichia coli, Proc. Natl. Acad. Sci. U.S.A.,* 71, 4713, 1974.
7. **Lee, J. C. and Anderson, R.,** Partial reassembly of yeast 60S ribosomal subunits *in vitro* following controlled dissociation under nondenaturing conditions, *Arch. Biochem. Biophys.,* 245, 248, 1986.
8. **Schlessinger, D.,** Ribosome formation in *Escherichia coli,* in *Ribosomes,* Nomura, M., Tissières, A., and Lengyel, P., Eds., Cold Spring Harbor Laboratory, Cold Spring Harbor, NY, 1974, 393.
9. **Guthrie, C., Nashimoto, H., and Nomura, M.,** Sturcture and function of *E. coli* ribosomes. VIII. Cold-sensitive mutants defective in ribosome assembly, *Genetics,* 384, 63, 1969.
10. **Nashimoto, H., Held, W., Kaltschmidt, I., and Nomura, M.,** Structure and function of bacterial ribosomes. XII. Accumulation of 21S particles by some cold-sensitive mutants of *Escherichia coli, J. Mol. Biol.,* 62, 121, 1971.
11. **Bryant, R. E. and Sypherd, P. S.,** Genetic analysis of cold-sensitive ribosome maturation mutants of *Escherichia coli, J. Bacteriol.,* 11, 1082, 1974.
12. **Pardo, D., Vola, C., and Rosset, R.,** Assembly of ribosomal subunits affected in a ribosomal mutant of *Escherichia coli* having an altered L22 protein, *Mol. Gen. Genet.,* 174, 53, 1979.
13. **Russel, P. J., Granville, R. R., and Tublitz, N. J.,** A cold-sensitive mutant of *Neurospora crassa* obtained using tritium-suicide enrichment that is conditionally defective in the biosynthesis of cytoplasmic ribosomes, *Exp. Mycol.,* 4, 23, 1980.
14. **Brimacombe, R. and Stiege, W.,** Structure and function of ribosomal RNA, *Biochem. J.,* 229, 1, 1985.
15. **Khan, M. S. N., Salim, M., and Maden, B. E. H.,** Extensive homologies between the methylated nucleotide sequences in several vertebrate rRNAs, *Biochem. J.,* 169, 531, 1978.
16. **Klootwijk, J., Van Den Bos, R. C., and Planta, R. J.,** Secondary methylation of yeast ribosomal RNA, *FEBS Lett.,* 27, 102, 1972.
17. **Maden, B. E. H. and Salim, M.,** The methylated nucleotide sequences in HeLa cell ribosomal RNA and its precursors, *J. Mol. Biol.,* 88, 133, 1974.
18. **Alix, J. H.,** Relationship between methylation and maturation of ribosomal RNA in prokaryotic and eukaryotic cells, in *Biological Methylation and Drug Design,* Borchardt, R.T., Creveling, C.R., and Ueland, R.M., Eds., Humana Press, Clifton, NJ, 1986, 175.
19. **Grummt, I.,** The effects of histidine starvation on the methylation of ribosomal RNA, *Eur. J.Biochem.,* 79, 133, 1977.
20. **Brand, R. C., Klootwijk, J., Van Steenbergen, R. J. M., De Kok, A. J., and Planta, R. J.,** Secondary methylation of yeast ribosomal precursor RNA, *Eur. J. Biochem.,* 75, 311, 1977.
21. **Fellner, P.,** Nucleotide sequence from specific areas of the 16S and 23S RNAs of *E. coli, Eur. J. Biochem.,* 11, 12, 1969.
22. **De Jonge, P., Klootwijk, J., and Planta, R. J.,** Sequence of the 3′ terminal 21 nucleotides of yeast 17S rRNA, *Nucleic Acids Res.,* 4, 3655, 1977.
23. **Choi, Y. C. and Busch, H.,** Modified nucleotides in T1 RNase oligonucleotides of 18S ribosomal RNA of the Novikoff hepatoma, *Biochemistry,* 17, 2551, 1978.
24. **Chelbi-Alix, M. K., Expert-Bezancon, A., Hayes, F., Alix, J. H., and Branlant, C.,** Properties of ribosomes and ribosomal RNAs synthesized by *Escherichia coli* grown in the presence of ethionine. Normal maturation of ribosomal RNA in the absence of methylation, *Eur. J. Biochem.,* 115, 627, 1981.
25. **Björk, G. R. and Isaksson, L. A.,** Isolation of mutants of *Escherichia coli* lacking 5-methyluracil in transfer ribonucleic acid or 1-methylguanine in ribosomal RNA, *J. Mol. Biol.,* 51, 83, 1970.
26. **Helser, T. L., Davies, J. E., and Dahlberg, J. E.,** Mechanism of kasugamycin resistance in *Escherichia coli, Nature (London),* 235, 6, 1972.
27. **Poldermans, B., Van Buul, C. P. J. J., and Van Knippenberg, P. H.,** Studies on the function of two adjacent N^G,N^G-dimethyladenosines near the 3′ end of 16S ribosomal RNA of *Escherichia coli.* II. The effect of the absence of the methyl groups on initiation of protein biosynthesis, *J. Biol. Chem.,* 254, 9090, 1979.

28. **Van Buul, C. P. J. J., Damm, J. B. L., and Van Knippenberg, P. H.,** Kasugamycin resistant mutants of *Bacillus stearothermophilus* lacking the enzyme for the methylation of two adjacent adenosines in 16S ribosomal RNA, *Mol. Gen. Genet.*, 189, 475, 1983.
29. **Thompson, J., Schmidt, F., and Cundliffe, E.,** Site of action of a ribosomal RNA methylase conferring resistance to thiostrepton, *J. Biol. Chem.*, 257, 7915, 1982.
30. **Skinner, R., Cundilffe, E., and Schmidt, F.J.,** Site of action of a ribosomal RNA methylase responsible for resistance to erythromycin and other antibiotics, *J. Biol. Chem.*, 258, 12702, 1983.
31. **Piendl, W. and Böck, A.,** Ribosomal resistance in the gentamicin producer organism *Micromonospora purpurea, Antimicrob. Agents Chemother.*, 22, 231, 1982.
32. **Thompson, J., Skeggs, P. A., and Cundliffe, E.,** Methylation of 16S ribosomal RNA and resistance to the aminoglycoside antibiotics gentamicin and kanamycin determined by DNA from the gentamicin-producer, *Micromonospora purpurea, Mol. Gen. Genet.*, 201, 168, 1985.
33. **Hardy, S. J., Kurland, C. G., Voynow, P., and Mora, G.,** The ribosomal proteins of *Escherichia coli*. I. Purification of the 30S ribosomal proteins, *Biochemistry*, 8, 2897, 1969.
34. **Kaltschmidt, E. and Wittmann, H. G.,** Ribosomal proteins. VII. Two dimensional gel electrophoresis for fingerprinting of ribosomal proteins, *Anal. Biochem.*, 36, 401, 1970.
35. **McConkey, E. H., Bielka, H., Gordon, J., Lastick, S. J., Lin, A., Ogata, K., Reboud, J. P., Traugh, J. A., Traut, R. R., Warner, J. R., Welfle, H., and Wool, I. G.,** Proposed uniform nomenclature for mammalian ribosomal proteins, *Mol. Gen. Genet.*, 169, 1, 1979.
36. **Bielka, H., Stahl, J., Bommer, U. A., Welfle, H., Noll, F., and Westermann, P.,** *The Eukaryotic Ribosome*, Bielka, H., Ed., Springer-Verlag, Heidelberg, 1982.
37. **Otaka, E. and Kobata, K.,** Yeast ribosomal proteins. I. Characterization of cytoplasmic ribosomal proteins by two dimensional gel electrophoresis, *Mol. Gen. Genet.*, 162, 259, 1978.
38. **Bollen, G. H. P. M., Mager, W. H., and Planta, R. J.,** High resolution mini-two-dimensional gel electrophoresisof yeast ribosomal proteins, *Mol. Biol. Rep.*, 8, 367, 1981.
39. **Otaka, E. and Osawa, S.,** Yeast ribosomal proteins. Correlation of several nomenclatures and proposal of a standard nomenclature, *Mol. Gen. Genet.*, 181, 176, 1981.
40. **Geisser, M., Tischendorf, G. W., Stoffler, G., and Wittmann, H. G.,** Immunological and electrophoretical comparison of ribosomal proteins from eight species belonging to Enterobacteriaceae, *Mol. Gen. Genet.*, 127, 11, 1973.
41. **Leader, D. P. and Coia, A. A.,** The acidic ribosomal phosphoprotein of eukaryotes and its relationship to ribosomal proteins L7 and L12 of *Escherichia coli, Biochem. J.*, 176, 569, 1978.
42. **Chooi, W. Y. and Otaka, E.,** Immunological evidence for structural homology between *Drosophila melanogaster* (S14), rabbit liver (S12), *Saccharomyces cerevisiae* (S25), *Bacillus subtilis* (S6) and *Escherichia coli* (S6) ribosomal proteins, *Mol. Cell. Biol.*, 4, 2535, 1984.
43. **Sudarickov, A. B. and Surguchov, A. P.,** Immunologically related proteins in cytoplasmic and mitochondrial ribisomes of yeast *Saccharomyces cerevisiae, Mol. Gen. Genet.*, 203, 316, 1986.
44. **Tanaka, T., Ishikawa, K., and Ogata, K.,** On the sequence homology of the ribosomal proteins, *Escherichia coli* S11, yeast r-protein 59 and Chinese Hamster S14, *FEBS Lett.*,202, 295, 1986.
45. **Stark, M. J. R., Cundliffe, E., Dijk, J., and Stoffler, G.,** Functional homology between *E. coli* ribosomal protein L11 and *B. megaterium* protein BM-L11, *Mol. Gen. Genet.*, 180, 11, 1980.
46. **Landers, T. A., Blumenthal, T., and Weber, K.,** Function and structure in ribonucleic acid phage Qβ ribonucleic acid replicase, *J. Biol. Chem.*, 249, 5801, 1974.
47. **Travers, A.,** Control of ribosomal RNA synthesis *in vitro, Nature (London)*, 244, 15, 1973.
48. **Jacobson, G. R. and Rosenbusch, J. P.,** Abundance and membrane association of elongation factor Tu in *E. coli, Nature (London)*, 261, 23, 1976.
49. **Amons, R., Pluijms, W., Roobol, K., and Moler, W.,** Sequence homology between EF-1α, the α-chain of elongation factor 1 from *Artemia salina* and elongation factor EF-Tu from *Escherichia coli, FEBS Lett.*, 153, 37, 1983.
50. **Terhorst, C., Möller, W., Laursen, R., and Wittmann-Liebold, B.,** The primary structure of an acidic protein from 50S ribosomes of *Escherichia coli* which is involved in GTP hydrolysis dependent on elongation factor G and T., *Eur. J. Biochem.*, 34, 138, 1973.
51. **Wittmann-Liebold, B. and Creuer, B.,** The primary structure of protein S5 from the small subunit of the *Escherichia coli* ribosome, *FEBS Lett.*, 95, 91, 1978.
52. **Yaguchi, M.,** Primary structure of protein S18 from the small *Escherichia coli* ribosomal subunit, *FEBS Lett.*, 59, 217, 1975.
53. **Isono, S. and Isono, K.,** Ribosomal protein modification in *Escherichia coli*. III. Studies of mutants lacking an acetylase activity specific for protein L12, *Mol. Gen. Genet.*,183, 473, 1981.
54. **Cumberlidge, A. G. and Isono, K.,** Ribosomal protein modification in *Escherichia coli*. I. A mutant lacking the N-terminal acetylation of protein S5 exhibits thermosensitivity, *J. Mol. Biol.*, 131, 169, 1979.

55. **Isono, K. and Isono, S.,** Ribosomal protein modification in *Escherichia coli.* II. Studies of a mutant lacking the N-terminal acetylation of protein S18, *Mol. Gen. Genet.,* 177, 645, 1980.
56. **Liew, C. C. and Yip, C. C.,** Acetylation of reticulocyte ribosomal proteins at time of protein biosynthesis, *Proc. Natl. Acad. Sci. U.S.A.,* 71, 2988, 1974.
57. **Reeh, S. and Pedersen, S.,** Post-translational modification of *Escherichia coli* ribosomal protein S6, *Mol. Gen. Genet.,* 173, 183, 1979.
58. **Hitz, H., Schäfer, D., and Wittmann-Liebold, B.,** Primary structure of ribosomal protein S6 from the wild type and a mutant of *E. coli, FEBS Lett.,* 56, 259, 1975.
59. **Gressner, A. M. and Wool, I. G.,** Influence of glucagon and cyclic adenosine 3′:5′-monophosphate on the phosphorylation of rat liver ribosomal protein S6, *J. Biol. Chem.,* 251, 1500, 1976.
60. **Gressner, A. M. and Wool, I. G.,** The stimulation of the phosphorylation of ribosomal protein S6 by cycloheximide and puromycin, *Biochem. Biophys. Res. Commun.,* 60, 1481, 1974.
61. **Kaerlein, M. and Horak, I.,** Phosphorylation of ribosomal proteins in HeLa cells infected with vaccinia virus, *Nature (London),* 259, 150, 1976.
62. **Scharf, K. D. and Nover, L.,** Heat shock induced alterations of ribosomal protein phosphorylation in plant cell cultures, *Cell,* 30, 427, 1982.
63. **Kruse, C., Johnson, S. P., and Warner, J. R.,** Phosphorylation of the yeast equivalent of ribosomal protein S6 is not essential for growth, *Proc. Natl. Acad. Sci. U.S.A.,* 82, 7515, 1985.
64. **Chen, R., Brosius, J., Wittmann-Liebold, B., and Schäfer, W.,** Occurrence of methylated amino acids as N-termini of proteins from *Escherichia coli* ribosomes, *J. Mol. Biol.,* 11, 173, 1977.
65. **Lederer, F., Alix, J. H., and Hayes, D.,** *N*-trimethylalanine, a novel blocking group, found in *E. coli* ribosome protein L11, *Biochem. Biophys. Res. Commun.,* 77, 470, 1977.
66. **Dognin, M. J. and Wittmann-Liebold, B.,** Identification of methylated amino acids during sequence analysis. Application to the *Escherichia coli* ribosomal protein L11, *Hoppe-Seyler's Z. Physiol. Chem.,* 361, 1697, 1980.
67. **Muranova, T. A., Muranov, A. V., Markova, L. F., and Ovchinnikov, Y. A.,** The primary structure of ribosomal protein L3 from *Escherichia coli* 70S ribosomes, *FEBS Lett.,* 96, 301, 1978.
68. **Arai, K., Clark, B. F. C., Duffy, L., Jones, M. D., Kaziro, Y., Laursen, R. A., L'Italien, J., Miller, D. L., Nagarkatti, S., Nakamura, S., Nielsen, K. M., Petersen, T. E., Takahashi, K., and Wade, M.,** Primary structure of elongation factor Tu from *Escherichia coli, Proc. Natl. Acad. Sci. U.S.A.,* 77, 1326, 1980.
69. **Chang, F. N., Chang, C. N., and Paik, W. K.,** Methylation of ribosomal proteins in *Escherichia coli, J. Bacteriol.,* 120, 651, 1974.
70. **Cannon, M., Schindler, D., and Davies, J.,** Methylation of proteins in 60S ribosomal subunits from *Saccharomyces cerevisiae, FEBS Lett.,* 75, 187, 1977.
71. **Hernandez, F., Cannon, M., and Davies, J.,** Methylation of proteins in 40S ribosomal subunits from *Saccharomyces cerevisiae, FEBS Lett.,* 89, 271, 1978.
72. **Kruiswijk, T., Kunst, A., Planta, R. J., and Mager, W. H.,** Modification of yeast ribosomal proteins. Methylation, *Biochem. J.,* 175, 221, 1978.
73. **Scolnik, P. A. and Eliceiri, G. L.,** Methylation sites in HeLa cell ribosomal proteins, *Eur. J. Biochem.,* 101, 93, 1979.
74. **Chang, F. N., Navickas, I. J., Au, C., and Budzilowicz, C.,** Identification of the methylated ribosomal proteins in HeLa cells and the fluctuation of methylation during the cell cycle, *Biochim. Biophys. Acta,* 518, 89, 1978.
75. **Cyrne, M. L., Pousada, C. R., and Hayes, D.,** Methylation of ribosomal proteins in *Tetrahymena pyriformis, Biochimie,* 63, 641, 1981.
76. **Chang, C. N. and Chang, F. N.,** Methylation of the ribosomal proteins in *Escherichia coli.* Nature and stoichiometry of the methylated amino acids in 50S ribosomal proteins, *Biochemistry,* 14, 468, 1975.
77. **Klagsbrun, M. and Furano, A. V.,** Methylated amino acids in the proteins of bacterial and mammalian cells, *Arch. Biochem. Biophys.,* 169, 529, 1975.
78. **Chang, F. N., Navickas, I. J., Chang, C. N., and Dancis, B. N.,** Methylation of ribosomal proteins in HeLa cells, *Arch. Biochem, Biophys.,* 172, 627, 1976.
79. **Coppard, N. J., Clark, B. F. C., and Cramer, F.,** Methylation of elongation factor 1α in mouse 3T3B and 3T3B/SV40 cells, *FEBS Lett.,* 164, 330, 1983.
80. **Alix, J. H., Hayes, D., Lontie, J. F., Colson, C., Glatigny, A., and Lederer, F.,** Methylated amino acids in ribosomal proteins from *Escherichia coli* treated with ethionine and from a mutant lacking methylation of protein L11, *Biochimie,* 61, 671, 1979.
81. **Alix, J. H. and Hayes, D.,** Properties of ribosomes and RNA synthesized by *Escherichia coli* grown in the presence of ethionine. III. Methylated proteins in 50S ribosomes of *E. coli* EA2, *J. Mol. Biol.,* 86, 139, 1974.
82. **Ferro-Luzzi Ames, G. and Niakido, K.,** *In vivo* methylation of prokaryotic elongation factor Tu, *J. Biol. Chem.,* 254, 9947, 1979.
83. **Nakamura, K. D. and Schlenk, F.,** Active transport of exogenous S-adenosylmethionine and related compounds into cells and vacuoles of *Saccharomyces cerevisiae, J. Bacteriol.,* 120, 482, 1974.

84. **Lhoest, J., Lobet, Y., Costers, E., and Colson, C.,** Methylated proteins and amino acids in the ribosomes of *Saccharomyces cerevisiae, Eur. J. Biochem.,* 141, 585, 1984.
85. **Kim, S., Lew, B., and Chang, F. N.,** Enzymatic methyl esterification of *Escherichia coli* ribosomal proteins, *J. Bacteriol.,* 130, 839, 1977.
86. **Cherest, H., Surdin-Kerjan, Y., Exinger, F., and Lacroute, F.,** S-adenosyl methionine requiring mutants in *Saccharomyces cerevisiae:* evidences for the existence of two methionine adenosyl transferases, *Mol. Gen. Genet.,* 163, 153, 1978.
87. **Chang, C. N. and Chang, F. N.,** Methylation of ribosomal proteins *in vitro, Nature (London),* 251, 731, 1974.
88. **Colson, C. and Smith, H. O.,** Genetics of ribosomal protein methylation in *Escherichia coli.* I. A mutant deficient in methylation of protein L11, *Mol. Gen. Genet.,* 154, 167, 1977.
89. **Lhoest, J. and Colson, C.,** Genetics of ribosomal protein methylation in *Escherichia coli.* II. A mutant lacking a new type of methylated amino acid, N^5-methylglutamine, in protein L3, *Mol. Gen. Genet.,* 154, 175, 1977.
90. **Jerez, C. and Weissbach, H.,** Methylation of newly synthesized ribosomal protein L11 in a DNA-directed *in vitro* system, *J. Biol. Chem.,* 255, 8706, 1980.
91. **Toledo, H. and Jerez, C. A.,** *In vitro* methylation of the elongation factor EF-Tu from *Escherichia coli, FEBS Lett.,* 193, 17, 1985.
92. **Chang, F.N.,** Temperature-dependent variation in the extent of methylation of ribosomal proteins L7 and L12 in *Escherichia coli, J. Bacteriol.,* 135, 1165, 1978.
93. **Chang, C. N., Schwartz, M., and Chang, F. N.,** Identification and characterization of a new methylated amino acid in ribosomal protein L33 of *Escherichia coli, Biochem. Biophys. Res. Commun.,* 73, 233, 1976.
94. **Chang, F. N. and Budzilowicz, C.,** Characterization of methylated neutral amino acid from *Escherichia coli* ribosomes, *J. Bacteriol.,* 131, 105, 1977.
95. **Chen, R. and Chen-Schmeisser, U.,** Isopeptide linkage between *N*-α-monomethylalanine and lysine in ribosomal protein S11 from *Escherichia coli, Proc. Natl. Acad. Sci. U.S.A.,* 74, 4905, 1977.
96. **Chang, J. Y.,** A novel Edman-type degradation: direct formation of the thiohydantoin ring in alkaline solution by reation of Edman-type reagents with *N*-monomethyl amino acids, *FEBS Lett.,* 91, 63, 1978.
97. **Kamp, R. and Wittmann-Liebold, B.,** Primary structure of protein S11 from *Escherichia coli* ribosomes, *FEBS Lett.,* 121, 117, 1980.
98. **Dognin, M. J. and Wittmann-Liebold, B.,** The primary structure of L11, the most heavily methylated protein from *Escherichia coli* ribosomes, *FEBS Lett.,* 84, 342, 1977.
99. **Brauer, D. and Wittmann-Liebold, B.,**The primary structure of the initiation factor IF-3ℓ from *Escherichia coli, FEBS Lett.,* 79, 269, 1977.
100. **Lhoest, J., Hespel, F., Lontie, J. F., Andrade, E., Digneffe, C., Colson, C., and Dabbs, E.,** Why is ribosomal protein L11 of *Escherichia coli* methylated?, in *Biochemistry of S-Adenosylmethionine and Related Compounds,* Usdin, E., Borchardt, R. T., and Creveling, C. R., Eds., Macmillan, London, 1982, 79.
101. **Cannon, M. and Cundliffe, E.,** Methylation of basic proteins in ribosomes from wild-type and thiostrepton-resistant strains of *Bacillus megaterium* and their electrophoretic analysis, *Eur. J.Biochem.,* 97, 541, 1979.
102. **Mardones, E., Amaro, A. M., and Jerez, C. A.,** Methylation of ribosomal proteins in *Bacillus subtilis, J. Bacteriol.,* 142, 355, 1980.
103. **Amaro, A. M. and Jerez, C. A.,** Methylation of ribosomal proteins in bacteria: evidence of conserved modification of the Eubacterial 50S subunit, *J. Bacteriol.,* 158, 84, 1984.
104. **Juan-Vidales, F., Sanches Madrid, F., Saenz-Robles, M. T., and Ballesta, J.P.G.,** Purification and characterization of two ribisomal proteins of *Saccharomyces cerevisiae.* Homologies with proteins from eukaryotic species and with bacterial protein EC L11, *Eur. J.Biochem.,* 136, 275, 1983.
105. **Locht, C., Beaudart, J. L., and Delcour, J.,** Partial purification and characterization of mRNA (guanine-7-) methyltransferase from the yeast *Saccharomyces cerevisiae, Eur. J. Biochem.,* 134, 117, 1983.
106. **Comb, D. G., Sarkar, N., and Pinzino, C. J.,** The methylation of lysine residues in protein, *J.Biol. Chem.,* 241, 1857, 1966.
107. **Hiatt, W. R., Garcia, R., Merrick, W. C., and Sypherd, P. S.,** Methylation of elongation factor 1α from the fungus *Mucor, Proc. Natl. Acad. Sci. U.S.A.,* 79, 3433, 1982.
108. **Fonzi, W. A., Katayama, C., Leathers, T., and Sypherd, P. S.,** Regulation of protein synthesis factor EF-1α in *Mucor racemosus, Mol. Cell. Biol.,* 5, 1100, 1985.
109. **Reporter, M.,** Methylation of basic residues in structural proteins, *Mech. Age Adv.,* 1, 114, 1973.
110. **Reporter, M.,** Methylation of structural and total ribosomal proteins of cultured muscle cells, *Fed. Proc.,* 33, 1584, 1974.
111. **Perisic, O. and Kanazir, D.,** *In vivo* methylation of rat liver ribosomal proteins, *Biochem. Soc. Trans. (England),* 9, 203, 1981.
112. **Goldenberg, C. J. and Eliceiri, G. L.,** Methylation of ribosomal proteins in HeLa cells, *Biochim. Biophys. Acta,* 479, 220, 1977.
113. **Amons, R., Pliujms, W., Roobol, K., and Möller, W.,** Sequence homology between EF-1α, the α-chain of elongation factor 1 from *Artemia salina* and elongation factor EF-Tu from *Escherichia coli, FEBS Lett.,* 153, 37, 1983.

114. **Kamp, R. M., Srinivasa, B. R., von Knoblauch, K., and Subramanian, A. R.,** Occurrence of a methylated protein in chloroplast ribosomes, *Biochemistry,* 26, 5866, 1987.
115. **Chang, F. N., Cohen, L. B., Navickas, I. J., and Chang, C. N.,** Purification and properties of a ribosomal protein methylase from *Escherichia coli* Q13, *Biochemistry,* 14, 4994, 1975.
116. **Colson, C. and Lhoest, J.,** Methylation of ribosomal proteins L3 and L11 in *Escherichia coli,* in *Biochemistry of S-Adenosylmethionine and Related Compounds,* Usdin, E., Borchardt, R. T., and Creveling, C. R., Eds., Macmillan, London, 1982, 79.
117. **Colson, C., Lhoest, J., and Urlings, C.,** Genetics of ribosomal protein methylation in *Escherichia coli.* III. Map position of two genes, *prm*A and *prm*B, governing methylation of proteins L11 and L3, *Mol. Gen. Genet.,* 169, 245, 1979.
118. **Kersten, H., Sandig, L., and Arnold, H. H.,** Tetrahydrofolate-dependent 5-methyluracil-tRNA transferase activity in *Bacillus subtilis, FEBS Lett.,* 55, 57, 1975.
119. **Durban, E., Nochumson, S., Kim, S., Paik, W. K., and Chan, S. K.,** Cytochrome *c*-specific protein-lysine methyltransferase from *Neurospora crassa, J. Biol. Chem.,* 253, 1427, 1978.
120. **Lhoest, J., Lobet, Y., and Colson, C.,** Characterization of a specific ribosomal protein methylase of *Saccharomyces cerevisiae, 15th FEBS Meet.,* (Abstr.), 251, 1983
121. **Lobet, Y., Lhoest, J., and Colson, C.,** manuscript in preparation.
122. **Lyon, E. S., McPhie, P., and Jakoby, W. B.,** Methinin: a peptide inhibitor of methylation, *Biochem. Biophys, Res. Commun.,* 108, 846, 1982.
123. **Hong, S. Y., Lee, H. W., Disa, S., Kim, S., and Paik, W. K.,** Studies on naturally occurring proteinous inhibitor for transmethylation reactions, *Eur. J. Biochem.,* 156, 79, 1986.
124. **Ballesta, J. P. G. and Vazquez, D.,** Activities of ribosomal cores deprived of proteins L7, L10, L11 and L12, *FEBS Lett.,* 48, 266, 1974.
125. **Howard, G. A. and Gordon, J.,** Peptidyl transferase activity of ribosomal particles lacking protein L11, *FEBS Lett.,* 48, 271, 1974.
126. **Cundliffe, E., Dixon, P., Stark, M., Stöffler, G., Ehrlich, R., Stöffler-Meilicke, M., and Cannon, M.,** Ribosomes in thiostrepton-resistant mutants of *Bacillus megaterium* lacking a single 50 S subunit protein, *J. Mol. Biol.,* 132, 235, 1979.
127. **Stöffler, G., Cundliffe, E., Stöffler-Meilicke, M., and Dabbs, E. R,** Mutants of *Escherichia coli* lacking ribosomal protein L11, *J. Biol. Chem.,* 255, 10517, 1980.
128. **Parker, J., Watson, R. J., and Friesen, J. D.,** A relaxed mutant with an altered ribosomal protein L11, *Mol. Gen. Genet.,* 144, 111, 1976.
129. **Stark, M. J. R. and Cundliffe, E.,** Requirement for ribosomal protein BM-L11 in stringent control of RNA synthesis in *Bacillus megaterium, Eur. J. Biochem.,* 102, 101, 1979.
130. **Röhl, R. and Nierhaus, K. H.,** Methyl groups of ribosomal protein L11 are not related to the synthesis of ppGpp, *Mol. Gen. Genet.,* 170, 187, 1979.
131. **Dabbs, E.R.,** Mutational alterations in 50 proteins of the *Escherichia coli* ribosome, *Mol. Gen. Genet.,* 165, 73, 1978.
132. **Pichon, J., Marvaldi, J., and Marchis-Mouren, G.,** The *"in vivo"* order of protein addition in the course of *Escherichia coli* 30S and 50S subunit biogenesis, *J. Mol. Biol.,* 96, 125, 1975.
133. **Fabian, U.,** Identification of proteins located in the neighborhood of the binding site for the elongation factor EF-Tu on the *Escherichia coli* ribosomes, *FEBS Lett.,* 71, 256, 1976.
134. **Lhoest, J. and Colson, C.,** Cold-sensitive ribosome assembly in an *Escherichia coli* mutant lacking a single methyl group in the ribosomal protein L3, *Eur. J. Biochem.,* 121, 33, 1981.
135. **Mangiarotti, G., Apriion, B., Schlessinger, D., and Silengo, L.,** Biosynthetic precursors of 30S and 50S ribosomal particles in *Escherichia coli, Biochemistry,* 7, 456, 1968.
136. **Miller, O. L. and Hamkalo, B. A.,** Visualization of RNA synthesis on chromosomes, *Int. Rev. Cytol.,* 33, 1, 1972.
137. **Lindahl, L.,** Two new ribosomal precursor particles in *E. coli, Nature (London),* 243, 170, 1973.
138. **Duncan, M. J. and Gorini, L.,** A ribonucleoprotein precursor of both the 30S and 50S ribosomal subunits of *Escherichia coli, Proc. Natl. Acad. Sci. U.S.A.,* 72, 1533, 1975.
139. **Splillmann, S., Dohme, F., and Nierhaus, K. H.,** Assembly *in vitro* of the 50S subunit from *Escherichia coli* ribosomes: proteins essential for the first heat dependent conformational change, *J. Mol. Biol.,* 115, 513, 1977.
140. **Branlant, C., Krol, A., Sriwidada, J., and Ebel, J. P.,** Characterization of ribonucleoprotein subparticles from 50S ribosomal subunits of *Escherichia coli, J. Mol. Biol.,* 116, 443, 1977.
141. **Chang, F. N. and Buszilowicz, C.,** Growth temperature dependent variation in the methylation of ribosomal proteins in *Escherichia coli,* in *Transmethylation,* Usdin, E., Borchardt, R. T., and Creveling, C. R., Eds., Elsevier/North Holland, 1979, 573
142. **Chang, F. N.,** Methylation of ribosomal proteins during ribosome assembly in *Escherichia coli, Mol. Gen. Genet.,* 183, 418, 1981.

Chapter 10

THE FUNCTION AND ENZYMOLOGY OF PROTEIN D-ASPARTYL/L-ISOASPARTYL METHYLTRANSFERASES IN EUKARYOTIC AND PROKARYOTIC CELLS

Irene M. Ota and Steven Clarke

TABLE OF CONTENTS

I. DIVERSITY OF PROTEIN CARBOXYL METHYLTRANSFERASES

Enzymes that catalyze the incorporation of methyl groups from *S*-adenosylmethionine into protein carboxyl groups have been found widely distributed in nature.[1,2] At this point, there appear to be three major classes of these protein methyltransferases that can be classified according to their methyl-acceptor specificity. One of these (type I) catalyzes the formation of glutamyl side chain methyl esters[3,4] from a set of specific L-glutamic acid residues on the receptors of chemotactic bacteria.[5,6] This glutamyl methyltransferase functions in the modulation of the output of these chemoreceptors.[7] The second class (type II) is much more widely distributed in both eukaryotic and prokaryotic cells and appears to methylate carboxyl groups on atypical amino acid residues which can be derived from both aspartyl and asparaginyl residues. These abnormal groups include D-aspartyl[8] and L-isoaspartyl residues.[9,10] No methyl-donor activity has been observed with this enzyme on substrates containing only normal L-aspartyl (or L-glutamyl) residues. The role of the type II enzyme appears to be in the recognition and methylation of abnormal or damaged proteins in the cell for possible repair or degradation reactions.[8-14] Finally, the existence of a third class of enzymes (type III) that appears to modify the alpha carboxyl group of the C-terminal cysteine residue of various membrane-related proteins has been inferred from the presence of methylated peptidyl yeast sex factors[15,16] and experimental evidence has been presented for such methylation of the mammalian *ras* oncogene products,[16a,16b] and other proteins. The differences between these three classes of enzymes are summarized in Table 1. The work in our laboratory has focused on the type II D-aspartyl/L-isoaspartyl methyltransferase, and we will concentrate on this enzyme in this review.

II. SUBSTRATE SPECIFICITY OF PROTEIN D-ASPARTYL/L-ISOASPARTYL METHYLTRANSFERASE

One of the most important clues to the function of type II protein carboxyl methyltransferase is its unusual substrate specificity. For example, the enzyme from mammalian erythrocyte cytosol transfers methyl groups from *S*-adenosylmethionine into base labile linkages on the carboxyl groups of a wide variety of cytosolic[17] and membrane proteins,[18-19] as well as proteins not found in these erythrocytes, such as ovalbumin and ribonuclease.[20,21] What is unusual is that each of these proteins is methylated at very low levels. For instance, only 3 out of every 6 million hemoglobin molecules[17] and 3 out of every 10,000 calmodulin molecules are methylated in intact cells.[22,23] Modification of such a small fraction of a protein should not have a significant effect on its cellular activity, and this enzyme is unlikely to function in a regulatory role similar to that of the bacterial L-glutamyl methyltransferase. Because a wide variety of substrates are methylated, this type II enzyme must recognize feature(s) common to all proteins. It now appears that there are two such features, and these are the presence of abnormal D-aspartyl and L-isoaspartyl residues. D-Aspartate was first recognized as a possible substrate for the enzyme when D-aspartyl β-methyl ester was recovered from carboxypeptidase Y digestions of enzymatically ^{3}H-methylated human erythrocyte membrane proteins.[8,17,24-27] In addition, L-isoaspartate was identified as a substrate when synthetic peptides containing L-isoaspartyl residues[9,11-12,14,26,28-35] (Table 2) and base-treated natural peptides[10] were found to be nearly stoichiometric methyl-accepting-substrates for the type II methyltransferase from a variety of tissues.

L-Isoaspartyl and D-aspartyl residues are not normally incorporated into proteins. They arise from relatively infrequent spontaneous degradation reactions at Asp and Asn residues and possible misincorporation reactions during protein synthesis (see below). All proteins would be expected to be subject to such reactions and this would account for the wide variety of substrates

TABLE 1
Classes of Protein Carboxyl Methyltransferases

	Type I (EC 2.1.1.24)	Type II (EC 2.1.1.77)	Type III (no EC number assigned)
Name	L-glutamyl protein methyltransferase	L-isoAspartyl/D-Aspartyl protein methyltransferase	C-terminal cysteinyl α-carboxyl protein methyltransferase
Residue(s) methylated	L-Glu	L-isoAsp, D-Asp	α-Carboxyl group of C-terminal L-Cys
Physiological substrates	Bacterial membrane chemoreceptors for Asp, Ser, dipeptides, etc.	Cytoskeletal proteins, band 3 anion transporter, many others	Yeast mating factors, ras oncogene products, lamin B (?), cGMP phosphodiesterase (?), G proteins (?)
Nonphysiological substrates	None detected	Many — ovalbumin, IgG, ribonuclease, etc.	?
Distribution of enzyme in nature	Limited to chemotactic bacteria	Ubiquitous in prokaryotic and eukaryotic cells	Mammalian cell lines, yeast, other cell types?
Functional role	Receptor modulation	Detection of abnormal aspartyl residues (possible participation in repair and degradation reactions)	Modulation of signal transduction pathways, membrane attachment?

TABLE 2
Synthetic Peptide Substrates for the Protein D-Aspartyl/L-Isoaspartyl Methyltransferase

L-Isoaspartyl Peptides	Km(μM)	Ref.
Lys-Ala-Ser-Ala-isoAsp-Leu-Ala-Lys-Tyr	0.4	32
Lys-Gln-Val-Val-isoAsp-Ser-Ala-Tyr-Glu-Val-Ile-Lys	0.55	31
Gly-Phe-Asp-Leu-isoAsp-Gly-Gly-Gly-Val-Gly	3.5	12
Trp-Met-isoAsp-Phe-NH_2	5	29
Lys-Val-Thr-Cys-Lys-isoAsp-Gly-Gln-Thr-Asn-Cys-Tyr-Gln-Ser-Lys	6.17	14
Val-Tyr-Pro-isoAsp-Gly-Ala	6.3	9
Leu-Met-isoAsp-Thr	+	33
Thr-Ser-isoAsp-Tyr-Ser-Lys-Tyr	+	33
Lys-Met-Lys-Asp-Thr-isoAsp-Ser-Glu-Glu-Glu-Ile-Arg	+	33[a]
Ala-Ala-isoAsp-Phe-NH_2	73	34
Tyr-Val-Ser-isoAsp-Gly-Asp-Gly	470	—[a]
isoAsp-Gly	(Not a substrate)	29
D-Aspartyl Peptides		
Val-Tyr-Pro-D-Asp-Gly-Ala	(Not a substrate)	9
Val-Tyr-Gly-D-Asp-Pro-Ala	(Not a substrate)	32
Lys-Ala-Ser-Ala-D-Asp-Leu-Ala-Lys-Tyr	(Not a substrate)	32
Trp-Met-D-Asp-Phe-NH_2	(Not a substrate)	11

[a] J. Lowenson and S. Clarke, unpublished results.

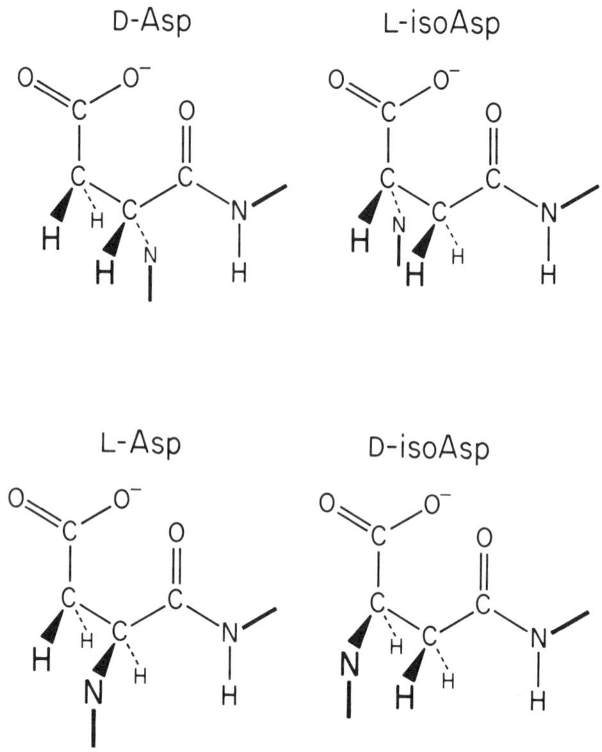

FIGURE 1. D-Asp and L-isoAsp are substrates for protein carboxyl methyltransferase type II. Protein carboxyl methyltransferase type II can methylate both D-aspartyl and L-isoaspartyl residues but not D-isoaspartyl and normal L-aspartyl residues in peptides and proteins. All four residues appear to be very similar except for the placement of the nitrogen atom in the peptide bond on the amino terminal side of the aspartyl residue (see text).

found for these enzymes. The slow rate of these degradation reactions can also account for the substoichiometry of the methylation reaction because these abnormal residues should be present at low levels compared to normal aspartyl residues.

L-Isoaspartyl and D-aspartyl residues may interfere with the activity of cellular proteins because the relative position of the peptide bond and side chain are altered. Thus, it may be advantageous for the cell to eliminate these residues. Accordingly, we have proposed that this enzyme may participate in the recognition of altered proteins for repair or degradation.[2,8]

The structure of racemized and isomerized aspartyl residues are shown in Figure 1. The ability of the enzyme to recognize both D-aspartyl and L-isoaspartyl residues may result from a requirement for the localization of the nitrogen atom in the peptide bond on the amino terminal side of this residue in either of two positions away from the face of the molecules depicted in Figure 1. When this nitrogen atom is localized on the face, as in L-aspartyl and D-isoaspartyl residues, no enzymatic reaction appears to occur although we cannot at this time exclude the participation of the D-isoaspartyl residue as a substrate.

There are several features of the substrate specificity of this enzyme that remain unexplained. First, although proteins containing D-aspartyl residues appear to be substrates in human erythrocyte membranes, short D-aspartyl containing-peptides are not substrates for the methyltransferase. Four different D-Asp-containing synthetic peptides have been tested, and all have failed to accept methyl groups (Table 2).[9,29,32] In an attempt to explain the lack of methylation of D-Asp-containing peptides, the possibility that the D-Asp β-methyl ester isolated from proteolytic digests of erythrocyte membrane proteins was a rearrangement product of an L-isoAsp α-methyl ester was tested. However, we were unable to generate D-methyl ester from a methylated L-isoAsp nonapeptide under a variety of conditions.[32] It appears that the lack of methylation of D-Asp containing peptides could be due to an absence of structural or sequence features that are necessary for recognition by the methyltransferase. (The requirements for L-isoAsp methylation are probably less stringent). Since erythrocyte membrane proteins yield D-Asp β-methyl ester upon proteolytic digestion, work in the laboratory is now focused on purifying and then fragmenting these proteins to identify the exact methylation sites.

Another problem with the elucidation of substrate specificty is that evidence for L-isoaspartyl methylation has not been obtained so far for proteins in intact cells. Because synthetic peptides containing L-isoaspartyl residues are generally good substrates for the methyltransferase, it is assumed that enzymatically methylated proteins should contain methylated isoaspartyl residues. The fact that L-aspartyl alpha-methyl esters have not been recovered from protease digestions of enzymatically methylated proteins can be explained by two properties of most proteases. While the isoaspartyl (β-peptide) bond is itself resistant to proteolysis, the alpha-methyl ester bond is readily hydrolyzed by these proteases. Thus, leucine aminopeptidase and carboxypeptidase Y digestions of several enzymatically methylated synthetic L-isoaspartyl containing peptides do not yield free L-aspartyl alpha-methyl ester, but rather isoaspartyl peptides and methanol (J. Lowenson, unpublished observations). To identify isoaspartyl esters in peptides and proteins by methods other than total enzymatic digestion, we have developed techniques to localize these sites in short peptide fragments and have been successful in demonstrating isoaspartyl methyl esters at specific positions in the peptide hormone glucagon[33] and in the protein calmodulin (see below).[33a,33b]

Although much of the work concerning the detailed substrate specificity of the protein D-aspartyl/L-isoaspartyl carboxyl methyltransferase has been done with the human erythrocyte and bovine brain enzymes, isozymes of protein carboxyl methyltransferases exhibiting the same wide substrate specificity and substoichiometric methylation have been found in tissues from a variety of eukaryotic organisms including human, bovine, horse, rabbit, rat, and *Torpedo* (Table 3) as well as in prokaryotic cells such as *E. coli*.[36] It has also been directly demonstrated that extracts from amphibian *Xenopus* oocytes,[30] PC-12 cells (derived from a rat adrenal medullary tumor),[28] and the bacteria *Salmonella typhimurium*[28] and *E. coli*[34] are capable of methylating L-isoaspartyl residues in synthetic peptides.

III. STRUCTURE OF METHYLTRANSFERASE ISOZYMES

The similarities among these type II methyltransferases extend to their physical properties as well. Each enzyme in this group appears to exist as a monomer and their molecular weights are similar (Table 3). Methyltransferase activity also generally exists in cells as more than one isoelectric species.

As significant quantities of purified human erythrocyte isozymes I and II and bovine erythrocyte isozymes I and II have become available, we have been able to examine more closely the structural similarities between these enzymes. The fragmentation patterns seen in Cleveland-Laemmli maps, in which *Staphylococcus aureus* V8 protease digested peptides of isozymes I and II from each species are separated by polyacrylamide gel electrophoresis in sodium dodecyl sulfate are nearly identical.[21] The profile of tryptic peptides of isozymes I and II from human erythrocyte and bovine erythrocyte isozyme II separated on reverse phase HPLC columns are also similar. The similarities seen by these techniques have been confirmed by sequence analysis of tryptic fragments of the methyltransferase from erythrocyte isozymes I and II and bovine erythrocyte isozyme II. In the tryptic peptides sequenced thus far, identity or probable identity has been found in 111 out of 112 residues between the human isozymes I and II and identities in 127 out of 134 residues between human and bovine isozyme IIs[21] (Table 4). These results suggest that the erythrocyte isozymes from both organisms may have nearly identical structures.

TABLE 3
Isozymes of Eukaryotic Protein Carboxyl Methyltransferases

Source	pI Major	pI Minor	Native MW	Polypeptide MW	Ref.
Human					
Erythrocyte	5.5, 6.6		25,000	28,000	21, 37
Bovine					
Brain	5.6, 6.5	5.7	27,000—28,000	24,300	38
Thymus	4.85		35,000		39
Horse					
Erythrocyte	5.6		25,400—26,000		40
Rabbit					
Brain	6.2	5.7, 5.9, 6.5			41
Muscle	6.2, 6.5				41
Heart	5.4, 6.5				41
Liver	5.8	5.6			41
Rat					
Testes	6.1, 7.35	6.7	25,000		42
Pituitary	6.1		25,000		42
Erythrocyte	4.9, 6.0	5.5	25,000		20
Torpedo ocellata	6.1, 6.4		29,000	29,000	43

TABLE 4
Differences in Amino Acid Sequences Among Bovine and Human Methyltransferase Tryptic Peptide Fragments

Comparison	Residues: Identical	Probably identical	Unknown	Probably different	Different	% Identity
Human I vs. human II	102	9	0	1	0	99.1
Human II vs. bovine II	119	8	2	1	4	94.8
Human I vs. bovine II	115	9	3	3	3	93.2

We have examined the methyltransferase isozymes from human erythrocytes to determine whether there are differences in their activities. These isozymes showed no obvious differences in the kinetics of the methylation of synthetic L-isoaspartyl containing peptides.[37] We have not been able to directly test whether both enzymes methylate D-aspartate similarly since no D-aspartyl-containing peptide has been methylated *in vitro* (see above). However, the fact that addition of synthetic L-isoaspartyl peptide to a methylation mixture containing erythrocyte membranes inhibits the production of D-Asp β-methyl ester suggests that the L-isoaspartyl methylating enzyme also methylates D-aspartyl sites.[26]

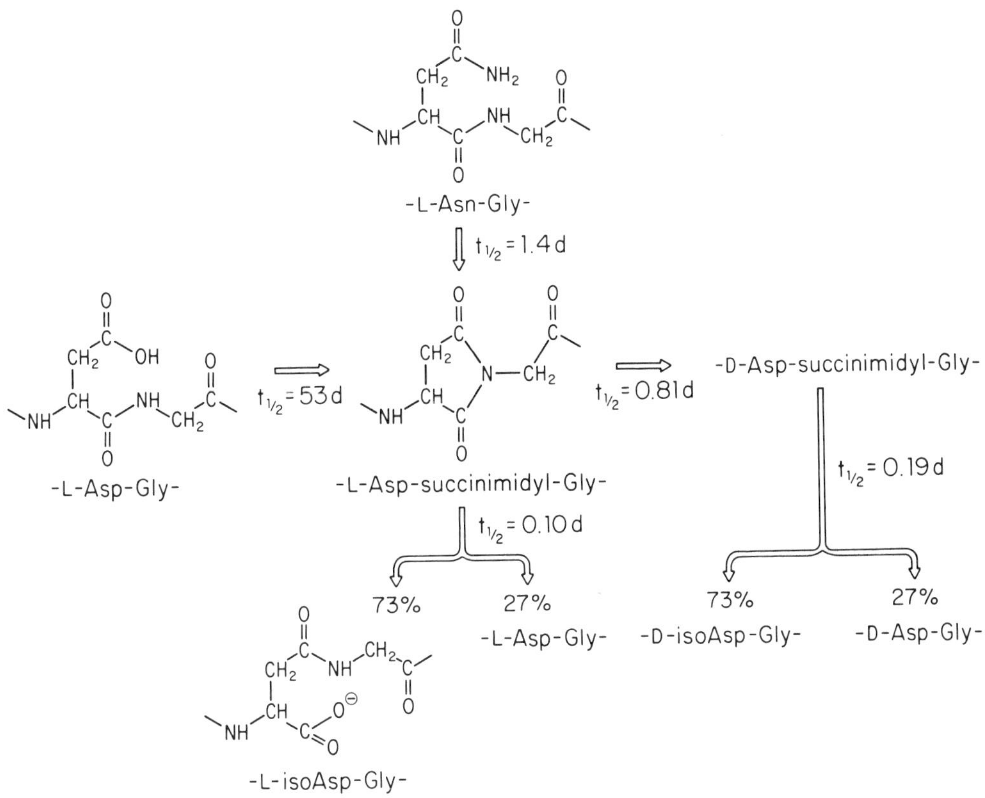

FIGURE 2. L-isoaspartyl and D-aspartyl residues arise from spontaneous chemical degradation reactions at L–asparaginyl and L-aspartyl residues in peptides and proteins. L-Asparaginyl and L-aspartyl residues are particularly labile amino acids that undergo an intramolecular reaction that results in the formation of L-succinimides. The formation of L-succinimides from these residues is particularly fast from Asn-Gly sequences. These succinimides can rapidly racemize to D-succinimides. The hydrolysis of these imides results in the formation of L-isoAsp, L-Asp, D–isoAsp, and D-Asp residues. Half-times are given for the peptide Val-Tyr-Pro-Asx-Gly-Ala.[46]

IV. ORIGIN OF SUBSTRATE SITES

Despite the highly accurate reactions of the enzymes that replicate DNA, transcribe RNA species, and translate messenger RNA into protein, and the fact that errors that do occur in these processes are often corrected by enzymatic proofreading mechanisms, almost all proteins appear to contain low levels of atypical amino acid residues. These include oxidized methionine residues,[44] glycated lysine residues,[45] as well as the racemized and isomerized aspartyl residues that are substrates for the D-aspartyl/L-isoaspartyl methyltransferase. Although the exact origin of these abnormal aspartyl residues is not clear, one possibility is that they are formed by spontaneous chemical processes during the aging of the protein. Even if "perfect" molecules are formed on ribosomes, they are immediately subject to a variety of spontaneous degradation reactions.

We have studied the spontaneous formation of L-isoaspartyl and D-aspartyl residues in short peptides containing either a single L-Asp or L-Asn residue.[46,46a] From the results of this work, the initial formation of an L-succinimide intermediate (Figure 2) could account for the bulk of the deamidation, racemization, and isomerization reactions. This intermediate racemizes to the D-succinimide at a rate approximately 30,000 faster than that expected for free aspartyl residues. Hydrolysis of the L- and D-succinimides generates L-Asp, L-isoAsp, D-Asp and D-isoAsp residues (Figure 2).

We have begun to examine the influence of peptide structure upon the rate of succinimide

TABLE 5
Rates of Degradation of Aspartyl and Asparaginyl Peptides under Physiological Conditions (pH 7.4, 37°C)[a]

Peptide	$t^1/2$ (d)	Products
Val-Tyr-Pro-Asn-Gly-Ala	1.4	73% isopeptide, 27% normal
Val-Tyr-Pro-Asn-Leu-Ala	70	64% isopeptide, 22% normal, 14% cleavage products[b]
Val-Tyr-Pro-Asn-Pro-Ala	106	Tetrapeptide cleavage products[b]
Val-Tyr-Pro-Asp-Gly-Ala	53	79% isopeptide, 21% normal

[a] Data taken from Reference 46.
[b] Cleavage to give a tetrapeptide probably occurred by the attack of the amide side chain nitrogen on the main peptide (see Reference 46).

formation. We found that the succinimide forms about 40 times more slowly from the synthetic hexapeptide Val-Tyr-Pro-Asp-Gly-Ala than from Val-Tyr-Pro-Asn-Gly-Ala.[46] This result is likely to reflect two competing factors. In the first place, the predominant form of aspartyl residues at neutral pH is the unprotonated carboxylate anion. This species does not participate in succinimide formation. In fact, given a pKa of an aspartyl residue of 3.4, only 1 in 10,000 groups will be protonated at pH 7.4 and will be potentially reactive for succinimide formation. Thus if the reactivity of the –COOH group was equal to that of a $-CONH_2$ group, we might expect succinimide formation from aspartyl residues to be about 10,000 times slower than from asparaginyl residues. The fact that succinimide formation is only 40 times slower for aspartyl residues in this peptide means that succinimide formation from the –COOH group must be faster than from the $-CONH_2$ group. The reason for this more rapid succinimide formation from the –COOH group versus the $-CONH_2$ group is not clear but might be expected because weaker bases (OH^- or H_2O) are generally better leaving groups than stronger bases (NH_2^- or NH_3). Thus, the small fraction of aspartyl residues in the protonated state under physiological conditions limits succinimide formation from aspartyl residues. We have also shown that replacement of the glycine residue in the Asn-containing hexapeptide with an alanine or leucine residue slows succinimide formation by factors of about 18 and 50, respectively.[46,46a] The large reduction seen with the peptide in which the glycine is replaced by an alanine residue suggests that factors in addition to the bulk of the residue following the aspartyl/asparaginyl residue may be important in the rate of succinimide formation. Glycine residues have a unique range of conformational flexibility as a result of lacking a beta-carbon atom, and Asp-Gly and Asn-Gly peptides may be able to assume conformations particularly suited for succinimide formation that other peptide pairs cannot (Table 5).

From these observations it was hoped that the aspartyl/asparaginyl residues most susceptible to degradation could be predicted in larger polypeptides. Based on the chemistry seen in the small peptide models, we would expect that Asn-Gly sites would represent major sites of degradation, Asn-nonGly and Asp-Gly would represent minor sites of degradation and that Asp-nonGly sites would be relatively resistant to succinimide formation. To test this idea, we mapped by protease digestions the methylation sites of glucagon, a 29 amino acid peptide hormone that contains three aspartyl residues and one Asn residue in the sequences AspSer, AspTyr, AspPhe, and AsnThr.[33] Based on the peptide studies described above, we predicted that the AsnThr site would be the major site of methylation by the type II D-aspartyl/L-isoaspartyl methyltransferase. We were surprised to find, however, that AspTyr was the origin for the major site and that there was only a minor site originating from AsnThr. Both of the methylated residues at these sites were determined to be L-isoaspartate. We confirmed that spontaneous processes were likely to be responsible for isoaspartate formation at this site because a similar result was obtained when a synthetic preparation of glucagon was analyzed for methylatable sites. We concluded that other factors such as acid/base catalysis by neighboring residues might facilitate imide formation and therefore isoaspartyl formation. For example, intramolecular interactions that result in the increase of the pKa of Asp-9 would be expected to increase the fraction of the protonated carboxyl group and thus the rate of succinimide formation. In at least one other case, however, an Asn-Gly sequence does appear to be a major site of isoaspartyl formation. This occurs in the peptide hormone ACTH, where the major site of methylation originates from the sequence Asn_{25}Gly.[9,10] These results suggest that it will probably not prove possible to predict methylatable sites in proteins based solely on the amino acid sequence.

The prediction of the origin of methylatable residues in proteins is made more complex by the fact they generally maintain strict three-dimensional conformations. The folding of the peptide backbone and positioning of the side chains may promote succinimide formation in some regions and inhibit it in others. Recently, this possibility was examined by looking at the structure of Asp-X and Asn-X sequences in 14 crystallized proteins.[47] It was found that most of these residues are protected from succinimide formation by excluding them from conformations where imide formation is favored. In fact, conformations that represent energy minimum states in protein folding are generally unfavorable for succinimide formation and this factor may account for the low rate of spontaneous damage to proteins by succinimide-linked reactions. These findings suggest that Asx residues that do degrade might be present in flexible or denatured regions of proteins that, like small peptides, lack conformational restraints. This appears to be true for calmodulin, in which we have recently elucidated the sites of methylation.

Calmodulin is a 17,000 Da calcium binding protein whose crystal structure is known to 2.2 Å.[48] Its structure has been described as a dumbbell with two globular calcium-binding domains, each binding two calcium ions, joined by a single long alpha helical segment. It has been suggested by Johnson et al. that the two AsnGly sequences in calmodulin, one in calcium-binding site II and the other in calcium-binding site III, might be the major sites of methylation in calmodulin.[49] However, we have found that native calmodulin appears to be methylated *in vitro* primarily at two other sites. Neither of these sites is in the calcium-binding domain and neither of them are derived from asparaginyl residues. In fact, both originate from Asp-nonGly sequences. One of these sites is in the eight turn alpha helical connector segment in the sequence Asp_{78}-Thr-Asp_{80}-Ser and the other site is at the N-terminus of the protein in the sequence Ala-Asp_2-Gln.[33a] The methylated residue at Asp_2 has been determined to be an L-isoaspartyl residue by comparison with the properties of a synthetic peptide corresponding to the sequence of the tryptic peptide, calmodulin 1-13, substituted with L-isoAsp at the second residue. The exact location and nature of the methylated residue in the Asp_{78}-Thr-Asp_{80}-Ser region is not clear, but there does not appear to be an L-isoaspartyl residue at the 78 position. We have not ruled out the possibility of a D-Asp site of methylation here. The formation of methylatable sites in calmodulin from aspartyl residues most likely proceeds via a succinimide intermediate. These may form most readily in the N-terminal and alpha helical connector regions because they are likely to be the most conformationally flexible regions of the molecule.[48] It is also possible that these sites are simply the most accessible to the methyltransferase.

These results are not consistent with earlier proposals that calmodulin is methylated primarily at Asn-Gly residues.[49,50] In both of these studies, calmodulin was first denatured by either base-treatment or incubation with EGTA to remove calcium ions. To investigate whether additional methylation sites may be exposed when calcium is removed we have methylated calmodulin in the calcium-free conformation in a medium containing EGTA. Because we find no additional sites it appears that Asp_2 and Asp_{78}/Asp_{80} are the origins of major methylatable sites in both forms of calmodulin.

It is also interesting that those regions that are expected to be rigid, such as the calcium-binding domains, do not appear to be methylated. These domains contain a large number of Asx residues, several of them in Asx–Gly sequences, and yet none of these residues appear to be methyl acceptors. Recently, Johnson et al. found that calmodulin preincubated under physiological conditions in the absence of calcium was a better methyl acceptor than calmodulin preincubated in the presence of calcium.[50] They suggested that the two AsnGly sequences, both present in calcium-binding domains, were more susceptible to degradation in the absence of calcium ion. We tested this idea directly with our mapping techniques and found that extended EGTA preincubation introduces new methylation sites in calcium binding domains II, III, and IV, as well as enhancing methylation at Asp_2 and in the $Asp_{78}ThrAsp_{80}Ser$ region. Since domain IV contains no AsnGly sequences and methylation was also enhanced at aspartyl residues it appears that these conditions do not specifically enhance methylation at AsnGly sequences. Interestingly, when calmodulin was preincubated in the presence of calcium, methylation was enhanced several-fold primarily at Asp_2 and at Asp_{78}/Asp_{80}. Taken as a whole, these results suggest that in the absence of calcium, calmodulin may be more conformationally flexible and more susceptible to succinimide formation.[336]

Another possible mechanism for the formation of L-isoaspartyl residues is for an error to occur during protein synthesis. If an aspartyl residue is linked by its tRNA through the side chain beta-carboxyl group instead of the alpha-carboxyl group, the result might be an L-isoaspartyl residue incorporated into the protein. We are particularly interested in this area both because there is some experimental evidence to support the methylation of nascent polypeptide chains,[51,52] and because such a role can explain previously poorly understood results. For example, the distribution of enzyme in cells and tissues roughly parallels protein synthesis capacity. With the exception of a recently described nuclear isozyme,[30] the bulk of the methyltransferase isozymes are located in the cytosolic compartment of the cell.[1,2] Furthermore, relatively high levels of methyltransferase activity have been reported in tissues engaged in protein and polypeptide secretion, such as the pituitary gland.[53] In such tissues, the rate of protein synthesis must be increased to provide for both cellular and secretory needs. A role for the methyltransferase in recognizing abnormal nascent polypeptide chains is also consistent with the presence of the D-aspartyl/L-isoaspartyl methyltransferase in cells such as bacteria that are capable of rapid protein turnover and synthesis.[28,34,36] On the other hand, O'Connor did not observe a change in the carboxyl methylation of frog oocyte proteins in the presence of protein synthesis inhibitors.[30] Thus, more work needs to be done to establish the precise relationship between protein synthesis and protein carboxyl methylation reactions.

V. WHAT HAPPENS AFTER METHYLATION?

Although the formation of D-aspartyl and L-isoaspartyl residues may not always seriously affect the cellular function of the protein, the cumulative effect of these, and other types of spontaneous damage, is expected to be significant. Therefore, the role of the D-aspartyl/L-isoaspartyl methyltransferase may be to mark damaged proteins for degradation and/or repair.[8] A degradation system might operate in a similar manner to the ubiquitination system where damaged proteins are enzymatically conjugated with ubiquitin and then degraded by ubiquitin-dependent proteases.[54] A repair system might use the energy of hydrolysis of the methyl ester to drive a reaction in which isopeptide bonds are converted to normal bonds or D-residues are converted to L-residues.

An important clue in understanding the fate of membrane protein carboxyl methyl esters in human erythrocytes was the observation that their hydrolysis rate in intact cells was comparable to that rate seen in isolated membranes.[55] These results indicated that there is no specific protein carboxyl methylesterase in human erythrocyte cytosol. Because the spontaneous rate of methyl ester hydrolysis via succinimide pathways in aspartyl peptides can be very rapid, these esters may be metabolized by non enzymatic processes.[19,56-57] If a succinimide intermediate is also involved in protein demethylation,* the regeneration of the succinimide after ester formation would now allow for repair because its hydrolysis generates normal aspartyl as well as isoaspartyl residues. Since the isoaspartyl residue would be recycled by the methyltransferase back to the ester (and thus the succinimide), the net result of these reactions would be the conversion of isoaspartyl residues to aspartyl residues. In several peptide systems, this is exactly what has been found to occur (Figures 3 and 4).[11,12,14] In this way, the succinimide would be a transient intermediate in both the degradation and repair reactions.

It also appears that proteins can be repaired as well as peptides. In recent experiments, it has been shown that calmodulin that has lost 82% of its activity by preincubation under physiological conditions for 28 d can have its activity partially restored (to 68% of the original activity) by incubation with *S*-adenosylmethionine and bovine brain protein carboxyl methyltransferase.[13] Since EGTA was present in the preincubation medium, it is likely that the calcium binding sites II, III, and IV, as well as the Asp_2 and $Asp_{78}ThrAsp_{80}Ser$ sites have become methyl acceptor sites. At this point it is unclear which residues must be methylated (and demethylated?) in order that activity can be recovered. Since calcium binding is critical for calmodulin activity, repair at these sites might be expected to restore activity. It is also possible that changes in the gross structure of calmodulin can reduce its activity. The formation of an L-isoaspartyl residue at the Asp_{78}/Asp_{80} region might alter the orientation of the two globular domains inhibiting the interaction with the target Ca^{++}-calmodulin-binding proteins. It would be interesting to test whether calmodulin that has been preincubated in the presence of calcium, where damage is limited primarily to the Asp_2 and Asp_{78}-Thr-Asp_{80}-Ser regions, loses activity and whether this activity can be restored by the methyltransferase.

One problem with this repair system is that L-isoaspartyl residues that have formed from asparagine residues cannot be fully repaired to asparagine but are "repaired" to L-aspartyl residues. It has been suggested, for instance, that calmodulin cannot regain 100% of its activity by methylation for this reason. Although this may be the case, it is also possible that some degraded residues in calmodulin are not accessible to the methyltransferase and therefore cannot be repaired. Another possibility is that not all L-isoaspartyl residues may be good substrates for the methyltransferase enzyme. Even the K_m values for small peptides vary greatly. As seen in Table 2, an L-isoaspartyl-containing nonapeptide has a very low K_m of 400 n*M* while a heptapeptide containing the sequence isoAspGlyAspGly derived from lysozyme has a K_m of 470 μ*M*. Another reason that repair may not be complete is that the succinimide can rapidly racemize to form a mixture of D- and L-succinimides. The hydrolysis of the D-succinimide results in the formation of an apparent nonsubstrate for the methyltransferase, D-isoaspartate, which may accumulate in cells. If the succinimide does not racemize, it may be that repair in proteins is more complete than in peptides because L-succinimides might preferentially open to L-aspartyl rather than L-isoaspartyl residues. The three-dimensional structure of the protein might be expected to favor the formation of the original L-Asp residue over the L-isoAsp residue.

It also appears that methylation may play a role in the removal of damaged proteins by degradative processes since the L-isoaspartyl containing hexapeptide is cleaved by proteolysis systems present in cytosolic extracts of human erythrocytes.[58] Although the hexapeptide with

* Although succinimides appear to be intermediates in the demethylation of several synthetic peptide esters, other chemical demethylation pathways may exist. For example, attack on the carbonyl carbon of the ester by a peptide bond oxygen can also occur to give an isoimide or an oxazalone derivative.[55] The detection of specific intermediates in the metabolism of cellular protein methyl esters may be technically difficult, but may be crucial in understanding the metabolism of these modified proteins.

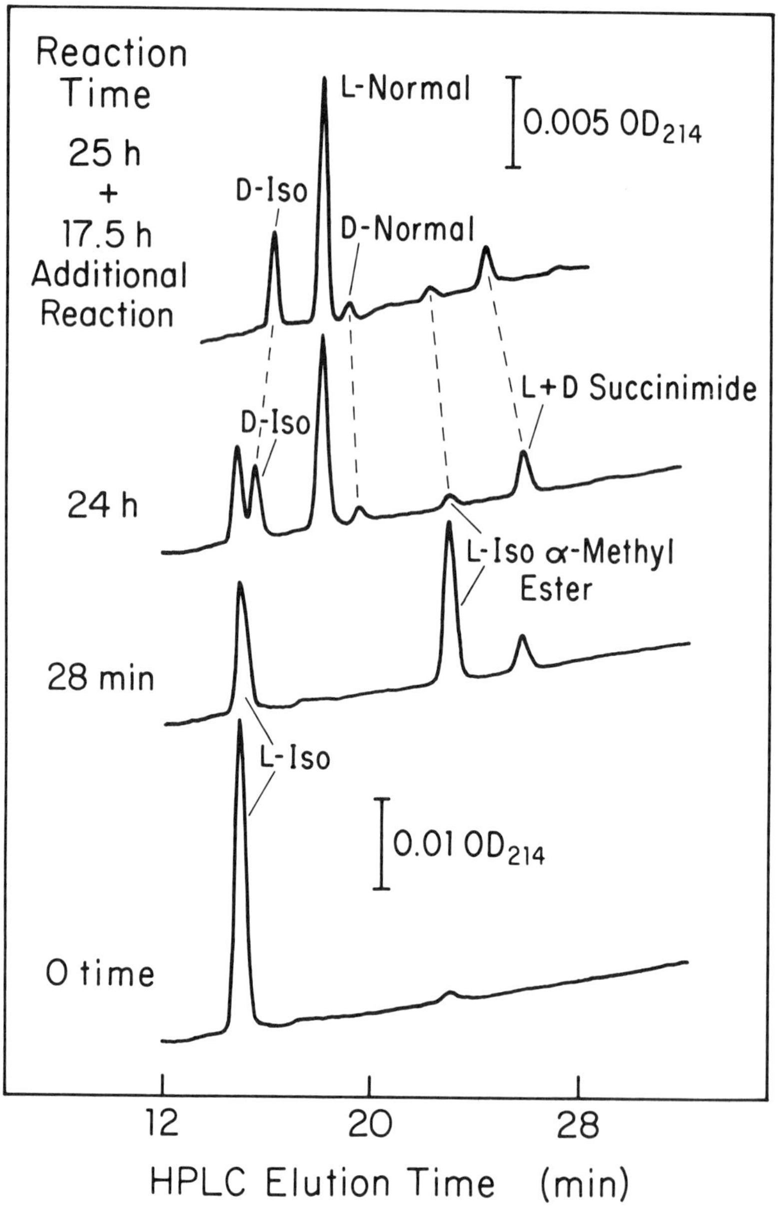

FIGURE 3. Time course of the repair of L-isoaspartyl residues to L-aspartyl residues in tetragastrin.[11] Iso-tetragastrin, Trp-Met-isoAsp-Phe-NH_2, was incubated with protein carboxyl methyltransferase type II and *S*-adenosylmethionine at 37°C and at various times aliquots were removed for analysis by reverse phase high performance liquid chromatography. After 24 h, a 50% conversion to the normal L-Asp containing peptide was achieved. Addition of more methyltransferase enzyme and *S*-adenosylmethionine and further incubation resulted in the depletion of the remaining L-isoaspartyl containing peptide. Repair was not complete because of the formation of D-isoAsp and D-Asp containing peptides that were not substrates for the methyltransferase. Although D-Asp appears to be a methyl acceptor in proteins it does not appear to be methylated in many peptides including tetragastrin.

a free N-terminus is rapidly degraded by proteases the *N*-acetylated peptide is stable and the reactions following methylation can be examined. The methylated peptide spontaneously forms a succinimide intermediate as described above, and both the L- and D-succinimides are

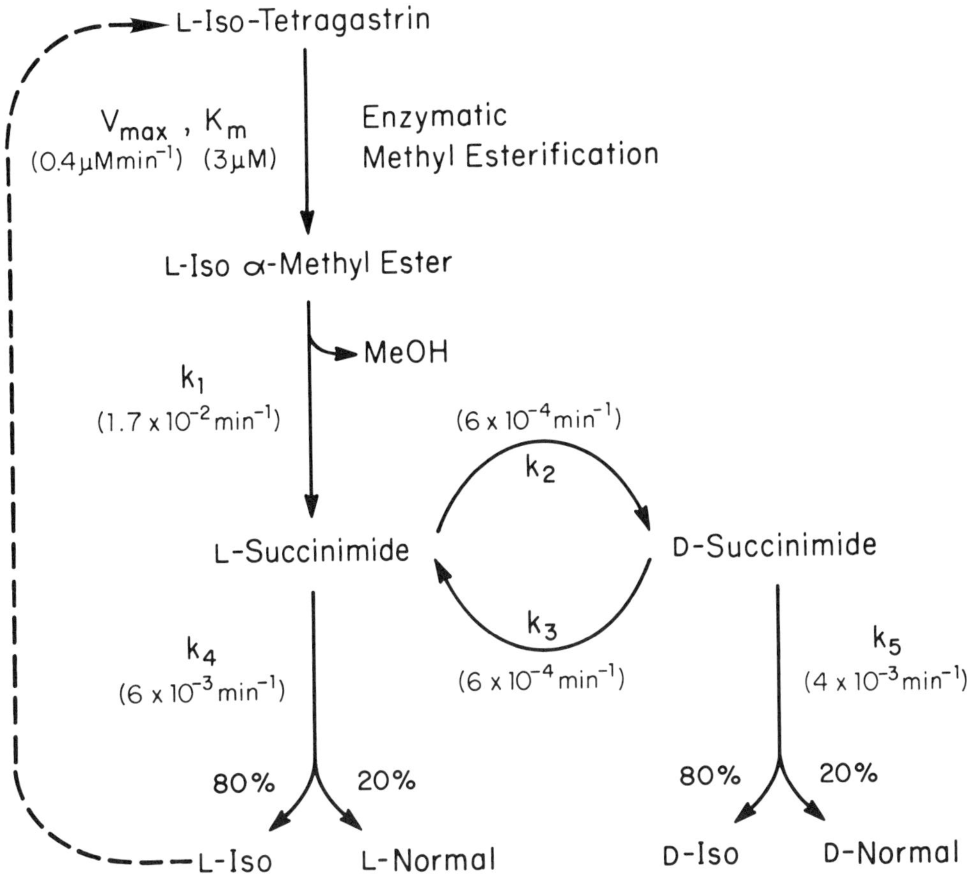

FIGURE 4. The enzymatic and nonenzymatic reactions involved in repairing L-isoaspartyl residues to L–aspartyl residues in tetragastrin.[11] The L-isoaspartyl residue in tetragastrin is enzymatically methylated to form L-isoAsp alpha-methyl ester. The methyl ester spontaneously hydrolyzes via an intramolecular reaction to form the L-succinimide. The succinimide then hydrolyzes to form 20% L-Asp and 80% L-isoAsp residues in this peptide. Repeated cycles of these enzymatic and nonenzymatic reactions over a 24-h period results in a 50% conversion to normal L-Asp containing tetragastrin (Figure 3). All of the L-isoAsp containing peptide could not be repaired to the normal L-Asp containing peptide because of the rapid racemization of the L-succinimide to the D-succinimide. The hydrolysis of the D-succinimide results in the formation of nonmethylatable D-isoaspartyl and D-aspartyl residue containing tetragastrins.

recognized by a post proline endopeptidase that clips the peptide after the prolyl residue. The rate of this reaction was tenfold greater than that for the corresponding peptide containing a normal Pro-Asp linkage. Significantly, the rate of cleavage of the D- or L-isoAsp containing-hexapeptide was 2- to 4.6-fold greater than that of the corresponding normal peptide. The facile cleavage of abnormal peptides by this enzyme suggests that proteins containing altered aspartyl residues may be targeted for cellular degradation.

Is it possible that proteolytic enzymes exist which may specifically recognize the original D-aspartyl beta methyl ester or the L-isoaspartyl alpha-methyl ester? Such enzymes might efficiently remove proteins containing these damaged residues from the cell. On the other hand, similar systems might function by recognizing unmethylated D-aspartyl or L-isoaspartyl residues directly and it is not clear why methyltransferases may be involved in such pathways. Further work is necessary to elucidate the pathway of the metabolism of these methylated proteins, particularly in nonerythroid cells.

REFERENCES

1. **Paik, W. K. and Kim, S.,** *Protein Methylation*, John Wiley & Sons, New York, 1980, chap. 12.
2. **Clarke, S.,** Protein carboxyl methyltransferases: two distinct classes of enzymes, *Annu. Rev. Biochem.*, 54, 479, 1985.
3. **Kleene, S. J., Hobson, A. C., and Adler, J.,** Isolation of glutamic acid methyl ester from an *Escherichia coli* membrane protein involved in chemotaxis, *J. Biol. Chem.*, 252, 3214, 1977.
4. **Van der Werf, P. and Koshland, D. E., Jr.,** Identification of a gamma-glutamyl methyl ester in bacterial membrane protein involved in chemotaxis, *J. Biol. Chem.*, 252, 2793, 1977.
5. **Clarke, S., Panasenko, S., Sparrow, K., and Koshland, D. E., Jr.,** *In vitro* methylation of bacterial chemotaxis proteins: characterization of protein methyltransferase activity in crude extracts of *Salmonella typhimurium*, *J. Supramol. Struct.*, 13, 315, 1980.
6. **Simms, S. A., Stock, A. M., and Stock, J. B.,** Purification and characterization of the S-adenosylmethionine:glutamyl methyltransferase that modifies membrane chemoreceptor proteins in bacteria, *J. Biol. Chem.*, 262, 8537, 1987.
7. **Stock, J. and Stock, A.,** What is the role of receptor methylation in bacterial chemotaxis?, *Trends Biochem. Sci.*, 12, 371, 1987.
8. **McFadden, P. N. and Clarke, S.,** Methylation at D-aspartyl residues in erythrocytes: possible step in the repair of aged membrane proteins, *Proc. Natl. Acad. Sci. U.S.A.*, 79, 2460, 1982.
9. **Murray, E. D., Jr. and Clarke, S.,** Synthetic peptide substrates for the erythrocyte protein carboxyl methyltransferase: detection of a new site of methylation at isomerized L-aspartyl residues, *J. Biol. Chem.*, 259, 10722, 1984.
10. **Aswad, D. W.,** Stoichiometric methylation of porcine adrenocorticotropin by protein carboxylmethyltransferase requires deamidation of asparagine 25: evidence for methylation at the alpha-carboxyl group of atypical L-isoaspartyl residues, *J. Biol. Chem.*, 259, 10714, 1984.
11. **McFadden, P. N. and Clarke, S.,** Conversion of isoaspartyl peptides to normal peptides: implications for the cellular repair of damaged proteins, *Proc. Natl. Acad. Sci. U.S.A.*, 84, 2595, 1987.
12. **Johnson, B. A., Murray, E. D., Jr., Clarke, S., Glass, D. B., and Aswad, D. W.,** Protein carboxyl methyltransferase facilitates conversion of atypical L-isoaspartyl peptides to normal L-aspartyl peptides, *J. Biol. Chem.*, 262, 5622, 1987.
13. **Johnson, B. A., Langmack, E. L., and Aswad, D. W.,** Partial repair of deamidation-damaged calmodulin by protein carboxyl methyltransferase, *J. Biol. Chem.*, 262, 12283, 1987.
14. **Galletti, P., Ciardiello, A., Ingrosso, D., Di Donato, A., and D'Alessio, G.,** Repair of isopeptide bonds by protein carboxyl *o*-methyltransferase: seminal ribonuclease as a model system, *Biochemistry*, 27, 1752, 1988.
15. **Ishibashi, Y., Sakagami, Y., Isogai, A., and Suzuki, A.,** Structures of tremerogens *A*-9291-I and *A*-9291-VIII: peptidyl sex hormones of *Tremella brasiliensis*, *Biochemistry*, 23, 1399, 1984.
16. **Clarke, S., Vogel, J. P., Deschenes, R. J., and Stock, J.,** Posttranslational modification of the Ha-ras oncogene protein: evidence for a third class of protein carboxyl methyltransferases, *Proc. Natl. Acad. Sci. U.S.A.*, 85, 4643, 1988.

16a. **Deschenes, R. J., Stimmel, J. B., Clarke, S., Stock, J., and Broach, J. R.,** *RAS* protein of *Saccharomyces cerevisiae* is methylesterified at its carboxyl terminus, *J. Biol. Chem.*, 264, 11865, 1989.

16b. **Ota, I. M. and Clarke, S.,** Enzymatic methylation of 23-29 kDa bovine retinol rod outer segment membrane proteins: evidence for methyl ester formation at carboxyl-terminal cysteinyl residues, *J. Biol. Chem.*, in press.

17. **O'Connor, C. M. and Clarke, S.,** Carboxyl methylation of cytosolic proteins in intact human erythrocytes: identification of numerous methyl-accepting proteins including hemoglobin and carbonic anhydrase, *J. Biol. Chem.*, 259, 2570, 1984.
18. **Freitag, C. and Clarke, S.,** Reversible methylation of cytoskeletal and membrane proteins in intact human erythrocytes, *J. Biol. Chem.*, 256, 6102, 1981.
19. **Terwilliger, T. C. and Clarke, S.,** Methylation of membrane proteins in human erythrocytes: identification and characterization of polypeptides methylated in lysed cells, *J. Biol. Chem.*, 256, 3067, 1981.
20. **Kim, S.,** S-Adenosylmethionine: protein-carboxyl methyltransferase from erythrocyte, *Arch. Biochem. Biophys.*, 161, 652, 1974.
21. **Gilbert, J. M., Fowler, A., Bleibaum, J., and Clarke, S.,** Purification of homologous protein carboxyl methyltransferases isozymes from human and bovine erythrocytes, *Biochemistry*, 27, 5227, 1988.
22. **Runte, L., Jurgensmeier, H.-L., Unger, C., and Soling, H. D.,** Calmodulin carboxylmethyl ester formation in intact human red cells and modulation of this reaction by divalent cations *in vitro*, *FEBS Lett.*, 147, 125, 1982.
23. **Brunauer, L. S. and Clarke, S.,** Methylation of calmodulin at carboxylic acid residues in erythrocytes: a non-regulatory covalent modification?, *Biochem. J.*, 236, 811, 1986.
24. **O'Connor, C. M. and Clarke, S.,** Methylation of erythrocyte membrane proteins at extracellular and intracellular D-aspartyl sites *in vitro*: saturation of intracellular sites *in vivo*, *J. Biol. Chem.*, 258, 8485, 1983.

25. **Clarke, S., McFadden, P. N., O'Connor, C. M., and Lou, L. L.**, Isolation of D-aspartic acid ß-methyl ester from erythrocyte carboxyl methylated proteins, *Methods Enzymol.*, 106, 331, 1984.
26. **O'Connor, C. M., Aswad, D. W., and Clarke, S.**, Mammalian brain and erythrocyte carboxyl methyltransferases are similar enzymes that recognize both D-aspartyl and L-isoaspartyl residues in structurally altered protein substrates, *Proc. Natl. Acad. Sci. U.S.A.*, 81, 7757, 1984.
27. **Lou, L. L. and Clarke, S.**, Enzymatic methylation of band 3 anion transporter in intact human erythrocytes, *Biochemistry*, 26, 52, 1987.
28. **O'Connor, C. M. and Clarke, S.**, Specific recognition of altered polypeptides by widely distributed methyltransferases, *Biochem. Biophys. Res. Commun.*, 132, 1144, 1985.
29. **McFadden, P. N. and Clarke, S.**, Chemical conversion of aspartyl peptides to isoaspartyl peptides: a method for generating new methyl-accepting substrates for the erythrocyte D-aspartyl/L-isoaspartyl protein methyltransferase, *J. Biol. Chem.*, 261, 11503, 1986.
30. **O'Connor, C. M.**, Regulation and subcellular distribution of a protein methyltransferase and its damaged aspartyl substrate sites in developing *Xenopus* oocytes, *J. Biol. Chem.*, 262, 10398, 1987.
31. **Aswad, D. W., Johnson, B. A., and Glass, D. B.**, Modification of synthetic peptides related to lactate dehydrogenase (231-242) by protein carboxyl methyltransferase and tyrosine protein kinase: effects of introducing an isopeptide bond between aspartic acid-235 and serine-236, *Biochemistry*, 26, 675, 1987.
32. **Lowenson, J. and Clarke, S.**, Protein carboxyl methyltransferase from human erythrocytes: substrate specificity with L-isoaspartyl and D-aspartyl-containing peptides and proteins, *Fed. Proc.*, 46, 2090, 1987.
33. **Ota, I. M., Ding, L., and Clarke, S.**, Methylation at specific altered aspartyl and asparaginyl residues in glucagon by the erythrocyte protein carboxyl methyltransferase, *J. Biol. Chem.*, 262, 8522, 1987.
33a. **Ota, I. M. and Clarke, S.**, Enzymatic methylation of L-isoaspartyl residues derived from aspartyl residues in affinity-purified calmodulin: the role of conformational flexibility in spontaneous isoaspartyl formation, *J. Biol. Chem.*, 264, 54, 1989.
33b. **Ota, I. M. and Clarke, S.**, Calcium affects the spontaneous degradation of aspartyl/asparginyl residues in calmodulin, *Biochemistry*, 28, 4020, 1989.
34. **Ding, L. and Clarke, S.**, Characterization of a L-isoaspartyl protein methyltransferase from *Escherichia coli*, *FASEB J.*, 2, A572, 1988.
35. **Aswad, D. W. and Johnson, B. A.**, The unusual substrate specificity of eukaryotic protein carboxyl methyltransferases, *Trends Biochem. Sci.*, 12, 155, 1987.
36. **Kim, S., Lew, B., and Chang, F. N.**, Enzymatic methyl esterification of *Escherichia coli* ribosomal proteins, *J. Bacteriol.*, 130, 839, 1977.
37. **Ota, I. M., Gilbert, J. M., and Clarke, S.**, Two major isozymes of the protein D-aspartyl/L-isoaspartyl methyltransferase from human erythrocytes, *Biochem. Biophys. Res. Commun.*, 151, 1136, 1988.
38. **Aswad, D. W. and Deight, E. A.**, Purification and characterization of two distinct isozymes of protein carboxymethylase from bovine brain, *J. Neurochem.*, 40, 1718, 1983.
39. **Kim, S. and Paik, W. K.**, Purification and properties of protein methylase II, *J. Biol. Chem.*, 245, 1806, 1970.
40. **Polastro, E. T., Deconinck, M. M., Devogel, M. R., Mailier, E. L., Looza, Y. B., Schnek, A. G., and Leonis, J.**, Purification and some molecular properties of protein methylase II from equine erythrocytes, *Biochem. Biophys. Res. Commun.*, 81, 920, 1978.
41. **Freitag, N. E.**, The distribution of isozymic forms of protein carboxylmethyltransferase in rabbit tissues, *J. Undergrad. Res. Univ. Calif. Irvine*, 13, 183, 1983.
42. **Cusan, L., Gordeladze, J. O., Andersen, D., and Hansson, V.**, Characterization of protein carboxyl methylase activities in rat testes: presence of testis specific charge isomer, *Arch. Androl.*, 7, 263, 1981.
43. **Haklai, R. and Kloog, Y.**, Purification and characterization of protein carboxyl methyltransferase from *Torpedo ocellata* electric organ, *Biochemistry*, 26, 4200, 1987.
44. **Brot, N. and Weissbach, H.**, Biochemistry and physiological role of methionine sulfoxide residues in proteins, *Arch. Biochem. Biophys.*, 223, 271, 1983.
45. **Vlassara, H., Brownlee, M., and Cerami, A.**, High-affinity-receptor-mediated uptake and degradation of glucose-modified proteins: a potential mechanism for the removal of senescent macromolecules, *Proc. Natl. Acad. Sci. U.S.A.*, 82, 5588, 1985.
46. **Geiger, T. and Clarke, S.**, Deamidation, isomerization, and racemization at asparaginyl and aspartyl residues in peptides: succinimide-linked reactions that contribute to protein degradation, *J. Biol. Chem.*, 262, 785, 1987.
46a. **Stephenson, R. C. and Clarke, S.**, Succinimide formation from aspartyl and asparaginyl peptides as a model for the spontaneous degradation of proteins, *J. Biol. Chem.*, 264, 6164, 1989.
47. **Clarke, S.**, Propensity for spontaneous succinimide formation from aspartyl and asparaginyl residues in cellular proteins, *Int. J. Peptide Protein Res.*, 30, 808, 1987.
48. **Babu, Y.S., Bugg, C. E., and Cook, W. J.**, Structure of calmodulin refined at 2.2 Å resolution, *J. Mol. Biol.*, 204, 191, 1988.

49. **Johnson, B. A., Freitag, N. E., and Aswad, D. W.,** Protein carboxyl methyltransferase selectively modifies an atypical form of calmodulin: evidence for methylation at deamidated asparagine residues, *J. Biol. Chem.*, 260, 10913, 1985.
50. **Johnson, B. A., Shirokawa, J. M., and Aswad, D. W.,** Deamidation of calmodulin at neutral and alkaline pH: quantitative relationships between ammonia loss and the susceptibility of calmodulin to modification by protein carboxyl methyltransferase, *Arch. Biochem. Biophys.*, 268, 276, 1989.
51. **Chen, J.-K. and Liss, M.,** Evidence of the carboxymethylation of nascent peptide chains on ribosomes, *Biochem. Biophys. Res. Comm.*, 84, 261, 1978.
52. **Nguyen, M. H., Harbour, D., and Gagnon, C.,** Secretory proteins from adrenal medullary cells are carboxyl-methylated *in vivo* and released under their methylated form by acetylcholine, *J. Neurochem.*, 49, 38, 1987.
53. **Diliberto, E. J., Jr. and Axelrod, J.,** Regional and subcellular distribution of carboxymethylase in brain and other tissues, *J. Neurochem.*, 26, 1159, 1976.
54. **Bachmair, A., Finley, D., and Varshavsky, A.,** *In vivo* half-life of a protein is a function of its amino-terminal residue, *Science*, 234, 179, 1986.
55. **Barber, J. R. and Clarke, S.,** Demethylation of protein carboxyl methyl esters: a nonenzymatic process in human erythrocytes?, *Biochemistry*, 24, 4867, 1985.
56. **Johnson, B. A. and Aswad, D. W.,** Enzymatic protein carboxyl methylation at physiological pH: cyclic imide formation explains rapid methyl turnover, *Biochemistry*, 24, 2581, 1985.
57. **Murray, E. D., Jr. and Clarke, S.,** Metabolism of a synthetic L-isoaspartyl-containing hexapeptide in erythrocyte extracts: enzymatic methyl esterification is followed by nonenzymatic succinimide formation, *J. Biol. Chem.*, 261, 306, 1986.
58. **Momand, J. and Clarke, S.,** Rapid degradation of D- and L-succinimide-containing peptides by a post-proline endopeptidase from human erythrocytes, *Biochemistry*, 26, 7798, 1987.

Chapter 11

IDENTITIES, ORIGINS, AND METABOLIC FATES OF THE SUBSTRATES FOR EUKARYOTIC PROTEIN CARBOXYL METHYLTRANSFERASES

Brett A. Johnson and Dana W. Aswad

TABLE OF CONTENTS

I. INTRODUCTION

Research on protein carboxyl methylation in eukaryotic cells has long been identified with the study of a single protein carboxyl methyltransferase (PCMT) that has been detected in and purified from a wide range of species and tissues.[1] Recent experiments have indicated that this enzyme modifies the alpha-carboxyl groups of L-isoaspartyl (beta-carboxyl-linked aspartyl) sites in peptide and protein substrates, and a functional role of methylation in the repair or proteolysis of the altered proteins has been proposed.[2,3] It therefore has been assigned a new Enzyme Commission number, EC 2.1.1.77, in order to distinguish it from the glutamyl methyltransferase, EC 2.1.1.24, that is involved in receptor regulation in chemotactic bacteria.[3] As yet, no methyltransferase activity directly comparable to EC 2.1.1.24 has been demonstrated in eukaryotic cells. However, there has been a recent report of a methyltransferase activity in eukaryotic cells which modifies the alpha-carboxyl group of C-terminal cysteine in certain protein substrates related to the *ras* oncogene protein.[4]

Because normal aspartyl and glutamyl residues do not appear to be substrates for eukaryotic PCMT (hereafter called simply PCMT), much recent work has shifted its focus away from the search for a regulatory function involving neutralization of negative charges. Instead, this research has been directed towards determining the major *in vivo* sources of the altered aspartyl methyl-accepting sites and the fates of these sites once methylated. In this chapter, we will describe the evidence that PCMT selectively modifies altered aspartate. We will then review the progress which has been made in elucidating the origins and destinies of the methyl-accepting sites.

II. THE SUBSTRATE SPECIFICITY OF PCMT

A. PCMT MODIFIES MANY PROTEINS WITH A LOW STOICHIOMETRY

PCMT catalyzes the methylation of a multiplicity of proteins, both *in vitro* and *in vivo*.[1] Many of the best *in vitro* substrates are probably not physiologically relevant methyl acceptors, being located in compartments separate from PCMT *in vivo*. For example, the substrates used most often to assay the enzyme, gelatin, ovalbumin, and gamma globulins, are obtained from extracellular or secretory sources, where PCMT seems to be absent.[1] The methylation of substrate proteins typically occurs with a very low stoichiometry, seldom exceeding 0.05 mol CH_3/mol even under optimal conditions.[3]

Another distinguishing characteristic of methylation by PCMT is the extreme lability of the methyl esters that are produced; these esters are 25 times less stable than the esters formed upon chemical methylation of carboxyl groups in the same proteins.[5] The lability of the methyl esters is responsible for their loss from proteins upon electrophoresis in alkaline sodium dodecyl sulfate gel systems. The recently described C-terminal cysteine methyltransferase apparently calayzes the formation of methyl esters which are much more stable than those formed by PCMT.[4] Thus, the few reports of *in vivo* carboxyl methylation of proteins using alkaline gel systems for detection, e.g., lamin B and cyclic GMP phosphodiesterase,[6,7] must be reevaluated to determine which methyltransferase was involved.

B. STOICHIOMETRIC METHYLATION OF L-ISOASPARTYL PEPTIDES

The broad protein substrate specificity of PCMT and the low stoichiometry of the reaction which it catalyzes prompted some workers to consider the possibility that PCMT recognizes a characteristic common to an altered subpopulation of each methyl-accepting protein. Evidence has now been presented which suggests that the major methylation site is L-isoaspartate, which is aspartate linked through its beta-carboxyl group to the next amino acid in the peptide backbone. Because L-isoAsp can be generated in proteins by known damage events, its methylation by PCMT might explain both the broad specificity and the low stoichiometry of the reaction.

TABLE 1
Methyl-Accepting Capacity of L-Val-L-Tyr-L-Pro-X-Gly-L-Ala Variants

Amino acid at X	Relative initial velocity[a]	mol CH_3/mol	Ref.
L-isoAsp	100	0.80, 0.91	8,9
L-Asp	4	0.01	8,9
D-isoAsp	18[b]	N D[c]	9
D-Asp	–2	N D	9
L-isoGlu	N D	0.00	2

[a] Methylation was by the type I isozyme of bovine brain PCMT, which modified the L-isoaspartyl peptide with a velocity of 10,700 pmol/min/mg.
[b] The methyl incorporation into this peptide was attributed to the presence of some L-isoaspartyl material.
[c] ND = not determined.

Proof of methylation of L-isoaspartate by PCMT comes from studies using synthetic peptide substrates containing altered aspartyl sequences. Table 1 shows the results which have been obtained for one of these peptides, L-Val-L-Tyr-L-Pro-X-Gly-L-Ala, which corresponds to a sequence present in adrenocorticotropin (ACTH) between residues 22 and 27. One variant of this sequence, in which L-isoaspartate occupied the fourth position, was found to be a stoichiometric substrate for both bovine brain and human erythrocyte PCMT.[8,9] Other analogs, containing L-Asp, D-Asp, D-isoAsp, or L-isoGlu in position four, were not substrates (Table 1).

A large number of distinct peptide sequences have now been investigated and have yielded similar results.[2,10] To date, whenever L-isoAsp is present in an internal position, the peptide has been found to be a stoichiometric substrate for PCMT, and most of the peptides are methylated with K_m between 0.4 and 5 μM.[2] L-isoAsp may not be a substrate when it is the N-terminal amino acid, as indicated by the lack of methylation of L-isoAspGly and L-isoAsp1-angiotensin II.[2] No stoichiometric methylation of peptides containing only D-Asp or normal L-Asp or L-Glu has yet been reported.

C. ISOLATION OF D-ASP BETA-METHYL ESTER FROM ERYTHROCYTE MEMBRANES

Clarke and colleagues have provided evidence that PCMT may also methylate the beta-carboxyl group of D-aspartate in erythrocyte membrane proteins. Carboxypeptidase Y digestions of erythrocyte membranes following incubation of intact erythrocytes with [methyl-^{3}H]-methionine *in vivo,*[11,12] or incubation of purified membranes with S-adenosyl-L-[methyl-^{3}H]-methionine and PCMT *in vitro,*[13,14] have been found to contain small amounts of radiolabeled material which is indistinguishable from D-Asp beta-methyl ester by a variety of criteria. The radioactivity in this material accounts for 4 to 15% of the radioactivity released by carboxypeptidase Y, and for about 0.1% of the radioactivity in the membranes used as the starting material for the digestion.[14]

A small amount of D-Asp beta-methyl ester has also been isolated from some fractions of erythrocyte cytosol in similar experiments, although other methylated fractions from the same source yield none of this material.[15] In an unpublished study performed in collaboration with Clarke's laboratory, we have failed to detect any D-Asp beta-methyl ester in bovine brain synaptic plasma membranes following *in vitro* methylation by purified PCMT. No L-Asp beta-methyl ester or L-Glu gamma-methyl ester has yet been isolated from any eukaryotic source.

As mentioned above, peptides containing D-Asp are not substrates for PCMT when the D-Asp is present at the same positions where L-isoAsp is methylated quantitatively.[2] This indicates that, if D-Asp is indeed methylated by PCMT, some specific surrounding sequence or structure must be necessary for the recognition of D-Asp by the enzyme. Flexible peptides should be able to achieve nearly every possible conformation when in solution, so that one would expect to see at least some degree of methylation if conformation were the limiting factor. On the other hand, the stoichiometric methylation of L-isoAsp in peptides of distinct sequence argues against the necessity for specific surrounding amino acids. Thus, it is difficult to reconcile the isolation of D-Asp beta-methyl ester with the lack of methylation of D-Asp in peptides. These concerns, as well as concern about the trace quantities of D-Asp beta-methyl ester isolated, make it important to eliminate possible mechanisms for the formation of D-Asp beta-methyl ester which do not require direct methylation of D-Asp by PCMT.

III. WHAT ARE THE MAJOR SOURCES OF ISOASPARTYL SEQUENCES?

A. *IN VITRO* SOURCES OF ISOASPARTATE

L-isoAsp can be generated in proteins from L-Asp and L-Asn through intramolecular reactions that proceed through cyclic imide intermediates. The formation of the cyclic imide involves a nucleophilic attack by the alpha nitrogen of the C-terminal neighbor amino acid on the side chain carbonyl of L-Asp or L-Asn, the latter being accompanied by deamidation. Subsequent hydrolysis of the five-membered ring produces a mixture of L-Asp and L-isoAsp, in the ratio of 1:3 to 1:4.[9,16-18] The cyclic imide is also prone to inversion of stereoconfiguration at the alpha carbon of the aspartate, followed by hydrolysis to produce D-isoAsp and D-Asp.[18-20] Methylation of L-isoAsp therefore selects for sequences which are prone to imide formation. Nonenzymatic demethylation also occurs through the imide intermediate, and the intramolecular catalysis can explain the extreme lability of the methyl esters formed by PCMT.[17]

The rates of these reactions at physiological pH and temperature have been studied recently using analogs of the ACTH (22-27) peptide.[20] L-Asn appears to be 40 times more rapidly damaged than L-Asp, and the final products of its deamidation are present in the ratio of 70 L-isoAsp: 19 L-Asp: 4 D-isoAsp: 1 D-Asp.[20] Other studies with model peptides have shown that cyclic imides and L-isoAsp form readily upon hydrolysis of synthetic beta-carboxyl esters of aspartate.[21,22] The rate of the rearrangement in dipeptides is dependent on the C-terminal neighbor, being most rapid when this amino acid has a polar, small side chain.[23] Of the amino acids tested, the most rapid imide formation was found when this amino acid was serine, threonine, or glycine, followed by asparagine, glutamine, alanine, and tryptophan, other amino acids giving much slower rates.[23] The rate of deamidation of asparagine in a variety of pentapeptides shows a similar dependence on the C-terminal neighbor of Asn, although the rate also varied with the nature of amino acids at other positions in the peptides.[24] In the limited number of peptides tested, asparagine appeared to deamidate more rapidly when followed by serine, histidine, and aspartate than when followed by alanine, and more rapidly when followed by alanine than when followed by threonine, proline, valine, leucine, or isoleucine.[24] Studies on variants of the asparagine-containing ACTH (22-27) peptide in which glycine was substituted with leucine and proline have confirmed that bulky and nonpolar amino acids decrease the rate of cyclic imide and L-isoAsp formation when they are present as the C-terminal neighbor of Asn.[20]

Evidence for L-isoAsp in purified proteins and peptides has been obtained by a variety of methods. The results from these studies are summarized in Table 2. One method involves exhaustive digestion of proteins by broad-specificity proteases, aminopeptidases, and/or carboxypeptidases, none of which is capable of cleaving the isoaspartyl linkage. These digestions liberate isoaspartyl dipeptides which can be identified using conventional chromatography,[25-27] or HPLC.[28] This method allows identification of the residues following L-isoAsp, but in most cases, it will not indicate whether the original amino acid was L-Asp or L-Asn unless the Asx-X sequence occurs only once in a polypeptide of known sequence. Because of the extensive manipulation entailed in these procedures, it is often difficult to determine whether isoAsp was present in the original protein or was generated in the peptides during analysis.

Another indicator of an isoaspartyl linkage is a drastic drop in the yield of Asp or Asn during

TABLE 2
Isoaspartate in Purified Proteins and Peptides

Polypeptide	Parent sequence	Method[a]	Ref.
Bovine calmodulin	Asx-Gly	A	28
Human and bovine collagen	Asx-Gly	A	25,26
Porcine pepsin	Asx-Gly	A	26
Chicken egg white lysozyme	Asx-Gly	A	25,26
Human fibrinogen and fibrin	Asx-Gly	A	25
Porcine ACTH	Asn^{25}-Gly^{26}	B	29,30
Smut fungus RNase U2	Asn^{23}-Gly^{24}	B	31
Human carbonic anhydrase C	Asn^{11}-Gly^{12} Asn^{56}-Gly^{57} Asn^{232}-Gly^{233}	B	32
Bovine seminal RNase	Asn^{67}-Gly^{68}	C	33
Bovine RNase A	Asn^{67}-Gly^{68} Asn^{44}-Thr^{45}	A	27
Porcine PHI-27	Asp^{3}-Gly^{4}	B	34
Human hemoglobin	Asp^{67}-Gly^{68b} Asx-Ala	A	26
Mouse epidermal growth factor	Asn^{1}-Ser^{2}	B	35
Porcine gastrin releasing peptide	Asn^{20}-His^{21}	B	36
Bovine/porcine glucagon	Asp^{9}-Tyr^{10} Asn^{28}-Thr^{29}	D	37

[a] Methods are: A, dipeptide in proteolytic digest; B, decreased yield of Asx during Edman degradation; C, internal alpha-carbon labeling; D, isolation of modified fragment after *in vitro* methylation. In many cases, especially those using methods B-D, additional data corroborates the existence of isoAsp. The reader is referred to the original references for details.

[b] Present in the beta-subunit.

Edman degradation of a purified polypeptide, often a proteolytic fragment, of known sequence. This occurs because the extra carbon introduced into the peptide backbone by the isoaspartyl linkage prevents the formation of the five-membered phenylthiohydantoin intermediate when the isoAsp is the N-terminal amino acid. Again, it is difficult to conclude with certainty whether the isoaspartyl linkage was present in the starting protein material or was generated during purification and/or analysis. A third method capitalizes on an acetic anhydride-catalyzed exchange of tritium from $^{3}H_2O$ to the alpha carbon of amino acids with free alpha carboxyl groups; the internal labeling that occurs during this procedure has been used to demonstrate isoAsp in a peptide derived from bovine seminal ribonuclease.[33]

In a recent study, glucagon was methylated with purified PCMT to the low stoichiometry of 0.004 mol CH_3/mol, and the methylated peptide was then digested with a variety of proteases in order to localize the methyl-accepting sites to small fragments which could be separated by reversed-phase HPLC.[37] Two sites of methylation were revealed, one in an L-isoAspTyr sequence produced by isomerization of Asp and the other at an L-isoAspThr sequence produced by deamidation of Asn. Depending on the protease, the recovery of methyl esters upon HPLC varied from 20 to 60%. Methylation at isoAspTyr accounted for 60% of the methyl groups recovered, whereas methylation at isoAspThr made up about 20%.[37] The relative recovery of methyl esters in such an analysis may not be representative of the relative amount of L-isoAsp originally present in the peptide, because methylation may have not been complete and because methyl esters of L-isoAsp in different sequences would be expected to have widely varying stabilities. In fact, those sequences most prone to cyclic imide and L-isoAsp formation might be expected to undergo the most rapid demethylation, as demethylation also proceeds through a cyclic imide intermediate.[17]

The sequences responsible for L-isoAsp formation in the limited list of proteins shown in Table 2 seem to confirm the general rules derived from model peptide studies: Asn is more prone to damage than Asp, and small, polar side chains are preferred in the C-terminal neighboring amino acid, Asn-Gly being the major source of L-isoAsp. Further studies on a wider range of proteins will of course be required before a strong conclusion can be made. It is probable that there are exceptions to the rule for individual proteins, because the Asn or Asp residues with the preferred C-terminal neighbors may lie in structures prohibiting imide formation, whereas other Asx sequences may have bond angles which favor formation of the imide.[38]

B. STOICHIOMETRIC METHYLATION OF PROTEINS BY PCMT?

Although most proteins are methylated by PCMT with a very low stoichiometry, there are several reports of proteins whose carboxyl methylation exceeds the typical < 0.05 mol CH_3/mol value. These include ACTH, calmodulin, the nicotinic acetylcholine receptor, and calcineurin. Because L-isoaspartyl methyl-accepting sites can be generated in proteins artifactually during purification, we have reevaluated the methyl-accepting capacity of each of these proteins.

1. Adrenocorticotropin

In an early study addressing the site of methylation of ACTH, a stoichiometry of 0.30 mol CH_3/mol was reported.[39] However, we observed a plateau of methylation at a stoichiometry of only 0.07 mol/mol using commercial porcine ACTH as substrate.[8] Addition of fresh ACTH after this plateau was reached resulted in a new burst of methylation representing 0.07 mol/mol modification of the additional ACTH, which indicated that a subpopulation of ACTH limited the reaction, not inactivation of PCMT or exhaustion of the radiolabeled methyl donor. After fractionation of the ACTH on cation-exchange chromatography, the methyl-accepting capacity was found in a peak corresponding to deamidated ACTH. This peak could be increased in proportion by incubation of the ACTH at alkaline pH (3 h in 0.1 *M* NH_4OH, 37°C), and the deamidated product accepted methyl groups to a stoichiometry of 0.70 mol/mol. No D-Asp was formed during the base-treatment. Thus, it appeared that methylation of L-isoAsp produced upon deamidation of ACTH could explain most of the methyl-accepting capacity of the peptide.

2. Calmodulin

Stoichiometries of 0.35 to 0.50 mol CH_3/mol have been reported for the methylation of calmodulin by purifed PCMT *in vitro*.[40,41] However, a number of laboratories have achieved far lower stoichiometries more representative of the < 0.05 mol/mol methylation observed for other proteins.[42-45] We showed that the low stoichiometry of the methylation was a result of the modification of a subpopulation of the calmodulin molecules, and that the subpopulation could be increased to 0.35 to 0.52 mol/mol upon brief alkaline treatment performed under the same conditions which caused deamidation of ACTH.[42] This alkaline treatment increases the amount of isoAspGly which can be recovered after extensive proteolytic digestion.[28] The increased methyl-accepting capacity upon incubation at alkaline pH is directly proportional to ammonia release (0.5 mol CH_3/mol NH_3), indicating that the methylation might be occurring at L-isoAsp produced upon deamidation of asparagine.[46] This hypothesis is further supported by the finding that the increases in methylation could be blocked by a prior chemical modification of side chain amides to prevent their deamidation.[42] The methyl-accepting subpopulation can also be increased by prolonged incubation at physiological pH,[46,47] and by exposure to elevated temperatures.[45,48] It is therefore likely that the early reports of near-stoichiometric methylation of calmodulin by PCMT reflected modification of an artifactually damaged subpopulation of molecules.

3. The Nicotinic Acetylcholine Receptor

The acetylcholine receptor protein isolated from the electric organ of *Torpedo californica* accepts methyl groups from erythrocyte PCMT with low stoichiometries ranging from 0.03 to 0.07 mol CH_3/mol of holoprotein,[49,50] but the *Torpedo* methyltransferase has been reported to methylate the receptor to 0.2 mol CH_3/mol of holoprotein.[51] This protein is comprised of five subunits, and all of the subunits accept methyl groups,[50,51] so that the actual stoichiometry at a single methyl-accepting site must be lower than that for the holoprotein. It also appears that a large amount of the methyl-accepting capacity *in vitro* is on the extracellular surface of the membrane, where PCMT is absent *in vivo*.[50]

Because the receptor used in those studies reporting moderate stoichiometries of methylation was isolated using procedures involving extraction at elevated pH,[50,51] and because alkaline conditions can encourage isoAsp formation through deamidation of Asn, we decided to evaluate the stoichiometry of the methylation using receptor protein purified by milder methods utilizing affinity chromatography techniques. The protein thus isolated was methylated at 2.5 μM concentration in a reaction containing 2.5 μM bovine brain PCMT and 200 μM radiolabeled *S*-adenosyl methionine, conditions giving maximal methylation of isoaspartyl peptides, deamidated ACTH, and deamidated calmodulin. We achieved a stoichiometry of only 0.06 mol CH_3/mol holoprotein under these conditions.[52] However, as shown in Figure 1, the methyl-accepting capacity could be increased markedly by prior incubation of the receptor protein at alkaline pH. Thus, it seems that PCMT is methylating a damaged subpopulation of this protein, as well.

4. Calcineurin

The calcium-dependent phosphatase calcineurin has been reported to accept methyl groups from purified PCMT to the very high stoichiometry of 2.0 mol CH_3/mol.[53] However, when we incubated calcineurin (a gift from Dr. Claude Klee, National Institutes of Health, Bethesda, MD) at a concentration of 5 μM in a pH 6, 30°C, 40 min reaction containing 200 μM *S*-adenosyl-L-[methyl-^{3}H]-methionine and 0 to 5 μM bovine brain PCMT, the methylation increased with PCMT concentration to a plateau between 2 and 5 μM. At the plateau, which is indicative of saturation of methyl-accepting sites,[8] the stoichiometry was 0.05 mol CH_3/mol. A prior incubation of the calcineurin under alkaline conditions (9 h in 0.1 M NH_4OH at 37°C) caused an increase in the methyl-accepting capacity (to 0.25 mol CH_3/mol). In the study reporting stoichiometric methylation,[53] the calculated stoichiometries were directly related to the ratio of PCMT/calcineurin, the 2 mol/mol stoichiometry requiring a ratio of 300:1. The fact that no saturation of methyl-accepting sites was observed suggests that the methyl incorporation may have actually occurred into endogenous substrates present in the enzyme preparation.

C. EVIDENCE FOR THE FORMATION OF ALTERED ASPARTATE *IN VIVO*

Although the near-stoichiometric methylation of purified proteins probably reflects a modification of artifactually damaged subpopulations, it is obvious that a wide range of proteins do serve as substrates for PCMT *in vivo*. Thus, it seems likely that altered aspartyl sequences are generated during the cellular lifetime of proteins. Possible mechanisms include the damage

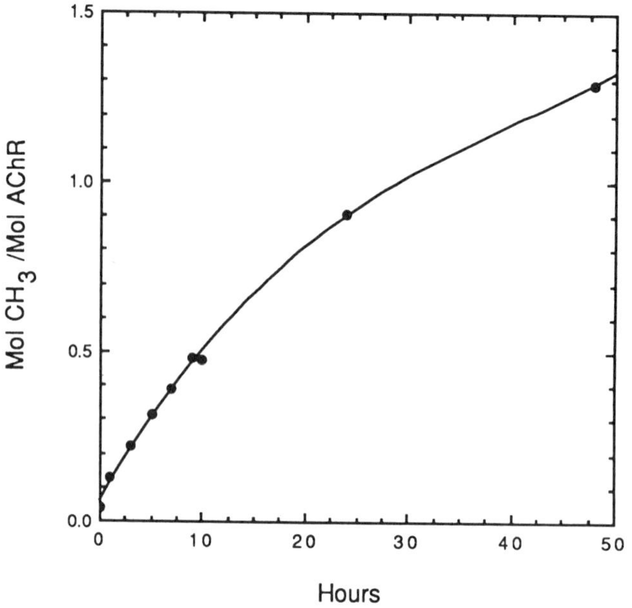

FIGURE 1. Alkaline treatment increases the methyl-accepting capacity of the *Torpedo californica* acetylcholine receptor. The isolated receptor protein (9.5 μ*M*) was incubated at 37°C for varying periods of time in 0.1 *M* NH_4OH. After lyophilization, it was methylated at a concentration of 2.5 μ*M* using 2.5 μM bovine brain PCMT and 200 μ*M* S-adenosyl-[*methyl*-^{3}H]-methionine. Methyl incorporation was determined after precipitation with trichloroacetic acid as described by Aswad. (From Aswad, D. W., *J. Biol. Chem.*, 259, 10654, 1984. With permission.)

reactions which have been observed *in vitro,* errors in protein synthesis,[9] and enzymatic production.

L-isoaspartyl di- and tripeptides have been identified in human urine,[25,54-58] and in the feces of rats and humans on antibiotic regimens.[59,60] The urinary isoaspartyl dipeptides are present in subjects who are fasting or adhering to protein-free diets, suggesting a metabolic, rather than dietary, origin.[55,56,58] In one study, the relative prevalence of urinary isoaspartyl dipeptides from a single fasting human was isoAspGly >> isoAspSer > isoAspThr, isoAspAla > isoAspAsn, isoAspGln > others,[56] which recalls the preference for a C-terminal neighboring amino acid with a small, polar side chain observed for isomerization of aspartyl beta-carboxyl esters.[23] Another study, which focused on the excretion of acidic isoaspartyl dipeptides from six cancer or fistula patients on protein-free diets, found isoAspGly > isoAspAla > isoAspAsp > isoAspGlu > isoAspSer > isoAspThr.[58] Hence, it is clear that isoAsp is generated *in vivo,* and it is possible that the *in vitro* sources discussed above, deamidation of asparagine and isomerization of aspartate, can account for its formation. However, an enzyme that can synthesize isoaspartyl dipeptides has been detected in rat kidney, and it may be responsible for the presence of certain dipeptides in urine.[61]

D. MODELED PROTEIN DAMAGE UNDER PHYSIOLOGICAL CONDITIONS

As discussed above, it is commonly very difficult to prove that L-isoAsp detected in proteins did not form during purification or analysis. On the other hand, analyses of purified proteins select for cases in which the L-isoAsp is not removed by possible *in vivo* processes. A likely function of PCMT is the removal of L-isoAsp by repair or proteolysis (see below), so that the most important *in vivo* sources of L-isoAsp methyl-accepting sites might not be detected in analyses of the methylation of purified proteins. (In this regard, it is interesting to note that most

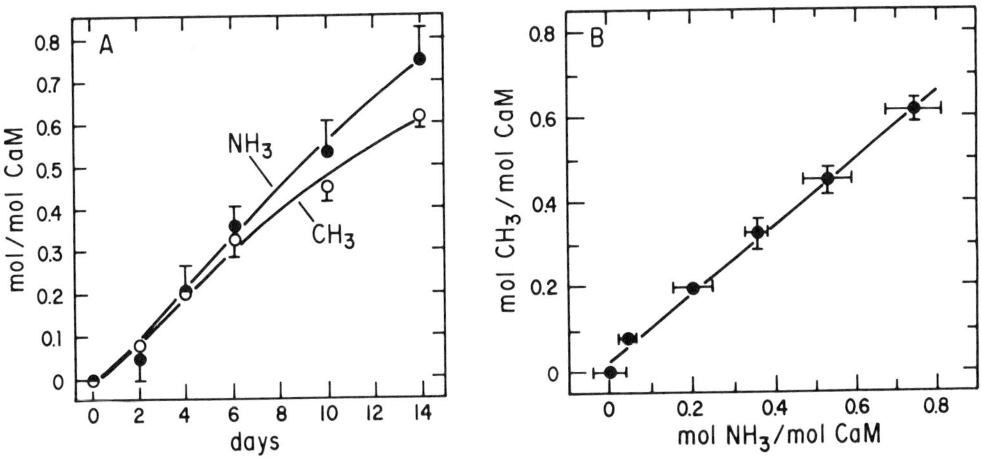

FIGURE 2. The methyl-accepting capacity of calmodulin is proportional to the amount of ammonia released upon incubation under physiological conditions. Bovine brain calmodulin (125 μ*M*) was incubated for varying periods of time at 37°C and pH 7.4 in 44 m*M* potassium Hepes, 1 m*M* EGTA. Separate samples were taken for measurement of free ammonia or for methylation by bovine brain PCMT.[46] Error bars denote sample standard deviations of triplicate assays. In panel B, the data from panel A are replotted in order to show the correlation between ammonia release and methyl incorporation. The line drawn is the result of a least-squares linear regression on the mean values, which gave a correlation coefficient of 0.997 and a slope of 0.79 ± 0.08 mol CH_3/mol NH_3 released.

of the proteins in which large amounts of L-isoAsp has been found (Table 2) are located in secretory or extracellular compartments, where PCMT is believed to be absent). One strategy which might be used to correct for both problems is to model the proposed damage by incubating a protein under physiological conditions. Comparison with the unincubated protein could then indicate probable sources of L-isoAsp formation *in vivo*.

We have begun to employ this strategy to analyze L-isoAsp formation in calmodulin, which has been found to be methylated to a small degree *in vivo*.[40,43,45,62] When calmodulin is incubated at pH 7.4, 37°C in the absence of calcium, it becomes deamidated as indicated by a loss of ammonia (Figure 2).[46] Methyl-accepting capacity increases proportionately (Figure 2), and ammonia release exceeds methylation, indicating that asparagine deamidation might account for most of the induced L-isoAsp.[46] Deamidation and increased methyl-accepting capacity are detectable within 2 d of incubation. Although the half-life of calmodulin is unknown, it seems likely that significant isoAsp formation could occur *in vivo*.

The increase in methyl-accepting capacity upon incubation at pH 7.4 and 37°C is unaffected by physiological concentrations of calcium (10^{-7} *M*), but higher calcium concentrations inhibit formation of methyl-accepting sites (Figure 3).[46] The biphasic shape of the calcium protection curve shown in Figure 3 suggests that there are at least two calcium binding sites, one of low affinity and one of high affinity, which are responsible for the inhibition of increased methylation. Calmodulin contains two AsnGly sequences, at positions 97 to 98 and 60 to 61, each of which is located in a separate calcium-binding domain.[63] From the studies on model peptides, these asparagines would be expected to be prone to imide and L-isoAsp formation. In preliminary studies, tryptic fragments containing these sequences from incubated calmodulin have been found to contain about half of the induced methyl-accepting sites.[64]

The chief problem with studies on modeled protein damage is in matching the *in vitro* incubation conditions with the appropriate *in vivo* environment for each protein of interest. Studies on deamidation of proteins *in vitro* have indicated that buffer composition and the presence of substrates, inhibitors, or allosteric effectors can have major effects on the rate of damage.[65] Also, the likelihood of significant L-isoAsp formation *in vivo* depends on the turnover rate of the protein: long-lived proteins would be expected to accumulate more altered L-isoAsp

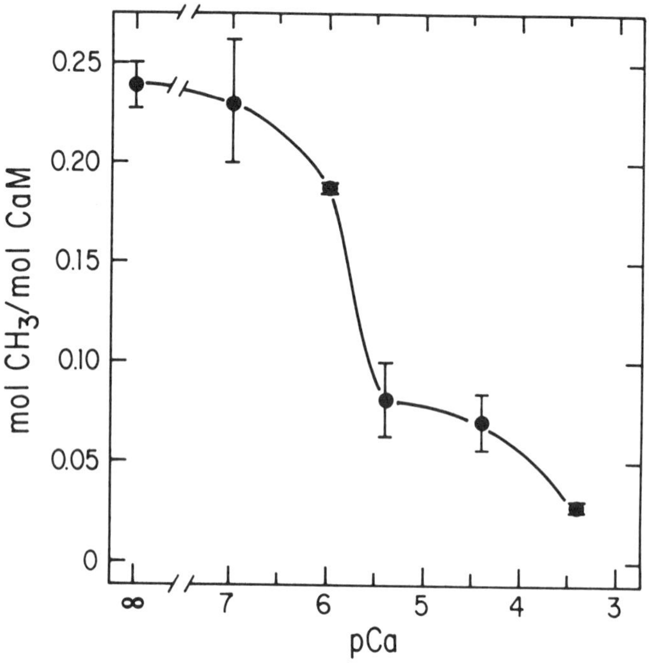

FIGURE 3. Effect of calcium concentration on the increase in methyl-accepting capacity of calmodulin upon incubation under physiological conditions. Calmodulin (54 μ*M*) was incubated for 14 d at pH 7.4, 37°C in 100 m*M* sodium Hepes, ionic strength adjusted to 0.16, with calcium buffers ranging in pCa from 3.4 to 7.[46] An additional incubation was performed in the presence of 10 m*M* EGTA and no added calcium (pCa = ∞). Sodium azide (0.2 mg/ml) was included in each incubation to prevent bacterial growth. After the incubation, the samples of calmodulin were dialyzed against a calcium-free buffer and then tested for methyl-accepting capacity.

than proteins which are degraded shortly after synthesis.

IV. WHAT ARE THE FATES OF ISOASPARTYL POLYPEPTIDES, AND WHAT IS THE ROLE OF METHYLATION?

A. CONVERSION OF L-ISOASP TO L-ASP BY PCMT

When peptides containing internal L-isoAsp are incubated with PCMT and S-adenosylmethionine at physiological pH and temperature, a majority of the L-isoAsp is converted to L-Asp.[18] An example, using the model peptide Lys-Gln-Val-Val-isoAsp-Ser-Ala-Tyr-Glu-Val-Ile-Lys, is shown in Figure 4. The conversion occurs because enzymatic methylation is followed by a rapid, nonenzymatic demethylation which proceeds through the cyclic imide intermediate.[17] Nonenzymatic hydrolysis of the cyclic imide produces a mixture of L-aspartyl and L-isoaspartyl peptides. The regenerated L-isoAsp is once again methylated by PCMT and the cycle continues. This conversion has now been observed for a total of five model peptides of varying sequence.[10,18,19] In no case is the conversion 100% effective, because inversion of stereoconfiguration occurs at the aspartyl alpha-carbon of the cyclic imide intermediate, yielding a significant amount of D-isoAsp, which is not a substrate for PCMT.

Conversion of L-isoAsp to L-Asp would give complete repair of a protein if the isoAsp were generated by isomerization of aspartate. If the isoAsp was formed upon deamidation of Asn, the replacement with Asp would result in an altered side chain, but the extra carbon in the peptide backbone of the isoAsp linkage would be removed, and this might allow the protein to achieve

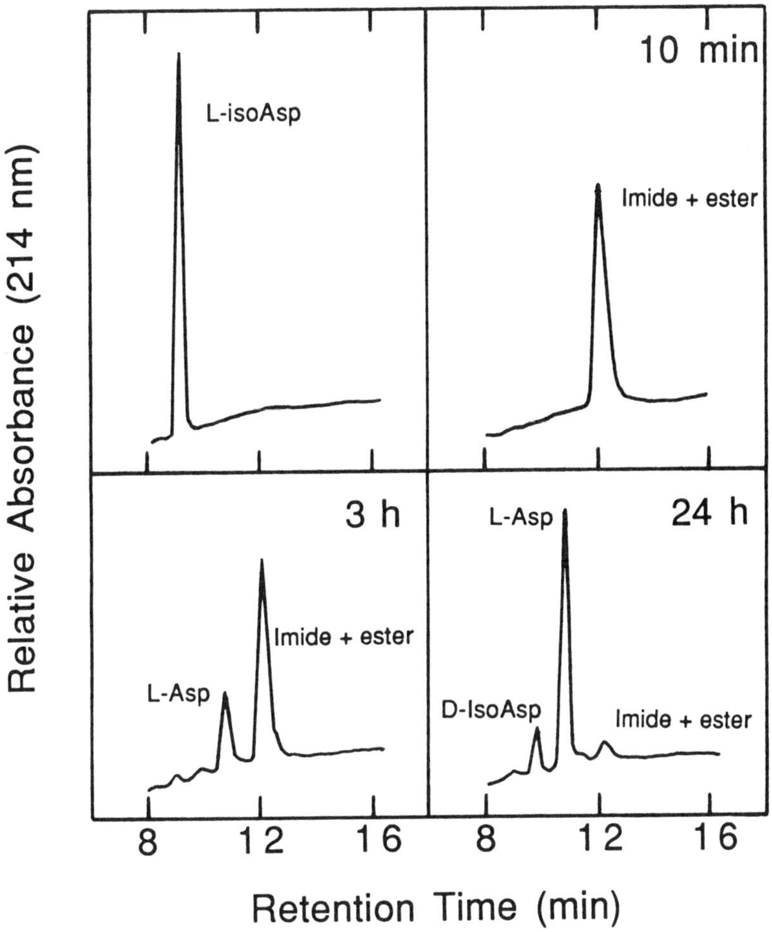

FIGURE 4. PCMT can convert an L-isoaspartyl peptide to an L-aspartyl peptide. The peptide Lys-Gln-Val-Val-isoAsp-Ser-Ala-Tyr-Glu-Val-Ile-Lys (6 μ*M*) was incubated at pH 7.4 and 37°C with 2.5 μ*M* bovine brain PCMT and 200 μ*M* S-adenosyl-L-methionine. The identity of the L-aspartyl product was determined by coelution with the synthetic peptide and by comparisons of proteolytic digests with those of the authentic peptide. (From Johnson, B. A., Murray, E. D., Jr., Clark, S., Glass, D. B., and Aswad, D. W., *J. Biol. Chem.*, 262, 5622, 1987. With permission.)

a more normal conformation and activity. We have tested the effect of the isoAsp-to-Asp conversion using deamidated calmodulin as substrate. Upon incubation at pH 7.4 and 37°C, methyl-accepting forms of calmodulin are generated which have a molecular radius greater than that of the native protein.[47] These forms were isolated using gel filtration chromatography and were found to be deficient in their ability to stimulate calcium-dependent phosphorylation in rat brain membranes. However, after incubation with PCMT and *S*-adenosylmethionine under conditions which result in the conversion of L-isoAsp to L-Asp in peptides, activity was restored (Figure 5). It therefore appears that the PCMT-dependent conversion of L-isoAsp to L-Asp in a protein might be of considerable functional significance.

Another possible role of the conversion of L-isoAsp to L-Asp is to allow facile cleavage of peptides by proteases involved in normal protein turnover. Most proteases are not capable of cleaving the isoaspartyl bond *in vitro*,[9,66] and conversion to L-Asp should remove this barrier to proteolysis.

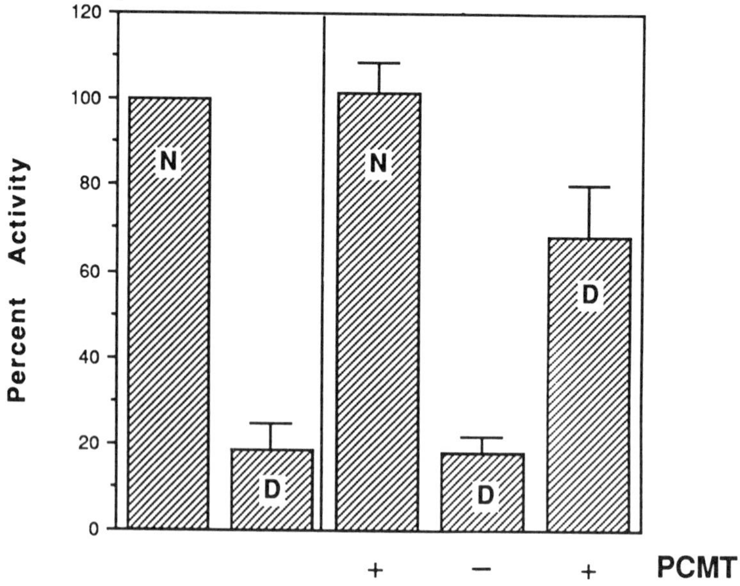

FIGURE 5. Functional repair of deamidated calmodulin by PCMT. Native calmodulin (N) or the high-M_r forms of calmodulin produced by a 28-d, pH 7.4, 37°C deamidation (D) were tested for their ability to stimulate the calcium-dependent phosphorylation of a 48,000 Da protein in rat brain membranes. In the left panel, activity was tested without additional incubation. In the right panel, the calmodulins were first incubated in a 48-h, pH 7.4, 37°C methylation reaction containing 200 μ*M* *S*-adenosyl-L-methionine in the presence or absence of bovine brain PCMT. Error bars denote standard deviations of triplicate activity assays. Further details are given in Reference 47.

B. POSSIBILITIES FOR MODIFICATION OF THE METHYL ESTER OR IMIDE INTERMEDIATE

Because of the slow rate of isoAsp-to-Asp conversion by PCMT and because of the inefficiency of the reaction, a number of other possible reactions involving modification of the methyl ester or imide intermediate have been proposed.[17] One possibility is the enzymatic addition of ammonia to the beta carbonyl of the aspartyl imide, which would result in the conversion of L-isoAsp to L-Asn. Such a reaction would lead to complete repair of a protein if Asn was the source of isoAsp, but it would be counterproductive if the isoAsp was produced by isomerization of L-Asp. Another possibility is direct proteolysis at the methyl ester or imide stages, or the addition of a marker molecule which would then signal proteolysis. The latter situation would be analogous to the stimulation of proteolysis which has been observed upon the enzymatic, covalent linkage of ubiquitin to denatured proteins.[67]

It is unclear which intermediate would be the most likely candidate for further modification. The methyl esters created by PCMT are very labile and may not survive long enough for enzymatic modification. On the other hand, most known reactions producing isoAsp require the formation of the cyclic imide, and it is unclear why any proposed modification of the imide would not occur at the imide produced upon deamidation of Asn or isomerization of Asp.

C. TESTS OF THE METABOLISM OF PEPTIDES CONTAINING INTERNAL L-ISOASP

To date, investigations into possible methylation-dependent fates of L-isoAsp have used model peptides incubated in cell extracts. For example, the ACTH (22-27) analog Val-Tyr-Pro-isoAsp-Gly-Ala has been incubated in erythrocyte cytosol, where it becomes methylated by PCMT and demethylated nonenzymatically through the imide intermediate.[68] In a further study on the metabolism of this peptide, it was labeled with a tritiated acetyl group at the N-terminus prior to incubation. The labeled peptide was rapidly cleaved by a postproline endopeptidase at the Pro-isoAsp bond, as were variants containing L-Asp or a cyclic imide.[69] Similar results have been obtained in our laboratory using analogs of the fluorescently labeled peptide dansyl-Gly-Phe-Asp-Leu-isoAsp-Gly-Gly-Gly-Val-Gly. When incubated in rat brain cytosol, methyl ester and imide formation are both evident, but the peptide is cleaved at the Leu-isoAsp bond, independently of methylation.[70] Analogs containing L-Asp or L-Asn in place of L-isoAsp are cleaved somewhat more rapidly.

In these two cases, N-terminal L-isoAsp peptides are created as a result of endoprotease activities. As discussed above, L-isoAsp may not be a substrate for PCMT in this position. There is evidence, however, for a protease activity in a variety of rat tissues that is capable of cleaving the isoaspartyl bond of certain di- and tripeptides possessing L–isoAsp as the N-terminal amino acid.[66] Thus, modification of L-isoAsp by PCMT might not have been necessary for the complete proteolysis of these two peptides.

We have also studied the metabolism of the peptide dansyl-Trp-Met-isoAsp-Phe-NH_2 in brain cytosol, with different results.[70] The aspartyl version of this peptide is destroyed rapidly by endogenous proteases, 80% of the original material disappearing within 5 min of incubation at pH 7.4 and 37°C. In contrast, the isoaspartyl version is very resistant to proteolysis, 76% of the original peptide remaining after 24 h of incubation in the presence of the methyltransferase inhibitor *S*-adenosylhomocysteine. When methylation is encouraged by the addition of *S*-adenosylmethionine, the methyl ester and imide both form and are quite stable, the imide persisting throughout a 24 h incubation.

These tests using model peptides have not revealed any evidence for specific modifications of the isoaspartyl methyl ester or cyclic imide. It is therefore possible that the PCMT-dependent L-isoAsp-to-L-Asp conversion is the relevant sequence of events *in vivo*. However, it is also possible that the cytosolic extracts used above are deficient in a cofactor or substrate required by other enzymes in the pathway, or that L-isoAsp must be in a specific sequence or structure to be recognized by the other enzymes. Further research using larger isoAsp-containing proteins and true *in vivo* systems is clearly needed to determine the contribution of PCMT to the metabolism of its substrates.

REFERENCES

1. **Paik, W. K. and Kim, S.,** *Protein Methylation,* John Wiley & Sons, New York, 1980.
2. **Aswad, D. W. and Johnson, B. A.,** The unusual substrate specificity of eukaryotic protein carboxyl methyltransferases, *Trends Biochem. Sci.,* 12, 155, 1987.
3. **Clarke, S.,** Protein carboxyl methyltransferases: two distinct classes of enzymes. *Annu.. Rev. Biochem.,* 54, 419, 1985.
4. **Clarke, S., Vogel, J. P., Deschenes, R. J., and Stock, J.,** The mammalian *ras* oncogene protein is modified by a new type of carboxyl methylation reaction, *Proc. Natl. Acad. Sci. U.S.A.,* 85, 4643, 1988.
5. **Kim, S. and Paik, W. K.,** Labile protein-methyl ester: comparison between chemically and enzymatically synthesized, *Experientia,* 32, 982, 1976.
6. **Chelsky, D., Olson, J. F., and Koshland, D. E., Jr.,** Cell-cycle dependent methyl esterification of lamin B, *J. Biol. Chem.,* 262, 4303, 1987.
7. **Swanson, R. J. and Applebury, M. L.,** Methylation of proteins in photoreceptor rod outer segments, *J. Biol. Chem.,* 258, 10599, 1983.
8. **Aswad, D. W.,** Stoichiometric methylation of porcine adrenocorticotropin by protein carboxyl methytransferase requires deamidation of asparagine-25. Evidence for methylation at the alpha-carboxyl group of atypical isopeptide-linked aspartyl residues, *J. Biol. Chem.,* 259, 10654, 1984.
9. **Murray, E. D., Jr. and Clarke, S.,** Synthetic peptide substrates for the erythrocyte protein carboxyl methyltransferase. Detection of a new site of methylation at isomerized L-aspartyl residues, *J. Biol. Chem.,* 259, 10662, 1984.

10. **Galletti, P., Cardiello, A., Ingrosso, D., DiDonato, A., and D'Alessio, G.,** Repair of isopeptide bonds by protein carboxyl O-methyltransferase: seminal ribonuclease as a model system, *Biochemistry,* 27, 1752, 1988.
11. **Janson, C. A. and Clarke, S.,** Identification of aspartic acid as a site of methylation in human erythrocyte membrane proteins, *J. Biol. Chem.,* 255, 11640, 1980.
12. **McFadden, P. N. and Clarke, S.,** Methylation at D-aspartyl residues in erythrocytes: possible step in the repair of aged membrane proteins, *Proc. Natl. Acad. Sci. U.S.A.,* 79, 2460, 1982.
13. **O'Connor, C. M. and Clarke, S.,** Methylation of erythrocyte membrane proteins at extracellular and intracellular D-aspartyl sites *in vitro, J. Biol. Chem.,* 258, 8485, 1983.
14. **O'Connor, C. M., Aswad, D.W., and Clarke, S.,** Mammalian brain and erythrocyte carboxyl methyltransferases are similar enzymes which recognize both D-aspartyl and isoaspartyl residues in structurally altered protein substrates, *Proc. Natl. Acad. Sci. U.S.A.,* 81, 7757, 1984.
15. **O'Connor, C. M. and Clarke, S.,** Carboxyl methylation of cytosolic proteins in intact human erythrocytes. Identification of numerous methyl-accepting proteins including hemoglobin and carbonic anhydrase, *J. Biol. Chem.,* 259, 2564, 1985.
16. **Bornstein, P. and Balian, G.,** Cleavage at Asn-Gly bonds with hydroxylamine, *Methods Enzymol.,* 47, 132, 1977.
17. **Johnson, B. A. and Aswad D. W.,** Enzymatic protein carboxyl methylation at physiological pH: cyclic imide formation explains rapid methyl turnover, *Biochemistry,* 24, 2581, 1985.
18. **Johnson, B. A., Murray, E. D., Jr., Clarke, S., Glass, D. B., and Aswad, D. W.,** Protein carboxyl methyltransferase facilitates conversion of atypical L-isoaspartyl peptides to normal L-aspartyl peptides, *J. Biol. Chem.,* 262, 5622, 1987.
19. **McFadden, P. N. and Clarke, S.,** Conversion of isoaspartyl peptides to normal peptides: implications for the cellular repair of damaged proteins, *Proc. Natl. Acad. Sci. U.S.A.,* 84, 2595, 1987.
20. **Geiger, T. and Clarke, S.,** Deamidation, isomerization, and racemization at asparaginyl and aspartyl residues in peptides. Succinimide-linked reactions that contribute to protein degradation, *J. Biol. Chem.,* 262, 785, 1987.
21. **Battersby, A. R. and Robinson, J. C.,** Studies on specific chemical fission of peptide links. I. The rearrangement of aspartyl and glutamyl peptides, *J. Chem. Soc.,* 259, 1955.
22. **DeTar, D. F., Gouge, M., Honsberg, W., and Honsberg, U.,** Sequence peptide polymers. I. Polymers based on aspartic acid and glycine, *J. Am. Chem. Soc.,* 988, 1961.
23. **Bodanszky, M. and Kwei, J. Z.,** Side reactions in peptide synthesis. VII. Sequence dependence in the formation of aminosuccinyl derivatives from beta-benzylaspartyl peptides, *Int. J. Peptide Protein Res.,* 12, 63, 1978.
24. **Robinson, A. B. and Rudd, C. J.,** Deamidation of glutaminyl and asparaginyl residues in peptides and proteins, *Curr. Top. Cell. Regul.,* 8, 247, 1974.
25. **Pisano, J. J., Prado, E., and Freedman, J.,** Beta-aspartylglycine in urine and enzymic hydrolyzates of proteins, *Arch. Biochem. Biophys.,* 117, 394, 1960.
26. **Haley, E. E., Corcoran, B. J., Dorer, F. E., and Buchanan, D. L.,** Beta-aspartyl peptides in enzymatic hydrolysates of protein, *Biochemistry,* 5, 3229, 1966.
27. **Haley, E. E. and Corcoran, B. J.,** Beta-aspartyl peptide formation from an amino acid sequence in ribonuclease, *Biochemistry,* 6, 2668, 1967.
28. **Aswad, D. W., Gleason, C. S., and Miller, P. G.,** Isolation of iso-aspartyl-glycine from deamidated calmodulin, *FASEB J.,* 3, A555, 1988.
29. **Gráf, L., Bajusz, S., Patthy, A., Barát, E., and Cseh, G.,** Revised amide location for porcine and human adrenocorticotropin hormone, *Acta Biochim. Biophys. Acad. Sci. Hung.,* 6, 415, 1971.
30. **Ekman, R., Norén, H., Håkanson, R., and Jörnvall, H.,** Novel variants of adrenocoticotrophic hormone in porcine anterior pituitary, *Regul. Pept.,* 8, 305, 1984.
31. **Kanaya, S. and Uchida, T.,** Comparison of the primary structures of ribonuclease U2 isoforms, *Biochem. J.,* 240, 163, 1986.
32. **Henderson, L. E., Henriksson, D., and Nyman, P. O.,** Primary structure of human carbonic anhydrase C, *J. Biol. Chem.,* 251, 5457, 1976.
33. **Di Donato, A. D., Galletti, P., and D'Alessio, G.,** Selective deamidation and enzymatic methylation of seminal ribonuclease, *Biochemistry,* 25, 8361, 1986.
34. **Tatemoto, K. and Mutt, V.,** Isolation and characterization of the intestinal peptide porcine PHI (PHI-27), a new member of the glucagon-secretin family, *Proc. Natl. Acad. Sci. U.S.A.,* 78, 6603, 1981.
35. **DiAugustine, R. P., Gibson, B. W., Aberth, W., Kelly, M., Ferrua, C. M., Tomooka, Y., Brown, C. F., and Walker, M.,** Evidence for isoaspartyl[1] (deamidated) forms of mouse epidermal growth factor, *Anal. Biochem.,* 159, 360, 1987.
36. **McDonald, T. J., Jornvall, H., Tatemoto, K., and Mutt, V.,** Identification and characterization of variant forms of the gastrin-releasing peptide (GRP), *FEBS Lett.,* 156, 349, 1983.
37. **Ota, I. M., Ding, L., and Clarke, S.,** Methylation at specific altered aspartyl and asparaginyl residues in glucagon by the erythrocyte protein carboxyl methyltransferase, *J. Biol. Chem.,* 262, 8522, 1987.

38. **Clarke, S.**, Propensity for spontaneous succinimide formation from aspartyl and asparaginyl residues in cellular proteins, *Int. J. Peptide Protein Res.*, 30, 808, 1987.
39. **Kim, S. and Li, C.H.**, Enzymatic methyl esterification of specific glutamyl residue in corticotrophin, *Proc. Natl. Acad. Sci. U.S.A.*, 76, 4255, 1979.
40. **Gagnon, C., Kelly, S., Manganiello, V., Vaughan, M., Odya, C., Strittmatter, W., Hoffman, A., and Hirata, F.**, Modification of calmodulin function by enzymatic carboxyl methylation, *Nature*, 291, 515, 1981.
41. **Gagnon, C.**, Enzymatic carboxyl methylation of calcium-binding proteins, *Can. J. Biochem. Cell Biol.* , 61, 921, 1983.
42. **Johnson, B. A., Freitag, N. E., and Aswad, D. W.**, Protein carboxyl methyltransferase selectively modifies an atypical form of calmodulin. Evidence for methylation at deamidated asparagine residues, *J. Biol. Chem.*, 260, 10913, 1985.
43. **Runte, L., Jürgensmeier, H.-L., Unger, C., and Söling, H. D.**, Calmodulin carboxylmethyl ester formation in intact human red cells and modulation of this reaction by divalent cations in vitro, *FEBS Lett.*, 147, 125, 1982.
44. **Billingsley, M.L., Kuhn, D., Velletri, P.A., Kincaid, R., and Lovenberg, W.**, Carboxylmethylation of phosphodiesterase attenuates its activation by Ca^{2+}-calmodulin, *J. Biol. Chem.*, 259, 6630, 1984.
45. **Brunauer, L. S. and Clarke, S.**, Methylation of calmodulin at carboxylic acid residues in erythrocytes. A non-regulatory covalent modification?, *Biochem. J.*, 236, 811, 1986.
46. **Johnson, B. A., Shirokawa, J. M., and Aswad, D. W.**, Deamidation of calmodulin at neutral and alkaline pH. Quantitative relationships between ammonia loss and the susceptibility of calmodulin to modification by protein carboxyl methyltransferase, *Arch. Biochem. Biophys.*, 268, 276, 1989.
47. **Johnson, B. A., Langmack, E. L., and Aswad, D. W.**, Partial repair of deamidation-damaged calmodulin by protein carboxyl methyltransferase, *J. Biol. Chem.*, 262, 12283, 1987.
48. **Murtaugh, T. J., Rowe, P. M., Vincent, P. L., Wright, L. S., and Siegel, F. L.**, Posttranslational modification of calmodulin, *Methods Enzymol.*, 102, 158, 1983.
49. **Kloog, Y., Flynn, D., Hoffman, A. R., and Axelrod, J.**, Enzymatic carboxymethylation of the nicotinic acetylcholine receptor, *Biochem. Biophys. Res. Commun.*, 97, 1474, 1980.
50. **Flynn, D. D., Kloog, Y., Potter, L. T., and Axlerod, J.**, Enzymatic methylation of the membrane-bound nicotinic acetylcholine receptor, *J. Biol. Chem.*, 257, 9447, 1982.
51. **Yee, A. S. and McNamee, M. G.**, Effects of carboxylmethylation by a purified *Torpedo californica* methylase on the functional properties of the acetylcholine receptor in reconstituted membranes, *Arch. Biochem. Biophys.*, 237, 349, 1985.
52. **Johnson, B. A.**, Modification of L-Isoaspartate by Protein Carboxyl Methyltransferase, Ph.D. dissertation, University of California, Irvine, 1986.
53. **Billingsley, M. L., Kincaid, R. L., and Lovenberg, W.**, Stoichiometric methylation of calcineurin by protein carboxyl O-methyltransferase and its effects on calmodulin-stimulated phosphatase activity, *Proc. Natl. Acad. Sci. U.S.A.*, 82, 5012, 1985.
54. **Kakimoto, Y. and Armstrong, M. D.**, Beta-L-aspartyl-L-histidine, a normal constituent of human urine, *J. Biol. Chem.*, 236, 3280, 1961.
55. **Buchanan, D. L., Haley, E. E., and Markiw, R. T.**, Occurrence of beta-aspartyl and gamma-glutamyl oligopeptides in human urine, *Biochemistry*, 1, 612, 1962.
56. **Dorer, F. E., Haley, E. E., and Buchanan, D. L.**, Quantitative studies of urinary beta-aspartyl oligopeptides, *Biochemistry*, 5, 3226, 1966.
57. **Lou, M. F.**, Isolation and identification of L-beta-aspartyl-L-lysine and L-gamma-glutamyl-L-ornithine from normal human urine, *Biochemistry*, 14, 2903, 1975.
58. **Tanaka, T. and Nakajima, T.**, Isolation and identification of urinary beta-aspartyl dipeptides and their concentrations in human urine, *J. Biochem.*, 84, 617, 1978.
59. **Welling, G. W. and Groen, G.**, Beta-aspartylglycine, a substance unique to caecal contents of germ-free and antibiotic-treated mice, *Biochem J.*, 175, 807, 1978.
60. **Welling, G. W.**, Comparison of methods for the determination of beta-aspartylglycine in fecal supernatants of leukemic patients treated with antimicrobial agents, *J. Chromatogr.*, 232, 55, 1982.
61. **Tanaka, T., Hirai, M., and Nakajima, T.**, Partial purification and characterization of an enzyme involved in the formation of beta-aspartyl dipeptides in rat kdney, *J. Biochem.*, 84, 1147, 1978.
62. **Vincent, P. L. and Siegel, F. L.**, Carboxylmethylation of calmodulin in cultured pituitary cells, *J. Neurochem.*, 49, 1613, 1987.
63. **Babu, Y. S., Sack, J. S., Greenhough, T. J., Bugg, C. E., Means, A. R., and Cook, W. J.**, Three-dimensional structure of calmodulin, *Nature*, 315, 31, 1985.
64. **Potter, S. M., Gleason, C. S., and Aswad, D. W.**, unpublished data.
65. **Yüksel, K. Ü. and Gracy, R. W.**, *In vitro* deamidation of human triosephosphate isomerase, *Arch. Biochem. Biophys.*, 242, 446.
66. **Dorer, F. E., Haley, E. E., and Buchanan, D. L.**, The hydrolysis of beta-aspartyl peptides by rat tissue, *Arch. Biochem. Biophys.*, 127, 490, 1968.

67. **Finely, D. and Varshavsky, A.,** The ubiquitin system: functions and mechanisms, *Trends Biochem. Sci.,* 10, 343, 1985.
68. **Murray, E. D., Jr. and Clarke, S.,** Metabolism of a synthetic L-isoaspartyl-containing hexapeptide in erythrocyte extracts. Enzymatic methyl esterification is followed by nonenzymatic succinimide formation, *J. Biol. Chem.,* 261, 306, 1986.
69. **Momand, J. and Clarke, S.,** Rapid degradation of D- and L-succinimide-containing peptides by a post-proline endopeptidase from human erythrocytes, *Biochemistry,* 26, 7798, 1987.
70. **Johnson, B. A., Terán, S., and Aswad, D. W.,** unpublished data.

Chapter 12

PROTEIN CARBOXYLMETHYLATION IN NERVOUS TISSUES

Melvin L. Billingsley

TABLE OF CONTENTS

I. INTRODUCTION

Protein *O*-carboxyl methyltransferase (EC 2.1.1.24; PCM) has been the subject of considerable investigation with respect to its potential as a mediator of signal transduction in excitable tissue. There are several reasons for suspecting a regulatory role of PCM in nervous system function; however, there are no definitive studies which demonstrate the exact functions of this unusual enzyme. One reason for suspecting a role in signal transduction stems from the numerous studies which have implicated reversable signaling via covalent posttranslational modifications of proteins such as phosphorylation/dephosphorylation reactions.[1] For example, there are calcium-stimulated protein kinases (e.g., protein kinase C[2] and calmodulin-dependent protein kinase II[3]) which are activated upon the regulated entry of calcium into the depolarized nerve terminal; specific protein substrates are stoichiometrically phosphorylated and their functional state is altered, leading to an alteration in cellular response.[4] Similarly, there are protein phosphatases which are responsive to depolarization, which will dephosphorylate substrates for protein kinases, leading to a cessation of phosphorylation-induced functional change.[5] Thus, protein phosphorylation has served as a prototype system for establishing paradigms and specific criteria for determining the potential roles of covalent protein modification in signal transduction.

Another reason for examining the role of PCM in nervous tissue is the description of the regulatory role of protein glutamyl carboxylmethylation in bacterial chemotaxis.[6-8] Extensive biochemical and genetic studies have demonstrated that carboxylmethylation and demethylation in bacteria are intimately linked to receptors which sense the chemotactic environment. Bacterial chemotaxic mechanisms have been used to describe and compare sensing, response and adaptation as it relates to neuronal signaling mechanisms.[9] Thus, exploration of potential roles for PCM in the mammalian nervous system have been guided to some extent by parallel studies of bacterial protein methylation.

A third reason for suspecting a functional role for PCM in nervous tissue is that PCM activity is highest in the brain.[10] Thus, even though this enzyme has a widespread tissue and phylogenetic distribution, considerable effort has been expended to determine its function in tissues with the highest enzymatic levels.

Finally, there are several reports which suggest that PCM activity can be influenced by drugs, hormones, and physiologic manipulations which affect neuronal function.[11] That a specific receptor can directly activate or influence PCM activity has yet to be demonstrated, however, and such associations may be shown to correlative rather than causal.

With regard to PCM function however, there are clear differences and disparities between bacterial and mammalian systems, between phosphorylation and methylation, and between cause and effect so as to leave open the question as to the ultimate role of this enzyme in physiologic functions. Several critical issues have revolved around (1) the rate of methylation, (2) the stoichiometry of methyl transfer, (3) the intramolecular site, (4) specific protein substrates of methylation, and (5) whether PCM is linked directly to a specific function. This chapter will attempt to focus on questions relating to localization, substrates, and function of PCM in nervous tissue, and will not address issues related to protein methylation in other tissues such as erythrocytes; comprehensive reviews on the subject of PCM function in other tissues have recently been published, and the reader is referred to these sources for further information.[12-14]

II. CHARACTERIZATION OF NEURONAL PCM

PCM has been purified from rat,[15] bovine,[16,17] and ox brain[18] and from the electric organ of *Torpedo*.[19] More recent purification schemes have centered around the use of *S*-adenosylhomocysteine (AdoHcy)-agarose affinity chromatography, coupled with more conventional

TABLE 1
Characterization of Cytosolic PCM from Neuronal Tissues

Tissue	Purification	kDa	pI	K_m (AdoMet)	K_i (AdoHcy)	Ref.
Pituitary	528	ND	ND	1.5 u*M*	N D	87
Calf brain	3000	25	6; 6.2	0.87 u*M*	0.9 u*M*	17
Ox brain	1339	35	ND	2.7 u*M*	N D	18
Bovine brain						
PCM I	1460	28	6.5	1.0 u*M*	N D	16
PCM II	1320	28	5.5; 5.6	1.4 u*M*	N D	
Rat brain	1500	27	ND	1.0 u*M*	0.5 u*M*	15
Torpedo electric organ	700	29	6.1; 6.4	1.5 u*M*	0.3 u*M*	19

Note: N D = not reported.

techniques. Table 1 compares the kinetic and molecular characteristics of neuronal PCM; there is general agreement that PCM is a monomer of 25 to 27 kDa, has a K_m for *S*-adenosylmethionine (AdoMet) of 1 to 5 u*M*, and has a K_i of 0.1 to 0.5 u*M* for AdoHcy. Amino acid composition analysis indicates that PCM is rich in acidic amino acid residues, and possesses one cysteine which may be near the active site of the enzyme.[15,19] Isoelectric focusing and ion-exchange chromatographic studies of purified brain enzymes have suggested that there are at least two isoforms which differ in charge; the basis for this difference is not known.[16]

Differences have been noted in terms of membrane-bound PCM activity. To date, several reports have suggested that in ox and rat brain, a significant fraction of PCM activity lies in membrane bound fractions. In ox brain, approximately 40% of the total homogenate activity was associated with membrane fractions; this activity could be partially solubilized using 0.1% Triton X-100.[18] The kinetics (K_m for AdoMet = 2.7 u*M*), pH optima (pH 6.2) and molecular weight (35 kDa) were similar for both soluble and membrane-bound forms. A detergent-extractable pool of PCM was also seen in synaptosomes prepared from rat brain;[20,21] although not fully characterized, the membrane-bound form of PCM lost activity following removal of detergent. Additional studies have suggested that membrane-bound forms of PCM are found in various neuronal cell types, and that the membrane-bound form may act on locally-generated substrates. Clearly, more work is needed to characterize the membrane-bound forms of this enzyme to determine whether differences exist in structure and function.

Our laboratory was interested in preparing polyclonal antibodies against brain PCM for purposes of immunologic characterizations and immunocytochemical mapping of PCM.[22] To this end, we prepared antibodies against purified bovine brain PCM. PCM was purified using DEAE-Trisacryl, AdoHcy-agarose and molecular sizing; the purified enzyme (corresponding to the isoform I of Aswad and Deight [16]) was homogeneous following SDS-PAGE. Bands of PCM were excised form the gel, mixed with complete Freund's adjuvant, and used to immunize New Zealand rabbits. Generally, reasonable antibody titers were obtained following the third boost of PCM. The antisera is usually purified to an IgG fraction using DEAE-Affigel Blue.

Antisera raised against bovine brain PCM was found, using Western blotting techniques, to recognize both isoforms of bovine and rat brain PCM, PCM from human and equine erythrocytes,[22] and PCM from PC-12 cells and other neuroblastoma lines (Billingsley, unpublished data). Thus, although some structural differences undoubtably exist between various isoforms and species, PCM antisera was crossreactive against PCM from a range of species and tissues. We have not yet determined whether there is immunologic crossreactivity against PCM from lower eukaryotic sources. More definitive cross-species comparisons will be possible when cDNA sequence information becomes available for this enzyme.

One unusual finding was that ^{3}H-methyl-AdoMet could be covalently crosslinked to bovine

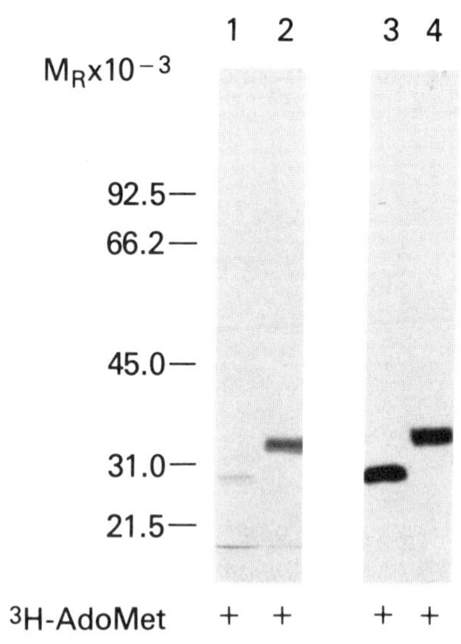

FIGURE 1. Labeling of purified PCM (lanes 1 and 3) or bovine adrenal PNMT (lanes 2 and 4) with ^{3}H-methyl AdoMet [60 Ci/mmol]. A total of 2 μg PCM or 6 μg PNMT were incubated with 5.0 μCi of ^{3}H-methyl-AdoMet (in 5 m*M* dithiothreitol, 0.1 M NaCl, 0.1 M HEPES buffer, pH 7.5) and irradiated with short wavelength UV light (1 cm distance) for 10 min at 4°C. Lanes 1 and 2 show the Coomassie blue protein staining pattern of the two enzymes, and lanes 3 and 4 show the autoradiograph of labeled proteins.

brain PCM; this crosslinking was not blocked by additions of dithiothreitol, but was inhibited by coincubation with AdoHcy; a similar phenomenon was observed with phenylethanolamine-*N*-methyltransferase (PNMT).[23] A representative experiment showing this type of covalent enzyme labeling is depicted in Figure 1. Although the stoichiometry of enzyme labeling was low (0.07 to 0.01 mol%), this technique was useful as a means on introducing a covalent tag near the presumed active site.

Thus, characterization of brain PCM has indicated strong immunologic crossreactivity and kinetic similarities between other tissue sources (e.g., erythrocyte) of PCM. However, as will be discussed, the localization of PCM in brain regions is characterized by regional variation in immunoreactive levels of PCM.

III. IMMUNOLOCALIZATION OF PCM IN NEURONS

Using the antisera against PCM, we have performed a series of immunocytochemical studies concerning the regional localization of PCM in brain. Sprague-Dawley rats were perfused with paraformaldehyde, their brains sectioned, and processed for immunocytochemistry using avidin-biotin linked methods as previously described.[22] Visualization of PCM immunoreactivity in these rats revealed a pattern of localization in the cell bodies of neurons; several examples of brain regions stained using PCM antisera are shown in Figure 2A to D. Several points are worth noting. First, PCM demarcated neuronal cell bodies throughout the CNS. However, there were regions which clearly expressed higher levels of immunoreactive PCM relative to one another. Most prominent were the substantia nigra, the locus coeruleus, where PCM was colocalized in tyrosine-hydroxylase containing neurons, and in the paraventricular nucleus of the hypothalamus.[24] Interestingly, the hypothalamic-pituitary system was the first tissue region investigated for "methanol formation" via enzymatic means.

On a subcellular level, PCM was localized in the cytoplasm and nuclei of the neuron; little staining was observed in terminal fields. Biochemical studies have shown PCM activity in synaptosomes;[10] however, it is often difficult to resolve synaptic concentrations of immunoreactive PCM at the light microscopic level. PCM immunoreactivity was concentrated over nuclei as well as cytoplasmic regions; this suggests that PCM may have a function in neuronal nuclei. As will be discussed later, PCM may modify nonhistone chromosomal proteins and nuclear envelope proteins.

Biochemical studies have indicated that significant levels of PCM activity are found in synaptosomal and cytosolic fractions throughout the brain. However, it is evident the PCM is distributed throughout the cell, with enzyme present in many compartments. Several experiments have suggested that PCM is transported down the axon to the nerve terminal region; these studies were performed on ligated peripheral nerves and suggested accumulation of PCM above the level of the block. However, unlike other proteins which are axonally transported, PCM was easily visualized using immunocytochemistry in neuronal cell bodies without the use of colchicine pretreatment to block axoplasmic transport. Thus, although some immunoreactive PCM clearly enters the axon and nerve terminal (see Figure 2C and D), there is no evidence to suggest that this protein is exclusively concentrated in synaptosomes.

Thus, immunolocalization of PCM in rat brain suggests that this enzyme may play a function in neuronal cell bodies, nuclei, and synaptic regions. Based on these studies, an examination of PCM function in hippocampus, paraventricular nucleus of the hypothalamus, and other regions with high levels of immunoreactive PCM may shed insight into the physiologic function of this enzyme.

IV. COLOCALIZATION OF PCM AND CALMODULIN-BINDING PROTEINS

One specific hypothesis that we have investigated is the possibility that PCM can carboxyl methylate calmodulin-regulated enzymes such as calcineurin (CN),[25] calmodulin-dependent protein kinase II,[26] and calmodulin-stimulated cyclic nucleotide phosphodiesterase (PDE).[27] In order to demonstrate that these enzymes are colocalized in the same brain regions, we have compared patterns of immunochemical distribution of CN, PDE, and PCM. One such comparison was performed in rat brain hippocampus, and is shown in Figure 3. PCM immunoreactivity was observed in the soma of all neurons in the hippocampus; similarly, CN immunoreactivity was localized in all hippocampal neurons. PDE immunoreactivity was only seen in specific subpopulations of neurons and was confined to the CA1-2 fields of the hippocampus. Of note is the observation that the subcellular pattern of labeling was different for each antigen—PCM was present in the cytoplasm and nuclei, CN was localized in punctate deposits in the cell soma, and PDE was localized in somatodendritic regions.[28] Thus, although PCM, CN, and PDE are colocalized in specific cells, they do not necessarily occupy the same subcellular compartments. Other studies have shown calmodulin-dependent protein kinase II to be localized in the CA fields of the hippocampus.[29] Recent studies have used immunocytochemistry to localize AdoHcy hydrolase (EC 3.3.1.1) in brain;[30] AdoHcy hydrolase immunoreactivity was localized in hippocampal CA fields and in neocortex and cerebellum. Subcellular localization was similar to that of PCM, with localization in neuronal soma and in nuclei. The colocalization of PCM and AdoHcy hydrolase in similar regions and subcellular compartments of the brain suggests that these areas are sites of active methyltransferase activity and are sensitive to the effects of inhibitors of AdoHcy hydrolase.

PCM activity has been studied during neuronal development in rat brain and superior cervical ganglia.[31-33] In general, these studies, utilizing enzyme activity measurements, indicated that PCM activity was present during neurogenesis and cellular migration, with activity levels generally reaching adult levels around puberty. The presence of PCM activity *in utero* suggests that the function of this enzyme is not limited to actions in fully differentiated neurons. Future

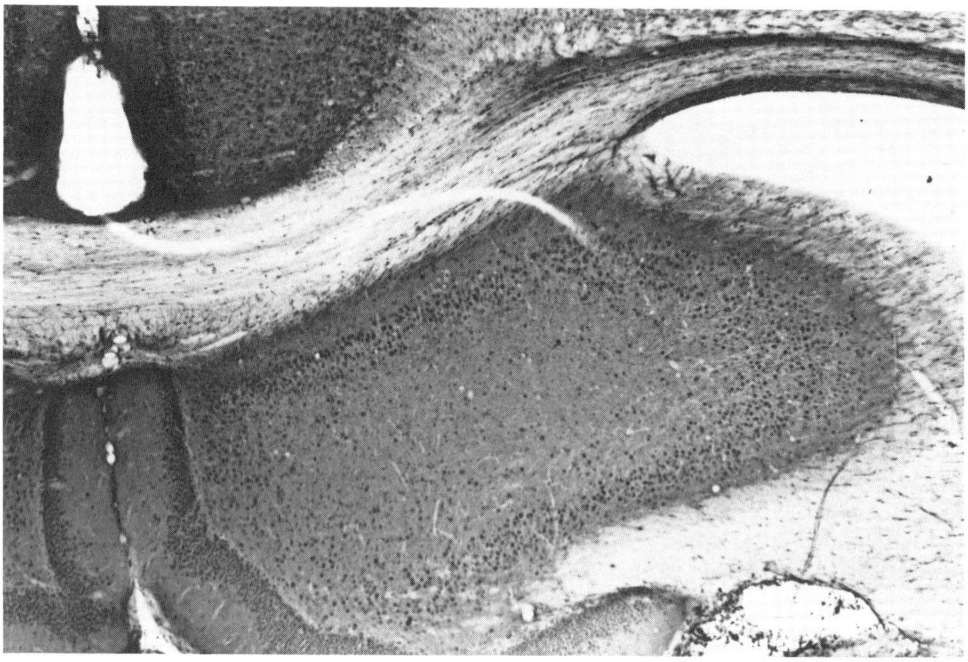

A

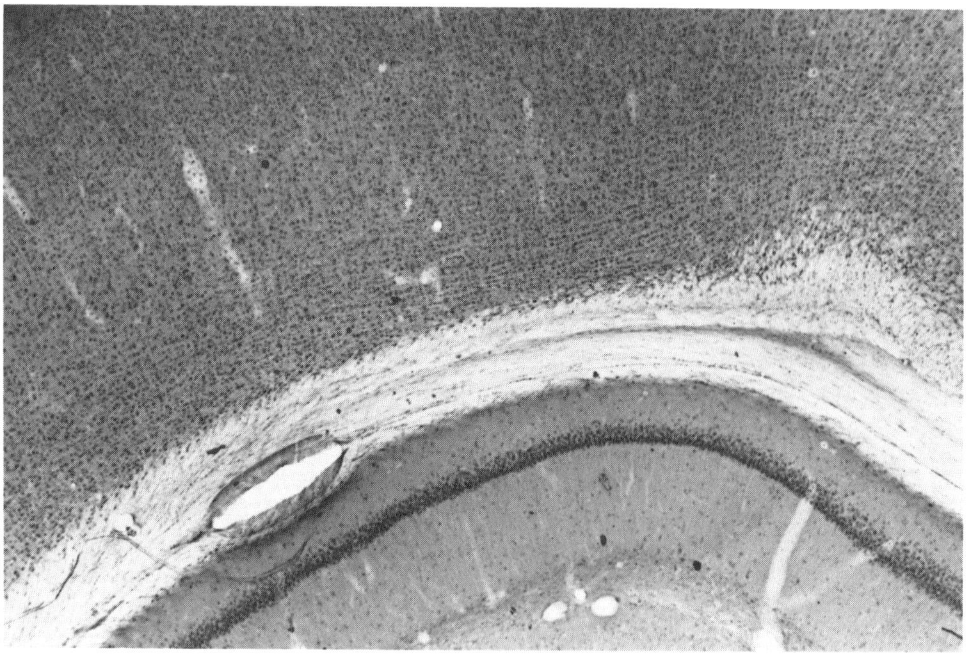

B

FIGURE 2. Immunolocalization of PCM in rat brain. (Panel A-) PCM immunoreactivity in the cell bodies of the head of the rat hippocampus and in the cingulate cortex. Note the relative absence of staining in the white matter (magnification, 90X) (Panel B-) PCM immunoreactivity in neocortex and hippocampus (90X). (Panel C-) PCM in neocortex, showing staining of pyramidal cell processes (180X) (Panel D-) Higher power view of neocortical pyramidal cells positive for PCM, showing apparent staining in dendrites (400X).

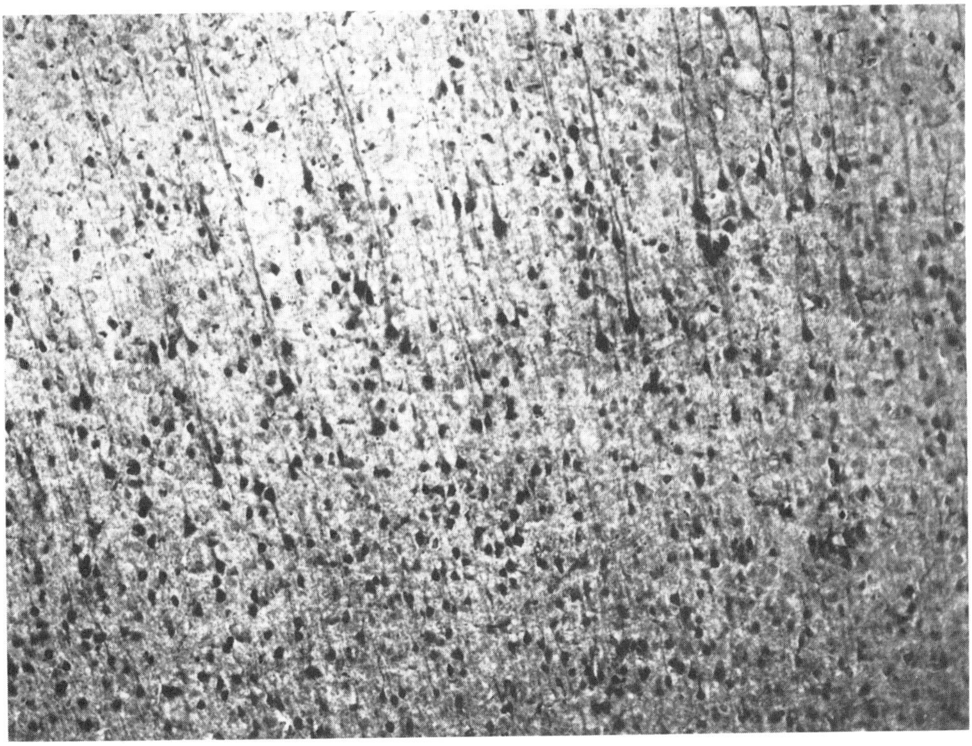

FIGURE 2C.

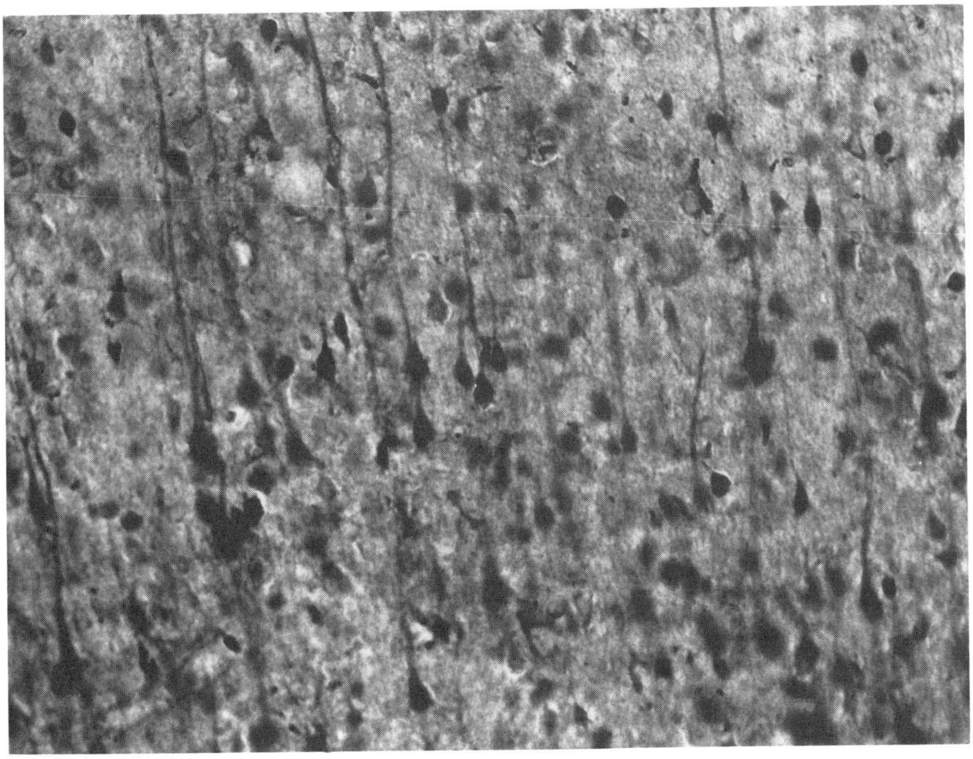

FIGURE 2D.

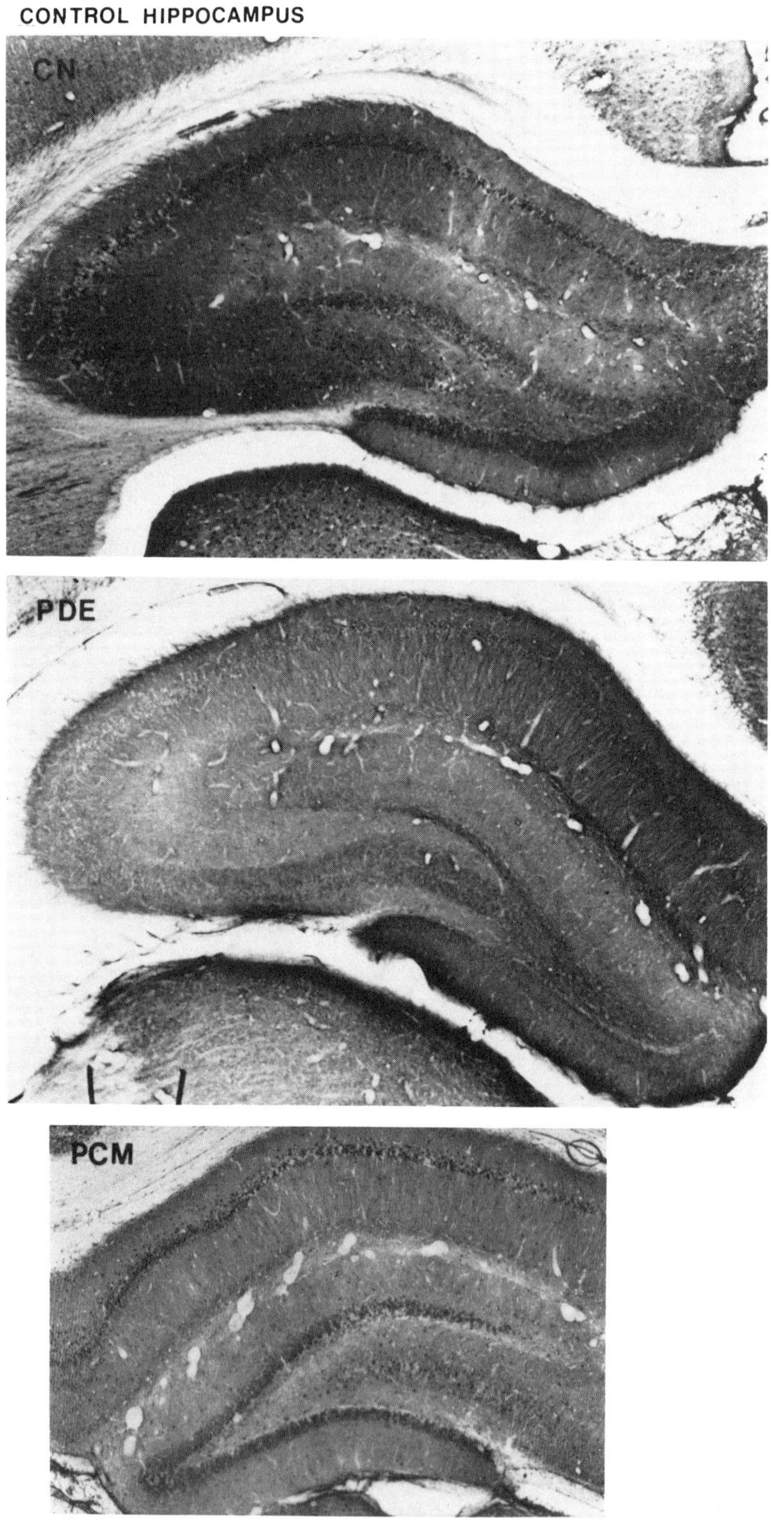

FIGURE 3. Comparison of immunolocalizations in rat hippocampus of calcineurin (CN); calmodulin-stimulated phosphodiesterase (PDE) and PCM. Note that CN and PCM are present in cell bodies throughout the cells of the hippocampus, while PDE is restricted to the CA1-2 fields. (80X).

TABLE 2
Putative Neuronal Substrates for PCM

Substrate	Functional effects	mol CH_3/mol	Ref.
Calmodulin	Decreased stimulation of target enzymes	0.01 to .50	27, 35, 40, 41
Nicotinic acetylcholine receptor	Decreased ion flux	0.01 to .20	44—46
Calmodulin-dependent enzymes	Decreased calmodulin stimulation	0.10 to 2.0	25—27, 88
cGMP-phosphodiesterase	?Enzyme activity	Not determined	47—49
Neurophysin	?Cotranslational; release of hormone	0.01 to.10	51—56
Chromogranin A	?Cotranslational; storage of transmitter?	0.26—1.3	50, 52, 91, 92
Isoaspartyl calmodulin	Repair, degradation	0.3 to 0.5	41
Synaptosomal proteins of 60, 48, 18 kDa	Synaptic activation?	0.01	37, 38
Synapsin	Synaptic vesicle change	0.01	16, 89
Myelin basic protein	Repair?	Not determined	89
Lamin B	Nuclear envelope function	0.3 to 0.7	60

immunochemical experiments to study the expression of PCM during development and aging are in progress.

V. SUBSTRATES FOR PCM IN NERVOUS TISSUE

A. CALMODULIN AND CALMODULIN-BINDING PROTEINS

Considerable effort has been expended to determine the "endogenous" substrates for PCM. Several problems have confounded the search for these substrates; most notably, stoichiometry of methylation; lability of the carboxyl methylated protein, range of substrate specificity, and functional change as a result of carboxyl methylation. Table 2 presents a list of substrates which have been reported to be carboxylmethylated by PCM; several of these studies employed purified PCM and purified substrate, while others were identified from homogenates or from *in vivo* labeling studies.

Initial studies in our lab focused on the possibility that calmodulin was carboxylmethylated, and that its ability to subsequently activate calmodulin-dependent enzymes was inhibited.[34] Calmodulin contains a high preponderance of acidic amino acid residues, many of which participate in the formation of calcium-binding pockets. Thus, it was hypothesized that enzymatic methylation of one or more residues would diminish calcium binding and activation of the model. However, rigorous analysis of the stoichiometry and kinetics of calmodulin methylation indicated that it was unlikely that its calcium-binding capacity was impaired since the stoichiometry of carboxyl methylation was 1 to 5%.[27] However, two possibilities could account for how conditions favoring carboxylmethylation could alter calmodulin function; first, some conditions used to purify calmodulin (boiling tissues for 5 min at pH 7 to 8) could have produced isoaspartyl adducts which were the site(s) of methylation or second, the calmodulin-binding protein itself was carboxyl methylated during the reaction with calmodulin.

In the initial studies of calmodulin methylation,[35] purified PCM and AdoMet were used to modify purified calmodulin; the entire contents of the reaction mixture were then added to a subsequent reaction containing the calmodulin-dependent phosphodiesterase and its reactants. This raised the possibility that the calmodulin-binding proteins (e.g., PDE) were themselves methylated.[36] We have explored this possibility and found that calmodulin-binding proteins could serve as methyl acceptors *in vitro;* however, these studies do not prove that calmodulin-binding proteins are regulated *in vivo* by carboxylmethylation. We have attempted to determine whether calmodulin-binding proteins are methylated to reasonable stoichiometries *in vivo* but have not found convincing evidence of such methylations to date. The only suggestion that calmodulin-binding proteins are methylated *in vivo* comes from studies of carboxyl methylation in synaptic structures; two laboratories have reported that a 60 kDa protein was the major methyl acceptor.[37,38] Numerous calmodulin binding proteins are concentrated in the synapse and have apparent molecular weights of 60 kDa; thus, it is possible that this methylated protein is a subunit of a calmodulin-binding protein. Conclusive demonstration that the carboxylmethylated 60 kDa synaptic protein is a calmodulin-binding protein remains to be demonstrated. Johnson and Aswad examined the topography of protein substrates in synaptic membrane, and concluded that most of the methylated proteins were found on the outside face of the vesicle.[39] One substrate, a 48 kDa protein, was resistant to proteolysis by trypsin, suggesting an inward facing component. The identity of this protein is not known.

A recent study using rapid HPLC separations of carboxylmethylated proteins reported that calmodulin was the major carboxylmethylated protein in cultured pituitary cells.[40] Because this study was performed on intact cells, calmodulin would not be extensively isomerized as could happen during a purification, raising the possibility that functional regulation could occur in cultured cells. Thus, the question remains open as to whether there is a functional role for methylation of calmodulin *in vivo.* Johnson and Aswad have suggested that some regulatory process may actually use transient deamidation, isoaspartyl formation and subsequent repair of the calmodulin to form a complete cycle.[41] These investigators have also shown that deamidated, isoaspartyl calmodulin is stoichiometrically methylated and that the methylation preceeds "repair" to a normal isoform; this finding raises several questions.[42]

First, is it necessary to repair isomerized calmodulin in a cell? Neurons and other cells have high concentrations of calmodulin (0.1 to 1.0 μM), which appears to be well in excess of calmodulin-binding protein concentrations in cells. Thus, a small concentration of isoaspartyl calmodulin may not have a significant impact on the functions of calcium-regulated processes. Alternatively, if deamidation-isoaspartyl formation was part of an important regulatory process, then repair would be necessary to maintain an adequate pool of functional calmodulin. It is possible that methylation could serve as a recognition signal for damaged proteins, and that calmodulin is simply more susceptible to isoaspartyl formation or damage. This latter possibility does not seem as likely since calmodulin can be exposed to rather stringent conditions (extremes of pH, temperature, chaotropes) and still exhibit functional binding and stimulation of target enzymes.[43]

B. NICOTINIC ACETYLCHOLINE RECEPTOR

Kloog and co-workers initially reported that the nicotinic acetylcholine receptor from *Torpedo californica* would be enzymatically carboxyl methylated by PCM from both erythrocytes and from electric organ.[44] These studies have been extended to show that PCM from electric organ is somewhat different than other isoforms, and that the γ and β subunits were selectively modified.[19] Yee and McNamee reported that all receptor subunits were carboxyl methylated, with preferential labelling of the α and γ subunits.[45] Further, the stoichiometry of methylation approached 20%. These authors reported that there was no effect of methylation on the interaction of the methylated receptor with its ligand, but did find a 20% inhibition of agonist-stimulated cation flux when the receptor was present in reconstituted membranes. Studies by

Nuske have suggested that the state of methylation of the acetylcholone receptor in synaptosomal preparations from *Torpedo* can be increased by cholinerigic effectors such as carbamylcholine, flavedil, and α-bungarotoxin.[46]

Taken together, these studies suggest that there is agonist-stimulated methylation of nicotinic acetylcholine receptor subunits, with a potential consequence of altering cation flux across the channel. This methylation can be accomplished by PCM intrinsic to and concentrated in *Torpedo* electroplax. However, there are several points which are unclear. First, the *Torpedo* enzyme is more sensitive to proteolytic degradation, is inhibited strongly by sulfhydryl-modification reagents, and is sensitive to copper ions;[18] it is possible that these structural differences between *Torpedo* and other enzyme sources are related to an altered function of the methyltransferase in this unique organ. Elucidation of the role of PCM in *Torpedo* should greatly facilitate understanding the potential role of PCM as a signal transduction enzyme or as a modifier of unusual amino acid modifications.

C. PHOSPHODIESTERASE METHYLATIONS

Several reports have indicated that cyclic nucleotide phosphodiesterases can be carboxyl methylated. The cGMP-phosphodiesterase (EC 3.1.4.17) in rod outer segments is intimately involved in the transduction of photons to neural impulses. Incubation of retinal preparations with ^{3}H-methyl-methionine resulted in the carboxylmethylation of proteins of 88, 61, and 21 to 26 kDa; the 88 kDa protein has been identified as cGMP phosphodiesterase.[47] More recently, one study has suggested that the level of phosphodiesterase methylation is greater during illumination, and that a protein inhibitor of phosphodiesterase decreased methylation.[48] These studies suggest that methylation may be involved in the modulation of retinal signal transduction. However, more stringent criteria concerning the stoichiometry and time course of methylation need to be performed in order to fully demonstrate a role for PCM in retinal signal transduction. We have observed that purified calmodulin-sensitive phosphodiesterase can be carboxylmethylated *in vitro* , but the relationship between these two phosphodiesterases and their ability to serve as substrate for PCM is not clear at this time. Coincidentally, inhibitors of cyclic nucleotide phosphodiesterases (isobutylmethyxanthine, papaverine, dipyradamole) have been reported to inhibit carboxyl methylation in intact platelets.[49]

D. SECRETORY PROTEINS: CHROMOGRANINS AND NEUROPHYSINS

Initial speculation on the putative function of PCM centered on its role in secretion. An early theory suggested that PCM could carboxylmethylate acidic proteins on the outside surface of secretory vesicles;[50] this methylation would have the net effect of reducing electrostatic barriers to vesicle-membrane fusion. This theory has not been widely accepted, due primarily to problems in substrate selectivity, amount and rate of methylation, and temporal and causal links between methylation and secretion. However, two recent reports have indicated that the dynamics of PCM action on neurophysin[51] and chromogranin[52] may suggest a role for PCM in the secretory process.

Neurophysins have been identified as the prime substrates in pituitary systems.[52-54] Neurophysin methylation has been observed both *in vivo* and *in vitro*. Manipulations of salt balance[55] and genetic absence of neurophysins in the Brattleboro rat[56] have been used to further suggest that neurophysin methylation occurs in concert with secretion. Recent work by Meydan et al. has suggested that carboxyl methylation occurs primarily on the pool of neurophysins which rapidly turn over as a consequence of release.[51] Since this pool, which comprises about 10% of the total neurophysin pool, also contains a high porportion of the newly synthesized neurophysin, it was suggested that recently synthesized neurophysins are the prime targets for methylation.

Similar results have been observed in adrenal chromaffin cells. Nguyen et al. reported that newly synthesized chromogranin A was preferentially carboxyl methylated and that methylated chromogranin A could be secreted into the media.[52] Carboxyl methylation of chromogranin was most likely cotranslational, and may be important for storage or functional release. Others have observed that carboxyl methylation is cotranslational,[57] and studies using inhibitors of protein synthesis have shown a drop in the extent of methylation following inhibition of *de novo* synthesis,[58] which further suggests that newly synthesized proteins are substrates for PCM. Although not compatible with age-related degradation and repair of proteins, it is possible that cotranslational methylation may "repair" errors in translation.

E. METHYLATION OF LAMIN B AND NUCLEAR PROTEINS

Lamin B is a protein of the nuclear envelope which is thought to participate in the attachment of intermediate filaments to the nucleus and in the nuclear pore complex.[59] During mitosis, the nuclear envelope is solubilized, and lamins, possibly as a consequence of phosphorylation, are also solubilized. Chelsky and co-workers recently reported that lamin B is methylated in murine lymphoma, neuroblastoma, HeLa cells, and in brain nuclei.[60] Lamin B was methylated in a cell-cycle-dependent manner, with demethylation occurring during mitosis. Radiolabeling and isoelectric focusing suggested that lamin B remains fully methylated. Since isolated brain nuclei were able to undergo methylation, these finding suggest that the level of lamin B methylation may be part of a series of control mechanisms which prevent cells from undergoing mitosis. Consistent with this finding is the fact that PCM levels are prominent in the nucleus of neurons, and that there appears to be light immunoreactivity around the nucleolus. Also of note is the finding that lamin B levels increase as embryonal carcinoma cells undergo retinoic acid-induced cellular differentiation.[61] However, the status of methylation of lamin B in these cells was not investigated.

One hypothesis which emerges from this line of research is that methylation is linked to the process of differentiation. Evidence from developing rat livers[62] and thymocytes[63] has suggested that nonhistone chromosomal proteins are methylated, and that their degree of methylation may play a role in development or differentiation. Studies of differentiating murine neuroblastoma cells have also suggested that PCM activity and substrates increase during morphologic differentiation.[64] Further, additional studies in murine neuroblastoma cells suggest that prolonged incubation with inhibitors of methylation can lead to a rebound increase in protein carboxyl methylation, suggesting that transcription of PCM can be increased by inhibition of methylation.[86]

Thus, a unifying hypothesis can be constructed concerning the role of methylation in brain nuclei. During development, lamin B formation would be predicted to remain stable shortly after the cessation of cytokinesis. Lamin B would then be maintained in a methylated state. Since some demethylation would be expected to occur spontaneously, PCM would serve to maintain the extent of methylation. Prolonged exposures to methylation inhibitors would then allow a fraction of the lamin B to become demethylated. This question can be experimentally approached using neuroblastoma cells which respond to differentiating agents, and by examining brain lamin methylation during development. We are currently pursuing such experiments in order to determine whether the function of methylation plays a role in the establishment of differentiation and cessation of mitosis in neuronal cells.

VI. PUTATIVE FUNCTIONS FOR PROTEIN CARBOXYLMETHYLATION

A. SECRETION

Initial hypotheses concerning the neuronal actions of PCM centered on its possible role in exocytosis and secretion; a more complete review of the role of PCM in secretion has been published, and the reader is referred to this source for more comprehensive reading.[65] Briefly, the finding that PCM was present in brain and was found in synaptosomal fractions led to speculation concerning a possible role for PCM in neurotransmitter release. One possible mechanism for PCM involvement in release processes could be through neutralization of negative charges on the surface of secretory granules. If the acidic character of intrinsic vesicle proteins presented a charge barrier, it is conceivable that carboxyl methylation of these charges might facilitate the interaction of the vesicle with the cytoplasmic face of the plasma membrane. However, the substrate specificity, kinetics, and stoichiometry of carboxyl methylation argue against a direct role for PCM in exocytosis.

Recent investigations have examined the possible role of PCM in dopamine receptor function. Initial studies suggested that PCM activity was stimulated by drugs which activated dopamine autoreceptors in both synaptosomes and striatal tissue slices.[66,67] Dopamine autoreceptor agonists increased PCM activity, an effect which was stereospecifically inhibited by autoreceptor antagonists. The stimulatory effect of dopamine agonists was not seen in tissues pretreated with the neurotoxicant 6-hydroxydopamine, suggesting that the effect was mediated by presynaptic dopamine neurons. Further, the high concentrations of immunoreactive PCM in substantia nigra support a possible role for PCM in dopaminergic neurons.

The use of inhibitors of methylation (homocysteine) coupled with studies of dopamine synthesis and release further suggested that PCM was coupled in some way to dopamine release, but not synthesis.[68] Studies on intact rats suggested that pretreatment with homocysteine prevented apomorphine-induced declines in dopamine metabolites, which is consistent with a possible role of PCM in dopaminergic neurons.[69] Thus, circumstantial evidence suggests that PCM could play a role in dopamine release. However, there are several caveats for these and other studies which employ inhibitors of methylation to demonstrate a role for PCM. First, methyltransferase inhibitors such as homocysteine thiolactone,[70] sinefungin,[71] 3-deazadenosine,[72] and neoplanacin A[73] allow AdoHcy to accumulate in the cell; this will inhibit a range of cellular methylation reactions, rendering it impossible to conclusively demonstrate that PCM was the “significant” methyltransferase involved the specific functional process affected by drug treatment. Second, it is unclear as to what substrates are involved in dopamine neurons, and no studies have shown a direct link between a carboxyl methylated protein and a change in the functional status of release. What is needed to answer some of these questions more directly is a specific pharmacologic inhibitor of PCM which can cross cell membranes. Thus, more tools are needed to conclusively demonstrate a role for PCM in the dopamine system.

B. EFFECTS OF METHYLATION INHIBITORS ON NERVE GROWTH FACTOR RESPONSES

Since methylation reactions have been implicated as potential signal transduction mechanisms, several studies have examined the effects of inhibitors of methylation on response to nerve growth factor (NGF) in NGF-sensitive cells. Initial studies using PC-12 cells indicated that NGF-induced neurite outgrowth, protein phosphorylation and morphologic changes were reversibly inhibited by incubating cells with either 5′-*S*-methyl adenosine, homocysteine thiolactone, and 3-deazadenosine.[74] At millimolar concentrations of these inhibitors, NGF-sensitive, but not epidermal growth factor-sensitive cell processes were inhibited. Although the concentrations used clearly inhibit methylations, it is not clear whether the effects of these inhibitors were mediated through inhibition of protein carboxyl methylation or through some other system. Similar findings were described using methylation inhibitors on calf adrenal chromaffin cells[75] and sympathetic neurons in culture;[76] the effects of methylation inhibitors were initially interpreted as evidence for a role of protein methylation in the signal transduction of NGF. However, in light of the findings that protein carboxyl methylation occurred primarily on isoaspartyl residues, the interpretation was revised to suggest that some other methylation process was responsible for the effects of inhibitors on NGF-induced processes. Thus, if more specific inhibitors of protein carboxyl methylation become available, a more definitive answer could be obtained concerning the effects of protein methylation in the transduction of NGF-mediated signals.

C. OTHER APPROACHES FOR STUDYING THE FUNCTIONAL ROLE OF PCM

Our laboratory has been interested in the use of biotinylated proteins as probes of protein-protein interactions.[77] Biotin is a vitamin with an exceptional affinity for the proteins avidin and streptavidin; the K_d is approximately 10^{-15} *M*. Thus, biotin-modified antibodies, lectins and proteins have been exploited as probes for identifying protein-protein interactions using avidin-linked chromogenic enzyme systems such as alkaline phosphatase or horseradish peroxidase.

One approach which has proven successful for identifying protein interactions is to subject a complex mixture of proteins containing putative sites of interaction to SDS-PAGE and subsequent electroblotting; the resolved, blotted proteins can then be incubated with the presumed interacting protein. This blot overlay approach has proven useful for identification of actin-binding proteins,[78] antibody-reactive proteins,[79] and calmodulin-binding proteins.[80] We have developed a sensitive, rapid method for detection of calmodulin-binding proteins following SDS-PAGE and blotting using biotinylated calmodulin; the details of this method have been recently described in several publications.[81-83]

Since biotinylation of proteins can take place under mild, nondenaturing conditions (unlike iodination) and the biotinylation reagents can be used to target specific amino acid residues such as lysine, cysteine, aspartate, glutamate, histidyl, and tyrosyl residues, modifications can be made with little or no change in functional activity of the protein. We have applied this approach to the question of the function of PCM and have asked "does PCM interact with specific proteins following SDS-PAGE, and can these proteins be detected using biotinylated PCM?" Since PCM is a protein methyltransferase, PCM, by definition, binds to protein substrates. The affinity of interaction is unclear, but in the case of isoaspartyl residues, the affinity has been estimated in the micromolar range, which is similar to some calmodulin-protein interactions. Thus, it is possible that PCM could interact with potential substrate proteins on nitrocellulose blots, and interact with an affinity that would allow detection.

A second consideration is whether the interaction is dependent on the presence of AdoMet, and is sensitive to AdoHcy. Finally, the interaction may be of reasonable affinity but the PCM-protein complex may rapidly dissociate following transfer of the methyl groups. In order to approach these questions, we have chosen to use the following strategy.

First, PCM was purified to apparent homogeneity from bovine brain; purified PCM (240 μg; 8.9 nmol) was dialyzed against phosphate-buffered saline (0.1 *M*; pH 7.0). In one case, PCM was incubated with 30 μl biotinyl-ε-aminocaproic acid *N*-hydroxysuccinimide ester (BRL or Pierce; 3.2 mg/100 μl dimethylformamide) containing 100 μ*M* AdoMet to protect the active site of the enzyme from biotinylation. The second reaction was carried out in the absence of AdoMet. After 2 h at 4°C, the reaction was dialyzed against excess phosphate buffer, and the dialysate made 20% with glycerol, aliquoted, and stored at -80°C until use. In order to demonstrate that PCM was successfully biotinylated, native and PCM biotinylated in the presence and absence of AdoMet were subjected to SDS-PAGE, transferred to nitrocellulose, and the nitrocellulose blot incubated with avidin-peroxidase complexes. As can be seen in Figure 4, PCM biotinylated in the presence (lanes d,e) and absence (lanes f,g) of AdoMet were visualized using avidin-chromogen systems; native PCM (lanes b,c) did not generate a signal on the blot. Biotinylation was further confirmed using slot-blot analysis and proteolytic digestions (Figure 5). Biotinylated PCMs (BioPCM-a; biotinylated with AdoMet; BioPCM b-biotinylated without AdoMet) were easily detected on slot blots; if avidin-color enzyme development was allowed to proceed overnight, as little as 10 ng of BioPCM could be detected. In one study, BioPCM-a was electrophoresed, the specific protein band excised, subjected to proteolytic digestion with *Staphylococcus* V8 protease, the digestion electrophoresed on a second 20% SDS-PAGE system and the separated peptides blotted on nitrocellulose. When the nitrocellulose blot was incubated with avidin-alkaline phosphatase chromogen systems, approximately ten distinct, biotin-containing bands were visualized, suggesting that the biotinylation occurred at multiple sites. PCM contains 14 lysines (amino acid content analysis), and thus, up to 15 distinct sites

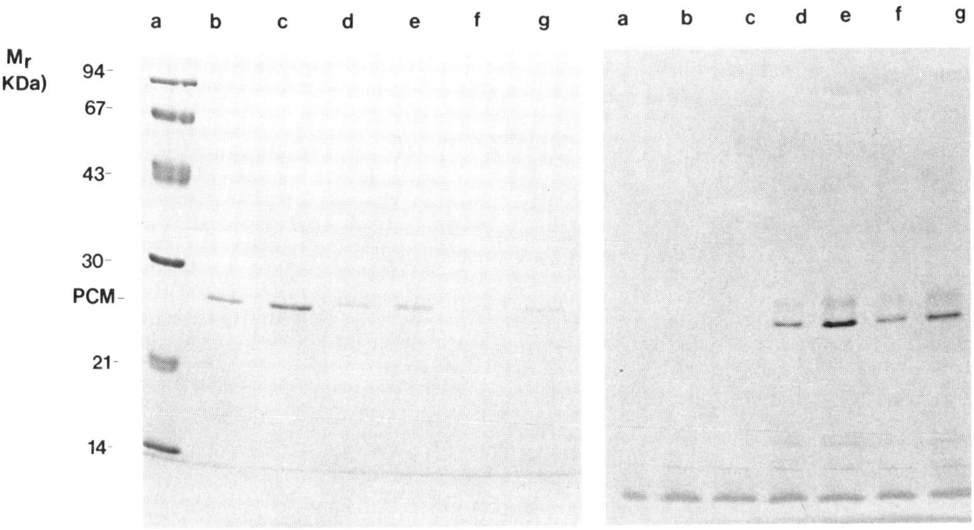

FIGURE 4. Biotinylation of bovine brain PCM. Purified PCM was biotinylated using biotinyl-*N*- hydroxysuccinimide ester as described in the text. PCM was biotinylated in the presence (lanes d, e) or absence (lanes f, g) of AdoMet; native PCM served as a control (lanes a, b). Following SDS-PAGE, the resolved proteins were transferred to nitrocellulose, and biotinylated PCM detected using avidin-based chromogenic enzymes. PCM was successfully biotinylated as judged by the reactivity on the Bio-Blot. PCM modified in the presence of AdoMet was used in subsequent studies.

could be biotinylated using the *N*-hydroxysuccinimide derivative of biotin. It is clear that the BioPCM preparation represents a population of biotinylated species, in accord with laws of mass action.

Our next objective was to determine whether BioPCM modified in the presence of AdoMet could be used as an overlay marker for PCM binding proteins separated by SDS-PAGE and blotted onto nitrocellulose. Initial studies suggested that in order to detect BioPCM, it was necessary to modify the traditional overlay approach by using rapid incubations in BioPCM (30 min) in the presence of AdoMet, followed by chemical fixation of the bound BioPCM to its putative binding proteins immobilized on nitrocellulose paper. Briefly, rat brain was homogenized, separated into crude cytosolic and membrane fractions using centrifugation (40,000 × *g*; 30 min) and the cytosol was chromatographed using DEAE-Trisacryl resin. Cytosolic proteins were eluted in a step gradient (0.05, 0.25, and 0.5 *M* NaCl washes) and approximately 100 μg of cytosolic and membrane fractions were subjected to SDS-PAGE and electroblotting. Care was taken not to boil the samples prior to electrophoresis for fear of creating conditions favoring racemization of proteins.

Two identical blots were made; one blot was incubated with biotinylated calmodulin as previously described in order to identify calmodulin binding proteins from rat brain, and the other blot was incubated with BioPCM as follows. Nonspecific protein binding sites on the nitrocellulose were blocked using 5% nonfat dry milk in 50 m*M* Tris-HCl, pH 7.4, containing 150 m*M* NaCl, 1 m*M* $CaCl_2$ (Buffer A). BioPCM (12 μg in 5 ml buffer A containing 100 μ*M* AdoMet) was incubated with the blot for 30 min at room temperature; the blot was then washed three times in buffer A with AdoMet for a total of 10 min. At this point, the blot was "fixed" for 1 min in 10% acetic acid/25% isopropanol; this was done to immobilize any BioPCM and prevent its desorption during subsequent washings. After fixation, the blot was washed 30 min in buffer A, and incubated with avidin-peroxidase complexes (Vector Labs); *p*-chloronaphthol was used as the peroxidase substrate.

Several proteins of 75 to 84 and 150 kDa appeared to bind BioPCM (Figure 6). These proteins were found in crude cytosolic, membrane, and 0.25 *M* NaCl DEAE-Trisacryl eluates, respec-

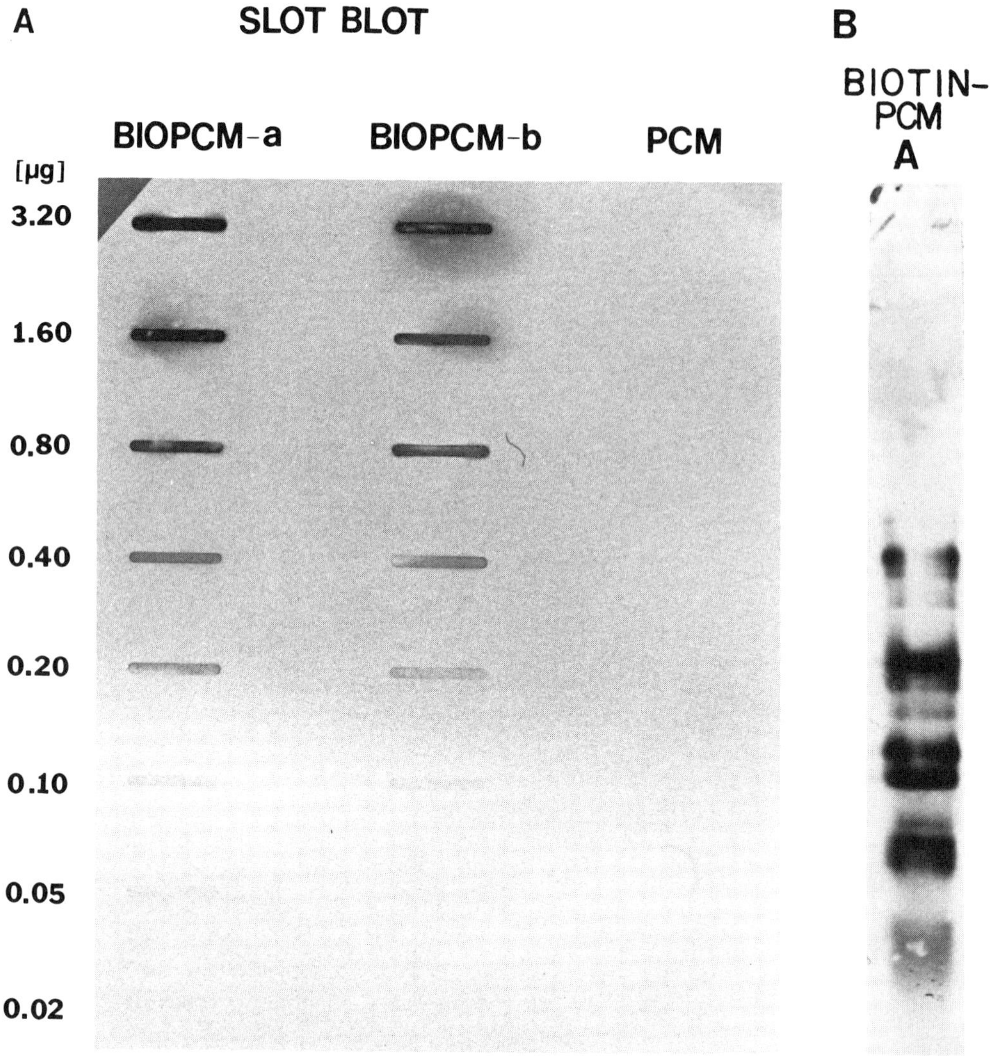

FIGURE 5. Detection of BioPCM on slot blots and after proteolytic degradation with *Staphylococcus* V8 protease. (Panel A) Biotinylated PCM (a, presence; b, absence of AdoMet) was immobilized on nitrocellulose slot blots, and incubated with avidin-alkaline phosphatase chromogen systems. Both derivatives were easily detected on the blot. (Panel B) Digestion of BioPCM-a with V8 protease. Following digestion, proteolytic fragments were separated on 20% SDS-PAGE, blotted, and the nitrocellulose incubated with avidin-alkaline phosphatase. At least 10 bands appeared following color development, suggesting that BioPCM was biotinylated in multiple sites.

tively (lanes 2, 3, and 5); these apparent PCM binding proteins were absent from the low salt and 0.5 *M* NaCl DEAE fractions (lanes 4, 6). This suggests, but does not prove, the BioPCM can interact with a specific series of proteins in brain fractions following SDS-PAGE, blotting, and overlay procedures. In contrast, a different range of proteins bound biotinylated calmodulin (BioCaM); most notable are the 52 and 62 kDa subunits of calmodulin-dependent protein kinase II (lanes 2 and 3) and the 61 kDa subunit of calcineurin (lane 5). From this experiment, it was clear that BioPCM did not bind to calmodulin-binding proteins with high affinity; similar results were obtained using native PCM and calmodulin-binding proteins immobilized on calmodulin-agarose columns (data not presented).

Several interpretations are possible for these experiments. First, there is no direct evidence that proteins binding to BioPCM were methylated. Initial studies using overlay carboxyl

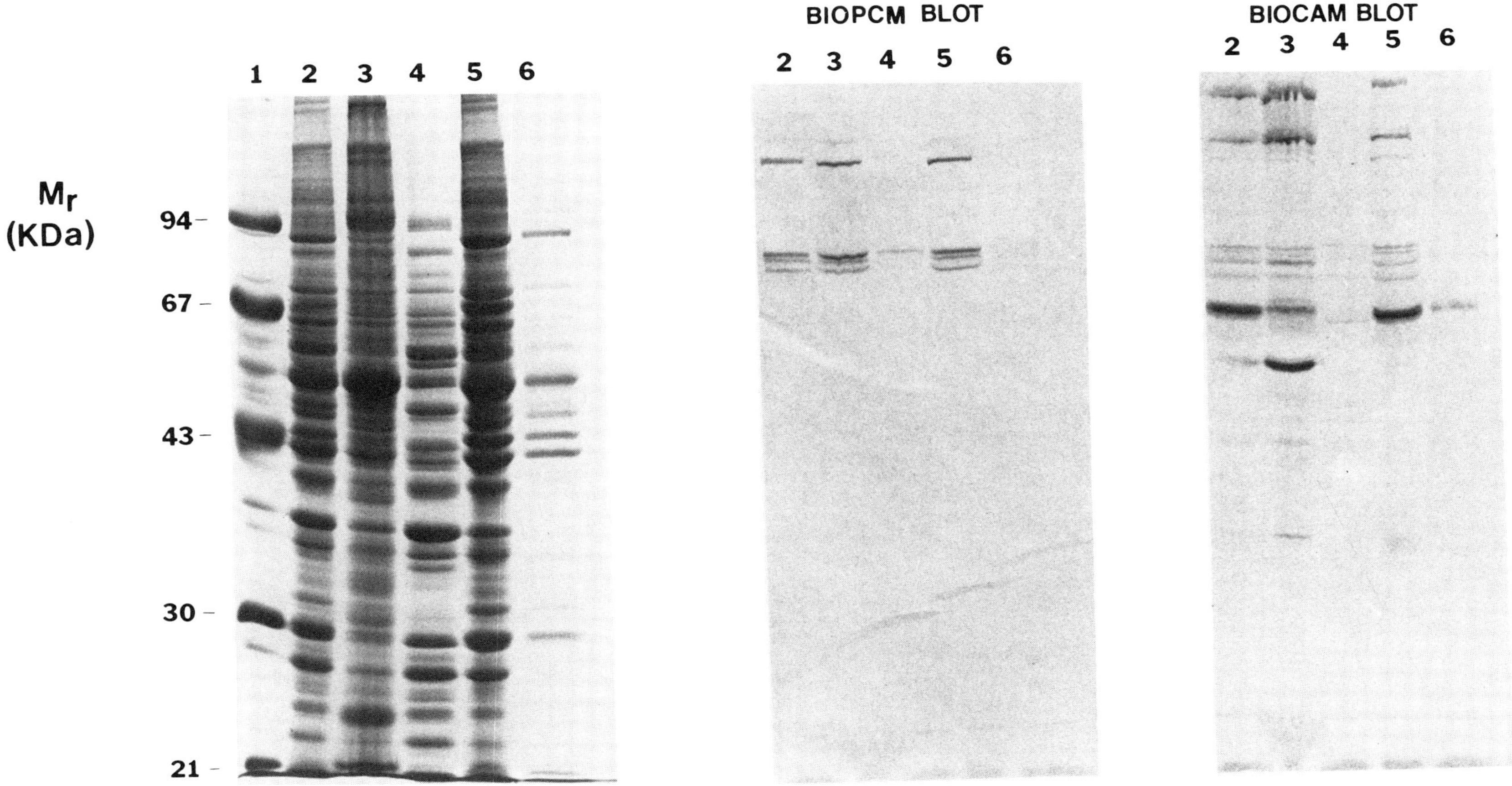

FIGURE 6. Detection of biotinylated calmodulin (BioCaM) or BioPCM binding proteins in fractions of rat brain. Rat brain was homogenized and separated into cytosolic (lane 2), membrane (lane 3), and cytosolic fractions from DEAE-Trisacryl chromatography (lane 4, effluent; lane 5,0.25 *M* NaCl eluate; lane 6, 0.5 *M* NaCl eluate). Following SDS-PAGE and blotting, identical nitrocellulose blots were incubated with either BioPCM or BioCaM, and the biotinylated proteins detected using avidin-peroxidase chromogens. Proteins of 75—84 kDa and 150 kDa bound BioPCM, while numerous proteins bound BioCaM (see text for description).

methylation reactions with ^{3}H-AdoMet were inconclusive. Second, it is possible that the binding of BioPCM is the result of biotinylation; however, it is more likely that biotinylation would interfere with protein binding per se. More evidence is needed to demonstrate that the binding is specific, in that it is inhibited by AdoHcy or other procedures that destroy BioPCM (boiling, low pH). It is also possible that PCM interacts with a range of proteins, but that the topology for interaction is destroyed by SDS-PAGE and conditions of blotting. All of these possibilities can be experimentally verified, and studies are in progress to explore each of these issues.

However, it is possible that specific proteins show a high affinity interaction with PCM, and that these will become apparent using this overlay technique. Because of the range of biotinylation reagents, it will be possible to create a series of BioPCMs with biotin attached to several different amino acid residues. One question is whether the binding of BioPCM is a consequence of its enzyme activity. By producing congeners that are modified such that enzymatic function is altered, one could dissociate the binding properties of PCM from the catalytic properties. It will also be possible to determine whether PCM binds with high affinity to isoaspartyl-containing proteins, or whether PCM has an affinity for lamin B in the nuclear membrane. It is also possible to inject BioPCM into cells, and to determine where the molecule binds by using avidin-based chromogen systems. Thus, approaching the questions of PCM function by examining the protein-protein interactions of PCM may shed light on the preferred substrates of this unusual enzyme.

D. PCM IN NEUROBLASTOMA CELLS

PCM activity has been detected in variants of the C1300 murine neuroblastoma cell line. Initial characterizations of C1300 cells grown *in vivo* have shown that the enzyme was present in all subcellular fractions, and that the enzyme was sensitive to inhibition by AdoHcy analogs.[84] When C1300 cells were cultured *in vitro,* similar activities were seen with PCM; however, there appeared to be a low concentration of methyl acceptor substrates.[85] These studies suggested that neuroblastoma cells would be good model systems for studying the effects of PCM on growing and differentiating cells. More recent experiments have shown that a 72-h exposure of murine neuroblastoma cells to adenosine dialdehyde, an inhibitor of AdoHcy hydrolase, produced a 350% increase in PCM activity; this effect was not seen with AdoHcy alone or sinefungin.[86] The increased PCM activity could be interpreted as an attempt by the cell to maintain some level of protein carboxyl methylation. Thus, future studies using inhibitors of methylation may prove useful in attempts to isolate the mRNA for PCM and to document changes in transcription as a result of drug treatment.

VII. SUMMARY

In this chapter, I have reviewed selected features of PCM function in the nervous system. It is clear that progress has been made on issues relating to enzyme purification, characterization, localization, and substrate specificity. However, there is still disagreement as to the physiologic role of this enzyme in brain and other tissue. Some of these questions may be answered using specific inhibitors of PCM, which do not currently exist. Other questions await the development of new ideas and experimental approaches to determine the functional role of PCM in the nervous system.

ACKNOWLEDGMENTS

I would like to thank Drs. R. L. Kincaid, C. D. Balaban, W. Lovenberg, D. M. Kuhn, M. Wolf, and R. H. Roth for scientific input over the years. The technical assistance of J. Kyle Krady and C. G. Hoover is greatly appreciated. Some of these studies were supported by PHS AG-06377 and a research grant from the International Life Sciences Institute Research Foundation.

REFERENCES

1. **Cohen, P.**, The role of protein phosphorylation in the normal control of enzyme activity, *Eur. J. Biochem.*, 151, 439, 1985.
2. **Nishizuka, Y.**, The role of protein kinase C in cell surface signal transduction and tumor promotion, *Nature*, 308, 693, 1984.
3. **Goldenring, J. R., Gonzalez, B., McGuire, J. S., Jr., and DeLorenzo, R. J.**, Purification and characterization of a calmodulin-dependent kinase from rat brain cytosol able to phosphorylate tubulin and microtubule-associated proteins, *J. Biol. Chem.*, 258, 12632, 1983.
4. **Nairn, A. C., Hemmings, H. C., Jr., and Greengard, P.**, Protein kinases in the brain, *Annu. Rev. Biochem.*, 54, 931, 1985.
5. **Stewart, A. A., Ingebritsen, T. S., and Cohen, P.**, The protein phosphatases involved in cellular regulation, *J. Biochem.*, 132, 289, 1983.
6. **Springer, M. S., Goy, M. F., and Adler, J.**, Protein methylation in behavioral control mechanisms and in signal transduction, *Nature*, 280, 279, 1979.
7. **Koshland, D. E., Jr.**, Biochemistry of sensing and adaptation in a simple bacterial system, *Annu. Rev. Biochem.*, 50, 765, 1981.
8. **Boyd, A. and Simon, M.**, Bacterial chemotaxis, *Annu. Rev. Physiol.*, 44, 501, 1982.
9. **Koshland, D. E., Jr.**, The bacterium as a model neuron, *Trends Neurosci.*, 6, 133, 1983.
10. **Diliberto, E. J., Jr. and Axelrod, J.**, Regional and subcellular distribution of protein carboxymethylase in brain and other tissues, *J. Neurochem.*, 26, 1159, 1976.
11. **Gagnon, C. and Heisler, S.**, Protein carboxylmethylation: role in exocytosis and chemotaxis, *Life Sci.*, 25, 993, 1979.
12. **Clarke, S.**, Protein carboxyl methyltransferase: two distinct classes of enzymes, *Annu. Rev. Biochem.*, 54, 479, 1985.
13. **Van Waarde, A.**, What is the function of protein carboxylmethylation?, *Comp. Biochem. Physiol.*, 86B, 423, 1987.
14. **O'Dea, R. F., Viveros, O. H., and Diliberto, E. J.**, Protein carboxymethylation: role in the regulation of cell function, *Biochem. Pharmacol.*, 30, 1163, 1981.
15. **Billingsley, M. L. and Lovenberg, W.**, Protein carboxylmethylation and nervous system function, *Neurochem. Int.*, 7, 575, 1985.
16. **Aswad, D. W. and Deight, B. A.**, Purification and characterization of two distinct isozymes of protein carboxylmethylase from bovine brain, *J. Neurochem.*, 40, 1718, 1983
17. **Kim, S.**, S-adenosylmethionine: protein carboxyl O-methyltransferase (protein methylase II), *Methods Enzymol.*, 106, 295, 1984.
18. **Iqbal, M. and Steenson, T.**, Purification of protein carboxymethylase from ox brain, *J. Neurochem.*, 27, 605, 1976.
19. **Haklai, R. and Kloog, Y.**, Purification and characterization of protein carboxyl methyltransferase from *Torpedo ocellata* electric organ, *Biochemistry*, 26, 4200, 1987.
20. **Brown, F. C.**, Protein carboxyl methylase isozymes in rat brain subcellular organelle, *Biochem. Pharmacol.*, 33, 2921, 1984.
21. **Sellinger, O. Z., Kramer, C. M., Fischer-Bovenkerk, C., and Adams, C. M.**, Characterization of a membrane-bound protein carboxylmethylation system in brain, *Neurochem. Int.*, 10, 155, 1987.
22. **Billingsley, M. L., Kim, S., and Kuhn, D. M.**, Immunohistochemical localization of protein-O-carboxylmethyltransferase in rat brain neurons, *Neuroscience*, 15, 159, 1985.
23. **Hurst, J. H., Billingsley, M. L., and Lovenberg, W.**, Photolabelling of methyltransferase enzymes with S-adenosylmethionine: effects of methyl acceptor substrates, *Biochem. Biophys. Res. Commun.*, 122, 499, 1984.
24. **Billingsley, M. L., Balaban, C. D., Berresheim, U., and Kuhn, D. M.**, Comparative studies on the distribution of protein-O-carboxyl methyltransferase and tyrosine hydroxylase by immunocytochemistry, *Neurochem. Int.*, 8, 255, 1986.
25. **Billingsley, M. L., Kincaid, R. L., and Lovenberg, W.**, Stoichiometric methylation of calcineurin by protein-O-carboxylmethyltransferase and its effects on calmodulin-stimulated phosphatase activity, *Proc. Natl. Acad. Sci. U.S.A.*, 85, 5612, 1985.
26. **Billingsley, M. L., Velletri, P. A., Lovenberg, W., Kuhn, D. M., Goldenring, J. R., and DeLorenzo, R. L.**, Is Ca^{2+}-calmodulin-dependent protein phosphorylation in rat brain modulated by carboxylmethylation?, *J. Neurochem.*, 44, 1442, 1985.
27. **Billingsley, M. L., Kuhn, D. M., Velletri, P. A., Kincaid, R. L., and Lovenberg, W.**, Carboxylmethylation of phosphodiesterase attenuates its activation by Ca^{2+}-calmodulin, *J. Biol. Chem.*, 259, 6630, 1984a.
28. **Kincaid, R. L., Balaban, C. D., and Billingsley, M. L.**, Differential localization of calmodulin-dependent enzyme in rat brain: evidence for selective expression of cyclic nucleotide phosphodiesterase in specific neurons, *Proc. Natl. Acad. Sci. U.S.A.*, 84, 1118, 1987.

29. **Ouimet, C. C., McGuinness, T. L., and Greengard, P.,** Immunocytochemical localization of calcium/calmodulin-dependent protein kinase II in rat brain, *Proc. Natl. Acad. Sci. U.S.A.*, 81, 5604, 1984.
30. **Patel, B. T. and Tudball, N.,** Localization of S-adenosylhomocysteine hydrolase and adenosine deaminase immunoreactivities in rat brain, *Brain Res.*, 370, 250, 1986.
31. **Paik, W. K. and Kim, S.,** Protein methylases during the development of rat brain, *Biochim. Biophys. Acta*, 313, 181, 1973.
32. **Clark, R. L., Venkatasubramanian, K., and Zimmerman, E. F.,** Prenatal and postnatal development of protein carboxylmethylation activity in the mouse, *Dev. Neurosci.*, 5, 465, 1982.
33. **Gilad, G., Gagnon, C., and Kopin, I. J.,** Protein carboxymethylation in the rat superior cervical ganglion during development and after axonal injury, *Brain Res.*, 183, 393, 1980.
34. **Billingsley, M. L., Velletri, P. A., Roth, R. H., and DeLorenzo, R. J.,** Carboxylmethylation of calmodulin inhibits calmodulin-dependent phosphorylation in rat brain membranes and cytosol, *J. Biol. Chem.*, 258, 5352, 1983.
35. **Gagnon, C., Kelly, S., Manganiello, V., Vaughn, M., Odya, C., Stritt-Matter, W., Hoffman, A., and Hirata, F.,** Modification of calmodulin function by enzymatic carboxymethylation, *Nature*, 291, 515, 1981.
36. **Runte, L., Jurgensmeier, H. L., Unger, C., and Soling, H. D.,** Calmodulin carboxyl methyl ester formation in intact human red blood cells and modulation of this reaction by divalent cations, *FEBS Lett.*, 147, 125, 1982.
37. **Miyake, M. and Innami, T.,** Protein carboxyl methylation in synaptic membrane of rat brain: the possible presence of adenosine-bound s-adenosyl homocysteine hydrolase in the membrane, *J. Neurochem.*, 49, 355, 1987.
38. **Billingsley, M. L. and Roth, R. H.,** Dopamine agonists stimulate protein carboxylmethylation in striatal synaptosomes, *J. Pharmacol. Exp. Ther.*, 223, 681, 1982.
39. **Johnson, B. A. and Aswad, D. W.,** Identification and topography of substrates for protein carboxyl methyltransferase in synaptic membrane and myelin-enriched fractions of bovine and rat brain, *J. Neurochem.*, 45, 1119, 1985.
40. **Vincent, P. L. and Siegel, F. L.,** Carboxylmethylation of calmodulin in cultured pituitary cells, *J. Neurochem.*, 49, 1613, 1987.
41. **Johnson, B. A., Freitag, N. E., and Aswad, D. W.,** Protein carboxyl methyltransferase selectively modifies an atypical form of calmodulin, *J. Biol. Chem.*, 260, 10913, 1985.
42. **Clarke, S. and O'Connor, C. M.,** Do eukaryotic carboxyl methyltransferases regulate protein function?, *Trends Biochem.*, 8, 391, 1983.
43. **Klee, C. B., Crouch, T. H., and Richman, P. G.,** Calmodulin, *Annu. Rev. Biochem.*, 49, 489, 1980.
44. **Kloog, Y., Flynn, D., Hoffman, A., and Axelrod, J.,** Enzymatic carboxymethylation of the nicotinic acetylcholine receptor, *Biochem. Biophys. Res. Commun.*, 97, 1474, 1980.
45. **Yee, A. S. and McNamee, M. G.,** Effects of carboxylmethylation by a purified *Torpedo californica* methylase on the functional properties of the acetylcholine receptor in reconstituted membranes, *Arch. Biochem. Biophys.*, 243, 349, 1985.
46. **Nuske, J. H.,** Protein methylase II in five taxa from three phyla, *Comp. Biochem. Physiol.*, 86B, 37, 1987.
47. **Swanson, R. J. and Applebury, M. L.,** Methylation of proteins in photoreceptor rod outer segments, *J. Biol. Chem.*, 258, 10599, 1983.
48. **Artemev, N. O. and Etingof, R. N.,** Methylation of 3′,5′-cGMP phosphodiesterase in the photoreceptor membranes of the retina, *Biochemistry (USSR)*, 52, 154, 1987.
49. **MacFarlane, D. G.,** Inhibitors of cyclic nucleotide phosphodiesterase inhibit protein carboxyl methylation in intact blood platelets, *J. Biol. Chem.*, 259, 1352, 1984.
50. **Diliberto, E. J., Viveros, O. H., and Axelrod, J.,** Subcellular distribution of protein carboxymethylase and its endogenous substrates in the adrenal medulla: possible role in excitation-secretion coupling, *Proc. Nat. Acad. Sci. U.S.A.*, 73, 4050, 1976.
51. **Meydan, N., Egozi, Y., and Kloog, Y.,** Enzymatic protein carboxylmethylation in rat posterior pituitary: neurophysins in rapid-turnover pool determine methyl accepting capacity, *J. Neurochem.*, 48, 208, 1987.
52. **Nguyen, M. H., Harbour, D., and Gagnon, C.,** Secretory proteins from adrenal medullary cells are carboxyl-methylated in vivo and released under their methylated form by acetylcholine, *J. Neurochem.*, 49, 38, 1987.
53. **Edgar, D. H. and Hope, D. B.,** Protein carboxyl methyl transferase of bovine posterior pituitary gland: neurophysin as a potential endogenous substrate, *J. Neurochem.*, 27, 949, 1976.
54. **Kloog, Y. and Saavedra, J. M.,** Protein carboxylmethylation in intact rat posterior pituitary *in vitro*, *J. Biol. Chem.*, 258, 7129, 1983.
55. **Gagnon, C., Axelrod, J., and Brownstein, M.,** Protein carboxylmethylation: effects of 2% sodium chloride administration on protein carboxymethylase and its endogenous substrates in rat posterior pituitary, *Life Sci.*, 22, 2155, 1978.
56. **Saavedra, S. M., Kloog, Y., Chevillard, C., and Fernandez-Pardal, J.,** High protein carboxymethylase activity and low endogenous methyl acceptor proteins in posterior pituitary were of rats lacking neurophysin-vasopressin (Brattleboro rats), *J. Neurochem.*, 41, 195, 1983.

57. **Chen, J. K. and Liss, M.**, Evidence of the carboxymethylation of nascent peptide chains on ribosomes, *Biochem. Biophys. Res. Commun.*, 84, 261, 1978.
58. **Chelsky, D., Ruskin, B., and Koshland, D. E., Jr.**, Methyl esterified proteins in a mammalian cell line, *Biochemistry,* 16, 972, 1985.
59. **Georgatos, S. D. and Blobel, G.**, Lamin B constitutes an intermediate filament attachment site at the nuclear envelope, *J. Cell Biol.*, 105, 117, 1987.
60. **Chelsky, D., Olson, J.F., and Koshland, D. E., Jr.**, Cell-cycle dependent methyl esterification of lamin B, *J. Biol. Chem.*, 262, 4303, 1987.
61. **Sidell, N. and Horn, R.**, Properties of human neuroblastoma cells following induction by retinoic acid, in *Advances in Neuroblastoma Res.*, Evans, A. E., Angio, G. D., and Seeger, R. C., Eds., Alan R. Liss, New York, 1985, 39.
62. **Fetters, H. A., Kelleher, J., and Duerre, J. A.**, Protein carboxyl methylation-demethylation in rat thymocytes, *Can. J. Biochem. Cell Biol.*, 63, 1112, 1985.
63. **Duerre, J. A. and Fetters, H. A.**, Protein carboxyl methylation-demethylation system in developing rat livers, *Biochemistry,* 24, 6848, 1985.
64. **Kloog, Y., Axelrod, J., and Spector, I.**, Protein carboxyl methylation increases in parallel with differentiation of neuroblastoma cells, *J. Neurochem.*, 40, 522, 1983.
65. **Diliberto, E. J., Jr.**, Protein-carboxyl methylation: putative role in exocytosis and in the cellular regulation of secretion and chemotaxis, in *Cellular Regulation of Secretion and Release,* Conn, P. M., Ed., Academic Press, New York, 1982.
66. **Billingsley, M. L., Galloway, M. P., and Roth, R. H.**, Possible role of protein carboxylmethylation in the autoreceptor-mediated regulation of dopamine neurons, in *Catecholamines,* Usdin, E., et al., Eds., Alan R. Liss, New York, 1984, 37.
67. **Wolf, M. E. and Roth, R. H.**, Dopamine autoreceptor stimulation increases protein carboxylmethylation in striatal slices, *J. Neurochem.*, 44, 291, 1985.
68. **Wolf, M. E. and Roth, R. H.**, *Dopamine Autoreceptors,* Alan R. Liss, New York, 1987.
69. **Saller, C. F. and Salama, A. I.**, Homocysteine prevents apomorphine-induced decreases in dopamine metabolites, *Brain Res.*, 360, 407, 1985.
70. **Gharib, A., Chabannes, B., Sarda, N., and Pacheco, H.**, *In vivo* elevation of mouse brain s-adenosyl-l-homocysteine after treatment with l-homocysteine, *J. Neurochem.*, 40, 1110, 1983.
71. **Borchardt, R. T., Kudnen, P, Huber, J. A., and Moorman, A.**, Inhibition of calf thymus and rat hypothalamic synaptosomal protein carboxyl methyltransferase by analogs of S-adenosyl homocysteine, *Mol. Pharmacol.*, 21, 181, 1986.
72. **Elden, L. G., Borchardt, R. T., and Rutledge, C. O.**, Characteristics of protein carboxyl methylation in the rat hypothalamus, *J. Neurochem.*, 38, 631, 1982.
73. **Keller, B. T. and Borchardt, R. T.**, Metabolism and mechanism of action of neoplanocin A—a potent inhibitor of S-adenosyl homocysteine hydrolase, in *Biological Methylation and Drug Design,* Borchardt, R. T., Creveling, C. R., and Neland, P. M., Eds., Humana Press, Clifton, NJ, 1986, 385.
74. **Seeley, J. P., Rukenstein, A., Connolly, J. L., and Greene, L. A.**, Differential inhibition of nerve growth factor and epidermal growth factor effects on the PC12 pheochromocytoma cell line, *J. Cell Biol.,* 98, 417, 1984.
75. **Acheson, A. and Thoenen, H.**, Both short-term and long-term effects of nerve growth factor on tyrosine hydroxylase in calf adrenal chromaffin cells are blocked by S-adenosyl homocysteine hydrolase inhibitors, *J. Neurochem.*, 48, 1416, 1987.
76. **Acheson, A., Vogl, W., Huttner, W., and Thoenen, H.**, Methyltransferase inhibitors block NGF-regulated survival and protein phosphorylation in sympathetic neurons, *EMBO J.*, 5, 2799, 1986.
77. **Billingsley, M. L., Pennypacker, K. R., Hoover, C. G., and Kincaid, R. L.**, Biotinylated proteins as probes of protein structure and protein-protein interactions, *Biotechniques,* 5, 22-31, 1987.
78. **Snabes, M. C., Boyd, A. G., and Bryan, J.**, Detection of actin binding proteins in human platelets by ^{125}I-actin overlay of polyacrylamide gels, *J. Cell Biol.*, 90, 809, 1981.
79. **Gershoni, J. and Palade, G. G.**, Protein blotting: some principles and applications, *Anal. Biochem.*, 131, 1, 1983.
80. **Glenney, J. R. and Weber, K.**, Calmodulin binding proteins of the microfilaments present in isolated brush borders and microvilli of intestinal epithelial cells, *J. Biol. Chem.*, 255, 10551, 1980.
81. **Billingsley, M. L., Pennypacker, K. R., Hoover, C. G., Brigati, D. J., and Kincaid, R. L.**, A rapid and sensitive method for detection and quantification of calcineurin and calmodulin binding proteins using biotinylated calmodulin, *Proc. Natl. Acad. Sci. U.S.A.*, 82, 7585, 1985.
82. **Kincaid, R. L., Billingsley, M. L., and Vaughan, M.**, Preparation of fluorescent, cross-linking and biotinylated calmodulin derivatives and their use in studies of calmodulin-activated phosphodiesterase and protein phosphatase, *Methods Enzymol.*, 159, 605, 1988.
83. **Billingsley, M.L., Polli, J. W., Pennypacker, K. R., and Kincaid, R. L.**, Use of biotinylated calmodulin derivatives for identification of calmodulin binding proteins, *Methods Enzymol.*, in press.

84. **O'Dea, R. F., Pons, G., Hansen, J. A., and Mirkin, B. L.,** Characterization of protein carboxyl-O-methyltransferase in the spontaneous *in vivo* murine C-1300 neuroblastoma, *Cancer Res.*, 42, 4433, 1982.
85. **O'Leary, M., Hansen, J., Mirkin, B. L., and O'Dea, R. F.,** Protein carboxyl-O-methyltransferase activity in cultured C-1300 neuroblastoma cells., *Biochem. Pharmacol.*, 32, 2339, 1983.
86. **O'Dea, R. F., Mirkin, B. L., Hogenkamp, H. P., and Barteni, D. M.,** Effect of adenosine analogs on protein carboxylmethyltransferase, s-adenosyl homocysteine hydrolase and ribonucleotide reductase activity in murine neuroblastoma cells, *Cancer Res.*, 47, 3656, 1987.
87. **Diliberto, E. J., Jr. and Axelrod, J.,** Characterization and substrate specificity of a protein carboxymethylase in the pituitary gland, *Proc. Natl. Acad. Sci. U.S.A.*, 71, 1701, 1974.
88. **Gagnon, C.,** Enzymatic carboxylmethylation of calcium binding proteins, *Can. J. Biochem. Cell Biol.*, 61, 921, 1983.
89. **Aswad, D. W. and Deight, E. A.,** Endogenous substrates for protein carboxylmethyltransferase in cytosolic fractions of brain, *J. Neurochem.*, 41, 1702, 1983.
90. **Gagnon, C., Viveros, O. H., Diliberto, E. J., Jr., and Axelrod, J.,** Enzymatic methylation of carboxyl groups of chromaffin granule membrane proteins, *J. Biol. Chem.*, 253, 3778, 1978.
91. **Gagnon, C., Veeraragavan, K., and Coulombe, R.,** Protein carboxyl methylation in adrenal medullary cells, *Cell Mol. Neurobiol.*, 8, 95, 1988.
92. **Veeraragavan, K., Coulombe, R., and Gagnon, C.,** Stoichiometric carbonyl methylation of chromogranins from bovine adrenal medullary cells, *Biochem. Biophys. Res. Commun.*, 152, 732, 1988.

Chapter 13

PROTEIN METHYLESTERASE FROM MAMMALS

C. Gagnon and K. Veeraragavan

TABLE OF CONTENTS

I. INTRODUCTION

Two decades after its discovery, the methylation of proteins on carboxyl groups under physiological conditions is still poorly understood. This posttranslational modification of proteins is catalyzed by the enzyme protein-carboxyl methylase (PCM). This ubiquitous enzyme esterifies free carboxyl groups on the side chains of proteins resulting in the neutralization of their negative charges.[1-4] The methyl esters formed can be hydrolyzed either enzymatically or nonenzymatically. The enyzmatic reaction is catalyzed by the enzyme protein methylesterase (PME) discovered about 10 years ago both in prokaryotes[5] and in eukaryotes.[6] The products of PME hydrolysis are methanol and unmethylated proteins (Figure 1). Spontaneous hydrolysis of protein-methyl esters also do occur but at a rate several orders of magnitude below that achieved by PME at PME's optimal pH.

In this chapter, the purification, properties, and substrate specificity of mammalian PME will be reviewed and compared to that of the bacterial enzyme.

II. PRESENCE OF PME IN MAMMALIAN TISSUES

The first evidence for the presence of PME in mammalian tissues dates back to 1979 when gelatin-methyl esters enzymatically synthesized from incubations of gelatin with *S*-adenosylmethionine and PCM, were hydrolyzed by kidney tissue homogenates.[6] The detection of this enzyme was made possible by the use of kidney homogenates as the source of enzyme and by selecting a pH around 5 to avoid spontaneous demethylation that occurs at neutral and alkaline pH. Under these acidic conditions, gelatin-methyl esters are greatly stabilized and spontaneous hydrolysis into methanol occurs only at a very slow rate. This instability of protein-methyl esters is not specific to gelatin-methyl esters since several other substrates have been shown to spontaneously hydrolyzed at neutral or alkaline pH.[7-10]

Whereas the level of PCM varies by a factor of 4 from one tissue to another, PME has a 26-fold variation. There seems to be an inverse correlation between the levels of PME and PCM activities in various tissues. High levels of PCM and low levels of PME have been detected in brain and testes whereas the reverse was observed in kidneys[11] (Table 1). The highest concentration of PME has been detected in kidneys. However, all tissues tested contain a certain amount of PME activity. Within the kidney, PME is preferentially concentrated in the cortex whereas PCM is more concentrated in the medulla. Within the cortex, PME is mainly localized in the proximal tubules, a site for several active transport systems[11] (Table 2).

At the subcellular level, PME appears to be predominantly localized in the cytosol. Little activity has been detected in the lysosomal or microsomal subcellular factions (Table 3). However the addition of nonionic detergent to kidney homogenate results in a 65% increase in enyzmatic activity (Table 1). In other tissues, the presence of a detergent increased up to eight-fold the level of PME activity.[6]

III. PURIFICATION OF KIDNEY PME

Kidney PME was purified 1200-fold by a series of chromatographic and dialysis steps with a yield of 28%.[12] First, kidneys were homogenized in 0.3 *M* sucrose. The cytosol was prepared by centrifuging the homogenate at 105,000 × *g* for 2 h. Proteins were precipitated with 65% ammonium sulfate. The resulting precipitated proteins were pelleted by centrifugation and resuspended in a low pH buffer (10 m*M*, sodium acetate, pH 4.35) containing 0.1 m*M* EDTA, 0.1 m*M* dithiothreitol (DTT) and 5% glycerol. The low pH and the presence of each component were essential to prevent loss of activity in subsequent steps. After dialysis against this low pH buffer and centrifugation to eliminate precipitated proteins, PME was loaded on a column of

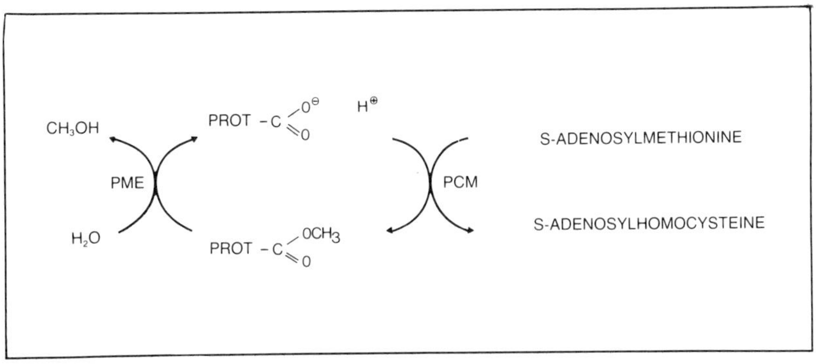

FIGURE 1. The protein-carboxyl methylating system. PCM: protein-carboxyl methylase; PME, protein methylesterase.

TABLE 1
Levels of Protein Methylesterase in Various Rat Tissue Homogenates

Tissues	Specific activity without detergent (pmol/min/mg protein)	Specific activity with detergent (pmol/min/mg protein)
Kidney	4.84	8.00
Liver	0.12	0.90
Adenohypophysis	0.09	0.36
Adrenal	0.08	0.59
Pancreas	0.18	0.36
Testis	0.12	0.55
Brain	0.03	0.17

Note: Gelatin-methyl esters were used as substrates for PME.

TABLE 2
Levels of PME and PCM Activities in Rat Kidney

	PME activity (pmol/min/mg protein)	PCM activity (pmol/min/mg protein)	PME/PCM
Pupilla	0.07	3.17	0.02
Medulla	0.48	1.68	0.28
Cortex	1.82	1.38	1.32
Distal tubules and glomeruli	0.87	2.25	0.35
Proximal and distal tubules	1.10	1.58	0.70
Proximal tubules	1.74	1.16	1.53

Note: These values are means of four to six animals. Gelatin and gelatin-methyl esters were used as substrates for PCM and PME, respectively.

Sephadex G-100. Active fractions were further purified by chromatofocusing on Polybuffer Exchanger 94 and adsorption chromatography on Matrex gel Green A. A summary of the various steps is presented in Table 4.

TABLE 3
Subcellular Distribution of PME Activity in Rat Kidney

Fraction	PME activity pmol/min	PME activity %	Specific activity (pmol/min/mg protein)
Homogenate	2300	100	7.8
Nuclei enriched	630	24	8.4
Mitochondria enriched	240	10	3.1
Lysosome enriched	170	7	3.9
Microsome enriched	160	7	4.2
Cytosol	1010	44	12.9

Note: The recovery of PME activity was 92%. Gelatin-methyl esters were used as substrates for PME.

TABLE 4
Purification of Protein Methylesterase from Rat Kidneys

Steps	Total activity (pmol/min)	Specific activity (pmol/min/mg protein)	Purification factor	Recovery (%)
Homogenate	4580	0.10	1.0	100
Cytosol	4250	0.330	3.3	93
Ammonium sulfate	4080	0.46	4.6	89
Dialysis pH 4.35	3960	1.13	11.4	86
Sephadex G-100	2860	6.30	63.6	62
Chromatofocusing	1930	42.80	432.0	42
Matrex gel Green A	1300	118.00	1190.0	28

Note: Ovalbumin-methyl esters were used as substrates for PME.

IV. PHYSICAL PROPERTIES OF PME

Analysis of the purified PME by polyacrylamide gel electrophoresis in denaturating conditions revealed the presence of a major band with a M_r of 31,000 and of a minor band with a M_r of 25,000. Attempts to separate these two proteins through a series of chromatographic steps were unsuccessful. However, following these chromatographies, ratios of 31,000/25,000 peptide varied from 0.2 to 5.0. A lowering of the ratio was always associated with a decrease in enzymatic specific activity of the preparation suggesting that the 31,000 peptide represented the active form of PME.[12] This value is consistent with the M_r attributed to PME by molecular sieving on Sephadex® G-100 column.[11] To determine whether both peptides were related, monoclonal antibodies against the purified PME preparation were developed. All monoclonals tested (up to ten) recognized both proteins. These results suggested that both peptides are immunologically related and that the 25,000 M_r peptide may represent a degradation product of the higher molecular weight peptide.

Once purified, PME is stable over a wide range of pH. In the presence of 0.1 m*M* DTT and 5% glycerol no significant loss of activity was observed between pH 2 and 8 whereas about 50% of the enzyme remains active at pH 1.0. On the other hand, PME was activated by 50% at pH 9.5 and 10, and all the enzymatic activity was lost at pH above 10.5.[12]

PME has an isoelectric point of 4.45 as measured by isoelectric focusing in polyacrylamide gels.[12] This is a reflection of the high content of Glx and Asx residues which represents up to 23% of all PME residues (Veeraragavan and Gagnon, unpublished results). Other preponderant residues include Ser (17%) and Gly (16%). Inhibition of PME activity by residue specific reagents indicated that a cysteine residue was very important for PME activity. Parachloro-

TABLE 5
Important Amino Acid Residues for PME Activity

Reagent	Molar ratio reagent/enzyme	PME activity %	Probable site of action
Mercuric chloride	100	0	Cys
Parachloromercuribenzoic acid	100	0	Cys
Tetranitromethane	100	0	Tyr
Diethylpyrocarbonate	100	0	His
1,2-cyclohexanedione	100	100	Arg
Acetylimidazole	100	100	Ser
2,4-pentanedione	1000	100	Lys

mercuribenzoic acid (PCMB) and mercuric chloride were very potent inhibitors of PME activity. Their effects were completely reversed by the addition of 2-mercaptoethanol. Diethylpyrocarbonate and tetranitromethane were also very effective inhibitors of PME activity. These two reagents react predominantly with His and Tyr residues, respectively, and to a minor extent to Cys residues. However, since PME free SH-groups were blocked with PCMB prior to the addition of the above reagents and unblocked thereafter with 2-mercaptoethanol, results with diethylpyrocarbonate and tetranitromethane indicate that some of the His and Tyr residues were essential for the hydrolysis of protein-methylesters and probably localized in or near the active site. Other residues such as Lys, Arg, and Ser do not appear to be important for the hydrolysis of protein-methyl esters (Table 5).

When incubated with ovalbumin-methyl esters as substrates, PME show an optimum pH of 4.0. Similar results are obtained when calmodulin-methyl esters were used as substrates.[12] On the other hand, the optimum pH was much broader when gelatin methyl esters were used. Under these conditions, the optimum pH ranged from 4.0 to 6.5 with 50% of PME activity still remaining at pH 8.0.[6] PME activity is not significantly affected by low concentrations of mono-divalent ions such as Na, K, Ca, and Mg. However, at concentrations above 0.2 *M*, divalent cations decreased PME activity. Up to 80% of PME activity was inhibited at 1 *M* $CaCl_2$.[12]

V. SUBSTRATE SPECIFICITY

Various protein-methyl ester substrates were incubated with PME to study the specificity of this enzyme. K_m values varied from 0.3 μM for histones to 64 μM for gamma globulin. Luteinizing hormone (LH), calmodulin, and adrenocorticotrophic hormone (ACTH) were among the best substrates for PME with K_ms of 1.2, 1.1, and 2.6 μM, respectively (Table 6). The K_{cat} also varied significantly from one substrate to another with the highest values observed for substrates that had the highest K_m (Table 6). K_{cat} values ranged from 20 to 500×10^{-6} s^{-1} whereas the K_{cat}/K_m ratios varied from 8 to 71 $M^{-1} \cdot s^{-1}$ for molecules under their original native form. There seems to be a direct relationship between K_m and K_{cat} values. Since alkaline deamidation of ACTH and calmodulin has been recently shown to increase the methyl acceptor capacity of these proteins and that under these conditions near stoichiometric carboxyl methylation was observed in the presence of an excess in PCM and *S*- adenosylmethionine,[10,13,14] methylated deamidated substrates were compared to native methylated calmodulin and ACTH. Deamidation of calmodulin prior to methylation by PCM marginally affected the K_m for PME. However, deamidation caused a sixfold increase in the K_{cat} (Table 7). On the other hand, deamidation of ACTH prior to methylation of PCM decreased the K_m value by a factor of 5 without modifying the K_{cat}. These results raised the questions as to whether methylation sites are different after deamidation and whether PCM considers these molecules (native vs. deamidated) as completely different molecules.

TABLE 6
Kinetic Parameters of Mammalian PME

Methylated substrate	K_m (μM)	$K_{cat} \times 10^6$ (s^{-1})
Histone	0.3	20
LH	1.2	65
FSH	3.2	92
Ovalbumin	7.6	397
GH	10.7	298
Prolactin	17.5	410
Bovine γ-globulin	64.0	500

TABLE 7
PME Kinetic Parameters for Native and Deamidated Substrates

Methylated substrate	K_m (mM)	$K_{cat} \times 10^6$ (s^{-1})
Native ACTH	2.6	117
Deamidated ACTH	0.7	107
Native calmodulin	1.1	37
Deamidated calmodulin	1.8	207

Note: ACTH and calmodulin were deamidated by alkaline treatment at pH 11.0 prior to being methylated by PCM. Native methylated and deamidated — methylated ACTH and calmodulin were used as substrates for PME.

TABLE 8
Inhibition of PME Activity by Protease Inhibitors

Inhibitor	Concentration (μM)	% inhibition of PME activity
Leupeptin	3.0	100
Chymostatin	24.0	100
α-1 Antitrypsin	2.7	10
STI	6.6	10
Antithrombin III	2.1	0
Aprotinin	17.6	0
LBTI	20.4	0
PMSF	657.0	0

VI. PROTEASE INHIBITORS AND PME ACTIVITY

Since esterases may also have proteolytic activities, a series of protease inhibitors were tested on PME activity as measured by the hydrolysis of ovalbumin-methyl esters. Antithrombin III, lima bean tryspin inhibitor (LBTI), aprotinin, soybean trypsin inhibitor (STI), α-1 antitrypsin, and phenylmethylsulfonylfluoride (PMSF) had no significant effect on PME activity. On the other hand, protease inhibitors with C-terminal aldehyde such as leupeptin and chymostatin were potent inhibitors (Table 8). Double recipricol plots of PME activity as a function of ovalbumin-methyl ester concentrations in the presence and absence of leupeptin and chymostatin indicated that these products acted as competitive inhibitors of the demethylation

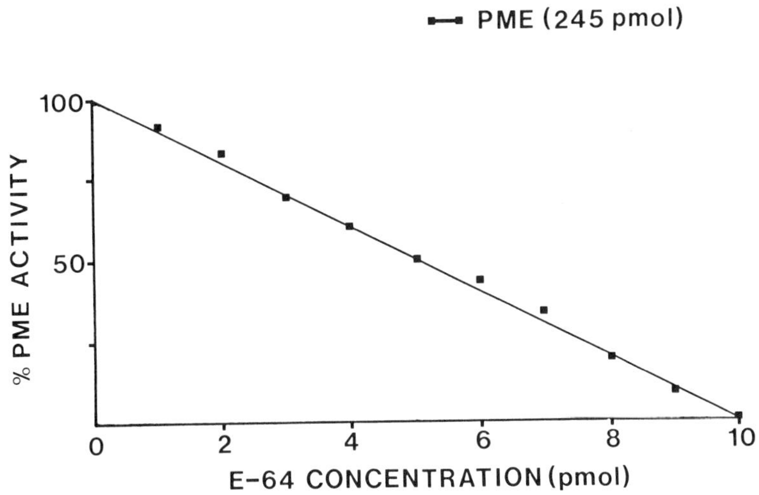

FIGURE 2. Inhibition of PME activity by the thiol protease reagent E-64. PME (245 pmol) was incubated with various concentrations of E-64 at 37°C for 5 min. Following this preincubation, ovalbumin-methyl esters were added and the PME activity remaining measured.

TABLE 9
Comparison between Bacterial and Mammalian PME

	Bacterial enzyme	Mammalian enzyme
M_r	37,000	31,000
Product formed	Methanol	Methanol
K_m	15 m*M*	0.3—64 m*M*
K_{cat}	270 $M^{-1} \cdot s^{-1}$	8—152 $M^{-1} \cdot s^{-1}$
Amino acids in the active site	Cys	Cys, Tyr, His
Function	Regulatory (chemotaxis)	Unknown

reaction.[15] Since these inhibitors are known to inhibit thiolproteases, these results suggest that a cysteine plays an important role in PME activity. This conclusion is in good agreement with results obtained with the two -SH blocking reagents discussed above. However, the most surprising finding of these experiments with leupeptin was the observation that 0.12 μ*M* leupeptin blocked 50% of the activity of 1.75 μ*M* PME. Since the enzyme used was more than 80% pure (e.g., the 31,000 M_r peptide), these results indicate that several PME molecules were inactive. To determine the exact number of active PME molecules in the PME preparation used, the latter was reacted with E-64, a cysteine protease inhibitor that inactivates enzymes by forming an irreversible thioether bond.[16,17] From the inhibition curve it was estimated that one molecule of E-64 inhibited the activity generated from 23 to 25 molecules of PME (Figure 2). This indicates that not more than one molecule out of 25 molecules was active. These results with protease inhibitors raise the question as to whether PME hydrolyzes peptide bonds. Preliminary results with synthetic peptides indicate that PME can hydrolyze peptide bonds in addition to methyl esters on proteins. However, the K_m for these peptides is two orders of magnitude above that of the peptide hormones LH and FSH.

VII. COMPARISON WITH BACTERIAL PME

In *Salmonella typhimurium* PME is encoded by the *che*B gene. This enzyme hydrolyzes methyl esters on methyl accepting chemotaxis proteins.[5,18] The bacterial PME has a molecular weight of 37,000, a value slightly higher than that of the mammalian enzyme (Table 9). Its K_m for the above proteins is 15 μM and a cysteine residue is essential for enzymatic activity.[19] In similarity with the mammalian enzyme, its turnover rate is very slow being 0.24 mol of methanol per mole of enzyme per minute. Like the mammalian PME, it appears that the bacterial enzyme predominantly exists under an inactive form or that the intrinsic activity of each molecule is very low. However, the poorly active 37,000 M_r peptide can be processed into a 21,000 M_r peptide through proteolysis. Limited proteolysis with trypsin results in a 15-fold activation of PME activity.[20]

VIII. PHYSIOLOGICAL SIGNIFICANCE OF THE MAMMALIAN PME

The physiological significance of PME will be understood only when the role of protein carboxyl methylation is established. At the present time, several experimental systems have been used to study PCM in live cells. The reader is referred to the numerous reviews published on the subject.[9,10,21-23] Other more recent *in vivo* systems where the methylation has been studied include those of Nguyen et al.[24] and Chelski et al.[25] In the first study, it was shown that methylation of chromogranin A, the major secretory protein in adrenal medullary cells occurs in intact cells. Methylated chromogranin A is stored in granules and can be released in the methylated form in the culture medium following cholinergic stimulation of adrenal cells.[24] In the second study, it was shown that the main methylated protein in mouse neuroblastoma cells is the nuclear envelope protein lamin B.[25] The methylation of this protein depends on the cell cycle and can reach, in live cells, the level of 1.3 mol of CH_3 per mole of protein. This stoichiometry is of a few orders of magnitude above the expected stoichiometry if the advocates of the "repair of abnormal Asp residue theory"[10,25] were correct. On the other hand, the repair theory is supported by numerous studies showing substoichiometric methylation of proteins in test tube reactions.[10,26] Thus, it is possible that PCM may turn out to have two types of physiological functions.

PME has been investigated in few physiological studies. In one of them, the enzyme has been shown to increase 7- to 20-fold in spermatozoa as they mature during epididymal transit.[27] In a second study PME has been shown to increase following stimulation of neutrophils by fmet-leu-phe, a chemotactic peptide.[28]

To play a role in a physiological system, PME must be able to rapidly hydrolyze protein-methyl esters. However, according to the enzymatic activity observed so far, there is doubt that this enzyme is sufficiently rapid. On the other hand, recent experimental evidence suggests that both bacterial and mammalian PME may exist under a partially active form. Whereas proteolysis activates the bacterial enzyme, the activation mechanism for PME is presently unknown. Nevertheless, a 15-fold activation like it has been observed for the bacterial enzyme is insufficient. With K_{cat} between 10^{-5} to $10^{-3} s^{-1}$, both enzymes are very slow catalysts. Under these special kinetics, PME must either be activated *in vivo* before fulfilling its biological function or be linked in a special way to its substrate. Further studies are required to elucidate how these slow enzymes are activated and how they fulfill their physiological roles.

REFERENCES

1. **Liss, M., Maxam, A. M., and Cuprak, L. J.,** Methylation of protein by calf spleen methylase: a new protein methylation reaction, *J. Biol. Chem.*, 244, 1617, 1969.
2. **Kim, S. and Paik, W. K.,** Purification and properties of protein methylase II, *J. Biol. Chem.*, 245, 1806, 1970.
3. **Kim, S. and Paik, W. K.,** Studies on the structural requirements of substrate protein for protein methylase II, *Biochemistry*, 10, 3141, 1971.
4. **Morin, A. M. and Liss, M.,** Evidence for a methylated protein intermediate in pituitary methanol formation, *Biochem. Biophys. Res. Commun.*, 52, 373, 1973.
5. **Stock, J. B. and Koshland, E. D., Jr.,** A protein methylesterase involved in bacterial sensing, *Proc. Natl. Acad. Sci. U.S.A.*, 75 , 3659, 1978.
6. **Gagnon, C.,** Presence of a protein methylesterase in mammalian tissues, *Biochem. Biophys. Res. Commun.*, 88, 847, 1979.
7. **Kim, S. and Paik, W. K.,** Labile protein-methyl esters: comparison between chemically and enzymatically synthesized, *Experientia*, 32, 982, 1976.
8. **Gagnon, C., Viveros, O. H., Diliberto, E. J., Jr., and Axelrod, J.,** Enzymatic methylation of carboxyl groups of chromaffin granule membrane proteins, *J. Biol. Chem.*, 253, 3778, 1978.
9. **Paik, W. K. and Kim, S.,** *Protein Methylation*, John Wiley & Sons, New York, 1980.
10. **Clarke, S.,** Protein carboxyl methyltransferase: two distinct classes of enzymes, *Annu. Rev. Biochem.*, 54, 479, 1985.
11. **Chene, L., Bourget, L., Vinay, P., and Gagnon, C.,** Preferential localization of a protein methylating demethylating system in proximal tubules of rat kidney, *Arch. Biochem. Biophys.*, 213, 299, 1982.
12. **Gagnon, C., Harbour, D., and Camato, R.,** Purification and characterization of protein methylesterase from rat kidney, *J. Biol. Chem.*, 259, 10215, 1984.
13. **Aswad, D. W.,** Stoichiometric methylation of porcine adrenocorticotropin by protein carboxyl methyl transferase requires deamidation of asparagine 25: evidence for the methylation at the carboxyl group of atypical L-isoaspartyl residues, *J. Biol. Chem.*, 259, 10714, 1984.
14. **Johnson, B. A., Freitag, N. E., and Aswad, D. W.,** Protein carboxyl methyltransferase selectively modifies an atypical form of calmodulin, *J. Biol. Chem.*, 260, 10913, 1985.
15. **Veeraragavan, K. and Gagnon, C.,** Leupeptin and chymostatin inhibit mammalian protein methylesterase activity, *Biochem. Biophys. Res. Commun.*, 142, 603, 1987.
16. **Hanada, K., Tamai, M., Ohmura, S., Samada, J., Seki, T., and Tanaka, I.,** Structure and synthesis of E-64, a new thiol protease inhibitor, *Agric. Biol. Chem.*, 42, 529, 1978.
17. **Hanada, K., Tamai, M., Yamagish, M., Ohmura, S., Samada, J., and Tanala, I.,** Isolation and characterization of E-64, a new thiol protease inhibitor, *Agric. Biol. Chem.*, 42, 523, 1978.
18. **Toews, M. L. and Adler, J.,** Methanol formation *in vivo* from methylated chemotaxis proteins in *Escherichia coli*, *J. Biol. Chem.*, 254, 1761, 1979.
19. **Snyder, M. A., Stock, J. B., and Koshland, D. E., Jr.,** The carboxylmethylase esterase of bacterial chemotaxis, *Methods Enzymol.*, 106, 321, 1984.
20. **Borczuk, A., Staub, A., and Stock, J.,** Demethylation of bacterial chemoreceptors is inhibited by attractant stimuli in the complete absence of the regulatory domain of the methylating enzyme, *Biochem. Biophys. Res. Commun.*, 141, 918, 1986.
21. **Gagnon, C. and Heisler, S.,** Protein carboxyl-methylation: role in exocytosis and chemotaxis, *Life Sci.*, 25, 993, 1979.
22. **O'Dea, R. F., Viceros, O. H., and Diliberto, E. J., Jr.,** Protein carboxylmethylation: role in the regulation of cell functions, *Biochem. Pharmacol.*, 30, 1163, 1981.
23. **Diliberto, E. J., Jr.,** Protein carboxyl methylation: putative role in exocytosis and in the cellular regulation of secretion and chemotaxis, in *Cellular Regulation of Secretion and Release*, Conn, P. M., Ed., Academic Press, New York, 1982, 147.
24. **Nguyen, M. H., Harbour, D., and Gagnon, C.,** Secretory proteins from adrenal medulla cells are carboxyl-methylated *in vivo* and released under their methylated form by acetylcholine, *J. Neurochem.*, 49, 38, 1987.
25. **Chelsky, D., Olson, J. F., and Koshland, D. E.,** Cell cycle-dependent methyl esterification of lamin B, *J. Biol. Chem.*, 262, 4303, 1987.
26. **Aswad, D. W. and Johnson, B. A.,** The unusual substrate specificity of eukaryotic protein carboxyl methyl transferase, *TIBS*, 12, 155, 1987.
27. **Gagnon, C., Harbour, D., de Lamirande, E., Bardin, C. W., and Dacheux, J. L.,** Sensitive assay detects protein methylesterase in spermatozoa: decrease in enzymatic activity during epididymal maturation, *Biol. Reprod.*, 30, 953, 1984.
28. **Venkatasubramanian, K., Hirata, F., Gagnon, C., Corcoran, B. A., O'Dea, R. F., Axelrod, J., and Schiffman, E.,** Protein methylesterase and leukocyte chemotaxis, *Mol. Immunol.*, 17, 201, 1980.

Chapter 14

CHEMOTACTIC METHYLATION IN *BACILLUS SUBTILIS*

George W. Ordal and David O. Nettleton

TABLE OF CONTENTS

I. INTRODUCTION

Chemotaxis is the process by which bacteria move to high concentrations of attractant or low concentrations of repellent. It is believed to be a primitive sensory-motor process that may well antedate the evolutionary separation of the archebacteria, eubacteria, and proeukaryotic superkingdoms. Active investigations of its mechanism have occurred in *Escherichia coli* and the close relative *Salmonella typhimurium, Bacillus subtilis, Spirochaetia,* and *Halobacterium halobium,* all phylogenetically distant organisms, and it may eventually be possible to hypothesize what the ancestral mechanism may have been like. Perhaps the best characterized aspect of chemotaxis has been the role of methylation, for the great majority of chemotactic baceteria possess methyl-accepting chemotaxis proteins (MCPs), which are methyl esterified. *E. coli* has four different MCPs,[1-4] and from *in vitro* methylation experiments[5] and partial proteolysis experiments,[6] it is apparent that *B. subtilis* has three different MCPs — H1, H2, and H3. However, in *B. subtilis* they function very differently than in *E. coli.* These differences are summarized in Table 1.

II. *IN VIVO* METHYLATION

A. SIMILARITIES *IN VIVO*

B. subtilis is a Gram-positive spore-forming bacterium and this is fundamentally different and phylogenetically distant from *E. coli,* a Gram-negative non-spore-forming bacterium. However, like *E. coli,* it is peritrichously flagellated and carries out chemotaxis by modulating durations of runs and tumbles.[7] Like *E. coli,* it has MCPs,[5] it is attracted to amino acids, sugars, and oxygen,[8-10] and it responds to addition of amino acid attractant by running (swimming smoothly) for a period of time and then returning to prestimulus random behavior (adaptation).[11-12]

One type of protocol that is typically used in analyzing biochemical events associated with chemotaxis is to suspend cells in buffer with chloramphenicol to inhibit protein synthesis, add [C^3H_3]-methionine, treat the cells with chemotactic effectors, and analyze the results in various ways. In the case of *E. coli,* when the effects of amino acid attractants are explored by fractionating the MCPs on SDS-polyacrylamide gels and fluorographing, addition of attractant causes a period of increasing methyl-esterification of MCPs.[13] The duration coincides with the period of running.[13] Thus, methylation of MCPs is assumed to somehow bring about adaptation. During this time, methanol, normally produced at a low level due to turnover of methyl groups on MCPs, is not made.[14]

B. DIFFERENCES *IN VIVO*

When the effects of amino acid attractants are explored using the same protocols in *B. subtilis,* very different results are obtained. First, there is an immediate change in the distribution of methyl groups among the three MCPs. Afterward, throughout the period of running and beyond, the distribution is completely stable (Figure 1).[15] Thus methylation changes are associated with excitation, not adaptation.

Second, there is net methyl transfer from H1 to H2. In *E. coli,* methyl transfers among MCPs is unknown. The evidence for methyl transfer from H1 to H2 was obtained by incubating the cells in nonradioactive methionine before addition of attractant. In this way the specific activity of methyl groups of *S*-adenosylmethionine (AdoMet), the proximate donor of methyl groups for the MCPs was reduced. Nevertheless, increase in methylation of H2 was observed. There was a very considerable decrease in methylation of H1, which was assumed to be the source of methyl groups for this methylation. Paradoxically, of the H1 MCPs that remain methylated, the degree of esterification becomes higher.[6]

This transfer may also occur for unstimulated cells *in vivo.* Addition of radioactive

TABLE 1
Different Roles of MCPs in *B. subtilis* and *E. coli*

B. subtilis	*E. coli*
On addition of attractant MCP methylation changes are immediate , and thus are probably connected with excitation (Figure 1)	MCP methlyation changes are gradual, throughout course of period of smooth swimming caused by attractant addition[13]
A period of enhanced flux of methyl groups through MCPs occurs on attractant addition. The flux is highest on addition of attractant but appears to continue through much or all of the period of smooth swimming	Turnover completely stops on attractant addition
Methyl groups are transferred to an unidentified carrier on addition of attractant, and they are transferred back on removal of attractant (Figure 2)	No such carrier is presumed to exist
No net change of number of methyl groups occurs on attractant addition, only a changed distribution[16]	Attractant addition causes increased methylation of the corresponding MCP[3]
Attractant causes transfer of methyl groups from MCP H1 to the MCP H2[6]	No inter-MCP transfer is known to occur
Repellents do not affect MCP methylation[18,19]	Repellents cause MCP demethylation[13]

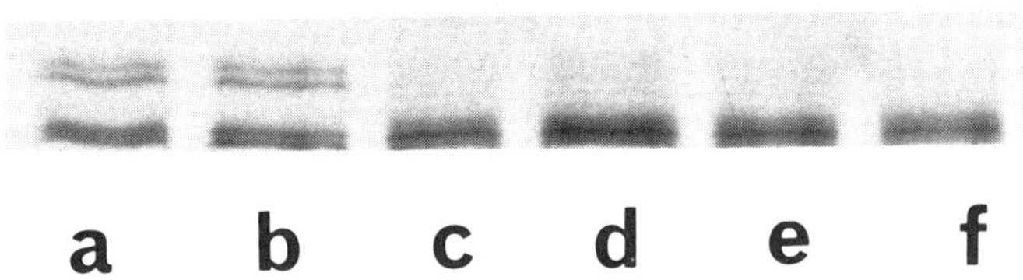

FIGURE 1. Time course of attractant-induced MCP methylation changes. Bacteria were radioactively labeled in absence of protein synthesis and exposed to 0.1 *M* aspartate (90% receptor occupancy) at the time noted by the arrow. Samples were fractionated by SDS polyacrylamide gel electrophoresis and fluorographed. Time (a) 30 s before addition; (b) 15 s before addition; (c) 5 s after addition; (d) 15 s after addition; (e) 30 s after addition; (f) 60 sec after addition. (From Thoelke, M., Parker, H., Ordal, E., and Ordal, G., *Biochemistry,* 27, 8453, 1988. With permission.)

methionine to cells in buffer shows that there is a lag for entry of methyl groups in H2, compared with H1 and H3, and subsequent addition of excess nonradioactive methionine shows immediate loss of methyl groups from H1 and H3 but continued increase in methylation of H2 for several minutes before decrease begins.[6] Thus, at least H2 is fundamentally different from *E. coli* MCPs, which are all receptors for attractants or for complexes of binding protein/attractant.

Third, by performing the same experiment and analyzing the total loss of methyl groups from the MCPs as the result of adding aspartate to cells previously labeled and then incubated in excess nonradioactive methionine, we found that attractant causes a flux of methyl groups through the MCPs. There is an immediate 50% loss of radioactive methyl groups when 0.1 *M* asparate is added (Figure 2).[15]

Fourth, by labeling the cells in low specific activity but abundant (10 μM) methionine so that they are labeled to a maximal (steady state) extent, we found that attractant caused no net change in number of methyl groups associated with the MCPs.[16] However, if labeling was carried out for only 15 s and attractant, even very low concentrations, or buffer was added and the cells were harvested only 5 s later, considerable increase in labeling was observed with attractant.[17] This is further evidence of the flux of methyl groups. It is also a very sensitive way of discerning the effect of attractants, for concentrations that have only scant effect on the distribution of methyl groups have a noticeable effect on labeling of MCPs by this protocol. This flux occurs not just right away, but later and possibly throughout the period of running caused by attractant.[17] Again,

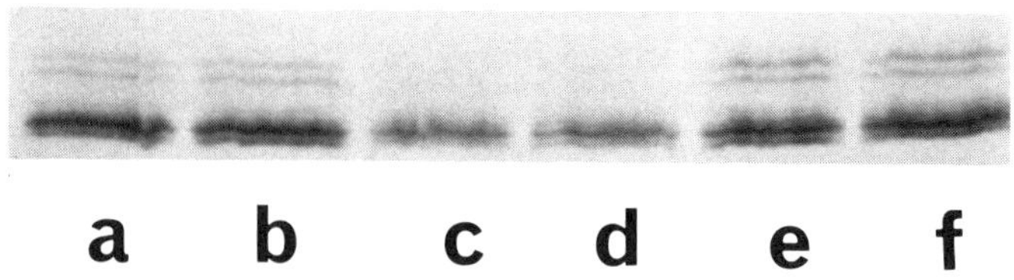

FIGURE 2. Effect of attractant addition and removal on MCP methylation. Bacteria were incubated in radioactive methionine. At 9 min excess (10 μ*M*) nonradioactive methionine was added. At 11 min 30 s 0.1 *M* aspartate was added and at 12 min 30 s was removed. Samples were taken at the following times: lanes 1, 2: 11 min; lanes 3, 4: 12 min; lanes 5, 6: 14 min. (From Thoelke, M., Kirby, J., and Ordal, G., *Biochemistry,* 28, 5585, 1989. With permission.)

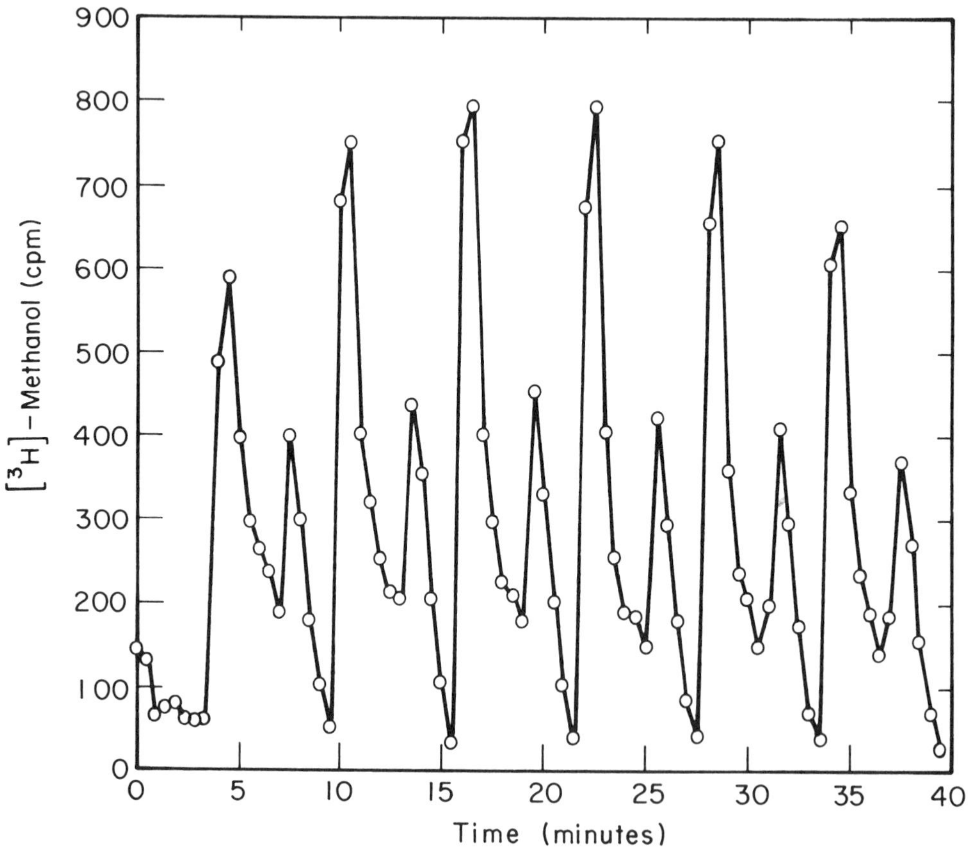

FIGURE 3. Time course of attractant-induced methanol production. Bacteria were labeled, given excess nonradioactive methionine, and incubated. Aliquots were removed at various times and the methanol quantitated. Aspartate was added at 4 min (arrow). Basla slopes were set equal and the data presented in relative cpm. Symbols: (Δ), 0.1 *M* aspartate, 90% receptor occupancy; (❑), 0.01 *M* aspartate, 50% receptor occupancy; (o), 4.3 m*M* aspartate, 30% receptor occupancy. (From Thoelke, M., Parker, H., et al., *op. cit.* With permission.)

by contract, in *E. coli,* MCP methylation increases, rather than stays constant, on addition of attractant, and the flux of methyl groups ("turnover") stops during the period of smooth swimming.[13,14]

Fifth, radioactive methyl groups are released at a high (maximal) rate throughout the course of the running period (Figure 3).[15] In *E. coli,* as stated above, methanol is not made during the comparable period.[14]

Sixth, all stimuli — addition or removal of attractant or repellent — produce methanol (Figure 4).[18-20] By contrast, in *E. coli,* only negative stimuli — addition of repellent or removal of attractant — produce it; positive stimuli block its production.[14,21] Very likely, the purpose of the methanol production is demethylation of some substance(s) to bring about adaptation. These experiements were performed by placing methylated cells in a flow cell, adding or removing stimuli, collecting effluent in a fraction collector, and quantitating the methanol. Unlike for *E. coli,* where the rate of methanol evolution falls rapidly, the rate of methanol evolution falls slowly, even when many repeated stimuli are given.[19] This implies that only a small fraction of the total number of methyl groups available are released as methanol from each stimulation.

Seventh, attractant causes immediate loss of radioactive methyl groups from labeled MCPs in presence of excess nonradioactive methionine. However, radioactive methanol is released throughout the course of the running period.[15] Therefore, the radioactive methyl groups must first go to an unidentified methyl carrier before their ultimate release as methanol. Were they to arise directly from the MCPs, they should have come off in a burst. No such carrier is believed to exist in *E. coli.*

Eighth, striking evidence for this carrier has come from experiments in which labeled cells are given excess nonradioactive methionine, the 0.1 *M* aspartate, and then are removed from the aspartate. As noted above, there is about 50% loss of radioactive methyl groups from the MCPs, but on removal of the aspartate, all the radioactive groups return (Figure 2).[19] On addition of aspartate, these radioactive methyl groups must have been transferred to this unidentified carrier, and then returned to the MCPs on removal of the aspartate. The nature of this carrier is unknown; however, nowhere on the SDS-polyacrylamide gel do the methyl groups lost from the MCPs arise.[19] Efforts to find this methylated carrier have not yet succeeded.

Further evidence for this carrier comes from repetitive stimulation of "steady state" labeled cells by 90% receptor occupancy (0.1 *M)* aspartate. That is, cells were labeled (at least 20') with low specific activity [C^3H_3]-methionine until MCPs were maximally labeled. They were placed in a flow cell and flushed with nonradioactive methionine for a few minutes until the background had become low. Then aspartate was added for several minutes, removed for several minutes, added, again, removed again, etc. In each instance, the ratio of radioactive methanol obtained from addition to removal was about 2.5. Interestingly, the maximal amount of radioactive methanol evolved on the *third* addition of aspartate, not the first (Figure 5). This result, incomprehensible if the MCPs are the proximate source of methanol, might be rationalized by assuming that methyl groups flow from MCPs to a methyl carrier and from there become methanol. If there are nonradioactive methyl groups on this carrier, added during growth, the specific activity of these groups will be lower than that of methyl groups on MCPs at the beginning of the experiment. However, mixing will occur due to additions and removals of aspartate. At the same time, of course, nonradioactive methyl groups originating from the added nonradioactive methionine will enter the pool of methyl groups on the MCPs via AdoMet during the experiment.

Finally, unlike *E. coli* where repellents affect MCP methylation by causing deesterification, in *B. subtilis* they do not directly affect methylation. For instance, adding chloropromazine to cells in presence or absence of aspartate, in presence or absence of excess nonradioactive methionine, does not affect MCP methylation.[18] Therefore, the signal from adding repellent is conveyed differently from the signal of attractant removal. As mentioned above, attractant removal leads both to methanol production and restoration of methyl groups previously lost from MCPs back to the MCPs.

It is interesting that methanol release occurs when attractant is added and when repellent is added. However, when both are added together at such concentrations that the behavior remains random, methanol production is not stimulated.[18] It seems that, as stated above, the purpose of methanol production is demethylation of substance(s) to bring about adaptation. Unless the behavior is nonrandom, these mechanisms do not operate.

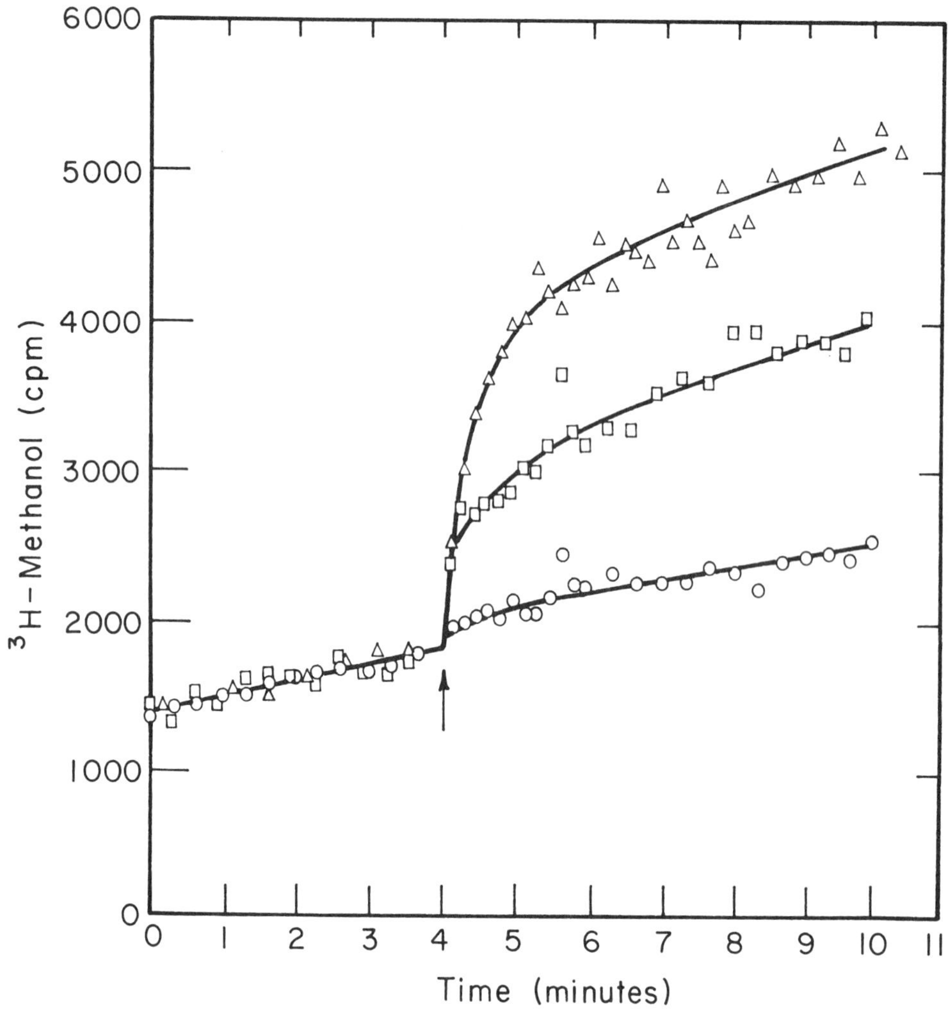

FIGURE 4. Flow assay showing effect of addition and removal of attractant and repellent. Bacteria were labeled, transferred to a flow cell, and buffer, attractant (23 m*M* aspartate, 70% receptor occupancy), or repellent (83 μ*M* chlorpromazine, the minimum concentration to prevent smooth swimming caused by 23 m*M* aspartate simultaneously). Reagents were added and removed at the times indicated by arrows. Upper panel, 0I1085 (*che*$^+$); lower panel, 0I1100, which is isogenic except for a *cheR* mutation, making the strain defective in chemotactic methyltransferase. (From Thoelke, M., Kirby, J., et al., *op. cit.* With permission.)

C. METHYLTRANSFERASE MUTANT

The importance of methylation on chemotaxis has been confirmed by examining mutants in the chemotactic methyltransferase. Furthermore, certain inferences about the nature of the chemotactic mechanism may be made from studies on these mutants. First, chemotaxis as measured in the capillary assay is very poor; yet the behavior of freely motile cells in indistinguishable from wild type. In this assay, a capillary containing attractant is placed in a suspension of cells. Attractant diffuses into the suspension creating a gradient, and bacteria travel up the gradient into the capillary. After 30 to 60 min the contents are plated and colonies counted the next day. The mutant shows only about 4% wild type chemotaxis for isoleucine and 0.5% wild type chemotaxis for mannitol.[22] In *E. coli,* the methyltransferase mutant only runs, never tumbles,[23] and hence cannot bias its motion, a prerequisite for showing chemotaxis so that capillary assays are meaningless. However, double mutants, lacking both methyltransferase and methylesterase, both run and tumble. For some attractants they show fairly good chemotaxis in

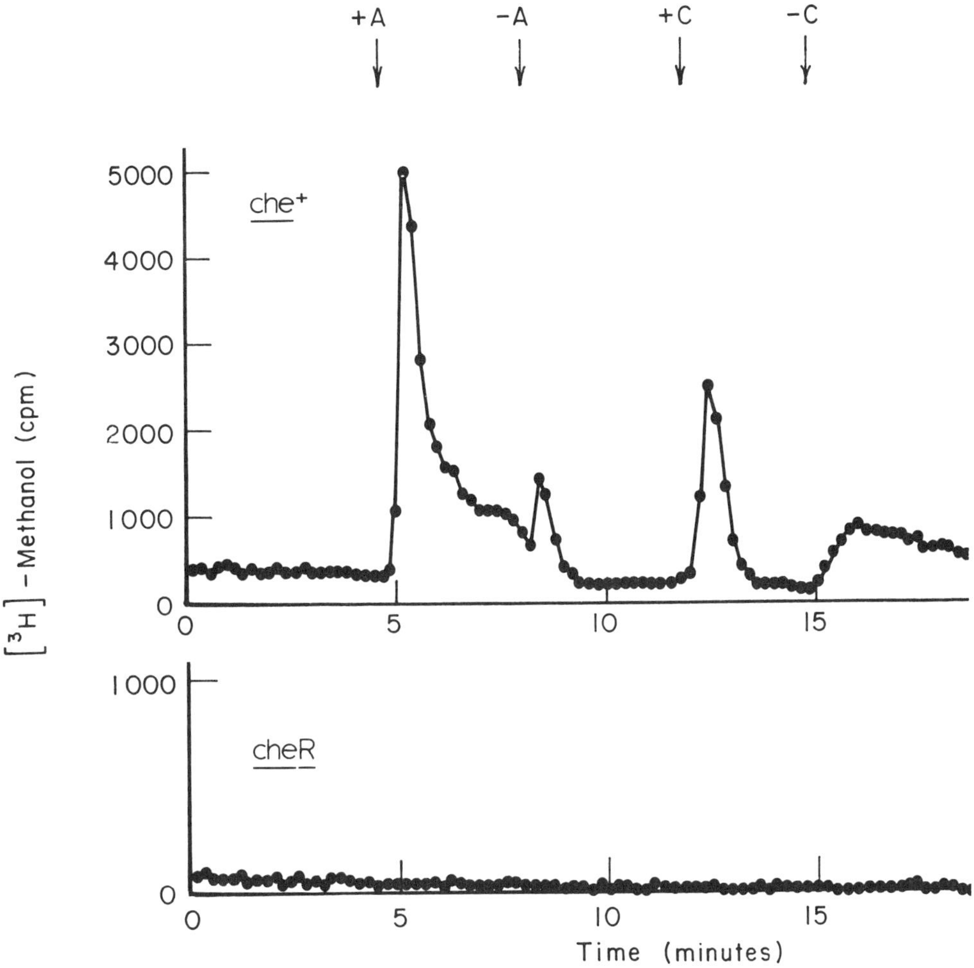

FIGURE 5. Flow assay showing effect of addition and removal of attractant on methanol evolution. Bacteria were radioactively labeled until labeling of MCPs reached a maximum (25 min) and then transferred to a flow cell. Buffer or 0.1 *M* aspartate was flowed by the cells and collected in a fraction collector. The resulting methanol was quantitated. (From Thoelke, M., Kirby, J., et al., *op. cit.* With permission.)

the capillary assay, and for others they do not.[24] Thus, for *B. subtilis,* methylated MCPs, presumably to enable methyl transfers, are vital for chemotaxis; in *E. coli,* other means seem to enable chemotaxis to take place.

The significance of the normal behavior of the unstimulated *cheR* mutant cells of *B. subtilis* is unclear. Since, unlike for *E. coli,* degree of methylation does not change on addition of attractant, it may not directly affect behavior, even when there is no methylation.

Furthermore, another rather unexpected difference between wild type and mutant is behavior when exposed to attractants and repellents. When amino acid attractant is added to wild type cells tethered to a coverslip, after most of their flagella were sheared off, they rotate counterclockwise (CCW) for a period, then briefly clockwise (CW) (the "overshoot"), then return to random behavior. Removal causes CW motion for about 60% as long, then a brief CCW overshoot, then return to random behavior. When amino acid attractant is added to the mutant, the bacteria rotate their flagella CCW punctuated by CW rotation, and there is no overshoot. The return to random behavior is slightly faster than for wild type. When attractant is removed, the mutant rotates its flagella CW punctuated by CCW rotations, and the return to random behavior is much longer than for wild type.[25] These results imply that the stimuli are not readily

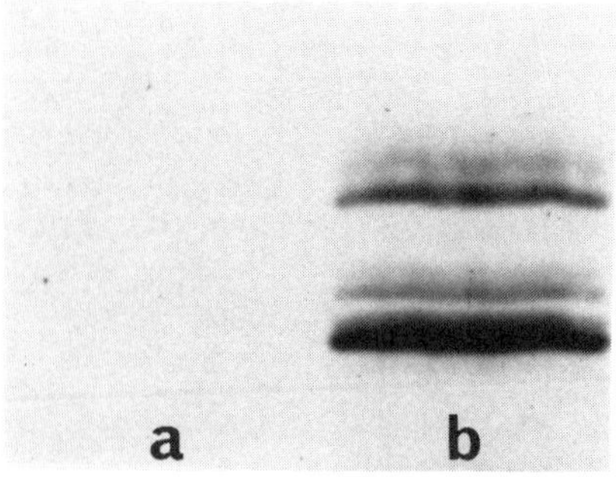

FIGURE 6. Methylation of MCPs by *B. subtilis* methyltransferase. Each assay contained 5 m*M* Ca^{++}, 1.5 μg membrane protein, 1.25 μC_i [C^3H_3]-AdoMet and either buffer (a) or methyltransferase (b) Methylation was stopped after 60 min and the products visualized by SDS-PAGE and fluorography.

"perceived". Perhaps the unmethylated MCPs do not register the fact of binding of attractant very well. Perhaps methyl transfers are involved in excitation. Perhaps some component of the signal transduction pathway works poorly in an unmethylated form. In addition, the fact that adaptation does occur without methanol release means that there must exist a methylation-independent system for bringing about return to random behavior as may be postulated for *E. coli*. However, this system is much more effective for adaptation to positive stimuli than to negative stimuli. Probably the reason for poor chemotaxis in the capillary assay is insensitivity to stimulation.

Finally, we placed labeled mutant cells in a flow cell and either attractant or repellent was added, then removed. No methanol was produced (Figure 4).[19] Therefore, all methanol originates from methyl groups on MCPs. However, as seen earlier, it appears to be released from different methyl-group carriers, rather than directly from MCPs.

III. *IN VITRO* METHYLATION

A. CHEMOTACTIC METHYLTRANSFERASE II

Methyltransferase catalyzes the transfer of methyl groups from AdoMet to MCPs, giving *S*-adenosylhomocysteine (AdoHcy) and methylated MCPs as products.[26] The methyl groups are present as esters of the γ-carboxyl of specific glutamic acid residues of the MCPs.[27,28] The progress of this reaction can be assayed by using radioactively tagged methyl donor, [C^3H_3]-AdoMet, and then monitoring the appearance of radioactivity on the MCPs (Figure 6).[26]

Methyltransferase from *B. subtilis* has been purified by Burgess-Cassler et al.[26] A lysate is formed from washed *B. subtilis* cells in a French press. Cell debris and membranes are then removed by centrifugation. The supernatant is then dialyzed and applied to a DEAE column, to which methyltransferase does not bind. The eluate from the DEAE column is then applied to a CM column, to which methyltransferase does bind. The methyltransferase is then eluted in a KCl gradient. The active fractions are pooled and subjected to ammonium sufate precipitation. The 60 to 80% precipitate is resuspended in buffer, dialyzed, and applied to an AdoHcy affinity column. Purified methyltransferase is then eluted in a gradient of NaC1. Table 2 summarizes the purification of methyltransferase II.[26]

The enzyme activity is followed during purification by filter assay.[26] This assay is performed by mixing 45 μg membrane protein (containing MCPs), 2 μCi [C^3H_3]-AdoMet, Mg^{++} to 5 m*M*, and methyltransferase activity in a reaction volume of about 50 μl. After incubation for 1 h at room temperature, the reaction is stopped by dilution with 5 ml of ice-cold buffer. The membrane proteins are then collected onto Millipore HAWP filters by vacuum filtration. The filters are then

TABLE 2
Purification of *B. subtilis* Methyltransferase

Step	Total protein (mg)	Total activity (units[a])	Specific activity (units/mg)	Fold-purification	Yield
1. Crude extract	919	1875	2.04	1.0 (–)	
2. Dialysis	550	3113	5.66	3.0 (1.0)	100[b]
3. DEAE-Bio-Gel A column	101	2747	27.2	13 (4.8)	88
4. CM-Bio-Gel A column	8.29	288	34.8	17 (6.2)	9.3
5. Ammonium sulfate fractionation	5.24	259	49.5	24 (8.6)	8.3
6. *S*-Adenosylhomo-cysteine affinity column	0.026	62	2380	1200 (420)	2.0

[a] Unit, picomoles of CH_3 transferred/min.
[b] Because of higher activity following dialysis, yield is designated "100" after dialysis step. (If purification is based on specific activity of dialyzed sample defined as 1.0, then values in parentheses apply.)

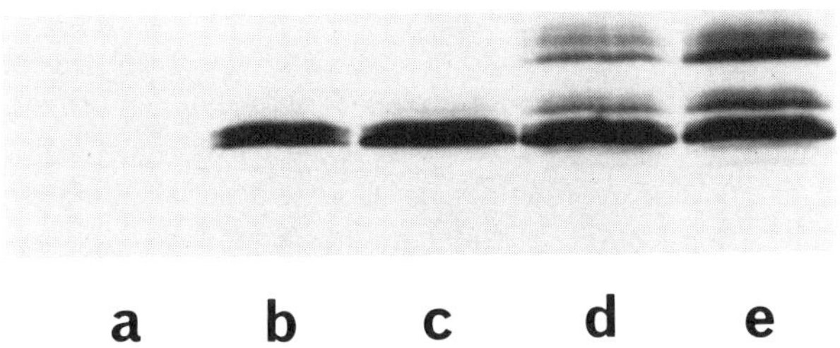

FIGURE 7. Effect of divalent cation concentration on methyltransferase activity. Methylations were carried out as described in the text, with the following modifications: (a) no enzyme, (b) no divalent cation, (c) 1 m*M* $MgCl_2$, (d) 3 m*M* $MgCl_2$, (e) 5 m*M* $MgCl_2$.

washed with buffer and the radioactivity associated with MCPs determined by liquid scintillation counting of the dried filter. Alternatively, the methylation reaction can be stopped with SDS and the MCPs separated on SDS-polyacrylamide gel electrophoresis (SDS-Page).[28]

Polyacrylamide gel electrophoresis of the purified enzyme indicates that the enzyme has a monomeric molecular weight of 30,000.[26] Gel filtration chromatography confirmed that the active species is a monomer. Methyltransferase was found to require divalent cation for activity, as either Mg^{++}, Ca^{++}, Sr^{++}, or Ba^{++},[26,29] although only Mg^{++} is present in sufficient quantity *in vivo* to promote methyltransferase activity. Furthermore, it was found that the relative degree of methylation of the various MCPs changed as the concentration of divalent cation also changed (Figure 7).[26] At lower levels of divalent cation, only the faster migrating species, H3, was found to be significantly methylated. As the concentration of divalent cation increased, the level of methylation of the other MCPs, H1 and H2, increased greatly relative to that of H3. At 5 m*M* divalent cation, the relative levels of methylation of the three MCPs approximated that found *in vivo*.

The K_m of methyltransferase for AdoMet was found to be about 5 μM.[26] AdoHcy, a product of the methyltransferase reaction, was found to inhibit the enzyme with a K_i of about 0.2 μM. Thus, the relatively stable AdoHcy could be used as the affinity substance in the affinity column for methyltransferase purification, instead of the relatively unstable AdoMet, because meth-

TABLE 3
Homology of *B. subtilis* and *E. coli/S. typhimurium* Methyltransferase

B. subtilis	*E. coli/S. typhimurium*
30,000 dalton monomer	31,000 dalton monomer
Monomeric in cell extract	Multimeric in cell extract, 38,000 daltons
$K_m = 5\ \mu M$[26]	$K_m = 12\ \mu M$
Inhibited by AdoHcy	Inhibited by AdoHcy
$K_I = 0.2\ \mu M$	
pH optimum 6.8—7	pH optimum 6.5—7
Requires divalent cation	No divalent cation requirement
Basic protein (pI > 7)	Basic protein (pI ≥ 9)
High substrate specificity	High substrate specificity
Little affected by attractant *in vitro*[5] (see below)	2× activated by attractant *in vitro*
Able to methylate *E. coli* MCPs[30]	Able to methylate *B. subtilis* MCPs[5]

yltransferase has a high affinity for it. Finally, methyltransferase was found to be most active at pH 6.9 at 20 to 25°C, with the activity decreasing rapidly as either pH or temperature was increased or decreased.

Methyltransferase was found to be highly specific in methylating MCPs.[30] The relative rate of methylation of other proteins was found to be less than 1% of the rate of the methylation of MCPs. The only exceptions to this are MCPs from *Escherichia coli* [30] and methylated membrane proteins from *Rhodospirillum rubum*.[31] Both of these proteins are methylated in response to chemotactic stimulus,[3,32,33] and at least the *E. coli* MCPs are also methylated to form carboxyl methyl esters on the glutamate residues.[34] The methyltransferase from *B. subtilis* was found to be functionally homologous to the methyltransferases from *E. coli* and *S. typhimurium* as described in Table 3.[30]

Methyltransferase was found to be active toward MCPs extracted from the *B. subtilis* membrane with the detergent Triton X-100.[35] The MCPs are solubilized in 1% Triton (v/v), and can be methylated by methyltransferase in the presence of the Triton (Figure 8),[35] with this methylation visualized by SDS-PAGE followed by fluorography.

B. CHEMOTACTIC METHYLESTERASE

Methylesterase from *B. subtilis* catalyzes the hydrolysis of glutamic acid methyl esters on the MCPs, giving rise to glutamate residues and methanol.[36] This reaction consumes one H_2O per hydrolysis. The reaction can then be written:

$$\text{MCP–CH}_2\text{COOCH}_3 + \text{H}_2\text{O} \xrightarrow[\text{esterase}]{\text{methyl-}} \text{MCP–CH}_2\text{COOH} + \text{HOCH}_3$$

The progress of this reaction can be monitored by mixing $[C^3H_3]$-methylated MCPs with methylesterase and assaying either the decrease in radioactivity associated with the MCPs or the formation of methanol.[20] The decrease in radioactivity of the MCPs is assayed by SDS-PAGE of the reaction mixture, followed by fluorography of the gel (Figure 9).[36] Alternatively, the formation of methanol is followed by placing the reaction mixture into the outer well of a Conway cell and water in the inner well. The Conway cell is then sealed and, after a suitable incubation period, the methanol is sampled from the inner well of the Conway cell.[20] The radioactivity of this sample is then determined by liquid scintillation counting.

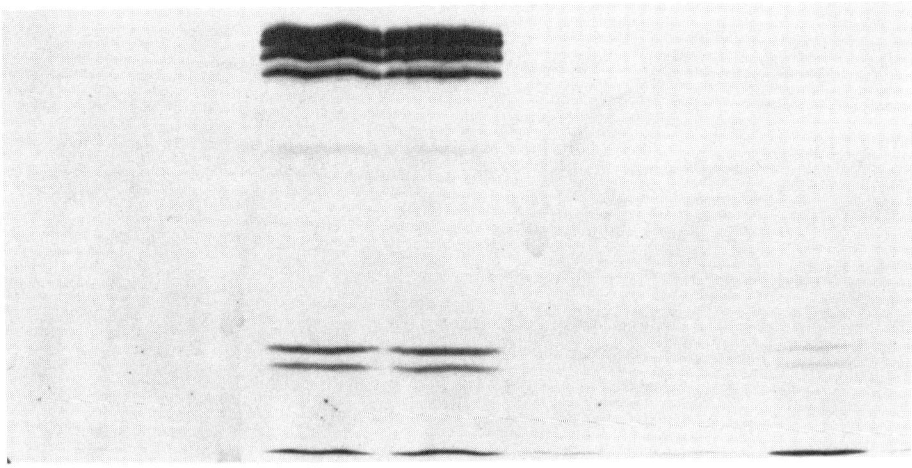

FIGURE 8. Methylation of solubilized MCPs. MCPs from *B. subtilis* were solubilized in 1% (w/v) Trition X-100, centrifuged and filtered to remove unsolubilized protein and membrane. Methylations were then carried out with: (a, b) extract from a methyltransferase defective mutant strain; (c, d) extract from a wild type strain; (e, f) buffer; or (g) no MCPs, wild type extract

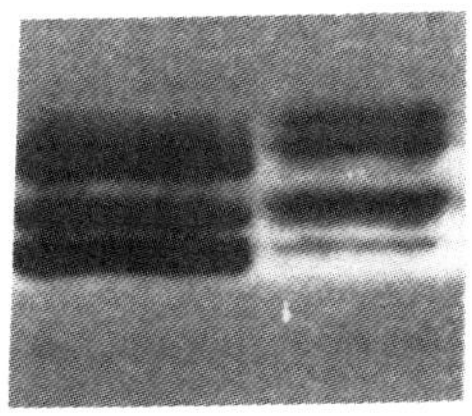

FIGURE 9. Effect of methylesterase on radiolabeled MCPs. Membrane from a methyltransferase mutant were methylated for 2 h, and then buffer (1) or 0.6 μg methylesterase (2) was added and incubation continued for an additional hour. MCPs were visualized by SDS-PAGE and fluorography. (From Goldman, D. and Ordal, G., *Biochemistry*, 23, 2600, 1984. With permission.)

Methylesterase is purified by a procedure similar to that for the purification of methyltransferase (see above).[26,36] The procedures are the same through forming a pool of active fractions from a CM column. The active pool of methylesterase from the CM column is then applied to a hydroxylapatite column and eluted with a phosphate gradient. The active fractions in this eluate are concentrated in an Amicon® concentrator (stirred cell) and then applied to a Bio Gel P-100 or P-60 column. Active fractions from this column, which represent the purified enzyme, are then pooled and stored for further use. The purification of methylesterase is summarized in Table 4.[36]

Methylesterase was found to be a monomer of 44,000 Da, as determined by both SDS-PAGE and size exclusion chromatography.[36] This enzyme also was found to require divalent cation for activity, with maximum activity at 1.1 m*M* Mg^{++}. Ca^{++}, Sr^{++}, and Ba^{++} can be substituted, but not Co^{++}, Ni^{++}, Mn^{++}, Fe^{++}, or Zn^{++}.[31] Maximum activity of methylesterase was observed at 28°C, with activity sharply decreasing with increasing temperature and more gradually decreasing

TABLE 4
Purification of *B. subtilis* Methylesterase

Step	Total protein (mg)	Total activity (pmol of CH_3OH produced/min)	Specific activity (pmol of CH_3OH produced min^{-1} mg^{-1})	Fold purification	Yield (%)
1. Dialyzed cytoplasm	658	10.72	0.016	1	100
2. CM-Bio-Gel A column	1.16	4.42	3.811	238	41
3. Bio-Gel HTP column	0.303	2.02	6.67	417	19
4. Bio-Gel P-60 column	0.0116	0.23	19.8	1237	2

with decreasing temperature. Methylesterase was virtually inactive at 40°C, a temperature at which chemotaxis is still very effective.[8] Perhaps protein-protein interactions in the cytoplasm of *B. subtilis* act to stabilize the enzyme, while such interactions are absent from the purified enzyme.

Methylesterase is most active at pH 7.5, with activity sharply declining with decreasing pH. As the pH is raised, however, the activity remains relatively stable, at least up to pH 9.5. Thus, the pH vs. activity profile resembles the titration of a protonatable group with a pKa of about 6.7, protonation of which renders the enzyme inactive. Histidine residues typically have a pKa in this range, and protonation of the ring nitrogen would make it less nucleophilic and, perhaps, less able to act as an intermediary in an S_N2 type reaction.[31]

Unlike methyltransferase, methylesterase is not sensitive to product inhibition.[31] Addition of an excess of either methanol or unmethylated MCPs has little or no effect on the hydrolysis of MCP methyl esters by methylesterase. Furthermore, methylesterase is not sensistive to the presence of ethanol, *n*--propanol, or *n*--butanol.

Purified methylesterase gradually loses activity when stored at 4°C.[36] This effect is diminished in the presence of 20% (v/v) glycerol. Furthermore, the activity of methylesterase is greatly enhanced by the presence of glycerol. Nearly a sixfold increase in activity was observed.[31, 36] That this is an activation and not merely a protection against denaturation is indicated by the observation that the ratio of activity in the presence and absence of glycerol at progressively shorter time points remains constant at about six.[31] If glycerol were merely stabilizing the enzyme, this ratio would approach one as the reaction time became very short. The unpurified enzyme also loses activity over time, even in the presence of glycerol. This is probably due to the action of residual proteases. Protease degradation is also indicated by the appearance of lower molecular weight methylesterase activities in the P-100 column eluate, especially when the earlier pools were deliberately allowed to stand before continuation of the purification procedure.[21] These lower molecular weight activities are found in discrete sizes of 22,500, 14,000, and 8000 Da, as determined by size exclusion chromatography.[31,36] That these products are digestion products and not separate gene products is indicated by their absence from carefully purified methylesterase and their inverse relationship to the activity of the 44,000 Da species. That is, as the activity of the 44,000 Da species decreases, the activity associated with the smaller fragments increases.[31] Therefore, the domain of methylesterase required for enzyme activity is small, and the segments connecting it to rest of the protein are susceptible to protease digestion. Trypsin must cleave within the domain required for activity, but the fragments freely associate with each other. This is indicated by the observation that methylesterase activity is undiminished by trypsin digestion only, but subsequent size exclusion chromatography of the digest results in complete loss of activity.[31] Examination of extracts of *B. subtilis* for higher molecular weight species has revealed no monomeric activity greater than 44,000 Da.[31,36]

TABLE 5
Homology of Methyleserases from *B. subtilis* and *E. coli/S. typhimurium*

B. subtilis	*E. coli/S. typhimurium*
44,000 m.w.[36]	38,000 m.w.[40]
Monomer is active,[36] but ssociation with other peptides observed[31]	Monomer is active, but association with other peptides is observed[40]
Requires divalent cation for activity	Does not requires divalent cation for activity
Activated by glycerol[31,36]	Activated by glycerol[31]
Inactivated by decreasing pH[31,36]	Inactivated by decreasing pH in phosphate buffer,[31] but not in MES buffer[40]
Basic protein (pI = 9.5)[31,41]	Basic protein (pI = 9.3)[40]
K_m = 9.8 n*M*[36]	K_m = 15 μ*M*[40]
Turnover number = 25/min[31]	Turnover number = 0.24/min[40]
Activated by attractant *in vivo*[20] and *in vitro*[36]	Inhibited by attractant *in vivo* [14] and *in vitro*[31]
Demethylates *E. coli* MCPs[5,31]	Demethylates *B. subtilis* MCPs[31]
Does not require divalent cation to demethylate *E. coli* MCPs[31]	Requires divalent cation to demethylate *B. subtilis* MCPs[31]

When a crude extract from *B. subtilis* is applied directly to a size exclusion column, two peaks of activity are eluted.[31] One activity is at the expected R^f for a 44,000 Da peptide, while the other elutes earlier, indicating a larger size. Examination of the larger sized activity revealed the presence of the 44,000 Da activity, among other proteins. Thus, the methylesterase enzyme must interact with other proteins in a concentrated extract and, probably, in the cytoplasm *in vivo* as well. This interaction with another protein is also indicated by examination of the eluate from a chromatofocusing column.[31] The purified methylesterase elutes at a relatively high pH, in a decreasing pH gradient, as expected of a peptide that does not bind to a positively charged column at neutral pH. However, when a crude extract from *B. subtilis* is applied to this column, two activities are eluted. One of these activities coelutes with the purified enzyme, while the other does not elute until a much lower pH is reached. These results are consistent with the methylesterase enzyme forming a relatively stable association with a specific other peptide, or peptides, which binds more tightly to the chromatofocusing column than does methlyesterase. Any role that these other proteins may have in regulating methylesterase activity remains, as yet, unknown.

Purified methylesterase has a relatively low turnover number, about 25/min, with respect to methylated MCPs.[31] Determination of the K_m of methylesterase for MCPs indicates that the methylesterase binds very tightly to the MCPs, with a K_m of about 9.8 n*M*.[36] Furthermore, methylesterase activity is very difficult to remove from membranes containing methylesterified MCPs, such as those isolated from wild type bacteria, but is easily removed from the membranes isolated from a mutant defective in methylation.[31,36] Again, this is consistent with the enzyme binding tightly and specifically to methylated MCPs. Therefore, the low rate of hydrolysis is not due to weak binding of enzyme to substrate. One explanation for the low rate of hydrolysis is that the methylesterase may have a different preferred acceptor for the methyl group than water,[18] but that dilution due to lysing the bacteria makes this other acceptor less available to the esterase, or that the presence of another protein at high concentration is required for maximum activity. The former is consistent with the *in vivo* observation that methyl groups are typically transferred to an intermediate before being released as methanol (see above). The nature of this intermediate is unknown, and this transfer has not yet been observed *in vitro*.

As indicated above, methylesterase binds tightly to methylated membranes.[31] This binding is specific for methylated MCPs, with unmethylated MCPs unable to bind significant amounts of enzyme. This binding to methylated MCPs is enhanced by, but does not require, Mg^{++}.[31] Thus,

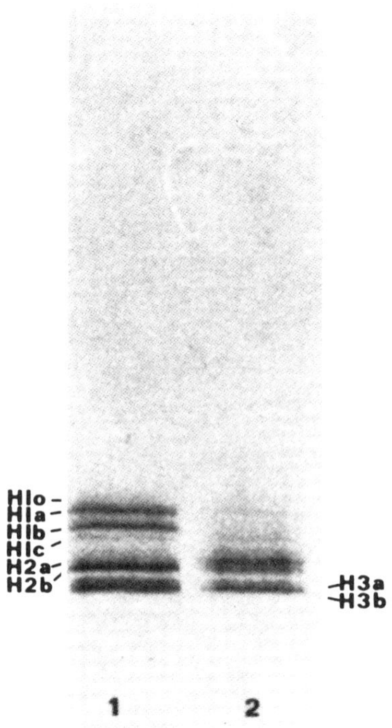

FIGURE 10. Effect of aspartate on *in vivo* methylation pattern of *B. subtilis* MCPs. *B. subtilis* was incubated with $[C^3H_3]$-methionine in the absence of protein synthesis. Buffer (1) or 0.1 *M* aspartate (2) was added, incubation stopped after 1 min and MCPs visualized by SDS-PAGE and fluorography. (From Goldman, D. and Ordal, G., *Biochemistry,* 23, 2600, 1984. With permission.)

while Mg^{++} may act as a salt bridge to stabilize methylesterase on the methylated MCP, its requirement for methylesterase activity must be due to other effects.

B. subtilis methylesterase is functionally homologous to the methylesterase from *E. coli.*[31] These homologies are summarized in Table 5.

C. EFFECTS OF ATTRACTANTS ON MCP METHYLATION *IN VITRO*

MCPs of *B. subtilis* fall into three groups, H1, H2, and H3, based on SDS-PAGE migration patterns (Figure 10).[5] Partial proteolysis indicates that the peptides of each group represent independent gene products.[6] The appearance of sub-bands within each group is explained by differing degree of methylation.[26] In *E. coli,* the increasing methylation of an MCP results in an increase in electrophoretic mobility.[37,38] Thus, the number of methyl groups on an MCP is indicated by its relative position on a polyacrylamide gel following electrophoresis. A similar situation exists for the *B. subtilis* MCPs, although the individual species do not resolve well enough to extract the quantitate the methylation level of each sub-band. Pulse-chase (Figure 11)[5] and demethylation (Figure 12)[5] studies *in vitro* with *B. subtilis* MCPs confirms that, as an MCP acquires more methyl groups, its electrophoretic mobility increases.[5] Thus, for MCP H 1, the sub-bands H1o, H1a, H1b, and H1c should represent one, two, three, and four methyl groups per MCP molecule, respectively.

In the methylation of MCPs from a methyltransferase mutant, which have no methyl groups on their MCPs, band H1a acquires label faster than band H1o (Figure 13).[5] This is probably the result of slow methylation of H1 to form H1o, and more rapid remethylation of H1o to form H1a. In this way, little H1o accumulates, while significant H1a is built up.

To investigate the effect of aspartate on methylation *in vitro,* unmethylated MCPs in

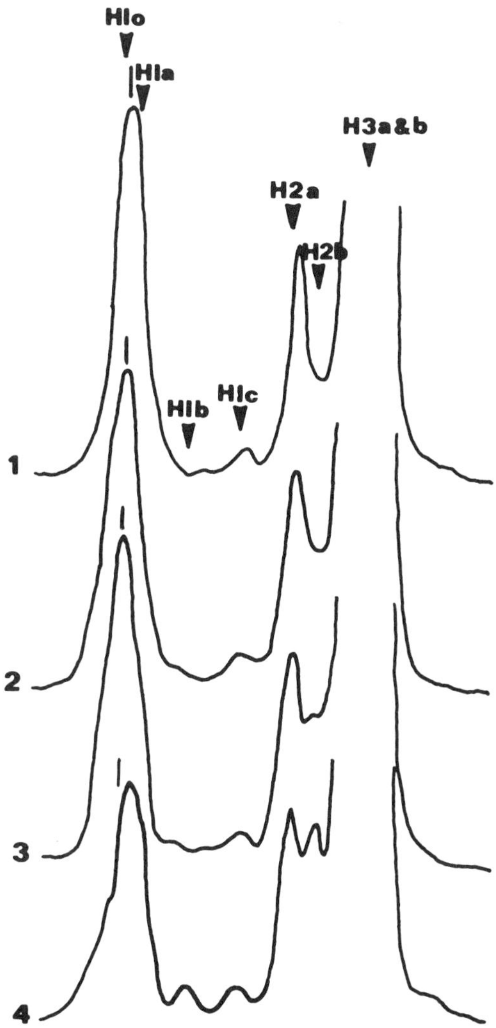

FIGURE 11. Pulse-chase methylation of MCPs isolated from a methylation defective mutant. Methyltransferase and {C^3H_3}-AdoMet were added to these MCPs. After 10 min incubation, an excess of nonradioactive AdoMet was added and reaction continued for another 10 (1), 20 (2), 40 (3) or 80 (4) min. MCPs were separated and visualized by SDS-PAGE and fluorography. The tracings are densitometer scans of the developed fluorogram. (From Goldman, D. and Ordal, G., *Biochemistry*, 23, 2600, 1984. With permission.)

membranes obtained from the *cheR* mutant were methylated in presence and absence of attractant. At early times there was no difference (Figure 13). At later times, there was inhibition of methylation of H1 and H2. However, H1c methylation was enhanced. H1c appears to be a more highly esterified form of H1 than H1b or H1a (see Figures 13 and 14). H3, on the other hand, was little affected by aspartate.

These results may be rationalized by assuming that aspartate causes inhibition of methylation of H1 but, for those H1s that become methylated, it causes more esterification. Indeed, this is what is observed *in vivo* (Figures 1 and 10). The lack of difference at early times may be understood by assuming that those MCPs were first methylated that had no aspartate bound to them. It is apparent that the mixture that contained aspartate then ran out of unbound MPCs

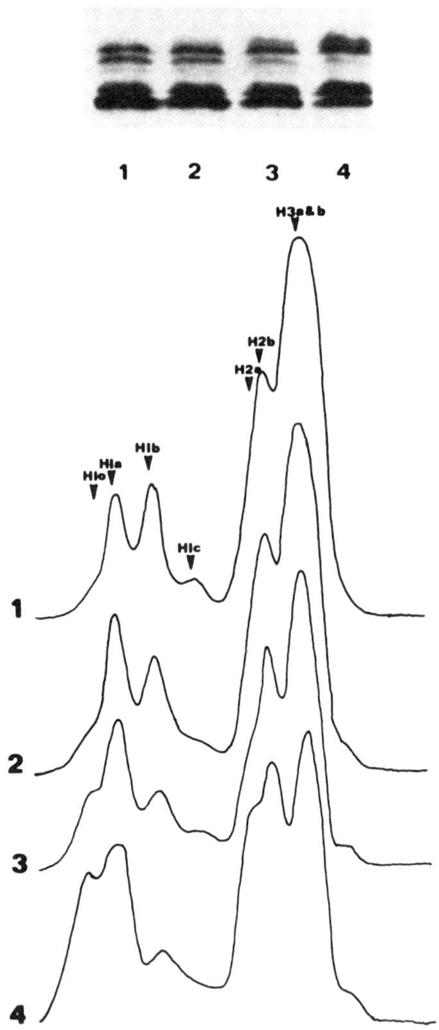

FIGURE 12. *In vitro* demethylation of [C^3H_3]-MCPs from wild type *B. subtilis*. Radiolabeled MCPs were mixed with methylesterase and incubated for 5 (1), 10 (2), 20 (3), or 40 (4) min. MCPs were separated and visualized by SDS-PAGE and fluorography. Tracings are densitometer scans of the fluorogram. (From Goldman, D. and Ordal, G., *Biochemistry*, 23, 2600, 1984. With permission.)

before 2 h (Figure 13) since at this time point there was less MCP methylation in the sample containing aspartate.

Another point of agreement between what is observed *in vitro* (Figure 13) and *in vivo* (Figure 1) is the lack of effect on methylation of H3. It is true that Figure 10 does not show apparent demethylation of H3 *in vivo*. However, it is likely that this was due to turnover in which nonradioactive methyl groups generated through internal proteolysis were "fluxed" into the MCPs (Figure 2). When excess radioactive methionine is present, there is little change in methylation of H3 *in vivo*.[16]

The major discrepancy between what is seen *in vitro* and *in vivo* is the methylation of H1, which is strongly inhibited *in vitro* (Figure 13) and enhanced *in vivo* (Figure 1).[6] In fact, there is methyl transfer from H1 to H2 *in vivo*.[6] This is, in fact, also observed *in vitro* (Figure 11). Here, following methylation using radioactive AdoMet, excess nonradioactive AdoMet was added

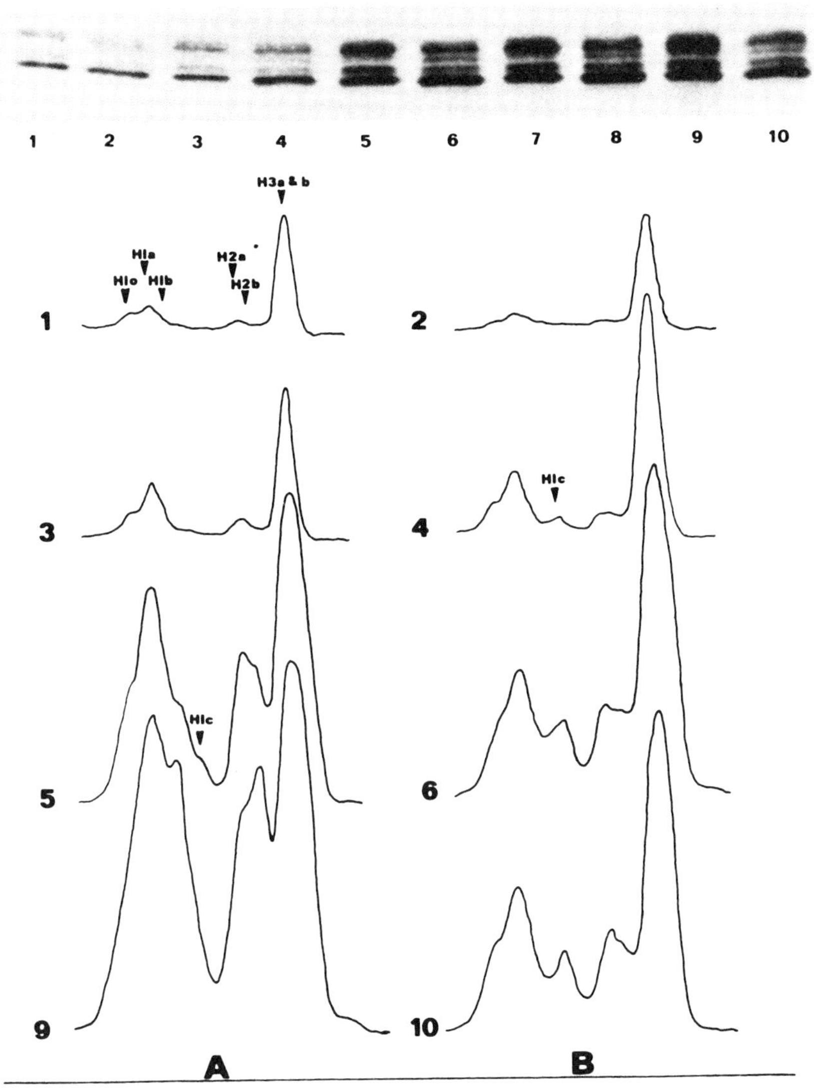

FIGURE 13. Effect of aspartate on *in vitro* MCP methylation. Membranes were isolated from methyltransferase defective *B. subtilis,* mixed with [C^3H_3]-AdoMet and methyltransferase in the presence (2,4,5,6,10;B) or absence (1,3,5,7,9;A) of 0.1 *M* aspartate. Methylation reactions were stopped after 19 (1,2); 45 (3,4); 120 (5,6); 165 (7,8), or 240 (9,10) min incubation. MCPs were separated and visualized by SDS-PAGE and fluorography (above). Tracings (below) are densitometer scans of the fluorogram. (From Goldman, D. and Ordal, G., *Biochemistry,* 23, 2600, 1984. With permission.)

and the incubation was continued. A decrease in apparent methylation of H1 and increase in apparent methylation of H2 occurred. Aspartate was not present in this incubation, and it is noteworthy that aspartate appears to inhibit a process that occurs fairly readily if we assume that, in absence of aspartate, H2 methylation occurs by transfer from H1. Thus, the effect of aspartate seems truly paradoxical: it enhances the methyl transfer *in vivo* but inhibits it *in vitro.* It must be that certain factors must be present *in vivo* that are scarce in the *in vitro* preparation.

Another significant difference betwen the methylation *in vitro* and *in vivo* is speed. Methylation changes occur immediatley, within 5 s *in vivo* but take hours *in vitro.* The basis of this difference is not understood. However, it may be that other components exist *in vivo* that may modify methyltransferase activity and, in any case, the much greater dilution of components in the *in vitro* incubation mixture contributes to the slower kinetics. In the future, using

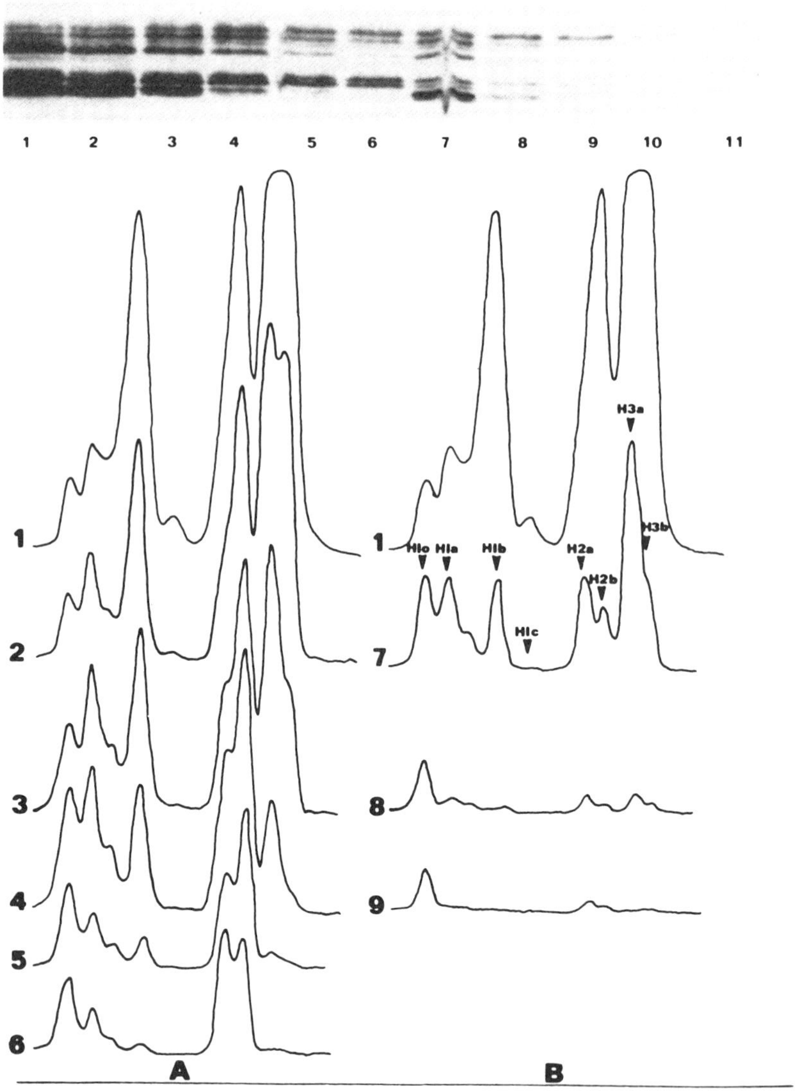

FIGURE 14. Effect of aspartate on *in vitro* demethylation of *B. subtilis* MCPs. [C^3H^3]-labeled MCPs were mixed with methylesterase in the absence (2—6) or presence (7—11) of 0.1 *M* aspartate. Reactions were stopped after 0 (1); 5 (2,7); 10 (3,8); 30 (4,9); 70 (5,10), or 120 (6,11) min incubation. MCPs were separated and visualized by SDS-PAGE and fluorography (above). Tracings are densitometer scans of lanes 1—9 of the fluorogram. (From Goldman, D. and Ordal, G., *Biochemistry,* 23, 2600, 1984. With permission.)

cloned methyltransferase, it may be possible to investigate the kinetics of methylation *in vivo* at very low enzyme concentrations.

In vitro, in the absence of stimulation, the preferred substrate for methylesterase appears to be the MCP H3, with H1 slightly more rapidly demethylated than H2 (Figure 14).[5] An analysis of this demethylation reveals that label is moving to the slower migrating, less methylated, species (H1c → H1b → H1a → H1o; H2b → H2a; and H3b → H3a). No evidence is available that any transfer of methyl groups from one MCP to another is facilitated by methylesterase.

To investigate the effect of aspartate on demethylation *in vitro,* MCPs in membranes obtained from the *cheR* mutant were first methylated using radioactive AdoMet. Then excess nonradioactive AdoMet, methylesterase, and buffer or aspartate were added. The purpose of the nonradioactive AdoMet was to render any further methylation of MCPs "silent". The effect of the aspartate, present at 90% receptor occupancy (0.1 *M*) was dramatic. There was an immense increase in the rate of demethylation of all MCPs (Figure 14). The only species that may have shown some resistance to aspartate-enhanced demethylation was H1o, which we believe to be the least methylated form of H1.

As mentioned above, this direct formation of methanol may not be physiological. Normally, the methyl groups may be transferred to another acceptor, and the methanol that arises might arise from some such carrier.[6,15,18,19] Evidence, described above, indicates that the MCPs are, indeed, not the direct source of methanol, but that aspartate induces a flux of methanol groups through all the MCPs (Figure 2).[6,19] Thus, it would seem reasonable to suppose that, *in vivo,* aspartate binds to each MCP, probably via a binding protein, to wreak its effects. *In vivo,* this leads to change in methyl esterification of MCPs and to other methyl transfers. *In vitro,* where some needed components may be scarce or absent, this leads to enhanced demethylation of all MCPs. However, it should be reemphasized that this enhanced demethylation may not be physiological. More experiments are needed to determine this, especially experiments to identify putative acceptors for methyl or methoxy groups that leave the MCPs.

REFERENCES

1. **Manson, M. D., Bland, V., Brado, G., and Higgens, C. F.,** Peptide chemotaxis in *E. coli* involves the tap signal transducer and the dipeptide permease, *Nature,* 321, 253, 1986.
2. **Mesibov, R. and Adler, J.,** Chemotaxis toward amino acids in *Escherichia coli, J. Bacteriol.,* 112, 315, 1972.
3. **Springer, M. S., Goy, M. F., and Adler, J.,** Sensory transduction in *Escherichia coli:* two complementary pathways of information processing that involve methylated proteins, *Proc. Natl. Acad. Sci. U.S.A.,* 74, 3312, 1977.
4. **Kondoh, H., Ball, C. B., and Adler, J.,** Identification of a methyl-accepting chemotaxis protein for ribose and galactose chemoreceptors of *Escheria coli, Proc. Natl. Acad. Sci. U.S.A.,* 76, 260, 1979.
5. **Goldman, D. J. and Ordal, G. W.,** *In vitro* methylation and demethylation of methyl-accepting chemotaxis proteins in *Bacillus subtilis, Biochemistry,* 23, 2600, 1984.
6. **Bedale, W. A., Nettleton, D. O., Sopata, C. S., Thoelke, M. S., and Ordal, G. W.,** Evidence for methyl-group transfer between the methyl-accepting chemotaxis proteins in *Bacillus subtilis, J. Bacteriol.,* 170, 223, 1988.
7. **Ordal, G. W.,** Chemotaxis away from uncouplers of oxidative phosphorylation in *Bacillus subtilis, Science,* 189, 802, 1975.
8. **Ordal, G. W. and Gibson, K. J.,** Chemotaxis toward amino acids by *Bacillus subtilis, J. Bacteriol.,* 129, 151, 1977.
9. **Ordal, G. W., Villani, D. P., and Rosendahl, M. S.,** Chemotaxis toward sugars by *Bacillus subtilis, J. Gen Microbiol.,* 115, 167, 1979.
10. **Baracchini, O. and Sherris, J. C.,** The chemotactic effect of oxygen on bacteria, *J. Pathol. Bacteriol.,* 152, 643, 1959.
11. **Ordal, G. W.,** Effect of methionine on chemotaxis by *Bacillus subtilis, J. Bacteriol.,* 125, 1005, 1976.
12. **Goldman, D. J. and Ordal, G. W.,** Sensory adaptation and deadaptation by *Bacillus subtilis, J. Bacteriol.,* 147, 267, 1981.
13. **Goy, M. F., Springer, M. S., and Adler, J.,** Sensory transduction in *Escherichia coli:* role of a protein methylation reaction in sensory adaptation, *Proc. Natl. Acad. Sci. U.S.A.,* 74, 4964, 1977.
14. **Toews, M. L., Goy, M. F., Springer, M. S., and Adler, J.,** Attractants and repellents control demethylation of methylated chemotaxis proteins in *Escherichia coli, Proc. Natl. Acad. Sci. U.S.A.,* 76, 5544, 1979.
15. **Thoelke, M. S., Parker, H. M., Ordal, E. A., and Ordal, G. W.,** Rapid attractant-induced changes in methylation of methyl-accepting chemotaxis proteins in *Bacillus subtilis,* submitted.
16. **Thoelke, M. S. and Ordal, G. W.,** unpublished.
17. **Casper, J. C., Thoelke, M. S., and Ordal, G. W.,** unpublished.
18. **Thoelke, M. S., Bedale, W. A., Nettleton, D. O., and Ordal, G. W.,** Evidence for an intermediate methyl-acceptor for chemotaxis in *Bacillus subtilis, J. Biol. Chem.* 262, 2811, 1987.
19. **Thoelke, M. S. and Ordal, G. W.,** Removing attractant causes methanol formation and non-S-adenosylmethionine-mediated MCP methylation in *Bacillus subtilis,* submitted.

20. **Goldman, D. J., Worobec, S. W., Siegel, R. B., Hecker, R. V. and Ordal, G. W.,** Chemotaxis in *Bacillus subtilis:* effects of attractants on the level of MCP methylation and the role of demethylation in the adaptation process, *Biochemistry* , 21, 915, 1982.
21. **Kehry, M. R., Doak, T. C., and Dahlquist, F. W.,** Stimulus-induced changes in methylesterase activity during chemotaxis in *Escherichia coli, J. Biol. Chem.*, 259, 11828, 1984.
22. **Ullah, A. H. J. and Ordal, G. W.,** *In vivo* and *in vitro* chemotactic methylation in *Bacillus subtilis, J. Bacteriol.*, 145, 958, 1981.
23. **Parkinson, J. S. and Revello, P. T.,** Sensory adaptation mutants of *E. coli, Cell,* 15, 1221, 1978.
24. **Stock, J., Borczuk, A., Chiou, F., and Burchenal, J. E. B.,** Compensatory mutants in receptor functions: a reevaluation of the role of methylation in bacterial chemotaxis, *Proc. Natl. Acad. Sci. U.S.A.*, 82, 8364, 1985.
25. **Ordal, G. W.,** Bacterial chemotaxis: biochemistry of behavior in a single cell, *Crit. Rev. Microbiol.*, 12, 95, 1985.
26. **Burgess-Cassler, A., Ullah, A. H. J., and Ordal, G. W.,** Purification and characterization of *Bacillus subtilis* methyl-accepting chemotaxis protein methyltransferase II, *J. Biol. Chem.*, 257, 8412, 1982.
27. **Ahlgren, J. A. and Ordal, G. W.,** Methyl-esterification of glutamic acid residues of methyl-accepting chemotaxis proteins in *Bacillus subtilis, Biochem. J.*, 213, 759, 1983.
28. **Ullah, A. H. J. and Ordal, G. W.,** Purification and characterization of methyl-accepting chemotaxis protein methyltransferase I in *Bacillus subtilis, Biochem J.*, 199, 795, 1981.
29. **Burgess-Cassler, A.,** Chemotaxis Methyltransferase II from *Bacillus subtilis,* Ph. D. thesis, University of Illinois, Urbana, 1983.
30. **Burgess-Cassler, A. and Ordal, G. W.,** Functional homology of *Bacillus subtilis* methyltransferase II and *Escherichia coli cheR* protein, *J. Biol. Chem.*, 257, 12835, 1982.
31. **Nettleton, D. O.,** Chemotactic Methylation in *Bacillus subtilis,* Ph.D. thesis, University of Illinois, Urbana, 1986.
32. **Goldman, D. G., Abbot, A., and Ordal, G. W.,** unpublished.
33. **Sockett, R. E., Armitage, J. P., and Evans, M. C. W.,** Methylation-independent and methylation-dependent chemotaxis in *Rhodobacter sphaeroides* and *Rhodospirillum rubrum, J. Bacteriol.*, 169, 5808, 1987.
34. **Kleene, S. J., Toews, M. L., and Adler, J.,** Isolation of glutamic acid methyl ester from an *Escherichia coli* membrane protein involved in chemotaxis, *J. Biol. Chem.*, 252, 3214, 1977.
35. **Ahlgren, J. A., Bedale, W. A., and Ordal, G. W.,** unpublished.
36. **Goldman, D. J., Nettleton, D. O., and Ordal, G. W.,** Purificiation and characterization of chemotactic methylesterase from *Bacillus subtilis, Biochemistry,* 23, 675, 1984.
37. **Chelsky, D. and Dahlquist, F. W.,** Structual studies of methyl-accepting chemotaxis proteins of *Escherichia coli:* evidence for multiple methylation sites, *Proc. Natl. Acad. Sci. U.S.A.*, 77, 2434, 1980.
38. **Engstrom, P. and Hazelbauer, G. L.,** Multiple methylation of methyl-accepting chemotaxis proteins during adaptation of *E. coli* to chemical stimuli, *Cell,* 20, 165, 1980.
39. **Stock, J. B., Clarke, S., and Koshland, D. E., Jr.,** The protein carboxymethyltransferase involved in *Escherichia coli* and *Salmonella typhimurium* chemotaxis, *Methods Enzymol.*, 106, 321, 1984.
40. **Snyder, M. A., Stock, J. B., and Koshland, D. E., Jr.,** Carboxymethyl esterase of bacterial chemotaxis, *Methods Enzymol.*, 106, 321, 1984.
41. **Fuhrer, D. and Ordal, G. W.,** unpublished.

Chapter 15

THE CHEMOTAXIS-SPECIFIC METHYLESTERASE OF ENTERIC BACTERIA

Richard C. Stewart and Frederick W. Dahlquist

TABLE OF CONTENTS

I. INTRODUCTION

During chemotaxis, bacteria such as *Escherichia coli* or *Salmonella typhimurium* detect temporal changes in the concentration of specific chemicals and behaviorally respond to these changes by modulation of the sense of rotation of the flagellar rotary motors. The ability of the bacteria to sense temporal gradients suggests that a mechanism exists for "remembering" the concentration of chemicals sensed in the recent past. This can be demonstrated by the sudden addition of a chemoattractant such as serine to a suspension of cells. The bacteria respond to this addition by counter-clockwise (CCW) rotation of their flagellar motors corresponding to a smooth swimming or run response. This response is maintained for a minute or so, and then the cells return to their prestimulus pattern of alternating clockwise (CW) and CCW rotation of their flagellar motors, corresponding to alternating periods of tumbling and smooth swimming. Thus the cells appear to behaviorally adapt to the temporal addition of chemoattractants. We imagine that the cells maintain a record or "memory" of their chemical environment over the recent past. If the current environment is detected to be "better" than the recorded one, smooth (CCW) swimming is generated. If the current environment is "worse" than the recorded one, tumbly (CW) swimming is generated, and the cells reorient to try a new random direction. The record is continually updated. It takes minutes to update the memory if a massive chemostimulus is applied.

The receptor proteins for one class of chemostimuli (the methylation dependent class) are sometimes also referred to as transducer proteins. These proteins are often involved in the detection of more than one chemostimulus. In some cases the transducer proteins directly bind the attractant molecules, while in other cases they interact with complexes of the chemoattractant and periplasmic binding proteins. The known transducer proteins in *E. coli* (Tsr, Tar, Trg, and Tap) and *S. typhimurium* are also referred to as methyl-accepting chemotaxis proteins (MCPs) because they are reversibly methylated[1-4] by a chemotaxis-specific methyltransferase (the *cheR* gene product). This enzyme catalyzes transfer of the activated methyl group from AdoMet to specific MCP glutamic acid γ-carboxylate groups in *E. coli*[5] and *S. typhimurium*.[6] The *cheB* gene product is the methylesterase that catalyzes hydrolysis of these methylesters,[7] producing methanol and regenerating the γ-carboxylates of the glutamic acid residues.[8]

In *E. coli, S. typhimurium*,[9-11] and *B. subtilis*[12,13] MCP methylation is thought to play an important role in behavioral adaptation, i.e., the return of swimming behavior to the prestimulus pattern of alternately running and tumbling.[14] Mutants deficient in CheR and/or CheB are defective with respect to adaptation and have extremely abnormal bias of flagellar rotation in their unstimulated swimming behaviors.[15-19] In *E. coli* and *S. typhimurium*, increased levels of MCP methylation are associated with adaptation to positive stimuli (addition of attractant or removal of repellent), while adaptation to negative stimuli is accompanied by decreased methylation.[20-26] The timecourse of the change in MCP methylation level approximates the timecourse for behavioral adaptation following such stimuli.[11,22] The four known MCPs serve as transducers for four different sets of stimuli. A given stimulus appears to alter mainly the methylation level of the MCP class that is responsible for detecting its concentration.[20-22] This section of this review covers what is known about the methylesterase, CheB, and the regulation of its activities. CheA and CheW are also discussed because they may be involved in this regulation. Although our focus is primarily on the enzymes from *E. coli* and *S. typhimurium*, some mention is made of the considerable progress which has been made in characterizing the methyltransferase and esterase from *B. subtilis* (see Reference 12 review). *B. subtilis* has MCPs that are methylated and demethylated in response to chemoeffectors. However, in this bacterium attractant stimuli cause decreased MCP methylation levels, and repellents cause increased levels of methylation.[13]

II. AN OVERVIEW OF THE METHYLESTERASE, CheB

The *cheB* gene product is thought to be a soluble cytoplasmic enzyme.[7,27-29] Subcellular fractionation of *E. coli* indicates that CheB is located in both the cytoplasm and the inner cytoplasmic membranes,[27] presumably reflecting the affinity of this cytosolic enzyme for the membrane-bound MCPs.[28] The DNA sequence of *cheB* from *E. coli*[30] predicts a corresponding amino acid sequence of 349 residues comprising a 37.5 kDa protein. In agreement with these predictions are the observed molecular weights of the *E. coli* (38 kDa; References 27, 31, 32) and *S. typhimurium* (37 kDa; References 29, 33) proteins and the amino acid content of the purified enzyme from *S. typhimurium*.[29] CheB catalyzes hydrolysis of the γ-glutamyl methyl esters formed by the methyltransferase, CheR; the products of the hydrolysis are methanol and the free γ-glutamyl carboxylates.[7,8] CheB is also responsible for the irreversible deamidation of two MCP glutamine residues, converting them to glutamates which are among the methylesterification sites in wild-type cells.[34-38] *In vitro* assays of the methylesterase activity of CheB (reviewed in Reference 28; see also Reference 29) enabled development of purification procedures for this enzyme.[28,39] A more straightforward and reproducible purification protocol has recently been reported by Simms et al.[29] In addition to the intact 37 kDa esterase, this procedure also yields a 21 kDa proteolytic fragment which represents the carboxy terminal F(3,5) of the uncleaved esterase and which has remarkably high activity (15-fold higher than that of the 37 kDa parent protein).[29] This finding may have significance when considering the regulation of methylesterase activity. Some of the kinetic and physical properties of the purified esterase have been determined.[28,29] The enzyme is readily inhibited by several thiol reagents,[28] perhaps indicating an essential role for one of the two cysteine residues[29,30] of CheB. Simms et al.[40] have further explored this possibility by using oligonucleotide-directed mutagenesis to convert these cysteine residues to alanines (individually). Cys^{207}→Ala CheB has normal esterase activity, while the Cys^{309}→Ala protein is completely inactive. In the sequence of CheB, the catalytically essential Cys^{309} residue follows what may be nucleotide-binding fold formed by the residues 277 to 307.[40] One exciting possibility is that this region is involved in regulating the activity of CheB (see below). Oligonucleotide-directed mutagenesis of a key element of the putative nucleotide-binding fold does result in inactive enzyme,[40] but no evidence supporting a regulatory role for this region of the protein has yet been reported.

The unstimulated swimming phenotype of *cheB* mutants is invariably tumbly.[7,17,41,42] This extreme CW bias of flagellar rotation can be altered by attractant stimuli, resulting in periods of smooth swimming.[17,41-43,52] The response thresholds for attractant stimuli are higher in *cheB* mutants than in wild-type cells.[42] However, CheB does not appear to be involved in generating the excitatory signal that links the transducers to the flagellar switch because *cheB* mutants exhibit normal response latencies to attractants.[43,51] By virtue of its role in determining methylation states of MCPs, CheB appears to be a part of the adaptation machinery. The MCPs in esterase deficient mutants have only about half of the methylation sites available in $cheB^+$ hosts (due to the loss of the sites generated by the CheB-catalyzed deamidation reaction; References 34 through 37). Methyl groups can be added to, but not removed from, the residual esterification sites. The resulting "overmethylation" apparently causes the nondeamidated MCPs to generate "tumble signal" constantly.[9-11,44] Stock et al.[43] have demonstrated that "tight" *cheB* mutants (no detectable esterase activity) are capable of some form of behavioral adaptation to stimuli of limited intensity.

III. REGULATION OF METHYLESTERASE ACTIVITY

A. RESPONSE TO POSITIVE AND NEGATIVE STIMULI

Several research groups have demonstrated that the activity of CheB is regulated in response to MCP-mediated stimuli.[45-48] By monitoring the disappearance of MCP methyl groups or the

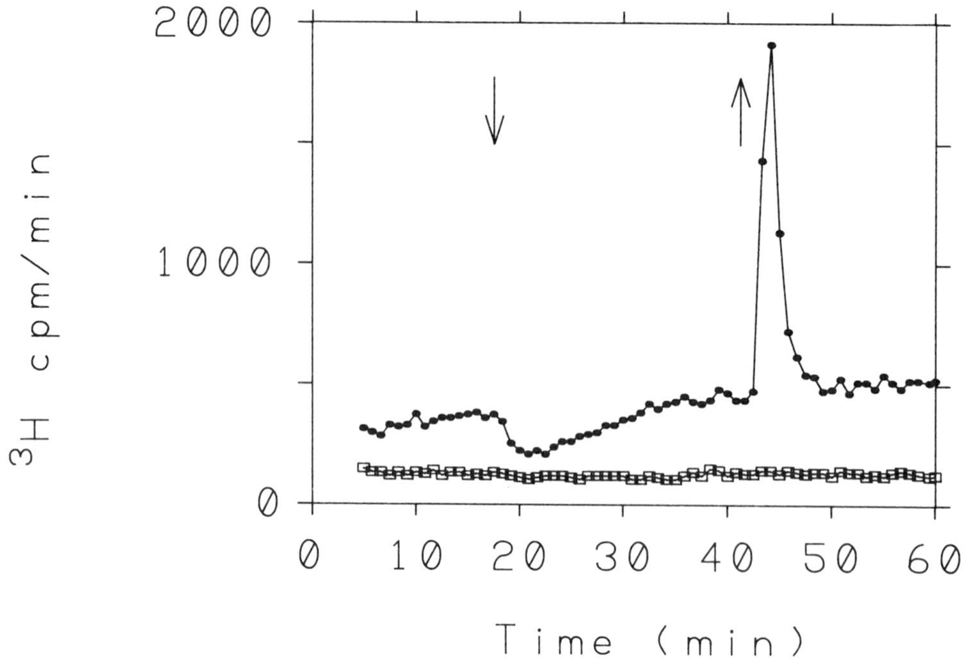

FIGURE 1. Response of steady state [^{3}H] methanol evolution from *E. coli* cells in response to 50 μ*M* serine addition (↓) and removal (↑). The experimental details are given in Kehry et al.[46] The closed circles represent data for wild type *E. coli* (strain RP487) while the open squares represent the response for a strain deleted for three (*tsr, tap, tar*) of the four MCPs (strain MS5228). Note the transient inhibition of methanol evolution seen on attractant addition and the transient activation seen on attractant removal with strain RP487. With MS5228 the level of methanol evolution is comparatively low and does not respond to chemotactic stimuli.

accumulation of methanol, Toews et al.[45] first demonstrated that the methylesterase is transiently inhibited by attractant stimuli; repellent stimuli (or removal of attractant) result in a transient increase in methylesterase activity.[45] Kehry et al.[46,47] have extended these studies by utilizing a flow assay that greatly facilitates monitoring of the *in vivo* methylesterase activity of CheB. This system basically involves pumping media over cells maintained on a filter; stimuli are presented to the cells simply by switching the inlet tubing from one source of media to another. The methyl groups on the MCPs in these cells have been prelabeled (using ^{3}H-methionine) prior to placing the cells on the filter, and esterase activity is monitored by collecting fractions of media flowing over the cells and determining the ^{3}H-methanol content of each. This is easily achieved by placing each fraction tube into a partially filled scintillation vial, capping the vial, and allowing vapor phase transfer to equilibrate the ^{3}H-methanol between the aqueous flow fraction and the scintillation cocktail in the larger vial. In "steady state flow" experiments, 3H-methionine is present in the flow medium at all times; typical results for wild-type and control cells are shown in Figure 1 for an experiment using serine addition and removal as stimuli. The transient decrease in methanol evolution (CheB activity) in response to serine addition is readily apparent, as is the dramatic increase in CheB activity when attractant is removed (a repellent stimulus). "Flow chase" experiments are performed in the same manner, except that only cold methionine is available to the cells in the flow medium following the initial prelabeling. Results from a typical flow chase experiment are shown in Figure 2. Again the esterase activity appears to decrease transiently following attractant addition and to increase transiently in response to attractant removal. It is gratifying to note that $t^1/2$ for an unstimulated flow chase (reflecting the net, weighted-average rate of deesterification of multiple methyl sites on each MCP; see Reference 46) agrees reasonably with the $t^1/2$ values for demethylation of individual sites as determined by Terwilliger et al.[49,50]

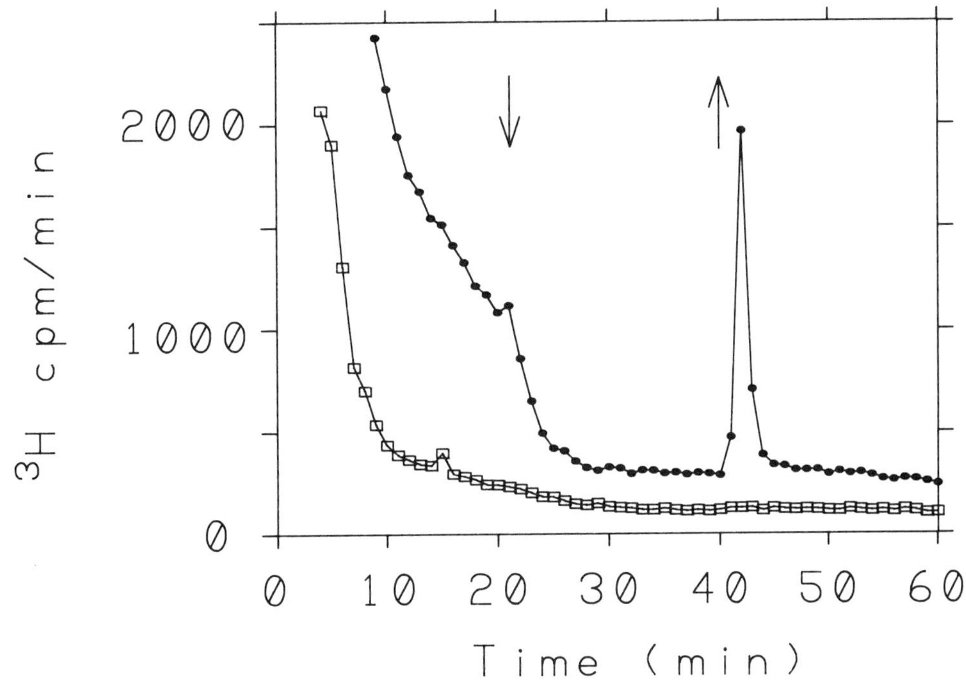

FIGURE 2. Response of [^{3}H] methanol production to chemotaxis stimuli in "flow chase" mode.[46] Serine (50 μ*M*) was added at the downward arrow and removed at the upward arrow. The data are shown for the same wild type and MCP deleted strains shown in Figure 1. Addition of attractant causes a transient decrease in the demethylation rate while removal of attractant causes a transient increase in demethylation rate.

There is evidence that the methylesterase activity may be regulated by the same signal generated by the transducers to communicate with the flagellar switch (or by the same mechanism responsible for signal generation). First, the response of the methylesterase activity to addition or removal of chemoeffectors always exhibits a characteristic biphasic timecourse: a very rapid change (faster than the mixing time of the flow apparatus) in one direction followed by a slower change (requiring 5 to 6 min) in the opposite direction that gradually restores the methylesterase activity to its prestimulus level. Although the mixing time of the flow apparatus (t^1/2~3.6 min) limits detailed determination of the kinetics of these changes, the timecourse of CheB activity change appears to mimic the sequence of behavioral events following stimulation of wild-type cells: a rapid excitatory response (latency ~0.2s; Reference 51) followed by adaptation to the prestimulus behavior (requires several minutes; Reference 52). Second, input from different classes of receptors is integrated in some manner prior to the event(s) that results in regulation of CheB activity,[47] a situation reminiscent of the integration that takes place in determining the swimming behavior of the cell.[14] And third, in *cheZ* mutants (which are known to be defective in excitatory signaling; Reference 51) the methylesterase activity responds to stimuli somewhat slower than is observed in *cheZ*$^+$ cells (Kehry et al., unpublished).

B. INVOLVEMENT OF OTHER CHEMOTAXIS GENES

Cells lacking most or all of the known flagellar and chemotaxis-associated proteins (*flaI* or *flaA* →*flbB* deletions; see Reference 53) of course show no methylesterase activity. When such cells are provided with CheR, CheB, and MCPs by expressing genes carried on plasmids, methylesterase activity is observable, but flow experiments show no regulation of this activity in response to chemotactic stimuli (Russell et al., unpublished results). Therefore, additional components of the chemotaxis sensory transduction machinery may be required for CheB regulation. Some progress has been made toward identifying those components. CheZ was a reasonable candidate because interspecific complementation studies suggest that CheZ interacts with CheB to form a complex that has some essential role in chemotaxis.[6] Furthermore, some *cheZ* mutants have low *in vitro* methylesterase activity.[39] However, CheZ is not required for methylesterase activity.[29,39] Flow experiments indicate that *CheZ* is not required for regulation of CheB per se, but rather for normal kinetics of regulation (Kehry et al., unpublished). CheY also does not appear to be involved in methylesterase regulation, as *cheY* mutants exhibit wild-type regulation patterns in flow experiments (Kehry et al., manuscript in preparation). However, *cheA* and *cheW* mutants, do not appear to be capable of regulating methylesterase activity in response to chemoeffectors, suggesting that the corresponding proteins may be involved in this regulation.

C. GLOBAL VS. LOCAL REGULATION

Can the regulation of methylesterase in response to chemotactic stimuli account for the stimulus-specific changes in methylation levels of corresponding MCPs? If inhibition of demethylation were primarily responsible for these changes, then one might expect interactions between the stimulated MCP and the methylesterase (e.g., accessibility) to directly determine the activity of the enzyme in response to stimuli (i.e., "local regulation"; References 1, 54, 55). However, CheB activity appears to be "globally regulated" in response to some parameter that reflects the integrated input from different classes of transducers.[47] This suggests that regulation of methylesterase activity is not primarily responsible for the stimulus-specific MCP methylation changes and predicts that this specificity results primarily from local regulation of the methyltransferase. In support of this prediction, Terwilliger et al.[49,50] have presented evidence that the increased level of Tar methylation following an aspartate stimulus is due primarily to increased rates of methylation at all four esterification sites, not to decreased rates of demethylation. The observation of stimulus-induced methylation changes in *in vitro* systems[54,55] suggests that this regulation of CheR may be at the local level.

Recent studies from Hazelbauer and co-workers[56] demonstrate that stimulation of mutant form of a Trg that lacks any functional methylation sites results in methyl group accumulation in other transducer species. This further strengthens the argument that, in addition to the "local" regulation of CheR, there is global inhibition of the methylesterase that facilitates the methyl group accumulation needed for adaptation to positive stimuli.

Studies of CheB proteolytic fragments[29] and of the *cheB* DNA sequence[57] may provide some insight into the molecular aspects of CheB regulation. When the 37 kDa mehtylesterase is proteolyzed to a limited extent, a highly active 21 kDa fragment is obtained.[29] This fragment is relatively resistant to further proteolysis and appears to be the carboxy-terminal three fifths of the intact 37 kDa protein. The specific activity the purified proteolytic fragment is fifteen times that of the parent protein, perhaps suggesting that the amino terminal segment of CheB plays some role in reversibly regulating the activity of the intact (37 kDa) enzyme. It remains to be seen if there is an *in vivo* role for proteolytic activation of the relatively low-activity 37 kDa enzyme.[29] Stock et al.[29] have determined that significant amino acid sequence homology exists between the entire CheY protein and the amino terminal region of CheB. Such homology raises the possibility that CheY and CheB interact with a common or similar substrate (e.g., MCPs) or perhaps with a common regulatory element.[29] As discussed below, *in vitro* experiments using purified CheB, CheY, and CheA have recently provided strong evidence that CheA is this common regulatory element and that regulation of the activities of CheB and CheY is achieved, at least in part, by phosphorylation of these proteins by an autophosphorylated form of CheA.[66-68]

D. ESTERASE IN OTHER BACTERIA

In *B. subtilis, in vitro* studies[58] have demonstrated that the activity of the purified methylesterase increases in response to MCP-mediated attractants; this result parallels *in vitro* results which have demonstrated that the decreased levels of MCP methylation observed in response to attractants is due to increased methylesterase activity.[13] The *B. subtilis* methylesterase has been purified and characterized;[59] it is a 41 kDa protein that requires a divalent cation for activity. This enzyme will utilize methylated MCPs from *E. coli* as substrates;[59] however, attractant stimuli have no effect on the methylesterase activity in this hybrid system.

IV. ROLE OF ChEA AND ChEW IN MODULATION OF METHYLESTERASE ACTIVITY

A. BEHAVIOR

Mutants in *cheA* or *cheW* have extreme CCW bias of flagellar rotation; they are smooth swimmers and seldom tumble.[5,17,38,41,60] Such mutants do respond transiently[38,61] to a strong pH repellent stimulus (benzoate) but do not appear to be capable of responding to other less potent CW stimuli.[38,60] Therefore, CheA and CheW do not appear to be required for CW flagellar rotation per se, although they seem to be essential for communicating MCP-chemoeffector binding events to the flagellar switch. The mapping of suppressors of *cheA* point mutants to genes encoding flagellar components[44] (and vice versa[62]) further supports such a signaling role for CheA.

CheA also appears to be involved in regulation of methylesterase activity in response to MCP-mediated stimuli. Springer and Zanolari[48] have found that several *cheA* point mutants and a *cheA* deletion mutant are defective in regulation of methylesterase activity in response to negative (repellent) stimuli *in vivo,* although regulation in response to positive stimuli is essentially normal. Work in the authors' laboratory has indicated that some *cheA* deletion mutants exhibit no regulation of CheB in response to positive or negative stimuli in flow chase experiments (Kehry et al., manuscript in preparation). These biochemical results, in conjunction with the previously described behavioral and genetic studies, suggest that CheA may have some role in signaling and that this signal (or some species affected by it) is also involved in regulation of CheB activity.

These observations suggest the pathway of information flow shown below:

$$\text{receptor-transducer} \xrightarrow[\text{CheW} \quad \text{CheZ}]{\text{CheA} \quad \text{CheY}} \text{flagella}$$

CheB

Here CheA and CheW are needed to interpret signals from the receptor-transducers. This signal is then communicated both to CheY and to CheB. The CheY information is used to regulate the sense of rotation of the bacterial flagella,[71-75] and the CheB information is used to regulate the methylation level of the receptor-transducer proteins.

B. BIOCHEMICAL ROLE

A partial *cheA* sequence and the entire *cheW* DNA sequence have been reported for the *E. coli* genes by Mutoh and Simon,[30] while Stock et al. have determined the entire sequences of the *cheA* and *cheW* genes of *S. typhimurium.*[69,70] Two polypeptides are encoded by the *cheA* locus of *E. coli;*[32,63,64] they are designated p[*cheA*]$_S$ (66 to 69 kDa) and p[*cheA*]$_L$ (76 to 78 kDa). The same coding sequence and reading frame are utilized for both CheA proteins, but the translation start site of p[*cheA*]$_L$ precedes that of p[*cheA*]$_S$ such that the larger protein has an additional 90 or so amino acids on its amino terminal end.[63] The two proteins have identical amino acid sequences beyond the first 90 residues of p[*cheA*]$_L$. Results of complementation studies suggest that the amino terminal portion of p[*cheA*]$_L$ has a function distinct from that shared by p[*cheA*]$_S$

and the remainder of p[*cheA*]$_L$.[63] The short form of CheA is found only in the cytoplasm of *E. coli,* while the longer form of CheA is found in both the cytoplasmic and inner membrane fractions.[27] So it is conceivable that the amino terminal region of p[*cheA*]$_L$ enables association of the longer protein with a different set of chemotaxis components from those which interact with the cytoplasmic pool of CheA.[44,63] Preliminary results from Parkinson's laboratory suggest that a construct which expresses only p[*cheA*]$_L$ is able to form normal swarms. This suggests that the short form of CheA may not be required for chemotaxis.

The function of CheW remains unknown. This 167 amino acid, 18 kDa protein[27,30,32,33,64,65] is found exclusively in the cytoplasm of *E. coli*.[27] The *S. typhimurium* protein has been purified to homogeneity by Stock and co-workers,[69] who found that CheW exists as a monodimer under nondenaturing conditions (confirming the results of cross-linking studies[65]). Analysis of the CheW amino acid sequence (predicted from the determined DNA sequences) by Stock et al.[69] indicates that residues 128 to 160 may form a purine nucleotide-binding site, an interesting possibility in view of the demonstrated requirement of ATP or derived nucleotides in chemotaxis.[76,77] Whether CheW actually binds nucleotides and the role of such an activity in chemotaxis have yet to be established. It seems likely that CheW has a function related to those of the other components of the *mocha* operon (*motA*, *motB*, and *cheA*), i.e., enabling or regulating motility. Some *cheW* mutants have overmethylated MCPs,[5] perhaps suggesting that the CheW protein is involved in regulation of CheB and/or CheR activity as well. Some role for CheW in generating the signal to which methylesterase responds is supported by the results of flow chase experiments that indicate that methylesterase activity is not regulated in response to negative chemotactic stimuli in *cheW* mutants (Kehry et al., manuscript in preparation).

V. ROLE OF PHOSPHORYLATION

Very recent results from Simon's laboratory[66-68] suggest that the phosphorylation of chemotaxis components plays an important role in the pathway of information flow in bacterial chemotaxis. Using purified proteins they have demonstrated that the CheA protein is autophosphorylated in the presence of ATP, and that phospho-CheA readily transfers its phosphoryl group to either CheY or CheB. These data and recent results from Taylor's laboratory[77] suggest that phosphorylated CheY could be the entity that signals CW rotation to the flagella in response to negative stimuli. By these arguments, the phosphorylated version of CheB would be the activated form of the esterase.[67]

If this view is correct, a number of important questions arise.

(1) How is CheB dephosphorylated?

To date there is no obvious candidate protein or gene product which might act to deactivate CheB by removal of its phosphate groups. Hess et al.[67] have demonstrated that CheB appears to spontaneously dephosphorylate, and this could provide a means of transiently activating CheB: CheA dependent phosphorylation followed by spontaneous dephosphorylation. In this view the activation of CheB appears to have some of the features of eukaryotic G protein. Rather than binding GTP to become activated, CheB is phosphorylated. The activated form of CheB or the G protein then decays in a first-order fashion with its own characteristic time constant.

(2) How is CheB inhibited?

There are two obvious ways that CheB could be transiently inhibited upon positive stimulation. In one case, the moderate activity observed in behaviorally adapted cells could reflect a low level of phosphorylated protein and further inhibition would then result from stimulated dephosphorylation of CheB. This would require some interaction with a factor which would have the properties of a CheB phosphatase which responds to the signaling state of the chemotaxis machinery. As stated above such a factor is currently unknown, and we feel this option is somewhat less likely than the one given below. In this case the inhibition of CheB by positive stimuli would involve a distinctly different event than phophorylation. We observe that mutant CheBs which are no longer activated by negative signals are still subject to inhibition by positive signals (Stewart and Dahlquist, submitted manuscript). This observation is consistent with the view that inhibition is not the absence of activation (but certainly doesn't prove the point). While it is premature to speculate too extensively on these points, one possibility is that the same event which might inhibit CheA phosphorylation during positive stimulation could also inhibit CheB activity. CheW could play such a role.

The above discussion should make clear that this is a very exciting time in the chemotaxis field and that the detailed understanding of the chemotaxis specific methylesterase will play an important role in delineating the pathway of information flow and feedback in this fascinating system.

REFERENCES

1. **Stock, J. B. and Koshland, D. E., Jr.,** Changing reactivity of receptor carboxyl groups during bacterial sensing, *J. Biol. Chem.*, 256, 10826, 1981.
2. **Kleen, S. J., Toews, M. L., and Adler, J.,** Isolation of glutamic acid methyl ester from an *Escherichia coli* membrane protein involved in chemotaxis, *J. Biol. Chem.*, 252, 3214, 1977.
3. **Van Der Werf, P. and Koshland, D. E., Jr.,** Identification of a γ-glutamyl methyl ester in bacterial membrane protein involved in chemotaxis, *J. Biol. Chem.*, 252, 2793, 1977.
4. **Kort, E. N., Goy, M. F., Larsen, S. H., and Adler, J.,** Methylation of a membrane protein involved in bacterial chemotaxis, *Proc. Natl. Acad. Sci. U.S.A.*, 72, 3939, 1975.
5. **Springer, W. R. and Koshland, D. E., Jr.,** Identification of a protein methyltransferase as the cheR gene product in the bacterial sensing system, *Proc. Natl. Acad. Sci. U.S.A.*, 74, 3659, 1977.
6. **DeFranco, A. L., Parkinson, J. S., and Koshland, D. E., Jr.,** Functional homology of chemotaxis genes in *Escherichia coli* and *Salmonella typhimurium*, *J. Bacteriol.*, 139, 107, 1979.
7. **Stock, J. B. and Koshland, D. E., Jr.,** A protein methylesterase involved in bacterial sensing, *Proc. Natl. Acad. Sci. U.S.A.*, 65, 3659, 1978.
8. **Toews, M. L. and Adler, J.,** Methanol formation *in vivo* from methylated chemotaxis proteins in *Escherichia coli*, *J. Biol. Chem.*, 254, 1761, 1979.
9. **Koshland, D. E., Jr.,** Biochemistry of sensing and adaptation in a simple bacterial system, *Annu. Rev. Biochem.*, 50, 765, 1981.
10. **Boyd, A. and Simon, M. I.,** Bacterial chemotaxis, *Annu. Rev. Physiol.*, 44, 501, 1982.
11. **Springer, M. S., Goy, M. F., and Adler, J.,** Protein methylation in behavioral control mechanisms and in signal transduction, *Nature*, 280, 279, 1979.
12. **Ordal, G. W. and Nettleton, D. O.,** Chemotaxis in *Bacillus subtilis*, in *The Microbiology of the Bacilli*, Academic Press, New York, 1983, 53.
13. **Goldman, D. J., Worobec, S. W., Siegel, R. B., Hecker, R. V., and Ordal, G. W.,** Chemotaxis in *Bacillus subtilis*: effects of attractants on the level of methylation of methyl-accepting chemotaxis proteins and the role of demethylation in the adaptation process, *Biochemistry*, 21, 915, 1982.
14. **Berg, H. C. and Tedesco, P. M.,** Transient response to chemotactic stimuli in *Escherichia coli*, *Proc. Natl. Acad. Sci. U.S.A.*, 72, 3325, 1975.
15. **Parkinson, J. S. and Revello, P. T.,** Sensory adaptation mutants of *E. coli*, *Cell*, 15, 1221, 1978.
16. **Rubik, B. A. and Koshland, D. E., Jr.,** Potentiation, desensitization, and inversion of response in bacterial sensing of chemical stimuli, *Proc. Natl. Acad. Sci. U.S.A.*, 75, 2820, 1978.
17. **Parkinson, J. S.,** cheA, cheB, and cheC genes of *E. coli* and their role in chemotaxis, *J. Bacteriol.*, 126, 758, 1976.
18. **Parkinson, J. S.,** Data processing by the chemotaxis machinery of *Escherichia coli*, *Nature*, 252, 317, 1974.
19. **Goy, M. F., Springer, M. S., and Adler, J.,** In search of the linkage between receptor and response: the role of a protein methylation reaction in bacterial chemotaxis, in *Taxis and Behavior, Receptors, and Recognition*, Vol. 5, Chapman and Hall, London, 1978, 1.
20. **Springer, M. S., Goy, M. F., and Adler, J.,** Sensory transduction in *Escherichia coli*: two complementary pathways of information processing that involve methylated proteins, *Proc. Natl. Acad. Sci. U.S.A.*, 74, 3312, 1977.
21. **Silverman, M. and Simon, M. I.,** Chemotaxis in *Escherichia coli*: methylation of che gene products, *Proc. Natl. Acad. Sci. U.S.A.*, 74, 3317, 1977.

22. **Goy, M. F., Springer, M. S., and Adler, J.**, Sensory transduction in *Escherichia coli:* role of protein methylation reaction in sensory adaptation, *Proc. Natl. Acad. Sci. U.S.A.*, 74, 4964, 1977.
23. **Chelsky, D. and Dahlquist, F. W.**, Methyl-accepting chemotaxis proteins of *Escherichia coli:* methylated at three sites in a single tryptic fragment, *Biochemistry*, 20, 977, 1981.
24. **Chelsky, D. and Dahlquist, F. W.**, Structural studies of methyl-accepting chemotaxis proteins of *Escherichia coli:* evidence for multiple methylation sites, *Proc. Natl. Acad. Sci. U.S.A.*, 77, 2434, 1980.
25. **DeFranco, A. L. and Koshland, D. E., Jr.**, Multiple methylation on processing of sensory signals during bacterial chemotaxis, *Proc. Natl. Acad. Sci. U.S.A.*, 77, 2429, 1980.
26. **Engström, P. and Hazelbauer, G. L.**, Multiple methylation of methyl-accepting chemotaxis proteins during adaptation of *E. coli* to chemical stimuli, *Cell*, 20, 165, 1980.
27. **Ridgway, H. F., Silverman, M., and Simon, M. I.**, Localization of proteins controlling motility and chemotaxis in *Escherichia coli, J. Bacteriol.*, 132, 657, 1977.
28. **Snyder, M. A., Stock, J. B., and Koshland, D. E., Jr.**, Carboxymethyl esterase of bacterial chemotaxis, *Methods Enzymol.*, 106, 321, 1984.
29. **Simms, S. A., Keane, M. G., and Stock, J.**, Multiple forms of the CheB methylesterase in bacterial chemosensing, *J. Biol. Chem.*, 260, 10160, 1985.
30. **Mutoh, N. and Simon, M. I.**, Nucleotide sequence corresponding to five chemotaxis genes in *Escherichia coli, J. Bacteriol.*, 165, 161, 1986.
31. **Matsumura, P., Silverman, M., and Simon, M. I.**, Synthesis of mot and che gene products of *Escherichia coli* programmed by hybrid ColE1 plasmids in minicells, *J. Bacteriol.*, 132, 996, 1977.
32. **Silverman, M. and Simon, M. I.**, Identification of polypeptides necessary for chemotaxis in *Escherichia coli, J. Bacteriol.*, 130, 1317, 1977.
33. **DeFranco, A. L. and Koshland, D. E., Jr.**, Molecular cloning of chemotaxis genes and overproduction of gene products in the bacterial sensing system, *J. Bacteriol.*, 147, 390, 1981.
34. **Kehry, M. R. and Dahlquist, F. W.**, Adaptation of bacterial chemotaxis: CheB dependent modification permits additional methylations of sensory transducer proteins, *Cell*, 29, 761, 1982.
35. **Kehry, M. R., Bond, M. W., Hunkapiller, M. W., and Dahlquist, F. W.**, Enzymatic deamidation of methyl-accepting chemotaxis proteins in *Escherichia coli* catalyzed by the *cheB* gene product, *Proc. Natl. Acad. Sci. U.S.A.*, 80, 3599, 1983.
36. **Terwilliger, T. C. and Koshland, D. E., Jr.**, Sites of methyl esterification and deamination on the aspartate receptor involved in chemotaxis, *J. Biol. Chem.*, 259, 7719, 1984.
37. **Sherris, D. and Parkinson, J. S.**, Posttranslational processing of methyl-accepting chemotaxis proteins in *Escherichia coli, Proc. Natl. Acad. Sci. U.S.A.*, 78, 6051, 1981.
38. **Parkinson, J. S. and Houts, S. E.**, Isolation and behavior of *Escherichia coli* deletion mutants lacking chemotaxis functions, *J. Bacteriol.*, 151, 106, 1982.
39. **Snyder, M. A. and Koshland, D. E., Jr.**, Identification of the esterase peptide and its interaction with the *cheZ* peptide in bacterial sensing, *Biochimie*, 63, 113, 1981.
40. **Simms, S. A., Cornman, E. W., Mottonen, J., and Stock, J.**, Active site of the enzyme which demethylates receptors during bacterial chemotaxis, *J. Biol. Chem.*, 262, 29, 1987.
41. **Parkinson, J. S.**, Complementation analysis and deletion mapping of *E. coli* mutants defective in chemotaxis, *J. Bacteriol.*, 135, 45, 1978.
42. **Yonekawa, H., Hayashi, H., and Parkinson, J. S.**, Requirement of the *cheB* function for sensory adaptation in *Escherichia coli, J. Bacteriol.*, 156, 1228, 1983.
43. **Stock, J., Kersulis, G., and Koshland, D. E., Jr.**, Neither methylating nor demethylating enzymes are required for bacterial chemotaxis, *Cell*, 42, 683, 1985.
44. **Parkinson, J. S.**, Genetics of bacterial chemotaxis, *Symp. Soc. Gen. Microbiol.*, 37, 265, 1981.
45. **Toews, M. L., Goy, M. F., Springer, M. S., and Adler, J.**, Attractants and repellents control demethylation of methylated chemotaxis proteins in *Escherichia coli, Proc. Natl. Acad. Sci. U.S.A.*, 76, 5544, 1978.
46. **Kehry, M. R., Doak, T. G., and Dahlquist, F. W.**, Stimulus-induced changes in methylesterase activity during chemotaxis in *Escherichia coli, J. Biol. Chem.*, 259, 11828, 1984.
47. **Kehry, M. R., Doak, T. G., and Dahlquist, F. W.**, Sensory adaptation in bacterial chemotaxis: regulation of methylesterase activity, *J. Bacteriol.*, 163, 983, 1985.
48. **Springer, M. S. and Zanolari, B.**, Sensory transduction in *Escherichia coli:* regulation of the demethylation rate by CheA protein, *Proc. Natl. Acad. Sci. U.S.A.*, 81, 5061, 1984.
49. **Terwilliger, T. C., Wang, J. Y., and Koshland, D. E., Jr.**, Kinetics of receptor modification. The multiply methylated aspartate receptors involved in bacterial chemotaxis, *J. Biol. Chem.*, 261, 10814, 1986.
50. **Terwilliger, T. C., Wang, J. Y., and Koshland, D. E., Jr.**, Surface structure recognized for covalent modification of the aspartate receptor in chemotaxis, *Proc. Natl. Acad. Sci. U.S.A.*, 83, 6707, 1986.
51. **Segall, J. E., Manson, M. D., and Berg, H. C.**, Signal processing times in bacterial chemotaxis, *Nature*, 296, 885, 1982.

52. **Block, S. M., Segall, J. E., and Berg, H. C.,** Adaptation kinetics in bacterial chemotaxis, *J. Bacteriol.,* 154, 312, 1983.
53. **Komeda, Y.,** Fusions of flagellar operons to lactose genes on a Mu *lac* bacteriophage, *J. Bacteriol.,* 150, 16, 1982.
54. **Wang, E. A. and Koshland, D. E., Jr.,** Receptor structure in the bacterial sensing system, *Proc. Natl. Acad. Sci. U.S.A.,* 77, 7157, 1980.
55. **Kleene, S. J., Hobson, A. C., and Adler, J.,** Attractants and repellents influence methylation and demethylation of methyl-accepting chemotaxis proteins in an extract of *Escherichia coli, Proc. Natl. Acad. Sci. U.S.A.,* 76, 6309, 1979.
56. **Bollinger, J., Park, C., Nowlin, D., and Hazelbauer, G. L.,** Adaptational crosstalk and the crucial role of methylation in chemotactic migration by *Escherichia coli,* submitted.
57. **Stock, A., Koshland, D. E., Jr, and Stock, J.,** Homologies between the *Salmonella typhimurium* CheY protein and proteins involved in the regulation of chemotaxis, membrane protein synthesis, and sporulation, *Proc. Natl. Acad. Sci. U.S.A.,* 82, 7989, 1985.
58. **Goldman, D. J. and Ordal, G. W.,** *In vitro* methylation and demethylation of methyl-accepting chemotaxis proteins in *Bacillus subtillis, Biochemistry,* 23, 2600, 1984.
59. **Goldman, D. J., Nettleton, D. O., and Ordal, G. W.,** Purification and characterization of chemotactic methylesterase from *Bacillus subtilis, Biochemistry,* 23, 675, 1983.
60. **Warrick, H. M., Taylor, B. L., and Koshland, D. E., Jr.,** Chemotactic mechanism of *Salmonella typhimurium:* preliminary mapping and characterization of mutants, *J. Bacteriol.,* 130, 223, 1977.
61. **Kihara, M. and Macnab, R. M.,** Cytoplasmic pH mediates pH taxis and weak-acid repellent taxis of bacteria, *J. Bacteriol.,* 145, 1209, 1981.
62. **Yamaguchi, S., Aizawa, S. -I., Kihara, M., Isomura, M., Jones, C. J., and Macnab, R. M.,** Genetic evidence for a switching and energy-transducing complex in the flagellar motor of *Salmonella typhimurium, J. Bacteriol.,* 168, 1172, 1986.
63. **Smith, R. A. and Parkinson, J. S.,** Overlapping genes at the cheA locus of *Escherichia coli, Proc. Natl. Acad. Sci. U.S.A.,* 77, 5370, 1980.
64. **Silverman, M., Matsumura, P., Hilmen, M., and Simon, M. I.,** Characterization of lambda *Escherichia coli* hybrids carrying chemotaxis genes, *J. Bacteriol.,* 130, 877, 1977.
65. **Chelsky, D. and Dahlquist, F. W.,** Chemotaxis in *Escherichia coli:* associations of protein components, *Biochemistry,* 19, 4633, 1980.
66. **Hess, J. F., Oosawa, K., Matsumura, P., and Simon, M. I.,** Protein phosphorylation is involved in bacterial chemotaxis, *Proc. Natl. Acad. Sci. U.S.A.,* 84, 7609, 1987.
67. **Hess, J. F., Oosawa, K., Kaplan, N., and Simon, M. I.,** Phosphorylation of three proteins in the signaling pathway of bacterial chemotaxis, *Cell,* 53, 79, 1988.
68. **Oosawa, K., Hess, J. F., and Simon, M. I.,** Mutants defective in bacterial chemotaxis show modified protein phosphorylation, *Cell,* 53, 89, 1988.
69. **Stock, A., Mottonen, J., Chen, T., and Stock, J.,** Identification of a possible nucleotide binding site in CheW, a protein required for sensory transduction in bacterial chemotaxis, *J. Biol. Chem.,* 262, 535, 1987.
70. **Stock, A., Chen., T., Welsh, D., and Stock, J.,** CheA protein, a central regulator of bacterial chemotaxis, belongs to a family of proteins that control gene expression in response to changing environmental conditions, *Proc. Natl. Acad. U.S.A.,* 85, 1403, 1988.
71. **Parkinson, J. S.,** *cheA, cheB,* and *cheC* genes of *Eescherichia coli* and their role in chemotaxis, *J. Bacteriol.,* 126, 45, 1976.
72. **Parkinson, J. S.,** Complementation amylasis and deletion mapping of *Escherichia coli* mutants defective in chemotaxis, *J. Bacteriol.,* 135, 45, 1978.
73. **Parkinson, J. S., Parker, S. R., Talbert, P. B., and Houts, S. E.,** Interactions between chemotaxis genes and flagellar genes, *J. Bacteriol.,* 155, 265, 1983.
74. **Clegg, D. O. and Koshland, D. E., Jr.,** The role of a signaling protein in bacterial sensing: behavioral effects of increased gene expression, *Proc. Natl. Acad. Sci. U.S.A.,* 81, 5056, 1984.
75. **Ravid, S., Matsumura, P., and Eisenbach, M.,** Restoration of flagellar clockwise rotation in bacterial envelopes by insertion of the chemotasxis protein CheY, *Proc. Natl. Acad. Sci. U.S.A.,* 83, 7157, 1986.
76. **Shioi, J., Galloway, R. J., Niwano, M., Chinnock, R. E., and Taylor, B. L.,** Requirement of ATP in bacterial chemotaxis, *J. Biol. Chem.,* 257, 7969, 1982.
77. **Smith, J. M., Roswell, E. H., Shioi, J., and Taylor, B. L.,** Identification of site of ATP requirement for signal transduction in bacterial chemotaxis, *J. Bacteriol.,* 170, 2698, 1988.

Chapter 16

ROLE OF PROTEIN CARBOXYL METHYLATION IN BACTERIAL CHEMOTAXIS

Jeff Stock

TABLE OF CONTENTS

I. INTRODUCTION

A. BACTERIAL CHEMOTAXIS

The motility mechanism in flagellated bacteria involves alternation between runs and tumbles (for recent reviews of bacterial chemotaxis and motility see Macnab, 1987[1,2]). In a constant environment cells generally run 1 to 2 s for a distance of about 30 μM (about 15 body lengths), and then tumble in place for about 0.1 to 0.2 s.[3] Bacteria swim by rotating their flagella.[4] Runs are associated with periods of constant rotation in a single direction while tumbles are caused by rotation reversals.[5] Tumbling serves to randomize the direction of each succeeding run so that the overall effect is a random walk. Chemotaxis results from a suppression of tumbles that occurs when cells detect increasing concentrations of attractants or decreasing concentrations of repellents.[3] Thus, if a bacterium senses that it is running toward attractants or away from repellents it will tend to continue on course, otherwise it will tumble and go off in a new direction. For *Escherichia coli* and *Salmonella typhimurium* the strongest attractants are nutrients such as aspartate and serine, while substances such as phenol, acetate, and indole act as repellents.

B. RECEPTOR STRUCTURE

Bacteria sense changes in the chemical composition of their surroundings through membrane receptor-transducer proteins.[2] In *E. coli* and *S. typhimurium* two major membrane receptors have been identified, one for aspartate, designated Tar; the other for serine, Tsr.[6,7] Additional, less abundant, transducer proteins mediate responses to ribose and galactose, Trg,[8] and peptides, Tap.[9] All of these proteins have similar primary sequences over their entire length of approximately 500 amino acids.[10-13] Each appears to be comprised of two domains: an N-terminal domain of approximately 200 residues embedded in the cytoplasmic membrane facing the periplasmic space, and a C-terminal cytoplasmic domain.[14-15] Stimulatory ligands bind to the N-terminal domain which acts in turn to regulate the signaling activity of the C-terminus. Within the C-terminal domain of each receptor-transducer protein there are at least four glutamic acid residues that are subject to methylation and demethylation.[16,17] A specific *S*-adenosylmethionine-dependent methyltransferase, the product of the *cheR* gene, methylates these residues;[18,19] and a specific methylesterase, the product of *cheB*, catalyzes the demethylation reaction.[20,21] The CheB protein is comprised of two domains: a C-terminal catalytic domain that functions to control the activity of the C-terminus.[21,22] In this review the role of the reactions catalyzed by CheR and CheB in sensory regulation will be examined.

C. SIGNAL TRANSDUCTION

Signal transduction in bacteria involves a network of phosphotransfer reactions where various forms of high energy phosphate are passed between cytoplasmic proteins.[22-25] A genetic analysis of the chemotaxis system indicates that signal transduction from the receptors to the flagellar motor requires the products of 4 Che genes: CheW, CheA, CheY, and CheZ.[2] All of these proteins have been purified and characterized (Table 1 and Figure 1). CheA is a kinase that phosphorylates CheY and phospho-CheY appears to bind to the flagella motor and cause tumbly behavior.[22-25] CheW[26] appears to provide a link between the receptors and CheA, and CheZ is a phosphatase that appears to act to control the lifetime of phospho-CheY.[24] In addition, the N-terminal regulatory domain of the CheB methylesterase is phosphorylated by the CheA kinase, and this modification acts to stimulate the demethylating activity of the C-terminal CheB catalytic domain.[22,24] Thus the CheB protein provides a link between the phosphorylation and methylation reactions involved in chemotaxis.

D. METHYLATION AND ADAPTATION

It has been generally accepted that chemoreceptor methylation is involved in adaptation.[17,27] This idea derived from an experimental paradigm for the chemotaxis response that was

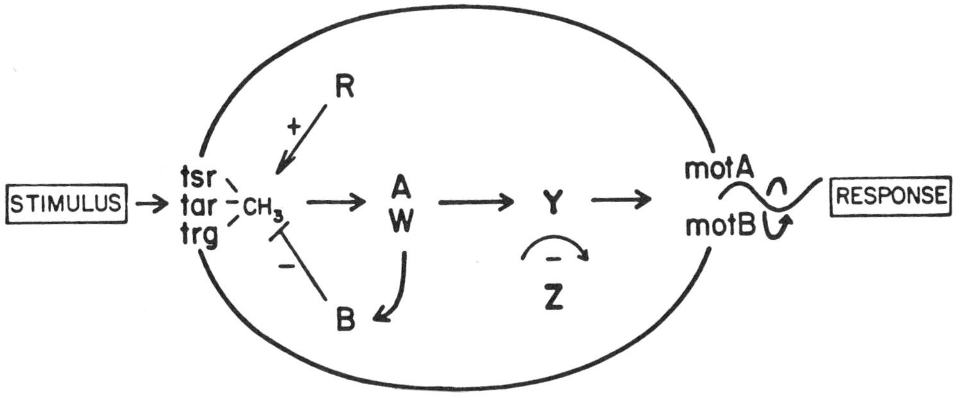

FIGURE 1. Signal transduction in bacterial chemotaxis. Stimuli interact with receptor-transducer proteins at the cell surface (Tar, Tsr, Trg). These proteins are methylated by an S-adenosylmethionine-dependent methyltransferase (R), and demethylated by a methylesterase (B). Information from the receptors regulates the activity of a protein kinase (A). The mechanism appears to require an accessory protein (W). The kinase phosphorylates both the esterase (B) and a response regulator (Y). Phosphorylated Y interacts with motor components to cause reversals. A phosphatase (Z) controls the liftime of phospho-Y.

TABLE 1
Components of the Chemotaxis Sensory Transduction System

Protein	Activity	Location	Molecular weight Native	Molecular weight Denatured	Ref.
Tar	Aspartate receptor	Transmembrane	120k	60k	7,14
Tsr	Serine receptor	Transmembrane	?	60k	10
CheR	Methyltransferase	Cytoplasm	32k	32k	19
CheB	Esterase/amidase	Cytoplasm	37k	37k	21
CheA	Protein kinase	Cytoplasm	290k	73k	23—25,78
CheW	?	Cytoplasm	36k	18k	26
CheY	Phosphorylated response regulator	Cytoplasm	14k	14k	22—25,79
CheZ	CheY phosphatase	Cytoplasm (flagellum)	>500k	24k	24,51

originally advanced by Macnab and Koshland in 1972.[28] Cells suddenly exposed to saturating concentrations of stimulatory ligands first undergo a change in swimming behavior (excitation), then after a few minutes prestimulus swimming behavior is restored despite the continued presence of the stimulus (adaptation). It is clear that changes in receptor methylation are an important component of the adaptation process with adaptation being operationally defined by the Macnab-Koshland stepwise response paradigm.[29-32] On the other hand, it is equally clear that mutant strains completely deficient in receptor methylation retain a significant ability to adapt their behavior on continued exposure to stimulatory ligands.[33-36] Moreover, it has not been possible to precisely define the relationship between swimming responses to large stepwise changes in stimulus concentration and the swimming behavior of cells actually migrating in chemical gradients.[27-39]

II. EFFECTS OF METHYLATION ON RECEPTOR ACTIVITY

Several distinct aspects of receptor activity have been assayed either in intact living cells or in reconstituted systems. These may be subdivided into three broad categories:

1. *Binding of stimulatory ligands or periplasmic binding proteins to the periplasmic domain.* Ligand binding can only be directly assayed *in vitro* using membrane preparations or purified receptors.[6,7]
2. *Interactions between the cytoplasmic domain and the methylating and demethylating enzymes.* Interactions with the CheR methyltransferase and the CheB methylesterase may be assayed with purified components *in vitro,* or in intact cells by following associated changes in methylation.[40,41]
3. *Effects on cellular motility.* Receptor effects on swimming behavior can only be assayed *in vivo* with most of the remaining chemotaxis components intact.[42,43]

The effects of changing levels of methylation on each of these activities are discussed separately below.

A. EFFECTS ON LIGAND BINDING

In the initial characterization of the aspartate and serine receptors it was shown that purified highly methylated receptors (receptors from *cheB* mutants) and unmethylated receptors (receptors from *cheR* mutants) show equivalent ligand binding properties.[6,7]

B. EFFECTS ON METHYLATION AND DEMETHYLATION

The evidence suggests that as far as the CheR methyltransferase is concerned, each methylation site is roughly equivalent.[44] There does not, for instance, appear to be an obligatory order of methylation where one glutamate must be methylated before the next is exposed. A somewhat different situation may pertain for the CheB methylesterase. Some methylated residues are hydrolyzed at an extremely rapid rate and some are extremely resistant to hydrolysis.[45,46] This produces a situation where the first residue hydrolyzed is generally the last residue to have been methylated. One might suppose that esterase activity would be affected by the level of receptor methylation, either because of changes in receptor conformation,[46] or because of feedback regulation from the signal transduction system.[22,48,49] There is no evidence concerning the effect of methylation on binding of transferase or esterase to the receptor. There are approximately 200 transferase monomers per cell,[19] approximately ten times as many esterase monomers (2000),[21] and about 10,000 receptor monomers.[6]

C. EFFECTS ON MOTILITY

In the absence of stimuli, mutants defective in the CheR methyltransfease run all the time,[18,31,32] and mutants defective in the CheB methylesterase tumble all the time[20,35,50] (note that the tumbly *cheZ* mutant whose behavior was investigated by Rubik and Koshland[50] was actually a *cheZ*$^+$ strain with a missense mutation in *cheB*[51]). Cells depleted of *S*-adenosylmethionine tend to run all the time just like CheR methyltransferase mutants.[52] Thus, the constant running phenotype of *cheR* mutants appears to be associated with lowered levels of receptor methylation rather than some other effect of CheR deficiency. Taken together the results suggest that low levels of receptor methylation produce a running phenotype while abnormally high levels cause tumbling.

Double mutants completely lacking both the CheR methyltransfease and the CheB methylesterase tend to behave like wild type strains[35,53] Each receptor contains at least four sites of glutamyl methylation.[16,17] Two of these are encoded as glutamine residues and two are encoded as glutamates. The glutamates can be methylated immediately by the CheR enzyme, but the glutamines must first be deamidated. The CheB esterase functions as an amidase to accomplish this, and thus provides two additional glutamates for the transferase.[54-56] The results with *cheRcheB* double mutants can be readily understood if it is assumed that an amide at a glutamate residue has the same effect on the receptors as a methyl ester.[35,38,39] Thus, *cheR* mutants run all the time because there are no methylesters and few amides, *cheB* mutants tumble all the time because most glutamates are either amidated or methyl esterified, and the double mutants run and tumble more or less like wild type because they have an intermediate level of modification, two glutamines and two glutamates.

III. MECHANISMS THAT REGULATE RECEPTOR METHYLATION

A. RECEPTOR CONFORMATION

Each type of receptor in a wild type cell exhibits a steady state level of methylation that varies with the degree of binding of its corresponding stimulatory ligands.[27,29,57] In *S. typhimurium*, for instance, saturating concentrations of serine and aspartate cause roughly a two-fold increase in the levels of methylation of the Tar and Tsr receptors.[45] Corresponding changes occur in all *che* mutants except *cheR* and *cheB*.[40] These changes in methylation appear to be the direct consequence of stimulus-induced changes in receptor conformation.[45] Studies with purified components indicate that the primary effect associated with attractant ligand binding is withdrawal of methylated residues from the esterase.[7,47] This conclusion fits nicely with the observed effects of attractants on methylation in intact cells.[45,46] The situation with the transferase is less clear. A cursory view of the effect of attractant stimuli on levels of methylation in intact cells would lead one to suppose that attractants cause conformational changes that expose unmethylated residues to the transferase. Studies with purified components indicate that attractant binding has little effect on transferase activity, however;[7,58] and recent studies with intact cells indicate that the transferase may be acting at its maximal velocity in the absence of attractant stimuli (Lupas and Stock, unpublished observations).

B. PHOSPHORYLATION OF THE ChEB METHYLESTERASE

Over the past few years attention has focused on a second locus of regulation, feedback activation of the esterase via the CheA protein. The first indication of an interaction between the esterase and CheA was the observation that *cheA* mutants showed much smaller repellent-induced increases in demethylation than corresponding *che*$^+$ strains.[48] It has subsequently been shown that the CheB protein has an N-terminal regulatory domain that is not required for esterase activity.[19] CheA is autophosporylated at a histidine residue,[22,23,59] and this group appears to be transferred to one or more aspartate residues in the esterase regulatory domain.[22-24] From studies with purified components, it is apparent that the phosphorylated form of the esterase is approximately ten-fold more active than the unphosphorylated enzyme.[22] It seems likely that repellent stimuli cause an increase in CheA kinase activity that leads to a dramatic activation of the esterase. It should be emphasized that this type of esterase regulation causes increased demethylation of all receptors.[41,48] This contrasts to the receptor-specific effects of attractants. Moreover, it has been shown that the latter mechanism is completely independent of the esterase regulatory domain since a fragment of CheB with this part entirely deleted appears to be regulated by attractants in the same way as the intact CheB protein.[47] The regulatory domain is, however, absolutely essential for CheA-dependent regulation of esterase activity both *in vivo* and with purified components (Lupas and Stock, unpublished results).

C. LEVELS OF *S*-ADENOSYLMETHIONINE

The *S*-adenosylmethionine (AdoMet) pool in *E. coli* and *S. typhimurium* appears to be remarkably unstable. For instance, as long as a methionine auxotroph is kept in saturating extracellular methionine the level of AdoMet is maintained at approximately 100 μM; but within seconds of removing exogenous sources of methionine the level drops to less than 3 μM.[52,60] Similar results are obtained with wild type cells grown on methionine.[52] These effects do not depend on the splitting of AdoMet to *S*-adenosylhomocysteine and methanol via the chemotaxis receptors, the CheR methyltransferase and the CheB methylesterase.[60] Chemotaxis does not, therefore, appear to play a major role as a pathway for AdoMet utilization in one-carbon metabolism.[16] Since the apparent K_m of the CheR methyltransferase is approximately 20 μM,[19] changes in the intracellular AdoMet pool could cause dramatic changes in rates of receptor methylation. AdoMet represents a convergence of three important end products of intermediary metabolism: aspartate-derived anabolism, serine-derived one-carbon metabolism, and ATP. Thus, while the major control mechanisms of CheB esterase activity appear to be the aspartate and serine receptors and ATP-dependent phosphotransfer reactions; the primary regulator of CheR methyltransferase activity may be the availability of the methyl donor, AdoMet, which in turn is controlled by the state of aspartate and serine based metabolism as well as the overall energy charge of the cell.

IV. CHEMOTAXIS IN THE ABSENCE OF RECEPTOR METHYLATION

A. RECEPTORS THAT ARE NOT METHYLATED TRANSDUCER PROTEINS

Soon after the discovery of methylated chemotaxis proteins (MCPs), it was generally assumed that they functioned as transducers relaying information from peripheral chemoreceptors across the membrane to signal transduction components in the cytoplasm.[57] This idea predominated because initial results indicated that there were only two MCPs, Tar and Tsr, both of which were associated with *che* genes, *cheM* and *cheD*, respectively;[61] while it was known that cells respond to over 20 different types of attractant and repellent stimuli.[62] The only receptors that had been identified were the periplasmic sugar binding proteins for galactose,[63] ribose,[64] and maltose.[65] Genetic, and later, biochemical evidence indicated that the maltose binding protein interacted with Tar.[57,66-68] It was also known that the galactose and ribose binding proteins interact with a common membrane component, designated Trg,[69] that was identified with a third, minor MCP in *E. coli* membranes.[8] Subsequent studies showed that Tar and Tsr act directly as aspartate and serine binding proteins,[6,7] and that Tsr also functions directly in thermotaxis[70] and pH taxis.[15,71] This still leaves numerous types of responses uncounted for, most notably, chemotaxis to hexoses and hexitols such as glucose, mannose, fructose, and mannitol,[72,73] and aerotaxis to oxygen.[74] In fact, it has been shown that cells respond to sugars and oxygen in the complete absence of MCP proteins.[34] Apparently the methylated proteins correspond to only one of several classes of membrane receptor proteins.

B. CHEMOTAXIS IN *cheRcheB* DOUBLE MUTANTS

Double mutants that completely lack both the CheR methyltransferase and the CheB methylesterase retain a significant ability to migrate in gradients of aspartate, serine, and ribose.[38] Thus, neither methylation nor demethylation is essential for chemotaxis. Apparently, the reason *cheR* and *cheB* point mutants are generally nonchemotactic is that the associated undermodified or overmodified receptors cause a disabling imbalance in steady state behavior so that the cells always run or always tumble. In the double mutant, the receptors are fixed at an intermediate level of modification (two glutamates and two glutamines) so their steady state behavior is similar to wild type.

The double mutant does not exhibit as efficient a response as wild type. Clearly, the methylating and demethylating enzymes make a substantial contribution to the overall response mechanism. By analyzing the difference between the double mutant and wild type one might hope to determine the role of the methylating and demethylating enzymes in chemotaxis. The results of such studies indicate that the methylation system is important in maintaining a balanced run-tumble switching frequency over a wide range of stimulus intensities. For instance, at saturating concentrations of aspartate or serine *cheRcheB* mutants tend to run all the time,[35] and under these conditions they are defective in chemotaxis.

V. POSSIBLE ROLES FOR METHYLATION IN CHEMOTAXIS

A. RECEPTOR SENSITIZATION AND DESENSITIZATION

Increased levels of methylation cause tumbly behavior and decreased levels suppress tumbles. These effects may be considered in terms of changes in receptor sensitivity. Thus, a cell with decreased levels of methylation is sensitized to attractant stimuli while a cell whose receptors are highly methylated is desensitized. Conditions that lower the level of intracellular AdoMet would be expected to lower the level of Tar and Tsr methylation and thereby sensitize the cell to serine and aspartate. Conversely, conditions that increase the level of AdoMet could have a desensitizing effect. Thus, methylation may function as a hunger/satiety signal where the AdoMet level provides a measure of the nutritional state of the cell.

B. BLOCKING THE REPELLENT RESPONSE

The only response that cells exhibit when they are actually undergoing chemotaxis is the suppression of tumbles induced by increasing attractant concentrations or decreasing repellent concentrations.[3] Decreasing attractant concentrations or increasing repellent concentrations have no effect unless the changes are extremely large and therefore irrelevant within the context of a cell actually migrating in a chemical gradient.[37,75] One can understand why it might by advantageous for a cell to be blind to negative stimuli be considering the environment of a bacterium that is rapidly growing on a rich substrate. The global concentration of nutrients, i.e., attractants, would decrease at an accelerated rate as the population density increased. A cell that responded to these decreases would tumble continuously, and be unable to swim productively. Global esterase activation via the CheB regulatory domain may function to prevent this type of debilitating negative response.

C. EXTENDING THE RANGE OF RECEPTOR SENSITIVITY

The receptors have a built in conformational mechanism to change their levels of methylation and thereby automatically adapt their sensitivity to counteract the effects of stimulatory ligands. The mechanism appears to be analogous to light and dark adaptation in the vertebrate visual system where rhodopsin phosphorylation appears to adjust the response of the rod photoreceptor system to varying background light intensities.[76,77]

IV. CONCLUSIONS

Over the past decade a wealth of information has been obtained concerning the molecular mechanism of sensory transduction in bacterial chemotaxis. Six cytoplasmic proteins are known to mediate the receptor signaling process. These interact with one another and with membrane receptors via a number of different covalent modification reactions including methylation, demethylation, deamidation, phosphorylation, and dephosphorylation. The methylation, demethylation, and deamidation reactions appear to function to control the sensitivity of a class of chemoreceptors in response to changing intra- and extracellular conditions. The recently identified phosphorylation and dephosphorylation reactions appear to be more directly associated with the temporal response mechanism that acts during chemotaxis to suppress tumbles when a cell senses it is moving toward a more favorable environment. The finding that the CheB methylesterase is phosphorylated and thereby activated by the CheA kinase provides a mechanism for feedback between signal transduction components and the receptors. Clearly methylation and phosphorylation reactions are coordinated and a detailed understanding of the role of one modification system must include an elucidation of the role of the other.

REFERENCES

1. **Macnab, R. M.,** Flagella, in *Escherichia coli* and *Salmonella typhimurium,* Vol. 1, Neidhardt, F. C., Ed., American Society for Microbiology, 1987, 70
2. **Macnab, R. M.,** Motility and chemotaxis, in *Escherichia coli* and *Salmonella typhimurium,* Vol. 1, Neidhardt, F. C., Ed., American Society for Microbiology, 1987, 732.
3. **Berg, H. C. and Brown, D. A.,** Chemotaxis in *Escherichia coli* analyzed by three-dimensional tracking, *Nature,* 239, 500, 1972.
4. **Silverman, M. and Simon, M.,** Flagellar rotation and the mechanism of bacterial motility, *Nature (London),* 249, 73, 1974.
5. **Macnab, R. M. and Ornston, M. K.,** Normal-to-curly flagellar transitions and their role in bacterial tumbling. Stabilization of an alternative quarternary structure by mechanical force, *J. Mol. Biol.,* 112, 1, 1977.
6. **Clarke, S. and Koshland, D. E., Jr.,** Membrane receptors for aspartate and serine in bacterial chemotaxis, *J. Biol. Chem.,* 254, 9695, 1979.
7. **Wang, E. A. and Koshland, D. E., Jr.,** Receptor structure in the bacterial sensing system, *Proc. Natl. Acad. Sci. U.S.A.,* 77, 7157, 1980.
8. **Kondoh, H., Ball, C. B., and Adler, A.,** Identification of a methyl-accepting chemotaxis protein for the ribose and galactose chemoreceptors of *Escherichia coli, Proc. Natl. Acad. Sci. U.S.A.,* 76, 260, 1979.
9. **Manson, M. D., Blank, V., Brade, G., and Higgins, C. F.,** Peptide chemotaxis in *E. coli* involves the Tap signal transducer and the dipeptide permease, *Nature,* 321, 253, 1986.
10. **Boyd, A., Kendall, K., and Simon, M. I.,** Structure of the serine chemoreceptor in *Escherichia coli, Nature,* 301, 623, 1983.
11. **Krikos, A., Mutoh, N., Boyd, A., and Simon, M. I.,** Sensory transducers of *E. coli* are composed of discrete structural and functional domains, *Cell,* 33, 615, 1983.
12. **Russo, A. F. and Koshland, D. E., Jr.,** Separation of signal transduction and adaptation functions of the aspartate receptor in bacterial sensing, *Science,* 220, 1016, 1983.
13. **Bollinger, J., Park, C., Harayama, S., and Hazelbauer, G. L.,** Structure of the Trg protein: homologies with and differences from other sensory transducers in *Escherichia coli, Proc. Natl. Acad. Sci. U.S.A.,* 81, 3287, 1984.
14. **Mowbray, S. L., Foster, D. L., and Koshland, D. E., Jr.,** Proteolytic fragments identified with domains of the aspartate chemoreceptor, *J. Biol. Chem.,* 260, 11711, 1985.
15. **Krikos, A., Conley, M. P., Boyd, A., Berg, H. C., and Simon, M. I.,** Chimeric chemosensory transducers of *Escherichia coli, Proc. Natl. Acad. Sci. U.S.A.,* 82, 1326, 1985.
16. **Stock, J. and Simms, S.,** Methylation, demethylation, and deamidation at glutamate residues in membrane chemoreceptor proteins, in *Post-translational Modifications of Proteins and Ageing,* Zappia, V., Ed., Plenum Press, 1988, 201.
17. **Stewart, R. C. and Dahlquist, F. W.,** Molecular components of bacterial chemotaxis, *Chem. Rev.,* 87, 997, 1987.
18. **Springer, W. R. and Koshland, D. E., Jr.,** Identification of a protein methyltransferase as the *cheR* gene product in the bacterial sensing system, *Proc. Natl. Acad. Sci. U.S.A.,* 74, 533, 1977.
19. **Simms, S. A., Stock, A. M., and Stock, J. B.,** Purification and characterization of the S-adenosylmethionine: glutamyl methyltransferase that modifies membrane chemoreceptor proteins in bacteria, *J. Biol. Chem.,* 262, 8537, 1987.
20. **Stock, J. B. and Koshland, D. E., Jr.,** A protein methylesterase involved in bacterial sensing, *Proc. Natl. Acad. Sci. U.S.A.,* 75, 3659, 1978.
21. **Simms, S., Keane, M., and Stock, J.,** Multiple forms of the CheB methylesterase in bacterial chemosensing, *J. Biol. Chem.,* 260, 10161, 1985.
22. **Stock, A., Wylie, D., Mottonen, J., Lupas, A., Schutt, C., and Stock, J.,** Phosphotransferase networks mediate stimulus-response coupling in bacteria, *C.S.H.S.Q.B.,* 53, 49, 1988.
23. **Wylie, D., Stock, A., Wong, C. -Y., and Stock, J.,** Sensory transduction in bacterial chemotaxis involves phosphotransfer between Che proteins, *Biochem Biophys. Res. Commun.,* 151, 891, 1988.
24. **Hess, J. F., Oosawa, K., Kaplan, N., and Simon, M. I.,** Phosphorylation of three proteins in the signaling pathway of bacterial chemotaxis, *Cell,* 53, 79, 1988.
25. **Ninfa, A. J., Ninfa, E. G., Lupas, A. N., Stock, A., Magasanik, B., and Stock, J.,** Crosstalk between bacterial chemotaxis signal transduction proteins and regulators of transcription of the Ntr regulon: evidence that nitrogen assimilation and chemotaxis are controlled by a common phosphotransfer mechanism, *Proc. Natl. Acad. Sci. U.S.A.,* 85, 5492, 1988.
26. **Stock, A., Mottonen, J., Chen, T., and Stock, J.,** Identification of a possible nucleotide binding site in CheW, a protein required for sensory transduction in bacterial chemotaxis, *J. Biol. Chem.,* 262, 535, 1987.
27. **Springer, M. S., Goy, M. F., and Adler, J.,** Protein methylation in behavioral control mechanisms and in signal transduction, *Nature,* 280, 279, 1979.

28. **Macnab, R. M. and Koshland, D. E., Jr.,** The gradient-sensing mechanism in bacterial chemotaxis, *Proc. Natl. Acad. Sci. U.S.A.*, 69, 2509, 1972.
29. **Springer, M. S., Goy, M. F., and Adler, J.,** Sensory transduction in *Escherichia coli:* a requirement for methionine in sensory adaptation, *Proc. Natl. Acad. Sci. U.S.A.*, 74, 183, 1977.
30. **Goy, M. F., Springer, M. S., and Adler, J.,** Sensory transduction in *Escherichia coli:* role of a protein methylation reaction in sensory adaptation, *Proc. Natl. Acad. Sci. U.S.A.*, 74, 4964, 1977.
31. **Parkinson, J. S. and Revello, P. T.,** Sensory adaptation mutants of *E. coli, Cell,* 15, 1221, 1978.
32. **Goy, M. F., Springer, M. S., and Adler, J.,** Failure of sensory adaptation in bacterial mutants that are defective in a protein methylation reaction, *Cell,* 15, 1231, 1978.
33. **Stock, J. B., Maderis, A. M., and Koshland, D. E., Jr.,** Bacterial chemotaxis in the absence of receptor carboxylmethylation, *Cell,* 27, 37, 1981.
34. **Niwano, M. and Taylor, B. L.,** Novel sensory adaptation mechanism in bacterial chemotaxis to oxygen and phosphotransferase substrates, *Proc. Natl. Acad. Sci. U.S.A.*, 79, 11, 1982.
35. **Stock, J., Kersulis, G., and Koshland, D. E., Jr.,** Neither methylating nor demethylating enzymes are required for bacterial chemotaxis, *Cell,* 42, 683, 1985.
36. **Segall, J. E., Block, S. M., and Berg, H. C.,** Temporal comparisons in bacterial chemotaxis, *Proc. Natl. Acad. Sci. U.S.A.*, 83, 8987, 1986.
37. **Brown, D. A. and Berg, H. C.,** Temporal stimulation of chemotaxis in *Escherichia coli, Proc. Natl. Acad. Sci. U.S.A.*, 71, 1388, 1974.
38. **Stock, J., Borczuk, A., Chiou, F., and Burchenal, J. E. B.,** Compensatory mutations in receptor function: a re-evaluation of the role of methylation in bacterial chemotaxis, *Proc. Natl. Acad. Sci. U.S.A.*, 82, 8364, 1985.
39. **Stock, J. and Stock, A.,** What is the role of receptor methylation in bacterial chemotaxis, *Trends Biochem. Sci.*, 12, 371, 1987.
40. **Stock, J. B., Clarke, S., and Koshland, D. E., Jr.,** The protein carboxyl methyltransferase involved in *Escherichia coli* and *Salmonella typhimurium* chemotaxis, *Methods Enzymol.*, 106, 310, 1984.
41. **Kehry, M. R., Doak, T. G., and Dahlquist, F. W.,** Stimulus-induced changes in methylesterase activity during chemotaxis in *Escherichia coli, J. Biol. Chem.*, 259, 11828, 1984.
42. **Clegg, D. O. and Koshland, D. E., Jr.,** The role of a signaling protein in bacterial sensing: behavioral effects of increased gene expression, *Proc. Natl. Acad. Sci. U.S.A.*, 81, 5056, 1984.
43. **Wolfe, A. J., Conley, M. P., Kramer, T. J., and Berg, H. C.,** Reconstitution of signaling in bacterial chemotaxis, *J. Bacteriol.*, 169, 1878, 1987.
44. **Terwilliger, T. C., Wong, J. Y., and Koshland, D. E., Jr.,** Kinetics of receptor modification. The multiply methylated aspartate receptors involved in bacterial chemotaxis, *J. Biol. Chem.*, 261, 10814, 1986.
45. **Stock, J. B. and Koshland, D. E., Jr.,** Changing reactivity of receptor carboxyl groups during bacterial sensing, *J. Biol. Chem.*, 256, 10826, 1981.
46. **Springer, M. S., Zanolari, B., and Pierzchala, P. A.,** Ordered methylation of the methyl-accepting chemotaxis proteins of *Escherichia coli, J. Biol. Chem.*, 257, 6861, 1982.
47. **Borczuk, A., Staub, A., and Stock, J.,** Demethylation of bacterial chemoreceptors is inhibited by attractant stimuli in the complete absence of the regulatory domain of the demethylating enzyme, *Biochem. Biophys. Res. Commun.*, 141, 918, 1986.
48. **Springer, M. S. and Zanolari, B.,** Sensory transduction in *Escherichia coli:* regulation of the demethylation rate by the CheA protein, *Proc Natl. Acad Sci. U.S.A.*, 81, 5061, 1984.
49. **Kehry, M. R., Doak, T. G., and Dahlquist, F. W.,** Sensory adaptation in bacterial chemotaxis: regulation of demethylation, *J. Bacteriol.*, 163, 983, 1985.
50. **Rubick, B. A. and Koshland, D. E., Jr.,** Potentiation, desensitization, and inversion of response in bacterial sensing of chemical stimuli, *Proc. Natl. Acad. Sci. U.S.A.*, 75, 2820, 1978.
51. **Stock, A. M. and Stock, J. B.,** Purification and characterization of the CheZ protein of bacterial chemotaxis, *J. Bacteriol.*, 169, 3301, 1987.
52. **Aswad, D. W. and Koshland, D. E., Jr.,** Evidence for an S-adenosylmethionine requirement in the chemotactic behavior of *Salmonella typhimurium, J. Mol. Biol.*, 97, 207, 1975.
53. **Segal, J. E., Manson, M. D., and Berg, H. C.,** Signal processing times in bacterial chemotaxis, *Nature,* 296, 855, 1982.
54. **Rollins, C. and Dahlquist, F. W.,** The methyl-accepting chemotaxis proteins of *E. coli:* a repellent-stimulated, covalent modification, distinct from methylation, *Cell,* 25, 333, 1981.
55. **Sherris, D. and Parkinson, J. S.,** Posttranslational processing of methyl-accepting chemotaxis proteins in *Escherichia coli, Proc. Natl. Acad. Sci. U.S.A.*, 78, 6051, 1981.
56. **Kehry, M. R., Bond, M. W., Hunkapiller, M. W., and Dahlquist, F. W.,** Enzymatic deamidation of methyl-accepting chemotaxis proteins of *Escherichia coli* catalyzed by the *cheB* gene product, *Proc. Natl. Acad. Sci. U.S.A.*, 80, 3599, 1983.
57. **Springer, M. S., Goy, M. F., and Adler, J.,** Sensory transduction in *Escherichia coli:* two complementary pathways of information processing that involve methylated proteins, *Proc. Natl. Acad. Sci. U.S.A.*, 74, 3312, 1977.

58. **Clarke, S., Sparrow, K., Panasenko, S., and Koshland, D. E., Jr.,** *In vitro* methylation of bacterial chemotaxis proteins: characterization of protein methyltransferase activity in crude extracts of *Salmonella typhimurium, J. Supramol. Struct.,* 13, 315, 1980.
59. **Hess, J. F., Oosawa, K., Matsumura, P., and Simon, M. I.,** Protein phosphorylation is involved in bacterial chemotaxis, *Proc. Natl. Acad. Sci. U.S.A.,* 84, 7609, 1987.
60. **Borczuk, A., Stock, A., and Stock, J.,** S-Adenosylmethionine may not be essential for signal transduction during bacterial chemotaxis, *J. Bacteriol.,* 169, 3295, 1987.
61. **Silverman, M. and Simon, M.,** Chemotaxis in *E. coli*: methylation of the che gene products, *Proc. Natl. Acad. Sci. U.S.A.,* 74, 3317, 1977.
62. **Adler, J.,** Chemotaxis in bacteria, *Annu. Rev. Biochem.,* 44, 341, 1975.
63. **Hazelbauer, G. L. and Adler, J.,** Role of the galactose binding protein in chemotaxis of *Escherichia coli* toward galactose, *Nature New Biol.,* 230, 101, 1971.
64. **Aksamit, R. R. and Koshland, D. E., Jr.,** Identification of the ribose binding protein as the receptor for ribose chemotaxis in *Salmonella typhimurium, Biochemistry,* 13, 4473, 1974.
65. **Hazelbauer, G. L.,** The maltose chemoreceptor of *Escherichia coli, J. Bacteriol.,* 122, 206, 1975.
66. **Koiwai, O. and Hayashi, H.,** Studies of bacterial chemotaxis. IV. Interaction of maltose receptor with a membrane-bound chemosensing component, *J. Biochem. (Tokyo),* 86, 27, 1979.
67. **Richarme, G.,** Interaction of the maltose-binding protein with membrane vesicles of *Escherichia coli, J. Bacteriol.,* 149, 662, 1982.
68. **Mowbray, S. L. and Koshland, D. E., Jr.,** Additive and independent responses in a single receptor: aspartate and maltose stimuli on the Tar protein, *Cell,* 50, 171, 1987.
69. **Strange, P. G. and Koshland, D. E., Jr.,** Receptor interactions in a signalling system: competition between ribose receptor and galactose receptor in the chemotaxis response, *Proc. Natl. Acad. Sci. U.S.A.,* 73, 762, 1976.
70. **Maeda, K. and Imae, Y.,** Thermosensory transduction in *Escherichia coli:* inhibition of the thermoresponse by L-serine, *Proc. Natl. Acad. Sci. U.S.A.,* 76, 91, 1979.
71. **Slonczewski, J. L., Macnab, R. M., Alger, J. R., and Castle, A. M.,** Effects of pH and repellent tactic stimuli on protein methylation levels in *Escherichia coli, J. Bacteriol.,* 152, 384, 1982.
72. **Adler, J. and Epstein, W.,** Phosphotransferase-system enzymes as chemoreceptors for certain sugars in *Escherichia coli* chemotaxis, *Proc. Natl. Acad. Sci. U.S.A.,* 71, 2895, 1974.
73. **Postma, P. W. and Lengler, J. W.,** Phosphoenolpyruvate: carbohydrate phosphotransferase system of bacteria, *Microbiol. Rev.,* 49, 232, 1985.
74. **Taylor, B. L.,** How do bacteria find the optimal concentration of oxygen?, *Trends Biochem. Sci.,* 8, 438, 1983.
75. **Block, S. M., Segall, J. E., and Berg, H. C.,** Adaptation kinetics in bacterial chemotaxis, *J. Bacteriol.,* 154, 312, 1983.
76. **Miller, J., Paulson, R., and Bownds, N. D.,** Control of light-activated phosphorylation in frog photoreceptor membranes, *Biochemistry,* 16, 2633, 1977.
77. **Wilden, U., Hall, S. W., and Kuhn, H.,** Phosphodiesterase activation by photoexcited rhodopsin is quenched when rhodopsin is phosphorylated and binds the intrinsic 48-kDa protein of rod outer segments, *Proc. Natl. Acad. Sci. U.S.A.,* 83, 1174, 1986.
78. **Stock, A., Chen, T., Welsh, D., and Stock, J.,** CheA protein, a central regulator of bacterial chemotaxis, belongs to a family of proteins that control gene expression in response to changing environmental conditions, *Proc. Natl. Acad. Sci. U.S.A.,* 85, 1403, 1988.
79. **Stock, A., Koshland, D. E., Jr., and Stock, J.,** Homologies between the *Salmonella typhimurium* CheY protein and proteins involved in the regulation of chemotaxis, membrane protein synthesis, and sporulation, *Proc. Natl. Acad. Sci. U.S.A.,* 82, 7989, 1985.

Chapter 17

SELF-METHYLATION BY SUICIDE DNA REPAIR ENZYMES

Bruce Demple

TABLE OF CONTENTS

I. OVERVIEW

The major mutagenic lesions formed in DNA by alkylating agents are repaired by a noncatalytic process in virtually all organisms. This repair is inducible by methylating agents in *Escherichia coli* and is mediated by suicidal enzymes, each of which transfers the methyl group from one premutagenic O^6-methylguanine in DNA irreversibly to one of its own cysteine residues. This inducible methyltransferase repair function resides in the 39-kDa Ada protein and can be isolated as the separated 19-kDa C-terminal fragment of Ada following mild proteolysis. A second suicidal transferase for methylations of the DNA phosphodiester backbone is located in the N-terminal half of the Ada protein, and its self-methylation renders Ada a potent transcriptional activator of the *ada* gene itself, along with several other genes that are inducible by alkylating agents. Proteins that function as suicide methyltransferases for O^6-methylguanine also exist in *E. coli* uninduced by alkylating agents, and are widely distributed in eukaryotic organisms, including mammalian cells. These self-inactivating "suizymes" are being exploited to understand the mechanism by which a protein recognizes rare damaged bases in DNA, and to establish the biological role of suicide repair in combatting mutagenesis and carcinogenesis by alkylating agents.

II. INTRODUCTION

Monofunctional alkylating agents include some of the most widespread and potent mutagens and carcinogens known, such as *N*-methyl-*N'*-nitro-*N*-nitrosoguanidine (MNNG), *N*-methyl-*N*-nitrosourea (MNU) and dimethylnitrosamine.[1] Mutagenesis by these agents is prevented in both bacteria and mammalian cells by unusual, self-inactivating repair proteins that transfer the alkyl groups of particular base adducts in alkylated DNA irreversibly to themselves in suicidal reactions. In *Escherichia coli,* a separate but analogous self-methylating transfer from methylated DNA phosphates converts the bacterial suicide repair protein Ada into a gene activator.[2,3] The mechanisms and specificities of these suicide transfer proteins account for their physiological effects and are the focus of this chapter.

A. DISCOVERY OF SUICIDE DNA REPAIR

The discovery of a DNA correction mechanism involving a self-inactivating repair protein depended crucially on the prior characterization of a novel inducible response to alkylation damage in *E. coli*.[4-6] This "adaptive response" was found to be strongly induced by agents such as MNNG, conferring high-level resistance in *E. coli* to the mutagenic and cell killing effects of such compounds.[4] The induced resistance to alkylation mutagenesis exhibited some unusual features,[6] the most prominent of which was its sharply limited capacity. For example, adapted cells could resist MNNG at 5 μg/ml indefinitely, but in 10 μg/ml they began accumulating mutations after only 10 min.[5] The saturable nature of the response is a direct result of the biochemical mechanism used to correct mutagenic damages in alkylated DNA.

MNNG and MNU induce a predominant single class of mutations in all the cases examined, transitions from G:C to A:T base pairs.[7] These reactive compounds alkylate DNA to form a variety of adducts, predominantly methylations of the guanine N7, the adenine N3, the guanine O^6 and phosphodiester oxygens, along with lesser amounts of methylated pyrimidine nitrogens and oxygens (Figure 1). Methylating chemicals that are less reactive with DNA (e.g., methyl methane sulfonate)* are far less efficient mutagens and form correspondingly less *O*-alkylation

* MNNG and methyl methane sulfonate are commonly described in the biochemical literature as reacting by S_N1 and S_N2 mechanisms, respectively. This distinction is probably incorrect (E. Loechler, personal communication), and we will describe the difference in terms of reactivities with DNA.

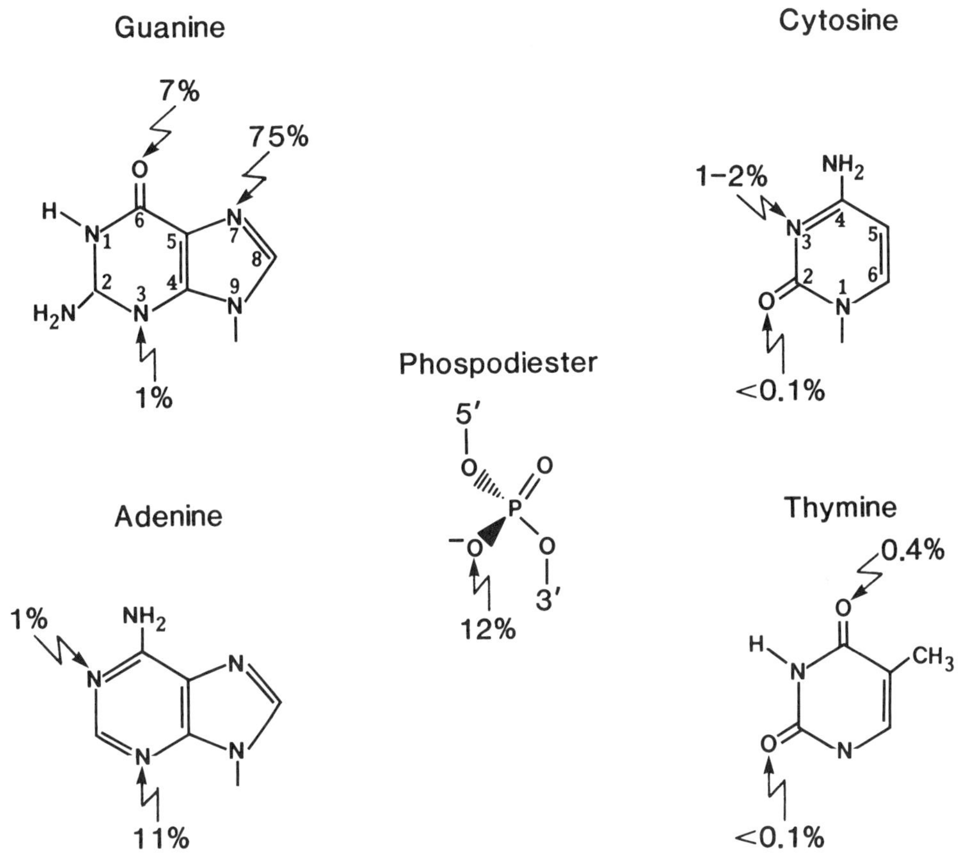

FIGURE 1. Important DNA alkylation sites for MNNG and MNU. The percentages given are independently averaged from the literature values (1) for the fraction of each of the products, and so do not sum to 100%. Additional minor products are not shown. Only the Sp stereoisomer of methyl phosphotriesters is shown, but the percentage indicated is for the total alkylation at DNA phosphates.

in DNA.[1] Loveless argued in 1969[8] that this fact, and the likely stabilization of O^6-methylguanine (O^{6}MeG) in the unusual enol (rather than keto) tautomeric form, could account for much of the mutagenesis by alkylating agents such as MNNG. Important support for this idea was later provided by the demonstration that the adaptive resistance to MNNG mutagenesis in *E. coli* correlated exactly with the induced cellular ability to remove O^{6}MeG from the chromosome.[9] This inducible repair was so efficient that most of the O^{6}MeG (several thousand molecules per cell) had been corrected during the alkylation challenge.[9] Most importantly, the repair capacity for O^{6}MeG exhibited exactly the saturable characteristic already seen for the adaptive resistance to mutagenesis. The correction of O^{6}MeG still proceeded after the initial rapid repair had been saturated, but was reduced to a rate of about 100 molecules/min/bacterium. The slow repair rate was abolished in preadapted cells that were subsequently treated with the protein synthesis inhibitor chloramphenicol, leading Cairns to hypothesize that each repair molecule acts only once.[10]

The development of an *in vitro* assay for O^{6}MeG correction immediately revealed an unusual biochemical feature: the labeled O^{6}MeG bases disappeared from DNA but did not appear in any acid-soluble form.[11] Thus, O^{6}MeG was not being removed as the free base, as a nucleotide or by demethylation to form methanol. This property of the reaction, at first baffling, was soon shown to result from the covalent transfer of the O^{6}MeG methyl group to a protein cysteine,[12] leading to the direct regeneration of a normal DNA guanine residue[13] (Figure 2). The methylated protein residue, *S*-methylcysteine, had been observed previously as a nonenzymatic protein

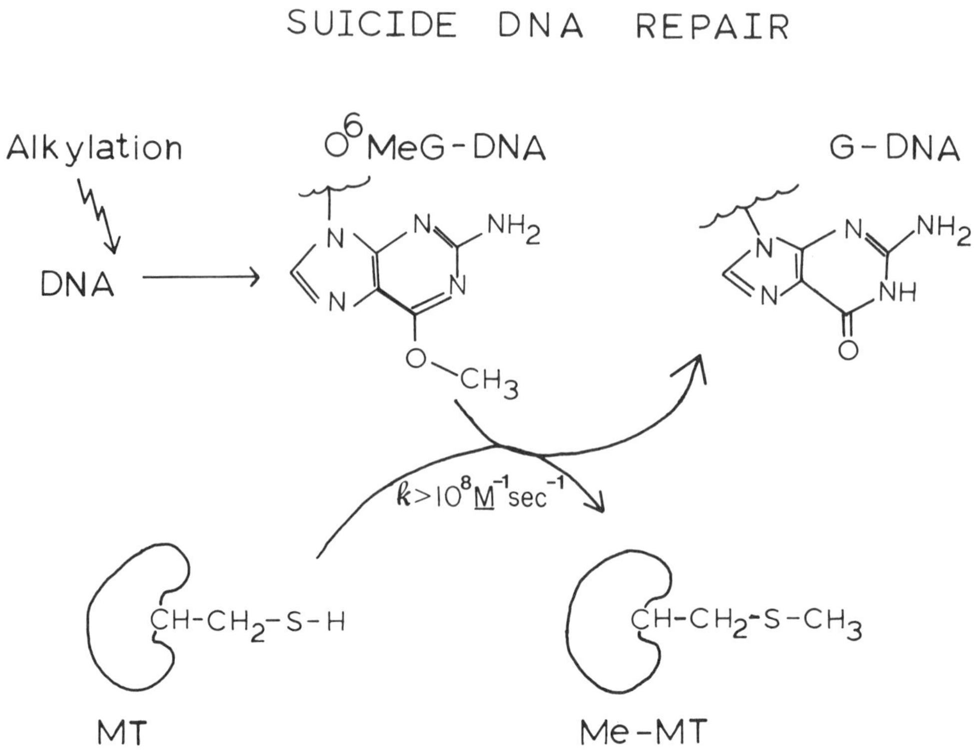

FIGURE 2. Suicide DNA repair to correct O^{6}MeG. An overall bimolecular reaction occurs in which the repair protein rapidly locates and transfers the guanine O^6 methyl group to one of its own cysteine residues. This leads to direct regeneration of normal guanine in DNA and the irreversible self-methylation of the protein itself.

product in animals treated with alkylating agents,[14] but never before as the product of an enzymatic reaction. It was suggested that the methylated cysteine and methyl transferring activity were in fact part of the same protein, with the transfer leading to irreversible inactivation of the transferase itself.[12]

III. PROPERTIES OF INDUCIBLE *E.COLI* O^{6}MEG-DNA METHYLTRANSFERASE

A. PHYSICAL PROPERTIES

The purification of the inducible *E. coli* repair protein, termed O^{6}MeG-DNA methyltransferase, revealed that a single polypeptide of M_r ~19,000 was sufficient for rapid and specific O^{6}MeG repair.[15] The apparent size of this polypeptide was unchanged upon methylation. Under reducing conditions, the highly purified transferase exhibited substantial heat stability, with a half-life of about 45 min at 95°C.[15] Consequently, it was unlikely that a mere acceptor of the O^{6}MeG methyl group had been purified, contaminated by a trace of a separate, enzymatic methyltransferase that happened to have the same thermostability. Instead, the methylated protein substrate and the transferring activity are one and the same molecule, each capable of scavenging just one methyl group from the guanine O^6 position (Figure 3). The 19-kDa protein contains six cysteines, one of which acts as the unique active residue for methyl transfer.[16] It is now known that this O^{6}MeG methyltransferase is actually the separated C-terminal domain of a 39-kDa protein, the primary product of the *ada* gene that regulates the adaptive response[16,17] (see Section IV). The Ada protein thus functions in both the antimutagenic repair of O^{6}MeG and in gene regulation.

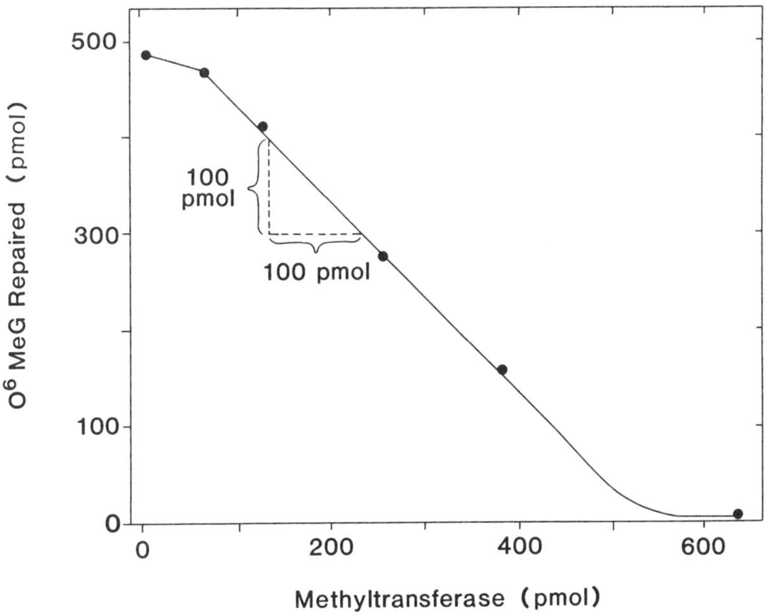

FIGURE 3. Stoichiometry of O[6]MeG repair by the purified 19-kDa O[6]MeG methyltransferase suizyme of *E. coli*. Only one methylated base is repaired for each methyltransferase molecule present. The amount of O[6]MeG in the DNA was determined by HPLC using fluorescence detection as described by Robins et al.[36] The concentration of the methyltransferase was determined by protein assay with Coomassie brilliant blue, by spectrophotometric measurement, and by amino acid analysis; all three determinations agreed to within 10%. (Taken from unpublished experiments by B. Demple and T. Lindahl.)

B. SUICIDE KINETICS

The self-methylating activity of the purified *E. coli* O[6]MeG-DNA methyltransferase does not exhibit any detectable cofactor requirement.[15] This irreversible, covalent transfer to an enzyme gives rise to the saturation kinetics that characterize O[6]MeG repair *in vivo*.[9] The maximum amount of methylated base repaired by the purified protein *in vitro* is strictly limited by the amount of active methyltransferase present in the reaction.[18] Just as in the case of the O[6]MeG repair in living cells,[10] there is no indication that active methyltransferase is regenerated *in vitro*.[18] It does not seem, therefore, that any additional protein component is missing from this repair system. Similarly, no specific activating or inhibiting effect could be detected for various small molecules, such as $MgCl_2$, ATP, or spermidine, while monovalent salts at concentrations above 50 m*M* decreased the reaction rate on duplex DNA nonspecifically.[18]

The consumption of the methyltransferase as a consequence of O[6]MeG repair means that the protein is not a true enzyme: the reaction is not catalytic. O[6]MeG methyltransferase does, however, satisfy the other major criteria for an enzyme. The protein acts in a highly specific fashion, lowering considerably the activation energy for an otherwise unfavorable reaction. The analogy has been drawn[18] between the suicidal methyltransferase reaction and convential covalent catalysis that terminates with hydrolysis of product from the enzyme. In the suicide reactions, this hydrolysis does not occur at a significant rate. These proteins that carry out stoichiometric reactions warrant a description that indicates their relation to true enzymes, and so we have called them suicide enzymes.[18] A useful shorthand term that retains this reference is "suizymes".

The *E. coli* methyltransferase exhibits very rapid kinetics when it acts upon O[6]MeG in duplex DNA at 37°C.[18] This fact necessitated the measurement of methyl transfer rates at lower temperatures (5 to 23°C) and extrapolation in order to determine an accurate rate for 37°C. At this temperature, 50% transfer is achieved in ≤1 s at a concentration of 10 n*M* each of methyltransferase and O^{6}MeG in DNA. The second-order rate constant for this reaction (≥10^8 $M^{-1}s^{-1}$) is rather faster than expected for a simple three-dimensional diffusion of the protein to the target site, indicating that the methyltransferase reaction might be augmented by facilitated diffusion of the protein along duplex DNA.[19] It is noteworthy that Cairns had previously calculated a similar repair rate *in vivo* of ~0.5 s, based on physiological studies.[20] The bacterial transferase is thus highly efficient at finding and acting upon its target damages in DNA (see Section VI. C).

Two mutant *E. coli* O^{6}MeG-DNA methyltransferases exhibited delayed reaction kinetics *in vitro*.[21] These proteins, the products of the *ada3* and *ada5* alleles,[17] scavenge O^{6}MeG methyl groups at rates that are 20- and 3000-fold slower, respectively, than the *ada*$^+$ gene product. The kinetic defects persisted in the isolated C-terminal 19-kDa O^{6}MeG methyltransferase domains of the Ada3 and Ada5 proteins, and thus are due to mutations in the 3′ half of the *ada* gene.[21] As expected, Southern analysis of the *ada3* and *ada5* genes revealed no detectable changes in their physical maps,[22] consistent with single amino acid substitutions causing the observed sluggishness of methyl group transfer. The mutation in the *ada3* gene has now been shown to be a single base pair change causing a single amino acid substitution in the C-terminal domain of the protein.[22] The mechanisms underlying the kinetic defects in the Ada3 and Ada5 proteins are unknown, but their elucidation may help illuminate the normal damage recognition and methyl transfer processes.

C. SUBSTRATE SPECIFICITY

The substrate specificity of the Ada O^{6}MeG methyltransferase of *E. coli* has been examined with respect to its relative preference for different methylated bases, its ability to act on small molecules, the effects of overall DNA structure, and the effects of sequence context on methyl transfer rates. The purified 19-kDa protein repairs not only O^{6}MeG, but also the corresponding pyrimidine analog O^4-methylthymine (O^{4}MeT) in DNA.[23,24] Other experiments indicated the presence in *E. coli* crude extracts of an inducible O^{4}MeT-DNA methyltransferase with M_r <25 kDa after self-methylation,[25] although this activity was not further identified. McCarthy et al.[26] had already demonstrated a probable *E. coli* methyltransferase for O^{4}MeT in crude extracts. The methylated species produced in this reaction had M_r 18 kDa, 20 kDa, and 39 kDa upon gel electrophoresis, along with a considerable amount of label migrating at the gel front. These results and recent *in vivo* studies[27] are all consistent with inducible O^{4}MeT repair mediated by the same suicidal methyltransferase that scavenges O^{6}MeG methyl groups (either the intact 39-kDa Ada protein or its 19-kDa O^{6}MeG methyltransferase domain). Although the presence of a separate inducible transferase for O^{4}MeT cannot be ruled out, there is no evidence for such a novel species; the multiple polypeptides observed by McCarthy et al.[26] are likely to have arisen from proteolysis occurring in the cell-free extracts. The O^2-methylated pyrimidines are not repaired by a methyltransferase, but rather are removed from DNA by an *N*-glycosylase.[23,25]

The purified *E. coli* O^{6}MeG methyltransferase fails to act on N-linked methyl groups of the DNA bases. These include N^6-methyladenine, which is a normal constituent of *E. coli* DNA, as well as the N7-methylguanine, N3-methyladenine, N1-methyladenine, and N3-methylthymine formed in DNA by alkylating agents.[15,18,23]

O^{6}MeG in duplex DNA is the best substrate for the *E. coli* Ada methyltransferase; single-stranded substrates are repaired about 1000-fold more slowly than duplex DNA.[18] Although the free O^{6}MeG base in large molar excess does not interfere detectably with the transferase activity upon methylated DNA,[18] there is some evidence that slow methyl transfer from free O^{6}MeG does occur. Yarosh et al.[28] reported inactivation of the purified *E. coli* methyltransferase during a long preincubation with O^{6}MeG, while the ribonucleoside O^6-methylguanosine was even more effective on a molar basis. From the lack of significant competition by the free base in the standard reaction, however, it may be calculated that the transfer rate from O^6MeG in DNA is $\geq 10^7$-fold more rapid than from the isolated base.[28]

In light of these data, it is not surprising that the purified bacterial transferase also acts on O^6MeG in synthetic oligonucleotides.[29-31] The repair rate increased with the length of the oligonucleotide substrate, with the transfer from a dodecanucleotide being about 50-fold faster than from a tetranucleotide.[31] It was not determined whether the self-complementary oligonucleotides used in these experiments were mostly in the single-stranded or in the duplex state, and such differences could certainly account for a significant amount of the difference in transfer rates.

The use of synthetic oligonucleotides has also allowed an assessment of the effect on the methyltransferase of the DNA sequence context. Small (less than fourfold) differences have been observed in various *in vitro* systems.[29-33] The *in vivo* significance of such sequence effects is unclear, although certain nucleotides in the genes of *E. coli*[7] and mammalian cells[34] are hotter targets for alkylation mutagenesis than are other sites. These mutation hot spots could be products of the individual contributions of DNA alkylation, repair, and the biological expression of the mutated genes.

The observation of methyltransferase action on the O^6MeG free base and the ribonucleoside show that the protein's activity is not restricted to DNA. Indeed, the purified *E. coli* enzyme also removed methyl groups from O^6MeG in transfer RNA *in vitro*.[35] The rate of this latter reaction relative to that on DNA was not reported.

The bacterial transferase repairs various O^6-alkylated guanine derivatives, but with rates that decrease with increasing size of the substituent. Ethyl groups are scavenged from O^6-alkylguanine 100-fold more slowly than are methyl groups,[18] while hydroxyethyl groups are transferred 300-fold more slowly than methyl groups.[36] These results are consistent with physiological observations that the adaptive response to alkylating agents becomes less effective with increasing size of the O^6-alkyl substituent.[27,37,38]

IV. OTHER SUICIDE REPAIR METHYLTRANSFERASES IN *E. COLI*

A. PHOSPHOTRIESTER METHYLTRANSFERASE

The discovery of the suicidal methyltransferase repairing O^6MeG prompted a search for analogous self-methylating activities directed against other lesions in alkylated DNA. An important result of this search was the observation that incubation of crude extracts from MNNG-adapted *E. coli* with a substrate containing methyl phosphotriesters (MeP) as the predominant *O*-methylated products led to the labeling of a 39-kDa polypeptide and some smaller species, none of which was identified further.[26] Protein *S*-methylcysteines were the products in this reaction, which also had the strange feature of achieving only 50% completion, i.e., about half of the phosphotriesters were refractory to repair. Weinfeld et al.[39] confirmed this limitation on MeP repair in bacterial crude extracts, and showed that only one of the two stereoisomers of MeP, the Sp configuration, was subject to repair. The existence of MeP repair by the transfer of methyl groups from the DNA phosphodiester backbone was thus established, and it was clear from earlier studies[15,18] (see also Figure 3) that the responsible activity was not the 19-kDa O^6MeG methyltransferase.

The O^6MeG-DNA methyltransferase purified originally[15] is actually an active fragment of the 39-kDa Ada protein. This realization was arrived at from two initially unrelated observations. First, antiserum directed against the purified 19-kDa protein cross-reacted strongly with a 39-kDa polypeptide (the size of the *ada* gene product).[40,41] Second, synthetic oligodeoxynucleotides predicted from the pure 19-kDa methyltransferase polypeptide sequence hybridized to the cloned *ada* gene.[41] The identification of the O^6MeG methyltransferase as part of the Ada protein was ultimately confirmed unambiguously by matching the C-terminal half of the polypeptide predicted from the *ada* DNA sequence with the extensive amino acid sequence already obtained from the purified 19-kDa methyltransferase.[16]

The 19-kDa suizyme that was originally purified[15] is the product of proteolysis that occurs upon cell lysis but hardly at all *in vivo*.[41] The protease responsible for this cleavage (between the Ada residues lysine-178 and glutamine-179) and another hydrolysis of Ada (between residues lysine-129 and alanine-130) is controlled by the *E. coli htpR* gene that regulates the bacterial heat-shock response, and the partly purified activity has been shown to mediate these same cleavages *in vitro* (L. Grossman, personal communication). Such proteolysis could provide a mechanism to destroy the irreversibly activated Ada protein that results from MeP repair and so switch off the adaptive response. The possible interplay of different stress-response systems that this idea implies could also apply to other damage-inducible regulons.

This assignment of an assayable activity to the Ada protein soon led to its purification in two laboratories as an intact, 39-kDa O^{6}MeG transferase.[43,44] The availability of the purified Ada protein revealed that it contains a suicide methyltransferase for MeP. The new activity repaired only one of the two MeP stereoisomers, and resided in a site in the protein distinct from the O^{6}MeG-repairing activity.[44] The Ada protein is therefore a doubly suicidal enzyme, with two independent methyltransferase activities. MeP is repaired by methyl transfer to a unique cysteine near the Ada N-terminus,[45] while the cysteine residue closest to the protein C-terminus is active as a methyltransferase for O^{6}MeG and O^{4}MeT.[16]

The independence of the Ada active sites for the *O*-methylated bases (cysteine-321) and MeP (cysteine-69) has recently been confirmed by protein engineering to modify these residues individually.[46] It is the self-methylation by the MeP methyltransferase that converts Ada protein into a potent transcriptional activator.[2,3,47] Although the self-alkylation at the Ada O^{6}MeG transferase site alone is not sufficient to confer this gene-activating function,[2] such a second methylation could have regulatory effects in addition to mediating antimutagenic repair. For example, the mutational changes that cause sluggish O^{6}MeG methyl transfer in the Ada3 and Ada5 proteins do not affect their respective rates of MeP repair, but both mutant proteins exhibit impaired ability as transcriptional activators.[21]

Cloned *E. coli* DNA fragments encoding various methyltransferase activities were isolated by Margison et al. using a direct assay for proteins that become labeled upon interaction with [^{3}H]MNU-treated DNA.[48] A subcloned fragment encoding MeP transferase activity directed the synthesis of a species with apparent M_r = 13,000 for the self-methylated protein.[48] Inspection of the restriction map of this fragment indicates that it encodes the first ~90 amino acids of Ada, which includes the MeP methyltransferase site.[45,46] Ada protein seems to be organized as two domains connected by an exposed hinge region, so that active MeP and O^{6}MeG methyltransferase fragments can be derived from the protein by gentle proteolysis *in vitro*.[45,47] This hinge region contains the main target for the endogenous protease of *E. coli*.[41,42]

B. CONSTITUTIVE O^{6}MEG METHYLTRANSFERASE DISTINCT FROM ADA

One of the cloned DNA fragments isolated by Margison et al.[48] encodes an ~18-kDa polypeptide that scavenged methyl groups from O^{6}MeG, but not from MeP. This DNA segment has now been examined in greater detail and shown to encode a new methyltransferase protein.[49] This protein has been purified as a 19-kDa polypeptide that cross-reacts only weakly antigenically with the 19-kDa Ada methyltransferase. The nucleotide sequence of the coding region of this genc, called *ogt* (for *O*^{6}Me*G*-DNA methyl<u>t</u>ransferase), encodes a 19,000-Da polypeptide that has limited amino acid sequence homology to the C-terminal half of the Ada protein, with the similarities confined largely to the C-terminal 93 residues of the two proteins.[49] Of particular interest is the observation that the Ogt protein contains a sequence very similar in composition and position to the Ada O^{6}MeG methyltransferase active site.[16] This sequence could contain the reactive cysteine, but the actual active site of Ogt has yet to be determined.[49] The possible significance of this limited sequence similarity is discussed below (see Section VI).

Others have independently found that *E. coli* contains a second, *ada*-independent O^{6}MeG-DNA methyltransferase. A 19-kDa polypeptide that becomes methylated upon interacting with O^{6}MeG-DNA was found in extracts either from a previously identified spontaneous *ada* mutant (BS23; Reference 40), now known to carry a deletion of the *ada* gene,[50] or from a strain in which the *ada* was deliberately replaced by an antibiotic resistance gene.[51] The *ada*-independent suizyme also appears to act on O^{4}MeT residues in DNA,[50] but this activity has not yet been reported for the purified Ogt protein.[49]

The new O^{6}MeG-DNA methyltransferase was not induced by alkylating agents and probably accounts for a significant (but as yet undetermined) fraction of the total activity directed against O^{6}MeG in nonadapted bacteria.[49-51] This constitutive methyltransferase may therefore serve to counteract occasional, low-level production of O^{6}MeG in the genome, with the Ada protein mobilized by induction in the face of a more substantial methylation challenge. The possible sources of constant, low-level alkylation in the bacterial environment constitute an interesting and challenging question for contemporary toxicology. One such source may reside in the cell itself. The major intracellular methyl donor *S*-adenosylmethionine alkylates DNA nonezymatically to form products like those caused by the weak mutagen methyl methane sulfonate.[52,53] Other, more potent natural alkylating mutagens could await discovery.

V. SUICIDE METHYLTRANSFERASES IN OTHER ORGANISMS

Proteins that become methylated upon interacting with DNA treated with MNNG, MNU, or the analogous ethylating compounds have now been reported from a wide variety of organisms. Although these proteins share some features with the *E. coli* Ada methyltransferase, there are also major differences in many cases. Some of these differences foreshadowed later discoveries in *E. coli,* such as the finding of a second methyltransferase for O^{6}MeG.

A. BACTERIAL METHYLTRANSFERASES

Evidence has been presented[54] for a rapid, saturable repair system that acts on O^{6}MeG in *Salmonella typhimurium*. The capacity of this repair is not inducible by MNNG, but otherwise has characteristics similar to the *E. coli* methyltransferase system. Assuming a suicidal reaction, it is estimated that *S. typhimurium* contains about 100 molecules of O^{6}MeG repair protein per cell (J. Guttenplan, personal communication) that is found as a 19-kDa methylated protein after incubation of cell extracts with O^{6}MeG-containing DNA (L. Samson, personal communication). The size of the unmethylated protein has not yet been determined.

In addition to the suicide methyltransferases of *E. coli,* proteins that apparently scavenge the O^{6}MeG methyl group exist in two Gram-positive organisms, *Micrococcus luteus*[56,57] and *Bacillus subtilis*.[58-60] Both organisms exhibit an adaptive response to alkylating agents[55-58] that includes increased synthesis of activities that remove methyl groups from O^{6}MeG and other *O*-methylated products in DNA. *B. subtilis* mutants have been isolated which fall into two categories, affected either in all aspects of the adaptive response (including an inducible methylpurine-DNA glycosylase), or only in the induction of O^{6}MeG repair activity.[59]

Both *M. luteus* and *B. subtilis* present a complicated complement of apparently self-methylating proteins. In crude extracts of MNNG-adapted *B. subtilis,* three polypeptides of M_r 20,000, 22,000, and 27,000 become labeled upon interaction with [^{3}H]methylated or [^{14}C]ethylated DNA.[60] Both the 20- and the 22-kDa species acquire the methyl groups from O^{6}MeG, but only the 22-kDa protein is induced by MNNG. The amount of label transferred from a [^{3}H]MNU-treated poly(dT) substrate to the 27-kDa species (which is also inducible by MNNG) was consistent with the fraction of MeP present.[60] This result indicates that the 27-kDa protein might be an MeP transferase, although it cannot be ruled out that the protein also takes up O^{4}MeT methyl groups. Such questions will be best answered by examining these *B. subtilis* proteins in purified form.

Several activities that repair *O*-alkylation damages have been at least partially purified from *M. luteus*.[56] The substrate specificities of these activities are distinct and nonoverlapping. A 31-kDa protein scavenges methyl groups from O^{6}MeG, a 22-kDa protein from O^{4}MeT, and a 13-kDa protein from MeP.[56] As seen in other systems, the repair of the *O*-methylated bases can go to completion, while only 50% of MeP are subject to repair. It is not known whether the MeP repair found in *M. luteus* is stereospecific for its substrate as found for the *E. coli* Ada system. The 31-kDa O^{6}MeG-repairing protein has been purified to apparent homogeneity and has some properties reminiscent of the inducible *E. coli* O^{6}MeG methyltransferase, including a strong preference for double-stranded over single-stranded DNA.[56] The *M. luteus* methyltransferase is thus one of the clearer examples of a suicide enzyme, since the homogeneous protein methylates itself; homogeneous protein preparations have not generally been obtained otherwise.

Although the methyltransferase systems observed in Gram-negative bacteria at first seemed more complicated than seen in *E. coli*, the existence of multiple methyltransferases for DNA *O*-alkylations now seems to be the rule for bacteria that possess an adaptive response. Separate constitutive and inducible repair proteins for O^{6}MeG have been detected in several cases, while the MeP repair activities seem to be confined to inducible species. It is important to note, however, that the possible effects of proteolysis in producing the various active polypeptides have not been assessed except for *E. coli*.[42]

B. EUKARYOTIC METHYLTRANSFERASES

Activities that repair O^{6}MeG have been detected in a variety of eukaryotes. Most workers have focused on mammalian cells, and their results are described in a separate section below. Despite an earlier negative report,[61] extracts of the baker's yeast *Saccharomyces cerevisiae* contain a small amount of a ~25-kDa polypeptide that is labeled during incubation with DNA containing O^6-[^{3}H]MeG (L. Samson, personal communication). A low level of methyltransferase activity for O^{6}MeG in *Drosophila melanogaster* has also been described,[62] as have activities in extracts of fish livers.[63] None of the responsible proteins has been significantly purified from these organisms, so unambiguous demonstrations of the suicide mechanism have not been made, and the sizes of the unmethylated proteins are unknown.

C. MAMMALIAN METHYLTRANSFERASES

Initial reports described activities from rodent[64,65] and human[66] liver tissue and in human lymphocytes[67] that repaired O^6-alkyl guanine by alkyl transfer to protein cysteine residues. A number of laboratories subsequently partly purified the O^{6}MeG-DNA methyltransferases to various extents from rat liver,[68,69] cultured human cells[70,71] or human placenta.[72,73] Some of these preparations[68,69,73] contained activities that were purified in excess of 1000-fold, but no physically homogeneous mammalian methyltransferase has yet been reported. Consequently, the direct demonstration of suicide methyltransferase activity has been elusive. Abundant circumstantial evidence nevertheless points to a repair mechanism for O^{6}MeG in mammalian cells that is indistinguishable from that employed by the bacterial Ada protein, i.e., suicidal self-methylation. For example, the apparent size of the active and methylated forms of the proteins from rat liver and human lymphoblasts are unchanged by methylation with M_r, respectively, of ~19,000[68] and ~22,000.[70] As suggested previously for the *E. coli* enzyme,[15] this indicates that the transferase and methylated sites reside in the same molecule. The repair reactions with the mammalian proteins form *S*-methylcysteine[64,67,70] or *S*-ethylcysteine[65] in amounts equal to the amount of O^6-alkylation that disappears.[64,66,69]

The general properties of these proteins acting on O^{6}MeG in DNA are similar to the *E. coli* Ada methyltransferase. No special cofactor requirements have been identified, while the mammalian activities are sensitive to thiol reagents such as $ZnCl_2$ or *p*-chloromercuribenzoate and protected by low molecular weight thiols such as dithiothreitol.[69,74] Both the rat[69] and human[70] methyltransferases exhibit the strong preference for double-stranded over single-stranded DNA substrates and for methyl over ethyl O^6-adducts as seen previously for the bacterial transferase.[18] One important difference between the prokaryotic and eukaryotic proteins is that the latter apparently do not repair O^4MeT at a significant rate[75] (although a separate O^4MeT-repairing activity might have escaped detection).

The amount of the O^6MeG repair activity varies among different animal tissues in a way that correlates with the resistance of the individual organ to alkylation carcinogenesis.[31,76] Chemically or virally transformed mammalian cells in culture exhibit two general phenotypes with respect to their sensitivities to alkylation toxicity and mutagenesis,[77] and to the *in vivo* repair of O^6MeG.[78] The alkylation-sensitive or repair-deficient cell lines, known respectively as Mer^- or Mex^- (in contrast to Mer^+, for Methylation repair, or Mex^+, for Methyl excision), are generally also deficient in or devoid of methyltransferase activity for O^6MeG.[70,72,79,80] These biochemical phenotypes are not discrete, since cell lines expressing various levels of O^6MeG methyltransferase have been identified,[81] and fusion of human Mex^- and Mex^+ lymphoblastoid cell lines yielded hybrids producing levels of the activity intermediate between the parents.[82] Moreover, the high frequency of Mer^- strains among transformed cells[83] virtually rules out direct mutation of the transferase structural gene as the cause of the repair deficiency. Instead, expression of O^6MeG-DNA methyltransferase in mammalian cells may be strongly regulated by an epigenetic mechanism (e.g., DNA cytosine methylation) that allows differential synthesis in various tissues and is disturbed upon cellular transformation.

VI. THE CHEMISTRY OF SUICIDE METHYLTRANSFERASES

The suicidal self-methylations described here have caught the attention of enzymologists. Aside from the possible regulatory functions of the methylated proteins, two key questions concern the chemical mechanisms of transfer of stable *O*-linked methyl groups to the polypeptides, and the mechanisms of target recognition in DNA. We will consider these questions in order, using the Ada O^6MeG methyltransferase as a paradigm.

A. ACTIVE SITES

The active sites of both the O^6MeG[16] and the MeP[45,46] methyltransferase of the *E. coli* Ada protein have been identified. The transfer from O^6MeG is to the C-terminal-most cysteine (residue 321) of Ada, as identified in the purified 19-kDa transferase[15] and located in the polypeptide translated from the *ada* DNA sequence.[16] It is noteworthy that the *S*-methylcysteine-containing sequence resembles the active site of thymidylate synthase (TS), in which the corresponding cysteine residue acts as a nucleophile to attack the six-carbon of dUMP:[84]

```
                                          321
                                      (S-methyl)
Ada-O6MeG    N...-Ala-Ile-Val-Ile-Pro-Cys-His-Arg-...C
                                   |   |   |
E. coli TS   N...-Met-Ala-Leu-Ala-Pro-Cys-His-Ala-...C
                                          146
```

The Pro-Cys-His triad seen in this comparison is preserved in nearly all TS proteins, while a sequence strongly similar to the Ada O^6MeG methyltransferase active site shown here is similarly positioned in the constitutive *ogt* methyltransferase of *E. coli*.[49] These similarities could reflect a common mechanism for generating the nucleophilic thiolate anion of cysteine by forming an ion pair with the adjacent histidine imidazolium. Such an ion pair is central to the hydrolytic mechanism of the thiol protease papain.[85,86] Indeed, Pro-Cys-His is statistically infrequent among sequenced proteins or predicted protein products, suggesting that this trio might fulfill a special enzymatic role.[16] The related model is consistent with the observed active site of the Ada MeP methyltransferase centered on cysteine-69, where the adjacent lysine could promote formation of the thiolate:[45]

69

(S–methyl)

Ada–MeP N . . .–Ala–Gly–Phe–Arg–Pro–*Cys*–Lys–Arg–. . . C

No active site sequence has yet been identified for any of the eukaryotic methyltransferases.

The conserved histidine of TS probably does not activate the adjacent cysteine thiol. At least one TS has the cysteine-proximal histidine replaced by a valine residue.[87] More importantly, the corresponding histidine residue in the crystallographic structure of *Lactobacillus casei* TS does not contact the active site cysteine.[88] In the *L. casei* enzyme, the cysteine is likely activated by an arginine located elsewhere in the polypeptide chain.[88] The proximity of such a (probably) cationic group would promote ionization of the nearby cysteine thiol to form the thiolate nucleophile. There are certainly basic residues other than the cysteine-proximal histidine in Ada[16] and Ogt[49] that could fulfill this cysteine-activating function.

The role of the other active site residues in the suicide methyltransferases is unknown, but some informed guesses can be made. For example, the active site proline probably constrains the conformation of its cysteine neighbor to position the thiolate for attack on its substrate. Each Ada methyltransferase and the putative Ogt active site also harbor a number of basic residues that could help anchor these proteins to DNA phosphates.

B. THE ROLE OF THIOLS

As mentioned above, the methyl groups of O^6MeG and MeP present a large energy barrier to demethylation.[89] The suicide methyltransferases must therefore employ strong nucleophiles to scavenge these groups. Chemical studies of model compounds indicated that thiolates (such as CH_3-S^-) have much greater "carbon basicity" (ability to attack nonactivated carbons, as in methanol) than do oxygen- or nitrogen-containing compounds.[90,91] Consequently, the central role of a cysteine residue in repair by suicidal methyl transfer is not surprising. The importance of the active cysteine in thymidylate synthase is underscored by a recent report that replacement of this cysteine with serine yielded a modified *E. coli* TS with a turnover rate 5000-fold lower than the normal enzyme.[92] Although the modified TS enzyme does display some activity, its rate is probably insufficient for normal metabolism. In the case of O^6MeG methyltransferase, rapid repair is essential to avert the fixation of mutations during replicationup6 18,20 and might only be accomplished with sufficient speed using an enzyme thiolate. The engineering of modified Ada proteins to replace the active site cysteines with serines or other amino acids will help address this point. However, the mutant Ada3 and Ada5 proteins (which exhibit slow repair kinetics) both retain cysteine at the transfer site.[21,22]

Two possible general mechanisms for the suicide methyltransferases may be proposed (Figure 4) based on the known three-dimensional structures and mechanisms of papain[85,86] and thymidylate synthase.[84,88] The active thiolate-imidazolium ion pair of papain is in equilibrium with an inactive unpaired form;[86] the two residues are separated by 133 other residues in the

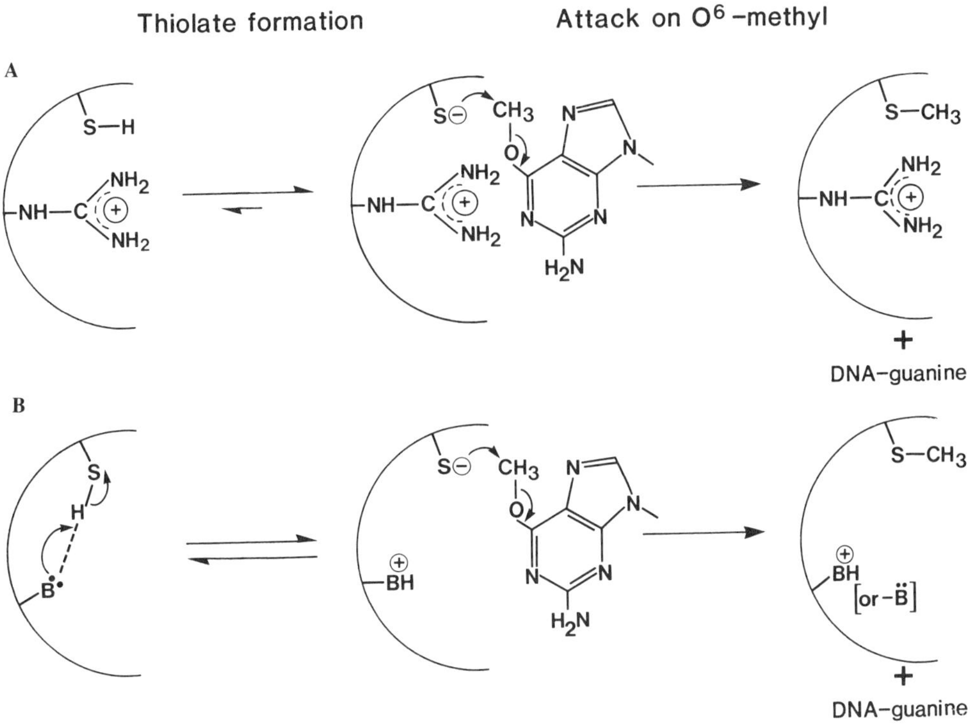

FIGURE 4. Proposed chemistry for O^{6}MeG repair by a suizyme. The possible source of the proton that returns to the guanine N1 position following demethylation is not indicated, but could be the proton that is thought to reside between the methylated guanine and the opposite cytosine (115) for the O^{6}MeG:C pair. (A) Based on the mechanism of thymidylate synthase.[84,88] (B) Based on the mechanism of papain.[85,86]

primary structure of the protein, but interact across a cleft in the folded protein.[85] The TS active site cysteine follows a proline residue (that breaks a small helical segment) and begins a stretch of antiparallel β sheet. That cysteine thiol interacts with an arginine side chain located 20 residues further toward the C-terminus of the β sheet.[88] The thiolate generated in this way attacks the unreactive 6-carbon of dUMP to initiate catalysis. In the proposed scenario for suicidal methyltransferases, the cysteine is activated as a nucleophile by a basic residue in the protein. If the active cysteine of the 19-kDa O^{6}MeG methyltransferase initiates a stretch of β sheet as in TS, the activating residue could be arginine-323, whose side chain would be *cis* to the cysteine sulfhydryl just two residues away.[16] An arginine residue is positioned similarly with respect to the active cysteine of Ada that scavenges MeP methyl groups[16,45] and in the probable active site of the putative *ogt* gene product.[49] The cysteine thiolate generated in this way could then directly attack the oxygen-linked alkyl groups in DNA, resulting in alkyl transfer to the protein cysteine (Figure 4). Stereochemical inversion of an isotopically labeled O^6-methyl group is predicted for this "in-line" mechanism. The protein product, *S*-methylcysteine, is extremely stable chemically,[12] and its formation may account, albeit on a trivial level, for the irreversible self-methylation seen with these repair proteins. Indeed, neither *E. coli* nor human O^{6}MeG methyltransferase was able to repair the thiol analog of O^{6}MeG in DNA, S^6-methylthioguanine,[93] a further indication of the stability of thioethers like *S*-methylcysteine.

Carboxylmethyl esters of glutamate or aspartate formed during a hypothetical methyltransferase reaction would be more easily reversed than *S*-methylcysteine,[94] but the oxygens of these residues lack sufficient nucleophilicity to attack the saturated methyl carbon of O^{6}MeG. The use of an enzyme serine to form *O*-methylserine ether in the protein might be just as suicidal as the use of a cysteine sulfhydryl, given the apparent stabilty to hydrolysis of the former residue.[95]

The role of cysteine in the suicidal methyltransferases is clearly an active one. Thus, terms such as "methyl accepting", which often appear in the literature, are misnomers that cast the key active residues of these proteins in a falsely passive light. Preferable and more accurate descriptions for these suizymes include "methyl scavenging", "methyl attacking", or "methyl transferring".

In the absence of DNA, the O^{6}MeG methyl-scavenging function of the Ada protein is not appreciably inhibited *in vitro* by MNU or by the sulfhydryl reagent iodoacetic acid, but the activity is highly sensitive to the more hydrophobic thiol reagent, *N*-ethylmaleimide (Demple and Lindahl, unpublished data). These observations are consistent with a methyltransferase structure in which the putative thiolate is poised at the active site but not directly exposed to solvent, coming in contact with its target methyl group only when the protein is bound to DNA. A further indication of the native reactivity of the active site thiols of Ada emerges from a recent report that methyl iodide induces the adaptive response in *E. coli* without causing substantial amounts of mutagenic DNA alkylation.[96] Methyl iodide may thus react directly with cysteine-69 of Ada to trigger it as a transcriptional activator, although it has not been ruled out that this compound forms large amounts of MeP in DNA accompanied by very little O^{6}MeG or O^{4}MeT.

It is unclear why the methyltransferase repair system evolved to use the protein itself as the site of methyl transfer, rather than some disposable, low molecular weight cofactor such as glutathione. It is also puzzling why an alternative mechanism of demethylation of O^{6}MeG did not evolve, such as the removal of methanol in an enzymatic process. For example, the enzyme adenosine deaminase is able to demethylate O^{6}MeG-nucleosides to form methanol in a catalytic reaction.[97] It may be that either of the above alternative mechanisms would not meet the requirements for speed in O^{6}MeG repair. Moreover, the guanine 6-methoxy group would be shielded by DNA structure from direct nucleophilic attack. In contrast, such factors may be less important for the repair of MeP in DNA, but in this case the covalent self-modification of the Ada MeP methyltransferase serves a crucial function: gene activation.[2,3,47]

C. SUBSTRATE RECOGNITION BY SUICIDE ENZYMES

A major outstanding problem in the field of DNA repair is the question of how repair proteins identify their target damages in the context of a large amount of undamaged DNA. The analogous question has been addressed for some proteins that bind to specific control[99,100] and restriction sites[101] in DNA, but similar structural studies of DNA repair proteins are just beginning. The suizymes present important opportunities to address this problem head-on for DNA repair.

In addition to O^{6}MeG, the 19-kDa domain of the *E. coli* Ada methyltransferase recognizes O^{4}MeT in DNA, albeit with a reduced efficiency.[23,24] It has been reported that the *ada*-independent O^{6}MeG methyltransferase in crude extracts of *E. coli* (presumably the *ogt* gene product[49]) also acts on O^{4}MeT,[50] although its relative preference for the purine compared to the pyrimidine adduct has not been measured. In other bacteria, the jobs of O^{6}MeG and O^{4}MeT repair appear to be divided among different proteins,[56,60] while the mammalian activities do not appear to repair O^{4}MeT either *in vitro* or *in vivo*.[75]

This ability of the *E. coli* Ada suizyme (and apparently the Ogt protein) to act on *O*-methylated derivatives of both a purine and a pyrimidine indicates that the minimal structure recognized by the transferases could be a pyrimidine ring presenting an *O*-methyl group in the major groove of DNA. However, recent structural studies of synthetic oligodeoxynucleotides containing O^{4}MeT indicate that the *O*-methyl group is directed toward the purine in the opposite strand.[102,103] Recent data also imply the the O^{6}MeG methoxy is directed into the center of the helix (D. Patel, personal communication), in contrast to earlier conclusions.[104,105] Although they form weakened base pairs with bases in the complementary strand of duplexes, neither

O^4MeT[102,103] nor O^6MeG[104,105] cause gross distortion of DNA structure, which remains in the B form overall. These modified bases do, however, alter the conformation of the phosphodiester at the modification site in all but the O^6MeG:T pair; the latter appears undistorted.[102-105] Since detailed studies of suicide repair rates on a set of defined complexes with different bases opposite O^6MeG or O^4MeT have not been reported, it is unclear to what extent the reported subtle conformational differences might affect repair. Perhaps the suicide methyltransferases scan DNA searching for modest conformational changes and then sample for the presence of a base-linked *O*-methyl group. On the other hand, it has been pointed out[106] that the observed mutational specificity of alkylating agents might be more a result of local DNA conformational effects on the incorporation of nucleotides by DNA polymerase than on the specificity of repair.

Since *O*-methylated bases are normally absent from undamaged DNA (except in some of the T-even phages, whose DNA may contain glucosylated 5-hydroxymethylcytosine residues), their recognition would supply the needed discrimination for the modified bases. Obviously the overall secondary structure of DNA does play a role, in that all of the enzymes reported show a strong preference for duplex over single-stranded DNA, and the rate of repair of synthetic oligonucleotides increases with increasing length of the substrate. Nevertheless, most reports agree that these same suicidal DNA methyltransferases are also capable of self-inactivation upon interaction with the O^6MeG free base.[28,107] Although self-methylation has not been unambiguously demonstrated as the cause of this inactivation, the HeLa O^6MeG transferase was reported to generate [^{3}H]guanine from O^6[^{3}H]MeG.[107]

The 19-kDa Ada methyltransferase can act on O^6MeG in tRNA,[35] a further indication of that protein's ability to recognize O^6-methylated guanine in a variety of structural contexts. It seems equally clear that reactions of the methyltransferase with substrates other than DNA (modified free bases, nucleosides, mononucleotides or RNA) do not constitute important reactions in living bacteria, since the number of O^6MeG repaired *in vivo*[9,10] and the number of methyltransferase molecules present in cell extracts[13,15] correlates almost exactly.

As mentioned earlier, both the bacterial and the mammalian alkyltransferases can carry out the transfer of larger groups from the guanine O^6 or the thymine O^4 positions, but at rates that decrease dramatically for ethyl and larger substituents. This difference could be due to steric problems, with the larger groups preventing access of the putative cysteine thiolate to the *O*-linked carbon. Alternatively, the ethyl and larger groups on the guanine O^6 or the thymine O^4 positions could adopt the *anti* conformation and be directed into the major groove in a way that impedes repair.

In the case of DNA damaged with 1,3-bis(2-chloroethyl)-1-nitrosourea, a cross-linking agent, intervention by the bacterial[36] or human[108] O^6MeG methyltransferase prevents cross-link formation between the strands of DNA, but may leave the protein itself covalently bound to the alkylated DNA.[108] Suicidal alkyl transfer in such cases may thus generate a new repair problem — the need to remove a DNA-protein cross-link.

Detailed understanding is beginning to emerge of the atomic interactions between various sequence-binding proteins and their specific target sequences in DNA.[99,101] Such information has not yet been obtained for DNA repair proteins that must largely ignore the sequence context of the damages they recognize. Some progress has been made for the *E. coli* photolyase that splits UV-induced pyrimidine dimers through the identification of the protein's contact sites in DNA using various footprinting techniques.[109] For O^6MeG methyltransferase such studies have not been reported and would likely be difficult with O^6MeG-containing substrates, since the normal suizyme carries out methyl transfer without any cofactors and even at low temperatures.[18] The use of modified substrates or altered proteins could obviate this problem, but would first require verification that the protein and DNA interact specifically.

Recently, we have obtained X-ray diffraction-quality crystals of the 19-kDa Ada O^6MeG methyltransferase domain.[110] The solution of the crystal structure of this paradigmatic protein will pave the way for elucidating the details of the damage-recognition mechanism of O^6MeG transferases, but a full understanding will also require co-crystals of the protein and the DNA substrate. Such work is underway.

The mechanism by which O^{6}MeG methyltransferase locates modified guanines in DNA is of great interest. A little information is available on this point from kinetic studies. As mentioned above, the 19-kDa Ada methyltransferase repairs O^{6}MeG in duplex DNA with a second-order rate constant $\geq 10^8\ M^{-1}s^{-1}$. This number indicates an efficient search and transfer process.[111] Such fast rates have been observed for other proteins acting at specific sites in DNA, such as bacterial gene regulatory proteins[111] and the *Eco*RI restriction endonuclease.[112] For these proteins, the search for the target site in DNA has been modeled by a facilitated transfer mechanism, in which the bound protein exchanges between domains of the DNA molecule[111] and perhaps also diffuses in one dimension along the DNA within domains.[19] Unfortunately, analogous kinetic data have not been reported for the Ada MeP or the Ogt O^{6}MeG methyltransferases.

VII. NATURAL ENZYME SUICIDE AS A CELLULAR DEFENSE

The notion that a whole protein molecule must be consumed to repair one lesion in DNA usually prompts the reaction that this is a lavish process that is inconsistent with the common view of cells as frugal spenders of energy. For the potently mutagenic lesion O^{6}MeG however, such apparent wastefulness is probably the by-product of the need to repair the lesion rapidly. This speed seems in turn to require the use of a strongly nucleophilic cysteine residue, leading to the formation of the hard-to-reverse *S*-methylcysteine in the protein. This investment of a relatively large amount of energy in a single repair cycle is not without a counterpart: it has been pointed out[47] that up to 1000 nucleotides may be removed and resynthesized in order to correct a single mismatch in the DNA of *E. coli.*[113] Moreover, cellular proteins normally turn over at various rates,[114] so that the occasional sacrifice of a small protein is not an undue burden.

The exploration of the control and mechanisms of suicide DNA repair have provided novel insights into cellular responses to mutagenic agents, gene control, and enzyme specificity. Continued new understanding is to be expected from this area of research, with the promise of future discoveries that may well have significance beyond the field of DNA repair.

ACKNOWLEDGMENTS

I am indebted to John Essigmann, Larry Hardy, Ed Loechler, and Leona Samson for their thoughtful and critical reading of the manuscript. I also wish to acknowledge Ed Loechler for numerous insightful discussions over the last several years about the structure of alkylated DNA and the mechanism of suicidal methyl transfer. Obviously, any errors and oversights that the reader might discover in this chapter are mine alone. Work in the author's laboratory was supported by start-up funds from the Department of Biochemistry and Molecular Biology, Harvard University, from the Clark Fund of Harvard University, and from the Dreyfus Foundation. B.D. is a Dreyfus Foundation Teacher-Scholar.

REFERENCES

1. **Singer, B.,** The chemical effects of nucleic acid alkylation and their relation to mutagenesis and carcinogenesis, *Prog. Nucleic Acids Res. Mol. Biol.*, 15, 219, 1975.
2. **Teo, I., Sedgwick, B., Kilpatrick, M. W., McCarthy, T. V., and Lindahl, T.,** The intracellular signal for induction of resistance to alkylating agents in *E. coli*, *Cell*, 45, 315, 1986.
3. **Nakabeppu, Y. and Sekiguchi, M.,** Regulatory mechanisms for induction of synthesis of repair enzymes in response to alkylating agents: Ada protein acts as a transcriptional regulator, *Proc. Natl. Acad. Sci. U.S.A.*, 83, 6297, 1986.

4. **Samson, L. and Cairns, J.,** A new pathway for DNA repair in *Escherichia coli, Nature,* 267, 281, 1977.
5. **Schendel, P. F., Defais, M., Jeggo, P., Samson, L., and Cairns, J.,** Pathways of mutagenesis and repair in *Escherichia coli* exposed to low levels of simple alkylating agents, *J. Bacteriol.,* 135, 466, 1978.
6. **Cairns, J., Robins, P., Sedgwick, B., and Talmud, P.,** The inducible repair of alkylated DNA, *Prog. Nucleic Acids Res. Mol. Biol.,* 26, 237, 1981.
7. **Coulondre, C. and Miller, J. H.,** Genetic studies of the *lac* Repressor IV: mutagenic specificity in the *lacI* gene of *Escherichia coli, J. Mol. Biol.,* 117, 577, 1977.
8. **Loveless, A.,** Possible relevance of O-6 alkylation of deoxyguanosine to the mutagenicity and carcinogenicity of nitrosamines and nitrosamides, *Nature,* 233, 206, 1969.
9. **Schendel, P. F. and Robins, P. E.,** Repair of O^6-methylguanine in adapted *Escherichia coli, Proc. Natl. Acad. Sci. U.S.A.,* 75, 6017, 1978.
10. **Robins, P. and Cairns, J.,** Quantitation of the adaptive response to alkylating agents, *Nature,* 280, 74, 1979.
11. **Karran, P., Lindahl, T., and Griffin, B.,** Adaptive response to alkylating agents involves alteration *in situ* of O^6-methylguanine residues in DNA, *Nature,* 280, 76, 1979.
12. **Olsson, M. and Lindahl, T.,** Repair of alkylated DNA in *Escherichia coli.* Methyl transfer from O^6-methylguanine to a protein cysteine residue, *J. Biol. Chem.,* 255, 10569, 1980.
13. **Foote, R. S., Mitra, S., and Pal, B. C.,** Demethylation of O^6-methylguanine in a synthetic DNA polymer by an inducible activity in *Escherichia coli, Biochem. Biophys. Res. Commun.,* 97, 654, 1980.
14. **Craddock, V. M.,** Reaction of the carcinogen dimethylnitrosamine with proteins and with thiol compounds in the intact animal, *Biochem. J.,* 94, 323, 1965.
15. **Demple, B., Jacobsson, A., Olsson, M., Robins, P., and Lindahl, T.,** Repair of alkylated DNA in *Escherichia coli.* Physical properties of O^6-methylguanine-DNA methyltransferase, *J. Biol. Chem.,* 257, 13776, 1982.
16. **Demple, B., Sedgwick, B., Robins, P., Totty, N., Waterfield, M. D., and Lindahl, T.,** Active site and complete sequence of the suicidal methyltransferase that counters alkylation mutagenesis, *Proc. Natl. Acad. Sci. U.S.A.,* 82, 2688, 1985.
17. **Jeggo, P.,** Isolation and characterization of *Escherichia coli* K-12 mutants unable to induce the adaptive response to simple alkylating agents, *J. Bacteriol.,* 139, 783, 1979.
18. **Lindahl, T., Demple, B., and Robins, P.,** Suicide inactivation of the *E. coli* O^6-methylguanine-DNA methyltransferase, *EMBO J.,* 1, 1359, 1982.
19. **Berg, O. G., Winter, R. B., and von Hippel, P. H.,** Diffusion-driven mechanisms of protein translocation on nucleic acids. I. Models and theory, *Biochemistry,* 20, 6929, 1981.
20. **Cairns, J.,** Efficiency of the adaptive response of *Escherichia coli* to alkylating agents, *Nature,* 286, 176, 1980.
21. **Demple, B.,** Mutant *Escherichia coli* Ada proteins simultaneously defective in the repair of O^6-methylguanine and in gene activation, *Nucleic Acids Res.,* 14, 5575, 1986.
22. **Abou-Zamzam, A. and Demple, B.,** manuscript in preparation.
23. **McCarthy, T. V., Karran, P., and Lindahl, T.,** Inducible repair of O–alkylated DNA pyrimidines in *Escherichia coli, EMBO J.,* 3, 545, 1984.
24. **Dolan, M. E., Scicchitano, D., Singer, B., and Pegg, A. E.,** Comparison of repair of methylated pyrimidines in poly(dT) by extracts from rat liver and *Escherichia coli, Biochem. Biophys. Res. Commun.,* 123, 324, 1984.
25. **Ahmmed, Z. and Laval, J.,** Enzymatic repair of O-alkylated thymidine residues in DNA: involvement of an O^4-methylthymine–DNA methyltransferase and a O^2-methylthymine DNA glycosylase, *Biochem. Biophys. Res. Commun.,* 120, 1, 1984.
26. **McCarthy, J. G., Edington, B. V., and Schendel, P. F.,** Inducible repair of phosphotriesters in *Escherichia coli, Proc. Natl. Acad. Sci. U.S.A.,* 80, 7380, 1983.
27. **Samson, L., Thomale, J., and Rajewsky, M.,** Alternative pathways for the *in vivo* repair of O^6-alkylguanine and O^4-alkylthymine in *E. coli:* the adaptive response and nucleotide excision repair, *EMBO J.,* in press.
28. **Yarosh, D. B., Hurst-Calderone, S., Babich, M. A., and Day, R. S., III,** Inactivation of O^6-methylguanine-DNA methyltransferase and sensitization of human tumor cells to killing by chloroethylnitrosourea by O^6-methylguanine as a free base, *Cancer Res.,* 46, 1663, 1986.
29. **Scicchitano, D., Jones, R. A., Kuzmich, S., Gaffney, B., Lasko, D. D., Essigmann, J. M., and Pegg, A. E.,** Repair of oligodeoxynucleotides containing O^6-methylguanine by O^6-alkylguanine-DNA alkyltransferase, *Carcinogenesis,* 7, 1383, 1986.
30. **Graves, R. J., Li, B. F. L., and Swann, P. F.,** Repair of synthetic oligonucleotides containing O^6-methylguanine, O^6-ethylguanine or O^4-methylthymine by O^6-alkyl-DNA alkyltransferase, in *Relevance of N-Nitroso Compounds to Human Cancer: Exposures and Mechanisms,* Bartsch, H., O'Neill, I.K., and Schulte-Hermann, Eds., International Agency for Research on Cancer, Lyons, 1987, 41.
31. **Dolan, M. E., Scicchitano, D., and Pegg, A. E.,** Use of oligodeoxynucleotides containing O^6-alkylguanine for the assay of O^6-alkylguanine-DNA alkyltransferase activity, *Cancer Res.,* 48, 1184, 1988.
32. **Boiteux, S., de Olivera, R. C., and Laval, J.,** The *Escherichia coli* O^6-methylguanine-DNA methyltransferase does not repair promutagenic O^6-methylguanine residues when present in Z-DNA, *J. Biol. Chem.,* 260, 8711, 1985.

33. **Topal, M. D., Eadie, J. S., and Conrad, M.,** O^6-methylguanine mutation and repair is nonuniform. Selection for DNA most interactive with O^6-methylguanine, *J. Biol. Chem.,* 261, 9879, 1986.
34. **Zarbl, H., Sukumar, S., Arthur, A. V., Martin-Zanca, D., and Barbacid, M.,** Direct mutagenesis of Ha-*ras*-1 oncogenes by *N*-nitroso-*N*-methylurea during initiation of mammary carcinogenesis in rats, *Nature,* 315, 382, 1985.
35. **Karran, P.,** Possible depletion of a DNA repair enzyme in human lymphoma cells by subversive repair, *Proc. Natl. Acad. Sci. U.S.A.,* 82, 5285, 1985.
36. **Robins, P., Harris, A. L., Goldsmith, I., and Lindahl, T.,** Cross-linking of DNA induced by chloroethylnitrosourea is prevented by O^6-methylguanine-DNA methyltransferase, *Nucl. Acids Res.,* 11, 7743, 1983.
37. **Sedgwick, B. and Lindahl, T.,** A common mechanism for repair of O^6-methylguanine and O^6-ethylguanine in DNA, *J. Mol. Biol.,* 154, 169, 1982.
38. **Todd, M. L. and Schendel, P. F.,** Repair and mutagenesis in *Escherichia coli* K-12 after exposure to various alkylnitrosoguanidines, *J. Bacteriol.,* 156, 6, 1983.
39. **Weinfeld, M., Drake, A. F., Saunders, J. K., and Paterson, M. C.,** Stereospecific removal of methyl phosphotriesters from DNA by an *Escherichia coli ada*$^+$ extract, *Nucleic Acids Res.,* 13, 7067, 1985.
40. **Sedgwick, B.,** Molecular cloning of a gene which regulates the adaptive response to alkylating agents in *Escherichia coli, Mol. Gen. Genet.,* 191, 466, 1983.
41. **Teo, I., Sedgwick, B., Demple, B., Li, B., and Lindahl, T.,** Induction of resistance to alkylating agents in *E. coli:* the *ada*$^+$ gene product serves both as a regulatory protein and as an enzyme for repair of mutagenic damage, *EMBO J.,* 3, 2151, 1984.
42. **Teo, I.,** Proteolytic processing of the Ada protein that repairs DNA O^6-methylguanine in *E. coli, Mut. Res.,* 183, 123, 1987.
43. **Nakabeppu, Y., Kondo, H., Kawabata, S., Iwanaga, S., and Sekiguchi, M.,** Purification and structure of the intact Ada regulatory protein of *Escherichia coli* K-12, O^6-methylguanine-DNA methyltransferase, *J. Biol. Chem.,* 260, 7281, 1985.
44. **McCarthy, T. V. and Lindahl, T.,** Methyl phosphotriesters in alkylated DNA are repaired by the Ada regulatory portein of *E. coli, Nucleic Acids Res.,* 13, 2683, 1985.
45. **Sedgwick, B., Robins, P., Totty, N., and Lindahl, T.,** Functional domains and methyl acceptor sites of the *Escherichia coli* Ada protein, *J. Biol. Chem.,* 263, 4430, 1988.
46. **Takano, K., Nakabeppu, Y., and Sekiguchi, M.,** Functional sites of the Ada regulatory protein of *Escherichia coli, J. Biol. Chem.,* 200, in press.
47. **Lindahl, T., Sedgwick, B., Sekiguchi, M., and Nakabeppu, Y.,** Regulation and expression of the adaptive response to alkylating agents, *Annu. Rev. Biochem.,* 57, 133, 1988.
48. **Margison, G. P., Cooper, D. P., and Brennand, J.,** Cloning of the *E. coli* O^6-methylguanine and methylphosphotriester methyltransferase gene using a functional DNA repair assay, *Nucleic Acids Res.,* 13, 1939, 1985.
49. **Potter, P. M., Wilkinson, M. C., Fitton, J., Carr, F. J., Brennand, J., Cooper, D. P., and Margison, G. P.,** Characterisation and nucleotide sequence of *ogt,* the O^6-alkylguanine-DNA-alkyltransferase gene of *E. coli, Nucleic Acids Res.,* 15, 9177, 1987.
50. **Rebeck, G. W., Coons, S., Carroll. P., and Samson, L.,** A new DNA methyltransferase repair enzyme in *E. coli, Proc. Natl. Acad. Sci. U.S.A.,* 85, 3039, 1988.
51. **Shevell, D., Abou-zamzam, A., Demple, B., and Walker, G. C.,** Construction of an *Escherichia coli ada* deletion by gene replacement in a *recD* strain reveals a second methyltransferase that repairs alkylated DNA, *J. Bacteriol.,* 170,3294, 1988.
52. **Rydberg, B. and Lindahl, T.,** Nonenzymatic methylation of DNA by the intracellular methyl group donor S-adenosyl-L-methionine is a potentially mutagenic reaction, *EMBO J.,* 1, 211, 1982.
53. **Barrows, L. R. and Magee, P. N.,** Nonenzymatic methylation of DNA by S-adenosylmethionine *in vitro, Carcinogenesis,* 3, 349, 1982.
54. **Guttenplan, J. B. and Milstein, S.,** Resistance of *Salmonella typhimurium* TA 1535 to O^6-guanine methylation and mutagenesis induced by low doses of N-methyl-N′-nitro-N-nitrosoguanidine: an apparent constitutive repair activity, *Carcinogenesis,* 3, 327, 1982.
55. **Ather, A., Ahmed, Z., and Riazuddin, S.,** Adaptive response of *Micrococcus luteus* to alkylating chemicals, *Nucleic Acids Res.,* 12, 2111, 1984.
56. **Riazuddin, S., Athar, A., and Sohail, A.,** Methyl transferases induced during chemical adaptation of *M. luteus, Nucleic Acids Res.,* 15, 9471, 1987.
57. **Hadden, C. T., Foote, R. S., and Mitra, S.,** Adaptive response of *Bacillus subtilis* to N-methyl-N′-nitro-N-nitrosoguanidine, *J. Bacteriol.,* 153, 756, 1983.
58. **Morohoshi, F. and Munakata, N.,** Adaptive response to simple alkylating agents in *Bacillus subtilis* cells, *Mut. Res.,* 110, 23, 1983.
59. **Morohoshi, F. and Munakata, N.,** Two classes of *Bacillus subtilis* mutants deficient in the adaptive response to simple alkylating agents, *Mol. Gen. Genet.,* 202, 200, 1986.

60. **Morohoshi, F. and Munakata, N.,** Multiple species of *Bacillus subtilis* DNA alkyltransferase involved in the adaptive response to simple alkylating agents, *J. Bacteriol.*, 169, 587, 1987.
61. **Maga, J. A. and McEntee, K.,** Response of *S. cerevisiae* to N-methyl-N′-nitro-N-nitrosoguanidine mutagenesis, survival, and *DDR* gene expression, *Mol. Gen. Genet.*, 200, 313, 1985.
62. **Green, D. A. and Deutsch, W. A.,** Repair of alkylated DNA: *Drosophila* have DNA methyltransferases but not DNA glycosylases, *Mol. Gen. Genet.*, 192, 322, 1983.
63. **Nakatsuru, Y., Nemoto, N., Nakagawa, K., Masahito, P., and Ishikawa, T.,** O^6-Methylguanine-DNA methyltransferase activity in liver from various fish species, *Carcinogenesis*, 8, 1123, 1987.
64. **Bogden, J. M., Eastman, A., and Bresnick, E.,** A system in mouse liver for the repair of O^6-methylguanine lesions in methylated DNA, *Nucleic Acids Res.*, 9, 3089, 1981.
65. **Mehta, J., Ludlum, D., Renard, A., and Verly, W. G.,** Repair of O^6-ethylguanine in DNA by a chromatin fraction from rat liver: transfer of the ethyl group to an acceptor protein, *Proc. Natl. Acad. Sci. U.S.A.*, 78, 6766, 1981.
66. **Pegg, A. E., Roberfroid, M., von Bahr, C., Foote, R. S., Mitra, S., Bresil, H., Likhachev, A., and Montesano, R.,** Removal of O^6-methylguanine from DNA by human liver fractions, *Proc. Natl. Acad. Sci. U.S.A.*, 79, 5162, 1982.
67. **Waldstein, E. A., Cao, E.-H., Miller, M. E., Cronkite, E. P., and Setlow, R. B.,** Extracts of chronic lymphocytic leukemia lymphocytes have a high level of DNA repair activity for O^6methylguanine, *Proc. Natl. Acad. Sci. U.S.A.*, 79, 4786, 1982.
68. **Hora, J. F., Eastman, A., and Bresnick, E.,** O^6-methylguanine methyltransferase in rat liver, *Biochemistry*, 22, 3759, 1983.
69. **Pegg, A. E., Wiest, L., Foote, R. S., Mitra. S., and Perry, W.,** Purification and properties of O^6-methylguanine-DNA transmethylase from rat liver, *J. Biol. Chem.*, 258, 2327, 1983.
70. **Harris, A. L., Karran, P., and Lindahl, T.,** O^6-methylguanine-DNA methyltransferase of human lymphoid cells: structural and kinetic properties and absence in repair-deficient cells, *Cancer Res.*, 43, 3247, 1983.
71. **Brent, T. P.,** Isolation and purification of O^6-alkylguanine-DNA alkyltransferase from human leukemic cells. Prevention of chloroethylnitrosourea-induced cross-links by purified enzyme, *Pharmacol. Ther.*, 31, 121, 1985.
72. **Yarosh, D. B., Rice, M., Day, R. S., III, Foote, R. S., and Mitra, S.,** O^6-Methylguanine-DNA methyltransferase in human cells, *Mut. Res.*, 131, 27, 1984.
73. **Boulden, A. M., Foote, R. S., Fleming, G. S., and Mitra, S.,** Purification and some properties of human DNA-O^6-methylguanine methyltransferase, *J. Biosci.*, 11, 215, 1987.
74. **Renard, A., Verly, W. G., Mehta, J. R., and Ludlum, D. B.,** Properties of the chromatin repair activity against O^6-ethylguanine lesions in DNA, *Eur. J. Biochem.*, 136, 461, 1983.
75. **Brent, T. P., Dolan, M. E., Fraenkel-Conrat, H., Hall, J., Karran, P., Laval, F., Margison, G. P., Montesano, R., Pegg, A. E., Potter, P. M., Singer, B., Swenberg, J. A., and Yarosh, D. B.,** Repair of O-alkylpyrimidines in mammalian cells: a present consensus, *Proc. Natl. Acad. Sci. U.S.A.*, 85, 1759, 1988.
76. **Goth, R. and Rajewsky, M. F.,** Persistence of O^6-ethylguanine in rat-brain DNA: correlation with nervous system-specific carcinogenesis by ethylnitrosourea, *Proc. Natl. Acad. Sci. U.S.A.*, 71, 639, 1974.
77. **Day, R. S., III and Ziolkowski, C.,** Human brain tumor strains with deficient host-cell reactivation of MNNG-damaged adenovirus 5, *Nature*, 279, 797, 1979.
78. **Sklar, R. and Strauss, B.,** Removal of O^6-methylguanine from DNA of normal and xeroderma pigmentosum-derived lymphoblastoid lines, *Nature*, 289, 417, 1981.
79. **Yarosh, D. B., Foote, R. S., Mitra, S., and Day, R. S., III,** Repair of O^6-methylguanine in DNA by demethylation is lacking in *Mer*$^-$ human tumor cell strains, *Carcinogenesis*, 4, 199, 1983.
80. **Foote, R. S., Pal, B. C., and Mitra, S.,** Quantitation of O^6-methylguanine-DNA methyltransferase in HeLa cells, *Mut. Res.*, 119, 221, 1983.
81. **Yarosh, D. B.,** The role of O^6-methylguanine-DNA methyltransferase in cell survival, mutagenesis and carcinogenesis, *Mut. Res.*, 145, 1, 1985.
82. **Ayres, K., Sklar, R., Larson, K., Lindgren, V., and Strauss, B.,** Regulation of the capacity for O^6-methylguanine removal from DNA in human lymphoblastoid cells studied by cell hybridization, *Mol. Cell. Biol.*, 2, 904, 1982.
83. **Day, R. S., III, Ziolkowski, C. H. J., Scudiero, D. A., Meyer, S. A., Lubiniecki, A. S., Girardi, A. J., Galloway, S. M., and Bynum, G. D.,** Defective repair of alkylated DNA by human tumor and SV40-transformed human cell strains, *Nature*, 288, 724, 1980.
84. **Maley, F., Belfort, M., and Maley, G.,** Probing the infrastructure of thymidylate synthase and deoxycytidylate deaminase, *Adv. Enzyme Regul.*, 22, 413, 1983.
85. **Drenth, J., Jansonius, J. N., Koekeok, R., Swen, H. M., and Wolthers, B. G.,** Structure of papain, *Nature*, 218, 929, 1968.
86. **Lewis, S. D., Johnson, F. A., and Schafer, J. A.,** Effect of cysteine-25 on the ionization of histidine-159 in papain as determined by proton nuclear magnetic resonance spectroscopy. Evidence for a His-159—Cys-25 ion pair and its possible role in catalysis, *Biochemistry*, 20, 48, 1981.

87. **Kenny, E., Atkinson, T., and Hartley, B. S.,** Nucleotide sequence of the thymidylate synthase gene (*thyP3*) from the *Bacillus subtilis* phage OT3, *Gene,* 34, 335, 1985.
88. **Hardy, L. W., Finer-Moore, J. S., Montfort, W. R., Jones, M. O., Santi, D. V., and Stroud, R. M.,** Atomic structure of thymidylate synthase: target for rational drug design, *Science,* 235, 448, 1987.
89. **Balsiger, R. W. and Montgomery, J. A.,** Synthesis of potential anticancer agents. XXV. Preparation of 6-alkoxy-2-aminopurines, *J. Org. Chem.,* 25, 1573, 1960.
90. **Hine, J. and Weimar J.,** Carbon basicity, *J. Am. Chem. Soc.,* 87, 3387, 1965.
91. **Jencks, W. P.,** *Catalysis in Chemistry and Enzymology,* McGraw-Hill, New York, 1969, 44.
92. **Dev, I. D., Yates, B. B., Leong, J., and Dallas, W. S.,** Functional role of cysteine-146 in *Escherichia coli* thymidylate synthase, *Proc. Natl. Acad. Sci. U.S.A.,* 85, 1472, 1988.
93. **Yarosh, D. B.,** Regulation and inhibition of O^6-methylguanine-DNA methyltransferase in human cells, in *Repair of DNA Lesions Introduced by N-Nitroso Compounds,* Myrnes, B. and Krokan, H., Eds., Norwegian University Press, Oslo, 1987, 135.
94. **Paik, W. K. and Kim, S.** *Protein Methylation,* John Wiley & Sons, New York, 1980, chap. 8.
95. **Sheid, B. and Pedrinian, L.,** DNA-dependent protein methylase activity in bull seminal plasma, *Biochemistry,* 14, 4357, 1975.
96. **Takahashi, K. and Kawazoe, Y.,** Methyl iodide, a potent inducer of the adaptive response without appreciable mutagenicity in *E. coli, Biochem. Biophys. Res. Commun.,* 144, 447, 1987.
97. **Wolfenden, R. V. and Kirsch, J. F.,** Enzymatic displacement of oxygen and sulfur from purines, *J. Am. Chem. Soc.,* 90, 6849, 1968.
98. **Wolfenden, R., Sharpless, T. K., and Allan, R.,** Substrate binding by adenosine deaminase. Specificity, pH dependence, and competition by mercurials, *J. Biol. Chem.,* 242, 977, 1967.
99. **Pabo, C. O. and Sauer, R. T.,** Protein-DNA recognition, *Annu. Rev. Biochem.,* 53, 923, 1984.
100. **Berg, J.,** Proposed structure for the zinc-binding domains from transcription factor IIIA and related proteins, *Proc. Natl. Acad. Sci. U.S.A.,* 85, 99, 1988.
101. **McClarin, J. A., Frederick, C. A., Wang, B.-C., Greene, P., Boyer, H., Grable, J., and Rosenberg, J. M.,** Structure of the DNA-EcoRI endonuclease recognition complex at 3 Å resolution, *Science,* 234, 1526, 1986.
102. **Kalnik, M. W., Kouchakdjian, M., Li, B. F. L., Swann, P. F., and Patel, D. J.,** Base pair mismatches and carcinogen-modified bases in DNA: an NMR study of A·C and A·O^4meT pairing in dodecanucleotide duplexes, *Biochemistry,* 27, 100, 1988.
103. **Kalnik, M. W., Kouchakdjian, M., Li, B. F. L., Swann, P. F., and Patel, D. J.,** Base pair mismatches and carcinogen-modified bases in DNA: an NMR study of G·T and G·O^4meT pairing in dodecanucleotide duplexes, *Biochemistry,* 27, 108, 1988.
104. **Patel, D. J., Shapiro, L., Kozlowski, S. A., Gaffney, B. L., and Jones, R. A.,** Structural studies of the O^6meG·C interaction in the d(C-G-C-G-A-A-T-T-C-O^6meG-C-G) duplex, *Biochemistry,* 25, 1027, 1986.
105. **Patel, D. J., Shapiro, L., Kozlowski, S. A., Gaffney, B. L., and Jones, R. A.,** Structural studies of the O^6meG·T interaction in the d(C-G-T-G-A-A-T-T-C-O^6meG-C-G) duplex, *Biochemistry,* 25, 1036, 1986.
106. **Basu, A. K. and Essigmann, J. M.,** Site-specifically modified oligodeoxynucleotides as probes for the structural and biological effects of DNA-damaging agents, *Chem. Res. Toxicol.,* 1, 1, 1988.
107. **Dolan, M. E., Morimoto, K., and Pegg, A. E.,** Reduction of O^6-alkylguanine-DNA alkyltransferase activity in HeLa cells treated with O^6-alkylguanines, *Cancer Res.,* 45, 6413, 1985.
108. **Brent, T. P., Smith, D. G., and Remack, J. S.,** Evidence that O^6-alkylguanine-DNA alkyltransferase becomes covalently bound to DNA containing 1,3-BIS(2-chloroethyl)-1-nitrosourea-induced precursors of interstrand cross-links, *Biochem. Biophys. Res. Commun.,* 142, 341, 1987.
109. **Husain, I., Sancar, G. B., Holbrook, S. R., and Sancar, A.,** Mechanism of damage recognition by *Escherichia coli* DNA photolyase, *J. Biol. Chem.,* 262, 13188, 1987.
110. **Moody, P. C. M. and Demple, B.,** Crystallization of O^6methylguanine-DNA methyltransferase from *Escherichia coli, J. Mol. Biol.,* 201, 200, 751, 1988.
111. **Fried, M. G. and Crothers, D. M.,** Kinetics and mechanism in the reaction of gene regulatory proteins with DNA, *J. Mol. Biol.,* 172, 263, 1984.
112. **Terry, B. J., Jack, W. E., and Modrich, P.,** Facilitated diffusion during catalysis by *Eco*RI endonuclease, *J. Biol. Chem.,* 260, 13130, 1985.
113. **Modrich, P.,** DNA mismatch correction, *Annu. Rev. Biochem.,* 56, 435, 1987.
114. **Rechsteiner, M.,** Ubiquitin-mediated pathways for intracellular proteolysis, *Annu. Rev. Cell Biol.,* 3, 1, 1987.
115. **Williams, L. D. and Shaw, B. R.,** Protonated base pairs explain the ambiguous pairing properties of O^6-methylguanine, *Proc. Natl. Acad. Sci. U.S.A.,* 84, 1779, 1987.

Chapter 18

NATURAL AND SYNTHETIC ANALOGS OF *S*-ADENOSYLHOMOCYSTEINE AND PROTEIN METHYLATION

F. Lawrence and M. Robert-Gero

TABLE OF CONTENTS

I. INTRODUCTION

Posttranslational modification of the amino acid residues of proteins is a well-established metabolic phenomenon. These modification reactions include methylation, acetylation, phosphorylation, thiolation, hydroxylation, ADP-ribosylation, nucleotidylation, glycosylation, carboxylation, and iodination, and concern a great variety of proteins. The biological significance of most of the above mentioned protein modifications is not yet fully understood.

Methylation of proteins has been studied during the last 20 years and several reviews have been published on the subject.[1-7] More than ten amino acid residues are susceptible to be methylated through three main linkages: the *N*-methylation of lysine, hydroxylysine, arginine, histidine, alanine, proline, phenylalanine, glutamine, and methionine, the *O*-methylation of glutamic and aspartic acids, and the *S*-methylation of methionine and cysteine.[4]

These methylation reactions are catalyzed by highly specific methyltransferases using *S*-adenosyl-L-methionine (AdoMet) as the methyl donor.[8]

A general reaction scheme illustrating the role of AdoMet in methylation reactions is given in Figure 1.

The methyl donor (AdoMet) is biosynthesized from ATP and methionine through the reaction catalyzed by ATP: L-methionine *S*-adenosyl transferase (EC 2.5.1.6).[9,10]

Transmethylation reactions involving AdoMet result in the formation of *S*-adenosyl-L-homocysteine (AdoHcy). AdoHcy was first characterized in 1954 by Cantoni and Scarano[11] as the product of enzymatic transmethylation from AdoMet and its structure was confirmed later by total synthesis by Baddiley and Jamieson.[12]

Each of the AdoMet-dependent methyltransferases has its own requirements for a methyl acceptor which are in general very specific.[4-7]

Besides the common requirement for AdoMet, each of these AdoMet-dependent methyltransferases also exhibits a sensitivity to inhibition by AdoHcy, one of the products of the transmethylation reaction. The product inhibition by AdoHcy and the enzymes involved in the metabolism of AdoHcy are part of a biological regulatory mechanism for AdoMet-dependent transmethylations.[13]

AdoHcy is metabolized by two pathways in yeast, plants, birds, and mammals. The first pathway is performed by the *S*-adenosyl-L-homocysteine-hydrolase (EC 3.3.1.1)[14] a NAD-dependent enzyme which catalyzes the reversible hydrolysis of AdoHcy to adenosine and L-homocysteine by an oxydation-reduction mechanism.[15-17] The second pathway for the metabolism of AdoHcy is catalyzed by L-amino acid oxidase (EC 1.4.3.2)[18-19] yielding *S*-adenosyl-γ-thio-α-ketobutyrate, ammonia, and H_2O_2.

In bacteria, the *S*-adenosylhomocysteine nucleosidase (EC 3.2.2.9) catalyzes the cleavage of the glycosyl linkage of AdoHcy to give adenine and 5′-ribosylhomocysteine.[20,21]

Several laboratories have attempted to take advantage of the inhibitory effect of AdoHcy for the design of specific inhibitors of AdoMet-dependent transmethylation. By systematically altering the chemical structure of AdoHcy and evaluating these analogs as inhibitors of various

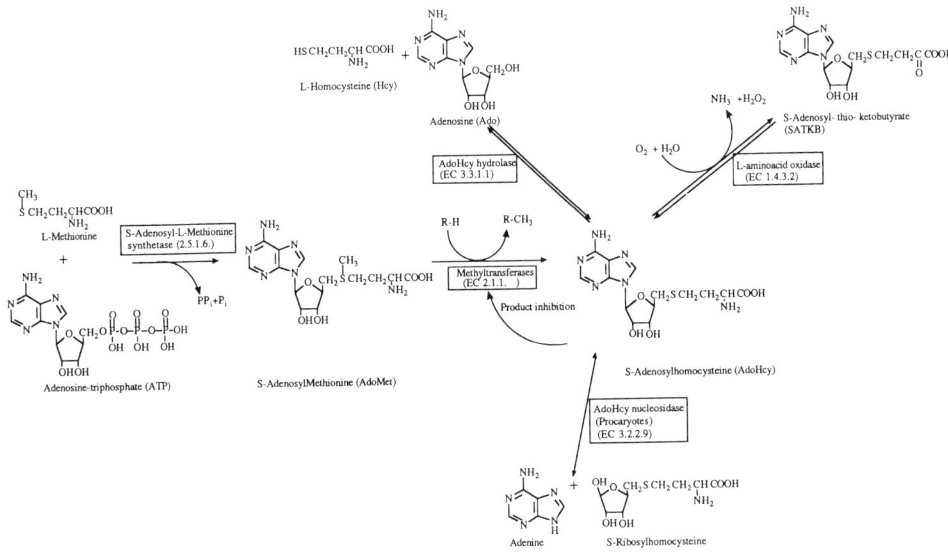

FIGURE 1. *S*-adenosylmethionine-dependent methyltransferases and *S*-adenosylhomocysteine metabolism.

methyltransferases, information is now available concerning the differences in the enzymatic binding requirements for AdoHcy.

Numerous analogs of AdoHcy and AdoMet have been synthesized to explore the structure-activity relationship of the various enzymes which recognize either AdoMet and/or AdoHcy.

II. ANALOGS OF *S*-ADENOSYL-L-HOMOCYSTEINE USED IN STUDIES ON PROTEIN METHYLATION

A general characteristic of AdoMet-dependent methyltransferases is their sensitivity to inhibition by AdoHcy, one of the products of the enzymatic methylation reaction.

AdoHcy is chemically more stable than AdoMet.[5] Its three dimensional X-ray crystal structure was established in 1981.[22]

The AdoHcy molecule consists of three parts (Figure 2A): a purine base (adenine), a pentose sugar (ribose) which with the adenine constitutes the adenosyl moiety, and a sulfur-containing amino acid 5′-side chain (L-homocysteine).

The names and formulas of the various AdoHcy analogs used for protein methylation studies are described in Figure 2, as well as the references concerning their discovery and/or their synthesis. The nomenclature of the analogues cited corresponds to that published by the authors. No attempt was made to uniformize it or to use the nomenclature of the UPAC.

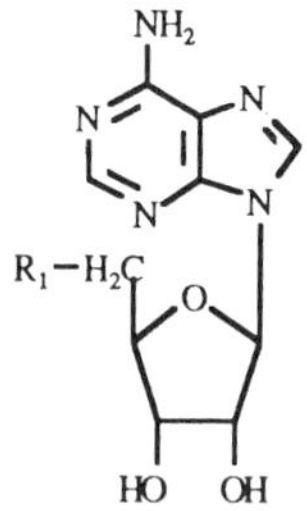

		R_1	Author and Year of synthesis/or finding.	Reference
1	5'-S -adenosyl-L-homocysteine	$-S-CH_2CH_2CH(NH_2)COOH$ (L)	Baddiley § Jamieson,1955,	12
2	5'-S-adenosyl-D-homocysteine	$-S-CH_2CH_2CH(NH_2)COOH$ (D)	Borchardt § Wu,1974,	24
3	5'-S-adenosyl-L-homocysteine sulfoxide	$-S(O)-CH_2CH_2CH(NH_2)COOH$ (L)	Borchardt § Wu,1974,	24
4	5'-S-adenosyl-L-homocysteine sulfone	$-S(O_2)-CH_2CH_2CH(NH_2)COOH$ (L)	Borchardt § Wu,1974,	24
5	Adenosine (Ado)	-OH	Davoll et al. 1948,	25
6	5'-S-methyl adenosine	$-S-CH_3$	Baddiley,1951,	26
7	5'-S-hydroxyethyl adenosine	$-S-CH_2CH_2OH$	Legraverend §Michelot,1976,	27
8	5'-O-(2 hydroxyethyl) adenosine	$-O-CH_2CH_2OH$	Coat §David,1970,	28
9	5'-S-carboxymethyl adenosine	$-S-CH_2COOH$	Gillet et al,1979,	29
10	5'-S-methylthiomethyl adenosine	$-S-CH_2SCH_3$	Legraverend et al,1977	30
11	5'-S-allyl adenosine	$-S-CH_2CH{=}CH_2$	Kuhn § Jahn,1965,	31
12	5'-S-(2-hydroxypropyl)adenosine	$-S-CH_2CH(OH)CH_3$	Vuilhorgne et al, 1978,	32
13	5'-S-(2,3-dihydroxypropyl) adenosine	$-S-CH_2CH(OH)CH_2OH$	Hildesheim et al, 1972,	33
14	5'-S-isopropyl adenosine	$-S-CH(CH_3)CH_3$	Kuhn § Jahn,1965,	31
15	5'-S-propyl adenosine	$-S-CH_2CH_2CH_3$	Kuhn § Jahn,1965,	31
16	5'-S-cysteamine adenosine	$-S-CH_2CH_2NH_2$	Hildesheim et al,1972,	33
17	5'-S-adenosyl-L-cysteine (AdoCys)	$-S-CH_2CH(NH_2)COOH$ (L)	Hildesheim et al,1971,	34
18	5'-S-adenosyl-D-cysteine	$-S-CH_2CH(NH_2)COOH$ (D)	Legraverend et al,1977,	30
19	5'-S-(1-methyl propyl) adenosine (IsoSIBA)	$-S-CH_2(CH_3)CH_2CH_3$	Legraverend et al,1977,	30
20	5'-S-(2-methyl propyl) adenosine (SIBA)	$-S-CH_2CH(CH_3)CH_3$	Hildesheim et al,1971,	34
21	5'-S-(2-methyl propyl)sulfinyl adenosine	$-S(O)-CH_2CH(CH_3)CH_3$	Blanchard, unpublished	
22	5'-N- (2-methyl propyl) adenosine	$-N-CH_2CH(CH_3)CH_3$	Legraverend et al,1977,	30
23	5'-S-(2-methyl allyl) adenosine	$-S-CH_2C(CH_3){=}CH_3$	Vuilhorgne et al, 1978,	32
24	5'-S-butyl adenosine	$-S-CH_2CH_2CH_2CH_3$	Hildesheim et al, 1972,	33
25	5'-S-adenosyl-4- thio-butyric acid (AdoTba)	$-S-CH_2CH_2CH_2COOH$	Borchardt § Wu,1974,	24
26	5'-S-adenosyl-3- thio-propylamine (AdoTpa)	$-S-CH_2CH_2CH_2NH_2$	Borchardt § Wu,1974,	24
27	5'-S-adenosyl- thio- ketobutyrate (SATKB)	$-S-CH_2CH_2C(=O)COOH$	Duerre et al, 1969,	35
28	5'-S-pentyl adenosine	$-S-CH_2CH_2CH_2CH_2CH_3$	Hildesheim et al, 1972,	33
29	5'-S-penicillamine	$-S-C(CH_3)_2CH(NH_2)COOH$	Gillet et al,1979,	29
30	5'-S-heptyl sulfinyl adenosine	$-S-(O)-(CH_2)_6CH_3$	Legraverend et al,1977,	30
31	5'-S-(β-D-glucopyranosyl) adenosine		Vuilhorgne et al, 1978,	32

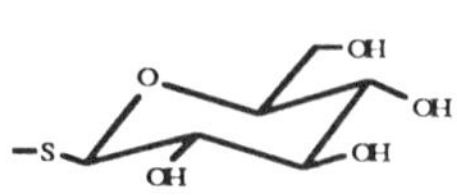

FIGURE 2. (A) Analogs modified in the 5′ side chain. (B) Analogs with modified base. (C) Analogs with modified sugar.

	R_1	Author and Year of synthesis/or finding.	Reference
32 5'-S-phenyl adenosine		Gillet et al,1979,	29
	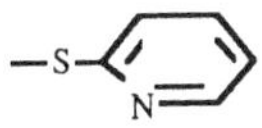		
33 5'-S-(2-pyridyl) adenosine		Nakagawa § Hata,1975,	36
	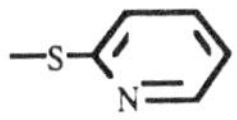		
34 5'-S-(3-carboxy-4-nitro) phenyl adenosine		Gillet et al, 1979,	29
	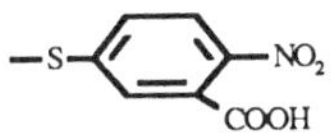		
35 Sinefungin	$-CH(NH_2)CH_2CH_2CH(NH_2)COOH$	Hamil § Hoehn,1973 Geze et al	37, 83
36 A9145C	$-CH(NH_2)CH_2CH_2CH(NH_2)COOH$	Nagarajan et al, 1977,	39
37 Sinefungin lactame		Blanchard etal, 1985,	40
	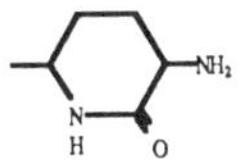		
38 5'chloro adenosine	-Cl	Jahn,1965,	41
39 5'-adenylic acid	$-PO_4H_2$	Baddiley § Todd,1947	42,
40 5'-S-adenosyl-L-ethionine (AdoEth)	$-S^+(CH_2CH_3)CH_2CH_2CH(NH_2)COOH$	Parks et al,1958,	43
41 5'-S-adenosyl-L-(2-amino-4-carbo-xymethylthio) butyrate	$-S^+(CH_2COOH)CH_2CH_2CH(NH_2)COOH$	Oliva et al,1980	44
42 5'-S-adenosyl-butylmethyl sulfonium	$-S^+(CH_3)CH_2CH_2CH_2CH_3$	Gillet et al,1979,	29
43 5'-S-adenosyl-L-(2-hydroxy-4-me-thylthio) butyrate	$-S^+(CH_3)CH_2CH_2CH(OH)COOH$	Zappia et al, 1969,	45
44 5'-S-adenosyl-(3-methyl-thio-propylamine	$-S^+(CH_3)CH_2CH_2CH_2NH_2$	Zappia et al, 1969,	45
45 5'-S-dimethylthioadenosine	$-S^+(CH_3)_2$	Oliva et al, 1980,	44

FIGURE 2 (continued)

A. SYNTHETIC ANALOGS

5′-Modified analogs of AdoHcy (Figure 2A)—These compounds can be obtained by two main chemical strategies: (1) by displacement of the *p*-toluene sulfonate group at the 5′ of 2′:3′-*O*-isopropylidene-5′-*O*-toluene-*p*-sulfonyladenosine, in presence of a thioalcoholate of formula RSH and dimethylformamide or liquid ammonia, known as Baddiley's method,[12] (2) by direct halogenation of the 5′ carbon of the adenosine, by thionyl chloride in hexamethylphosphorotriamide, followed by thiolation of the 5′-deoxy-5′-chloroadenosine obtained, with a thioalcoholate RSH in an aqueous solution of sodium hydroxide (Kikugawa's method).[23]

Base-modified analogs of AdoHcy (Figure 2B)—The strategy used to obtain base-modified analogs consists in coupling a base modified nucleoside with homocysteine or a thiolate.

***N*⁶-methyladenosine derivatives (n° 57, 58, 67)**—The first fully characterized methyladenosine derivatives were reported by Jones and Robins.[46] They used *N,N*-dimethylacetamide as the methylating agent at room temperature with excess methyl *p*-toluenesulfonate. Under these conditions 1-methyladenosine is isolated as the tosylate which by treatment with aqueous sodium hydroxide was readily converted to N^6-methyladenosine. Analogs n° 57, 58, and 67 are prepared using the method described by Baddiley and Jamieson.[12]

N^6-ethylAdoHcy (n° 68)—N^6-ethyladenosine is prepared by adding ethylamine to 6-chloro-2′-3′-5′ triacetyl 9-β-D-ribofuranosyl purine in dimethyl-formamid as described by Legraverend and Michelot 1976.[27] The final product is prepared by the Kikugawa method.[23]

N^6-carboxymethyl analogs (n° 59, 60)—These derivatives are prepared by treating ethyl-N^6-carboxymethyl 2′-3′-*O*-isopropylidene-5′-*O*-tosyl adenosine in tetrahydrofuran with sodium and two equivalents of the corresponding thiol.[32]

N^6-benzoyl-5′-deoxy-5-*S*-(2-methylpropyl)-5′-thioadenosine (n°61)—This is prepared by adding dropwise a threefold excess of distilled benzoylchloride to a solution of 5′-deoxy-5′-*S*-(2-methylpropyl)-5′-thioadenosine in pyridine at 0°.[30]

3-deazaAdoHcy (n° 70)—3-deazaadenosine is prepared according to Rousseau et al. by treating 4-chloro-1(β-D-ribofuranosyl)imidazo[4,5-c]pyridine at 100° in anhydrous hydrazine under N_2. The resulting 4-hydrazino derivative is then reduced using W-5 Raney nickel.[47] The recrystallized 3-deazaadenosine is converted to the corresponding 2′,3′-isopropylidene derivative as described by Mizuno[48] followed by the conversion to the 5′-tosylate. The 5′-tosylate is condensed with *S*-benzyl-L-homocysteine in sodium and liquid NH_3.[49]

***S*-tubercidylhomocysteine (n° 71)**—This molecule, described by Coward et al.[50] is prepared by converting natural tubercidin to the 2′,3′-isopropylidene and then to the 5′-*O*-tosyl derivative, followed by the condensation with *S*-benzyl-DL-homocysteine in liquid NH_3, and sodium.

2-azaadenosyl-L-homocysteine (n° 72)—2-Azaadenosine is prepared according to the procedure of Montgomery and Thomas.[51] The oxidation of adenosine with hydrogen peroxide in glacial acetic acid yields adenosine-1-*N*-oxide. The reaction of this intermediate with benzyl bromide followed by treatment with absolute methanol saturated with ammoniac affords 5-amino-4-(*N*-benzyloxy)carboxamidine-1-β-D-ribofuranosylimidazole which upon reduction with Raney nickel and cyclization by treatment with sodium nitrite in glacial acetic acid yields the desired 2-azaadenosine. The final product is prepared as described by Borchardt et al.[52]

8-azaadenosyl-L-homocysteine (n° 73)—Borchardt et al.[53] prepared 8-azaadenosine as described by Montgomery et al.[54] by condensation of 2,3,5-tri-*O*-acetyl-D-ribofuranosyl chloride with 6-nonamido-8-azapurine in benzene in the presence of Linde molecular sieves. The 2′,3′,5′-triacetyl-6-nonamido-8-azaadenosine after purification and hydrolysis with methanolic ammonia was converted to the 5′-chloro derivative, followed by condensation with L-homocysteine in sodium and liquid ammonia.[52,55]

Replacement of adenosine by inosine (n° 62, 66 and 69)—5′-Deoxy-5′-*S*-(isobutyl)-5′-thio inosine (n° 62) was synthesized by Vuilhorgne et al.[32] by condensation of 5′-deoxy-5′-

B. ANALOGUES WITH MODIFIED BASE

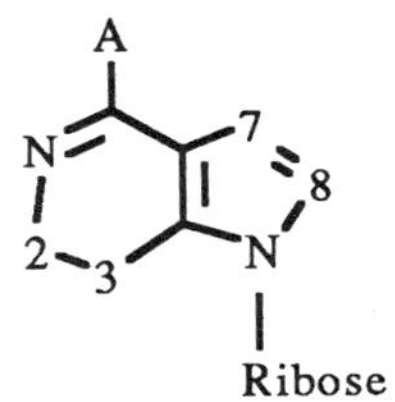

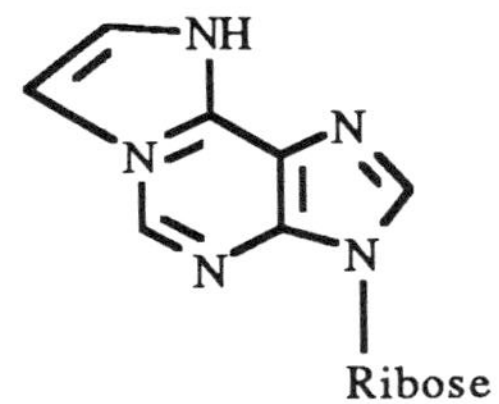

51

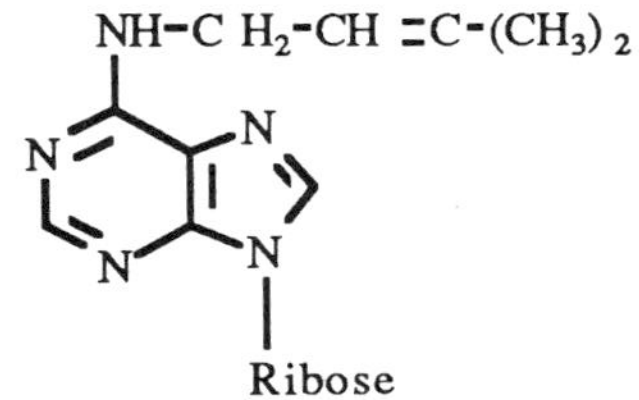

53

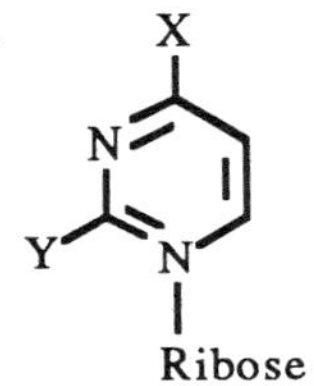

		A	2	3	7	8	Author and Year of synthesis/or finding	Reference
5	Adenosine	$-NH_2$	CH	N	N	CH	Davoll et al, 1948,	25
46	3-deazaadenosine	$-NH_2$	CH	CH	N	CH	Rousseau et al,1966,	47
47	N^6methyladenosine	$-NHCH_3$	CH	N	N	CH	Jones § Robins,1963,	46
48	Toyocamycine	$-NH_2$	CH	N	C-CN	CH	Tolman et al, 1968,	57
49	Tubercidine	$-NH_2$	CH	N	CH	CH	Tolman et al,1969,	58
50	8-Bromoadenosine	$-NH_2$	CH	N	N	CBr	Reist et al,1967,	59
51	Ethenoadenosine	§Fig.	CH	N	N	CH	Kochetkov et al, 1971,	60
52	Inosine (Ino)	OH	CH	N	N	CH	Kalkar,1947,	61
53	Isopentenyladenosine	§Fig.	CH	N	N	CH	Hall et al,1976,	62
56	Guanosine	-OH	CNH	N	H	CH	Davoll,1958,	63
57	N^6methylSIBA	$-NHCH_3$	CH	N	N	CH	Vuilhorgne et al,1978,	32
58	N^6methylisoSIBA	$-NHCH_3$	CH	N	N	CH	Vuilhorgne et al,1978,	32
59	N^6carboxymethyl SIBA	$-NHCH_2COOH$	CH	N	N	CH	Vuilhorgne et al,1978,	32

FIGURE 2B.

60	N^6carboxymethyl allyladenosine	-$NHCH_2COOH$	CH	N	N	CH	Vuilhorgne et al,1978,	32
61	N^6benzoylSIBA	-$NHCOC_6H_5$	CH	N	N	CH	Legraverend et al 1977,	30
62	5'-S(2methylpropyl inosine (SIBI)	-OH	CH	N	N	CH	Vuilhorgne et al,1978,	32
66	S-inosyl-L(2-hydroxy-4-methyl)thio butyrate	OH	CH	N	N	CH	Zappia et al, 1969,	45
67	N^6methylAdoHcy	-$NHCH_3$	CH	N	N	CH	Borchardt et al,1976	53
68	N^6ethylAdoHcy	-$NHCH_2CH_3$	CH	H	H	CH	Legraverend §Michelot,1976,	27
69	Inosylhomocysteine (InoHcy)	OH	CH	N	N	CH	Borchardt et al,1974,	49
70	3-deazaAdoHcy	NH_2	CH	CH	N	CH	Borchardt et al, 1974,	49
71	Tubercidinylhomo-cysteine	NH_2	CH	N	CH	CH	Coward et al, 1974,	50
72	2-azaAdoHcy	NH_2	NH	N	N	CH	Borchardt et al,1982,	52
73	8-azaAdoHcy	NH_2	CH	N	N	N	Borchardt et al,1976,	53
		X	Y					
54	Cytidine	NH_2	OH				Howard et al,1947,	64
55	Uridine	OH	OH				Loring § Ploeser,1949,	65
63	5'-S(2methylpropyl cytidine	NH_2	OH				Vuilhorgne et al,1978,	32
64	5'-S(2methylpropyl uridine	OH	OH				Vuilhorgne et al,1978,	32
74	Cytosylhomocysteine	NH_2	OH				Borchardt et al,1974,	49
75	Uridylhomocysteine	OH	OH				Hildesheim et al,1971,	34
76	S-ribosylhomocysteine	no base					Duerre et al,1970,	56
77	S-ribosylhomocysteine sulfoxide	no base					Duerre et al,1970,	56

FIGURE 2B continued.

chloro inosine in dimethylformamide under nitrogen and sodium hydride with 2-methyl-1-propanethiol. 5′-*S*-inosyl-L-homocysteine (n° 69) is prepared by reacting *S*-benzyl-L-homocysteine in liquid NH_3 and Na with 2′,3′-isopropylideneinosine 5′-tosylate.[49] *S*-inosyl-L-(2-hydroxy-4-methylthio)butyric acid (n° 66) was prepared by Zappia et al.[45] by double deamination of *S*-adenosylmethionine with *3 M* $NaNO_2$.

Replacement of adenosine by cytosine and uridine (n° 63, 64, 74, 75)—Analogs n° 63 and 64 are prepared by reaction of 4 mmol of 5′-deoxy-5′-chlorocytidine in aqueous *2 N* NaOH with two equivalents of 2-methyl-1-propanethiol at 80° for 4 h.. The two thionucleosides were separated after neutralization and chromatography.[32] *S*-cytidyl-L-homocysteine (n° 74) is prepared by condensation of the 5′-tosylate of 2′,3′-isopropylidenecytidine with *S*-benzyl-L-homocysteine in Na and liquid NH_3 as described for analog *n*°69.[49] 5′-*S*-uridyl-(D,L)-homocysteine (n° 75) was synthesized in a similar way by Hildesheim et al.[34] by condensation of 2′,3′-*O*-isopropylidene 5′-*O*-*p*-toluenesulfonyluridine with sodium homocysteinate.

Synthetic analogs with a modified sugar moiety (Figure 2C)—The preparation of 2′-deoxynucleosides (78) can be achieved either via reduction of 2′ halogenated compounds or desulfurization of the 2′-thio derivatives. 3′-Deoxynucleosides (80) can be prepared by the same methods, namely, by catalytic or radical induced reduction of 3′-halonucleosides or desulfurization of the 3′-thio derivatives.[66]

The monomethylation of the 2′- or the 3′-hydroxyl function of adenosine (79 and 81) is performed by the use of catalytic amounts of stannous chloride dihydrate and diazomethane.[67]

•Analogs 86, 87, and 88 are prepared by periodate oxidation followed by reduction with $NaBH_4$ in water.[32] The compounds formed were cyclized by treatment with sodium methoxide in dimethylformamide to yield the pseudo nucleoside. After tosylation the compounds were reacted with 2-methyl-propane-1-thiol.[32]

•Analog 89 was prepared by direct periodate oxidation of AdoHcy in water. Reduction by $NaBH_4$ yielded the analog 90.[71]

B. NATURAL ANALOGS

Several AdoHcy analogs are found in nature. Some are metabolic products found within most cells, others are excreted by microorganisms and have antibiotic properties.

Methylthioadenosine (n° 6)—Methylthioadenosine (MTA) is an ubiquitous nucleoside biosynthesized from AdoMet via several metabolic pathways in mammalian cells.[78] It normally does not accumulate (its intracellular concentration is lower than 5 μM),[79] but is rather rapidly degraded to 5′-methylthioribose-one-phosphate and adenine by MTA phosphorylase, or excreted by some MTA-phosphorylase deficient cell lines.[80] For a review see Reference 81.

•***Sinefungin*** (n° 35) is a nucleoside antibiotic isolated from cultures of *Streptomyces griseolus*[37] and *Streptomyces incarnatus*. This molecule is composed of an adenosine linked by a carbon-carbon bond to an ornithine residue.[39] Its chemical synthesis has been undertaken in various laboratories.[38,82-85]

•***A9145C*** (n° 36) was isolated in the same conditions than sinefungin. Its structure is similar to sinefungin except that it has a double bond at the $C_{4'}$-$C_{5'}$ position of ribose.[39]

•***S-adenosyl γ-thio-α-ketobutyrate*** **(SATKB)** (n° 27) is a product of AdoHcy metabolism excreted in rat urine.[19]

AdoMet analogs can also be formed *in situ* when methionine analogs are substrates of AdoMet synthetase, (EC 2.5.1.6) (ex.: ethionine) or when adenosine analogs which have been previously converted into triphosphate derivatives by the adenosine kinases are substrates of the AdoMet synthetase (ex.: neplanocin A, cordycepin). Similarly, AdoHcy analogs can be obtained *in situ* when adenosine analogs are substrates of AdoHcy hydrolase (see review Ueland[86]) or when the AdoMet analog is a substrate of an AdoMet-dependent methyltransferase (ex.: 3-deazaAdoMet; NpcMet).

C. ANALOGUES WITH MODIFIED SUGAR

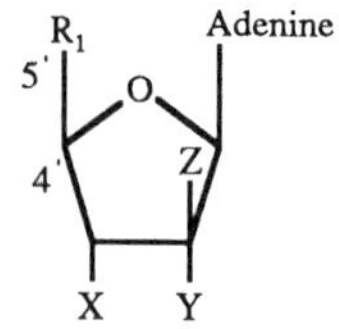

		X	Y	Z	R_1	Author and Year of synthesis/or finding	Reference
5	Adenosine	OH	OH	H	$-CH_2OH$	Davoll et al,1948,	25
36	A9145C	OH	OH	H	$=CHCH(NH_2)(CH_2)_2CH(NH_2)COOH$	Nagarajan et al, 1977,	39
78	2'-deoxyadenosine	OH	H	H	$-CH_2OH$	Robins et al, 1976,	66
79	2'-O-methyladeno-sine	OH	OCH_3	H	$-CH_2OH$	Robins et al,1974,	67
80	3'-deoxyadenosine (Cordycepin)	H	OH	H	$-CH_2OH$	Cunningham et al,1951 Walton et al, 1964,	68 69
81	3'-O-methyladeno-sine	OCH_3	OH	H	$-CH_2OH$	Robins et al,1974,	67
82	Arabinosyl adenine	OH	H	OH	$-CH_2OH$	Lee et al,1960,	70
83	Adenine	no sugar moiety					
84	SIBAra	OH	H	OH	$-CH_2SCH_2CH(CH_3)CH_3$	Blanchard unpublished	
85	SIBA 2'(3') phosphite				$-CH_2SCH_2CH(CH_3)CH_3$	Legraverend et al,1977,	30

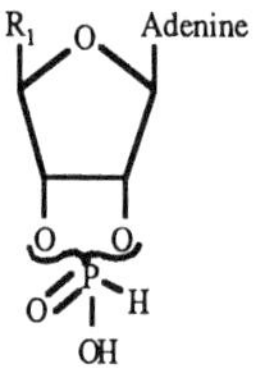

86 SIBALDO-SO $-CH_2S(O)CH_2CH(CH_3)CH_3$ Vuilhorgne et al,1978, 32

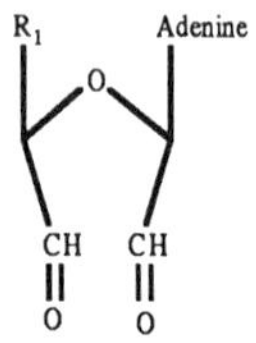

87 SIBAO $-CH_2SCH_2CH(CH_3)CH_3$ Vuilhorgne et al,1978, 32

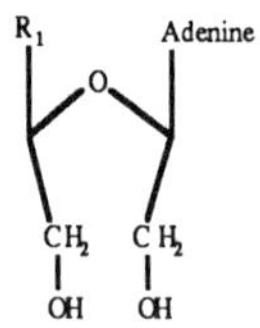

88 SIBAD $-CH_2SCH_2CH(CH_3)CH_3$ Vuilhorgne et al,1978, 32

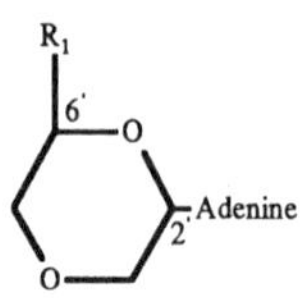

FIGURE 2C.

89 AdoHcy dialdehyde — $-CH_2SCH_2CH_2CH(NH_2)COOH$ — Borchardt et al. 1978, 71

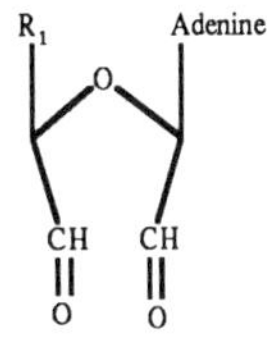

90 2',3' acyclic AdoHcy — $-CH_2SCH_2CH_2CH(NH_2)COOH$ — Borchardt et al. 1978, 71

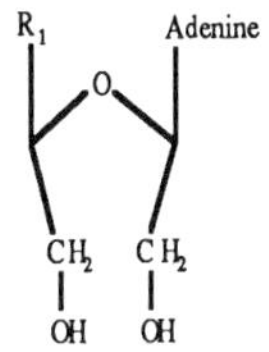

91 Neplanocin(Npc) — $-CH_2OH$ — Yaginuma et al. 1981, 72

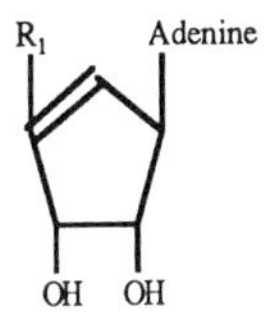

92 S-Neplanocin-
methionine (NpcMet) — $-CH_2S^+(CH_3)CH_2CH_2CH(NH_2)COOH$ — Keller & Borchardt,1984,73

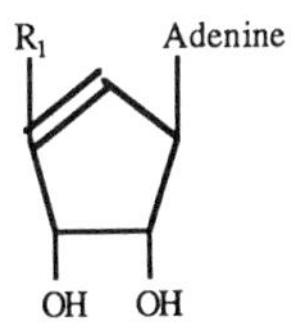

D. ANALOGUES WITH -C-C- BOND WITH SUGAR

93 Formycin — Hori et al,1964, 74

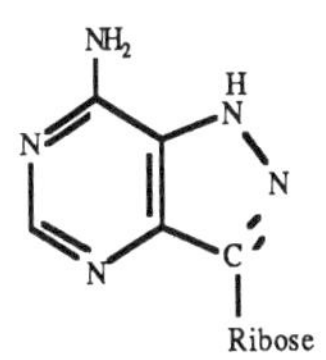

94 Pyrazofurin — Farkas et al. 1972, 75

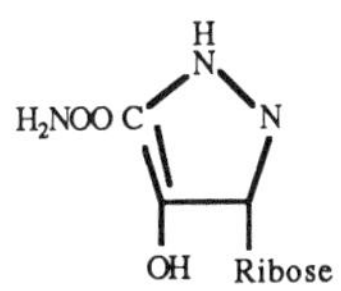

FIGURE 2C continued.

95 Minimycin Kusakabe et al, 1972, 76

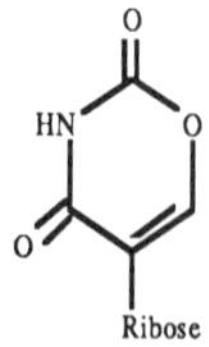

96 Showdomycin Kalvoda et al, 1970, 77

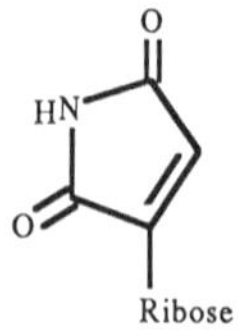

FIGURE 2C continued.

III. INHIBITION OF PROTEINE METHYLASES *IN VITRO* BY *S*-ADENOSYL HOMOCYSTEINE ANALOGS

A. METHYLATION OF THE GUANIDINO GROUP OF ARGININE RESIDUES

The *S*-adenosylmethionine:protein-arginine-*N*-methyltransferase (EC 2.1.1.23) is responsible for the methylation of the guanidino group of arginine residues of proteins. It is currently named protein methylase I since it was the first methylating enzyme identified.[87]

The existence of three methylated arginine derivatives in proteins is well established.[87,88] These include N^G-monomethylarginine, N^G,N^G-dimethylarginine (asymmetric), and N^G,N'^G-dimethylarginine (symmetric). They are widely distributed in proteins ranging from those of Actinomytes to mammals.[4]

1. Apparent Kinetic Constants of Protein Methylase I for AdoMet and AdoHcy *In Vitro*

The K_m values for *S*-adenosyl-L-methionine for various protein methylases I described are in the range of 2.5 to 40 μ*M* (Table 1). Since the intracellular concentration of *S*-adenosyl-L-methionine in organisms varies between 10 and 100 μ*M*, [4,89,90] the concentration of this methyl donor does not seem to be a limiting factor for enzymatic methylation *in vivo*.

S-adenosyl-L-homocysteine (n° 1), the demethylated product of *S*-adenosyl-L-methionine, was found to inhibit these enzymes in a competitive manner with respect to *S*-adenosylmethionine. The K_i values for *S*-adenosyl-L-homocysteine ranging between 1.5 to 12 μ*M* are well above the intracellular concentration of 0.1 μ*M*.[4,91]

The fact that the K_m values for *S*-adenosyl-L-methionine and the K_i values for *S*-adenosyl-L-homocysteine are not too different suggests that these two components of the methylation reaction have similar affinity for protein methylase I. However in all cases reported the K_i values for AdoHcy were slightly lower than the K_m values for AdoMet (Table 1).

2. Inhibition of Protein Methylase I by Various Analogs of *S*-adenosyl-L-Homocysteine (AdoHcy)

In order to characterize the AdoHcy binding site on the protein methylase I, various analogs of AdoHcy were evaluated for their abilities to inhibit this transmethylation *in vitro*. The AdoHcy analogs used for this study had modifications in either the amino acid, base, or sugar portion of this molecule (Figure 2A—C).

a. Analogs with Modified 5′-Side Chain

The chirality of the asymmetric carbon of AdoHcy is not absolutely critical for the binding to protein methylase I, since *S*-adenosyl-D-homocysteine (n° 2) exhibits inhibitory activity: the K_i values for D- and L-isomers of AdoHcy are, respectively, 3.3 and 2.0 μ*M* for the enzyme from Krebs II ascite cells[92] and 45 and 8 μ*M* for the enzyme from chick embryo fibroblasts;[93] furthermore the L- and D-isomers of *S*-adenosylcysteine (n° 17, 18) are both equally recognized by these enzymes. Similar results are observed for the cytochrome specific-arginine methylase from *Euglena gracilis* which is inhibited by the two isomers of AdoHcy (at 50 μ*M* 24 and 75% inhibition is observed for the D- and L-isomers, respectively),[94] and for the histone arginine methylase from wheat germ.[95]

With few exceptions [*S*-adenosyl-D-homocysteine (n° 2), *S*-adenosylethionine (n° 40), sinefungin (n° 35) and A9145C (n° 36)] modification of the 5′-side chain of the AdoHcy molecule is accompanied by an important decrease of the inhibitory activity *in vitro*. Thus, for the best of them, the apparent K_i values are 30- to 70-fold higher than those with *S*-adenosyl-L-homocysteine. Even *S*-adenosyl-L-cysteine (n° 17) which is the most closely related to *S*-adenosyl-L-homocysteine with only a one carbon shorter side chain displays a drastic reduction in inhibitory activity *in vitro*. Most of the molecules tested exhibit affinities of the same order

TABLE 1
Comparison of the Kinetic Constant of Various AdoHcy Analogs for Protein-Methylase I from Various Sources

	Sources of enzyme and references												
											Leishmania		
Compound	**Calf thymus 101**	**Rat liver 100**	**MEF 96**	**MEF Py[v] 96**	**Mouse macrophages 97**	**Krebs II ascites 92**	**CEF 93/98·**	**CEF RSV 98**	**Wheat germ 95**	***Euglena gracilis* 94**	***donovani* 99**	***tropica* 99**	***enrietti* 99**
K_m (μ*M*)													
AdoMet	2.0	4.4	15	14	14	2.5	38-12·	24.7	5.7	40	24	14	60
K_i (μ*M*) Analogs													
A. With modified 5′-side chain													
1 AdoHcy(L)	—	1.2	3.5	4.7	1.5	2.0	8/1.5·	4.9	1.05	11.8 (75%)[b]	3.0	19	2
2 AdoHcy(D)	—	—	—	—	—	3.3	45	—	(27%)[a]	—(24%)[b]	—		
5 Adenosine (= Ado)	—	—	—	—	—	500	2700	—	1.58	—(2%)[b]			
6 5′-*S*-methyl-Ado	—	—	—	—	—	50	396	—	(25%)[a]	—(11%)[b]			
7 5′-*S*-hydroxy-ethyl-Ado	—	—	—	—	—	—	223	—	(32%)[a]				
8 5′-*O*-(2-hy-droxyethyl)-Ado	—	—	—	—	—	—	∞						
10 5′-*S*-methyl-thiomethyl-Ado	—	—	—	—	—	50 (90%)[c]	320						
11 5′-*S*-allyl-Ado	—	—	—	—	—	66	421						
12 5′-*S*-(2-hy-droxypropyl)-Ado	—	—	—	—	—	—	794						
13 5′-*S*-(2,3-di-hydroxy-propyl)-Ado	—	—	—	—	—	—	1700						

15	5′-*S*-propyl-Ado	—	—	—	—	—	46	240						
16	5′-*S*-cysteamine Ado	—	—	—	—	—	30	40						
17	AdoCys (L)	—	—	—	—	—	150	1660	—	—	(2%)[b]			
18	AdoCys (D)	—	—	—	—	—	130	1056						
19	5′-*S*-(1-methyl-propyl)-Ado	—	—	—	—	—	40	307						
20	5′-*S*-(2-methyl-propyl)-Ado (SIBA)	—	—	255	117	—	100 (88%)	635/182[e]	395					
21	5′-*S*-(2-methyl-propyl)sul-finyl–Ado	—	—	—	—	—	330	1600						
22	5′-*N*-(2-methyl-propyl)-Ado	—	—	—	—	—	(20%)[e]	∞						
23	5′-*S*-(2-methyl-allyl)-Ado	—	—	—	—	—	—	410						
24	5′-*S*-butyl-Ado	—	—	—	—	—	40	400	—	(31%)[a]				
28	5′-*S*-pentyl-Ado	—	—	—	—	—	—	405						
30	5′-*S*-heptyl sulfinyl-Ado	—	—	—	—	—	—	1247						
31	5′-*S*-(β-D-gluco-pyranosyl)-Ado	—	—	—	—	—	—	∞						
33	5′-*S*-(2-pyridyl) Ado	—	—	—	—	—	66	1285						
35	Sinefungin	—	—	0.43	0.46	9.0	—	3.5·	11.1	1.67	(48%)[b]	145	57	120
36	A9145C	—	—	—	—	—	—	0.5·	—	0.023	(11%)[b]	∞	∞	500
37	Sinefungin lactame	—	—	—	—	—	—	—	—	—	—	253	∞	1370
38	5′-chloro-ado	—	—	—	—	—	85	—						
39	$AMP_{5'}$	—	—	—	—	—	(0%)[e]	—						
40	AdoEth-(L)	5.0	10											

TABLE 1 (continued)
Comparison of the Kinetic Constant of Various AdoHcy Analogs for Protein-Methylase I from Various Sources

Compound	Sources of enzyme and references												
	Calf thymus 101	Rat liver 100	MEF 96	MEF Py[v] 96	Mouse macrophages 97	Krebs II ascites 92	CEF 93/98	CEF RSV 98	Wheat germ 95	*Euglena gracilis* 94	*Leishmania donovani* 99	*Leishmania tropica* 99	*Leishmania enrietti* 99
B. With modified base													
5 (Adenosine)	—	—	—	—	—	500 (60%)[e]	2700		1.58	(2%)[b]			
46 3-Deazaadenosine	—	—	—	—	—	—	—	—	(24%)[c]				
47 N⁶-methyladenosine	—	—	—	—	—	(29%)[e]							
48 Toyocamycine	—	—	—	—	—	330 (62%)[e]							
49 Tubercidin	—	(72%)[d]	—	—	—	(25%)[e]							
50 8-Bromoadenosine	—	—	—	—	—	(37%)[e]							
51 Ethenoadenosine	—	—	—	—	—	(30%)[e]							
53 Isopentenyl-Ado	—	(79%)[d]											
54 Cytidine	—	—	—	—	—	(0%)[e]	—	—	(0%)[c]				
55 Uridine	—	—	—	—	—	(0%)[e]	—	—	(0%)[c]				
56 Guanosine	—	—	—	—	—	(0%)[e]			(0%)[c]				
57 N⁶-methyl–SIBA	—	—	—	—	—	—	2700						
58 N⁶-methyliso–SIBA	—	—	—	—	—	—	1660						
59 N⁶-carboxymethyl–SIBA	—	—	—	—	—	—	∞						
60 N⁶-carboxymethyl-5′-allyl-Ado	—	—	—	—	—	—	∞						

61 N^6-benzoyl–SIBA	—	—	—	—	—	—	2100						
62 5′-*S*-(2-methyl-propyl)–inosine	—	—	—	—	—	—	∞						
63 5′-*S*-(2-methyl-propyl)–cytidine	—	—	—	—	—	—	∞						
64 5′-*S*-(2-methyl-propyl)–uridine	—	—	—	—	—	—	∞						
66 5′-*S*-inosyl (2-hydroxy-4-methylthio-butyrate)	—	—	—	—	—	—	—		1.2	(6%)[b]			
C. With modified sugar													
5 Adenosine	—	—	—	—	—	500 (60%)[d]	2700	—	1.58	(2%)[b]			
36 A9145C	—	—	—	—	—	—	0.5·	——	0.023	(11%)[b]	∞	∞	500
78 2′-Deoxy-adenosine	—	—	—	—	—	(15%)[d]	—	—	(17%)[c]	(10%)[b]			
79 2′-*O*-methyl-adenosine	—	—	—	—	—	(3%)[d]							
80 3′-Deoxy-adenosine–(cordycepin)	—	—	—	—	—	—0	∞	—	(0%)[c]	(16%)[b]			
81 3′-*O*-methyl-adenosine	—	—	—	—	—	660 (23%)[d]							
82 Arabinosyl-adenine	—	—	—	—	—	(13%)[d]	—	—	(0%)[c]	(12%)[b]			
83 Adenine	—	(16%)[d]	—	—	—	(30%)[d]	—	—	(22%)[c]				
84 SIBAra	—	—	—	—	—	—	∞						
85 SIBA 2′3′ phosphite	—	—	—	—	—	—	1960						

TABLE 1 (continued)
Comparison of the Kinetic Constant of Various AdoHcy Analogs for Protein-Methylase I from Various Sources

	Sources of enzyme and references												
	Calf thymus	Rat liver	MEF	MEF Py	Mouse macrophages	Krebs II ascites	CEF	CEF RSV	Wheat germ	*Euglena gracilis*	*Leishmania donovani*	*Leishmania tropica*	*Leishmania enrietti*
Compound	**101**	**100**	**96**	**96**	**97**	**92**	**93/98**[v]	**98**	**95**	**94**	**99**	**99**	**99**
86 SIBALDO SO	—	—	—	—	—	—	471						
87 SIBAO	—	—	—	—	—	—	∞						
88 SIBAD	—	—	—	—	—	—	∞						
D. C–C BOND													
93 Formycin	—	—	—	—	—	—	—		(0)[c]				
94 Pyrazofurin	—	—	—	—	—	—	—		(0)[c]				
95 Minimycin	—	—	—	—	—				(0)[c]				
96 Showdomycin	—	—	—	—	—				(0)[c]				

Note: The K_i values are expressed in μ*M*; the sign ∞ is used when inhibitory activity could not be detected according to the authors. Values in brackets represent the percentage of inhibition of the methylation reaction observed *in vitro* using: [a] 40 μ*M* AdoHcy analog; [b] 50 μ*M*; [c] 100 μ*M*; [d] 1 m*M*; [e] 2 m*M*. The analogs are numbered as in Figures 1A, B, C, and D. [v] denotes virus.

of magnitude irrespective of the length (methyl to pentyl, n° 6, 15, 24, 28) and shape of their side chains which may be linear, branched, or even bear an aromatic ring.

Modifications at the level of the sulfur atom reduce or abolish the inhibitory activity except for sinefungin and its derivatives; oxidation of the sulfur to sulfoxide (S→O) reduces the inhibition; the K_i values are increased from 635 to 1600 μ*M* and from 100 to 300 μ*M* for SIBA (n° 20) and SIBA-SO (n° 21) with respect to avian[93] and mammalian[92] enzymes. The replacement of the sulfur atom by a nitrogen (n° 22) or by an oxygen (n° 8) atom abolishes the inhibitory activity,[92,93] whereas the replacement of the sulfur atom by an amino-substituted methylene group (>$CH\text{-}NH_2$) as in the antifungal antibiotic sinefungin (n° 35) alters slightly the inhibitory activity: the ratio of the K_i value of sinefungin to the K_i value of AdoHcy ranges between 1.6 and 2.2 for most of the enzymes tested, except those of *Leishmania donovani* and *L. enrietti*[99]

Among the sulfonium analogs of AdoHcy, *S*-adenosyl-L-ethionine (n° 40) is a good substrate for protein methylase I from rat liver[100] and thus an inhibitor of AdoMet utilization. Cory et al. found that *S*-adenosyl-L-ethionine inhibited protein methylase I activity by about 50% at 80 μ*M*.[101]

Adenosine itself (n° 5) is inhibitory,[92] indicating that the nucleoside is important for the binding. It should be noted that for protein methylase I from Krebs II ascite, chloroadenosine (n° 38) is a more powerful inhibitor than adenosine itself.[92]

b. Analogs with a Modified Base

The 6-amino group of adenine plays a crucial role (n° 62, 59, 60) although some monosubstitutions by methyl (n° 57, 58, 61), or etheno groups (51) can be tolerated by the enzymes from chick embryo fibroblasts and Krebs II ascite cells.[92,93]

Bases other than adenine are not recognized: thus compounds n° 62, 63, 64 are inactive toward the enzyme from chick embyro[93] and compounds n° 54, 55, 56 toward those of Krebs II ascite cells and wheat germ.[92,95]

The nitrogen atom at position 7 also seems to play an important role in recognition as it cannot be replaced by a carbon atom as in tubercidin (n° 49).[92] If however this carbon bears a –CN substituent [toyocamycin, (n° 48)] most of the activity is recovered.[92]

The position 8 appears to be much less important for recognition since a major alteration like substitution by a bromide atom (n° 50) conferred a limited loss of inhibition.[92]

Among the sulfonium analogs of AdoHcy, *S*-inosyl-L-(2-hydroxy-4-methyl-thio)-butyrate (n° 66) is a very efficient inhibitor (K_i 1.2 μ*M*) of protein methylase I from wheat germ.[95] However at 50 μ*M* this product inhibits only by 6% the enzyme from *Euglena*.[94]

c. Analogs with a Modified Sugar

Any modification of the ribose ring except its 3′-*O*-methylation (n° 81) is accompanied by a drastic reduction in inhibitory activity.[92] Even the complete loss of ribose is less deleterious, as adenine (n° 83) still has a significant activity.[92,95]

Loss of the 2′-hydroxyl (n° 78), its methylation (n° 79) or its change to arabino configuration (n° 82) retain part of the inhibitory activity.[92] More critical is the loss of the 3′-hydroxyl group (cordycepin n° 80) or oxidation (n° 87, 88) which result in a loss of activity for the enzyme from vertebrate[92,93] and from wheat germ.[95]

However, it seems that the enzyme from *Euglena* behaves differently since cordycepin (n° 80) and 9-β-D-arabinofuranosyladenine (n° 82) are more inhibitory than adenosine when tested at 50 μ*M*.[94]

The nature of the link at the 4′,5′-position (single or double bond) influences the binding. Comparison of the K_i values for sinefungin (n° 35) and for the 9145C derivative (n° 36) shows that the presence of the double bond decreases the inhibitory power of the molecule toward protein methylase I from *Euglena*[94] and *Leishmania*[99] whereas it is increased for the enzyme from chick embryo fibroblasts[98] and that of wheat germ.[95]

Formycin (n° 93), pyrazolin (n° 94), minimycin (n° 95), and showdomycin (n° 96), which have a C–C–bond instead of C–N, do not have any effect on the activity of wheat germ enzyme, at 100 μ*M*.[95]

d. Conclusion

Inhibition of protein methylase I by *S*-adenosyl-L-homocysteine analogs greatly depends on the integrity of the adenine and homocysteine moieties of the molecules. The integrity of the ribose moiety is essential only with mammalian and avian enzymes. In all cases the chirality of the asymmetric carbon in the side chain is not an absolute requirement for activity but the sulfur atom appears important, sinefungin being an exception. The comparison of the crystalline structure of AdoHcy and sinefungin might provide an explanation.

B. METHYLATION OF FREE CARBOXYL GROUPS OF PROTEINS

The *S*-adenosyl methionine: protein carboxyl *O*-methyltransferases (EC 2.1.2.24) catalyze the posttranslational AdoMet-dependent methylation of the carboxyl side chains of glutamyl and/or aspartyl residues of cellular proteins resulting in the formation of labile protein methylesters.

These enzymes are also named protein methylases II by Paik and Kim.[102]

There are at least two classes of these enzymes which play two very different physiological roles.[4,103,104] The first class catalyzes the methylation of carboxylic acid residues of specific proteins; the second recognizes structurally altered carboxylic acids on a very large variety of proteins.

In prokaryotes, protein methylase II regulates chemotaxis by methylating and demethylating specific membrane proteins which modify flagellar rotation. In eukaryotes protein carboxy-methylase is involved in the regulation of leukocyte and slime mold chemotaxis, cellular secretion, racemization, repair of aging proteins, phototransduction, and cellular growth and development.

Three types of methylated carboxylic acid derivatives in proteins have been found: L-glutamyl, D-aspartyl, and L-isoaspartyl residues.

1. Apparent Kinetic Constants of Protein Methylase II for AdoMet and AdoHcy *In Vitro*

The K_m values for AdoMet for various protein methylases II described so far are in the range of 1 to 12 μ*M* and the K_i for the product inhibitor AdoHcy varies from 0.1 to 1.8 μ*M* (Table 2).

Reports on the enzymatic activity in presence of AdoHcy structural analogs are relatively recent.[29,44]

2. Inhibition of Protein Methylase II by Various Analogs of *S*-Adenosyl-L-Homocysteine (AdoHcy)

a. Analogs with Modified 5′-Side Chain

The chirality of the asymmetric carbon of AdoHcy is crucial for the recognition by protein methylase II, *S*-adenosyl-D-homocysteine (n° 2) being inactive toward calf brain enzyme[44] and a weak inhibitor toward calf thymus enzyme.[52]

The removal of one carbon on the 5′-side chain (*S*-adenosyl-L-cysteine n° 17) leads to an inactive molecule.[44,52] Among the AdoHcy analogs modified in the amino acid moiety only 5′-methylthioadenosine (n° 6) is endowed with a significant inhibitory power (K_i 41 μ*M*) on calf brain enzyme. Isobutylthioadenosine (n° 20) did not exert any inhibitory effect on protein methylase II from various origin.[44,105,106]

Modification at the level of the sulfur atom typified by the conversion of *S*-AdoHcy to *S*-AdoHcy sulfoxide (n° 3) results in a decrease of the recognition by the enzyme from human erythrocytes the K_i values being 0.1 and 7 μ*M*, respectively, for L-AdoHcy and L-AdoHcy

TABLE 2
Comparison of the Kinetic Constant of Various AdoHcy Analogs for Protein-Methylase II from Various Sources

Compound	Human erythrocyte 29/106·	Calf thymus 52*/107	Calf brain 44	Bovine adrenal 107	Murine neuroblastoma 108	Mouse pancreas 112	Mouse cell 113	Chick embryo 105	Wheat germ 106	*Leishmania* 109
	Sources of enzyme and references									
K_m (μ*M*) AdoMet	—	0.97·	0.87	—	3.2	0.75	4.2	1.5	5.0	1.0
K_i (μ*M*) Analogs										
A. With modified 5′-side chain										
1 AdoHcy(L)	0.1/1.25·	1.03·*	0.65 (90%)[b]	0.30	0.64/1.67	—	—	0.7	1.5	0.8/1.16
2 AdoHcy(D)	—	34.10*	(0%)[b]							
3 AdoHcySO(L)	7	1.55*	—							
4 $AdoHcySO_2$(L)	—	1.39*	—	—						
5 Ado	3500	1200·	—	—						
6 5′-*S*-methyl-Ado	—	—	41 (22%)[c]							
7 5′-*S*-hydroxy-ethyl-Ado	3000	—	(0%)[e]							
9 5′-*S*-carboxy-methyl-Ado	>1600	—								
14 5′-*S*-isopropyl-Ado	>300	—								
16 5′-*S*-cyste-amine-Ado	500	—								
17 AdoCys(L)	—	∞	(0%)[d]							
20 5′-S(2-methyl-propyl)Ado (SIBA)	—	—	(0%)[e]	—	—	—	—	>6000	(0%)[g]	
24 5′-*S*-butyl-Ado	>6000	—	(0%)[e]							
25 AdoTba	—	∞*								
26 AdoTpa	—	∞*								

TABLE 2 (continued)
Comparison of the Kinetic Constant of Various AdoHcy Analogs for Protein-Methylase II from Various Sources

Compound	Human erythrocyte 29/106·	Calf thymus 52·/107	Calf brain 44	Bovine adrenal 107	Murine neuroblastoma 108	Mouse pancreas 112	Mouse cell 113	Chick embryo 105	Wheat germ 106	*Leishmania* 109
	Sources of enzyme and references									
29 5′-*S*-penicillamine	1600	—								
32 5′-*S*-phenyl-Ado	>1100	—								
33 5′-*S*-(2 pyridyl) Ado	>800									
34 5′-*S*-(3 carboxy-4 nitro)phenyl-Ado	15	—								
35 Sinefungin	2.6·	0.22	—	0.5	0.47/0.97	—		0.90	0.40	1.8/62.4
36 A9145C	0.4	0.02	—	0.04	0.04/0.05	—		0.08	0.10	
38 5′-chloro-Ado	650		—		—					
40 AdoEth	—	—	—	—	—	13				
41 5′-*S*-Ado-L-(2-amino-4-carboxy methylthio)–butyrate	—	—	(50%)[a]		—					
42 5′-*S*-methyl-butyl-Ado	38	—	—							
43 5′-*S*-L-(2-hydroxy-4-methylthio)-butyrate	—	—	(0%)[e]		—					
44 5′-*S*-(3-methyl-thiopropyl-amine)-Ado	—	—	0%[e]	—			—	—	—	
45 5′-*S*-dimethyl-Ado	—	—	35%[e]	—	—	—	—	—		

B. With modified base										
46 3-deaza-Ado	—	—	—	—	—	—	—	—	(0%)[h]	
67 N[6]-methyl-AdoHcy	—	19.9*								
68 *N*[6]-ethyl-AdoHcy	—	∞								
69 InoHcy	—	∞	(7.5%)[c]							
70 C[3]-AdoHcy	—	∞								
71 C[7]-AdoHcy (TubHcy)	—	14.5*								
72 2-azaAdoHcy	—	42.5*								
73 8-azaAdoHcy	—	93.9*								
74 CytHcy	—	∞								
75 UriHcy	—	∞								
C. With modified sugar										
36 A9145C	0.4	0.02	—	004	004/005	—	—	0.08	010	—
89 AdoHcy dial-dehyde		∞								
90 2′,3′-Acyclic AdoHcy	—	∞								
92 NpcMet	—	—	—	—			(50%)[f] 122 Km			

Note: The K_i values are expressed in μ*M*; The sign ∞ is used when inhibitory activity could not be detected according to the authors. Values in brackets represent the percentage of inhibition obtained with the following concentrations of AdoHcy analogs: [a] 5 μ*M*; [b] 10 μ*M*; [c] 20 μ*M*; [d] 50 μ*M*; [e]: 100 μ*M*; [f] 205 μ*M*; [g] 500 μ*M*., [h] 1000 μ*M*. The analogs are numbered as in Figure 1A, B, and C.

sulfoxide.[29] Whereas in the case of the calf thymus enzyme, analogs with oxidized sulfur atom (n° 3 and 4) show appreciable inhibitory activity.[52]

The replacement of the sulfur atom of AdoHcy by an amino substituted methylene group as in sinefungin (n° 35) increases the affinity toward the enzymes from calf thymus,[107] wheat germ,[106] and murine C1300 neuroblastoma cells.[108] The inhibitory potency of sinefungin toward the enzymes from chick embryo,[105] bovine adrenal,[107] human erythrocytes,[29] and leishmanial enzymes[109] from susceptible strains is slightly lower than that of AdoHcy. In the case of leishmanial enzyme from low susceptibility strains the K_i for sinefungin is higher (55 against 2 μM in susceptible cells).[109]

Among various sulfonium compounds tested for their inhibitory effect of protein methylase II from calf brain[44] only *S*-adenosyl-L-(2-amino-4-carboxy-methylthio)butyrate (n° 41) and 5′-dimethyl-thioadenosine (n° 45) were inhibitory in the concentration range (up to 100 μM) used: the former was a very effective inhibitor (I_{50} = 5 μM), the latter a weak one (I_{35} = 100 μM). 5′-dimethylthioadenosine (n° 45) is not a methyl donor in the carboxymethyl esterification of corticotropin;[44] 5′-*S*-adenosyl-3-methylthiopropylamine (n° 44), a product of *S*-adenosylmethionine metabolism involved in polyamine biosynthesis,[110] is inactive either as substrate or inhibitor of the calf brain enzyme whereas *S*-adenosyl-L-(2-hydroxy-4-methylthio)butyrate (n° 43) has a moderate activity as methyl donor but was inactive as inhibitor (up to 100 μM).[44] These results indicate that the 1-carboxyl and 2-amino groups of the methionine moiety are essential for the binding of the methyl donor to the enzyme and/or to the substrate peptide.

The 5′-*S*-methylbutyl sulfonium derivative (n° 42) is a better inhibitor of protein carboxyl methyltransferase from human erythrocyte than the 5′-*S*-butyl derivative (n° 24) (K_i 38 μM and >6000 μM, respectively). The methyl donor capacity of the former molecule was not measured.[29]

AdoEth is a competitive inhibitor of protein carboxymethylase from mouse pancreas, with a lower affinity than AdoHcy.[112]

Adenosine (n° 5) itself is slightly inhibitory. As for protein methylase I, 5′-chloroadenosine (n° 38) is a better inhibitor than adenosine of the enzyme from human erythrocytes.[29]

b. Analogs with a Modified Base

S-inosyl-L-homocysteine (n° 69), the 6-deaminated analog of AdoHcy, gave negligible or no inhibition of protein methylase.[44,52] This loss of inhibitory effect with the deaminated derivative indicates the importance of the 6-aminopurine ring as binding site on the enzyme. However, small alkyl substituent at the 6-amino group (e.g., N^6-methyl-AdoHcy, n° 67) can be tolerated whereas a larger substituent at N^6, such as an ethyl group (n° 68) results in total loss of inhibitory activity.[52] Of the base modified analogs tested on calf thymus enzyme only tubercidinyl L-homocysteine, (n° 71), 8-aza-AdoHcy, (n° 73), and 2-azaAdoHcy (n° 72) exhibited inhibitory effects,[52] but all were less active than AdoHcy itself. More drastic structural changes in the base moiety, such as the replacement of the adenine ring by a pyrimidine base (UriHcy n° 75, CytHcy n° 74) result in a complete loss of inhibitory activity.[52]

c. Analogs with a Modified Sugar Moiety

A9145C (n° 36) a sinefungin analog is, in all cases tested, more effective than sinefungin (n° 35): thus for human erythrocyte and wheat germ[105] the K_i for A9145C is four- to sixfold lower than that for sinefungin, whereas for the enzymes from bovine adrenals,[107] murine neuroblastoma cells,[108] and chick embryo,[105] it is one order of magnitude lower than that of sinefungin.

AdoHcy dialdehyde (n° 89), which is an affinity labeling reagent for the histamine *N*-methyltransferase[71] is inactive as an inhibitor of the protein methylase II from calf thymus, as well as 2′,3′-acyclic AdoHcy (n° 90).[52]

Neplanocylmethionine [NpcMet (n° 92)] a carbocyclic AdoMet analog, in which the ribose moiety is replaced by a cyclopentene ring, is a weak inhibitor of protein carboxymethylase from L cells.[113] The IC_{50} value is 205 μM, or 20-fold higher than the concentration of AdoMet used in the assay. Studies with [*methyl*-^{3}H]*S*-neplanocylmethionine indicate that this analog has little substrate activity: the K_m value was approximately 30-fold higher than that of AdoMet.

d. Conclusion

The specificity of protein carboxyl methyltransferases appears to be rather strict compared to the other methyltransferases. The structural features of primary importance (in binding the amino acid portion of AdoHcy to the enzyme) include: (1) the chirality of the amino acid asymmetric carbon atom; (2) the terminal amino group; (3) the terminal carboxyl group, and (4) the three carbon atom distance between the sulfur atom and the terminal amino and carboxyl groups. The inhibitory activity of sulfur-modified analogs (AdoHcy sulfoxide, AdoHcy sulfone, sinefungin, and A9145C) indicate that protein-carboxy methylases can accomodate structural changes around the sulfur atom of AdoHcy. Minor changes in the adenine ring of AdoHcy produce a significant reduction in enzyme affinity. Furthermore the results obtained with compound (89) and (90) suggest that the rigidity of the ribose ring of AdoHcy contributes to its enzymatic binding.

C. METHYLATION OF THE ε-AMINO GROUP OF LYSINE RESIDUES

The *S*-adenosylmethionine:protein lysine-*N*-methyltransferases (EC 2.1.1.43) catalyze the methylation of the ε-amino group of lysine residues of proteins. It is also known as protein methylase III.[114]

Three methylated derivatives of lysine are found in proteins: these include ε-*N*-mono-methyl-, ε-*N*-dimethyl-, and ε-*N*-trimethyllysine. They occur in a wide variety of proteins from many species of organisms.[4]

1. Apparent Kinetic Constants of Protein Methylase III for AdoMet and AdoHcy *In Vitro*

The K_m values for AdoMet for various protein methylases III are in the range of 3 to 130 μ*M* (Table 3); as for most of the AdoMet dependent methyltransferases, the concentration of AdoMet does not seem a limiting factor *in vivo*.

The K_i values of protein methylase III for AdoHcy are more homogeneous than those of protein methylase I and II and range between 2 to 5.9 μ*M*. These values are well above the intracellular concentration of this molecule.[4] The ratio of K_i for AdoHcy to $K_{m\,for}$ AdoMet varies from 0.04 to 0.47, the lowest values are observed for the enzymes from *Leishmania*, *Saccharomyces*, and *Neurospora crassa*.

2. Inhibition of Protein Methylase III by Various Analogs of *S*-Adenosyl-L-Homocysteine (AdoHcy)

a. Analogs with a Modified 5′-Side Chain

The D-isomer of AdoHcy inhibits the protein methylases III from wheat germ,[115] rat brain,[116,117] and *Euglena gracilis*.[118] These results indicate that the chirality of the asymmetric carbon of AdoHcy is not absolutely critical for recognition by the enzyme. In this respect protein methylase III behaves as protein methylase I[92,93] but differently from protein methylase II.[44]

With one exception, modification of the 5′-side chain of the AdoHcy molecule is accompanied by an important decrease or loss of inhibitory activity *in vitro*. Thus, 5′-*S*-adenosyl-L-cysteine (n° 17) inhibits the histone methylase "VB" from *Euglena gracilis* (30% inhibition at a concentration of 50 μ*M*) but had no effect on the histone methylase "VA".[118] 5′-*S*-isobutylthioadenosine (SIBA n° 20) has a weak or no inhibitory activity on protein methylase III from Krebs ascites cells,[92] chick embryo fibroblasts,[93] and *Euglena gracilis*.[118] The other analogs tested so far, such as methylthioadenosine (n° 6), (hydroxyethyl)thioadenosine (n° 7), *n*-butylthioadenosine (n° 24) and *S*-adenosyl-γ-thio-α-ketobutyrate (n° 27) are inactive.

The state of oxidation of the sulfur atom of AdoHcy is important, *S*-adenosyl-L-homocysteine sulfoxide (n° 3) being inactive at 15 μ*M* on the enzyme from rat brain.[116,117]

TABLE 3
Comparison of the Kinetic Constant of Various AdoHcy Analogs for Protein-Methylase III from Various Sources

Compound	Rat spleen 122	Rat liver 119	Rat brain 117	Rat brain nuclei 116	Krebs ascites 120	CEF 98/938· 121·	CEF RSV 98	Wheat germ 115	*Euglena gracilis* 118	*Leishmania* d**	*Leishmania* t** 99	*Leishmania* e**
	Sources of enzyme and references											
K_m (μM) AdoMet K_i (μM) Analogs	—	3.0	—	12.5	—	59/3.1* 130·	101 177·	47.6	27^A—34^B	120	37	50
A. With modified 5′-side chain												
1 AdoHcy (L)	(70%)^d	12.0	i	5.9	4.4	5.5	5.8	(73%)^b	$15^A/1.6^B$ $(80^A/99^B)^c$	4.8	3.0	4.9
2 AdoHcy (D)	—	—	i	i	—	—	—	(33%)^b	$(46^A/87^B)$	—	—	—
3 AdoHcySO(L)	—	—	— (0%)^a	(0%)	—	—						
5 Adenosine (= Ado)	(0%i)^d	—	— (0%)^a	(0%)	—	—	—	—	— $(5^A/0^B)^{cc}$			
6 5′-*S*-methyl-Ado		—	(0%)^a	(0%)	—	—	—	(0%)^b	$(4^A/8^B)^c$			
7 5′-*S*-hydroxy-ethyl-Ado	—	—	—	—	—	—	—	(0%)^b				
17 AdoCys(L)	—	—	—	—	—	—	—	—	—$(3^A/38^B)^c$			
20 5′-S(2-methyl-propyl)-Ado (SIBA)	—	—	—	—	—	1328 3348·	3117 1984·	—	$(0^A/4^B)^c$			
24 5′-S-butyl- Ado	—	—	—	—	—	—	—	(0%)^b				
27 *S*-adenosyl-γ-thio-α-keto-butyrate	—	—	— (0%)^a	— (0%)	—	—						
35 Sinefungin	—	—	—	—	—	4.9	7.1	—	$19^A/1.5^B$ $(66^A/94^B)^c$	1.8	47	73
36 A9145C	—	—	—	—	—	3.9	—	—	— $(0^A$—$26^B)^c$	∞	—	∞
40 AdoEth(L)	(25%)^d	170	—	—	133	i	—	—	$40^A/44^B$ $(34^A/56^B)^c$			
B. With modified base												
49 Tubercidin	(0%)^e											

53 Isopentenyl-Ado	(0%)[e]												
69 InoHcy(L)	—	—	—	—	—	—	—	(0%)[b]					
76 *S*-ribosyl-homocys-teine	—	—	(0%)[a]	(0%)									
77 *S*-ribosyl-homocys-teinesulfoxide	—	—	(0%)[a]										
C. With modified sugar													
36 A9145C	—	—	—	—	—	3.9	—	—		($O^A/26^B$)[c]	∞	—	∞
80 Cordycepin	—	—	—	—	—	—	—	—	—	($5^A/16^B$)[c]			
82 Arabinosyl-adenine	—	—	—	—	—	—	—	—	—	($4^A/O^B$)[c]			
83 Adenine	(0%)[f]	—	(0%)[a]	(0%)									

Note: The K_i values are expressed in μ*M*. When no kinetic values are given, the data are expressed in percentage of inhibition (%) observed *in vitro* with a given concentration of AdoHcy analog: [a] 15 μ*M;* [b] 40 μ*M;* [c] 50 μ*M;* [d] 100—130 μ*M;* [e] 500—600 μ*M;* [f] 3000 μ*M.* [i] when a product is mentionned to inhibit, without any values.
** d = *L. donovani;* t = *L. tropica;* e = *L. enrietti.*
The analogs are numbered as in Figure 1A, B, and C.

The replacement of the sulfur atom by an amino-substituted methylene group (sinefungin n° 35) does not abolish the inhibitory activity; in fact besides AdoHcy, only sinefungin and A9145C (n° 36) are good competitive inhibitors of protein methylase III *in vitro*.[98,99,115,118]

S-adenosyl-L-ethionine (n° 40) is the only sulfonium analog tested on protein methylase III: it appears that, in contrast to the observation on protein methylase I which utilizes AdoMet and AdoEth with comparable efficiency, AdoEth inhibits noncompetitively the utilization of AdoMet by protein methylase III from rat liver;[119] the K_m value of AdoMet is 3 μM while the K_i value for AdoEth is 170 μM, thus obliterating the biochemical significance of this analogue on the *in vivo* enzymatic methylation of the ε-amino group of histone lysine residues. However, in the case of *Euglena gracilis*, AdoEth inhibits both histone H 1-specific methylases in a competitive manner with respect to *S*-adenosylmethionine, the K_i values being three to four times higher than those of *S*-adenosylhomocysteine,[118] while for the enzyme from Krebs II ascite tumor cells the K_i value of AdoEth (133 μM) is 30 times higher than that for AdoMet.[120] Greenaway and Levine found that AdoEth inhibits the protein methylase III from chick embryo nuclei.[121]

Adenosine has no effect on the activity of protein methylase III from various sources.[116-118,122]

b. Analogs with a Modified Base

The replacement of the 6-amino group of the adenine moiety of AdoHcy by a hydroxyl group (InoHcy n° 69), as well as the absence of adenine ring (*S*-ribosylhomocysteine n° 76) abolish the inhibitory activity toward the protein methylase III from wheat germ, indicating the importance of this part of the molecule.[115]

c. Analogs with a Modified Sugar

A9145C (n° 36) is as active as sinefungin (n° 35) towards protein methylase III from chick embryo fibroblast[98] suggesting that the nature of the 4′-5′-bond is not important for recognition by protein methylase III.

d. Conclusion

These results indicate that the inhibition by AdoHcy analogs is strongly dependent on the integrity of both the adenine and homocysteine portions of the molecule, the exceptions being sinefungin and A9145C.

D. METHYLATION OF HISTIDINE RESIDUES

3-*N*-methylhistidine is found in the contractile proteins, actine and myosine.

The first report on partial purification and characterization of *S*-adenosylmethionine: protein-histidine *N*-methyltransferase appeared only recently.[123] As for the other AdoMet-dependent methyltransferases this enzyme is inhibited by AdoHcy, but no results are yet available on structural requirements for this inhibition.

E. METHYLATION OF THE SULFUR ATOM OF METHIONINE RESIDUES

A cytochrome *c*-specific enzyme which methylates methionine residues has been reported in *Euglena gracilis*.[94]

1. Apparent Kinetic Constants of Methionine-*S*-Methylase for AdoMet and AdoHcy *In Vitro*

The K_m value for AdoMet for this enzyme is 16.6 μM (Table 4). As for the other AdoMet-dependent transmethylases, AdoHcy is a competitive inhibitor with respect to AdoMet, with a K_i value of 8.13 μM.[94]

TABLE 4
Inhibition by AdoHcy Analogs of Methionine *S*-Methylase

Compound	Source of enzyme and references *Euglena gracilis*[94]	
A. With modified 5′-sidechain		
1. AdoHcy(L)	(88.5%)[a]	8.13 μ*M*
2. AdoHcy(D)	(27.4%)[a]	—
5. Adenosine(Ado)	(0.8%)[a]	—
6. 54-*S*-methyl-Ado	(6.0%)[a]	—
17. AdoCys(L)	(0.6%)[a]	—
35. Sinefungin	(52.1%)[a]	—
36. A9145C	(9.8%)[a]	—
40. AdoEth(L)	(15.3%)[a]	—
B. With modified base		
66. 5′-*S*-inosyl(2-hydroxy-4-methyl-thiobutyrate)	(4.5%)[a]	—
C. With modified sugar		
5. Adenosine	(0.8%)[a]	—
36. A9145C	(9.8%)[a]	—
78. 2′-deoxyadenosine	(7.0%)[a]	—
80. 3′-deoxyadenosine	(9.5%)[a]	—
82. Arabinosyladenine	(10.4%)[a]	—

[a] Values in brackets represent the percentage of inhibition observed with 50 μ*M* AdoHcy analogs. The analogs are numbered as in Figure 1A, B, and C.

2. Inhibition of Methionine-*S*-Methylase by Various Analogs of *S*-Adenosyl-L-Homocysteine

Various AdoHcy analogs were tested for their inhibitory activity toward this enzyme (Table 4).[94] Only D- and L-isomers of AdoHcy and sinefungin are significantly inhibitory. All the other analogs show very little inhibitory effect, indicating that the observed inhibition of AdoHcy is strongly dependent on the integrity of both the adenine and homocysteine portions of the molecule.

IV. EFFECT OF ADOHCY ANALOGS ON PROTEIN METHYLATION *IN VIVO*

Based on the metabolic pathways of AdoMet and AdoHcy, there are several approaches to design *in vivo* inhibitors of AdoMet-dependent methyltransferases:

1. Molecules which act directly on a specific methyltransferase (i.e., analogs of the methyl acceptor substrate, the methylated product, AdoHcy or AdoMet)
2. Molecules that act indirectly by inhibiting AdoMet biosynthesis or AdoHcy metabolism.

The approach of altering the activity of adenosyltransferase has been explored. It resulted in the decrease of the intracellular level of AdoMet leading to an inhibition of all AdoMet-dependent enzymes as well as polyamine synthesis. Analogs of AdoMet have been examined as possible inhibitors of various AdoMet-dependent methyltransferases. The use of these analogs as such has two disadvantages: their chemical instability, and problems associated with cellular transport. However, a variation of this approach is the use of substrate analogs for this enzyme directly in cell culture allowing the formation of AdoMet analogs *in situ* (ex.: ethionine, neplanocin....).

The approach of altering AdoHcy metabolism by changing the activity of AdoHcy hydrolase (EC 3.3.1.1) has been widely studied; the results being either the increase of the intracellular concentration of AdoHcy when this enzyme is inhibited or the formation of analogs of AdoHcy *in situ*.

A third approach has been explored: evaluation of the activity of exogenous AdoHcy analogs.

In this section the term *in vivo* will be used in a wide sense and will include studies carried out on animals or in organs and cell cultures.

Only few studies dealing with the biological effects of methyltransferases inhibitors describe the effect of AdoHcy analogs on protein methylation *in vivo*.

In order to evaluate a potential selectivity of the effect of AdoHcy analogs we will summarize the various results according to the compounds studied and not according to the methylase.

Several compounds have been shown to inhibit the methylation of proteins *in vivo*. They include molecules like sodium butyrate[124] inhibiting arginine methylation, or molecules which are either inhibitor (cycloleucine)[125] or alternate substrate of AdoMet synthetase (ethionine).[126] Following administration of ethionine, alteration in the methylation pattern in rat liver histones has been observed. Thus, the methylation of arginine and lysine was inhibited by 89 and 50%, respectively, after a 0.5 mg ethionine/g body weight administration.[127] *In vivo* ethionine is processed to AdoEth, which ethylates histones, specifically on arginine residues.[128] Gilliland et al. have shown that ethionine ingestion reduces pancreatic carboxymethylase activity.[112]

Intact erythrocytes cultured with adenosine and homocysteine, and thus having a higher intracellular pool of AdoHcy, have a lower amount of carboxymethylation.[129] Similar results have been obtained in brain; adenosine and homocysteine treatment decreases the incorporation of labile and stable methyl groups into brain proteins in response of the elevation of AdoHcy level.[130]

3-Deazaadenosine (C^3Ado) (n° 46) treatment leads to both accumulation of AdoHcy and accumulation of C^3AdoHcy which results, in turn, in the inhibition of various cellular transmethylation reactions in mouse lymphocytes,[131] and decreases the protein carboxymethylations in intact rat posterior pituitary lobes.[132]

Sinefungin (n° 35) has been shown to inhibit *in vivo* methylation to various extents, according to the methylase tested and the organism studied. In *Leishmania donovani* cultured in the presence of 0.26 μ*M* sinefungin the methylation of the arginine and lysine residues of the proteins found in the 12,000 *g* supernatant fraction was not affected.[97] However, results were different for the proteins extracted from the 12,000 *g* pellet. Although a slight inhibition was observed for the N^G-N^G-dimethylarginine (6%), more significant values were obtained for the methylation of the dicarboxylic amino acids and lysine: 52% inhibition for methanol production, 26 and 16% inhibition for monomethyllysine and dimethyllysine, respectively. The formation of trimethyllysine and *N*-methylhistidine was not inhibited. (Lawrence et al. in preparation). These results are in good agreement with K_i values obtained *in vitro;* only methylase III of *L. donovani* had a high affinity for sinefungin (K_i 1.8 μ*M*),[97] and protein carboxymethylase of *L. mexicana* has been efficiently inhibited by this nucleoside.[109]

The methylation of methyl-accepting chemotaxis proteins (MCPs) is inhibited by the addition of sinefungin (n° 35) or A9145C (n° 36) in the culture of *E. coli* made permeable to AdoMet related molecules by EGTA treatment: 50% inhibition occurs at 10^{-8} *M* sinefungin, 10^{-9}*M* A9145C and 10^{-5} *M* AdoHcy.[133] The ability of several AdoHcy analogs to inhibit methylation of endogenous methyl acceptor proteins (MAPs) by endogenous protein carboxymethylase in rat hypothalamic synaptosomes has been tested.[52] AdoHcySO (n° 3), AdoHcySO_2 (n° 4) and 2-azaAdoHcy (n° 72) produced significant inhibition of protein carboxymethylation in this system. The natural analogs sinefungin (n° 35) and A9145C (n° 36) were inactive as inhibitors of protein methylase II in intact synaptosomes but active in lysed synaptosomes, suggesting problems of transport of these inhibitors within these cells.

Brain cells treated with 30 μ*M* sinefungin produce vacuolated myelin as observed under cycloleucine treatment, while the myelin basic protein (arginine) methyltransferase is inhibited by 60%.[134]

Among the AdoHcy analogs tested *in vivo,* SIBA (n° 20) inhibits slightly the overall methylation of the Polyoma virus proteins. Lysine and histidine methylations are unaffected in empty capsids and in complete virions, whereas the methyl incorporation in the N^G,N^G-dimethylarginine is three times lower in the virus particles produced by SIBA-treated cells.[96] In herpes virus-infected cells, the methylation of total proteins was slightly affected by SIBA.[135] Methyl esterification of human erythrocyte membrane proteins *in vivo* is inhibited by 5′MTA (n° 6), SIBA (n° 20) and adenine (n° 83).[136] This *in vivo* effect of SIBA could be due to the action of adenine produced by a phosphorolytic cleavage of this molecule.[137] The lack of any inhibition by SIBA in phosphate depleted cells supports this view, and is in agreement with *in vitro* data on protein carboxymethylase.[44]

In vivo S-tubercidinyl homocysteine (n° 70), a physiologically stable AdoHcy analog,[138] is a potent inhibitor of protein methylase II involved in chemotaxis.[139]

V. CONCLUSIONS

The great variety of natural and synthetic analogs of AdoHcy allowed to shed light on the specific binding requirements of the three most studied protein methylases:*S*-adenosyl-methionine:protein-arginine-*N*-methyltransferase (PM I), *S*-adenosylmethionine:protein-carboxyl-*O*-methyltransferase (PM II) and *S*-adenosylmethionine:protein-lysine-*N*-methyl-transferase (PM III). Systematic studies using analogs modified on the base, sugar, and amino acid moieties suggest that PM II has the most strict structural requirements with respect to the effectors. It is more difficult to evaluate the results obtained with AdoHcy analogs using whole cells or organs. However, in most cases, when an analog showed good affinity for an enzyme *in vitro,* its effect *in vivo* could also be demonstrated.

In the future, efforts should be made to perform similar systematic work with the less studied protein methylases, namely with the enzymes catalyzing the *S*-methylation of methionine, the *N*-methylation of histidine, and the enzymes recently reviewed by Stock et al.,[7] catalyzing the methylation of *N*-terminal α-amino groups of methionine, alanine, phenylalanine, and proline in specific proteins. The increasing number of protein methylases described in the last 5 years should stimulate further research to synthesize new types of nucleosides which should uncover subtle structural differences in the active site of these enzymes.

ACKNOWLEDGMENT

This chapter is dedicated to the memory of Professor Edgar Lederer.

REFERENCES

1. **Paik, W. K. and Kim, S.,** Protein methylation: chemical, enzymological, and biological significance, in *Advances in Enzymology,* Vol. 42, John Wiley & Sons, New York, 1975.
2. **Cantoni, G. L.,** Biological methylation: selected aspects, *Annu. Rev. Biochem.,* 44, 435, 1975.
3. **Byvoet, P. and Baxter, C. S.,** *Chromosomal Proteins and their Roles in the Regulation of Gene Expression,* Academic Press, New York, 1975, 127.
4. **Paik, W. K. and Kim, S.,** *Protein Methylation,* Meister A., Ed., John Wiley & Sons, New York, 1980.
5. **Paik, W. K. and Kim, S.,** *The Enzymology of Post-Translational Modification of Proteins,* Vol. 2, Freedman R. and Hawkins H. C., Eds., Academic Press, London, 1985.

6. **Paik, W. K., Kim, S., and Lee, H. W.,** Enzymology of protein methylation: recent development, *Biological Methylation and Drug Design,* Borchardt, R. T., Creveling, C. R., and Ueland, P. M., Eds., Humana Press, Clifton, NJ, 1986, 15.
7. **Stock, A., Clarke, S., Clarke, C., and Stock, J.,** N-Terminal methylation of proteins: structure, function and specificity, *FEBS Lett.,* 220, 8, 1987.
8. **Cantoni, G. L.,** The nature of the active methyl donor formed enzymatically from L-methionine and adenosinetriphosphate, *J. Am. Chem. Soc.,* 74, 2942, 1952.
9. **Cantoni, G. L.,** Activation of methionine for transmethylation, *J. Biol. Chem.,* 189, 745, 1951.
10. **Schlenk, F.,** The biosynthesis of S-adenosylmethionine by yeast cells, in *Transmethylation,* Elsevier/ North Holland, New York, 1979, 3.
11. **Cantoni, G. L. and Scarano, E.,** The formation of S-adenosylhomocysteine in enzymatic transmethylation reactions, *J. Am. Chem. Soc.,* 76, 4744, 1954.
12. **Baddiley, J. and Jamieson, G. A.,** Synthesis of S-(5′-deoxyadenosine-5′)-homocysteine, a product from enzymatic methylations involving "active methionine", *J. Chem. Soc.,* 5941, 1085, 1955.
13. **Deguchi, T. and Barchas, J.,** Inhibition of transmethylations of biogenic amines by S-adenosylhomocysteine, *J. Biol. Chem.,* 246, 3175, 1971.
14. **de la Haba, G. and Cantoni, G. L.,** The enzymatic synthesis of S-adenosylhomocysteine from adenosine and homocysteine, *J. Biol. Chem.,* 234, 603, 1959.
15. **Palmer, J. L. and Abeles, R. H.,** Mechanism for enzymatic thioether formation, *J. Biol. Chem.,* 251, 5817, 1976.
16. **Richards, H. H., Chiang, P. K., and Cantoni, G. L.,** Adenosylhomocysteine hydrolase crystallization of the purified enzyme and its properties, *J. Biol. Chem.,* 253, 4476, 1978.
17. **Palmer J. L. and Abeles R. H.,** The mechanism of action of S-adenosylhomocysteinase, *J. Biol. Chem.,* 254, 1217, 1979.
18. **Duerre, J. A. and Walker, R. D.,** Metabolism of adenosylhomocysteine, in *The Biochemistry of Adenosylmethionine,* Salvatore, F., Borek, E., Zappia, V., Williams-Ashman, H.G., and Schlenk, F., Eds., Columbia University Press, New York, 1977, 43.
19. **Miller, C. H. and Duerre, J. A.,** Oxidative deamination of S-adenosyl-L-homocysteine by rat kidney L-aminoacid oxidase, *J. Biol. Chem.,* 244, 4272, 1969.
20. **Duerre, J. A.,** Hydrolytic nucleosidase acting on S-adenosyl homocysteine and on 5′-methylthioadenosine, *J. Biol. Chem.,* 237, 3737, 1962.
21. **Walker, R. D. and Duerre, J. A.,** S-adenosylhomocysteine metabolism in various species, *Can. J. Biochem.,* 53, 312, 1975.
22. **Ishida, T., Morimoto, H., and Inoue, M.,** Three dimensional X-ray crystal structure of S-adenosyl-L-homocysteine, *J. Chem. Soc. Chem. Commun.,* 671, 1981.
23. **Kikugawa, K. and Ichino, M.,** Direct halogenation of sugar moiety of nucleosides, *Tetrahedron Lett.,* 87, 1971.
24. **Borchardt, R. T and Wu, Y. S.,** Potential inhibitors of S-adenosyl methionine-dependent methyltransferases. I. Modification of the aminoacid portion of S-adenosylhomocysteine, *J. Med. Chem.,* 17, 862, 1974.
25. **Davoll, J., Lythgoe, B., and Todd, A. R.,** Experiments on the synthesis of purine nucleosides. XIX. A synthesis of adenosine, *J.Chem.Soc.,* 987, 1948.
26. **Baddiley, J.,** Adenine 5′-deoxy-5′-methylthiopentoside (adenine thiomethyl pentoside) a proof of structure and synthesis, *J. Chem. Soc.,* 1348, 1951.
27. **Legraverend, M. and Michelot, R.,** Nouvelles synthèses d'analogues de la S-adenosylhomocystèine et de la S-adenosylmèthionine, *Biochimie,* 58, 723, 1976.
28. **Coat, J. P. and David, S.,** Préparation et couplage sur des polypeptides de 5′-O-carboxyméthyl-ribonucléosides, *Carbohydr. Res.,* 12, 335 1970.
29. **Gillet, L., Looze, Y., Deconnick, M., and Leonis, J.,** Binding capacities of various analogues of S-adenosyl-L-homocysteine to protein methyltransferase II from human erythrocytes, *Experientia,* 35, 1007, 1979.
30. **Legraverend, M., Ibanez, S., Blanchard, P., Enouf, E., Lawrence, F., Robert-Gero, M., and Lederer, E.,** Structure-activity relationship of synthetic S-adenosyl-homocysteine analogues: effect on oncogenic virus-induced cell transformation and on methylation of macromolecules, *Eur. J. Med. Chem.,* 12, 105, 1977.
31. **Kuhn, R. and Jahn, W.,** Vom adenosin abgeleitete thioäther und S-oxide, *Chem. Ber.,* 98, 1699, 1965.
32. **Vuilhorgne, M., Blanchard, P., Hedgecock, C. J. R., Lawrence, F., Robert-Gero, M., and Lederer, E.,** New synthetic S-adenosylhomocysteine analogues with oncostatic and antiviral activity, *Heterocycles,* 11, 495, 1978.
33. **Hildesheim, J., Hildesheim, R., and Lederer, E.,** Nouvelles synthèses d'analogues de la S-adenosylhomocystéine, inhibiteurs potentiels des méthyltransférases, *Biochimie,* 54, 431, 1972.
34. **Hildesheim, J., Hildesheim, R., and Lederer, E.,** Synthèses d'inhibiteurs des méthyl-transférases: analogues de la S-adénosyl homocystéine, *Biochimie,* 53, 1067, 1971.
35. **Duerre, J. A., Miller, C. H., and Reams, G. G.,** Metabolism of S-adenosyl-L-homocysteine *in vivo* by the rat, *J. Biol. Chem.,* 244, 107, 1969.

36. **Nakagawa, I. and Hata, T.**, A convenient method for the synthesis of 5′-S-alkylthio-5′-deoxyribonucleosides, *Tetrahedron Lett.*, 1409, 1975.
37. **Hamil, R. L. and Hoehn, M. M.**, A9145, a new adenine-containing antifungal antibiotic. I. Discovery and isolation, *J. Antibiot.*, 26, 463, 1973.
38. **Mock, G. A. and Moffat, J. G.**, An approach to the total synthesis of sinefungin, *Nucleic Acids Res.*, 10, 6223, 1982.
39. **Nagarajan, R., Chao, B., Dorman, D. E., Nash, S. M., Occolowitz, J. L., and Shabel, A.**, Sinefungin (A9145), an adenine containing antifungal antibiotic: structure elucidation of sinefungin and related metabolites, *17th Intersci. Conf. Antimicrob. Ag. Chemother.*, Abstr. n° 50, 1977.
40. **Blanchard, P., Dodic, N., Fourrey, J. L., Géze, M., Lawrence, F., Malina, H., Paolantonacci, P., Vedel, M., Tempéte, C., Robert-Gero, M., and Lederer, E.**, Sinefungin and derivatives: synthesis, biosynthesis and molecular target studies in *Leishmania,* in *Biological Methylation and Drug Design,* Borchardt, R. T., Creveling, C. R., and Ueland, P. M., Eds., Humana Press, Clifton, NJ, 1986, 435.
41. **Jahn, W.**, Synthese 5′-substituierter adenosinderivate, *Chem. Ber.*, 98, 1705, 1965.
42. **Baddiley, J. and Todd, A. R.**, Nucleotides. I. Muscle adenylic acid and adenosine diphosphate, *J. Chem. Soc.*, 648, 1947.
43. **Parks, L. W.**, S-adenosylethionine and ethionine inhibition, *J. Biol. Chem.*, 232, 169, 1958.
44. **Oliva, A., Galletti, P., Zappia, V., Paik, W. K., and Kim, S.**, Studies on substrate specificity of S-adenosylmethionine: protein-carboxyl methyltransferase from calf brain, *Eur. J. Biochem.*, 104, 595, 1980.
45. **Zappia, V., Zydek-Cwick, R., and Schlenk, F.**, The specificity of S-adenosylmethionine derivatives in methyltransfer reactions, *J. Biol. Chem.*, 244, 4499, 1969.
46. **Jones, J. W. and Robins, R. K.**, Purine nucleosides. III. Methylation studies of certain naturally occurring purine nucleosides, *J. Am. Chem. Soc.*, 85, 193, 1963.
47. **Rousseau, R. J., Townsend, L. B., and Robins, R. K.**, The synthesis of 4-amino-β-D-ribofuranosylimidazo[4,5-C] pyridine (3-deazaadenosine) and related nucleotides, *Biochemistry,* 5, 756, 1966.
48. **Mizuno, Y., Tazawa, S., and Kageura, K.**, Synthetic studies of potential antimetabolites. XII. Synthesis of 4-substituted 1-(β-D-ribofuranosyl)-1-H-imidazo[4,5-C] pyridines, *Chem. Pharmacol. Bull.*, 16, 2011, 1968.
49. **Borchardt, R. T., Huber, J. A., and Wu, Y. S.**, Potential inhibitors of S-adenosylmethionine-dependent methyltransferases. II. Modification of the base portion of S-adenosylhomocysteine, *J. Med. Chem.*, 17, 868, 1974.
50. **Coward, J. K., Bussolotti, D. L., and Chang, C. D.**, Analogs of S-adenosyl homocysteine as potential inhibitors of biological transmethylation. Inhibition of several methylases by S-tubercidinyl homocysteine, *J. Med. Chem.*, 17, 1286, 1974.
51. **Montgomery, J. A. and Thomas, H. J.**, Nucleosides of 2-azapurines and certain ring analogs, *J. Med. Chem.*, 15, 182, 1972.
52. **Borchardt, R. T., Kuonen, D., Huber, J. A., and Moorman, A.**, Inhibition of calf thymus and rat hypothalamic synaptosomal protein-carboxymethyltransferase by analogues of S-adenosylhomocysteine, *Mol. Pharmacol.*, 21, 181, 1982.
53. **Borchardt, R. T., Huber, J. A., and Wu, Y. S.**, Potential inhibitors of S-adenosylmethionine-dependent methyltransferases. IV. Further modifications of the amino acid and base portions of S-adenosyl-L-homocysteine, *J. Med. Chem.*, 19, 1094, 1976.
54. **Montgomery, J. A., Thomas, H. J., and Clayton, S. J.**, A convenient synthesis of 8-azaadenosine, *J. Heterocycl. Chem.*, 7, 215, 1970.
55. **Borchardt, R. T., Huber, J. A., and Wu, Y. S.**, A convenient preparation of S-adenosylhomocysteine and related compounds, *J. Org. Chem.*, 41, 565, 1976.
56. **Duerre, J. A., Salisbury, L., and Miller, C. H.**, Preparation and characterization of sulfoxides of S-adenosylhomocysteine and S-ribosyl-L-homocysteine, *Anal. Biochem.*, 35, 505, 1970.
57. **Tolman, R. L., Robins, R. K., and Townsend, L. B.**, Pyrrolo[2,3-d]pyrimidine nucleoside antibiotics. Total synthesis and structure of toyocamycin unamycin B, vengicide antibiotic E-212 and sangivamycin (BA.9012), *J. Am. Chem. Soc.*, 90, 524, 1968.
58. **Tolman, R. L., Robins, R. K., and Townsend, L. B.**, Pyrrolopyrimidine nucleosides. III. The total synthesis of toyocamycin, sangivamycin, tubercidinyl, and related derivatives, *J. Am. Chem. Soc.*, 91, 2102, 1969.
59. **Reist, E. J., Calkins, D. F., Fisher, L. V., and Goodman, L.**, The synthesis and reactions of some 8-substituted purine nucleosides, *J. Org. Chem.*, 33, 1600, 1967.
60. **Kochetkov, N. K., Shibaev, V. N., and Kost, A. A.**, New reaction of adenine and cytosine derivatives potentially useful for nucleic acids modification, *Tetrahedron Lett.*, 22, 1993, 1971.
61. **Kalckar, H. M.**, Differential spectrophotometry of purine compounds by means of specific enzymes—determination of hydroxypurine compounds, *J. Biol. Chem.*, 167, 429, 1947.
62. **Hall, R. H., Robins, M. J., Stasiuk, L., and Thedford, R.**, Isolation of N^6(γ,γ-dimethylallyl)adenosine from soluble ribonucleic acid, *J. Am. Chem. Soc.*, 88, 2614, 1966.
63. **Davoll, J.**, The synthesis of the V-triazolo[d]pyrimidine analogues of adenosine, inosine, guanosine and xanthosine, and a new synthesis of guanosine, *J. Chem. Soc.*, 1593, 1958.

64. **Howard, G. A., Lythgoe, B., and Todd, A. R.,** A synthesis of cytidine, *J. Chem. Soc.*, 1052, 1947.
65. **Loring, H. S. and Ploeser, J. T.,** The deamination of cytidine in acid solution and the preparation of uridine and cytidine by acid hydrolysis of yeast nucleic acid, *J. Biol. Chem.*, 178, 439, 1949.
66. **Robins, M. J., Mengel, R., Jones, R. A., and Fouron, Y.,** Nucleic acid related compounds. XXII. Transformation of ribonucleoside 2′,3′-O-ortho esters into halo, deoxy, and epoxy sugar nucleosides using acyl halides. Mechanism and structure of products, *J. Am. Chem. Soc.*, 98, 8204, 1976.
67. **Robins, M. J., Naik, S. R., and Lee, A. S. K.,** Nucleic acid related compounds. XII The facile and high yield stannous chloride catalyzed monomethylation of the cis-glycol system of nucleosides by diazomethane, *J. Org. Chem.*, 39, 1891, 1974.
68. **Cunningham, K. G., Hutchinson, S. A., Manson, W., and Spring, F. S.,** Cordycepin a metabolic product from cultures of Cordyceps militaris. I. Isolation and characterization, *J. Chem. Soc.*, 2299, 1951.
69. **Walton, E., Nutt, R. F., Jenkins, S. R., and Holly, F. W.,** 3′-deoxynucleosides. I. A synthesis of 3′-deoxyadenosine, *J. Am. Chem. Soc.*, 86, 2952, 1964.
70. **Lee, W. W., Benitez, A., Goodman, L., and Baker, B. R.,** Potential anticancer agents. XL. Synthesis of the β-anomer of 9-(D-arabinofuranosyl)-adenine, *J. Am. Chem. Soc.*, 82, 2648, 1960.
71. **Borchardt, R. T., Wu, Y. S., and Wu, B. S.,** Affinity labeling of histamine N-methyltransferase by 2′,3′,dialdehyde derivatives of S-adenosylhomocysteine and S-adenosylmethionine: kinetic of inactivation, *Biochemistry*, 17, 4145, 1978.
72. **Yaginuma, S., Muto, N., Tsujino, M., Sudate, Y., Hayash, M., and Otani, M.,** Studies on neplanocin A, new antitumor antibiotic. I. Producing organism, isolation and characterization, *J. Antibiot.*, 34, 359, 1981.
73. **Keller, B. T. and Borchardt, R. T.,** Metabolic conversion of neplanocin A to S-neplanocylmethionine by mouse L929 cells, *Biochem. Biophys. Res. Commun.*, 120, 131, 1984.
74. **Hori, M., Ito, E., Takita, T., Koyama, G., Takeuchi, T., and Hamao, U.,** A new antibiotic formycin, *J.Antibiot.*, 17A, 96, 1964.
75. **Farkas, J., Flegelova, Z., and Sorm, F.,** Synthesis of pyrazomycin, *Tetrahedron Lett.*, 22, 2279, 1972.
76. **Kusakabe, Y., Nagatsu, J., Shibuya, M., Kawaguchi, O., Hirose, C., and Shirato, S.,** Minimycin, a new antibiotic, *J. Antibiot.*, 25, 44, 1972.
77. **Kalvoda, L., Farkas, J., and Sorm, F.,** Synthesis of showdomycin, *Tetrahedron Lett.*, 26, 2297, 1970.
78. **Ferro, A. J.,** Function and metabolism of 5′-methylthioadenosine, in *Transmethylation*, Usdin, E., Borchardt, R. T., and Creveling, C. R., Eds., Elsevier/North Holland, New York, 1979, 117.
79. **Pegg, A. E., Borchardt, R. T., and Coward, J. K.,** Effects of inhibitors of spermidine and spermine synthesis on polyamine concentrations and growth of transformed mouse fibroblasts, *Biochem. J.*, 194, 79, 1981.
80. **Kamatani, N. and Corson, D. A.,** Abnormal regulation of methylthioadenosine and polyamine metabolism in methylthioadenosine phosphorylase-deficient human leukemic cell lines, *Cancer Res.*, 40, 4178, 1980.
81. **Schlenk, F.,** Methylthioadenosine, *Adv. Enzymol.*, 54, 195, 1983.
82. **Lyga, J. W. and Secrist, J. A., III,** Synthesis of chain-extended and C-6′ functionalized precursors of the nucleoside antibiotic sinefungin, *J. Org. Chem.*, 48, 1982, 1983.
83. **Gèze, M., Blanchard, P., Fourrey, J. L., and Robert-Gero, M.,** Synthesis of sinefungin and its C-6′ epimer, *J. Am. Chem. Soc.*, 105, 7638, 1983.
84. **Moorman, A. R., Martin, T., and Borchardt, R. T.,** Addition of 1,nitroalkanes to methyl 2,3-O-isopropylidene-β-D-ribo-penta dialdo-1,4-furanoside and N6-benzoyl-2′,3′-O-isopropylideneadenosine-5′-aldehyde, *Carbohydr. Res.*, 113, 233, 1983.
85. **Mizuno, Y., Tsuchida, K., and Tampo, H.,** Studies directed toward the total synthesis of sinefungin. Synthesis of 4-(5′-deoxyuridin-5′-yl)-4-nitro-butyronitrile, 4-(5′-deoxyadenosin-5′-yl)-4-nitro-butyramide and closely related nucleosides, *Chem. Pharmacol. Bull.*, 32, 2915, 1984.
86. **Ueland, P. M.,** Pharmacological and biochemical aspects of S-adenosylhomocysteine and S-adenosylhomocysteine hydrolase, *Pharmacol. Rev.*, 34, 223, 1982.
87. **Paik, W. K. and Kim, S.,** Protein methylase I, purification and properties of the enzyme, *J. Biol. Chem.*, 243, 2108, 1968
88. **Kakimoto, Y. and Akazawa, S.,** Isolation and identification of N^G,N^G- and N^G,N'^G-dimethylarginine, N^ε-mono, di, and trimethyllysine, and glucosyl-galactosyl- and galactosyl-δ-hydroxylysine from human urine, *J. Biol. Chem.*, 245, 5751, 1970.
89. **Hoffman, D. R., Cornatzer, W. E., and Duerre, J. A.,** Relationship between tissue levels of S-adenosylmethionine, S-adenosylhomocysteine, and transmethylation reactions, *Can. J. Biochem.*, 57, 56, 1979.
90. **Baldessarini, R. J. and Kopin, J.,** S-adenosylmethionine in brain and other tissues, *J. Neurochem.*, 13, 769, 1966.
91. **Salvatore, F., Utili, R., and Zappia, V.,** Quantitative analysis of S-adenosylmethionine and S-adenosylhomocysteine in animal tissues, *Anal. Biochem.*, 41, 16, 1971.
92. **Casellas, P. and Jeanteur, P.,** Protein methylation in animal cells; inhibition of S-adenosyl-L-methionine: protein (arginine)N-methyltransferase by analogs of S-adenosyl-L-homocysteine, *Biochim. Biophys. Acta*, 519, 255, 1978.

93. **Enouf, J., Lawrence, F., Tempete, C., Robert-Gero, M., and Lederer, E.,** Relationship between inhibition of protein methylase I and inhibition of Rous sarcoma virus-induced cell transformation, *Cancer Res.*, 39, 4497, 1979.
94. **Farooqui, J. Z., Tuck, M., and Paik, W. K.,** Purification and characterization of enzymes from *Euglena gracilis* that methylate methionine and arginine residues of cytochrome C, *J. Biol. Chem.*, 260, 537, 1985.
95. **Gupta, A., Jensen, D., Kim, S., and Paik, W. K.,** Histone-specific protein-arginine methyltransferase from wheat germ, *J. Biol. Chem.*, 257, 9677, 1982.
96. **Raies, A.,** Etude comparative du mécanisme et du site d'action de la 5′-désoxy-5′-S-isobutylthioadenosine sur la réplication du virus polyome et la production du virus du sarcome de Rous, *Thèse de doctorat es Sciences naturelles, Orsay*, 1983.
97. **Paolantonacci, P., Lawrence, F., Lederer, E., and Robert-Gero, M.,** Protein methylation and protein methylases in *Leishmania donovani* and *Leishmania tropica* promastigotes, *Mol. Biochem. Parasitol.*, 21, 47, 1986.
98. **Vedel, M., Lawrence, F., Robert-Gero, M., and Lederer, E.,** The antifungal antibiotic sinefungin as a very active inhibitor of methyltransferases and of the transformation of chick embryo fibroblasts by Rous sarcoma virus, *Biochem. Biophys. Res. Commun.*, 85, 371, 1978.
99. **Paolantonacci, P., Lawrence, F., and Robert-Gero, M.,** Differential effect of sinefungin and its analogs on the multiplication of three *Leishmania* species, *Antimicrob. Agents Chemother.*, 28, 528, 1985.
100. **Baxter, C. S. and Byvoet, P.,** Effect of carcinogens and other agents on histone methylation by a histone arginine methyltransferase purified from rat liver cytoplasm, *Cancer Res.*, 34, 1418, 1974.
101. **Cory, M., Henry, D. W., Taylor, D. L., and Koskela, K. J.,** Inhibitors of histone methylation, *Chem. Biol. Interact.*, 9, 253, 1974.
102. **Kim, S. and Paik, W. K.,** Purification and properties of protein methylase II, *J. Biol. Chem.*, 245, 1806, 1970.
103. **Clarke, S.,** Protein carboxyl methyltransferases: two distinct classes of enzymes, *Annu. Rev. Biochem.*, 54, 479, 1985.
104. **Van Waarde, A.,** What is the function of protein carboxyl methylation?, *Comp. Biochem. Physiol.*, 86B, 423, 1987.
105. **Pierré, A. and Robert-Gero, M.,** Selective methylation of Rous sarcoma virus glycoproteins by protein methylase II, *FEBS Lett.*, 113, 115, 1980.
106. **Trivedi, L., Gupta, A., Paik, W. K., and Kim, S.,** Purification and properties of protein methylase II from wheat germ, *Eur. J. Biochem.*, 128, 349, 1982.
107. **Borchardt, R. T., Eiden, L. E., Wu, B., and Rutledge C. O.,** Sinefungin: a potent inhibitor of S-adenosylmethionine: protein-O-methyltransferase, *Biochem. Biophys. Res. Commun.*, 89, 919, 1979.
108. **O'Dea, R. F., Pons, G., Hansen, J. A., and Mirkin, B. L.,** Characterization of protein-carboxyl O-methyltransferase in the spontaneous *in vivo* murine C-1300 neuroblastoma, *Cancer Res.*, 42, 4433, 1982.
109. **Avila, J. L. and Avila, A.,** Correlation of sinefungin susceptibility and drug-affinity for protein carboxymethyltransferase activity in American *Leishmania* species, *Mol. Biochem. Parasitol.*, 26, 69, 1987.
110. **Zappia, V., Carteni-Farina, M., and Porcelli, M.,** Biochemical and chemical aspects of decarboxylated S-adenosylmethionine, in *Transmethylation,* Usdin, E., Borchardt, R. T., and Creveling, C. E., Eds., Elsevier/North Holland, Amsterdam, 1979, 95.
111. **Jamaluddin, M., Kim, S., and Paik, W. K.,** Studies on the kinetic mechanism of S-adenosylmethionine: protein O-methyltransferase of calf thymus, *Biochemistry,* 14, 694, 1975.
112. **Gilliland, E. L., Turner, N., and Steer, M. L.,** The effects of ethionine administration and choline deficiency on protein carboxymethylase activity in mouse pancreas, *Biochim. Biophys. Acta,* 672, 280, 1981.
113. **Keller, B. T., Clark, R. S., Pegg, A. E., and Borchardt, R. T.,** Purification and characterization of some metabolic effects of S-neplanocylmethionine, *Mol. Pharmacol.*, 28, 364, 1985.
114. **Paik, W. K. and Kim, S.,** Solubilization and partial characterization of protein methylase III from calf thymus nuclei, *J. Biol. Chem.*, 245, 6010, 1970.
115. **Di Maria, P., Kim, S., and Paik, W. K.,** Cytochrome c specific methylase from wheat germ, *Biochemistry,* 21, 1036, 1982.
116. **Duerre, J. A., Wallwork, J. C., Quick, D. P., and Ford, K. M.,** *In vitro* studies on the methylation of histone in rat brain nuclei, *J. Biol. Chem.*, 252, 5981, 1977.
117. **Wallwork, J. C., Quick, D. P., and Duerre, J. A.,** Properties of soluble rat brain histone lysine methyltransferase, *J. Biol. Chem.*, 252, 5977, 1977.
118. **Tuck, M. T., Farooqui, J. Z., and Paik, W. K.,** Two histone H1-specific protein-lysine N-methyltransferases from *Euglena gracilis:* purification and characterization, *J. Biol. Chem.*, 260, 7114, 1985.
119. **Baxter, C. S. and Byvoet, P.,** Effect of carcinogens and other agents on histone methylation in rat liver nuclei by endogeneous histone lysine methyltransferase, *Cancer Res.*, 34, 1424, 1974.
120. **Burdon, R. H. and Garven, E. V.,** Enzymatic modification of chromosomal macromolecules: formation of histone ε-N-trimethyllysine by a soluble chromatin enzyme, *Biochim. Biophys. Acta,* 232, 371, 1971.

121. **Greenaway, P. J. and Levine, D.,** Identification of a soluble protein methylase in chicken embryo nuclei, *Biochim. Biophys. Acta,* 350, 374,1974.
122. **Baxter, C. S. and Byvoet, P.,** *In vitro* inhibition of histone methylation, *Proc. Am. Assoc. Cancer Res.,* 98 (abstr. n° 389), 1973.
123. **Vijayasarathy, C. and Rao, B. S. N.,** Partial purification and characterization of S-adenosylmethionine: protein-histidine N-methyltransferase from rabbit skeletal muscle, *Biochim. Biophys. Acta,* 923, 156, 1987.
124. **Boffa, L. C., Gruss, R. J., and Allfrey, V. G.,** Manifold effects of sodium butyrate on nuclear function: selective and reversible inhibition of phosphorylation of histone H1 and H2A and impaired methylation of lysine and arginine residues in nuclear protein fractions, *J. Biol. Chem.,* 256, 9612, 1981.
125. **Small, D. H., Carnegie, P. R., and Anderson, R. Mc.D.,** Cycloleucine-induced vacuolization of myelin is associated with inhibition of protein methylation., *Neurosci. Lett.,* 21, 287, 1981.
126. **Parks, L. W.,** S-adenosylethionine and ethionine inhibition, *J.Biol.Chem.,* 232, 169, 1958.
127. **Cox, R. and Tuck, M. T.,** Alteration of methylation patterns in rat liver histones following administration of ethionine, a liver carcinogen, *Cancer Res.,* 41, 1253, 1981.
128. **Friedman, M., Schull, K. H., and Farber, E.,** Highly selective *in vivo* ethylation of rat liver nuclear proteins by ethionine, *Biochem. Biophys. Res. Commun.,* 34, 857, 1969.
129. **Barber, J. R. and Clarke, S.,** Inhibition of protein carboxylmethylation by S-adenosyl-L-homocysteine in intact erythrocytes, *J. Biol. Chem.,* 259, 7115, 1984.
130. **Schatz, R. A., Wilens, T. E., and Sellinger, O. Z.,** Decreased *in vivo* protein and phospholipid methylation after *in vivo* elevation of brain S-adenosyl-homocysteine, *Biochem. Biophys. Res. Commun.,* 98, 1097, 1981.
131. **Zimmerman, T. P., Iannone, M., and Wolberg, G.,** 3-deazaadenosine, S-adenosylhomocysteine hydrolase-independent mechanism of action in mouse lymphocytes, *J. Biol. Chem.,* 259, 1122, 1984.
132. **Kloog, Y. and Saavedra,** Protein carboxylmethylation in intact rat posterior pituitary lobes *in vitro, J. Biol. Chem.,* 258, 7129, 1983.
133. **Rollins, C. M. and Dahlquist, F. W.,** Methylation of chemotaxis specific proteins in *Escherichia coli* cells permeable to S-adenosylmethionine, *Biochemistry,* 19, 4627, 1980.
134. **Amur, S. G., Shanker, G., Cochran, J. M., Veo, H. S., and Pieringer, R. A.,** Correlation between inhibition of myelin basic protein arginine methyltransferase by sinefungin and lack of compact myelin formation in cultures of cerebral cells from embryonic mice, *J. Neurosci. Res.,* 16, 367, 1986.
135. **Jacquemont, B. and Huppert, J.,** Inhibition of viral RNA methylation in *Herpes simplex* virus type 1-infected cells by 5′-S-isobutyladenosine, *J. Virol.,* 22, 160, 1977.
136. **Galletti, P., Oliva, A., Manna, C., Della Ragione, F., and Carteni-Farina, M.,** Effect of 5′-methylthioadenosine on *in vivo* methyl esterification of human erythrocyte membrane proteins, *FEBS Lett.,* 126, 236, 1981.
137. **Carteni-Farina, M., Della Ragione, F., Ragosta, G., Oliva, A., and Zappia, V.,** Studies on the metabolism of 5′-isobutylthioadenosine (SIBA): phosphorolytic cleavage by methylthioadenosinephosphorylase, *FEBS Lett.,* 104, 266, 1979.
138. **Crooks, P. A., Dreyer, R. N., and Coward, J. K.,** Metabolism of S-adenosylhomocysteine and S-tubercidinylhomocysteine in neuroblastoma cells, *Biochemistry,* 18, 2601, 1979.
139. **Coward, J. K. and Crooks, P. A.,** *In vitro* and *in vivo* effects of S-tubercidinylhomocysteine: a physiologically stable methylase inhibitor, in *Transmethylation,* Usdin, E., Borchardt, R. T., and Creveling, C. R., Eds., Elsevier/North Holland, New York, 1979, 215.

Chapter 19

UPDATE ON THE SYNTHESIS OF *N*-METHYL AMINO ACIDS

N. Leo Benoiton

TABLE OF CONTENTS

I. INTRODUCTION

The methods which have been used to synthesize the N^{ω}-alkyldiamino acids were reviewed in detail in 1978.[1] The compounds of major interest amongst this group are N^{ε}-methyllysine, N^{ε},N^{ε}-dimethyllysine, and N^{ε},N^{ε},N^{ε}-trimethyllysine and the corresponding ornithines. Since 1978, except for two papers on the synthesis of N^{ε},N^{ε},N^{ε}-trimethyllysine,[2,3] few publications have provided improvements in the methods described. This chapter presents an update on the preparation of these compounds. However, since the last review,[1] it has become apparent that posttranslational modification of proteins also occurs at the NH_2-terminal amino group.[4] The synthesis of N^{α}-methyl-, N^{α},N^{α}-dimethyl-, and N^{α},N^{α},N^{α}-trimethylamino acids involves essentially the same chemistry as that employed for the synthesis of the N^{ω}-methylated compounds except that the whole process is simpler. There is not a second reactive functional group which must be protected. Since the synthesis of N^{α}-methylamino acids is of interest primarily for purposes other than those related to the central theme of this monograph, and since all of the methods are reviewed in relation to the N^{ω}-methylated diaminoacids, the question of N^{α}-methylation has not been addressed in separate sections. All compounds and derivatives that have been crystallized or characterized by some criterion besides chromatography have been tabulated.

The abbreviations used are in accordance with the rules and recommendation of IUPAC-IUB Commission on Biochemical Nomenclature.[5] When not indicated, the amino acid symbol represents the L isomer. For derivatives, a substituent appearing above or immediately to the right of the symbol indicates substitution at the side chain functional group. A substituent separated by a dash and appearing at the left of the symbol is located at the α-amino group; at the right, is at the carboxy group. Symbols for the N^{ε}-alkylamino acids and abbreviations for the protecting groups are as follows: Lys(Me), N^{ε}-methyl-L-lysine; Lys(Me_2), N^{ε},N^{ε}-dimethyl-L-lysine; Lys(Me_3^+), N^{ε},N^{ε},N^{ε}-trimethyl-L-lysine; Boc, *tert*-butoxycarbonyl; Bz, benzoyl; Bzl, benzyl; Pht, phthaloyl; Tos, *p*-toluenesulfonyl; Trt, triphenylmethyl; Z, benzyloxycarbonyl. The counterion, or the salt form in which products were isolated, are usually not indicated in the equations given.

II. N^{ε}-METHYLLYSINE AND DERIVATIVES

The first synthesis of N^{ε}-methyllysine was carried out in 1930 by amination of *N*-benzoyl-$^{\varepsilon}$-methylaminocaproic acid (Equation 1).[6] After deprotection, the resulting racemic hydrochloride salt failed to crystallize, so the product was characterized as the picrate and the picrolonate. In 1944, a group studying the utilization of N^{ε}-methyllysine by the rat prepared the compound in the same manner, except that the starting material was cyclohexanone oxime. The intermediate ε-aminocaproic acid was in turn converted into ε-methylaminocaproic acid through the *p*-toluenesulfonyl derivative (Equation 2).[7] The intriguing observation was made that acid hydrolysis of H-DL-Lys(Me,Bz)-OH gave a mixture of N^{ε}-methyllysine and lysine. Deprotection by refluxing with 2 *N* barium hydroxide gave the product as the picrate in 77% yield, from which the crystalline monohydrochloride was readily obtained. The absence of lysine in the product was established using the lysine decarboxylase from *Bacterium cadaveris*, which is known not to attack the N^{ε}-methyllysine. N^{ε}-methyl-DL-lysine was again prepared 15 years later as a reference compound in connection with studies on the specificity of proteases using methylated proteins as substrates.[8] The same method was used except for the choice of hydrogen bromide/acetic acid at 75°C for the detosylation and thionyl chloride/iodine instead of bromine for the halogenation. The latter combination was selected because it was found that bromination according to the original synthesis caused some demethylation. Paper chromatography in *n*-BuOH-AcOH-$H_2$0 (6:1:3; upper layer) for 14 d allowed a good separation of N^{ε}-methyllysine from lysine for analytical purposes.

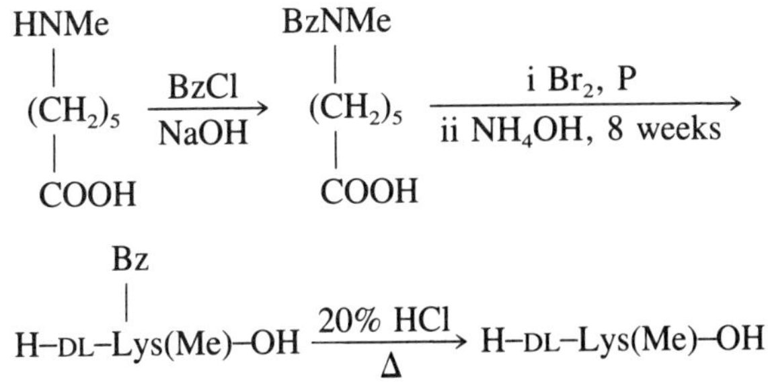

(1)

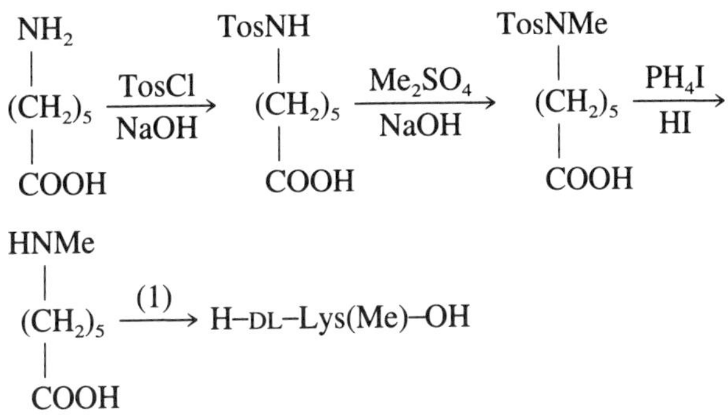

(2)

A synthesis of N^{ε}-methyl-DL-lysine unrelated to the approach described above was included in a description of a general method for preparing N^{ω}-alkyllysines (Equation 3).[9] This route involved amination of δ-bromobutylhydantoin with methylamine followed by hydrolysis. However, the intermediate for methyllysine was difficult to purify. Hydrolysis with 48% hydrobromic acid gave a much lower yield than with barium hydroxide. The analytical data are presented for the monohydrobromide, but there is no indication of how this salt was obtained.

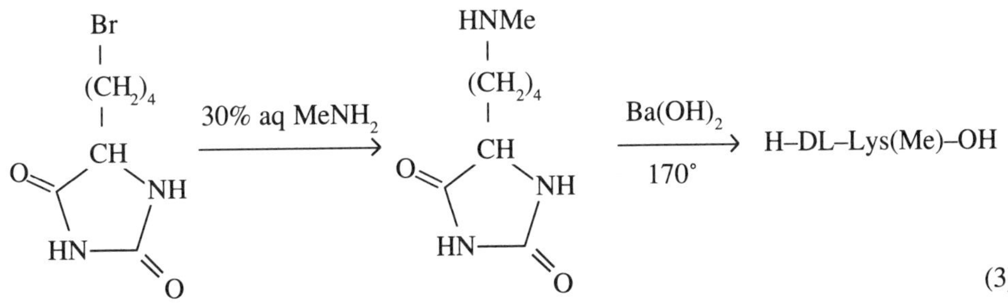

(3)

In 1963, Drs. Woon Ki Paik and Sangduk Kim, then of this department, and I were interested in studying the enzyme responsible for the utilization of N^{ε}-methyllysine by the rat.[10] For this study, the L isomer was desired, and since none of the above synthesis could be used to obtain the enantiomers without effecting a resolution, we had to find another synthetic route to N^{ε}-methyllysine. Optically active N^{α}-methyllysine had previously been prepared by methylation of Tos-Lys(Bz)-OH with methyl iodide[11] as well as N^{α}-methylornithine from the corresponding derivative,[12] so we set out to prepare N^{ε}-methyl-L-lysine in the analogous fashion (Equation 4).[13]

It should be pointed out that the introduction of a single methyl group on nitrogen through the *N*-tosyl derivative, using dimethyl sulfate, had been carried out as far back as 1919.[14] Our synthesis had two serious shortcomings; neither of the tosyl intermediates could be crystallized, and the methylation was incomplete. As a result, the product was contaminated with up to 10% of lysine. The pure product was nevertheless obtained by destroying the lysine with lysine decarboxylase. It is now obvious that our choice of methyl iodide as methylating agent was a bad one (Equation 9).

$$\text{H–Lys–OH} \xrightarrow[\text{iii } H_2S,\ H^+]{\text{i } CuCO_3 \quad \text{ii TosCl, NaOH}} \text{H–}\overset{\text{Tos}}{\overset{|}{\text{Lys}}}\text{–OH} \xrightarrow[\text{NaOH}]{\text{BzCl}}$$

$$\text{Bz–}\overset{\text{Tos}}{\overset{|}{\text{Lys}}}\text{–OH} \xrightarrow[70^\circ]{\text{MeI, NaOH}} \text{Bz–}\overset{\text{Tos}}{\overset{|}{\text{Lys}}}\text{(Me)–OH} \xrightarrow[\text{ii Dowex 50; } NH_4OH]{\text{i 48\% HBr}} \text{H–Lys(Me)–OH} \qquad (4)$$

No attempt was made, however, to improve this synthesis because at this time the feasibility of another approach became apparent. This new one was derived from a report on a new synthesis of N^α-methylamino acids[15] (exemplified in Equation 5), a method also applied to the synthesis of N^α-methyllysine using the benzyloxycarbonyl group for N^ε-protection. During the following decade, this was the method of choice for preparing *N*-methylamino acids; however, it is now recognized that the compounds prepared by this method are generally slightly racemized (0.3 to 5% D isomer; more for valine) and that some racemization can occur during both of the alkylation steps of the synthesis.[16] The synthesis of optically pure N^α-benzylamino acids can be best achieved using sodium cyanoborohydride at pH 6 instead of sodium borohydride.[16] My synthesis of N^ε-methyl-L-lysine[13,17] was carried out as shown in Equation 6. The product crystallizes readily as the monohydrochloride and contains no lysine. The D enantiomer has also been prepared and its optical purity established at >99.6% using L-amino acid oxidase.[18] Since the alkylation was effected on the N^ε-amino group, racemization as alluded to above does not enter into the picture. The N^ε-methyl-L-lysine that is commercially available has been prepared by this method. The synthesis of the starting N^α-benzyloxycarbonyllysine is based on well-known methods.[19,20] The product is not always free of N^ε-benzyloxycarbonyl isomer, but this impurity can be removed using a cupric carbonate-alumina column.[21,22]

$$\text{H–Leu–OH} \xrightarrow[\text{NaBH}_4 \text{ or } H_2,\ \text{Pd–C}]{\text{PhCHO, NaOH}} \text{Bzl–Leu–OH} \xrightarrow[HCO_2H,\ \Delta]{\text{38\% aq. } CH_2O}$$

$$\text{Bzl–MeLeu–OH} \xrightarrow[\text{Pd–C}]{H_2} \text{H–MeLeu–OH} \qquad (5)$$

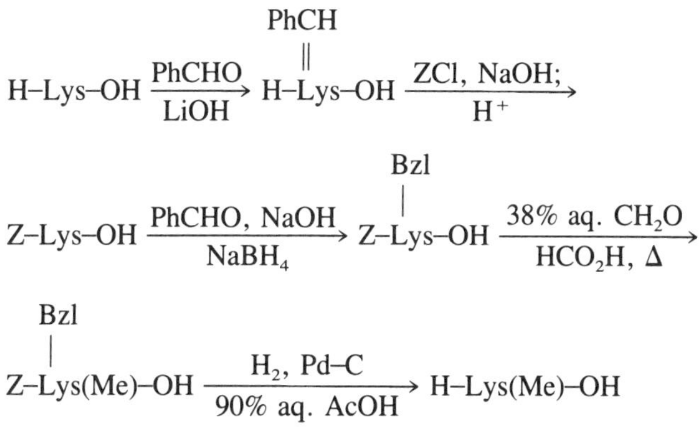

(6)

The next synthetic route to N^{ε}-methyllysine arose from a new synthesis of N^{α}-methylamino acid derivatives devised in our own laboratory (Equation 7).[23]

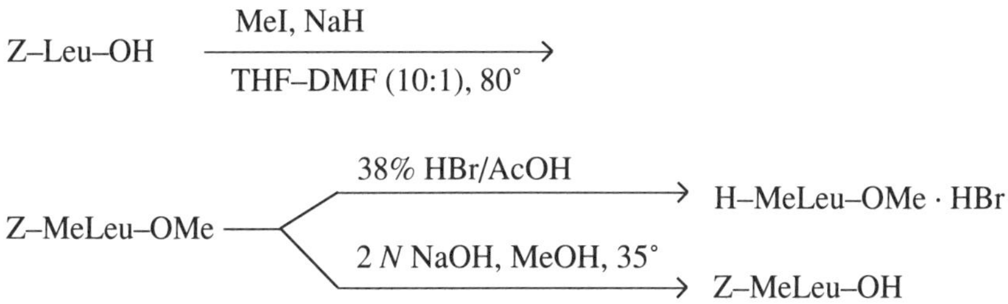

(7)

The methylation has since been simplified.[24,25] The reaction is carried out in tetrahydrofuran at room temperature, and the product of the methylation is the free carboxylic acid instead of the ester. Both N^{α}-benzyloxycarbonyl- and N^{α}-butoxycarbonyl-N^{α}-methylamino acids are accessible by this route, and it has been established that the products are optically pure.[16] It is now the method of choice for the preparation of N^{α}-methylamino acids and these derivatives and the method used by all commercial houses which are selling these products. According to this procedure, the ε-benzyloxycarbamido group of lysine was methylated using methyl iodide and sodium hydride, the N^{α}-amino group being protected against methylation by the trityl group (Equation 8).[26] Selective removal of the latter by hot acetic acid provided the pure product as the methyl ester, still carrying the benzyloxycarbonyl group at the N^{ε} position. This is a most useful intermediate for synthetic work in peptide chemistry and one which readily can be deprotected to give the methylamino acid.

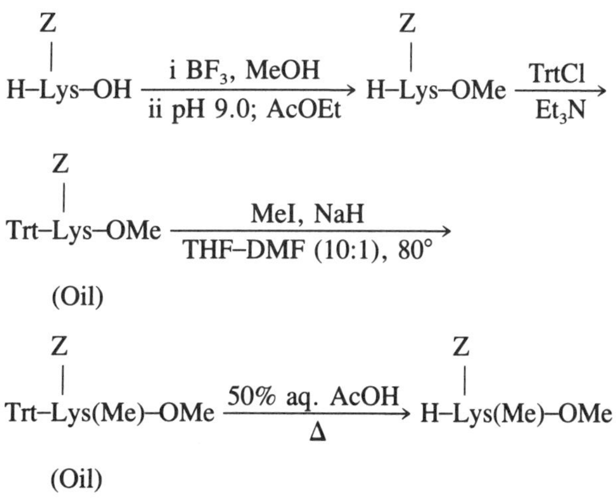

(8)

In this synthesis, the N^{α},N^{ε}-disubstituted starting material was the methyl ester because this was the best way of obtaining the required derivatized lysine. A new synthesis of methyl N^{ε}-benzyloxycarbonyllysinate was devised in the process.[27] That the trityl group would provide protection against the methylation of an N^{α}-amino group had been established using *N*-tritylleucine as a model compound. Another possible candidate, the phthaloyl group, failed to serve the purpose[26] since the phthalimido group is partially cleaved by sodium hydride even under the milder conditions of methylation.[28] It is likely that the Ox-protecting group (incorporation of the amino group into a 4,5-diphenyl-4-oxazolin-2-one ring)[29] would provide adequate protection against N^{α}-methylation. However, prudence dictates that the free carboxylic acid and not the ester should be used as starting material in this case, since N^{α},N^{α}-disubstituted amino acid esters are prone to racemize in alkali.[30] We have recently established that the milder conditions of methylation will also completely methylate an ω-benzyloxycarbamido group using *N*-benzyloxycarbonyl-γ-aminobutyric acid.[28]

My synthesis of N^{ε}-methyllysine from the N^{ε}-tosyl derivative has recently been superseded by a contribution that is characterized in particular by the enviable yields reported.[31] This work formed part of a study on the conformation of poly(N^{ε}-methyllysine) and related polymers. As shown in Equation 9, these workers used dimethylsulfate[14] instead of methyl iodide for the methylation, crystallized the two intermediate tosyl derivatives, and employed hydrochloric acid for final deprotection. It must be noted, however, that characterization of the product included R_f values for thin layer chromatography in two solvent systems that do not distinguish between N^{ε}-methyllysine and lysine. In addition, there is no reason to question the optical purity of this product, but the statement that the compound could not be evaluated by use of L-amino acid oxidase because N^{ε}-methyllysine is not a substrate for this enzyme is false.

$$\text{H–}\overset{\text{Tos}}{\overset{|}{\text{Lys}}}\text{–OH} \xrightarrow[\text{NaOH}]{\text{BzCl}} \underset{(99\%)}{\text{Bz–}\overset{\text{Tos}}{\overset{|}{\text{Lys}}}\text{–OH}} \xrightarrow[\text{33\% NaOH, 50°}]{\text{Me}_2\text{SO}_4}$$

$$\underset{(97\%)}{\text{Bz–}\overset{\text{Tos}}{\overset{|}{\text{Lys}}}\text{(Me)–OH}} \xrightarrow[100°]{12\ N\ \text{HCl}} \underset{(66\%)}{\text{H–Lys(Me)–OH} \cdot \text{HCl}} \qquad (9)$$

The success achieved in methylating the ω-tosylamido grou permitted evaluation of the idea to prepare the corresponding N^{α}-butoxycarbonyl derivative by the same route (Equation 10) since the butoxycarbonyl group is much easier to remove by acidolysis than the benzoyl and tosyl groups. This selectivity of cleavage would make it a derivative useful for peptide synthesis. The compound was obtained without difficulty and characterized as the dicyclohexylammonium salt.[28] A portion, after deprotection with sodium in ammonia followed by trifluoroacetic acid, gave only one amino acid that coincided with authentic N^{ε}-methyllysine on the amino acid analyzer. Though crystallization of the methyl-amino acid was not attempted, it should be possible after purification using resin[13] or after cleavage of the derivative by acid.[6,8] This procedure therefore provides an additional route to N^{ε}-methyl-L-lysine.

$$\text{H–}\overset{\text{Tos}}{\overset{|}{\text{Lys}}}\text{–OH} \xrightarrow[\text{NaOH}]{\text{BocN}_3} \underset{\text{(Oil)}}{\text{Boc–}\overset{\text{Tos}}{\overset{|}{\text{Lys}}}\text{–OH}} \xrightarrow[\text{33\% NaOH, 50°}]{\text{Me}_2\text{SO}_4} \underset{\text{(Oil)}}{\text{Boc–}\overset{\text{Tos}}{\overset{|}{\text{Lys}}}\text{(Me)–OH}} \qquad (10)$$

TABLE 1
Derivatives of N^{ε}-Methyllysine

Compound	Melting point (°C)	Method[a]	Ref.
H-Lys(Me)-OH		4	13
H-Lys(Me)-OH·HCl	237—239	6	13,17
H-Lys(Me)-OH·HCl	240	9	31
H-Lys(Me)-OH[b]	230	4	13
H-DL-Lys(Me)-OH·HCl·H_2O		2	7
H-DL-Lys(Me)-OH·HBr	234—235	3	9
H-DL-Lys(Me)-OH[b]	227	1	6—8
H-DL-Lys(Me)-OH·H_2O[c]	228	1	6
H-Lys(Me)-OEt·2HCl	145		13
H-DL-Lys(Me,Bz)-OH	225—227	1	6—8
H-Lys(Me,Z)-OH	210—212	11	6
H-Lys(Me,Z)-OH	215—217		31
H-Lys(MeZ)-OMe·HCl	113—115[d]	8	26
Lys(Me,Z)NCA[e]	83	17	31,33
Bz-Lys(Me,Z)-OH	Oil	11	33
Bz-Lys(Me,Tos)-OH	118	9	31
Bz-Lys(Me,Tos)-OH	Oil	4	13
Boc-Lys(Me,Z)-OH	Oil	11	28
Boc-Lys(Me,Tos)-OH[f]	60—62	10	28
Bz-Lys(Me)-OMe·HCl	Oil	11	33
Bz-Lys(Me)-NH_2·HCl	199—200	11	32,33
Bz-Lys(Me,Z)-OMe	Oil	11	33
Bz-Lys(Me,Z)-NH_2	114—115		32
Trt-Lys(Me,Z)-OMe	Oil	8	26

[a] Refers to equation in the text.
[b] Picrate.
[c] Picrolonate.
[d] Unpublished data.
[e] *N*-carboxyanhydride
[f] *N,N*-dicyclohexylammonium salt.

Adapted from Benoiton, N. L., *Chemistry and Biochemistry of Amino Acids, Peptides and Proteins,* Vol. 5, Weinstein, B., Ed., Marcel Dekker, New York, 1978, 163. With permission.

Besides the few derivatives originating from the synthesis of the methyl-amino acid, the derivatives of N^{ε}-methyl-L-lysine described in the literature were prepared in our laboratory, mostly for the purpose of establishing the susceptibility of appropriate derivatives to the action of the basic amino acid-directed proteases, trypsin and carboxypeptidase B. These derivatives are primarily esters and amides, many of them benzoylated at the α position. All were prepared by standard procedures as shown in Equation 11, which comes from References 26, 28, 32, and 33. Derivatives of N^{ε}-monoalkyllysines are compiled in Table 1.

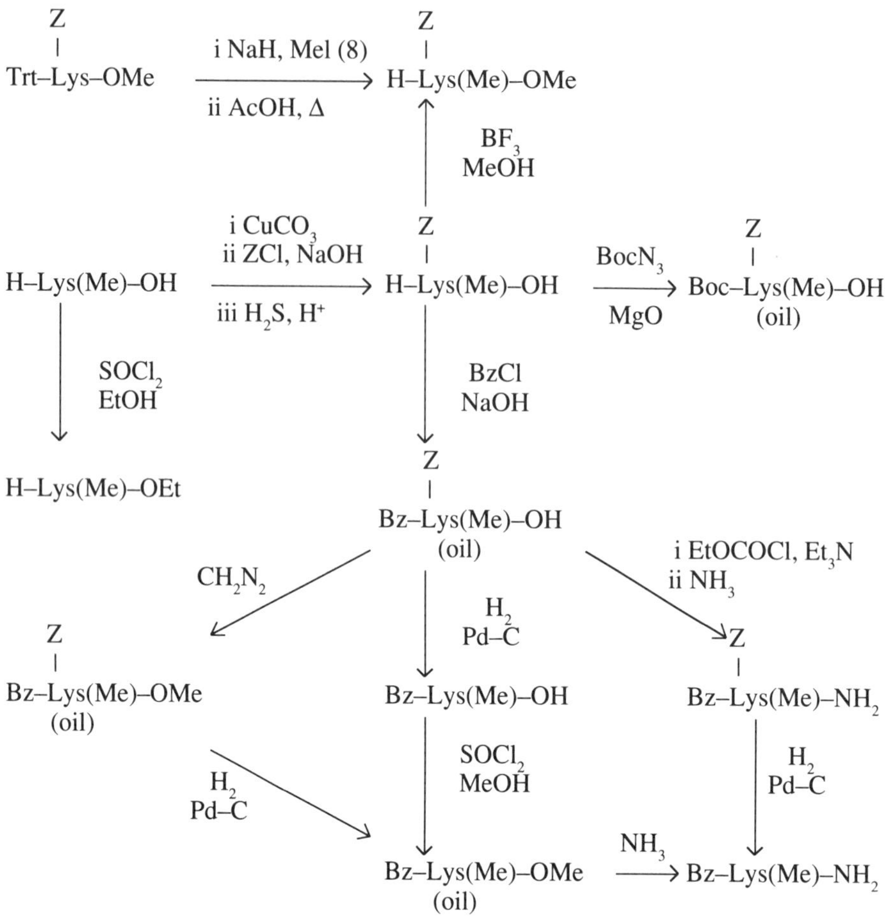

(11)

III. N^{δ}-METHYLORNITHINE

A synthesis of N^{δ}-methyl-DL-ornithine was described in the literature in 1922, 8 years before that of N^{ε}-methyl-DL-lysine, as part of a study on the origin of creatine in animals.[34] This synthesis (shown in Equation 12), actually represented the first for mono-*N*-alkyldiamino acids which are based on alkylation of the tosylamido group of the derivatized parent acid. Two subsequent synthesis of the L isomer appeared much later along with those of N^{ε}-methyl-L-lysine.[13,31] These were minor variations of the original method.

$$\text{Bz–DL–}\underset{|}{\overset{\text{Bz}}{\text{Orn}}}\text{–OH} \xrightarrow{\text{Ba(OH)}_2} \text{Bz–DL–Orn–OH} \xrightarrow[\text{NaOH}]{\text{TosCl}}$$

$$\text{Bz–DL–}\overset{\text{Tos}}{\text{Orn}}\text{–OH} \xrightarrow[\text{NaOH, 23°}]{\text{Me}_2\text{SO}_4} \text{Bz–DL–}\overset{\text{Tos}}{\text{Orn}}\text{(Me)–OH} \xrightarrow[55°]{\text{HI, PH}_4\text{I}}$$

$$\text{Bz–DL–Orn(Me)–OH} \xrightarrow[\Delta]{\text{20\% HCl}} \text{H–DL–Orn(Me)–OH} \qquad (12)$$

TABLE 2
Derivatives of N^{ε}-Methylornithine

Compound	Melting point (°C)	Method[a]	Ref.
H-Orn(Me)-OH·HCl		4	13
H-Orn(Me)-OH·HCl	252	9	31
H-DL-Orn(Me)-OH·2HCl	157	12	34
H-DL-Orn(Me)-OH[b]	206	12	34
H-Orn(Me,Z)-OH	212		31
H-DL-Orn(Me,Tos)-OH	245	12	34
H-DL-Orn(Me,Tos)-OH·HCl	228	12	34
Orn(Me,Z)NCA[c]	Oil	17	31
Bz-DL-Orn(Me)-OH	215	12	34
Bz-Orn(Me,Tos)-OH	163	4	13
Bz-Orn(Me,Tos)-OH	160	9	31
Bz-DL-Orn(Me,Tos)-OH	188—189	12	34

[a] Refers to equation in the text.
[b] Chloroplatinate.
[c] *N*-Carboxyanhydride.

Adapted from Benoiton, N. L., *Chemistry and Biochemistry of Amino Acids, Peptides and Proteins,* Vol. 5, Weinstein, B., Ed., Marcel Dekker, New York, 1978, 163. With permission.

In the original work, DL-ornithine was prepared by the condensation of malonic ester with an alkyl halide followed by hydrolysis, after which the product was isolated as the *N,N′*-dibenzoyl derivative (ornithuric acid). This was selectively hydrolyzed with barium hydroxide, the more soluble N^{α}-benzoyl isomer being obtained by fractional crystallization. Subsequent reactions were as indicated.

My synthesis of the L isomer[13] was carried out as described for N^{ε}-methyl-L-lysine (Equation 4), except that the two intermediates crystallized readily. N^{δ}-Methyl-L-ornithine has also been prepared essentially as described in Equation 10[31] with the same high yields. The high yield obtainable for a methylation using dimethyl sulfate had been pointed out in the original paper.[14] There is some discrepancy between the physical data reported by the two groups with respect to the melting points of intermediates and specific rotations. N^{δ}-monoalkylornithine derivatives are compiled in Table 2.

IV. N^{ε},N^{ε}-DIMETHYLLYSINE

The first synthesis of an N^{ω},N^{ω}-dimethyldiamino acid made use of the general reductive alkylation procedure which involves catalytic hydrogenation in the presence of aqueous formaldehyde,[35,36] Both N^{α},N^{α}-dimethyl-DL-lysine and N^{ε},N^{ε}-dimethyl-DL-lysine were prepared in this manner in 1959 with the benzoyl group used for amino protection.[8] Subsequent syntheses of the L isomer were minor modifications of this work. The sequence of reactions is described in Equation 13. I repeated this synthesis later with the L isomer[13] starting, however, with N^{ε}-benzyloxycarbonyl-L-lysine obtained through the copper salt, a method available since 1943.[37] Debenzoylation was effected with hot acid instead of barium hydroxide, and a similar difficulty with the purification of the product was encountered. In fact, I finally prepared the N^{ε},N^{ε}-dimethyl-L-lysine from the N^{α}-acetyl derivative which is much easier to hydrolyze. This same N^{α}-acetyl-N^{ε},N^{ε}-dimethyl-L-lysine was also made during a study on the reductive alkylation of proteins using sodium borohydride as a reducing agent.[38] However, no details were given as to how the compound was isolated from the buffered salt-containing reaction mixture, nor was evidence of adequate characterization presented.

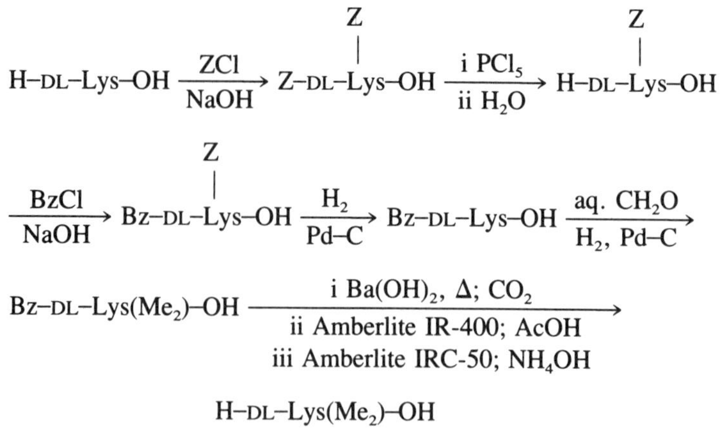

(13)

A completely different synthetic approach was that used in 1969 by a group who had discovered N^ε,N^ε,N^ε-trimethyllysine in histones the previous year.[39] These workers prepared several tritium-labeled *N*-methyllysines from products obtained by classical malonic ester type condensations. Their synthesis of cold N^ε,N^ε-dimethyl-DL-lysine (Equation 14) involved the condensation of *N*-benzyl-protected γ-dimethylaminobutyl bromide with diethyl acetamidomalonate in excess, followed by catalytic hydrogenation and acid hydrolysis in that order. The product was purified and freed of glycine by elution chromatography on a 3-cm^2 resin column. [^{3}H-4,5]-N^ε,N^ε-dimethyl-DL-lysine was prepared similarly using the C-2,3-unsaturated butyl bromide as starting material, effecting the reduction with tritium gas, followed by purification with an amino acid analyzer. In this case, norleucine was formed as a major byproduct due to –C–N≡ fission during the saturation step. In the same study, the [^{3}H-methyl]-labeled amino acid was also prepared as shown below (Equation 15) using the amino acid analyzer for isolation of the product.

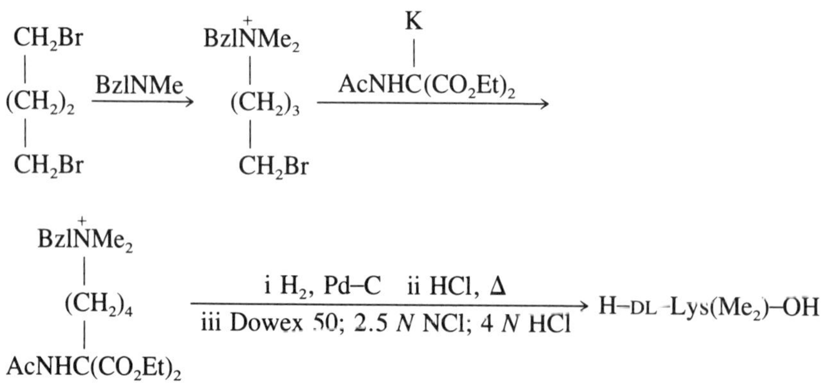

(14)

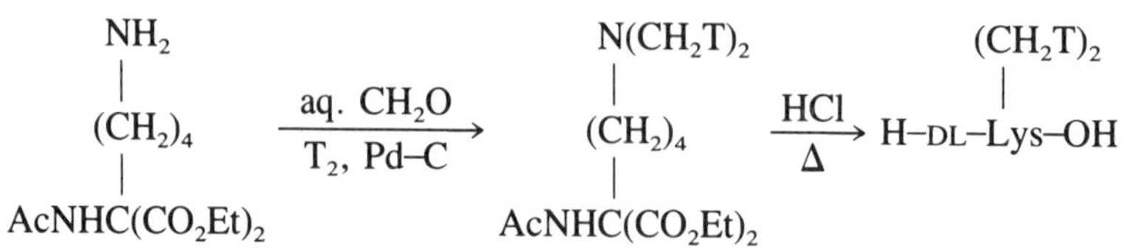

(15)

Dr. F.M. Chen in my laboratory discovered a new method of synthesizing *N*,*N*-dialkylamino acids as a result of the fortuitous choice of reaction conditions for a catalytic hydrogenation. Instead of using aqueous alcohol made slightly acidic as the solvent for removing an *N*-benzyloxycarbonyl group, he used absolute ethanol, with the astounding result that the pertinent amino group was not only deprotected but also diethylated in the process, as shown in Equation 16.[40] It transpired that he had not completely removed atmospheric oxygen from the mixture, having evacuated the hydrogenation apparatus only once before introducing the hydrogen. The oxygen remaining was sufficient to oxidize some of the ethanol to acetaldehyde, which served as carbonyl component for the well-known dialkylation reaction.[35,36]

$$\text{Bz–Lys–NHPh} \xleftarrow[\text{aq. EtOH}]{\text{H}_2\text{, Pd–C}} \text{Bz–}\overset{\overset{\text{Z}}{|}}{\text{Lys}}\text{–NHPh} \xrightarrow[\text{abs. EtOH}]{\text{air, H}_2\text{, Pd–C}} \text{Bz–Lys(Et}_2\text{)–NHPh} \qquad (16)$$

Further investigation established that in the presence of palladium-on-charcoal catalyst and air, methanol also is readily converted to formaldehyde, and the solution obtained can be used instead of aqueous formaldehyde for reductive methylations.[40,41]

Using methanolic formaldehyde solution as the source of carbonyl compound, we have prepared the N^{α}-butoxycarbonyl derivatives of dimethyllysine and dimethylornithine from the corresponding N^{ω}-benzyloxycarbonyl derivatives (Equation 17)[41] and found it much easier to crystallize the products than when aqueous formaldehyde had been used. This is because the methanolic formaldehyde solution leaves no residue of insoluble paraformaldehyde after evaporation as does an aqueous formaldehyde solution. Difficulties with purification after reductive alkylations using the latter have been reported.[13,42] The first of these compounds had been prepared previously by the conventional procedure from the same starting material. Since the starting materials are commercially available and the products can be deprotected by mild acidolysis, this is the simplest method of obtaining these N^{ω},N^{ω}-dimethylamino acids. It is worth pointing out also that heat, higher than atmospheric pressure or pH adjustments are not necessary for carrying out these or similar reductive alkylations.[41]

$$\text{Boc–}\overset{\overset{\text{Z}}{|}}{\text{Xxx}}\text{–OH} \xrightarrow[\text{H}_2\text{, Pd–C}]{\text{CH}_2\text{O–MeOH}} \text{Boc–Xxx(Me}_2\text{)–OH}$$

$$\text{Xxx} = \text{Lys, Orn} \qquad (17)$$

A recent synthesis of N^{ε},N^{ε}-dimethyl-L-lysine involved a modified Clarke-Eschweiler methylation of N^{α}-benzyloxycarbonyl-L-lysine using sodium borohydride to reduce the Schiff base produced by the addition of aqueous formaldehyde (Equation 18).[3] The protecting group was removed by catalytic hydrogenation, and the product was purified by ion-exchange chromatography. The authors were unable to get a crystalline dihydrochloride. This may be an additional example of the deleterious effect of paraformaldehyde[13,42] mentioned above

$$\text{Z–Lys–OH} \xrightarrow[\text{ii aq. HCl, pH 5}]{\text{i aq. CH}_2\text{O, NaBH}_4} \text{Z–Lys(Me}_2\text{)–OH} \xrightarrow[\text{Pd–C}]{\text{H}_2} \text{H–Lys(Me}_2\text{)–OH} \qquad (18)$$

Derivatives of N^{ε},N^{ε}-dimethyllysine are compiled in Table 3. Some of these were prepared from the corresponding dimethylated derivative, but they could also be prepared by the simultaneous deprotection and reductive alkylation of the corresponding N^{ε}-benzyloxycarbonyl derivative as in Equation 17.

TABLE 3
Derivatives of N^{ε},N^{ε}-Dimethyllysine

Compound	Melting point (°C)	Ref.
H-Lys(Me_2)-OH·HCl		13
H-DL-Lys(Me_2)-OH	243—245	8
H-DL-Lys(Me_2)-OH·2HCl	176	39
H-DL-Lys(Me_2)-OH[a]	183—184	8
H-Lys(Me_2)-OEt·2HCl	Oil	3,33
Ac-Lys(Me_2)-OH·$2H_2O$	196—198	13
Bz-Lys(Me_2)-OH	209	13
Bz-DL-Lys(Me_2)-OH	207—210	8
Boc-Lys(Me_2)-OH	156—158	41
Bz-Lys(Me_2)-OMe·HCl	Oil	33
Bz-Lys(Me_2)-NH_2·HCl·H_20	155—157	33
Boc-Orn(Me_2)-OH	169—170	41

[a] Picrate.

Adapted from Benoiton, N. L., *Chemistry and Biochemistry of Amino Acids, Peptides and Proteins*, Vol. 5, Weinstein, B., Ed., Marcel Dekker, New York, 1978, 163. With permission.

V. N^{ε},N^{ε},N^{ε}-TRIMETHYLLYSINE

N^{ε},N^{ε},N^{ε}-trimethyllysine was first described in the literature in 1930 as part of a study designed to gain information on the manner in which amino acids were linked together to form proteins. With this objective in mind, casein was methylated using dimethyl sulfate, the product was hydrolyzed, the sulfate ion was removed with barium, and then on the supposition that an animal would not utilize a modified amino acid, the mixture was fed to a dog.[43] After considerable effort, N^{ε},N^{ε},n^{ε}-trimethyllysine was isolated from the urine and converted to the dichloroaurate salt. Its identity was established by comparison with the gold salts of synthetic α- and ε-betaines of lysine.[44] This experiment demonstrated that the ε-amino group of lysine in casein was not bound. The synthesis of the N^{ε},N^{ε},N^{ε}-trimethyl-DL-lysine involved the methylation of N^{α}-benzoyl-DL-lysine using dimethyl sulfate followed by acid hydrolysis (Equation 19). Points of interest are that the starting material was obtained from the dibenzoylamino acid by selective hydrolysis with barium hydroxide, with the surprisingly high yield of 85%, as had been done for the synthesis of N^{α}-methylornithine several years earlier (Equation 12).[34] In contrast, debenzoylation by acid of N^{α}-benzoyl-N^{α},N^{α},N^{α}-trimethylornithine was found to take place more easily than the debenzoylation of N^{α}-benzoyl-N^{ε},N^{ε},N^{ε}-trimethyllysine. It also appears from this work that the gold salts of the racemic and optically active ε-betaines of lysine have the same melting points. We prepared N^{α}-benzoyl-N^{ε},N^{ε},N^{ε}-trimethyl-L-lysine from N^{α}-benzoyl-L-lysine by the same method much later, isolating the compound by adsorption on Dowex 50, followed by elution with ammonium hydroxide.[33] We were unaware of the above work at the time.

$$\text{Bz–DL–}\overset{\text{Bz}}{\overset{|}{\text{Lys}}}\text{–OH} \xrightarrow{\text{Ba(OH)}_2} \text{Bz–DL–Lys–OH} \xrightarrow[\text{Ba(OH)}_2]{\text{Me}_2\text{SO}_4}$$

$$\text{Bz–DL–Lys(Me}_3^+\text{)–O}^- \xrightarrow[\text{36 h}]{6\ N\ \text{HCl},\ \Delta} \text{H–DL–Lys(Me}_3^+\text{)–O}^- \tag{19}$$

Not until 30 years after the original work did this derivative again appear on the scene. Syntheses of the three possible betaines of lysine were described in a French patent in 1959.[45] These compounds were described as foaming agents, useful for the preparation of nontoxic emulsions in foods and medicines. The synthesis of the N^{ε}-derivative, which is applicable for the racemate only, is shown in Equation 20. The novelty here was that the N^{α}-amino group was introduced into the molecule in the protected form using dibenzylamine. After methylation, the benzyl groups were removed by hydrogenation over palladium black at elevated temperature.

$$\begin{array}{c}\text{HNBz}\\ |\\ (\text{CH}_2)_4\\ |\\ \text{BrCHCO}_2\text{H}\end{array} \xrightarrow[\Delta]{\text{HN(Bzl)}_2} \text{(Bzl)}_2\text{–}\overset{\text{Bz}}{\overset{|}{\text{Lys}}}\text{–OH} \xrightarrow[\Delta]{8\ N\ \text{HCl}} \text{(Bzl)}_2\text{–Lys–OH}$$

$$\xrightarrow[\text{ii 12 } N \text{ HCl, } \Delta\text{; Ba}^{2+}]{\text{i Me}_2\text{SO}_4\text{, KOH; 100°}} \text{(Bzl)}_2\text{–Lys(Me}_3^+\text{)–O}^- \xrightarrow[70°]{\text{H}_2\text{, Pd–C}} \text{H–DL–Lys(Me}_3^+\text{)–O}^- \tag{20}$$

The next syntheses were carried out in 1964 in Japan by a group that had isolated a new Dragendorf-positive amino acid from an aqueous extract of *Laminaria augustata*.[46] This new amino acid, called laminine, was established to be $N^{\varepsilon},N^{\varepsilon},N^{\varepsilon}$-trimethyl-L-lysine by comparison of its oxalic acid salt with that of synthetic material prepared by the two methods described below (Equations 21 and 22).[47] Both these methods included a purification of the methylamino acid product by displacement chromatography on a cation-exchange resin, followed by anion removal with a second resin treatment. The salt crystallized from ethanol-water as the dioxalate hemihydrate. The crystalline oxalates of the corresponding ornithine and diaminobutyric acid derivatives were also prepared by the method described in Equation 21. In addition, racemic laminine was prepared by total synthesis from diethyl acetamidomalonate (Equation 23). Synthesis by the same method, followed by purification of the product on a column of silica gel or by thin-layer chromatography, was described 4 years later,[48] as well as an isolation of the L isomer from hydrolyzed casein which had been methylated.[49] About the same time, N^{ε}-trimethyl-L-lysine was isolated from seeds of *Reseda luteola* L.[50]

$$\text{H–Lys–OH} \xrightarrow[\substack{\text{iii Amberlite IR-120; NH}_4\text{OH}\\ \text{iv Amberlite IRA-410}}]{\substack{\text{i CuCO}_3\\ \text{ii Me}_2\text{SO}_4\text{, NaOH}}} \text{H–Lys(Me}_3^+\text{)–O}^- \tag{21}$$

$$\text{Ac–Lys–OH} \xrightarrow[\substack{\text{iii 12 } N \text{ HCl, } \Delta\\ \text{iv Amberlite IRA-410}}]{\substack{\text{i Me}_2\text{SO}_4\text{, NaOH}\\ \text{ii Amberlite IR-120; NH}_4\text{OH}}} \text{H–Lys(Me}_3^+\text{)–O}^- \tag{22}$$

$$\mathrm{Cl{-}(CH_2)_4{-}Br} \xrightarrow{\mathrm{AcNHC(CO_2Et)_2\ (Na)}} \mathrm{Cl{-}(CH_2)_4{-}AcNHC(CO_2Et)_2} \xrightarrow[\Delta]{6\,N\ \mathrm{HCl}}$$

$$\mathrm{Cl{-}(CH_2)_4{-}NH_2{-}CH{-}CO_2H} \xrightarrow[\text{ii Amberlite IRA-410}]{\text{i aq. } \mathrm{Me_3N},\ 115^\circ} \mathrm{H{-}DL{-}Lys(Me_3^+){-}O^-} \qquad (23)$$

$N^\varepsilon,N^\varepsilon,N^\varepsilon$-trimethyl-L-lysine was prepared in the process of making the N^α-benzyloxycarbonyl derivative for further synthetic work on peptides related to angiotensin.[51] An earlier synthetic approach (Equation 22) was used but the whole of the ammonium hydroxide eluate from the resin column was collected (Equation 24). As a consequence, the product which crystallized, in good yield, contained about 5% of lysine. This was removed by conversion to the ether-soluble dibenzyloxycarbonyl derivative. However, this reaction also was incomplete, requiring purification of the desired N^α-benzyloxycarbonyl derivative on a cellulose column. The free trimethylamino acid was crystallized as the dichloride salt after hydrogenolytic deprotection.

$$\mathrm{Ac{-}Lys{-}OH} \xrightarrow[\substack{\text{iii } \mathrm{Ba(OH)_2} \\ \text{iv Amberlite IR-120; } \mathrm{NH_4OH}}]{\substack{\text{i } \mathrm{Me_2SO_4},\ \mathrm{Ba(OH)_2} \\ \text{ii } 4\,N\ \mathrm{H_2SO_4},\ \Delta}}$$

$$\begin{matrix}\mathrm{H{-}Lys(Me_3^+){-}O^-} \\ + \\ \mathrm{H{-}Lys{-}OH}\end{matrix} \xrightarrow[\mathrm{NaOH}]{\mathrm{ZCl}} \begin{matrix}\mathrm{Z{-}Lys(Me_3^+){-}O^-} \\ + \\ \mathrm{Z{-}Lys(Z){-}OH}\end{matrix} \xrightarrow[\text{ii chromatography on cellulose}]{\text{i ether extraction}}$$

$$\mathrm{Z{-}Lys(Me_3^+){-}O^-} \xrightarrow[\mathrm{Pd{-}C}]{\mathrm{H_2}} \mathrm{H{-}Lys(Me_3^+){-}O^-} \qquad (24)$$

A synthesis of racemic laminine also was described by another group after they had discovered it in histones.[52] The method involved the same condensation reaction as had been used for the synthesis of the dimethylamino acid except that the amine was trimethylamine (Equation 25).[39] The product again was purified by passage through an ion-exchange resin column. As for the dimethylamino acid, [^{3}H-4,5]-$N^\varepsilon,N^\varepsilon,N^\varepsilon$-trimethyl-DL-lysine was prepared using the appropriate C-2,3-unsaturated butyl bromide. Norleucine was again formed as byproduct during the saturation step. The product was purified by paper chromatography. [^{3}H-*methyl*]-$N^\varepsilon,N^\varepsilon,N^\varepsilon$-trimethyl-DL-lysine was prepared from the dimethylamino acid by quaternization using methyl iodide in ethanol in a sealed tube at 50°C, followed by purification with the analyzer (Equation 26).[39] On the basis of our work described below, this reaction would have been incomplete.

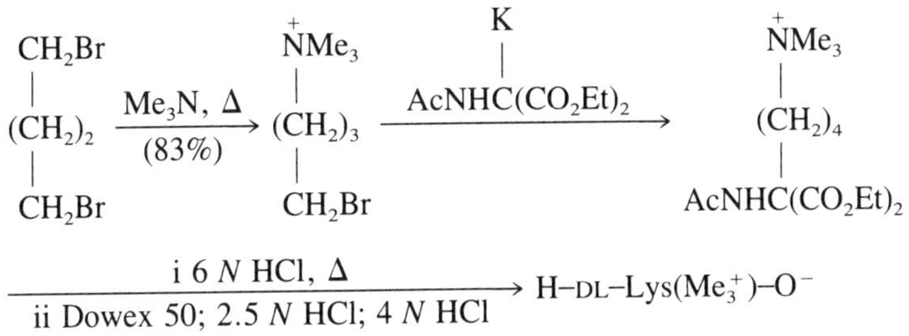

(25)

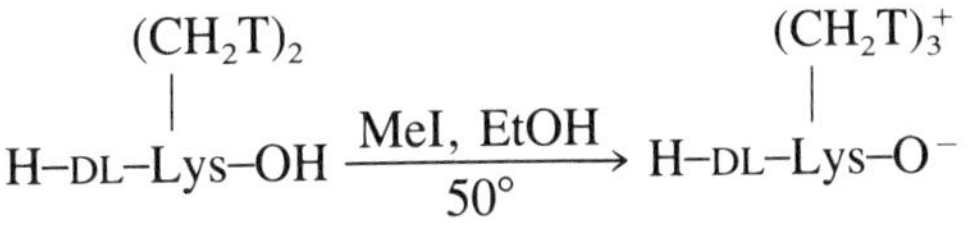

(26)

An additional synthesis from N^{α}-acetyl-L-lysine was reported by a group who discovered the amino acid in cytochrome *c*.[53] These authors, however, used methyl iodide in refluxing methanol containing sodium hydroxide for the methylation followed by hydrolysis under unnecessarily rigorous conditions. No adequate characterization was ever published. Two other groups studying the biosynthesis of carnitine from N^{ε},N^{ε},N^{ε}-trimethyllysine have prepared the [^{3}H-*methyl*]-labeled amino acid by minor variations of the above methods, using barium hydroxide as base and aqueous methanol as solvent at ambient temperature.[54,55] These were reaction conditions adopted from previous work on carnitine. The product was purified in both cases using ion-exchange resins.

A different route to the optically active trimethylamino acid arose from our studies on *N*-methylation using sodium hydride-methyl iodide. The finding that an amino group was quaternized by the reagent and that an N^{α}-tritylamino group resisted methylation (see Equation 8) allowed the preparation of an N^{ε},N^{ε},N^{ε}-trimethyl-L-lysine derivative from the appropriate starting material[26] as shown in Equation 27. The chemistry involved in making the starting material represented the first report of the selective cleavage of an N^{ω}-benzyloxycarbonyl group in the presence of an α-tritylamino group.

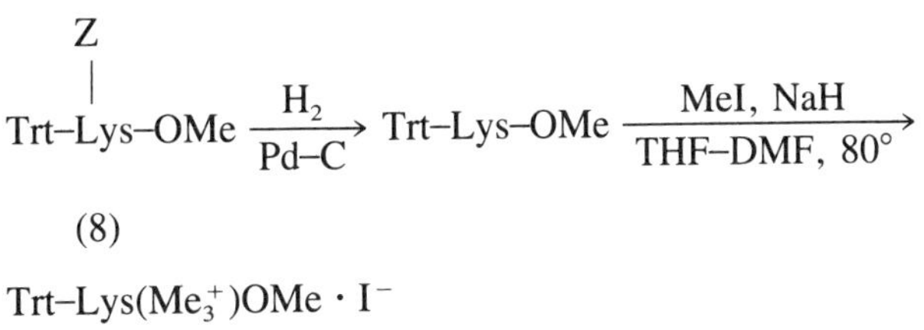

(27)

During the decade following the discovery of N^{ε}-methyllysine in histones,[56] I received many requests for samples of the three N^{ε}-methyllysines for use as reference compounds. And at this time, Moore, in our laboratory, was trying to make Bz-Gly-Lys(Me_3)-0$^-$ for studies on the specificity of carboxypeptidase B. He methylated the parent peptide by the method used previously for the synthesis of N^{ε},N^{ε},N^{ε}-trimethyllysine,[33] and to our surprise and dismay, he found that the benzamido group also had undergone methylation to the extent of 6 to 10%, giving Bz-MeGly-Lys(Me_3^+)-0$^-$ as a byproduct.[57] He therefore had to resort to some other method of methylation. In exploring other methods, we found that it was very easy to prepare a mixture of the three N^{ε}-methyllysines suitable for use as a reference mixture for chromatography (see Equation 28). The overmethylation observed by Moore worried me for some time because it meant that the earlier reaction[33] also might have led to some overmethylation. Fortunately, this now seems unlikely since we were able to show that *N*-benzoylleucine did not undergo methylation under the same conditions.[28] Benzamido methylation by dimethyl sulfate and barium hydroxide is therefore probably restricted to the benzoylglycyl residue.

$$\text{Bz–Lys–OMe} \underset{\text{OH}^-}{\overset{\text{a}}{\rightleftarrows}} \underset{7\%}{\text{Bz–}\overset{\text{H}^+}{\overset{|}{\text{Lys}}}\text{–O}^-} \xrightarrow{\text{a}}$$

$$\underset{7\%}{\text{Bz–}\overset{\text{H}^+}{\overset{|}{\text{Lys}}}\text{(Me)–O}^-} \xrightarrow{\text{a}} \underset{11\%}{\text{Bz–}\overset{\text{H}^+}{\overset{|}{\text{Lys}}}\text{(Me}_2\text{)–O}^-} \xrightarrow{\text{a}} \underset{75\%}{\text{Bz–Lys(Me}_3^+\text{)–O}^-}$$

$$\text{a} = \text{aq. CH}_2\text{N}_2 \qquad (28)$$

The reagent used by Moore to circumvent the overmethylation problem was diazomethane. Though used routinely for the esterification of carboxy groups, it is not so well known that in aqueous solution, diazomethane methylates amino groups to the quaternary state, converting amino acids into betaines.[58,59] The reaction is slow because much of the reagent is decomposed by hydrolysis. Initially, the carboxy group of the amino acid is partially esterified, but it is regenerated by hydrolysis resulting from the increase in pH brought about by the methylation. N^{α}-protected lysines are converted in sequence to the monomethyl-, dimethyl-, and trimethylamino derivatives, the amount of each depending on the time of reaction. The results indicated were obtained after 24 h (Equation 28).[21] The method is not practical for preparing the pure quaternized amino acid, but it is a simple method of acquiring a mixture of the N^{ε}-methyllysines for reference purposes. Because of the slow reaction, we used the corresponding dimethylated derivative as starting material to prepare Bz-Gly-Lys(Me_3^+)-O^- by this method.[37] Only 2% of unreacted material remained. Despite its serious shortcomings, the use of diazomethane in aqueous solution for quaternization should not be disregarded because the reaction is such a clean one; the only byproducts are methanol and nitrogen.

The reader who ponders over the above will realize that no single method for trimethylating an amino group has had preference because none of the methods is entirely satisfactory. The reaction is either incomplete, or unusually vigorous conditions leading to side reactions are required. The obvious deficiencies prompted us to search for other methods. The applicability of a general quaternization reaction described in 1970[60,61] to amino acid derivatives was examined.[40,62] This method involves reaction of the amine with methyl iodide in dimethylformamide in the presence of an organic base with a higher pK_a' as hydrohalide acceptor (Equation 29). The highly hindered and weakly nucleophilic 1,2,2,6,6-pentamethylpiperidine is the base of choice. The hydroiodide formed is separated from the desired product by extraction into hot acetone. It was found that the reaction proceeded smoothly for protected derivatives (X = Bz; Y = NHPh) and small peptides (X = Bz-Phe; Y-OMe). However, it was incomplete for derivatives containing a free carboxy group (X = Boc; Y=OH), and furthermore, some of the base became quaternized as well. These complications arise because these derivatives exist as zwitterions, which resist complete methylation because of the protonated amino group. The reaction therefore is of some use for preparing derivatives, but is not superior to previous methods for making trimethyllysine.

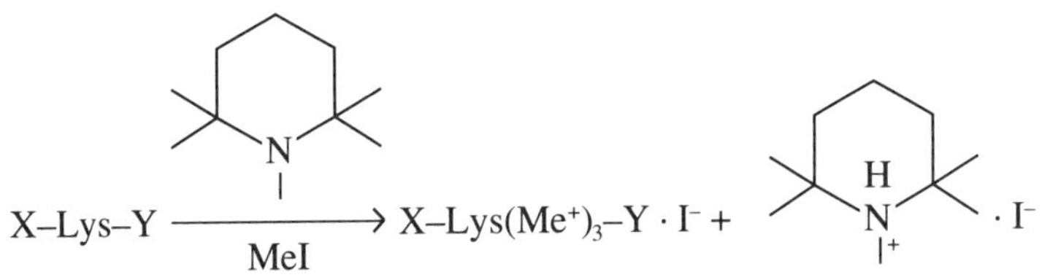

(29)

We then looked at the possibility of quaternization simply by the addition of methyl iodide to a dimethylamino group, since methiodide formation is a known reaction. Again, this reaction was complete for protected derivatives but incomplete for zwitterionic compounds (Equation 30).[62] However, the reaction could be induced to go to completion by the addition of solid potassium bicarbonate. The obvious thing to do next was to try the same thing on a free amino group and to our delight, quaternization occurred without difficulty (Equation 31). The reaction was studied in detail with respect to R and found to be specific and generally applicable.[40,62] The reagent does not attack hydroxy groups and esterifies free carboxy groups only to a small extent (0 to 8%). The latter presents no obstacle because the esters are readily separated from the zwitterionic products by extraction into chloroform. The other attractive feature is that the reagent is so mild that it can be used by quaternizing amino groups in derivatives containing sensitive protecting group, and also in peptides. The N^{α}-benzyloxycarbonyl and N^{α}-butoxycarbonyl derivatives of N^{ε},N^{ε},N^{ε}-trimethyl-L-lysine have been prepared (Equations 32 and 33). The former is a crystalline hydrate, the latter an amorphous powder. Since the starting materials are commercially available and the protecting groups can be readily removed by catalytic hydrogenolysis and mild acidolysis, respectively, this method provides a convenient source for the free trimethylamino acid. This route seemed to be the method of choice for trimethylating an amino group. The N^{α}-butoxycarbonyl derivative of N^{δ},N^{δ},N^{δ}-trimethyl-L-ornithine was also similarly prepared.

$$\text{X–Lys(Me}_2\text{)–Y} \xrightarrow[\text{MeOH}]{\text{MeI}} \text{X–Lys(Me}_3^+\text{)–Y} \cdot \text{I}^-$$

$$\text{only if Y} \neq \text{OH} \tag{30}$$

$$\text{RNH}_2 \xrightarrow[\text{MeOH, 24 h}]{\text{MeI, KHCO}_3} \text{R}\overset{+}{\text{N}}\text{(Me}_3\text{)I}^- \tag{31}$$

$$\text{Z–Lys–OH} \xrightarrow[\text{ii Dowex 50; NH}_4\text{OH}]{\text{i MeI, KHCO}_3\text{, MeOH}} \text{Z–Lys(Me}_3^+\text{)–O}^- \tag{32}$$

It was soon realized, however, that the method would only be of value to others if complete details of the synthesis were available. We thus designed a synthesis of N^{ε},N^{ε},N^{ε}-trimethyl-L-lysine from N^{α}-*tert*-butoxycarbonyl-N^{ε}-benzyloxycarbonyl-L-lysine in which the product is crystallized as the dioxalate salt (Equation 34). Considerable originality was introduced in order to purify and isolate the product.

$$\text{Boc–}\overset{\text{Z}\atop|}{\text{Lys}}\text{–OH} \xrightarrow[\text{Pd–C}]{H_2} \text{Boc–Lys–OH} \xrightarrow[\text{MeOH, 12 h}]{\text{MeI, NaHCO}_3}$$

$$\xrightarrow[\substack{\text{iii filtration}\\ \text{iv concentration, 80°}}]{\substack{\text{i filtration}\\ \text{ii oxalic acid}}} \text{H–Lys(Me}_3^+\text{)–O}^- \quad (34)$$

A solution of N^{α}-Boc-N^{ε}-Z-L-lysine (12 g; 31.5 mmol) in 300 ml of methanol-water (19:1) was stirred overnight in a hydrogen atmosphere in the presence of 1 g of 5% palladium-on-charcoal. The catalyst was removed by filtration through Celite and the filtrate was evaporated to dryness to remove the water. Methyl iodide (10 ml; 0.14 mmol) and potassium hydrogen carbonate (10 g; 0.10 mmol) were added to the residue (N^{α}-Boc-lysine) dissolved in 250 ml of methanol and the mixture was stirred for 10 to 12 h at room temperature (gas was released at 100 ml/h). The addition of the methyl iodide and $KHCO_3$ and stirring was repeated twice more. A 5-ml aliquot was examined by proton nuclear magnetic resonance spectroscopy in D_2O after removal of the solvent. The absence of the (CH_2)-N-peak of dimethyllysine at $\delta = 2.87$ ppm (relative to sodium 2,2-dimethyl-2-silapentane-5-sulfonate as internal standard) indicated that the reaction had gone to completion. The (CH_3)-N^+– peak of trimethyllysine was at $\delta = 3.08$ ppm. The mixture was filtered ($KHC0_3$ and KI) and the filtrate was evaporated to dryness. The residue was dissolved in 100 ml of ethanol, any insoluble material was removed by filtration, and the filtrate was again evaporated to dryness. Oxalic acid (30 g) in 200 ml of ethanol was added to the residue. The mixture was stirred, cooled, and filtered to remove final traces of K^+ as potassium oxalate. The ethanol was removed by evaporation, 100 ml of water was added, and the solution was concentrated to 80°C until oxalic acid precipitated out. This removed I^- as hydriodic acid. The addition of water and concentration was repeated three to seven times until the distillate was free of I^-, as evidenced by testing with silver nitrate. The heating removed the Boc group and also hydrolyzed any ester produced during the *N*-alkylation.

The trimethyllysine dioxalate was crystallized by adding 500 ml of ethanol and a seed crystal to the cooled solution. The crystals were collected and washed with ethanol. A second crop could sometimes be obtained from the mother liquor. Recrystallization was effected by dissolving the compound in the minimum amount of water and adding ethanol to turbidity, followed by oxalic acid (2 g). The yield, m.p. 124 to 126°C, was 8 to 10 g (70 to 80%). Molecular formula: $C_9H_{22}N_20_2^{2+}\cdot C_4H_20_8^{2-}\cdot 1/2H_20$ (377.35). Poor yields indicated incomplete desalting or iodide removal. The mother liquor could be treated as follows as an alternative for inorganic ion removal, but only after any protecting groups had been eliminated. The ethanol was removed by evaporation, Dowex-50 H^+ (20 to 50 mesh) was added to the residue dissolved in water, and the mixture was shaken for 1 h. The resin was filtered off, washed with water, and then suspended in 7.5 *N* ammonium hydroxide for 15 min. After filtering, the eluate was evaporated to dryness, the residue was dissolved in water, oxalic acid in ethanol was added, and crystals were obtained.

The first step in the synthesis can be avoided by purchasing the N^{α}-Boc-L-lysine, but this doubles the cost of the starting material. The main difficulty in obtaining pure betaines such as trimethyllysine, resides not in the quaternization reaction but in separation of the product from inorganic ions, all of which are soluble in water at all pHs. Conventional trimethylations are done in aqueous barium hydroxide (as base), followed by acid hydrolysis and desalting with an ion-exchange resin. In our procedure, the methanol as solvent allows for removal of a large part of the potassium and iodide ions as potassium iodide by filtration of the reaction mixture. Replacement of the solvent by ethanol and addition of excess oxalic acid allows for removal of the remaining K^+ as insoluble potassium oxalate. The oxalic acid also displaces the remaining

TABLE 4
Derivatives of N^ω,N^ω,N^ω-Trimethyldiamino Acids

Compound	Melting point (°C)	Method[a]	Ref.
H-Lys(Me_3^+)-O^-·2HCl	232	24	51
H-Lys(Me_3^+)-O^-·2HCl	dec.	35	3
H-Lys(Me_3^+)-O^-·2(CO_2H)$_2$·1/2$H_2$0	124—126	21,22,34	2,46,47
H-Lys(Me_3^+)–O^-[b]	225—226		63
H-Lys(Me_3^+)–O^-·HCl	240	20	45
H-DL-Lys(Me_3^+–O^-·2HCl		19	44
H-DL-Lys(Me_3^+)–O^-·2HCl	260—261	25	39
H-DL-Lys(Me_3^+)–O^-·Au_2Cl_8(CO_2H)	187—188		43,44
H-DL-Lys(Me_3^+)–O^-·2(CO_2H)$_2$·1/2H_2O	129	23	47
H-DL-Lys(Me_3^+)-O^-[c]	275—280	20	45
H-Lys(Me_3^+)-OMe[d]	142—143		47
H-Lys(Me_3^+)-NH_2[d]	118—119		47
Bz-Lys(Me_3^+)-O^-	240—241		33
Bz-Lys(Me_3^+)-O^-·$AuCl_4$	159—160		33
(Bzl)$_2$-DL-Lys(Me_3^+)-0$^-$·2HCl	170		45
Boc-Lys(Me_3^+)-O^-			33
Z-Lys(Me_3^+)-O^-·HCl	157—158	24	51
Z-Lys(Me_3^+)-0$^-$·H_2O	199—201	32	62
Trt-Lys(Me_3^+)-OMe·I^-	201—202		26
H-Orn(Me_3^+)-0$^-$·2(CO_2H)$_2$·H_2O	111—113	21	47
Boc-Orn(Me_3^+)-O^-		33	62
H-A_2bu(Me_3^+)-O^-·(CO_2H)$_2$	181	21	47

[a] Refers to equation in the text.
[b] Di-*p*-hydroxyazogenzene-*p*′-sulfonate.
[c] Diflavianate.
[d] Dipicrate.

Adapted from Benoiton, N. L., in *Chemistry and Biochemistry of Amino Acids, Peptides and Proteins,* Vol. 5, Weinstein, B., Ed., Marcel Dekker, New York, 1978, 163. With permission.

iodide from the betaine, thus rendering it volatile. This approach for preparing and purifying betaines should be generally applicable to other betaines which can be crystallized as oxalate salts.

At about the same time, another synthesis of N^ε,N^ε,N^ε-trimethyl-L-lysine in large amounts was published. It involves quaternization of N^α-benzyloxycarbonyl-N^ε,N^ε-dimethyl-L-lysine using *O*-methyl-*N*,*N*′-dicyclohexylisourea followed by removal of the protecting group by hydrogenolysis (Equation 35). The product was obtained in 57% yield as the dihydrochloride, but with a melting point which is unreliable for characterization purposes. A dioxolate salt prepared and crystallized as described in the literature had a melting point 20°C higher than the literature value. Aqueous formaldehyde had been used to prepare the starting material, and the disadvantages of this have been alluded to above.

$$\text{Z–Lys(Me}_2\text{)–OH} \xrightarrow[\substack{\text{ii filtration} \\ \text{iii HCl, evaporation}}]{\substack{\text{C}_6\text{H}_{11}\text{NH–C(OMe)=NC}_6\text{H}_{11} \\ \text{MeOH, 6 d}}} \text{Z–Lys(Me}_3^+\text{)–O}^-$$

$$\xrightarrow[\text{ii HCl}]{\text{i H}_2\text{, Pd–C}} \text{H–Lys(Me}_3^+\text{)–O}^- \qquad (35)$$

The salts and derivatives of the trimethyldiamino acids which have been described are compiled in Table 4.

REFERENCES

1. **Benoiton, N. L.,** N^{ω}-Alkyldiamino acids: chemistry and properties, in *Chemistry and Biochemistry of Amino Acids, Peptides and Proteins,* Vol. 5, Weinstein, B., Ed., Marcel Dekker, New York, 1978, 163.
2. **Chen, F. M. F. and Benoiton, N. L.,** A synthesis of N^6,N^6,N^6-trimethyl-L-lysine dioxalate in gram amounts, *Biochem. Cell Biol.,* 64, 182, 1986.
3. **Hertzberg, T., Kühler, T., and Nilsson, M.,** Preparation of (S)-5-Amino-5-carboxy-*N,N,N*-trimethyl-1-pentaneaminium chloride (L-lysinebetaine hydrochloride), *Acta Chem. Scand.,* 840, 387, 1986.
4. **Stock, A., Clarke, S., Clarke, C., and Stock, J.,** N-Terminal methylation of proteins: structure, function and specificity, *FEBS Lett.,* 220, 8, 1987.
5. IUPAC-IUB Joint Commission on Biochemical Nomenclature (JCBN) Nomenclature and symbolism for amino acids and peptides recommendations, 1983, *Eur. J. Biochem.,* 138, 9, 1984.
6. **Enger, R. and Steib, H.,** α- and ε-Monomethyllysine, *Z. Physiol. Chem.,* 191, 97, 1930.
7. **Neuberger, A. and Sanger, F.,** The availability of ε-acetyl-D-lysine and ε-methyl-DL-lysine for growth, *Biochem. J.,* 38, 125, 1944.
8. **Poduska, K.,** Preparation and chromatographic behaviour of some *N*-methyl derivatives of DL-lysine, *Collect. Czech. Chem. Commun.,* 24, 1025, 1959.
9. **Babineau, L. M. and Berlinguet, L.,** Synthèse de dérivés de la lysine, *Can. J. Chem.,* 40, 1626, 1962.
10. **Kim, S., Benoiton, L., and Paik, W. K.,** On the metabolism of ε-methyl-L-lysine by rat kidney homogenate, *Biochim. Biophys. Acta,* 71, 745, 1963.
11. **Izymiya, N. and Ota, S.,** Inversion of ε-benzoyl-L-lysine, *J. Chem. Soc. (Jpn.),* 72, 445, 1951.
12. **Izymiya, N.,** On the walden inversion of amino acids, *J. Chem. Soc. (Jpn.),* 72, 550, 1951.
13. **Benoiton, N. L.,** Synthesis of ε-N-methyl-L-lysine and related compounds, *Can. J. Chem.,* 42, 2043, 1964.
14. **Thomas, K. and Goerne, M. G. H.,** Weitere Untersuchungen über die Herkunft des Kreatins, *Z. Physiol. Chem.,* 104, 73, 1919.
15. **Quitt, P., Hellerback, J., and Vogler, K.,** Die Synthese optisch aktiver *N*-monomethyl-aminosäuren, *Helv. Chim. Acta,* 46, 327, 1963.
16. **Cheung, S. T. and Benoiton, N. L.,** *N*-Methylamino acids in peptide synthesis. VII. Studies on the enantiomeric purity of *N*-methylamino acids prepared by various procedures, *Can. J. Chem.,* 55, 916, 1977.
17. **Benoiton, L. and Berlinguet, L.,** ε-N-methyl-L-lysine monohydrochloride, *Biochem. Prep.,* 11, 80, 1966.
18. **Kim, S., Benoiton, L., and Paik, W. K.,** Purification and properties of ε-N-methyl-L-lysine: 0_2 oxidoreductase, *J. Biol. Chem.,* 239, 3790, 1964.
19. **Bezas, B. and Zervas, L.,** On the peptides of l-lysine, *J. Am. Chem. Soc.,* 83, 719, 1961.
20. **Wünsch, E. and Zwick, A.,** Zur Synthese des Glucagon. IV. Darstellung der Sequenz 12—15, *Chem. Ber.,* 97, 3305, 1964.
21. **Benoiton, N. L., Demayo, R. E., Moore, G. J., and Coggins, J. R.,** A modified synthesis of α-N-carbobenzoxy-L-lysine and the preparation and analysis of mixtures of ε-N-methyllysines, *Can. J. Biochem.,* 49, 1292, 1971.
22. **Savrda, J. and Bricas, E.,** Synthesis of peptides related to the N-terminal sequence of bovine trypsinogen. I. Synthesis of protected peptides, *Bull. Soc. Chim. Fr.,* 2423, 1968.
23. **Coggins, J. R. and Benoiton, N. L.,** Synthesis of *N*-methylamino acid derivatives from amino acid derivatives using sodium hydride/methyl iodide, *Can. J. Chem.,* 49, 1968, 1971.
24. **McDermott, J. R. and Benoiton, N. L.,** *N*-Methylamino acids in peptide synthesis. II. A new synthesis of *N*-benzyloxycarbonyl, *N*-methylamino acids, *Can. J. Chem.,* 51, 1915, 1973.
25. **Cheung, S. T. and Benoiton, N. L.,** *N*-Methylamino acids in peptide synthesis. V. The synthesis of *N-tert*-butoxycarbonyl, *N*-methylamino acids by *N*-methylation, *Can. J. Chem.,* 55, 906, 1977.
26. **Benoiton, N. L. and Coggins, J. R.,** Methylation as a route to peptide intermediates of *N*-methylamino acids and ε-*N*-methyllysines, in *Progress in Peptide Research,* Lande, S., Ed., Gordon & Breach, New York, 1972, 145.
27. **Coggins, J., Demayo, R., and Benoiton, N. L.,** Esterification of ε-*N*-carbobenzoxy-L-lysine with boron trifluoride-alcohol, *Can. J. Chem.,* 48, 385, 1970.
28. **Benoiton, N. L. and Kuroda, K.,** unpublished data, 1976.
29. **Sheehan, J. C. and Guziec, F. S.,** Amino group protection in peptide synthesis. The 4,5-diphenyl-4-oxazoline-2-one group, *J. Org. Chem.,* 38, 3034, 1973.
30. **McDermott, J. R. and Benoiton, N. L.,** *N*-Methylamino acids in peptide synthesis. III. Racemization during saponification and acidolysis, *Can. J. Chem.,* 51, 2555, 1973.
31. **Yamamoto, H. and Yang, J. T.,** Syntheses and conformational studies of poly-(N^{ε}-methyl-L-lysine), poly(N^{δ}-methyl-L-ornithine), poly(N^{δ}-ethyl-L-ornithine) and their carbobenzoxy derivatives, *Biopolymers,* 13, 1093, 1974.
32. **Benoiton, N. L. and Deneault, J.,** The hydrolysis of two ε-*N*-methyl-L-lysine derivatives by trypsin, *Biochim. Biophys. Acta,* 113, 613, 1966.

33. **Seely, J. H. and Benoiton, N. L.,** Effect of *N*-methylation and chain length on kinetic constants of trypsin substrates, *Can. J. Biochem.*, 48, 1122, 1970.
34. **Thomas, K., Kapfhammer, J., and Flaschenträger, B.,** Über δ-Methylornithin und δ-methylarginin. Zur Frage Nach der Herkunft des Kreatins, *Z. Physiol. Chem.*, 124, 75, 1922.
35. **Emerson, W. S.,** Preparation of amines by reductive alkylation, *Org. Reactions*, 4, 174, 1948.
36. **Bowman, R. E. and Stroud, H. H.,** N-Substituted amino-acids. I. A new method of preparation of dimethyl-amino-acids, *J. Chem. Soc.*, 1342, 1950.
37. **Neuberger, A. and Sanger, F.,** The availability of acetyl derivatives of lysine for growth, *Biochem. J.*, 37, 515, 1943.
38. **Means, G. E. and Feeney, R. E.,** Reductive alkylation of amino groups in proteins, *Biochemistry*, 7, 2192, 1968.
39. **Hempel, K., Lange, H. W., and Birkofer, L.,** Synthese tritium-markierter, N-methylierter Lysine, *Z. Physiol. Chem.*, 350, 867, 1969.
40. **Benoiton, N. L. and Chen, F. M. F.,** The synthesis of amino acid derivatives and peptides containing *N,N*-dimethylamino and *N,N,N*-trimethylamino groups, in *Peptides 1976*, Loffet, A., Ed., Presse Universitaire de Bruxelles, Brussels, 1976, 149.
41. **Chen, F. M. F. and Benoiton, N. L.,** Reductive *N,N*-dimethylation of amino-acid and peptide derivatives using methanol as the carbonyl source, *Can. J. Biochem.*, 56, 150, 1978.
42. **Bowman, R. E.,** N-Substituted amino-acids. II. The reductive alkylation of amino-acids, *J. Chem. Soc.*, 1346, 1950.
43. **Kapfhammer, J.,** Isolierung von ε-Monogetain des Lysin aus methyliertem Casein, *Z. Physiol. Chem.*, 191, 112, 1930.
44. **Enger, R. and Halle, F.,** Monobetaine von Lysin, *Z. Physiol. Chem.*, 191, 103, 1930.
45. **Goffinet, B. and Amiard, G.,** α,ε-Diaminocaproic acid betaines, French Patent 1,176,117, 1959; *Chem. Abstr.*, 55, 19865a, 1961.
46. **Takemoto, T., Daigo, K., and Takagi, N.,** Hypotensive constituents of marine algae. I. A new basic amino acid (laminine) and the other basic constituents isolated from *Laminaria augustata*, *Yakugaku Zassi*, 84, 1176, 1964.
47. **Takemoto, T., Daigo, K., and Takagi, N.,** Studies on the hyposensitive constituents of marine algae. II. Synthesis of laminine and related compounds, *Yakugaku Zasshi*, 84, 1180, 1964.
48. **Puskás, J. and Tyihák, E.,** Synthesis of DL-2-amino-6-trimethylaminocaproic acid betaine (ε-*N*-trimethyllysine), Periodica Polytechnica (Budapest), *Chem. Eng. Series*, 13, 261, 1969.
49. **Tyihák, E. and Puskás, J.,** Isolation of *N*-trimethyllysine from protein hydrolyzates and cell saps, *Die Pharmazie*, 1, 69, 1970.
50. **Larsen, P. O.,** N^6-Trimethyl-L-lysine betaine from seeds of *Reseda luteola* L, *Acta. Chem. Scand.*, 22, 1369, 1968.
51. **Havinga, E. and Schattenkerk, C.,** Polypeptides. V. Properties of the arginyl residue of des-arginyl^{1}isoleucyl^5-angiotensin II essential to pressor activity, *Tetrahedron*, Suppl. 8, (part I), 313, 1966.
52. **Hempel, K., Lange, H. W., and Birkofer, L.,** ε-*N*-Trimethyllysin, eine neue Aminosäure in Histonen, *Naturwissenschaften*, 55, 37, 1968.
53. **DeLange, R. J., Glazer, A. N., and Smith, E. L.,** Presence and location of an unusual amino acid, ε-*N*-trimethyllysine, in cytochrome *c* of wheat germ and *Neurospora*, *J. Biol. Chem.*, 244, 1385, 1969.
54. **Haigler, H. T. and Broquist, H. P.,** Carnitine synthesis in rat tissue slices, *Biochem. Biophys. Res. Commun.*, 56, 676, 1974.
55. **Hochalter, J. B. and Henderson, L. M.,** Carnitine biosynthesis: the formation of glycine from carbons 1 and 2 of 6-*N*-trimethyl-L-lysine, *Biochem. Biophys. Res. Commun.*, 70, 364, 1976.
56. **Murray, K.,** The occurrence of ε-*N*-methyl lysine in histones, *Biochemistry*, 3, 10, 1964.
57. **Moore, G. J. and Benoiton, N. L.,** Effect of $N^ε$-alkylation of the substrate on the hydrolysis of benzoylglycyl-L-lysine by carboxypeptidase B, *Can. J. Biochem.*, 53, 1145, 1975.
58. **Kuhn, R. and Brydówna,** Action of diazomethane on amino acids, *Berichte*, 70, 1333, 1937.
59. **Kuhn, R. and Ruelis, H. W.,** Uber die Umsetzung von Diazomethane mit Zwitter-Ionen und anorganischen Salzen, *Chem. Ber.*, 83, 420, 1950.
60. **Sommer, H. Z. and Jackson, L. L.,** Alkylation of amines. A new method for the synthesis of quaternary ammonium compounds from primary and secondary amines, *J. Org. Chem.*, 35, 1558, 1970.
61. **Sommer, H. Z., Lipp, H. I., and Jackson, L. L.,** Alkylation of amines. A general exhaustive alkylation method for the synthesis of quaternary ammonium compounds, *J. Org. Chem.*, 36, 824, 1971.
62. **Chen, F. M. F. and Benoiton, N. L.,** A new method of quaternizing amines and its use in amino acid and peptide chemistry, *Can. J. Chem.*, 54, 3310, 1976.
63. **Kakimoto, Y. and Akazawa, S.,** Isolation and identification of N^G,N^G- and N^G,N'^G-dimethylarginine, $N^ε$-mono, di-, and trimethyllysine, and glucosyl-galactosyl- and galactosyl-δ-hydroxylysine from human urine, *J. Biol. Chem.*, 245, 5751, 1970.

Chapter 20

BIOLOGICAL EFFECTS OF METHYLATED AMINO ACIDS

Ernö Tyihák, Béla Szende, and Lajos Trézl

TABLE OF CONTENTS

I. INTRODUCTION

It is becoming increasingly evident that the posttranslational enzymatic modification of various proteins influences the primary and secondary metabolism (e.g., cell proliferation, differentiation, signal systems).[1-3] Among these protein-modifying reactions, the methylation reaction is of particular importance. Methylation of proteins results primarily in the formation of very stable *N*-methyl derivatives of basic amino acids such as L-lysine, L-arginine, and L-histidine, or less stable carboxyl-methyl ester derivatives of dicarboxylic amino acids such as aspartyl and glutamyl residues.[4,5] The stable methylated basic aminno acids as ε-*N*-methylated lysines, *N*^G-methylated arginines, and 3-*N*-methylhistidine also occur in free form, possibly resulting from enzymatic hydrolysis of methylated proteins *in vivo* (see also Chapter 1 in this series).[6]

It is known that L-lysine and L-arginine are in indispensable requirement for normal growth of microorganisms,[7] plants,[8,9] and animals.[10] Furthermore, an antagonistic role between these two basic amino acids has also been observed in different organisms,[11,12] especially when considering their effects on animal tumors. Oral or peritoneal administration of a large dosage of either L-lysine or L-arginine has been found to produce a highly significant change in the growth of various transplantable animal tumors in mice:[13,14] L-lysine strongly inhibited the tumor growth when L-arginine promoted the growth of tumor. This fundamental relationship suggests that ε-*N*-methylated lysines and N^G-methylated (and N^G-hydroxymethylated) arginines deserve a distinguished place among the methylated amino acids.

It is thought that the biological effects of methylated amino acids are derived from their methyl groups which are transferred to the native amino acids in posttranslational enzymatic reactions by use of *S*-adenosyl-L-methionine (AdoMet) as a methyl donor. Thus, metabolism of the methyl groups in an organism is a key to understanding the mechanism of action of methylated amino acids.

II. BIOLOGICAL EFFECTS OF ε-*N*-METHYLATED LYSINES

A. BIOLOGICAL EFFECTS OF ε-*N*-TRIMETHYL-L-LYSINE (TML) ON HUMAN AND ANIMAL TEST SYSTEMS

The first experimental data on the biological effect of TML was found in its blood-pressure

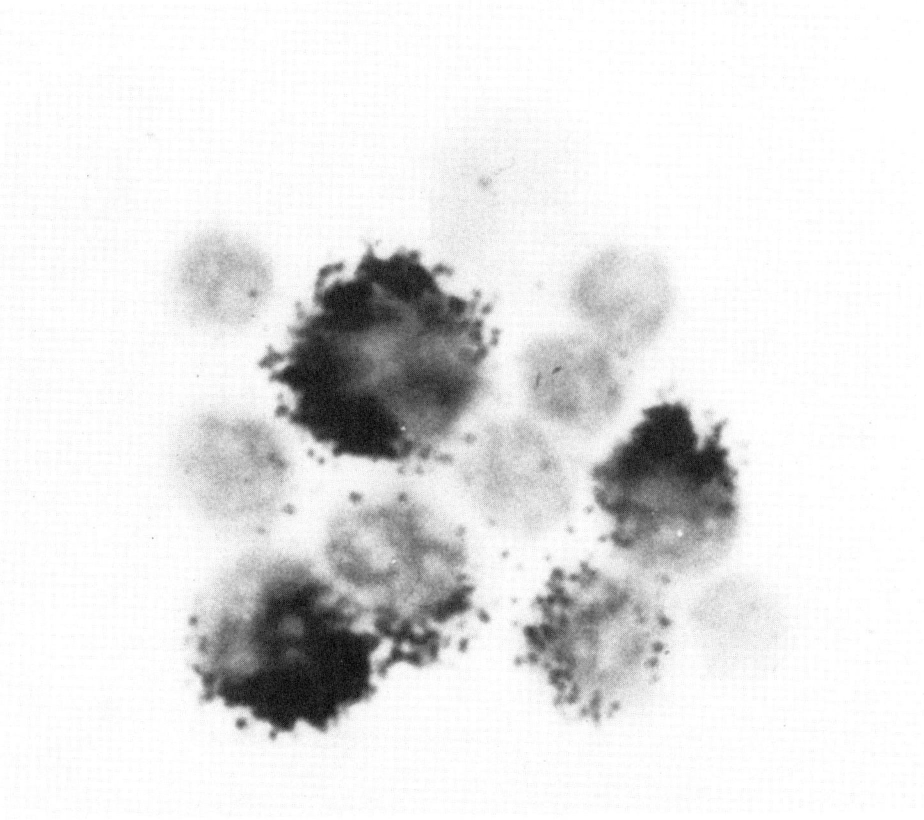

FIGURE 1. Human peripheral blood lymphocyte culture treated for 48 h with 100 μg/ml TML and labeled for 1 h with [^{3}H]TdR. Five out of 11 cells contain grains (300×).

reducing activity.[15] Subsequently, a series of studies on the biological effects of TML was performed in Hungarian institutes confirming the cell proliferation stimulating activity of this compound.[16]

1. Blood-Pressure Reducing Effect of TML

Laminine, a compound isolated from a marine algae *Laminaria angustata,* has been shown to be effective in reducing blood pressure and was later proven to be identical to TML.[15] Since its toxicity is small in comparison with other blood pressure reducing quaternary ammonium compounds, it can be administered in higher dosages.[16] However, because of a moderate activity TML is not used in clinical practice.

2. Blastic Transformation of Human T Lymphocytes[17,18]

Ten μg/ml TML hydrochloride or 100 μg/ml TML glutamate caused the blastic transformation of human T lymphocytes in *in vitro* culture. The transformation was proven by light and electron microscopical morphological studies (Figure 1) by measuring the incorporation of [^{3}H]TdR into the lymphocytes (Figure 2) and by establishing DNA and histone histograms (Figures 3 and 4). The degree of transformation was identical to that caused by phytohemagglutinin (PHA). Simultaneous treatment with PHA and TML did not result in blastic transformation, but TML—given 1 h after PHA—increased the transforming effect of PHA (Figure 5).

3. Effect on *In Vitro* Cultured Mouse Embryonic Fibroblasts

A cell line, derived from CBA mouse embryo fibroblasts, was treated with 10 to 100 μg/ml

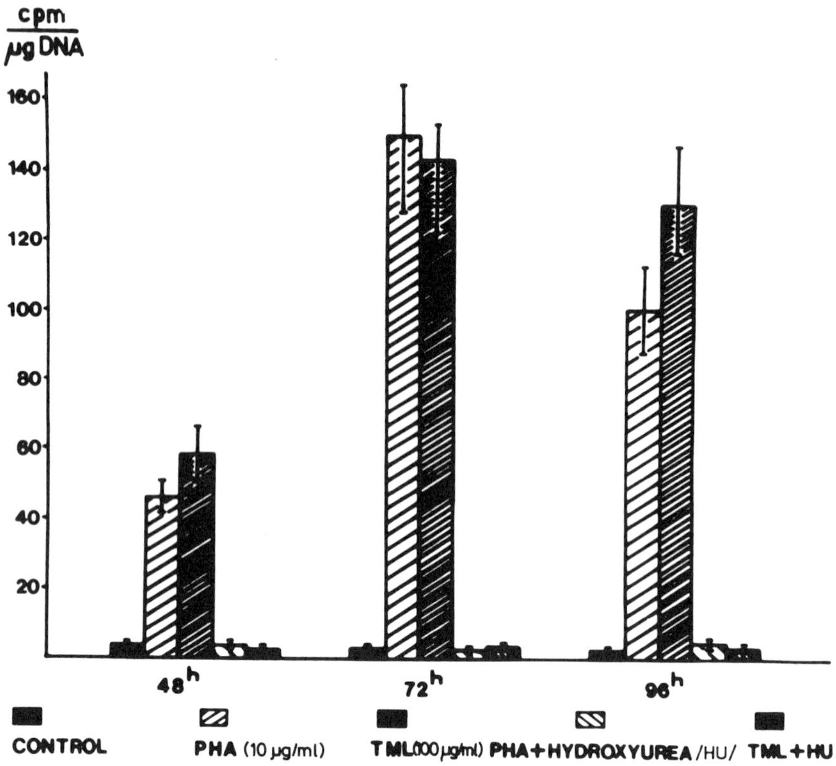

FIGURE 2. [^{3}H]thymidine incorporation into human T lymphocyte cultures after TML, PHA, and hydroxyurea (HU) treatment. (From Suba, Z. S., Szende, B., Takáts, J., Lapis, K., and Elek, G., *Neoplasma,* 27, 11, 1980. With permission.)

TML. This treatment was ineffective when given in the exponential growth phase, but caused an additional cell division in the stationary phase (Figure 6).

4. Effect on [^{3}H]TdR Incorporation into Mouse Thymus, Spleen, Bone Marrow, Jejunum, and Liver[19]

A single dose of TML at 100 mg/kg i.p. resulted in a moderate increase of the incorporation of [^{3}H]TdR into the thymus, spleen, bone marrow (Figure 7) and jejunum of healthy CBA mice.

Whole body irradiation of healthy CBA mice (900 rad) is characterized by partial necrosis and subsequent regeneration of the jejunal mucosa. TML treatment (100 mg/kg, i.p.), administered simultaneously with the irradiation, caused a significantly higher degree of regeneration, measured by both [^{3}H]TdR incorporation (Figure 8) and by histological studies. The [^{3}H]TdR incorporation into the liver cells of CBA inbred mice was enhanced by TML treatment (100 mg/kg, as a single i.p. treatment). This enhancement was followed by liquid scintillation (Figure 9) and by autoradiography. A difference was observed in the TML effect between males and females: females responded more quickly and to a more measurable extent than males.

5. Effect of TML in Splenic Colony Assay[19]

Inbred DBA/2 mice were irradiated with 900 rad and 2×10^4 nucleated bone marrow cells were injected into the tail vein of the animals 24 h later. Simultaneously, 20 mg/kg TML was given i.p. to the assay animals. Splenic colonies were counted 7 d after the irradiation. It was observed that the average number of splenic colonies doubled in the treated group when compared with untreated controls. The effect of TML was less noticeable when the donors were treated (Figure 10).

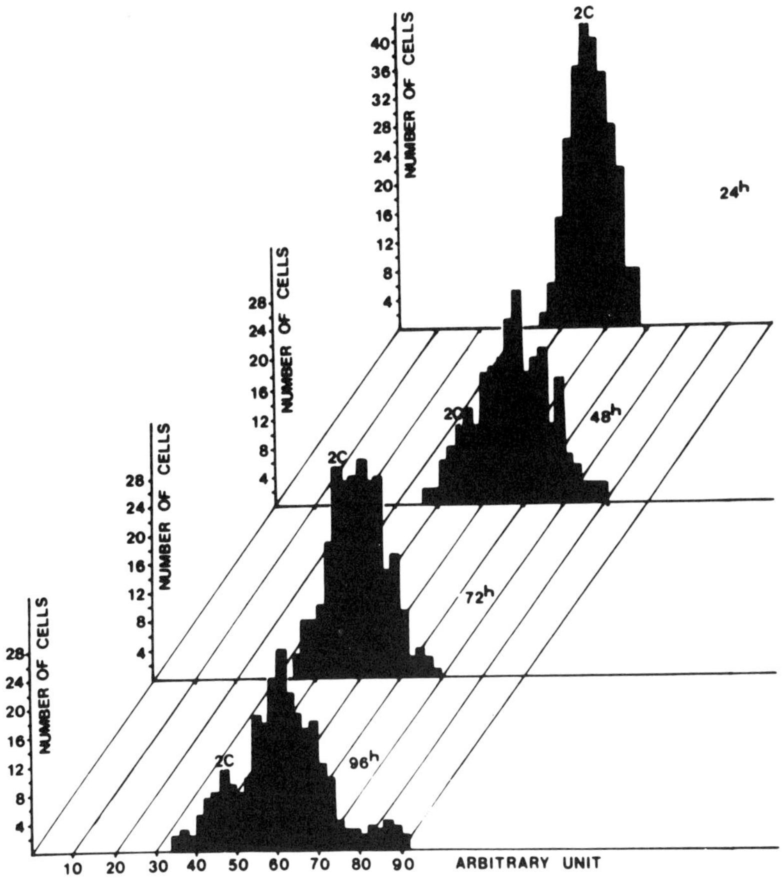

FIGURE 3. Relative DNA content distribution in human lymphocyte cultures after TML treatment. (From Suba, Z. S., Szende, B., Takáts, J., Lapis, K., and Elek, G., *Neoplasma*, 27, 11, 1980. With permission.)

6. Effect of TML on Anemia Caused by NK/Ly Ascites Tumor, Lykurim, and Whole Body Irradiation

TML treatment (100 mg/kg i.p. three times on consecutive days) did not influence the red or white blood cell counts in healthy DBA/2 inbred mice. However, when hemopoiesis was damaged in some way, TML prevented the decrease of erythrocyte or leukocyte numbers.

a. Effect of TML Treatment in Hemolytic Anemia Caused by NK/Ly Ascites Tumor

The ascites lymphoma described by Németh and Kellner results in a severe hemolytic anemia in the later phase of its growth.[20] Repeated i.p. treatment with 100 mg/kg TML moderated the terminal anemia and leukopenia of NK/Ly ascites tumor bearing animals.

b. Effect of TML Treatment on White and Red Blood Cell Count of Swiss Mice Treated with Lykurim

The alkylating agent Lykurim causes a drop in red blood cell count 17 d after the i.p. injection of Lykurim (30 mg/kg) in Swiss mice. Granulocyte count also shows a drop by 17 d after a transient increase. TML, given simultaneously in a single 100 mg/kg i.p. dose with Lykurim, prevented the drop of red blood cell and granulocyte number, moreover, resulted in an increase in red blood cells. Pre- or posttreatment with TML was less effective in the case of granulocytes, but was more effective in the case of red blood cells.

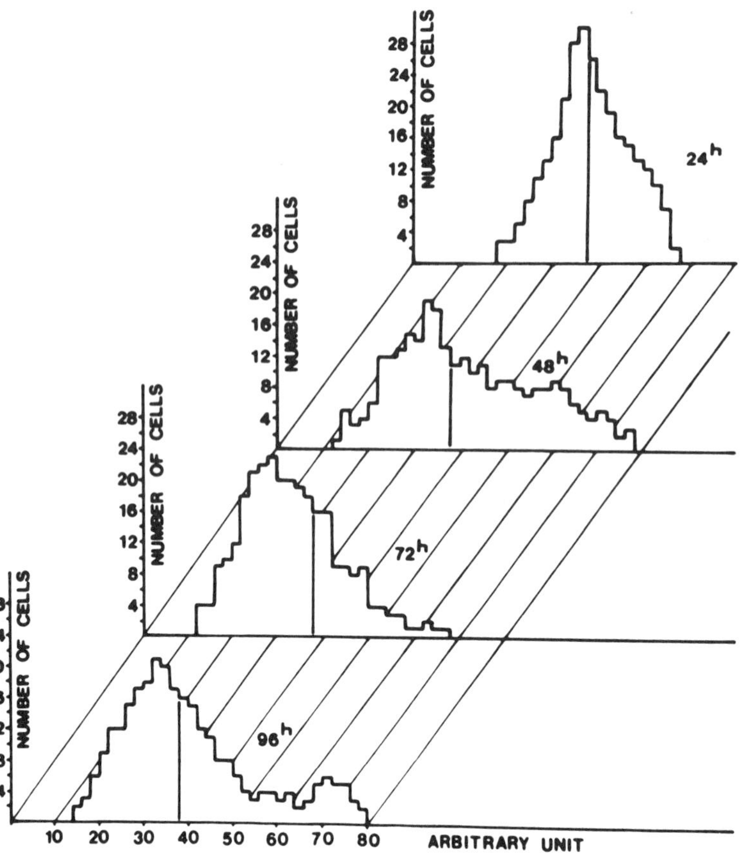

FIGURE 4. Relative histone content distribution in human lymphocyte cultures after TML treatment. (From Suba, Z. S., Szende, B., Takáts, J., Lapis, K., and Elek, G., *Neoplasma,* 27, 11, 1980. With permission.)

c. *Effect of TML on Hematocrit Values of Irradiated DBA/2 Mice*[21]

Whole body of inbred DBA/2 mice was irradiated with 500 rad, together with TML administered (100 mg/kg as a single i.p. treatment) 24 h before, 24 h after irradiation, and simultaneously with irradiation, respectively. Following the irradiation, hematocrit values were determined weekly. Three weeks after the irradiation, TML given simultaneously with 500 rad irradiation, prevented the drop in hematocrit value observed in the first 2 weeks after irradiation.

7. Effect of TML on Humoral and Cellular Immune Response

As TML proved to be mitogenic in *in vitro* lymphocyte cultures, the effect of this compound on the immune system was of great interest. Thus, inbred female CBA mice were irradiated with 400 rad, and 1 d later SRBC suspension was injected as antigen. TML treatment was initiated 24 h after the antigen stimulus, and 100 mg/kg TML (i.p.) was administered every 4th day for 5 d. Levamisole and *E. coli* endotoxine treatment (5×3 mg/kg i.p.) were applied as positive control. Irradiation decreased the hemagglutinin titer when compared to nonirradiated, untreated controls; however, both TML and Levamisole treatment prevented this irradiation-induced decrease (Figure 11). On the other hand, the cellular immune response, i.e., the late type hypersensitivity determined by the foot-pad swelling test, decreased by repeated TML (100 mg/kg i.p.) treatment of both irradiated (400 rad) or nonirradiated CBA mice (Figure 12).

In *in vitro* studies, spontaneous lymphocyte mediated cytotoxicity (SLMC), antibody-

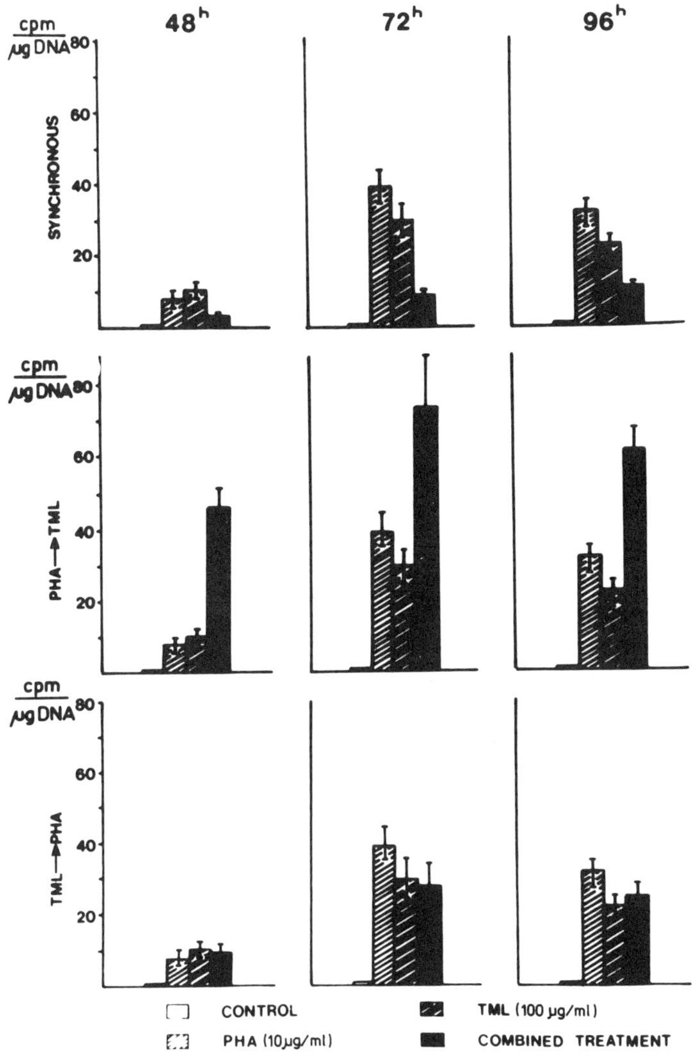

FIGURE 5. Effect of combined PHA-TML treatment on human lymphocyte cultures.

dependent cytotoxicity (ADCC), and chicken red blood cell (CRBC) ADCC were determined using lymphocytes isolated from the peripheral blood of healthy persons. When TML was incubated together with the target and effector cells in 200, 400, or 800 µg/ml concentrations, a dose-dependent decrease of cytotoxicity was observed in all three systems treated with TML (Figure 13).

8. Studies on the Mode of Action of TML[22,23]

Paik and Kim[4] had demonstrated the presence of an enzyme, ε-alkyllysinase (ε-alkyl-L-lysine:oxygen oxidoreductase; EC 1.5.3.4) in rat kidney, which dealkylates free MML and DML. They later observed another enzyme from rat kidney which demethylates an enzymatically labeled [*methyl*-^{14}C]histone, in which one of the reaction products has been identified as formaldehyde. However, all available evidence indicates that these two enzymes are identical.[24] It is especially of interest that the ε-alkyllysinase activity is practically absent in fast-growing Novikoff hepatoma where the enzyme which methylates the ε-amino group of protein-bound L-

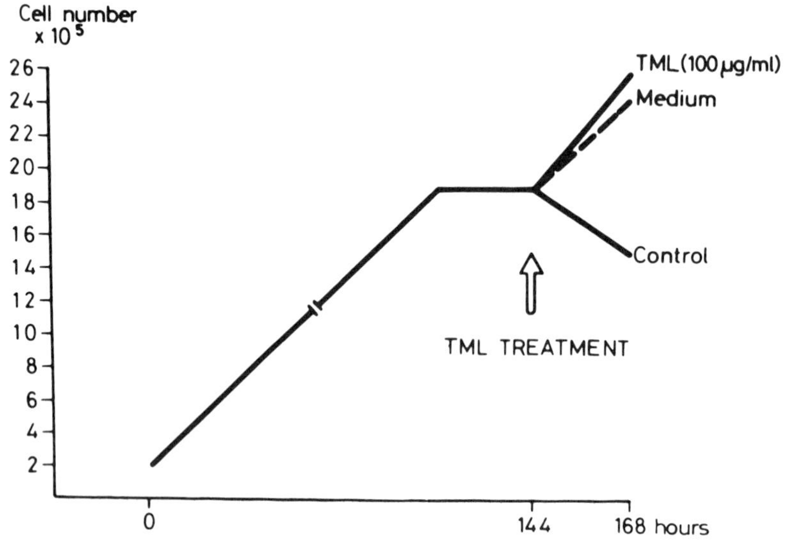

FIGURE 6. Effect of medium change or TML treatment on the proliferation of a CBA_{T6T6} mouse fibroblast line.

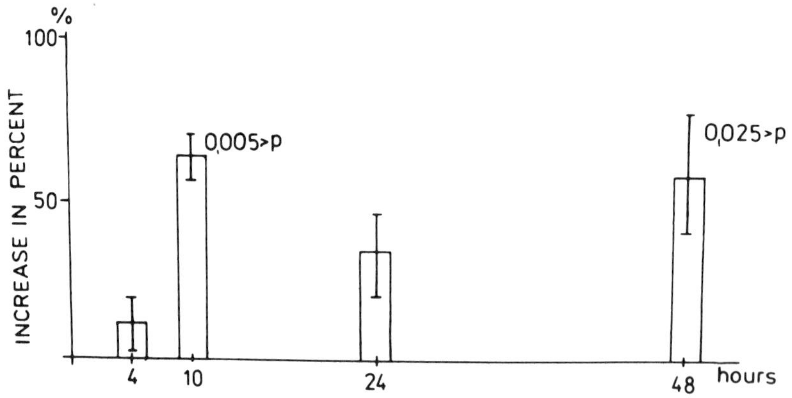

FIGURE 7. The effect of *in vivo* TML treatment (100 mg/kg i.p.) on the incorporation of [^{3}H]TdR into the bone marrow of mice.

lysine (protein methylase III; AdoMet: protein-lysine *N*-methyltransferase; EC 2.1.1.43) was found to be highly active.[25]

We have observed in lymphocyte cultures treated with TML that the formaldehyde level of the culture medium increased by 300 to 900% as early as 10 min after treatment and persisted for 120 min, then decreased to the control level. Since, according to Novogrodsky,[26] the formation of aldehyde groups on the cell surface is enough to initiate cell division, we have to take into consideration the role of the formaldehyde molecule and its reaction products when studying the cell proliferation inducing effect of TML. Thus, the distribution of radioactivity in NK/Ly ascites tumor cells subsequent to the administration of [^{3}H]TML[27] was studied using thin layer chromatography. It was found that the radioactivity was mostly found in the macromolecule-free fraction of the hyaloplasma. Further, most of the TML undergoes a rapid metabolic process and the major end product is of a ninhydrin negative and Dragendorff positive material.

Subsequently, an increasing proportion of radioactivity was found associated with macromolecules such as H1 and H4 histones as well as with the bases of nucleic acid, primarily adenine

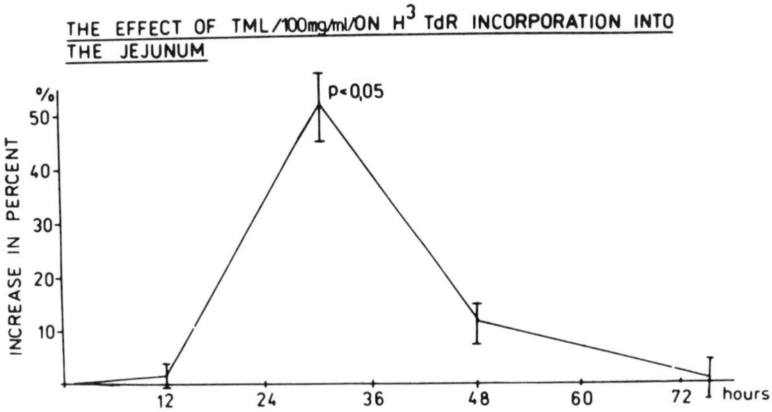

FIGURE 8. The effect of *in vivo* TML treatment (100 mg/kg i.p.) on [^{3}H]TdR incorporation into the jejunum of mice. (From Lapis, K., Szende, B., Kovács, L., and Simon, K., *Exp. Pathol.*, 24, 97, 1983. With permission.)

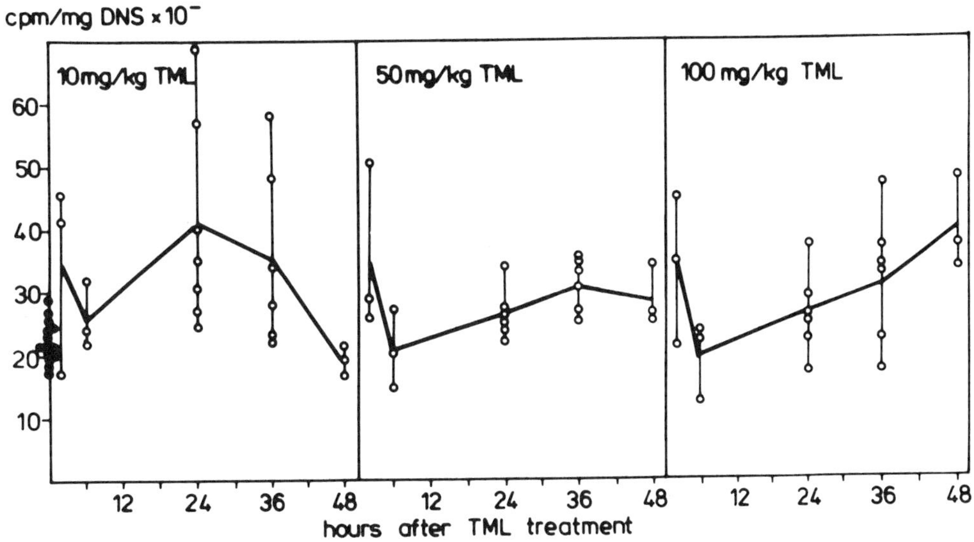

FIGURE 9. Incorporation of [^{3}H]TdR into liver cells of CBA_{T6T6} (1 μCi/g body weight [^{3}H]thymidine-30′).

(Figure 14). It appeared, therefore, that the methylation of histones and nucleic acids occurs a few hours after TML treatment. The prevalency of adenine methylation seems to indicate that, while methylation of guanine usually leads to malignant transformation, methylation of adenine may play a role in normal cell proliferation.

According to biochemical studies carried out on TML treated NK/Ly ascites tumor cells, increased RNA synthesis takes place 1 h after the treatment and it lasts for 18 h, decreasing to control level by 48 h; DNA synthesis starts to increase 18 h after TML treatment and the increase stops at 48 h, meanwhile, DNA polymerase activity increases as well. The increase of mitotic index occurs 24 h after TML treatment (Figure 15). It is of interest that after oral administration of TML, this compound first concentrates in the liver, later in the kidney and a considerable amount can be detected in the urine. The urine of normal persons and tumor bearing patients also contains a low amount of TML.[28] It should be noted that TML does not penetrate the blood-liquid-barrier.

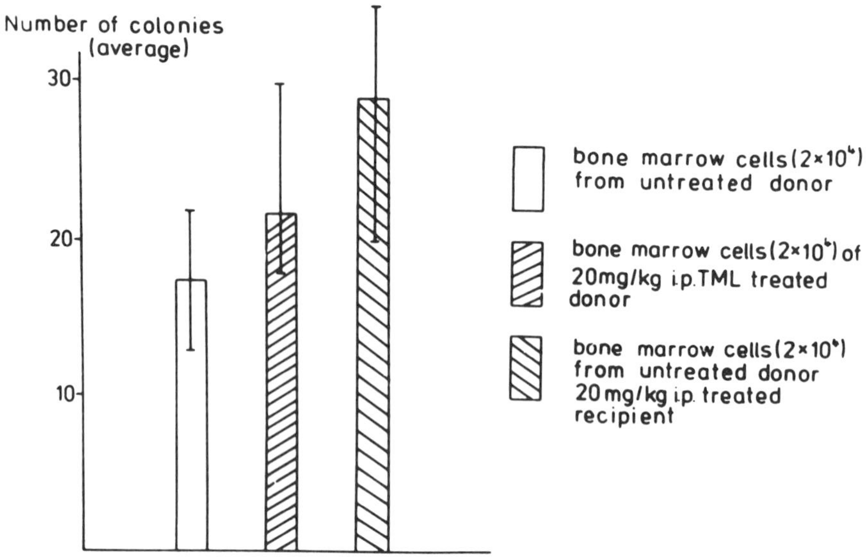

FIGURE 10. The number of splenic colonies of irradiated (900 rads) BDF_1 mice with and without TML treatment. (From Lapis, K., Szende, B., Kovács, L., and Simon, K., *Exp. Pathol.*, 24, 97, 1983. With permission.)

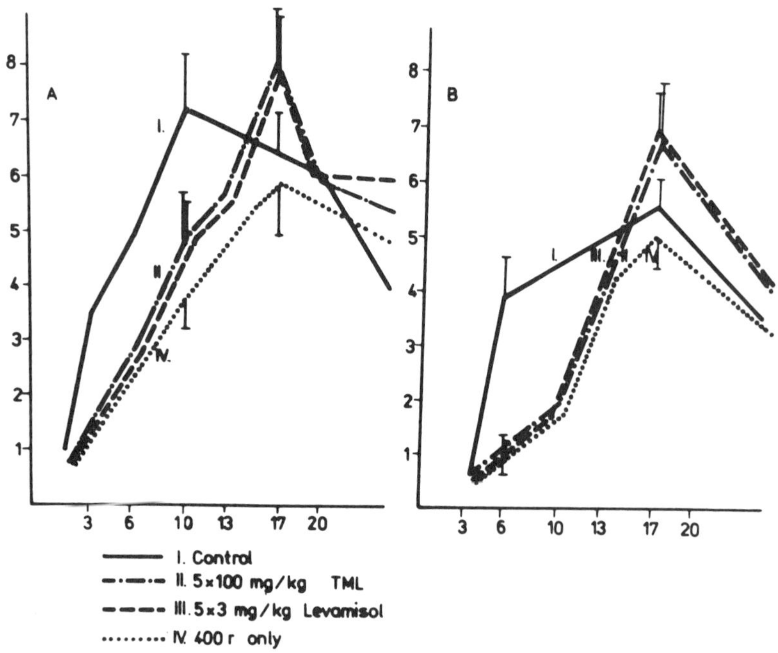

FIGURE 11. The effect of *in vivo* TML and Levamisol treatment on the humoral immune response (hemagglutinin titers) of irradiated (400 rads) mice. (A) Mercaptoethanol sensitive titers. (B) Mercaptoethanol resistant titers. (From Lapis, K., Szende, B., Kovács, L., and Simon, K., *Exp. Pathol.*, 24, 97, 1983. With permission.)

9. Studies on the Toxicity of TML

Sprague-Dawley rats were treated with 100, 300, and 1000 mg/kg TML (daily, via gastric tube) for 90 or 120 d when the animals were sacrificed. None of the TML doses caused toxic

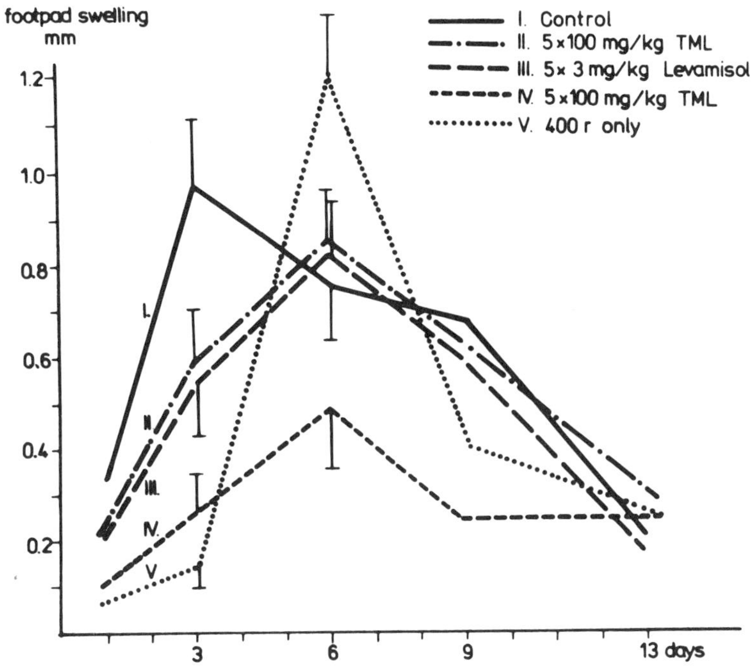

FIGURE 12. The effect of prolonged TML and Levamisol treatment on the cellular immune-response of irradiated (400 rads) mice, measured by foot-pad swelling. (From Elek, G., Láng, I., Szende, B., and Lapis, K., *Agents Actions,* 12, 503, 1982. With permission.)

death. The lower doses of 100 mg/kg or 300 mg/kg did not result in the alteration of laboratory parameters, however, with the 1000 mg/kg dose, a moderate but significant alteration in hemoglobin, red blood cell, serum protein, serum cholesterol, and serum lipid level was observed. At the same time, the weights of the thymus, adrenal gland, and ovary decreased moderately when compared to the control. Histologically, mild parenchymal degeneration of the liver was found.

a. Effect of TML on Polyamine Level in Various Organs

It was possible that some of the toxic or even nontoxic effects of TML could be attributed to an increased polyamine level produced by high doses of TML. Thus, a relatively high dose of TML (200 mg/kg i.p.) was administered to Sprague-Dawley rats as a single treatment. Rats were sacrificed 24 h later and putrescine, spermidine, spermine level and spermidine/spermine ratio were determined in the liver and kidney. A significant elevation of spermidine level and a slight elevation of spermidine/spermine ratio were observed in the liver. When 100 mg/kg TML was given i.p. four times on consecutive days, no alternation in spermidine or spermine level was observed in the liver 24 h after the last treatment. However, putrescine level went up from 9.2 $\pm$3.4 mmol/g liver weight to 17.0$\pm$7.4 mmol/g liver weight. This significant elevation may also be responsible in some extent for stimulation of liver cell proliferation by TML treatment. No significant alterations in the kidney were found regarding the polyamine level.

10. The Application of TML in Combination Chemotherapy of Animal Tumors[29]

The stimulatory effect of TML on the regeneration of artificially damaged cell systems, such as bone marrow and intestinal mucosa, led to an examination of the combination of TML and various cytostatics in the chemotherapy of experimental transplantable animal tumors.

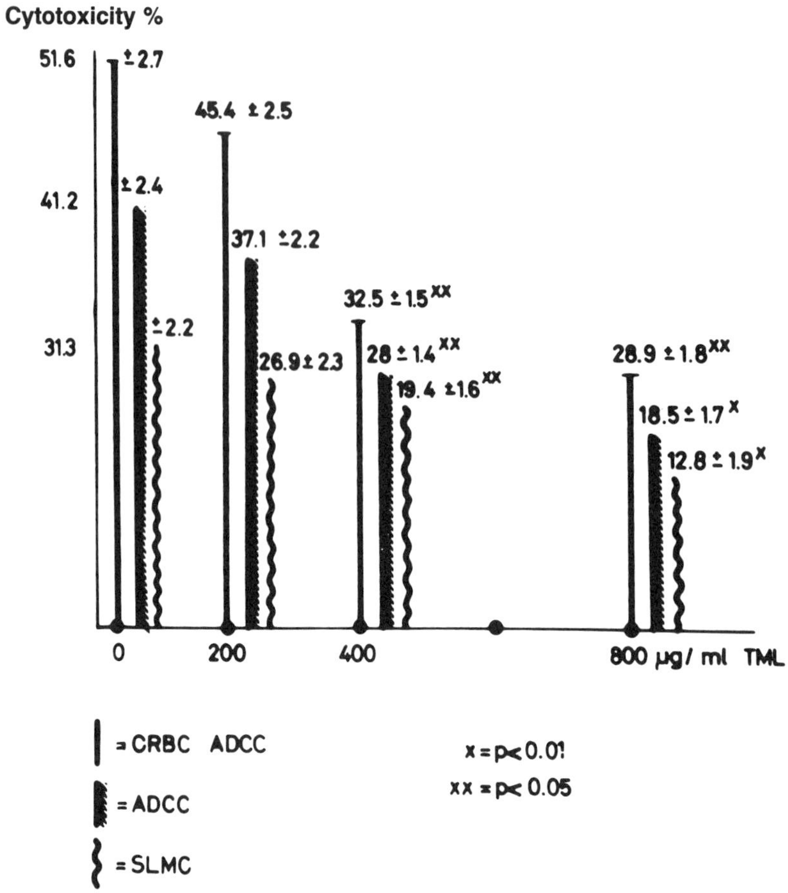

FIGURE 13. The effect of TML on cellular immune response *in vitro*. CRBC = chicken red blood cell; ADCC = antibody dependent cytotoxicity; SLMS = spontaneous leukocyte mediated cytotoxicity. (From Elek, G., Láng, I., Szende, B., and Lapis, K., *Agents Actions*, 12, 503, 1982. With permission.)

a. Effect of TML Treatment, Combined with Various Cytostatic Agents on the Survival of Healthy Mice and Chickens

Inbred DBA/2 and Swiss random mice were injected with the LD_{50} of cyclophosphamide, vincristine, and Adriamycin as a single i.p. treatment. TML (100 mg/kg) was also administered as a single i.p. injection either as a simultaneous treatment with the cytostatic agents or as pretreatment (24 h before the cytostatic agent) or as posttreatment (24 h after the cytostatic agent). Simultaneous treatment with TML decreased or prevented the toxic effect of all types of cytostatic drugs used, resulting in the decrease of deaths. Pre- or posttreatment did not show this effect (Figure 17 a,b,c). In chickens, vincristine treatment (5 mg/kg i.p.) causes severe neurotoxicity and death within 1 to 3 d TML (100 mg/kg i.p. simultaneously with vincristine) prevented the death of 5 animals out of 15.

b. Effect of TML Treatment, Combined with Cytostatic Agents on Tumor-Bearing Mice

The TML-cyclophosphamide combination was used in experiments with the solid form of Sarcoma 180 (Crocker). L_{-1210} tumor was the test object for the combination chemotherapy with TML + vincristine, and TML + Adriamycin. The simultaneously applied TML treatment decreased the toxic effect and resulted in increased survival of tumor-bearing animals when combined with vincristine or adriamycin to L_{-1210} tumor bearing, and with cyclophosphamide to

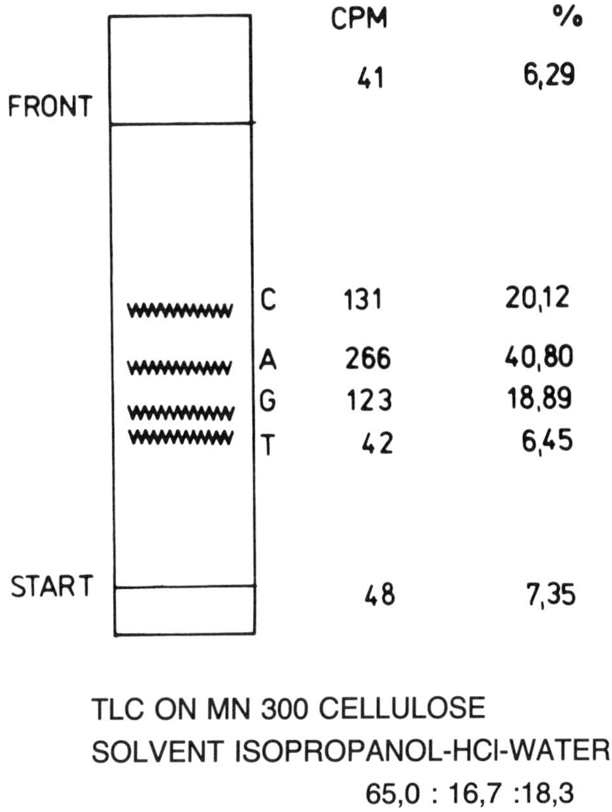

FIGURE 14. Binding of [^{3}H]TML to various bases of DNA. (From Elek, G., Láng, I., Szende, B., and Lapis, K., *Agents Actions,* 12, 503, 1982. With permission.)

S-180 tumor-bearing animals (Table 1). An increased regeneration of the bone marrow of mice receiving combination therapy was observed when compared with the animals treated only with the cytostatic drugs.

B. BIOLOGICAL EFFECTS OF ε-*N*-METHYLATED LYSINES ON PLANT TEST SYSTEMS

1. Treatment of Tobacco Plant Tissue Cultures by ε-*N*-Methylated and ε-*N*-Formylated Lysines

It has been established that ε-*N*-dimethyl-L-lysine (DML) is the major N^{ε}-methylated lysine derivative in the histones.[30-32]

Table 2 demonstrates the effect of DML on the growth of tobacco tissue culture.[33] It can be seen that DML induced significant promotion of tissue growth while a considerable decrease of cell number can be observed.

Table 3 illustrates the effect of DML on the histone fractions of tobacco tissues. The treatment decreased the amount of H1 histone fraction in proportion with the concentration of DML applied.

Table 4 summarizes the statistical evaluation of the effect of L-lysine, ε-*N*-methylated lysine and N^{ε}-formyl-L-lysine on the growth of tobacco tissue cultures.[34] It is seen in these results that MML induces a cell proliferation-retarding effect while TML and L-lysine exhibit a promoting effect. It should also be pointed out that MML produced practically a similar effect in all cases and that the effect of TML and other derivatives depended on the age and type of cultures ("well-

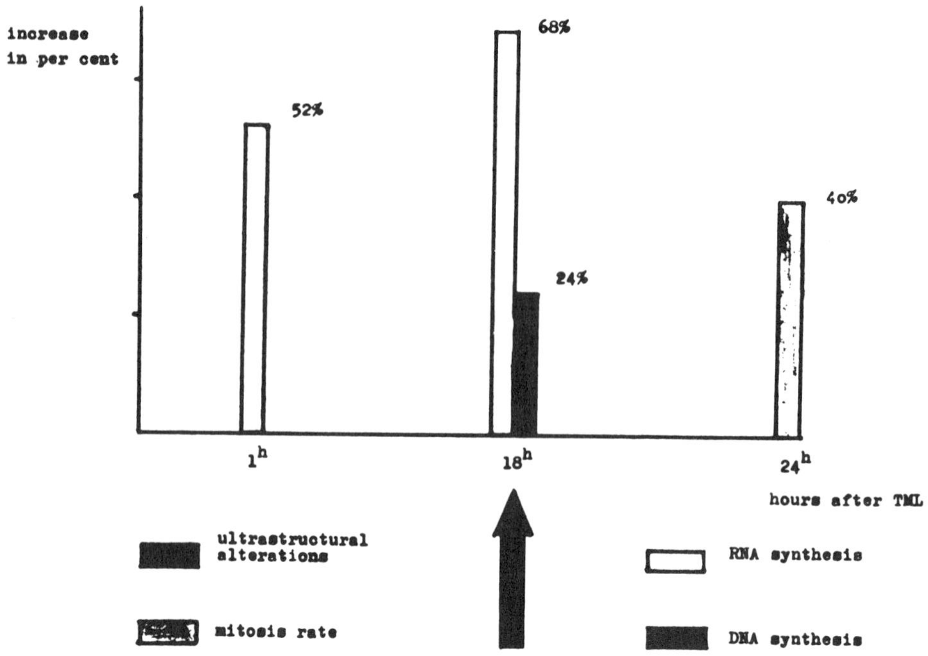

FIGURE 15. The sequence of increase in RNA, DNA synthesis, mitotic rate, and ultrastructural alterations after *in vivo* TML treatment of mice.

growing" or "poor-growing"). FL always generates a moderate effect on the plant cell proliferation. It is quite likely that these seemingly anomalous effects are due to the endogenous level of the substances examined; there might be a characteristic equilibrium between the methylating and the demethylating enzymes in plant tissue cultures, and this balance can be positively or negatively modified with administration of methylated basic amino acids.

2. Effect of TML on the Growth of Germinating Poppy Plants[35]

The seeds of "Kék-DUNA" poppy *(Papaver somniferum)* sort were germinated on filter paper in Petri dishes containing different concentrations of TML-HCl for treatment. The data were evaluated with *t* test, and the results are summarized in Table 5. It can be concluded that TML generates a cell proliferation-promoting effect at all concentrations examined but the optimum value can be observed at 10 ppm concentration. It should also be noted that TML glutamate produced similar activity, but a moderate activity can be demonstrated in the case of TML dioxalate. Similar results have also been obtained with TML treatment in the case of other plant species (e.g., wheat, barley, flax, hemp, and bean). It follows from these results that the application of TML and related compounds in agricultural practice offers an excellent prospect.

3. Effect of TML on Photosynthetic CO_2 Fixation, on Protein Synthesis, and on the Decomposition of the Chlorophyll[36]

Figure 16 demonstrates the effect of TML on photosynthetic CO_2 fixation and on protein synthesis in leaves of bean plant subjected to a proper spraying with solutions of different TML concentration. The application of TML could also prevent the decomposition of chlorophyll (Figure 17). The increase in photosynthetic CO_2 fixation and in protein synthesis and the preventive action on chlorophyll decomposition might be the underlying mechanism for the function of TML in plant cell proliferation and plant crop production.

TABLE 1
Mean Surviving Time

				Survivors		
	0 + 0	0	0	0 + 0	0	0
Control	7.27	6.77	7.77	—	—	—
2.0 mg/kg vincristine	8.5	8.2	8.8	8	4	4
	$0.15 < p < 0.20$	$0.20 < p < 0.25$	$0.25 < p < 0.30$			
2.0 mg/kg vincristine	11.1	10.8	11.2	14	6	8
+ 100 mg/kg TML	$0.005 < p < 0.025$	$0.25 < p < 0.05$	$0.005 < p < 0.01$			
2.5 mg/kg vincristine	7.8	7.4	8.2	6	3	3
	$0.30 < p < 0.35$	$0.35 < p < 0.40$	$0.35 < p < 0.40$			
2.5 mg/kg vincristine	11.6	10.5	12.7	16	6	10
+ 100 mg/kg TML	$p < 0.0005$	$0.0125 < p < 0.025$	$p < 0.0005$			
3.0 mg/kg vincristine	10.05	10.1	10.0	10	4	6
	$0.05 < p < 0.01$	$0.025 < p < 0.05$	$0.05 < p < 0.01$			
3.0 mg/kg vincristine	11.45	11.7	11.2	16	7	9
+ 100 mg/kg TML	$p < 0.0005$	$0.0025 < p < 0.005$	$0.0005 < p < 0.0025$			
3.2 mg/kg vincristine	—	—	10.2	—	—	13
			$0.20 < p < 0.25$			
3.2 mg/kg vincristine	—	—	12.8	—	20	—
+ 100 mg/kg TML			$0.0005 < p < 0.0025$			
3.5 mg/kg vincristine	10.45	9.7	10.7	11	5	6
	$p < 0.0005$	$0.025 < p < 0.05$	$0.0005 < p < 0.0025$			
3.5 mg/kg vincristine	11.25	10.3	11.4	15	8	9
+ 100 mg/kg TML	$0.005 < p < 0.0025$	$0.05 < p < 0.10$	$0.0005 < p < 0.0025$			
4.0 mg/kg vincristine	6.4	4.8	8.2	4	—	4
	—	—	$0.35 < p < 0.30$			
4.0 mg/kg vincristine	9.8	10.0	8.4	10	6	4
	$0.005 < p < 0.01$	$0.025 < p < 0.05$	$0.25 < p < 0.30$			

From Tyihák, E., Maróti, M., and Vágujfalvi, D., *Bot. Közlem,* 58, 85, 1971 (in Hungarian); and Tyihák, E., Maróti, M., Vágujfalvi, D., Bajusz, S., and Patthy, A., *Experientia,* 31, 818, 1975. With permission.

4. Application of TML and Other Preparations Containing ε-*N*-Methylated Lysines in Agricultural Practice

Spontaneous formation of ε-*N*-methylated and ε-*N*-formylated L-lysine derivatives in the reaction between L-lysine and formaldehyde had been observed earlier.[37] As described above, TML has been shown to be biologically active, increasing cell proliferation in human and animal test systems,[17-19] and similar effects by the reaction products of L-lysine and formaldehyde (optimally produced in an aqueous solution, termed formetol) could also be observed in a number of different plant series.[38,39]

Plant growth and increase in crop yield could be demonstrated after proper spraying of plants with solutions containing N^{ε}-methylated lysines in small scale as well as productional scale experiments. An average increase of 8% (from 6 to 20%) in the crop yield of eight different plant species has been achieved in large scale productional experiments during the last 5 years.[40]

5. Induction of Plant Disease Resistance by TML

According to Schönbeck et al.[41] there is a resistance potential in every plant depending on its genetic constitution which can be activated by appropriate inducers. It seems that disease resistance is not a static but an elastic feature of plants. Induced resistance appears to be only effective against obligate biotrophic pathogens which form haustoria.

Recently, it was observed that the pre-treatment of bean plants with TML induces a resistance to *Uromyces phaseoli* (Table 6). As shown in the table, there is a time- and dose-dependent

TABLE 2
The Effect of N^{ε}-Dimethyl-L-Lysine (DML) on the Growth of Tobacco Tissue Cultures

Concentration in the standard culture medium (mg/l)	End weight (g)	Daily growth (mg)	Relative growth	Cell number (average) (× 10^4 /g)	Longitudinal size of cells (μm)
50	8.3578	132	41.6	98.0	22—528
10	8.4223	133	42.0	114.5	—
1	8.1688	129	40.7	129.0	—
Control	5.8349	92	29.0	139.5	22—319

Note: Initial weight: 200—200 mg; temperature: 28(±2)°C; incubation time: 49 d: Average values of 20-20 samples (4 parallel with 5-5 samples).

From Tyihák, E., Maróti, M., and Vágujfalvi, D., *Bot. Közlem.*, 58, 85, 1971 (in Hungarian); and Tyihák, E., Maróti, M., and Ráthony, Z., *Bot. Közlem.*, 61, 149, 1974 (in Hungarian). With permission.

TABLE 3
The Effect of ε-*N*-Dimethyl-L-Lysine (DML) on the Histone Fractions of Tobacco Tissue Cultures

Concentration in the standard culture medium (mg/l)	Histone fraction (Johns) mg/100 g fresh tissue		
	H1	H2B + H3	H2B
50	0.30	2.25	0.93
10	0.48	2.05	1.10
1	0.65	2.00	0.96
Control	1.06	1.85	1.13

TABLE 4
Statistical Evaluation of the Effect of L-lysine and ε-*N*-Methylated Lysines as well as ε-*N*-Formyl-L-Lysine on the Growth of Tobacco Tissue Cultures

Treatment	*n*		S	t (control)	t (MML)
Control	10	26.31	2.14		
MML	9	22.87	1.72	1.4	
DML	10	27.96	1.66	0.66	$2.03_{P5\%}$
TML	9	30.72	1.75	$1.76_{P10\%}$	$3.08_{P1\%}$
FL	10	30.13	1.78	1.56	$2.92_{P1\%}$
Lysine	10	32.29	1.77	$2.45_{P5\%}$	$3.85_{P0.1\%}$

Note: Conditions: Initial weight: 200—200 mg; incubation time: 97 d; temperature: 28(±2)°C; 10, 0—10.0 mg amino acid/l standard culture medium.

induced resistance using TML as a natural inducer. The dose-dependent induced resistance is especially interesting because there is perhaps a "pharmacological" dose-dependent induced resistance (10^{-3} to 10^{-5} *M* TML) and a "physiological" dose-dependent induced resistance (10^{-8} to 10^{-10} *M* TML) in bean plants.[42] These investigations on induced resistance offer the possibility of obtaining more information about the mechanism of compatibility which operates in complex host-parasite relationships.

TABLE 5
The Effect of TML on the Growth of Germinating Poppy Plants

Treatment mg/l	n	(cm)	t (control)
100	130	2.71	$-6.58_{P0.1\%}$
10	100	2.89	$-8.20_{P0.1\%}$
1	96	2.58	$-4.4_{P0.1\%}$
Control	97	2.23	

Note: For explanation, see text.

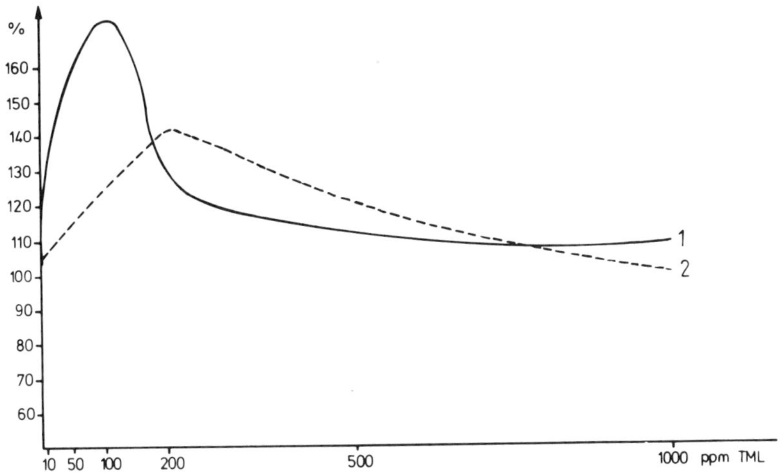

FIGURE 16. Effect of TML on photosynthetic CO_2 fixation and on protein synthesis occurring in leaves of Pinto bean plant. Photosynthesis (on basis of measurement of chlorophyl). 2. Protein synthesis (measurement of incorporation of ^{14}C glycine by Packard liquid scintillator). (From Jeney, A., Gyapay, G., Lapis, K., Szende, B., Bursics, L., and Tyihák, E., *Biochem. Pharmacol.*, 29, 2729, 1980. With permission.)

C. BIOLOGICAL EFFECTS OF ε-*N*-METHYLATED LYSINES ON MICROBIOLOGICAL TEST SYSTEMS

1. Cell Proliferation-Promoting Effect of TML on Obligate Biotrophic Pathogens

It has very recently been shown that a number of colonies of *Uromyces phaseoli* on bean plants treated with TML- and other ε-*N*-methylated lysines increased considerably if the time period between the pretreatment and the inoculation was short (e.g., Table 6, first column). A similar cell proliferation-promoting effect was also observed in the relationship between barley and powdery mildew.[42] In this host-parasite relationship, it is possible that there exists a competition for the TML molecules, which are available in larger amount, between the pathogens immediately after the treatment, and the plant tissue trying to use them in its defense mechanisms.

III. BIOLOGICAL EFFECTS OF N^G-METHYLATED AND N^G-HYDROXYMETHYLATED ARGININES

Guanidino-methylated derivatives of L-arginine (N^G-monomethyl-L-arginine, MMA; N^G,N^G-dimethyl-L-arginine, DMA; N^G,N'^G-dimethyl-L-arginine, DM′A) have been found in some proteins,[43,44] as well as in free state in various tissues and biological fluids.[45,56]

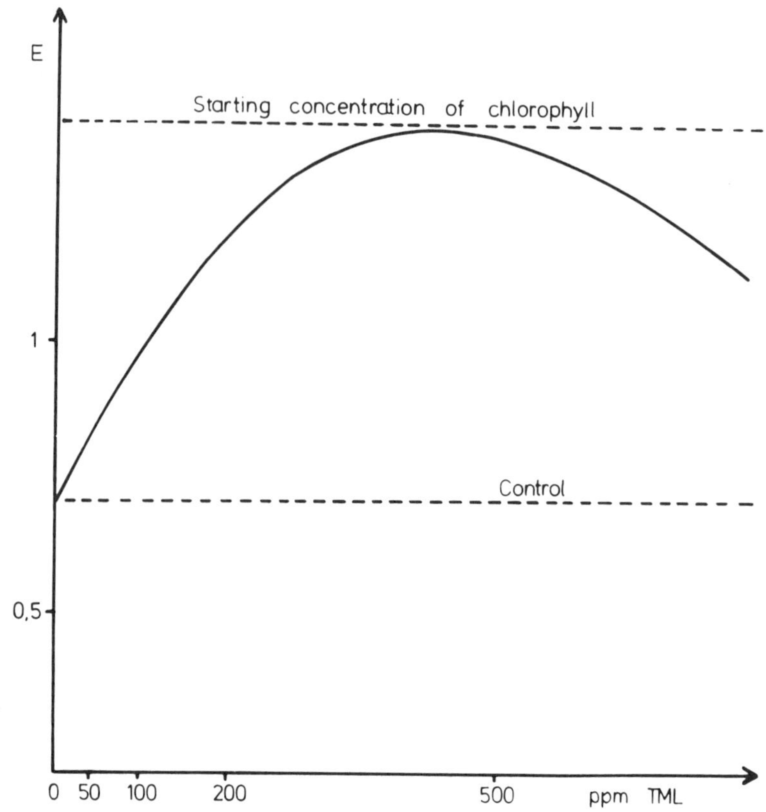

FIGURE 17. Effect of TML on the decomposition of the chlorophyll. Plants were grown in growth chamber at 20°C in darkness for 4 d; E = specific extinction. (From Szende, B., Jeney, A., Benedeczky, I., and Lapis, K., *Adv. Tumor Prev. Detect.*, 3, 122, 1976. With permission.)

TABLE 6
Effect of ε-*N*-Trimethyl-L-Lysine (TML) Pre-Treatment of "Saxa" Bean Plants on the Colony Number of *Uromyces phaseoli*

Concentration of TML in dist. water (mol/l)	Time of inoculation after pretreatment in days					
	1		6		8	
		%		%		%
10^{-3}	17.5	91.1	4.1	48.2	6.0	75.6
10^{-4}	27.1	141.1	4.6	51.1	6.3	79.1
10^{-5}	32.5	169.3	8.2	96.5	4.5	57.2
10^{-6}	26.6	138.5	7.1	83.5	6.1	76.8
10^{-7}	29.4	153.1	7.3	85.9	4.8	61.3
10^{-8}	20.6	107.1	5.5	64.7	5.8	73.6
10^{-9}	26.5	138.0	3.1	36.5	4.7	59.5
10^{-10}	27.2	141.6	3.9	45.8	5.1	64.0
Control (H_2O)	19.2	100.0	8.5	100.0	7.9	100.0

Note: Colony number/leaf surface of cm^2/mean value of 36 measurements.

More recently, it has been shown that the L-lysine and L-arginine behave antagonistically in the reactions with formaldehyde:[47,48] L-lysine can be methylated and can also be formylated[49] in

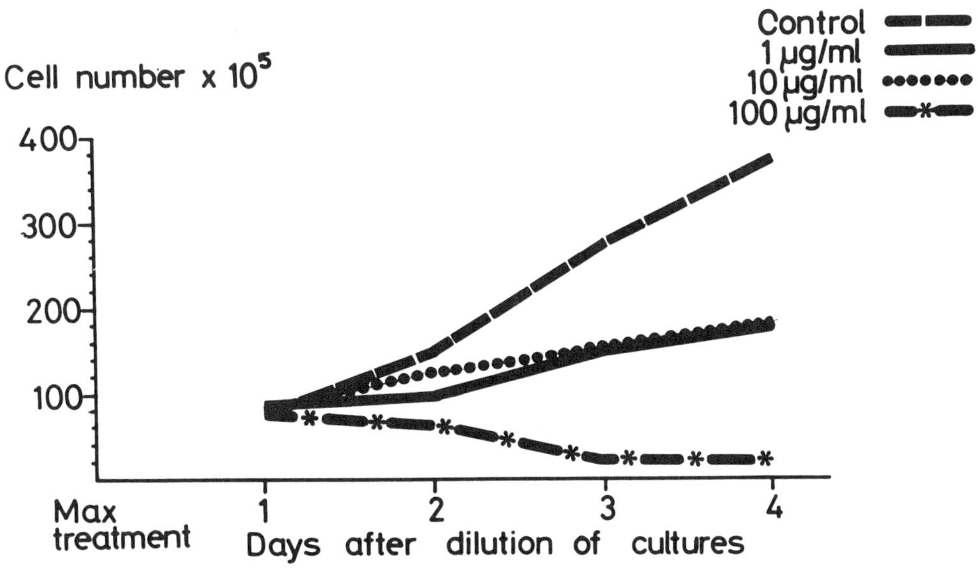

FIGURE 18. Effect of MAX treatment on the proliferation of K-562 cells *in vitro*. (From Szende, B., Lapis, K., and Simon, K., *Neoplasma,* 29, 427, 1982. With permission.)

this spontaneous reaction,[37,50] whereas L-arginine can only be hydroxymethylated by formaldehyde. These hydroxymethylated arginines are relatively stable molecules and can be isolated from the reaction mixture.[48,51,52] Our preliminary studies indicated that such molecules occur in human blood and urine,[52] and it has been hypothesized that the guanidine group of L-arginine of enzyme protein could be hydroxymethylated[53] in enzymatic transmethylation reactions. There is only a limited amount of investigation dealing with the biological effects of these special molecules.

A. BIOLOGICAL EFFECTS OF N^G-HYDROXYMETHYLATED ARGININES ON ANIMAL TEST SYSTEM

1. Growth-Retarding Effect of N^G-Hydroxymethylated Arginines on Animal Tumors

In vitro, a 100 µg/ml mixture of N^G-hydroxymethyl-L-arginines (MAX) caused a total inhibition of proliferation of K-562 cells (Figure 18), while 10 µg/ml and 100 µg/ml MAX treatment with P-388 cell culture resulted in a total inhibition of proliferation and 1 µg/ml MAX was ineffective (Figure 19). In an *in vivo* experiment, 400 mg/kg daily i.p. treatment (started 24 h after tumor inoculation and followed for 10 d) caused a complete inhibition of the growth of Ehrlich ascites tumor.

C_{57}Bl mice, inoculated with Lewis lung tumor (LLT), i.m. and made tumor-free by amputation of the tumorous leg 10 d after tumor inoculation, were treated seven times with 400 to 400 mg/kg MAX i.p. from the 11th d after inoculation, were sacrificed and examined for lung metastasis. An inhibitory effect of MAX treatment was observed in both metastasis number and average volume of the lung nodules (Figure 20). The tumor-growth inhibitory effect of MAX may have some therapeutical value in tumor chemotherapy.

B. BIOLOGICAL EFFECTS OF N^G-METHYLATED ARGININES ON PLANT TEST SYSTEMS

1. Growth-Retarding Effect of Guanidino-Methylated Arginines on the Tobacco Tissue Cultures[54]

The test material used was a secondary callus tissue isolated from tobacco (*Nicotiana tabacum* L.) stem. The tissue consisted of a yellowish-green cell population in intensive growth

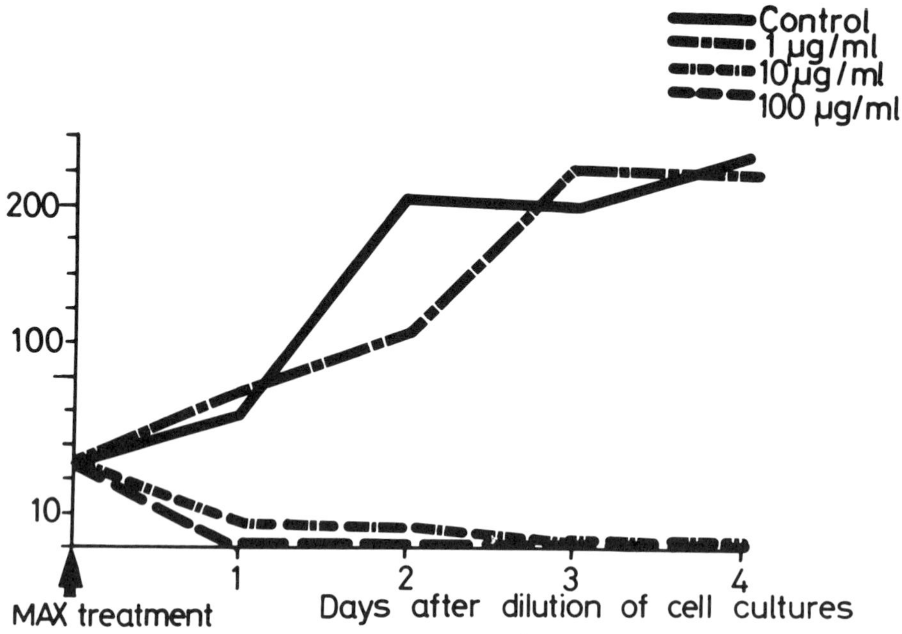

FIGURE 19. Effect of MAX treatment on the proliferation of P-388 cells *in vitro*. (From Szende, B., Lapis, K., and Simon, K., *Neoplasma,* 29, 427, 1982. With permission.)

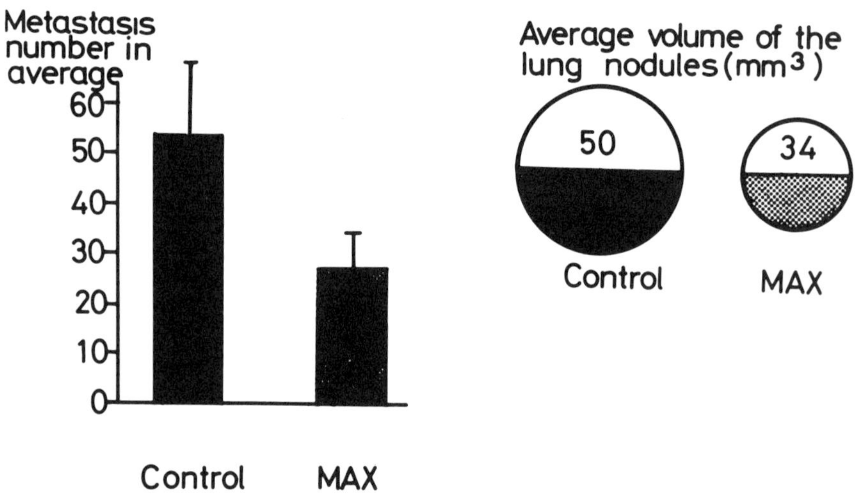

FIGURE 20. Effect of *in vivo* MAX treatment on the lung metastasis number of C_{57}Bl mice inoculated with Lewis lung tumor i.m. and amputated after 10 d. Treatment from the 11th d after inoculation seven times, 400-400 mg/kg/day, i.p.).

which did not show organ formation on standard culture medium and only some tissue differentiation. The different amino acids were applied at a concentration of 10.0 to 100.0 mg/ml agar-agar culture medium. In the Erlenmeyer dishes, each piece of tissue was placed on 50 ml culture medium with an initial weight of 200 mg. The tissue grew for 62 d in a thermostat of 28 ± 2°C with a natural alternation of day and night.

As shown in Table 7, considerable growth inhibition by MMA, DM′A, and DMA at concentrations of 10 and 100 ppm after 62 d of culturing was observed. After 62 d, the added L-arginine, particularly its N^G-methylated derivatives, can be shown in alcoholic extracts of

TABLE 7
Effect of L-Arginine and its Guanidino-Methylated Derivatives on the Growth of Tobacco Tissue Cultures

Treatment	Final weight after 62 d (g)	Daily growth (mg)	Growth inhibition (%)
Control	16.38	204	0
A-10	16.22	261	0
A-100	8.49	137	$49_{(P1\%)}$
MMA-10	9.89	159	$40_{(P1\%)}$
MMA-100	2.32	37	$87_{(P0.1\%)}$
DM′A-10	10.23	160	$38_{(P1\%)}$
DM′A-100	3.84	62	$77_{(P0.5\%)}$
DMA-10	10.62	170	$36_{(P1\%)}$
DMA-100	3.10	44	$79_{(P0.1\%)}$

Note: Signs and abbreviations: A, L-arginine; MMA, N^G-monomethyl-L-arginine; DM′A, N^G,N'^G-dimethyl-L-arginine; DMA, N^G,N^G-dimethyl-L-arginine; 10, 10 mg amino acid/1 culture liquid; 100, 100 mg amino acid/1 culture liquid.

TABLE 8
Effect of N^G-Monomethyl-L-Arginine Acetate (MMA) Pretreatment of "Saxa" Bean Plants on the Colony Number of *Uromyces phaseoli*

Concentration of MMA in dist. water (mg/l)	Time of inoculation after pretreatment in days: 1		6		8	
		%		%		%
10^{-3}	19.4	112.8	12.4	86.7	20.1	82.4
10^{-4}	14.4	83.7	9.9	69.2	25.5	104.5
10^{-5}	19.4	112.8	10.1	70.6	24.6	100.8
10^{-6}	19.2	111.6	7.6	53.1	18.7	76.6
10^{-7}	19.0	110.5	8.5	59.4	17.4	71.3
10^{-8}	18.8	109.3	14.3	100.0	14.4	59.0
10^{-9}	17.8	103.5	10.3	72.0	13.7	56.1
Control	17.2	100.0	14.3	100.0	24.4	100.0

Note: Colony number/leaf surface of cm^2 (mean value of 36 measurements).

culture medium and in the tissue extracts by ion exchange thin-layer chromatographic methods. The three guanidino-methylated arginines indicate the following order of enrichment in the tissue compared to the control: control < arginine < MMA < DM′A < DMA. It is probable that the tobacco tissue can not demethylate the two dimethyl-L-arginines.

2. Induction of Plant Disease Resistance by MMA[42]

Resistance against *Uromyces phaseoli* was induced in bean by treating the plants with MMA acetate (Table 8). It can be seen that a cell proliferation-promoting effect was not detectable in the case of short inoculation time. It is probable that the metabolism of methyl group of MMA plays a role in the induction of resistance because several days are necessary for the activating plant tissue. It should be noted that the methyl linkage of MMA appears to be less stable.

IV. BIOLOGICAL EFFECT OF LEUCINE-METHYLESTER

It is well known that some compounds such as ammonium chloride, chloroquine, and *O*-methylesters of various amino acids such as L-Leu-*O*Me are lysosomotrophic. L-Leu-*O*Me causes lysosomal disruption and rapid death of human monocytes at concentrations of 1 to 10 m*M* at 37°C.[55] In addition,[56] *O*-methylesters of various amino acids impair the *in vitro* function of monocytes, macrophages, and natural killer cells within a short period of time. The incorporation of [^{14}C]labeled amino acids into mouse peritoneal macrophages decreases following treatment with amino-acid-*O*-methylesters and the stereoisomers of these compounds, depending on concentration and time of exposure. Macrophages were rapidly destroyed by treatment with *O*-methylesters of several amino acids but the *O*-methylester-L-leucine proved to be the most effective compound.

Cytological changes of the macrophages were also followed. An excessive vacuolization occurred in the cytoplasm 20 to 40 min after *in vitro* treatment (Figure 21). Most of the macrophages were necrotic 1 h after treatment. The activity of endogenous peroxidase, nonspecific esterase and acidic phosphatase increased in the first 5 min and markedly decreased at 50 min after the treatment. According to ultrastructural studies, it is very likely that the formation of phagolysosomes is inhibited and cells are killed because of membrane damage.

V. CONCLUSIONS

The methylated amino acids, mainly ε-*N*-methylated lysines and N^G-methylated (and N^G-hydroxymethylated) arginines exhibit different biological effects on various living systems after exogenous administration. It is assumed that these amino acids can also manifest similar effects endogenously in different regions of cells and tissues. The amount of these compounds, their localization and the ratio between themselves may be important factors in determining the rhythm of cell cycle, differentiation, resistance potential against biotic and abiotic stresses, etc.

These methylated amino acids also occur as peptide bound. The exposure of chicken embryonic fibroblasts to sodium arsenite or to heat shock results in methylating specific polypeptides in which the major methylated lysine is TML and the major arginine derivative is MMA.[57] Additionally, during heat shock in *Drosophila* cells, the methylation pattern of core histones changed dramatically.[58]

The mechanism by which the free methylated amino acids manifest the aforementioned various effects is not clear at present. The demethylation of the methylated amino acids may be a basic phenomenon in this mechanism. During this process formaldehyde and demethylated compound can be formed as it has been previously demonstrated, e.g., in the case of MML and DML.[4,24] More recently, it is shown that the formation of formaldehyde from AdoMet is linked to an enzymic transmethylation.[59] The generalization of this new finding may lead to the confirmation of a formaldehyde cycle[53] which will facilitate elucidation of certain perplexing phenomena in this field, too.

Finally, the biological systems include specific enzymes for the elimination of methyl groups in another way such as hydrolysis of MMA or direct conversion of DMA to citrulline[60-62] and it is probable that competition exists between the original molecules and their methylated derivatives as well.

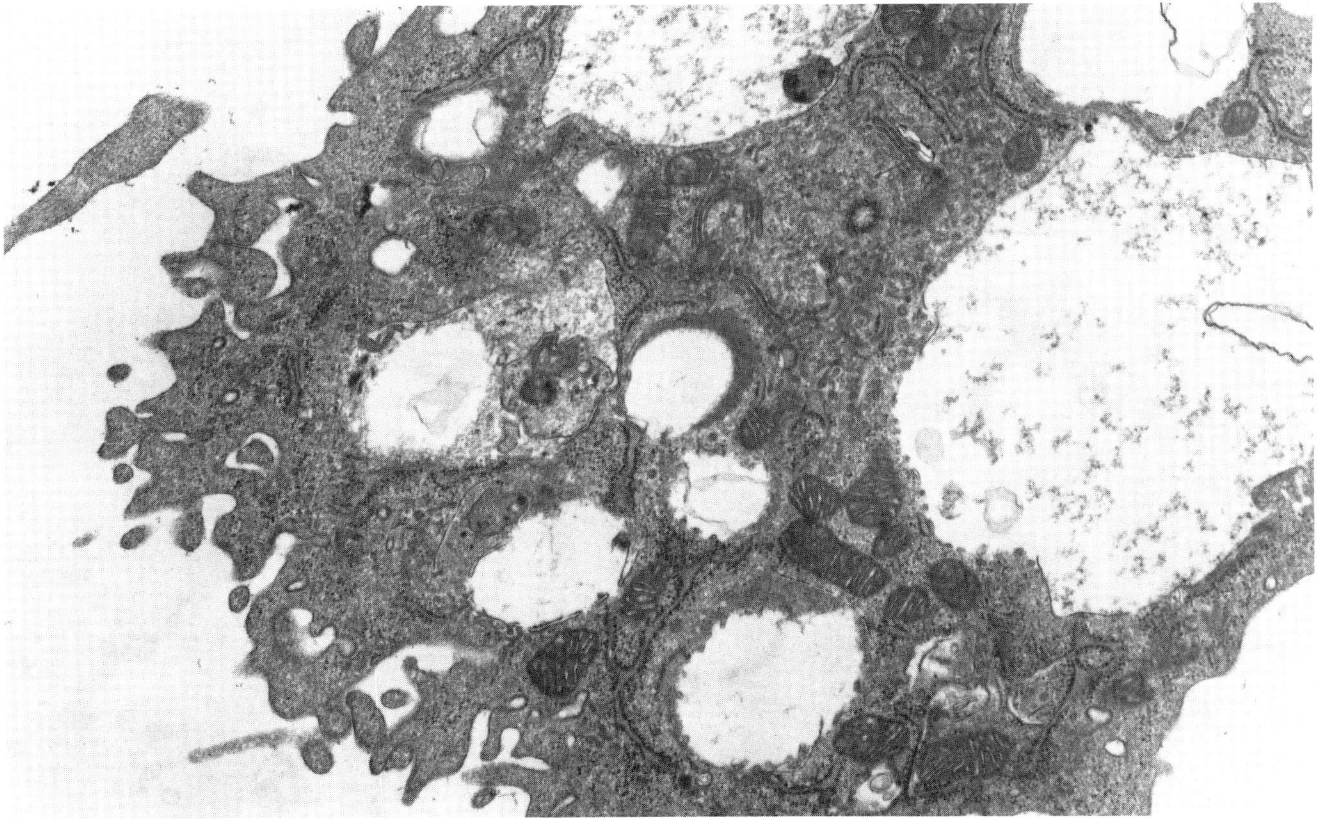

FIGURE 21. Mouse peritoneal macrophages treated *in vitro* with L-leu-*O*Me (5 m*M* for 40 min). The cytoplasm shows vacuolization (14,440×).

REFERENCES

1. **Francis, G. L. and Ballard, F. J.,** Enzyme inactivation via disulphide-thiol exchange as catalysed by a rat liver membrane protein, *Biochem. J.,* 186, 581, 1980.
2. **Wold, F.,** *In vivo* chemical modification of proteins (post-translational modification), *Annu. Rev. Biochem.,* 50, 783, 1981.
3. **Freedman, R. B. and Howkins, H. C., Eds.,** *The Enzymology of Post-Translational Modification of Proteins,* Vol. 2, Academic Press, New York, 1985.
4. **Paik, W. K. and Kim, S.,** *Protein Methylation, Biochemistry: A Series of Monographs,* Vol. 1, John Wiley & Sons, New York, 1980.
5. **Usdin, E. and Borchardt, C., Eds.,** *Biochemistry of S-Adenosyl-Methionine and Related Compounds,* Macmillan, London, 1982.
6. **Paik, W. K. and Kim, S.,** Protein methylation. Enzymatic methylation of proteins after translation may take part in control of biological activities of proteins, *Science,* 174, 114, 1971.
7. **Inglis, V. B. M.,** Requirement for the replication of herpes virus, *J. Gen. Virol.,* 3, 9, 1968.
8. **Nickell, L. G. and Maretzki, A.,** Growth of suspension cultures of sugarcane cells in chemically defined media, *Physiol. Plant.,* 22, 117, 1969.
9. **Furuhashi, K. and Yatazawa, M.,** Amino acids as nitrogen sources for the growth of rice callus tissue, *Plant Cell Physiol.,* 11, 559, 1970.
10. **Rogers, Q. R., Chen, D. M. M. Y., and Harper, A. E.,** Importance of dispensable amino acids for maximal growth in the rat, *Proc. Soc. Exp. Biol. Med.,* 134, 517, 1970.
11. **Jones, J. D.,** Lysine-arginine antagonism in the chick, *J. Nutr.,* 84, 313, 1964.
12. **Lerch, B. and Stegemann, H.,** Variabilität der freien Aminosäuren in Zuckerrüben-Blättern, insbesondere Abhängigkeit des Glutaminsäure- und Asparaginsäure-Gehalts von der Belichtung, *Landw. Forschg.,* 23, 181, 1970.
13. **Szende, B.,** The Effect of Amino Acids on Tumor Growth, Dissertation, Budapest, 1973.
14. **Tyihák, E., Szende, B., and Lapis, K.,** Biological significance of methylated derivatives of lysine and arginine, *Life Sci.,* 20, 385, 1977.
15. **Takemoto, T., Daigo, K., and Takagi, N.,** Studies on the hypotensive constituents of marine algae. I. New basic amino acid "laminine" and the other basic constituents isolated from *Laminaria angustata, Yakugaku Zasshi,* 84, 1176, 1964.
16. **Szende, B., Tyihák, E., Kopper, L., and Lapis, K.,** The tumor growth promoting effect of ε-*N*-trimethyllysine, *Neoplasma,* 17, 433, 1970.
17. **Stotz, G., Szende, B., Lapis, K., and Tyihák, E.,** The effect of ε-*N*-DL-trimethyl-lysine on human lymphocyte cultures, *Exp. Pathol.,* 9, 317, 1974.
18. **Suba, Zs., Szende, B., Takáts, J., Lapis, K., and Elek, G.,** ε-*N*-Trimethyllysine: a natural cell component with mitogenic activity, *Neoplasma,* 27, 11, 1980.
19. **Lapis, K., Szende, B., Kovács, L., and Simon, K.,** The effect of ε-*N*-trimethyllysine (TML) treatment on the regeneration of bone marrow and intestinal mucosa in mice, *Exp. Pathol.,* 24, 97, 1983.
20. **Lapis, K., Davies, A. J. S., and Cross, A. M.,** Anaemia associated with the NK/lymphoma in mice, *Br. J. Cancer,* 16, 770, 1962.
21. **Elek, G., Láng, I., Szende, B., and Lapis, K.,** The effect of ε-*N*-trimethyllysine (TML) on the humoral and cellular immune response, *Agents Actions,* 12, 503, 1982.
22. **Szende, B., Jeney, A., Benedeczky, I., and Lapis, K.,** Investigation of the mode of action of ε-amino-trimethyllysine, *Adv. Tumor Prev. Detect.,* 3, 122, 1976,.
23. **Jeney, A., Gyapay, G., Lapis, K., Szende, B., Bursics, L., and Tyihák, E.,** Methylation-like reaction of [^{3}H-methyl]-N-5-trimethyllysine (TML) to chromatin components, *Biochem. Pharmacol.,* 29, 2729, 1980.
24. **Paik, W. K. and Kim, S.,** ε-Alkyllysinase. New assay method, purification and biological significance, *Arch. Biochem. Biophys.,* 165, 369, 1974.
25. **Paik, W. K., Lee, H. W., and Lawson, D.,** Age-dependency of various protein methylases, *Exp. Gerontol.,* 6, 271, 1971.
26. **Novogrodsky, A.,** A chemical approach for the study of lymphocyte activation, in *Mitogens in Immunobiology,* Oppenheim, J. J. and Rosenstreich, D. L., Eds., Academic Press, New York, 1976, 43.
27. **Lapis, K., Jeney, A., Tyihák, E., Kopper, L., Szarvas, T., Bursics, L., Ujhelyi, E., and Szende, B.,** Studies on uptake of *N*-ε-^{3}H-trimethyllysine in the organs of mice with NK/Ly ascites tumors, *Acta Biochem. Biophys. Acad. Sci. Hung.,* 13, 47, 1978.
28. **Tyihák, E., Patthy, A., Ferenczi, S., Eckhardt, S., Kralovánszky, J., Lapis, K., and Szende, B.,** Methods for the study of methylated basic amino acids in the urine of healthy and cancerous people, *Orvostud.,* 27, 532, 1975 (in Hungarian).

29. **Szende, B., Lapis, K., and Simon, K.**, Combined effect of cytostatic drugs and ε-*N*-L-trimethyllysine in healthy and transplantable tumor bearing mice, *Neoplasma*, 29, 427, 1982.
30. **Murray, K.**, The occurrence of ε-*N*-methyl-lysine in histones, *Biochemistry*, 3, 10, 1964.
31. **Paik, W. K. and Kim, S.**, ε-*N*-Dimethyllysine in histones, *Biochem. Biophys. Res. Commun.*, 27, 479, 1967.
32. **Hempel, K., Lange, H. W., and Birkofer, L.**, ε-*N*-Trimethyllysine, eine neue Aminosäure in Histonen, *Naturwiss.*, 55, 37, 1968.
33. **Tyihák, E., Maróti, M., and Vágujfalvi, D.**, Effect of ε-*N*-dimethyllysine on tobacco tissue cultures, *Bot. Közlem.*, 58, 85, 1971 (in Hungarian).
34. **Tyihák, E., Maróti, M., and Ráthonyi, Z.**, The effect of L-lysine and N^{ε}-methylated lysines and N^{ε}-formyl-L-lysine on the growth of the tobacco tissue cultures, *Bot. Közlem.*, 61, 149, 1974 (in Hungarian).
35. **Tyihák, E.**, unpublished data, 1983.
36. **Szarvas, T.**, unpublished data, 1986.
37. **Tyihák, E., Trézl, L., and Rusznák, I.**, Spontaneous N^{ε}-methylation of L-lysine by formaldehyde, *Pharmazie*, 35, 18, 1980.
38. **Kovács, G., Tyihák, E., Rusznák, I., Trézl, L., Földesi, D., Szabó, B., Bódi, I., Császár, Sz., Szopko, M., and Gombár, M.**, Hungarian Patent 182.677, 1984; and e.g., U.S. Patent 4.532.214, 1985.
39. **Tyihák, E., Trézl, L., Rusznák, I., Kovács, G., Szabó, B., Császár, Sz., Gombár, M., and Földesi, D.**, Hungarian Patent 190.357, 1985.
40. **Rusznák, I., Szarvas, T., Gombár, M., and Terbe, I.**, New results in the application of Formetol, in *Proc. 2nd Int. Conf. The Role of Formaldehyde in Biological Systems*, Tyihák, E. and Gullner, G., Eds., SOTE Press, Budapest, 1987, 219.
41. **Schönbeck, F., Dehne, H.-W., and Beicht, W.**, Activation of unspecific resistance mechanisms in plants, *J. Plant Dis. Prot.*, 87, 654, 1980.
42. **Tyihák, E. and Schönbeck, F.**, unpublished data, 1987—1988.
43. **Paik, W. K. and Kim, S.**, ω-*N*-Methylarginine in histones, *Biochem. Biophys. Res. Commun.*, 40, 224, 1970.
44. **Baldwin, G. S. and Carnegie, P. R.**, Specific enzymic methylation of an arginine in the experimental allergic encephalomyelitis protein from human myelin, *Science*, 171, 579, 1971.
45. **Kakimoto, Y. and Akazawa, S.**, Isolation and identification of N^G,N^G- and N^G,N'^G-dimethylarginine, N^{ε}-mono-, di- and trimethyl-lysine, and glycosyl- and galactosyl-hydroxylysine from human urine, *J. Biol. Chem.*, 245, 5751, 1970.
46. **Tyihák, E. and Patthy, A.**, On the chemical nature of "promine" and "retine", *Acta Agronom. Acad. Sci. Hung.*, 22, 445, 1973.
47. **Tyihák, E., Rusznák, I., and Trézl, L.**, Antagonistic behaviour in spontaneous reactions of L-lysine and L-arginine with formaldehyde and their biological significance, in *Proc. Hung. Annu. Meet. Biochem.*, Miskolc, Hungary, 1975; *Chem. Abstr.*, 85, 118, 182, 1976.
48. **Trézl, L., Rusznák, I., Tyihák, E., Szarvas, T., Csiba, A., and Szende, B.**, Antagonistic behaviour in spontaneous reactions of L-lysine and L-arginine with formaldehyde and their biological significance, in *Proc. 2nd Int. Conf. The Role of Formaldehyde in Biological Systems*, Tyihák, E. and Gullner, G., Eds., SOTE Press, Budapest, 1987, 147.
49. **Tyihák, E., Trézl, L., and Kolonits, P.**, The isolation of N^{ε}-formyl-L-lysine from the reaction between formaldehyde and L-lysine and its identification by OPLC and NMR spectroscopy, *J. Pharm. Med. Anal.*, 3, 343, 1985.
50. **Trézl, L., Rusznák, I., Tyihák, E., Szarvas, T., and Szende, B.**, Spontaneous N^{ε}-methylation and N^{ε}-formylation reactions between L-lysine and formaldehyde inhibited by L-ascorbic acid, *Biochem. J.*, 214, 289, 1983.
51. **Csiba, A., Trézl, L., Tyihák, E., Gráber, H., Vári, E., Téglás, G., and Rusznák, I.**, Assumed role of L-arginine in mobilization of endogenous formaldehyde, *Acta Physiol. Acad. Sci. Hung.*, 59, 35, 1982.
52. **Csiba, A., Trézl, L., Tyihák, E., Szarvas, T., and Rusznák, I.**, N^G-Hydroxymethyl-L-arginines: new serum and urine components: their isolation and characterization by ion-exchange TLC, in *Proc. Int. Symp. TLC with Special Emphasis on OPLC*, Szeged, Hungary, 1984, Tyihák, E., Ed., Labor MIM, Budapest, 1986, 146.
53. **Tyihák, E.**, Is there a formaldehyde cycle in biological systems?, in *Proc. 2nd Int. Conf. The Role of Formaldehyde in Biological Systems*, Tyihák, E. and Gullner, G., Eds., SOTE Press, Budapest, 1987, 137.
54. **Tyihák, E., Maróti, M., Vágujfalvi, D., Bajusz, S., and Patthy, A.**, The growth-retarding effects of guanidino-methylated arginines on the tobacco tissue cultures, *Experientia*, 31, 818, 1975.
55. **Thiele, D. L. and Lipsky, P. E.**, Modulation of human natural killer cell function by L-leucine methyl ester: monocyte-dependent depletion from human peripheral blood mononuclear cells, *J. Immunol.*, 134, 786, 1985.
56. **Csuka, I., Szende, B., Antoni, F., and Lapis, K.**, Effect of *O*-methylesters of amino acids on macrophages, in *Proc. 2nd Int. Conf. The Role of Formaldehyde in Biological Systems*, Tyihák, E. and Gullner, G., Eds., SOTE Press, Budapest, 1987, 27.

57. **Chung, C., Lazarides, E., O'Connor, C. M., and Clarke, S.,** Methylation of chicken fibroblast heat shock proteins at lysyl and arginyl residues, *J. Biol. Chem.*, 257, 8356, 1982.
58. **Camato, R. and Tanguay, R. M.,** Changes in the methylation pattern of core histones during heat shock in *Drosophila* cells, *EMBO J.*, 1, 1529, 1982.
59. **Huszti, S. and Tyihák, E.,** Formation of formaldehyde from S-adenosyl-L-[*methyl*-^{3}H]methionine during enzymic transmethylation of histamine, *FEBS Lett.*, 209, 362, 1986.
60. **Paik, W. K., Abou-Gharbia, M., Swern, D., Lotlikar, P. D., and Kim, S.,** Metabolism of N^G-monomethyl-L-arginine, *Can. J. Biochem. Cell Biol.*, 61, 850, 1983.
61. **Ogawa, T., Kimoto, M., Watanabe, H., and Sasaoka, K.,** Metabolism of N^G,N^G- and N^G,N'^G-dimethylarginine in rats, *Arch. Biochem. Biophys.*, 252, 526, 1987.
62. **Ogawa, T., Kimoto, M., and Sasaoka, K.,** Occurrence of a new enzyme catalyzing the direct conversion of N^G,N^G-dimethyl-L-arginine to L-citrulline in rats, *Biochem. Biophys. Res. Commun.*, 148, 671, 1987.

Chapter 21

METHYLATED POLYPEPTIDE MODELS

Jake Bello

TABLE OF CONTENTS

I. INTRODUCTION

The study of polypeptides of simple composition has been of great importance in establishing basic aspects of protein structure, even though natural proteins are far more complex in composition and sequence. The study of methylated poly(L-lysine)(PLL) and other methylated polypeptides may contribute to a better understanding of the effects of methylation on peptide conformation, and interactions with solvents, membranes, peptides, and nucleic acids.

Among the questions which readily come to mind are the following:

1. How does methylation affect the formation and stability of the several conformations which a polypeptide chain may assume?
2. How does methylation affect intra- and interchain interactions?
3. How does methylation affect hydrophobic interactions and coulombic interactions?
4. How does methylation affect interactions with nucleic acids, membranes, and polysaccharides?
5. How does methylation affect internal dynamics?

Some information is now available on some of these questions, but not all.

An important question is: how will methylation of amino or guanidino (more accurately, ammonium or guanidinium) groups affect interactions with anionic groups, especially phosphodiester groups of nucleic acids, but also carboxylate of proteins, phosphate, and sulfate esters of posttranslationally modified proteins, and perhaps anionic groups of charged polysaccharides and lipds?

Baxter and Byvoet[1] began the study of this question with a ^{13}C-NMR investigation of N^{ε}-methylated lysine and methylated 1-aminohexane. Methylation caused a shift of the ^{13}C resonances of the ε and methyl carbons of lysine, as well as that of the 1-CH_2 carbon of aminohexane. ^{14}N spectra of methyl-, dimethyl-, trimethyl-, ethyl-, diethyl-, and triethylamine also showed progressive downfield shifts with increasing alkylation. Baxter and Byvoet suggested that since the shifts reflect decreased charge densities at side-chain nitrogen and next-neighbor carbon atoms, methylation would be expected to enhance binding to DNA. This conclusion ignores several factors. First, if the charge density decreases on some atoms, it will increase elsewhere. Second, the number of H-atoms available for H-bonding to oxygen of phosphodiester would decrease; third, the steric effects would be different, in particular, the intercharge distance between phosphate and alkylammonium would increase (for the trimethyl certainly); and fourth, hydration would change, and with it would come a change in ΔG of extrusion of solvent on interaction of protein with DNA.

Byvoet and Baxter[2] studied the elution of histones from chromatin by increasing concentrations of HCl, protamine, PLL, or deoxycholate, finding that the most highly methylated histones required the highest concentration of eluting agent, suggesting that methylation promotes binding to DNA. This result is apparently the opposite of the results on dissociation of PLL and Poly (N^{ε},N^{ε},N^{ε}-trimethyl-L-lysine)(PTMLL) from DNA, to be described in more detail below. However, as Byvoet and Baxter noted, it is not possible to conclude that methylated histones bind better than do the same histones in the unmethylated state.

Most of the research to be discussed has been carried out on polypeptides with methylated lysine residues. The field of methylated arginine and methylated histidine is essentially untouched. The published work in this field is not yet abundant, leaving many questions largely open. Much of the data cited here are from work recently done in this laboratory, and not yet published in detail.

II. SYNTHESIS OF METHYLATED POLYPEPTIDES

Seely and Benoiton[3] prepared poly (N^{ε}-methyl-L-lysine)(PMMLL) from the *N*-carboxy anhydride of N^{ε}-methyl-N^{ε}-carbobenzyloxy-L-lysine followed by removal of the blocking group with HBr in acetic acid. PMMLL and Poly (N^{δ}-ethyl-L-ornithine)(PMELO) were prepared by Yamamoto and Yang,[4] from the *N*-carboxy anhydride of N^{ε}-methyl-N^{ε}-carbobenzyloxy-L-lysine, or N^{δ}-ethyl-N^{δ}-carbobenzyloxy-L-ornithine. The original papers should be consulted for details of the synthesis of the monomers. Poly (N^{ε},N^{ε}-dimethyl-L-lysine)(PDMLL) is readily prepared from PLL by reductive alkylation using formaldehyde and $NaBH_4$.[5,6] It may be preferable to use $NaCNBH_3$, following the recommendation of Jentoft and Dearborn[7] to recrystallize the $NaCNBH_3$. A preparation of this type was reported to give a ^1H NMR spectrum expected for PDMLL.[8] PDMLL has also been prepared by reaction of PLL with formaldehyde and formic acid.[3] PTMLL and Poly (N^{δ},N^{δ},N^{δ}-trimethyl-L-ornithine)(PTMLO) have been easily prepared by treatment of PLL and Poly (L-ornithine)(PLO) with dimethyl sulfate at pH 9 to 10.[9] The PTMLL product gives the expected ^1H NMR spectrum and, after acid hydrolysis, a single spot on thin-layer chromatography, corresponding to N^{ε},N^{ε},N^{ε}-trimethyllysine. In both of these procedures, the reaction mixture is dialyzed to remove by-products, and then dialyzed against a salt or acid of the desired counterion. As an aside, acid hydrolysis of PTMLL and PTMLO is a very convenient way to prepare trimethyl-L-lysine (TMLL) and trimethyl-L-ornithine (TMLO); the excess HCl is removed by repeated evaporation and dissolutions in water.

PMMLL cannot be made by direct methylation of PLL, as the reaction cannot be stopped at the stage of monomethylation. An example of a relatively low conversion (30%) of lysine residues to monomethyl lysine residues has been published; in this example the proportion of dimethyl- and trimethyllysine residues was small.[8]

A copolymer of dimethylarginine and ornithine was prepared from PLO by reaction with dimethylcyanamide. About 60 to 70% of the amino groups of PLO were converted to N^{ω},N^{ω}-dimethylguanidino groups.[8]

III. METHYLATED PLLS AS ENZYME SUBSTRATES

The hydrolysis of methylated PLLs has been studied with trypsin and carboxypeptidase B (CPB). Trypsin catalyzes hydrolysis at internal lysine or arginine residues, while carboxypeptidase B catalyzes hydrolysis of C-terminal lysine (or arginine). Seely and Benoiton[3] found that porcine CPB hydrolyzes PMMLL at about one third the rate of PLL, and PDMLL at about 1/50 of the rate of PLL, in the initial stage. After 1 h, the rate dropped and the release of monomer proceeded linearly, rather than parabolically. PMMLL was hydrolyzed to small peptides, probably in the range of dimers to pentamers, as for PLL. Seely and Benoiton[10] also investigated monomeric substrates and found that *N*-methyl-lysine ethylester and N^{α}-benzoyl-N^{ε}-methyllysine ethyl ester are substrates for trypsin, with k_{cat} 1 to 2 orders of magnitude smaller than for the corresponding unmethylated lysine. The dimethyl and trimethyllysine esters were not substrates.

Paik and Kim[6] reductively methylated pancreatic RNase, PLL, and a histone, and tested the action of trypsin on these substrates. Control PLL and methylated PLL (56% dimethyl, 13% monomethyl, and 32% unmethylated residues) were hydrolyzed by trypsin at the same rate. Similar results were obtained for control and methylated histone and for ribonuclease. These results differ from those of Benoiton and Deneault[11] and of Gorecki and Shalitin[12] on methylated lysine esters. Paik and Kim suggest that the difference may reside in differences in the mechanisms of peptide and ester cleavage of trypsin.

IV. CONFORMATIONAL PROPERTIES OF MONOMETHYLATED PLL AND PLO AND MONOETHYLATED PLO

The first conformational study of a methylated PLL was done by Yamamoto and Yang[4] who prepared the monomethyl polymer, poly (N^ε-methyl-L-lysine), PMMLL, from the Z-blocked N^ε-L-methyl-lysine-*N*-carboxy anhydride. By similar procedures they synthesized poly (N^ε-methyl-L-ornithine), (PMMLO), and the corresponding N^ε-ethylornithine polymer, PMELO. At pH 6.1 all three polypeptides, like PLL, are in the random coil conformation. (We shall here accept the convention that what is called the random coil for PLL is just that, and shall consider later the question of its being in the extended helix proposed by Krimm and associates.) At pH 12.4 PMMLL and PMELO were found by optical rotary dispersion (ORD) to be largely α-helical. But PMMLL, on the basis of CD, was incompletely helical, about 75%, while PMELO was close to being completely helical. PMMLO was α-helical to a lesser extent, about 45 to 50%, but addition of methanol (to 1:1 v/v) enhanced the helicity markedly, as is usually the case for polypeptides in solvents of relatively low polarity. From titration curves at several temperatures (5 to 35°) they calculated ΔG at 25° for the coil-to-helix transition to be about –60 cal/residue mole for PMMLL and for PMMLO, compared with –90 for PLL (noting that ΔG values of –80 and –140 have been reported by others). It appears that ΔG of helix formation in uncharged polymer is less negative for the methylated PLL than for the parent. For PMMLL and PMMLO ΔH values were, respectively, –370 and –540 cal/residue mole, compared with –505 to –880 reported for PLL by several investigators. Entropy changes were –1.1 and –1.6 eu/residue mole for PMLL and PMELO, respectively. Values of ΔS for PLL have been reported variously at –1.2 to –2.7 eu/residue mole. The thermodynamic quantities are not strikingly different for PMMLO and PMELO from those of PLL. In summary, Yamamoto and Yang found that monomethylation did not enhance helix formation, but that monoethylation (in comparing PLO, PMMLO, and PMELO) did.

Yamamoto and Yang[13] then studied the thermal helix→β transition for uncharged PMMLL and PMELO at elevated temperature. At 60° both polymers go over to the β-structure, with $[\theta]^{217}$ value close to that of PLL under similar conditions of pH and temperature. Yamamoto and Yang[13] also observed that the helicity of PMLO is very little changed between 25 and 50°, but for PLO the helicity goes from 50% to completion over the same temperature range. Not so for uncharged PMMLO, which on heating is converted to a state approximating the random coil.

Yamamoto and Yang also studied the rate of α → β conversion. In addition to the effects of pH (i.e., completeness of deprotonation) and temperature, the rate is dependent on polypeptide concentration, as expected for an intermolecular association reaction (although at low concentration, intramolecular β-structure can form). The rate of α → β is greater for PMELO than for PMMLL, which are isomers, and both are slower than for PLL. On the basis of only these three, conclusions are uncertain, but, tentatively, it appears that alkylation of $–NH_2$ reduces the α → β rate, but that a larger alkyl group has the lesser slowing effect.

Tseng and Yang[14] studied polymers of lysine homologs, with 5, 6, or 7 CH_2 groups in the side chain, to compare with the 4 CH_2 groups of lysine. While the NH_3^+ groups are not alkylated, the effect of a more hydrophobic side chain could be studied. They found that the longer the side chain the more readily formed and stable the α-helix at high pH, while the α → β transition at high pH and high temperature is inhibited. Also, sodium dodecyl sulfate ($NaDodSO_4$) converts all three homologs to α-helix. Thus, the effect of hydrophobic groups in the side chain is not clear cut, and depends on which side of the amino group the hydrophobicity is increased.

V. CONFORMATIONAL PROPERTIES OF PTMLL AND PTMLO

A. EFFECTS OF TEMPERATURE

In our laboratory we decided to study first the properties of completely methylated PLL and

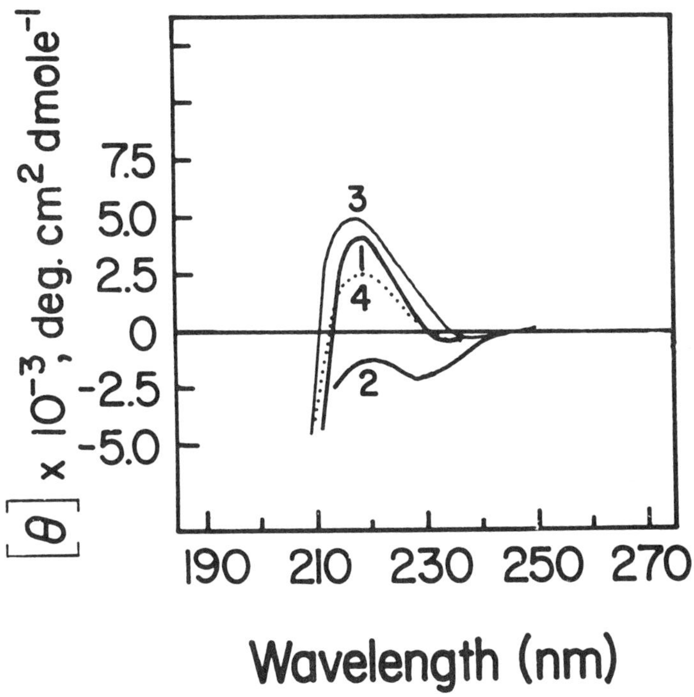

FIGURE 1. CD spectra of $(Lys)_n$ and $[Lys(Me_3)]_n$ at neutral pH: 1. $(Lys)_n$, 25°C; 2. $(Lys)_n$, 68°C; 3. $[Lys(Me_3)]_n$, 25°C; $[Lys(Me_3)]_n$, 68°C. (From Granados, E. N. and Bello, J., *Biopolymers*, 18, 1479, 1979. With permission.)

PLO in order to observe the extreme effects of methylation. These were prepared by the action of excess dimethyl sulfate on PLL and PLO at pH about 9.5 followed by dialysis (at which stage a choice of counterion may readily be made).[9]

Conformational studies[9] by circular dichroism (CD) of PTMLL·$HClO_4$ show it to be a random coil at 25° in H_2O, similar to PLL. However, $[\theta]_{218}$ is larger for PTMLL than for PLL. The 218 nm extremum is considered indicative of the random coil form. PTMLL and PTMLO at high pH remain random coils because their quaternary ammonium groups are not titratable. The larger $[\theta]_{218}$ for PTMLL over PLL suggests that PTMLL is more random than PLL, or PTMLL is more ordered than is PLL. These are contradictory alternatives, which depend on whether or not protonated PLL is a random coil, a question to be discussed below. However, PLL at 68° undergoes a substantial decrease in $[\theta]_{218}$ while $[\theta]_{218}$ of PTMLL decreases by half as much (Figure 1).[9] The decrease in $[\theta]_{218}$ for PLL is accompanied by the appearance of negative ellipticity around 230 nm. It is not clear if this represents disordering of the putative extended helix or the onset of order.[15,16] There has been a controversy as to the state of PLL in the completely protonated form. It had been accepted that protonated PLL is a random coil, and its CD spectrum has been accepted as a standard for random coils. However, Tiffany and Krimm[17] proposed that the CD spectrum is that of a locally organized structure, an extended helix. This proposal was based on the resemblance of the CD spectrum to that of helical $[Pro]_n$, differences from the CD spectra of denatured proteins, on the temperature dependence of the CD spectrum, on Raman and infrared spectra, and on theoretical ideas. Some researchers have supported the extended helix proposal,[18,19] but it has not met with broad acceptance, and some rebuttals have been published.[20-22] The difference between PLL and PTMLL in their response to elevated temperature may have a bearing on this question. The greater resistance of PTMLL to CD changes induced by elevated temperature suggests a more stable conformation, one which may be more rigid and resistant to randomization. An alternative is that both are random coils at 25°,

TABLE 1
Conformations of Methylated PLL and PLO

Polypeptide	pH 6	pH 11	$NaDodSO_4$	$NaClO_4$ (or $Ca[ClO_4]_2$)
PLL	r.c.	α-Helix	—	α-Helix
PMMLL[a]	r.c.	α-Helix	—	—
PDMLL[b]	r.c.	α-Helix	α-Helix, 50%	α-Helix
PTMLL[c]	r.c.	r.c.	r.c.	α-Helix
PLO	r.c.	Partial α-helix[d]	α-Helix[d]	Partial helix[e]
PTMLO	r.c.	r.c.	r.c.	α-Helix
$(Lys^{67},Ala^{33})_n$	r.c.	—	—	α-Helix
$(TMLys^{67},Ala^{33})_n$	Partial helix[f]	—	—[g]	α-Helix
$(Lys^{50},Ala^{50})_n$	α-Helix	—	α-Helix, 75%	α-Helix
$(TMLys^{50},Ala^{50})_n$	Partial helix	—	α-Helix, 50%	α-Helix

Note: r.c. = random coil.

[a] Yamamoto and Yang.[4,13]
[b] Bello et al.[8]
[c] Bello et al.[8]
[d] Grourke and Gibbs.[23]
[e] Precipitates on heating in 1 *M* $Ca(ClO_4)_2$.
[f] Helicity dependent on concentration, see text.
[g] See text.

but that PLL undergoes a partial transition to some structured form at 68°, while PTMLL is more resistant to this change. The data available do not resolve the question, but further work, using different techniques, on alkylated PLLs may help to do so. In this review I shall refer to protonated PLL and to methylated polypeptides with similar CD spectra as random coils, in the absence of definitive evidence of a different conformation, except in connection with discussion of the coil/extended helix controversy.

B. DODECYL SULFATE DETERGENT

$NaDodSO_4$ converts PLL to a β-structure,[23] but has no effect on the PTMLL CD spectrum.[9] This does not result from the absence of an interaction, since PTMLL in 0.005, 0.025 and 0.47 M $NaDodSO_4$ shows greatly increased light scattering over PTMLL in H_2O, or over $NaDodSO_4$ without PTMLL (unpublished data from this laboratory). PLO in $NaDodSO_4$ is α-helical to at least 60%, but PTMLO is a random coil.[9] PTMLO is a random coil in the protonated form. In $NaDodSO_4$ the positive $[\theta]_{218}$ falls to about one third of the value in H_2O.[9] Again we have the uncertainty of interpretation; either random coil is being converted, in part at least, to an ordered structure, or the extended helix is being randomized. But α-helix is not formed to any significant extent. These differences suggest that methylation of proteins might affect their interactions with lipids. A summary of conformations of PLL, PLO, PTMLL, PTMLO, and PDMLL is given in Table 1.

C. HELIX INDUCTION BY PERCHLORATE ION

Perchlorate salts of various cations (Na^+, Ca^{++}, H^+, etc.) convert protonated PLL to the α-helix.[19,22,24,25] The effect of ClO_4^- on the conformation of PTMLL, PDMLL, and PTMLO has been studied.[8,9] $Ca(ClO_4)_2$, 0.5 *M*, at 25° caused some turbidity in PTMLL and generated a CD spectrum which appeared to be that of the α-helix distorted by turbidity, as suggested by its resemblance to the spectrum of helical poly(glutamic acid) which is aggregated. PTMLO in 0.5 *M* $Ca(ClO_4)_2$ showed an α-helical CD spectrum, truncated on the short wave length side, and of twice the magnitude, at 223 nm, as that of PLO.

A more detailed study of PTMLL in $NaClO_4$ has now been made,[26] which showed that PTMLL adopts the α-helical conformation at about 3% of the ClO_4^- concentration required for PLL, namely, for 6×10^{-4} *M* (residue) of PTMLL at O°, $1{\cdot}2 \times 10^{-3}$ *M* $NaClO_4$ vs. 40×10^{-3} *M* for 8×10^{-4} *M* PLL. (These results were on PTMLL and PLL of M_r 40,000, with ClO_4^- introduced as the counterion by dialysis.) These concentrations represent ClO_4^- to peptide residue ratios of 2 and 51 for PTMLL and PLL, respectively. At so low a ClO_4^- concentration the helix-generating cause must be an interaction between the ClO_4^- and the peptide, and not an effect of ions on water structure. A direct binding effect is supported by the linearity of log K vs. log $[ClO_4^-]$, where K = [fraction helical]/[fraction random], although some effect on water structure in the immediate vicinity of the cationic side chain group and the ClO_4^- ion is possible. The equilibrium constant for binding of ClO_4^- to a quaternary ammonium resin, Dowex-2, is considerably larger than for Cl^- binding (Belle, cited in Reference 27). But if PTMLL or PLL is brought to 50% helix with $NaClO_4$, addition of Cl^- to half the concentration of ClO_4^- present reduces the helix content by half. At NaClO4 above 1 *M* helix content gradually falls, under the general denaturing influence of ClO_4^-. As the $NaClO_4$ concentration is raised above about 0.01 *M*, turbidity develops in PTMLL, which reaches a plateau over the range from about 0.1 *M* to 1 *M* $NaClO_4$, above which turbidity falls, reaching baseline at about 6 *M* $NaClO_4$. Even at 6 *M* $NaClO_4$ both PLL and PTMLL retain much helicity. When the PTMLL is lightly dansylated, it is found that the fluorescence intensity runs parallel to the turbidity. Since $NaClO_4$ does not affect the fluorescence of dansylated PLL in the helix-inducing range of $NaClO_4$ concentration, it appears that the fluorescence increase for dansyl-PTMLL arises from aggregation.

When random coil PLL is converted to α-helix at high pH, the fluorescence of the dansyl label is markedly affected in parallel to the conformational change.[28] But the absence of a fluorescence change for helix induction in PLL by $NaClO_4$ suggests that at high pH the fluorescence effect may not be a direct effect of helicity, but an indirect effect of helicity on the side chains.

The major decrease in the concentration of ClO_4^- required for helicity in PTMLL compared with PLL, indicates an important effect of added methyl groups, since the charge remains the same. In a protein, of course, the number of methyl groups will be relatively small, and usually not closely spaced. Yet, the results suggest that in proteins important local effects on conformational stability and on interactions between peptide chains might arise from methylation. When the concentration is high enough, PTMLL precipitates in $NaClO_4$ solution, as was observed during dialysis of the methylation reaction mixture against about 0.1 *M* $NaClO_4$ in the cold. PLL does not precipitate under these conditions. The lower solubility of PTMLL indicates an enhancement of interchain interactions arising from methylation.

A comparison was also made as to the concentration of polypeptide required for spontaneous helix formation with only stoichiometric ClO_4^- counterion and no added $NaClO_4$. For PLL·$HClO_4$ and PTMLL·$HClO_4$ of M_r 40,000 to 50,000 the concentrations for 50% α-helix formation are, respectively, 0.14 and 0.003 *M*, and for M_r 500,000 these are, respectively, 0.12 and 0.002 *M*. Thus, the ratios of $C_{50\%}$ (the concentration of peptide required for 50% helix) are 47:1 and 60:1 for the low and high molecular weight polypeptides, respectively. This emphasizes further the marked promotion of helix formation resulting from methylation. In addition it is seen that for PTMLL·$HClO_4$ there is a somewhat greater dependence on the molecular weight of the polypeptide than for PLL. (A detailed description of this work is in preparation.)

D. EFFECTS OF GUANIDINIUM AND SODIUM CHLORIDES ON PTMLL AND PDMLL CONFORMATION

1. Guanidinium Chloride

Tiffany and Krimm[29] found that 0.66 *M* guanidinium chloride (GdmCl) lowers $[\theta]_{218}$ of PLL, and that higher concentrations of GdmCl increase $[\theta]_{218}$; in 5.7 *M* GdmCl $[\theta]_{218}$ is 18% above the value in water. We found a similar effect, but with only a 10% increase in $[\theta]_{218}$ at 5.6 *M* GdmCl.[8] With PTMLL the initial decrease is smaller and the subsequent increase considerably larger, about 50% (Figure 2). PDMLL gives an intermediate result. The addition of the effect

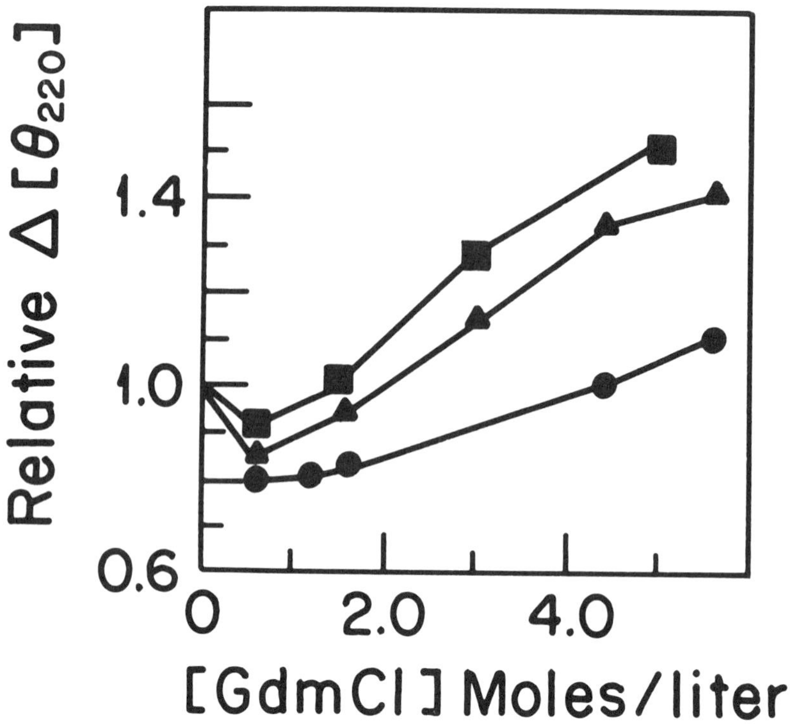

FIGURE 2. Effect of GdmCl on $[\theta]_{220}$ of PLL, PDMLL and PTMLL ●, PLL ▲, PDMLL ■, PTMLL. Peptide concentrations were 6.9×10^{-3} *M* in residues. (From Bello, J., Granados, E. N., Lewinski, S., Bello, H. R., and Trueheart, T., *J. Biomol. Struct. Dynamics*, 2, 899, 1985. With permission.)

of elevated temperature (to 90°) to that of GdmCl, results in a progressive decrease in $[\theta]_{218}$ for PLL, similar to the effect of elevated temperature on PLL in water, with appearance of a negative extremum above 230 nm, which becomes deeper and shifts to shorter wavelength with increasing temperature (Figure 3). The 230 to 238 nm extremum appears in calculated spectra16 for mixtures of random coil and α-helical conformations, based on charged PLL being a random coil and on the well-established helical CD spectrum of uncharged PLL. However, it is not precluded that the high temperature CD spectrum of charged PLL (in water or in strong GdmCl) represents an approach to that of disordered natural proteins, which do not show positive 218 nm bands. The idea that the extremum at 230 to 238 nm is that of partial α-helical character has been disputed. For PTMLL in 4.4 *M* GdmCl (Figure 3) the effect of elevated temperature is weaker, as is the effect of temperature in the absence of GdmCl (noted earlier). For PDMLL in GdmCl the temperature effect is intermediate between PLL and PTMLL. Abundant data on GdmCl indicate that at 6 *M* it randomizes non-crosslinked proteins, and model compound results indicate that aqueous GdmCl increases the solubility of both polar and apolar parts of proteins,[30] which readily explains the randomizing effect of GdmCl. Thus, The GdmCl data for PLL appear to support the idea that protonated PLL is a random coil in water, which first takes on some order in dilute GdmCl (screening of charge repulsion), and becomes disordered again, or more disordered, in concentrated GdmCl. If concentrated GdmCl disorders PTMLL to a state similar to that of PLL, then perhaps the conformation of PTMLL in the absence of GdmCl is not totally disordered, but has a significant element of order, which PLL may have to a smaller degree. The bulkier methylated side chains might contribute to greater rigidity. However, Tiffany and Krimm[29] have proposed that PLL in concentrated GdmCl is an extended helix; if so, then PTMLL and PDMLL may be more ordered, perhaps more rigid (since rigidity contributes to enhanced ellipticity[31]), than is PLL. A point in favor of the suggestion of Tiffany and Krimm is

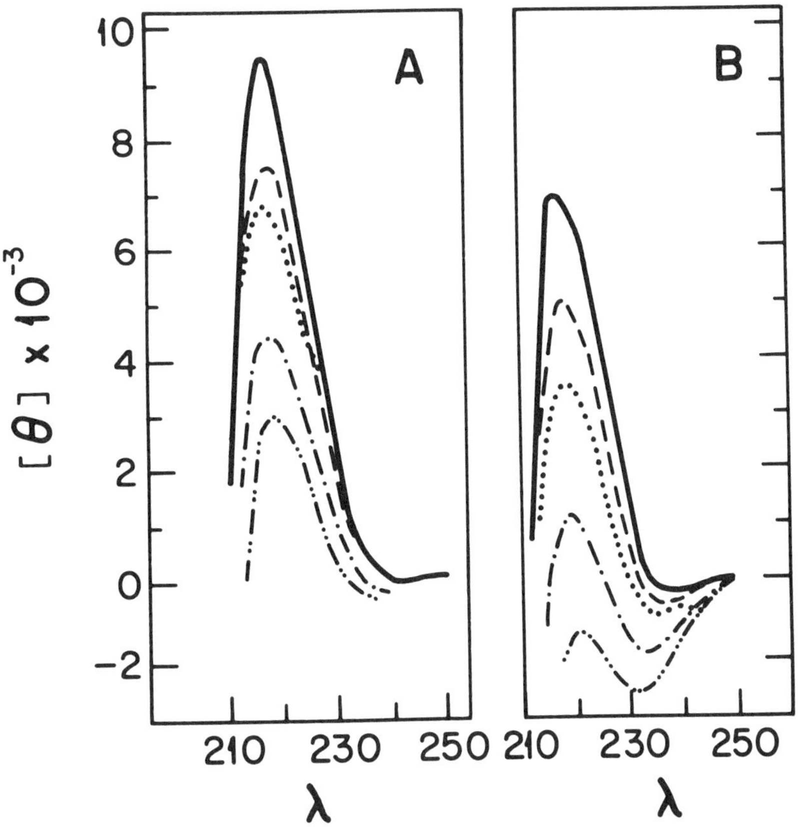

FIGURE 3. Effect of temperature on CD spectra of PTMLL and PLL in 4.4 *M* GdmCl. Temperatures from top to bottom are: 25, 40, 50, 70, and 90°C. PDMLL gives an intermediate result. (From Bello, J., Granados, E. N., Lewinski, S., Bello, H. R., and Trueheart, T., *J. Biomol. Struct. Dynamics*, 2, 899, 1985. With permission.)

that at elevated temperature $[\theta]_{218}$ decreases for PLL, PDMLL, and PTMLL, but less so for the latter two, with PDMLL being the intermediate case. The decrease in $[\theta]_{218}$ at elevated temperature is easily accepted as an indicator of disordering, in conformity with the general disordering capability of GdmCl and hot water. Binding of Gdm^+ to peptide groups has been suggested as the cause of formation of the extended helix.[17,29] On the other hand, binding to peptide groups has been proposed as a mechanism of disruption of native protein conformation,[32] which results, as noted, in a CD spectrum different from that of PLL, etc. in GdmCl. But the $[\theta]_{218}$ data appear to require some induction of order by GdmCl, rather than mere absence of disordering. Although we are not in a position to decide on the random coil vs. extended helix question, it is plain that methylation affects the conformation. It may be that further study of methylated PLL (and of other methylated polypeptides) by other methods may contribute to understanding this fundamental question of peptide conformation.

2. Sodium Chloride

The magnitude of $[\theta]_{218}$ for the polypeptides in zero or low salt at room temperature increases in the order PLL < PDMLL < PTMLL (Figure 4). This can be rationalized in terms of increasing extent or rigidity of the extended helix if we consider the possibility that addition of methyl groups may result in some restriction of sidechain mobility, which may contribute to restricted mainchain mobility or enhancement of hydrophobic interactions. On adding NaCl to a concentration of several molar, $[\theta]_{218}$ decreases, the fractional decreases being in the order PLL

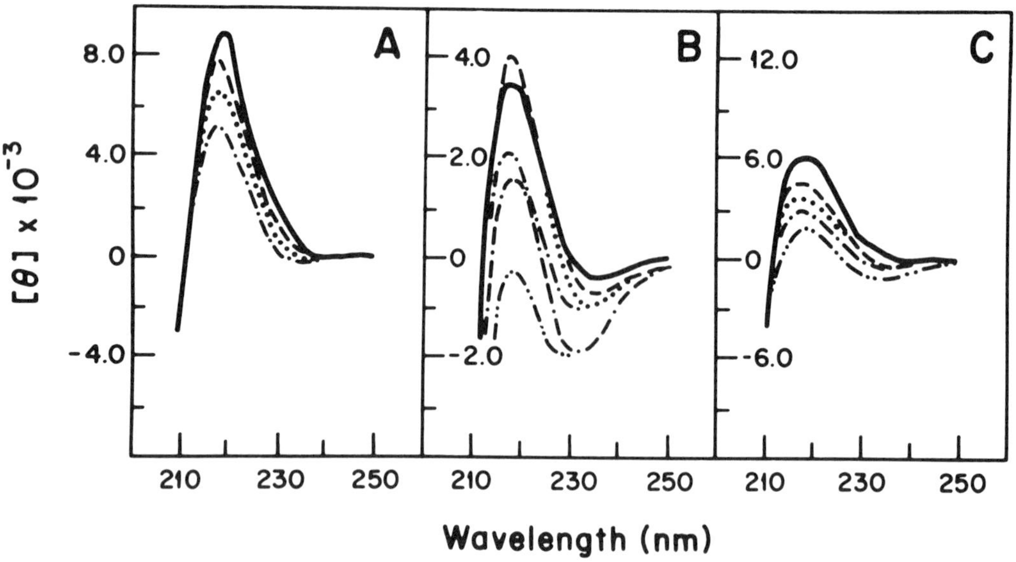

FIGURE 4. Effect of NaCl concentration on CD spectra of PLL, PDMLL and PTMLL. For A (PTMLL), the [NaCl] is 0, 0.25, 2.0 and 4.90 *M* from top to bottom. For B (PLL), [NaCl] is 0.4, 0.2, 0.8 (coincides with 0.4 below 225 nm), 1.6, 2.4 and 3.2 *M*. For C (PDMLL), [NaCl] is 0.2 to 0.4, 0.8, 1.6, 2.4 and 3.2 *M* from top to bottom. Note reversal of direction of [θ] for 0.4 *M* NaCl and PLL. Note differences in ordinate scales. The buffer was $2 \times 10^{-4} M$ sodium phosphate, pH 7.9, with $1 \times 10^{-5} M$ EDTA. (From Bello, J., Granados, E. N., Lewinski, S., Bello, H. R., and Trueheart, T., *J. Biomol. Struct. Dynamics*, 2, 899, 1985. With permission.)

> PDMLL > PTMLL, but being about –4000 to –5000 $cm^2\ dmol^{-1}$ in all cases.[8] The decrease in $[\theta]_{218}$ accompanies a change in spectral shape, particularly in PLL, less so in PDMLL, and still less in PTMLL. This change is similar to that observed with elevated temperature in the absence of salt, i.e., a decrease in $[\theta]_{218}$ and the appearance of a 230 nm extremum, discussed above in connection with temperature effects. It is interesting that NaCl has an effect similar to that of elevated temperature, i.e., a reduction of $[\theta]_{218}$, while GdmCl increase $[\theta]_{218}$ (after an initial decrease). Since high concentrations of NaCl and GdmCl have opposite effects on the CD spectrum of PLL, PDMLL, and PTMLL, they cannot be operating by the same mechanism. If one effect arises from charge screening, the other may operate by a binding mechanism or they may operate by different binding mechanisms.

E. POLY(N^ε,N^ε-DIMETHYL-L-LYSINE)

Some work has been done on the conformational properties of PDMLL, but not nearly as much as on PTMLL. Some results have already been mentioned above in connection with GdmCl, NaCl, and temperature. PDMLL is a random coil in the absence of helix-inducing ions. $Ca(ClO_4)_2$ at 0.5 *M* converts it to (probably) the α-helix, with turbidity, based on the resemblance of the CD spectrum to that of PTMLL under similar conditions.[8] Other concentrations of ClO_4^- have not been studied. $NaDodSO_4$, 0.05 *M*, generates partial α-helical character, perhaps 50%, intermediate between PLL and PTMLL.[8] At pH 11 PDMLL is an α-helix, and on heating it precipitates. The precipitated PDMLL may be in the β-structure by analogy with PLL and PMMLL.

VI. METHYLATED $(LYS,ALA)_N$

In a step toward a more natural sequence of amino acids, random sequence copolymers of lysine and alanine were methylated. These had the composition $(Lys^{50},Ala^{50})_n$ and $(Lys^{67},Ala^{33})_n$, and their derivatives were trimethylated on the lysine residues. $(Lys^{67},Ala^{33})_n$ in the absence of added salt (with only stoichiometric ClO_4^- counterion) is a random coil at 0°, at a concentration of 0.7 m*M* in residues, and 20% helical at 16 m*M*. At 0.7 m*M* the methylated polymer is a random coil, but it is 50% helical at 4.4 m*M*. Thus, methylation promotes helicity, as it does for PTMLL vs. PLL. $(Lys^{50}, Ala^{50})_n$ is a stronger helix former than is $(Lys^{67},Ala^{33})_n$, being 75% helical at 0.1 m*M* at 0° (lower concentrations were not studied). But with $(Lys^{50},Ala^{50})_n$ methylation weakens the helical tendency, resulting in 50% helix at 2.2 m*M*. Thus, as the lysine proportion decreases, methylation of the lysine residues diminishes the helix-forming tendency, at least with alanine as the comonomer.

The effect of added $NaClO_4$ to these polymers was that at 0°, polymer at 0.7 m*M* in peptide residues, 50% helix for $(Lys^{67},Ala^{33})_n$ required 7.5 m*M* $NaClO_4$, while $(TMLys^{67},Ala^{33})_n$ required 1 m*M* $NaClO_4$, a ratio of 7.5 compared with 34 for PLL vs. PTMLL. $(TMLys^{50},Ala^{50})_n$ required 0.04 m*M* $NaClO_4$ for 50% helix; but no comparison with the unmethylated (Lys^{67},Ala^{33}) could be made because this polymer is already largely helical in the absence of $NaClO_4$. Thus, for the 1:1 copolymer the unmethylated one has stronger helix-forming tendency than the methylated polymer. Again, the alanine residues promote helicity, since less ClO_4^- is required for $(TMLys^{50},Ala^{50})_n$ than for $(TMLys^{67},Ala^{33})_n$.

As noted above, PLL in $NaDodSO_4$ has a β structure and PTMLL the random coil. The 2:1 copolymer of lysine and alanine has a β structure in $NaDodSO_4$, while the trimethylated derivative has some structure with $[\theta]_{222} = -6000$ cm² dmol⁻¹, but it is not clear what the partial structure is. The copolymer $(Lys^{50},Ala^{50})_n$ is 75% helical in $NaDodSO_4$, and the methylated polymer is about 50% helical. Kubota et al.[33] reported nearly 100% helix for $(Lys^{50},Ala^{50})_n$ in 0.025 *M* $NaDodSO_4$, and a mixture of α and β for a 3:1 Lys,Ala copolymer. These results indicate that trimethylation tends to inhibit induction by $NaDodSO_4$ of ordered structure in PLL and in copolymers of lysine with alanine, the inhibitory effect diminishing with dilution of the trimethyllysine residues among the alanine residues.

Light scattering measurements on $(Lys^{50},Ala^{50})_n$ and $(TMLys^{50},Ala^{50})_n$ in 0.025 *M* and 0.43 *M* $NaDodSO_4$, show that all scatter weakly, and the latter more weakly (by 50%) than the former. But for $(Lys^{67},Ala^{33})_n$ and $(TMLys^{67},Ala^{33})_n$ in 0.025 and 0.43 *M* $NaDodSO_4$, unmethylated polymer scatters strongly, while the methylated polymer scatters about 10% as strongly as the former. Thus the results for PLL vs. PTMLL, and $(Lys^{X},Ala^{100-X})_n$ vs. $(TMLys^{X},Ala^{100-X})_n$ show that methylation affects interaction with an amphiphile. The detailed results on lysine-alanine copolymers are being prepared for publication.

The results for the series PLL, $(Lys^{67},Ala^{33})_n$, $(Lys^{50}, Ala^{50})_n$ and their completely methylated derivatives show that the effects of methylation on conformation are not yet predictable.

VII. METHYLATED MELITTIN

The foregoing work on conformational effects was done on polymers composed entirely of methylated lysine (or ornithine) residues or simple copolymers of lysine and alanine. Will there be significant effects with peptides containing a smaller fraction of methylated residues? We have made some preliminary studies of the 26-residue bee venom peptide melittin, trimethylated on its four NH_2 groups. Melittin was selected because it readily undergoes a random → helix transconformation and because it contains a sequence of four positive charges, lys-arg-lys-arg (residues 21 to 24), as well as a terminal α-amino and a lysine residue at position 7. There are no negative charges. The sequence of four charges bears a relationship to the continuous charge sequence of PLL, but the fact that only two of these are lysine residues takes it further from the artificial state of PLL, and means that if methylation should exert an effect on conformation the result would be more relevant to peptides of interest. Amino acid analysis confirmed the methylation of the glycine and 2.93 lysine residues. Melittin is converted by many salts from a random coil monomer to a largely α-helical tetramer.[34,35] Methylated melittin requires about twice the NaCl concentration for 50% of the ultimate helicity as does melittin. But with $NaClO_4$, the salt concentrations are about equal. This was somewhat obscured by the appearance of considerable turbidity in the methylated melittin at 0.2 *M* $NaClO_4$, but little in native melittin. Thus, relative to the NaCl effect, methylated melittin is affected twice as strongly by $NaClO_4$ as is melittin. It appears, therefore, that in this peptide with four residues out of 26 methylated (compared to all residues in PTMLL), methylation still influences the helix-forming tendency. Thus, while only 4 of its 26 residues are methylated, a distinct effect of methylation is seen, although much smaller than for PTMLL vs. PLL. Also the appearance of turbidity in methylated melittin in $NaClO_4$ recalls the similar effect for PTMLL. A similar comparison between NaCl and $NaClO_4$ acting on PLL and PTMLL is less clear because NaCl induces very little helix even at very high NaCl concentration, and the interpretation of the CD spectrum is uncertain because the apparent helical contribution is small relative to that of the random coil or extended helix contribution. NaCl does affect the CD spectrum of PLL and PTMLL, the latter less than the former, as noted earlier. Thus, methylation of PLL and of melittin has parallel effects. A detailed account of these results is being prepared for publication.

VIII. INTERACTIONS WITH POLYNUCLEOTIDES

A. DISSOCIATION BY SALT

PTMLL, like PLL, binds to polynucleotides. Both polypeptides bind preferentially to A,T-polymers over G,C-polymers in dilute salt, and while PLL binds preferentially to $(dA\text{-}dT)_n \cdot (dA\text{-}dT)_n$ over $(dA)_n \cdot (dT)_n$, PTMLL shows no preferential binding between these two.[36,37] PTMLL is dissociated from polynucleotides by salt at half the NaCl or $MgCl_2$ concentration needed to dissociate PLL.[38] (These experiments were done with lightly dansylated PLL and PTMLL. Complexation with DNA increases the fluorescence intensity; dissociation reduces it to baseline.) This is true for calf thymus (CT) DNA as well as for $(dG\text{-}dC)_n \cdot (dG\text{-}dC)_n$, $(dG)_n \cdot (dC)_n$, $(dA\text{-}dT)_n \cdot (dA\text{-}dT)_n$ and $(dA)_n \cdot (dT)_n$. The same relationship holds for native and for thermally denatured and cooled calf thymus DNA, and for polypeptides of M_r 30,000 and 3400, although lower [NaCl] is needed to dissociate both PLL and PTMLL of the lower molecular weight. For M_r 30,000 PLL and PTMLL complexed with CT DNA the NaCl or KCl concentrations needed for dissociation are about 1.2 and 0.6 *M*, respectively; for M_r 3400 the corresponding values are about 0.6 and 0.4 *M*. In the case of 3400 M_r PLL and PTMLL, the [NaCl] needed to reduce the fluorescence by 50% is 0.45 and 0.2 *M*, respectively. The latter is near physiological salt concentration (ignoring the fact that a Na salt was used, not K); changes in cellular salt concentration may differentially affect binding of methylated and nonmethylated residues. As the number of methyl groups per lysine residue is lowered the [NaCl] required to dissociate the polypeptide from CT-DNA increases linearly.[8]

The dissociation by salt is independent of temperature for both PLL and PTMLL, indicating that binding is athermic for both. Calorimetric measurements of binding of PTMLL to CT DNA were shown by Sturtevant to be athermic.[38] PLL binding to DNA was already known to be nearly athermic,[39,40] as is that of histone H1.[41] Binding appears to be driven by ΔS of solvent extrusion.[39] It might be anticipated that the cationic trimethylamino group, being larger than the NH_3^+ would cover more of the DNA than would the latter, and hence result in more loss of water of hydration. However the bulkier Me_3N group may not permit the side chain to approach as closely, resulting in less solvent extrusion. Differential solvation effects are difficult to disentangle. Therefore, we shall fall back on the more obvious explanations of weaker binding of PTMLL to DNA, namely, absence of H-bonding, more diffuse positive charge, greater distance between charge centers, and, perhaps, steric hindrance to binding.

We have noted that displacing PTMLL from various polynucleotides requires about half the salt concentration needed with PLL. We have also examined the effect of using as the dissociating salt tetramethylammonium chloride (TMACl), a cation which resembles the trimethylated side chain of PTMLL. Shapiro et al.[42] showed that TMA^+ binds more tightly to AT-rich DNA than to GC-rich DNA. We have found that PTMLL-DNA is dissociated at about 1.1 *M* TMACl while PLL-DNA is not dissociated even at 3 *M* TMACl. The result for PLL-DNA with TMACl is in agreement with the result found by Shapiro et al. for PLL-DNA in 2 *M* TMACl. They found that in 2 *M* TMACl PLL is displaced from AT- to GC-rich regions. TMA^+ is a weaker binder to DNA than are Na^+ and other inorganic cations, and since with Na^+ both PLL and PTMLL are completely dissociated from DNA (albeit at different ionic strengths), while with TMA^+ a qualitative difference is seen (i.e., complete dissociation of PTMLL-DNA, but not dissociation of PLL-DNA), we may consider that methylation of protein lysyl residues may cause a marked change in protein-DNA binding in the presence of competitive ions other than Na^+, K^+, Mg^{+2}, etc., i.e., competitors weaker than the latter ions. Conversely, in the presence of very strongly binding competitors the difference may be diminished. (Detailed results on TMACl are being prepared for publication.)

B. THERMAL DENATURATION OF PTMLL-POLYNUCLEOTIDE COMPLEXES

It is well known that binding of PLL to polynucleotides raises Tm of the latter. It was shown by Gabbay[44] that as the number of methyl substituents on some diamines increased, Tm of complexes of these with $(I)_n{\cdot}(C)_n$ and $(A)_n{\cdot}(U)_n$ increased. Similar results were found by Orosz and Wetmur[45] for complexes with alkyltrimethylammonium salts. We found a similar effect for CT-DNA in a series of methylammonium chlorides.[38] It was therefore of interest to see if methylated polypeptides do likewise. It was found that PTMLL and PTMLO raise Tm of CT-DNA and of $(dA\text{-}dT)_n{\cdot}(dA\text{-}dT)_n$ more than PLL and PLO do.[38] PTMLO and PTMLL raise Tm of CT-DNA by about 7 to 10° more than do PLO and PLL, depending on the peptide and salt concentration. With $(dA\text{-}dT)_n{\cdot}(dA\text{-}dT)_n$ the PTMLL complex melts 13° higher than the PLL complex. These increases in Tm are in contrast to the effects of the series NH_4^+ to $(CH_3)_4N^+$ on Tm of CT-DNA, in which Tm decreases with increasing methylation. The reason for increased Tm with PTMLL and PTMLO may reside in a larger reduction in the energy of the native state, or a larger increase in the energy of the denatured state of DNA. The underlying causes of either effect are to be found in the conformational entropies of the two polymers (polynucleotide and polypeptide) in the two states and in the energetics of solvation. A choice among the various factors is not yet possible.

Complexes of $(dA\text{-}dT)_n{\cdot}(dA\text{-}dT)_n$ with PTMLL, PDMLL, and 30% monomethylated PLL show linearly increased Tm with increased methylation, except for the step from PLL to 30% PMMLL, for which a jump of 3° occurs, or 9° per methyl group, in contrast to an increase of 1.9°/Me from 0.3 methyl to three methyl groups per residue (Figure 5).[8] This result is of some interest since it shows the possibility of a significant effect on DNA thermodynamics at a relatively low degree of methylation, closer to physiological conditions than are the cases of PDMLL and PTMLL.

The difference in the effect on Tm between methylated small amines and PTMLL and PTMLO may lie in differences in conformational ΔS on denaturation. The ΔS of the polypeptide residue on melting the DNA is likely to be greater than for a simple amine. There may also be solvation differences. Also, we have to compare the relative stabilizations of native and denatured DNA for four ligands, i.e., NH_3^+ and, say, Me_4N^+ vs. PLL and PTMLL. The situation is very complex.

The slope of the melting profile of DNA-PTMLL is smaller than for DNA-PLL, indicating a smaller ΔH. A smaller ΔH and a higher Tm result in a smaller ΔS of melting. The ΔS of melting arises from conformational change in DNA, conformational change in polypeptide, and change in solvation.

Tm measurements have also been made for a variety of synthetic double-stranded polynucleotides (ribo-, deoxyribo-, and ribo-deoxyribo-duplexes) complexed with PLL, PDMLL, and PTMLL (Table 2).[8] A complex set of effects is observed, namely, (1) increased Tm with increased methylation; (2) decreased Tm with increased methylation; (3) little effect on Tm. One

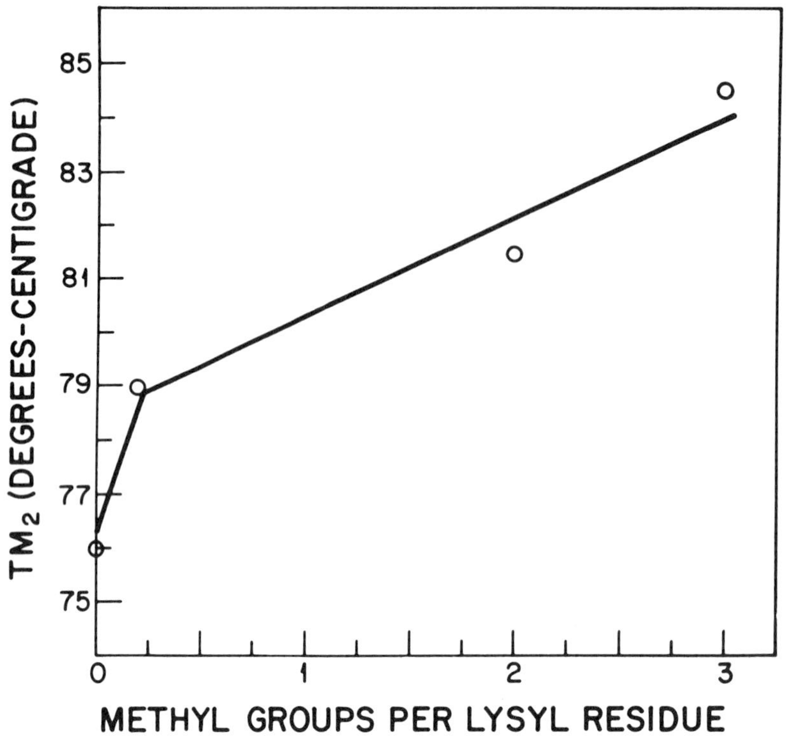

FIGURE 5. Tm of $(dAdT)_n \cdot (dAdT)_n$ complexed with methylated $[Lys]_n$. Peptide and nucleotide concentrations were 2×10^{-5} and 4×10^{-5} *M* in residues, respectively, in 0.01 *M* NaCl, 2.5×10^{-4} EDTA, pH 7.0. (From Bello, J., Granados, E. N., Lewinski, S., Bello, H. R., and Trueheart, T., *J. Biomol. Struct. Dynamics,* 2, 899, 1985. With permission.)

TABLE 2
Melting Temperatures of Complexes

Polynucleotide	PLL	PDMLL	PTMLL
$(dIdC)_n \cdot (dIdC)_n$	68.2	71.8	75.6
$(dI)_n \cdot (dC)_n$	72.3	54.7	55.3
$(dI)_n \cdot (rC)_n$	69.3	54.2	54.6
$(rI)_n \cdot (rC)_n$	82.5	70.5	70.3
$(rI)_n \cdot (dI)_n$	61.8	58.4	60.6
$(rI)_n \cdot (rC_m)_n$	83.8	64.2	64.5
$(dAdT)_n \cdot (dAdT)_n$	79.2	82.7	87.2
$(dAdU)_n \cdot (dAdU)_n$	69.4	74.5	79.1
$(dA)_n \cdot (dT)_n$	>98	>94.6	96.4
$(rA)_n \cdot (dT)_n$	71.1	75.8	77.4
$(dA)_n \cdot (rU)_n$	79.7	78.0	74.5
$(rA)_n \cdot (rU)_n$	81.0	71.5	66.2
$(rA)_n \cdot (rU_m)_n$	81.0	78.2	76.7
$(rA_m)_n \cdot (rU)_n$	76.7	76.0	68.2
$(rA_m)_n \cdot (rU_m)_n$	71.0	84.5	81.8
		Poly(Arg)	Poly(DMA)
$(dI)_n \cdot (dC)_n$	71.0	74.1	

From Bello, J., Granados, E. N., Lewinski, S., Bello, H. R., and Trueheart, T., *J. Biomol. Struct. Dynamics,* 2, 899, 1985. With permission.

TABLE 3
Viscosities of Polypeptide-CT DNA Complexes

Polypeptide	r[a]	NaCl (mol/l)	Viscosity[b]
$(Lys)_n$	0.25	2.0	1.03
$(Lys)_n$	0.25	0.1	0.77
$[Lys(Me_3)]_n$	0.25	2.0	1.06
$[Lys(Me_3)]_n$	0.25	0.1	0.74
$(Lys)_n$	0.50	2.0	1.12
$(Lys)_n$	0.50	0.1	0.30
$[Lys(Me_3)]_n$	0.50	2.0	1.12
$[Lys(Me_3)]_n$	0.50	0.1	0.60
$(Lys)_n$	0.75	2.0	1.03
$(Lys)_n$	0.75	0.1	0.10
$[Lys(Me_3)]_n$	0.75	2.0	1.03
$[Lys(Me_3)]_n$	0.75	0.1	0.31

[a] Ratio of amino acid residues to nucleotide residues.
[b] Viscosities of the complexes relative to the viscosity of DNA at the same ionic strength. All solutions contained 2.5×10^{-4} *M* EDTA.

result which shows some systematic effect is that for alternating purine-pyrimidine polynucleotides Tm increases with increasing methylation. All of the homopolymer duplexes so far studied give lower Tms with PDMLL and PTMLL, with the exception of $(rA)_n{\cdot}(dT)_n$ and $(rAm)_n{\cdot}(rUm)_n$.

One experiment with $(Arg)_n$ and $(DMArg^{70},Orn^{30})_n$ complexed with $(dI)_n{\cdot}(dC)_n$ showed that Tm is higher for the latter than for $(Arg)_n$, opposite to the effect of dimethylation of PLL.[8] $(DMArg^{70},Orn^{30})_n$ was made by reaction of $(Orn)_n$ with dimethylcyanamide to convert ornithine to dimethylarginine residues. Perhaps methylation of arginine residues will have effects different from methylation of lysine; additional data are certainly required to test this idea. Since about one third of the residues were ornithine, $(Arg)_n$ and $(DMArg^{70},Orn^{30})_n$ are not strictly comparable. It is possible that the inverted Tm effect arises from the ornithine residues. However, the effect of $(Orn)_n$ on $(dI)_n{\cdot}(dC)_n$ has not yet been studied.

Both PLL and PTMLL bind preferentially to AT-rich DNA over CG-rich DNA, at low salt concentration. By the use of melting profiles it was found that PLL preferentially binds $(dA)_n{\cdot}(dT)_n$ over $(dA\text{-}dT)_n{\cdot}(dA\text{-}dT)_n$, while PTMLL shows no preference.[37] This raises the possibility that methylation may alter relative binding strengths to local sequences of DNA.

C. VISCOSITY OF DNA-PTMLL COMPLEXES

Since histones have the function of packaging DNA compactly, we have examined the viscosity (an indicator of compactness) of complexes of PTMLL and PLL with CT-DNA.[38] At low ratio ($r = 0.25$) of amino acid residue to nucleotide residue, there is little difference in viscosity (Table 3) at 0.1 *M* NaCl, but at $r = 0.5$ and 0.75 the viscosity of the PTMLL complex is two to three times as large as for PLL, suggestive of a less compact packing. The potential relevance to histone-DNA structure is obvious. More detailed study, using less highly methylated PLL as well as methylated $(Arg)_n$, might be useful.

D. CIRCULAR DICHROISM OF PTMLL COMPLEXES OF POLYNUCLEOTIDES

The CD spectra for complexes of PLL with polynucleotides have been studied extensively; the literature is too voluminous to review here. Our interest is in comparison of the effects of PLL and methylated PLL on polynucleotides.[37] So far almost all of the work has been on PTMLL, and with most polynucleotides PTMLL and PLL produce somewhat different CD spectra. The greatest difference observed was for PTMLL-$(dG\text{-}dC)_n\cdot(dG\text{-}dC)_n$ vs. PLL-$(dG\text{-}dC)_n\cdot(dG\text{-}dC)_n$. The latter gives a spectrum of large negative ellipticity, termed a ψ-spectrum, which arises from differential light scattering by chirally organized aggregates. PTMLL-$(dG\text{-}dC)_n\cdot(dG\text{-}dC)_n$ gives positive ellipticity in 0.1 to 0.3 *M* NaCl, but negative ellipticity at 0.4 *M* NaCl. Work in progress by Iijima in this laboratory shows that the appearance of the positive ellipticity is dependent on the conditions of molecular weight, mode of mixing, etc. This work has given positive ellipticities much smaller than those of Granados and Bello;[37] these new results may arise from a nonaggregated complex of A-form DNA. PDMLL-$(dG\text{-}dC)_n\cdot(dG\text{-}dC)_n$ and PDMLL-$(dA\text{-}dT)_n\cdot(dA\text{-}dT)_n$ give only ψ CD spectra at any NaCl concentration short of dissociation (unpublished, H. K. Iijima, this laboratory). The fact that CD differences are observed between PLL and PTMLL complexes indicates that methylation of the polypeptide can change the conformation of the DNA.

IX. SUMMARY

The main results of the research described here are as follows:

1. Methylation of PLL or PLO results in changes in conformational properties, leading in some environments to increased ordering, in others to decreased ordering.
2. Methylation of copolymers results in enhancement or inhibition of ordered structure depending on the proportion of lysine residues.
3. Methylation of one natural peptide with a relatively small proportion of lysine residues resulted in effects similar to those of the PLL → PTMLL conversion, but, as might be expected, to a lesser extent.
4. Methylation of PLL results in changes in interactions with polynucleotides in the following properties: strength of binding, Tm, compactness, base preference, and geometry.
5. The effects of methylation are not yet predictable.

All of the work described has dealt with what are presumed to be equilibrium states, except for some experiments by Yamamoto and Yang on rates of thermal transitions for PMMLL, PMELO, and PMMLO. Dynamic aspects, including rates of intermolecular interactions and internal dynamics, are almost unexplored.

REFERENCES

1. **Baxter, C. S. and Byvoet, P.,** CMR studies of protein modification. Progressive decrease in charge density at the α-amino function of lysine with increasing methyl substitution, *Biochem. Biophys. Res. Commun.*, 64, 514, 1975.
2. **Byvoet, P. and Baxter, C. S.,** Histone methylation, a functional enigma, *Chromosomal Proteins and Their Role in the Regulation of Gene Expression*, Stein, G. S. and Kleinsmith, L. J., Eds., Academic Press, New York, 1975, 127.
3. **Seely, J. H. and Benoiton, N. L.,** The carboxypeptidase B-catalyzed hydrolysis of poly-ε-*N*-methyl-L-lysine and poly-ε-*N*-ε-*N*-dimethyl-L-lysine, *Biochem. Biophys. Res. Commun.*, 37, 771, 1969.
4. **Yamamoto, H. and Yang, J. T.,** Synthesis and conformational studies of poly(N^{ε}-methyl-L-lysine), poly(N^{δ}-methyl-L-ornithine), poly(N^{ε}-ethyl-L-ornithine), and their carbobenzoxy derivatives, *Biopolymers*, 13, 1093, 1974.
5. **Means, G. E. and Feeney, R. E.,** Reductive alkylation of amino groups in proteins, *Biochemistry*, 7, 2192, 1968.
6. **Paik, W. K. and Kim, S.,** Effect of methylation on susceptibility of protein to proteolytic enzymes, *Biochemistry*, 11, 2589, 1972.

7. **Jentoft, N. and Dearborn, D. G.,** Labeling of proteins by reductive methylation using sodium cyanoborohydride, *J. Biol. Chem.*, 254, 4359, 1979.
8. **Bello, J., Granados, E. N., Lewinski, S., Bello, H. R., and Trueheart, T.,** Methylated poly(L-lysine): conformational effects and interactions with polynucleotides, *J. Biomol. Struct. Dynamics*, 2, 899, 1985.
9. **Granados, E. N. and Bello, J.,** Alkylated poly(amino acids). I. Conformational properties of poly(N^{ε}-trimethyl-L-lysine) and poly(N^{ε}-trimethyl-L-ornithine, *Biopolymers*, 18, 1479, 1979.
10. **Seely, J. H. and Benoiton, N. L.,** Effect of N-methylation and chain length on kinetic constants of trypsin substrates. ε-*N*-methyllysine and homolysine derivatives as substrates, *Can. J. Biochem.*, 48, 1122, 1970.
11. **Benoiton, L. and Deneault, J.,** The hydrolysis of two ε-*N*-methyl-L-lysine derivatives by trypsin, *Biochim. Biophys. Acta*, 113, 613, 1966.
12. **Gorecki, M. and Shalitin, Y.,** Non cationic substrates of trypsin, *Biochem. Biophys. Res. Commun.*, 29, 189, 1967.
13. **Yamamoto, H. and Yang, J. T.,** The thermal induced helix-β transition of poly(N^{ε}-methyl-L-lysine) and poly(N^{δ}-ethyl-L-ornithine) in aqueous solution, *Biopolymers*, 13, 1109, 1974.
14. **Tseng, Y.-W. and Yang, J. T.,** Conformation of poly(L-lysine) homologs in solution, *Biopolymers*, 16, 921, 1977.
15. **Rippon, W. B. and Hiltner, W. A.,** The 225—240 nm circular dichroism band in disordered and charged polypeptides, *Macromolecules*, 6, 282, 1973.
16. **Meyer, Y. P.,** The pH-induced helix-coil transition of poly-L-lysine and poly-L-glutamic acid and the 238 μm dichroic band, *Macromolecules*, 2, 624, 1969.
17. **Tiffany, M. L. and Krimm, S.,** Circular dichroism of the "random" polypeptide chain, *Biopolymers*, 8, 347, 1969.
18. **Hiltner, W. A., Hopfinger, A. J., and Walton, A. G.,** Helix-coil controversy for polyamino acids, *J. Am. Chem. Soc.*, 94, 4324, 1972.
19. **Painter, P. C. and Koenig, J. L.,** The solution conformation of poly(L-lysine) A raman and infrared spectroscopic study, *Biopolymers*, 15, 229, 1976.
20. **Epand, R. M., Wheeler, G. E., and Moscarello, M. A.,** Circular dichroism and protein magnetic resonance studies of random chain poly-L-lysine, *Biopolymers*, 13, 359, 1974.
21. **Balasubramanian, D.,** Critque of the interpretation of the circular dichroism of unordered polypeptides and proteins, *Biopolymers*, 13, 407, 1974.
22. **Dearborn, D. G. and Wetlaufer, D. B.,** Circular dichroism of putative unordered polypeptides and proteins, *Biochem. Biophys. Res. Commun.*, 39, 314, 1970.
23. **Grourke, M. J. and Gibbs, J. H.,** Transition from random coil to α-helix induced by sodium dodecyl sulfate, *Biopolymers*, 5, 586, 1967.
24. **Peggion, E., Cosani, A., Terbojevich, M., and Borin, G.,** Conformational studies on polypeptides. The effect of sodium perchlorate on the conformation of poly-L-lysine and of random copolymers of L-lysine and L-phenylalanine, *Biopolymers*, 11, 633, 1972.
25. **Cassim, J. Y. and Yang, J. T.,** Critical comparison of the experimental optical activity of helical polypeptides and the predictions of the molecular exciton model, *Biopolymers*, 9, 1475, 1970.
26. **Bello, J.,** Formation and stability of helical poly(N^{ε},N^{ε},N^{ε}-trimethyl-L-lysine) in sodium perchlorate solution, *Biopolymers*, in press.
27. **Peterson, S.,** Anion exchange processes, *Ann. N.Y. Acad. Sci.*, 57, 144, 1954.
28. **Gill, T. J., Lodoulis, C. T., Kunz, H. W., and King, M. F.,** Studies of intramolecular transitions of polypeptides by fluorescence techniques, *Biochemistry*, 11, 2644, 1972.
29. **Tiffany, M. L. and Krimm, S.,** Extended conformations of polypeptides and proteins in urea and guanidine hydrochloride, *Biopolymers*, 12, 575, 1973.
30. **Tanford, C.,** Protein denaturation, *Adv. Protein Chem.*, 24, 1, 1970.
31. **Goodman, M., Toniolo, C., and Falcetta, J.,** Conformational aspects of polypeptide structure. XXX. Rotary properties of cyclic and bicyclic amides. Restricted and rigid model compounds for peptide chromophores, *J. Am. Chem. Soc.*, 91, 1816, 1969.
32. **Bello, J., Haas, D., and Bello, H. R.,** Interactions of protein-denaturing salts with model amides, *Biochemistry*, 8, 2539, 1969.
33. **Kubota, S., Ikeda, K., and Yang, J. T.,** Conformation of sequential and random copolypeptides of lysine and alanine in sodium dodecyl sulfate solution, *Biopolymers*, 22, 2219, 1983.
34. **Bello, J., Bello, H. R., and Granados, E. N.,** Conformation and aggregation of melittin: dependence on pH and concentration, *Biochemistry*, 21, 461, 1982.
35. **Tatham, A. S., Hider, R. C., and Dicke, A. F.,** The effects of counterions on melittin aggregation, *Biochem. J.*, 21, 683, 1983.
36. **Leng, M. and Felsenfeld, G.,** The preferential interactions of polylysine and polyarginine with specific base sequences in DNA, *Proc. Natl. Acad. Sci. U.S.A.*, 56, 1325, 1966.

37. **Granados, E. N. and Bello, J.,** Interactions of poly(N^{ε},N^{ε},N^{ε}-trimethyllysine and poly(lysine) with polynucleotides: circular dichroism and A-T sequence selectivity, *Biochemistry,* 20, 476, 1981.
38. **Granados, E. N. and Bello, J.,** Interactions of poly(Nε,Nε,Nε-trimethyllysine and poly(N^{δ},N^{δ},N^{δ}-trimethylornithine) with polynucleotides: salt dissociation and thermal denuration, *Biochemistry,* 19, 3227, 1980.
39. **Giancotti, V., Cesaro, A., and Crescenzi, V.,** Heat of interaction of DNA with polylysine and trilysine, *Biopolymers,* 14, 675, 1975.
40. **Ross, P. D. and Schapiro, J. T.,** Heat of interaction of DNA with polylysine, spermine and $M6^{++}$, *Biopolymers,* 13, 415, 1974.
41. **Bradbury, E. M., Chapman, G. E., Danby, S. E., Hartman, P. G., and Riches, P. L.,** Studies on the role and mode of operation of the very lysine-rich histone Hl(Fl) in eukaryote chromatin. *Eur. J. Biochem.,* 57, 521, 1975.
42. **Shapiro, J. T., Leng, M., and Felsenfeld, G.,** Deoxyribonucleic acid-polylysine complexes. Structure and nucleotide specificity, *Biochemistry,* 8, 3219, 1969.
43. **Shapiro, J. T., Stannard, B. S., and Felsenfeld, G.,** The binding of small cations to deoxyribonucleic acid. Nucleotide specificity, *Biochemistry,* 8, 3233, 1969.
44. **Gabbay, E.,** Topography of nucleic acid helices in solution. I. The nonidentity of polyadenylic-polyuridylic and polyinosine-polycytidylic acid helices, *Biochemistry,* 5, 3036, 1966.
45. **Orosz, J. M. and Wetmur, J. G. G.,** DNA melting temperature and renaturation rates in concentrated alkylammonium salt solutions, *Biopolymers,*16, 1183, 1977.

Chapter 22

NONENZYMATIC PROTEIN METHYLATION AND ITS BIOLOGICAL SIGNIFICANCE

Lajos Trézl, Ernö Tyihák, and Prabhakar D. Lotlikar

TABLE OF CONTENTS

I. INTRODUCTION

Synthesis of monomethylated amino acids was first reported by Volhard in 1862 (cited in Reference 1). This synthesis was achieved by condensing α-halogenated fatty acids with aqueous solution of methylamine at 120 to 130°C. Later, Fischer and co-workers devised many new methods to produce monomethylated amino acids (cited in Reference 1). Novak[2] used dimethylsulfate to hypermethylate glycine to produce betaine.

In spite of the fact that the methylated amino acids were recognized very early in biological systems (Liebig isolated the methylated glycine that is the sarcosine in 1847), the possibility of the chemical methylation to occur emerged only in the last few decades.

Nonenzymatic chemical methylation in biological systems is prevalent predominantly with the methylating capacity of chemical carcinogenic compounds which directly can methylate amino acids, peptides, proteins, DNA, RNA, etc. However, it has been reported that *S*-adenosyl-L-methionine (AdoMet), a biological methyl donor, may also serve as a direct methylating agent for proteins[3] and DNA[4,5] via a nonenzymatic process. Nevertheless, AdoMet is recognized as a key active methyl donor for the enzymatic methylation in biological systems; this fundamental discovery originated from Cantoni in 1951, who enzymatically methylated nicotinamide with AdoMet (cited in References 6 and 7). This discovery opened a new era in biochemistry.

The other possibilities for the methylation were restricted to carcinogenic nitroso compounds such as alkylnitrosamines. The active direct methylating species, the methyl cation (CH_3^+) generated via oxidative dealkylation can methylate proteins, DNA and RNA. The original work of Magee et al.[8-10] supported this concept. Magee et al. had already demonstrated in 1962 that dimethylnitrosamine (DMN) could methylate proteins and DNA, but no one had supposed and proven that formaldehyde liberated from DMN, simultaneously with methyl cation (CH_3^+) could also directly methylate amino acids, proteins, DNA etc.

Magee and Farber[10] had suggested however that the liberated formaldehyde from DMN may participate in enzymatic methylating processes involving proteins and DNA methylation only after its incorporation into the C_1-pool. Turberville and Craddock[11] provided some evidence that the liberated formaldehyde from DMN may be involved in the direct methylating processes, especially in the methylation of lysine residues in proteins, but they were unable to provide any mechanism about the chemical processes of methylation.

In 1973, Reis and Tunks[12] treated casein with formaldehyde and identified the ε-*N*-methylated-L-lysine in the hydrolysate of casein; this compound however was absent in the hydrolysate of untreated casein. In spite of their important results, they did not recognize the direct methylating capacity of formaldehyde.

Johansson and Tjälve[13] demonstrated the important role of formaldehyde liberated from [^{14}C]DMN in toxic and carcinogenic processes in mice with autoradiographic investigations, but they could not provide any information about the chemical nature of the liberated formaldehyde.

Hungarian scientists recognized precisely that formaldehyde is itself a direct, spontaneous methylating and simultaneous formylating agent which can methylate nonenzymatically amino acids, proteins, nucleic acid bases, etc. and elucidated the correct mechanism of the reaction. Tyihák et al.[14] demonstrated in 1975 that L-lysine, but not L-arginine, could be directly methylated by formaldehyde.

It is important to note that there are a lot of exogenous (e.g., DMN[9,10]) and endogenous (e.g., L-methionine[15-17]) formaldehyde precursors from which formaldehyde could be liberated. This, Mackenzie et al.[15] observed in 1947 that L-methionine may also serve as a formaldehyde precursor in *in vivo* systems. Kim et al.[18] proved that alkyllysinase (EC 1.5.3.4) demethylates ε-*N*-methyl-L-lysine liberating formaldehyde. Hungarian research workers[19-22] found that the demethylase enzyme activity increased in TMV infected hypersensitive tobaccos and increased the formaldehyde formation from L-[$^{14}CH_3$]methionine. Tyihák et al.[23] proved that the formaldehyde level increased in plants after heat shock.

The above important data indicate that formaldehyde could be liberated from a large number of precursors and that this liberated formaldehyde may be involved in nonenzymatic direct protein methylation processes similar to the reactive direct methylating species. Thus, it is extremely important to elucidate their role in nonenzymatic protein methylation.

II. DIRECT NONENZYMATICALLY ACTING METHYLATING AGENTS WHICH HAVE BIOLOGICAL SIGNIFICANCE

In organic synthesis there are many chemicals which are used as direct methylating agents for *N*-, *S*-, *O*-methylation in SN_1, SN_2 or nucleophilic addition mechanisms. Some of these chemicals have been widely used either in laboratories or in industries to produce numerous different compounds, drugs, pesticides, detergents, etc.; for example:

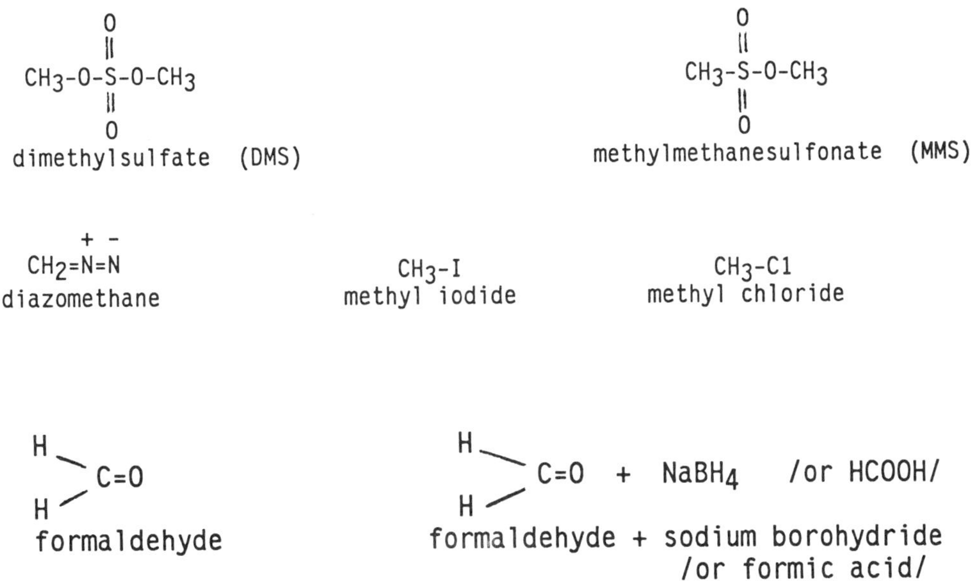

Since these compounds[24-27] can directly methylate amino acids, proteins, DNA, RNA, etc., they may be potential health hazards for people who are exposed to these compounds in laboratories and factories. These compounds may also be dangerous for the environment since they are potentially carcinogenic.

Many methylating agents are precarcinogenic and need to be activated into ultimate carcinogens. The mechanisms whereby the simple direct methylating species are generated from secondary carcinogens have been clearly defined: they involve either spontaneous or enzyme-catalyzed processes.

Methyl phenyltriazole[25] (MPT) is a typical example of the former (nonenzymatically catalyzed processes) and is transformed to give methyl cation, a direct methylating species, as a result of either acidic or basic hydrolysis, for example:

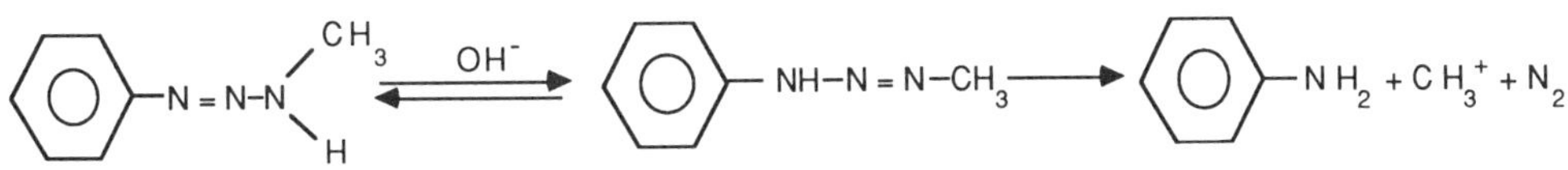

The nitrosoureas, nitrosoguanidines, and nitrosourethanes also generate the same methylating species (CH_3^+) as a result of hydrolysis. As a consequence, a more rapid breakdown pathway catalyzed by thiol groups may well be responsible for the primary mechanism of methylation by MNNG in cells.[25]

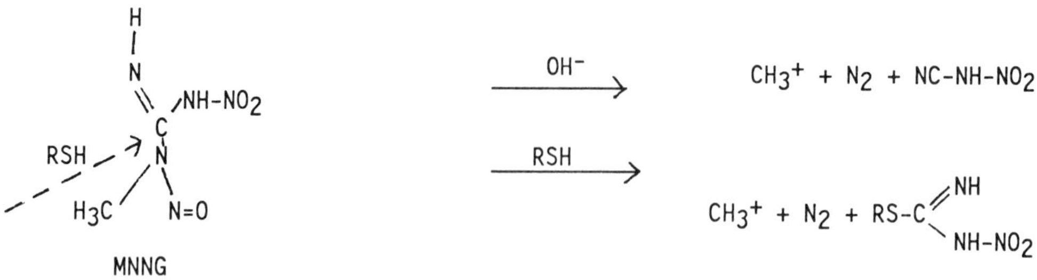

N-nitroso-*N*-methylurea (MNU) is a typical representative of this class of compounds: it can undergo disproportionate decomposition very rapidly and spontaneously at physiological conditions, ph 7.4 and at 37°C, in 0.1 *M* phosphate buffer, generating CH_3^+.[28]

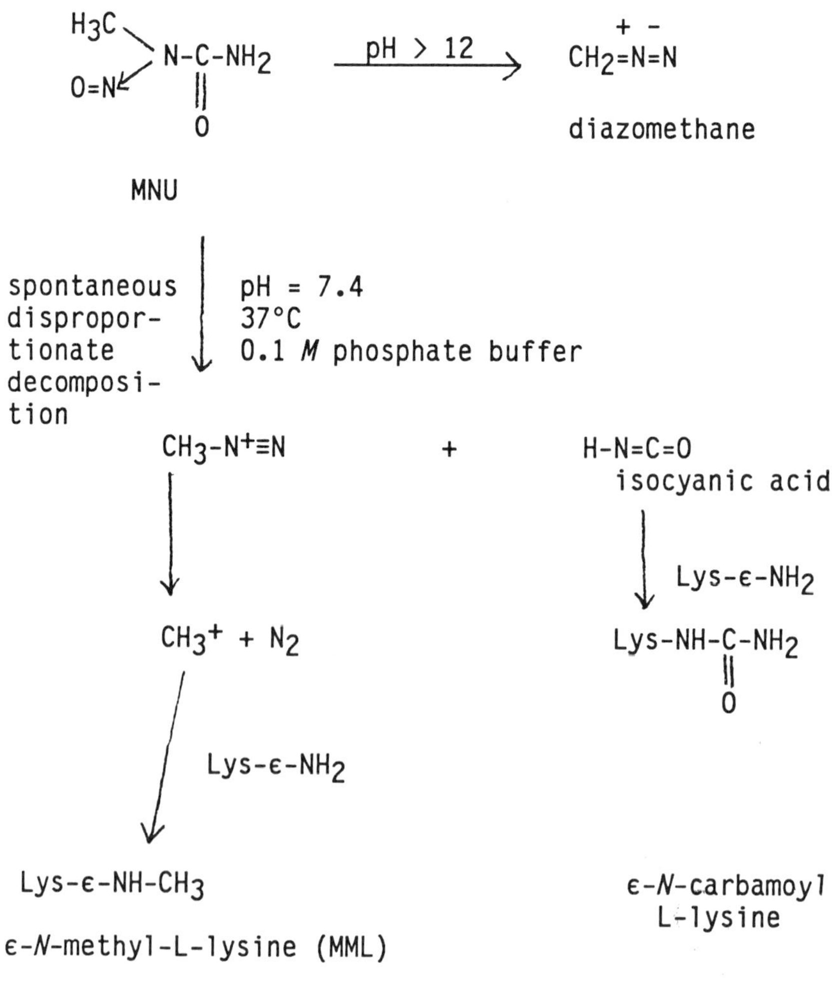

MNU can directly attack and methylate the ε-amino group of L-lysine, and the guanidine group of L-arginine residues in cytochrome *c* protein[29] and can also methylate rapidly *in vitro* the –SH groups of naturally occurring thiol compounds such as L-cysteine, L-homocysteine, and glutathione at physiological conditions without any enzymatic activation.[30] The rapidity of the methylation of –SH groups with MNU is about ten times higher than the methylation of the ε-NH_2 group of L-lysine.[30]

$$R\text{-}SH + CH_3^+ \xrightarrow{K_1} R\text{-}S\text{-}CH_3$$

$$Lys\text{-}\epsilon\text{-}NH_2 + CH_3^+ \xrightarrow{K_2} Lys\text{-}NH\text{-}CH_3$$

$$K_1/K_2 \sim 10$$

Some of the chemicals of this class have been widely used in organic chemical laboratories as methylating compounds, notably the diazomethane precursors,[25] MNU, and *N*-nitroso-*N*-methyl-*p*-toluenesulfonamide:

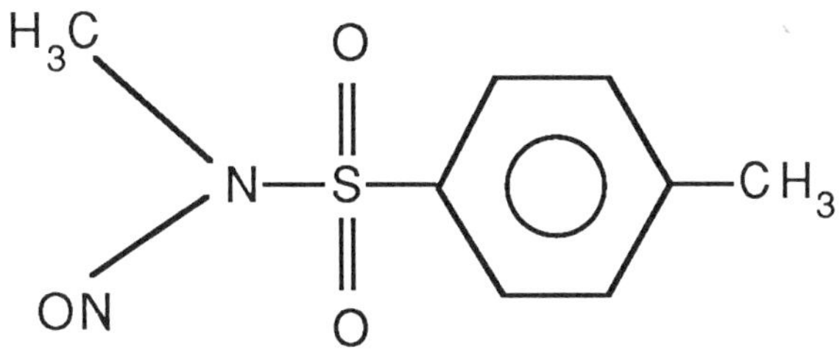

The latter has been shown to cause lung tumors in mice while the former can lead to the production of skin tumors as a result of topical application to mice. Some of these compounds are widely used in medical and biological research laboratories as model compounds to initiate tumors in animals.

The powerful methylating capacity of these compounds indicates that these agents which are strong health hazards can easily attack the macromolecular cellular components, especially proteins, DNA, and RNA.

A. FORMALDEHYDE IS A POTENTIALLY DIRECT METHYLATING AGENT IN BIOLOGICAL SYSTEMS

It is necessary to deal with the role of formaldehyde in more detail because up until now formaldehyde had not been accepted as a direct methylating compound in biological systems, but only accepted as a reductive methylating reagent together with $NaBH_4$ or concentrated HCOOH which had been widely used in organic chemistry.

On the other hand, there is a strong relationship between direct methylating species (CH_3^+) and formaldehyde because CH_3^+ can react instantaneously with tissue water forming methanol which will be oxidized to formaldehyde *in vivo* conditions.[24]

$$CH_3^+ + HOH \xrightarrow{\text{fast}} CH_3OH \xrightarrow{\textit{in vivo}} CH_2O$$

Furthermore, formaldehyde could be liberated, not only from alkylnitrosamines, but also from many exogenous and endogenous precursors in biological systems. This formaldehyde which is liberated through different pathways can directly methylate amino acids, peptides, proteins, DNA, RNA, etc. It is important to review the historical background of the formaldehyde reactions which occurred during the last 100 years so that the significance of the direct spontaneous methylation reaction of formaldehyde could be appreciated.

Formaldehyde, discovered by Hofmann (cited in Reference 31) in 1868 and first manufactured in 1889, has become a chemical of outstanding importance and an indispensable ingredient in the production of industrial and consumer products. Formaldehyde

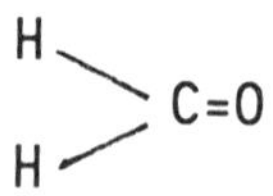

is the simplest and very reactive aliphatic aldehyde molecule which is widely used as a methylating agent in organic chemistry for almost 100 years.

The historical background of the most important methylation reactions of formaldehyde will provide evidence that these reactions take place under extreme conditions or with specific catalysts which might not be adaptable for biological systems.

Some of the manipulatively simple methylation procedures require heating together with a primary or secondary amine, formaldehyde, and formic acid at a temperature during which carbon dioxide is generated. The overall reaction appears to be

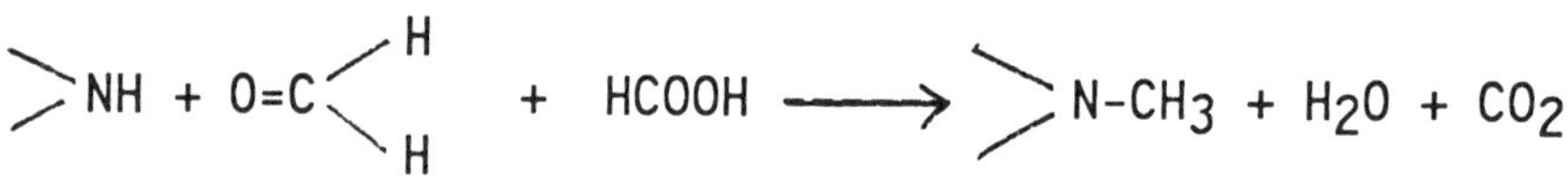

The method is credited to Leuckart,[32] who in 1885, alkylated ammonia by heating carbonyl compounds with ammonium formate. Wallach[33] subsequently showed that in the presence of formic acid, alkylation of an amine by carbonyl compound proceeds at lower temperatures. Plöchl[34] trimethylated ammonium sulfate by heating it with formaldehyde in 1888, and Eschweiler[35] in 1905 methylated ammonia (NH_3) and aliphatic amines with formaldehyde at 120 to 160°C under high pressure.

Hess[36] synthesized hygrin from hydramine base with formaldehyde and HCl in 1913.

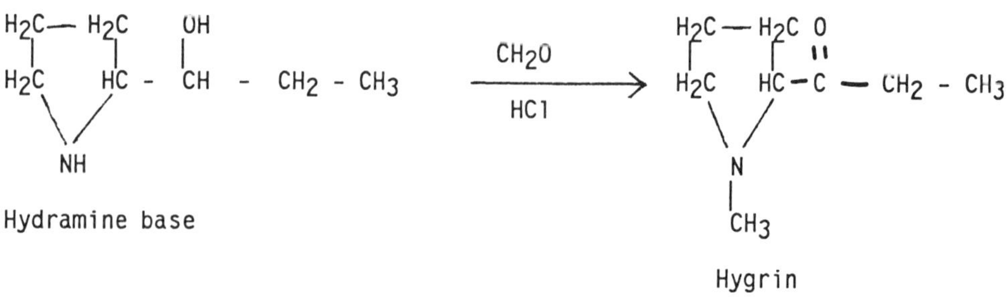

It is very interesting that the reaction does occur quantitatively. Clarke et al.[37] methylated monobasic amino acids with formaldehyde and concentrated formic acid at 100°C, refluxing for

many hours.

Staple and Wagner[38] reinvestigated the mechanism of Wallach alkylation reaction and assumed that the Schiff base is an intermediate structure in the reaction. Lindeke et al.[39] reevaluated the Eschweiler-Clarke reactions with the specific method of deutero methylation with CD_2O and D-COOH in 1976. They verified that the formic acid is itself the hydride anion H^- or D^-donor in the reductive methylation reaction.

In 1968, Means and Feneey[40] reductively methylated the free amino groups in proteins such as lysozyme, insulin, ribonuclease, and α-chymotrypsin, with formaldehyde and sodium borohydride ($NaBH_4$). They showed that 77% of the ε-amino group of L-lysine residues in bovine zinc insulin had been methylated while its two α-amino groups, those of glycine and phenylalanine, were only 63 and 20% methylated, respectively. They assumed the mechanism of reductive methylation reaction to form Schiff base as an intermediate in the reactions.

The historical perusal of the most important methylation reactions proved very clearly that the previously mentioned authors applied extreme conditions in the direct chemical methylation reactions such as high temperature, high pressure, concentrated formic acid, HCl and $NaBH_4$, which might not be adaptable for biological systems; no one in the previous 100 years had suggested that formaldehyde itself can spontaneously methylate NH_3, amines, amino acids, or proteins under physiological conditions.

The formation of labile *N*-hydroxymethyl groups, ($N\text{-}CH_2OH$), was only supposed to occur between amino acid residues and formaldehyde, but they were not isolated by anyone although the formation of cross-links and the endomethylene bridge in the case of L-cysteine and L-histidine were shown under physiological conditions.

The basic reaction, so-called "formol titration", between amino acids and formaldehyde was first published by Sörensen[41] in 1908, and this concept has remained in many textbooks even today:

$$^{-}OOC - R - \overset{+}{N}H_3 + OH^- \rightleftharpoons {}^{-}OOC - R - NH_2 + H_2O$$

$$^{-}OOC - R - NH_2 + CH_2O \rightleftharpoons {}^{-}OOC - R - NH\text{-}CH_2OH$$

$$^{-}OOC - R - \overset{+}{N}H_3 + OH^- + CH_2O \rightleftharpoons {}^{-}OOC - R - NH - CH_2OH + H_2O$$

The above mentioned statements were well demonstrated by Kitamoto and Maeda[42] in 1980. The authors reevaluated the reactions of formaldehyde with amino acids at low concentrations using modern analytical methods (MS, NMR, IR, etc.). They verified the formation of labile *N*-hydroxymethyl groups, but the stability of these groups except for sensitivity to acid hydrolysis was different in lysine as compared to arginine. On the basis of their studies, they concluded that neither tyrosine nor lysine yielded stable products to an appreciable extent even at a tenfold molar excess of formaldehyde. Though the ε-NH_2 group of lysine might interact with formaldehyde, the product was too labile for amino acid analysis as was the case for reaction at α–NH_2 position. They found no evidence of the so-called Schiff-base ($-N{=}CH_2$) formation.

B. THE MECHANISM OF THE SPONTANEOUS METHYLATION AND FORMYLATION REACTIONS BY FORMALDEHYDE, ANTAGONISTIC BEHAVIOR BETWEEN L-LYSINE AND L-ARGININE AGAINST FORMALDEHYDE AND THEIR BIOLOGICAL SIGNIFICANCE

In spite of the fact that the research on the reaction of formaldehyde with amino acids, proteins, aliphatic amines and nucleic acid bases was carried out in Hungary for more than a decade, we (Trézl and Tyihák) have obtained many new significant results.[14,43-68] However, until now, these studies were not recognized by the scientific world, except by a few French scientists[69,70] who verified our results in 1985.

Hemminki[71] described the formation of *N*-2-methylguanosine in the reaction between formaldehyde and guanosine, but the author did not recognize the significance of this reaction and could not explain the mechanism of the reaction. It is therefore necessary to deal in detail with the reactions and the mechanism of the methylation reactions of formaldehyde and to elucidate its important role in nonenzymatic protein methylation. It will be shown below that many *in vivo* experiments such as the results of Turberville and Craddock[11] cannot be explained without the knowledge of our results.

During the last 10 years, a large research team from several research institutes has been organized in Hungary (the Hungarian research team). This Hungarian research team has obtained many important basically new results and has recognized very clearly that formaldehyde is not only a spontaneous direct methylating agent but also a simultaneously formylating agent for amino acids, aliphatic amines, and bases in nucleic acids etc., especially for L-lysine at physiological conditions (37°C, pH 7.4; in 0.1 *M* phosphate buffer).

It was shown that the two basic amino acids (L-lysine and L-arginine) behave antagonistically in the reactions with formaldehyde. L-lysine could readily be methylated and formylated at room temperature whereas L-arginine could only be hydroxymethylated with formaldehyde on the guanidino group. The investigation of spontaneous reaction in model systems is very important because these reactions can occur in biological systems as well.

The methylated derivatives of L-lysine and hydroxymethyl derivatives of L-arginine are biologically active compounds which occur in biological systems and as such, exert an influence on cell proliferation. The methylated derivatives of L-lysine,[63] especially ε-*N*-trimethyl-L-lysine (TML) increase[64-66] cell proliferation whereas the N^G-hydroxymethyl derivatives of L-arginine[55-59,61] have an inhibitory effect. A review report and summary of the results of the Hungarian research team will now be presented.

During the potentiometric titration of L-lysine with formaldehyde,[49,52] the pH of the aqueous system decreased in accordance with what could be expected on the basis of the Sörensen formol titration[41] at room temperature. After the neutralization there was no further change in the pH of the aqueous solution with further increase in formaldehyde concentration. If, however, the reaction mixture was examined by OPLC method[52] (over pressure layer chromatography developed by E. Tyihák), ninhydrin positive spots at the same positions as the spots of reference samples of ε-*N*-formyl-L-lysine (F-lys), L-lysine (Lys), ε-*N*-methyl-L-lysine (MML), ε-*N*-dimethyl-L-lysine (DML) and ε-*N*-trimethyl-L-lysine (TML) could be observed after the first minute. The amounts of these derivatives increased with time, but approximately only one third of the Lys was modified after a few days. The pH had an important influence on the composition of the reaction mixture. The amount of F-Lys-like substance was small in the reaction mixture at pH 7; in addition, other unknown substances were present on the chromatogram at the spot of the F-Lys-like substances. If the pH of the reaction mixture was adjusted to the basic range, for example, pH 9, only one substance was present on the chromatogram at the position of F-Lys. Figure 1 illustrates a good separation obtained with the OPLC method of F-Lys from the ε-*N*-methylated compounds (MML, DML, TML) which were formed in different reaction mixtures of L-lysine with formaldehyde.

^{1}H-NMR[50] and [^{13}C]NMR[52] studies yielded confirmatory evidence that the structure of the F-Lys-like substance isolated from the reaction mixture was indeed identical with that of the authentic sample of F-Lys. Mass spectral studies confirmed the formation of the ε-*N*-methyl group.[47] MML could be characterized by an intense fragment ion at m/e 44 ($CH_2{=}^+NH\text{-}CH_3$). MML was isolated from the reaction mixture giving both similar mass number and similar melting point range to the authentic MML (SIGMA sample), M.P. 248 to 249°C.

Figure 2 shows the ^{1}H-NMR spectra of α-*N*-acetyl-L-lysine-methylamide and its reaction with formaldehyde respectively; α-*N*-acetyl-L-lysine-methylamide may serve as a model compound for L-lysine[50] in peptide linkage. This figure also indicates the sharp singlet lines

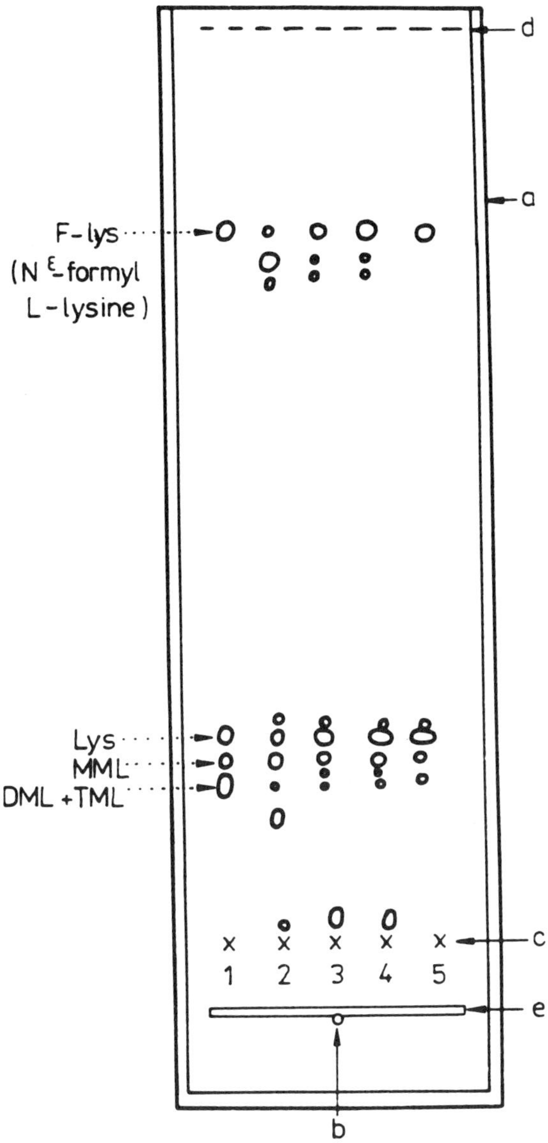

FIGURE 1. Separation of reaction mixture between L-lysine and formaldehyde by OPLC (overpressured layer chromatography).[53] a. Precoated plates of Ionpres™6, with impregnated edges. b. Solvent inlet point, 0.9% (w/v) sodium chloride solution. c. Start point. d. Front line. e. Solvent directing channel.

1. Standard compounds.
2. Reaction mixture of L-lysine with formaldehyde (1 g L-lysine free base was dissolved in 50 ml of water + 5 ml 36% w/v formaldehyde solution, pH was adjusted to 7 with NaOH solution) 15 min, 20°C.
3. Same as 2 except incubated for 1 h at 20°C, pH = 7
4. Same as 2 except incubated for 12 h at 20°C, pH = 7
5. Same as 4 except pH was adjusted to 9 with NaOH solution.

characteristic of ε-*N*-methyl groups (2.8 ppm) and ε-*N*-formyl groups (8.1 ppm), which were formed simultaneously in the reaction mixture.

The [^{13}C]NMR spectrum of the isolated F-Lys from the reaction mixture is shown in Figure

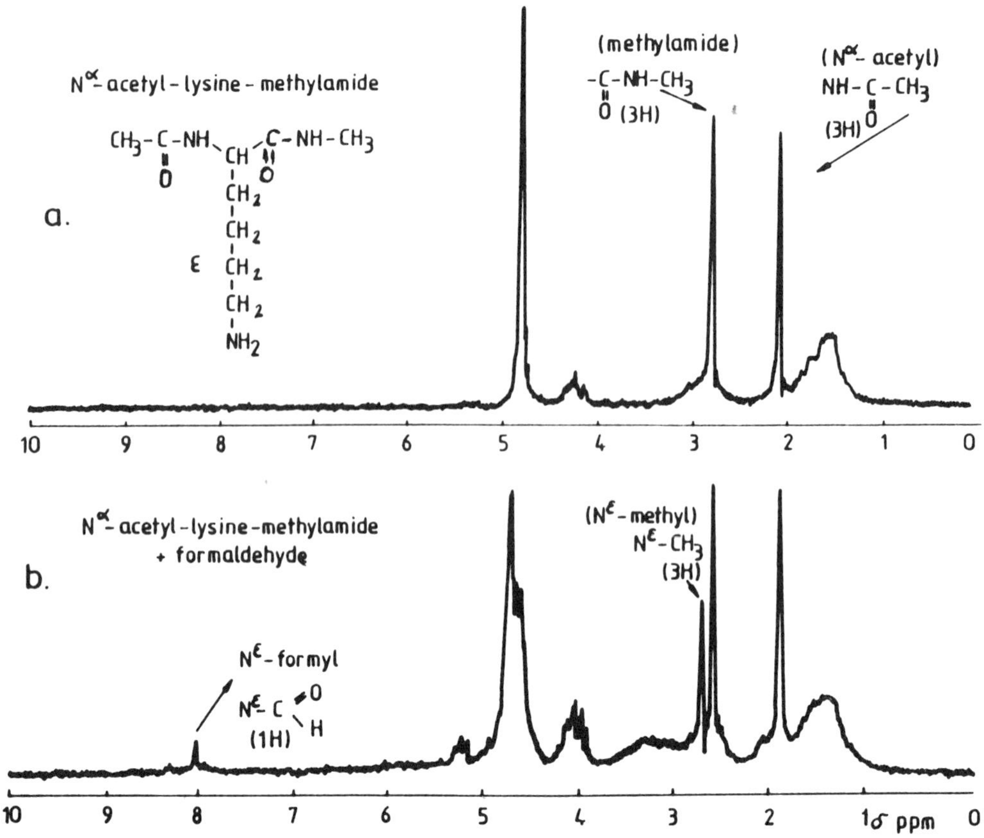

FIGURE 2. ^{1}H-NMR spectra of (a) *N*-α-acetyl-L-lysine-methylamide in D_2O; (b) reaction mixture of *N*-α-acetyl-L-lysine-methylamide with formaldehyde in D_2O at 37°C, after 12 h, 0.15 g *N*-α-acetyl-L-lysine-methylamide was dissolved in 1 ml formaldehyde solution, 60 mg paraformaldehyde was dissolved in boiling D_2O to 1 ml.

3. The chemical shifts in ppm were identical with the shifts of the authentic synthesized sample.[52] According to the results, the formaldehyde is capable not only of methylating, but also of formylating the ε-NH_2 group of L-lysine. It is very interesting that the yield of F-Lys is limited to only 4 to 6% in this reaction which is much less than the stoichiometric amount.

What is the explanation for the simultaneous formation of ε-*N*-methyl and ε-*N*-formyl groups? What is the mechanism of the reaction?

According to the authors' hypothesis based on previous data, the mechanism of the reaction could be explained in the following way[52,53,61] (see Scheme 1).

First, in the reaction, an additional compound I could be formed between L-lysine and formaldehyde. This very labile, inseparable compound, ε-*N*-hydroxymethyl-L-lysine, may be stabilized by resonance to an iminium cation II (no Schiff base) or the reaction may be reversed to form the initial compounds. The next step is the formation of the ε-*N*-methyl group III (ε-*N*-monomethyl-L-lysine, MML) by the reduction of the iminium cation group with a further formaldehyde molecule, since formaldehyde is itself a hydride anion donor H^- (e.g., instead of $NaBH_4$ or HCOOH) in the reduction mechanism. Simultaneously with this reduction step, formyl cation group IV could also be formed from formaldehyde. This formyl cation is very reactive and can react very fast with water or the ε-NH_2 group of another L-lysine molecule. Since

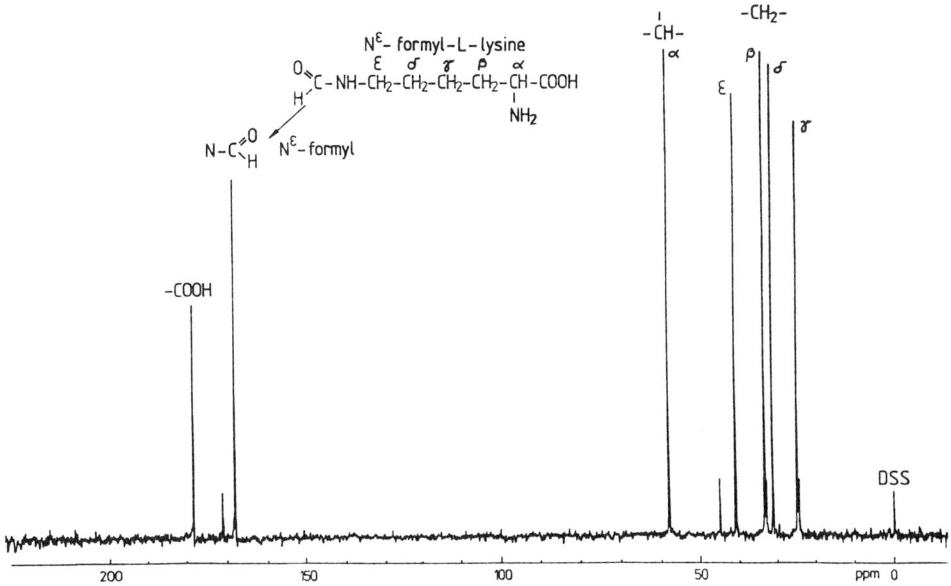

FIGURE 3. ^{13}C-NMR spectrum of N^{ε}-formyl-L-lysine isolated from the reaction mixture of L-lysine and formaldehyde (1 g L-lysine free base was dissolved in 50 ml of water + 5 ml 36% w/v formaldehyde, 12 h, 20°C, pH = 9).

water is always in excess in the reaction mixture, the amount of formic acid V is always higher than that of F-Lys VI. This explains why the small amount of F-Lys is formed in the reaction between formaldehyde and L-lysine.

The amount of F-Lys depends on the pH of the reaction mixture. A basic pH (e.g., pH 9 in borate buffer) is more favorable for the formation of F-Lys than is an acidic or a neutral pH. This shows that in a basic medium there are many more free ε-NH_2 groups which can react with the formyl cation group. This is in good agreement with the described mechanism.

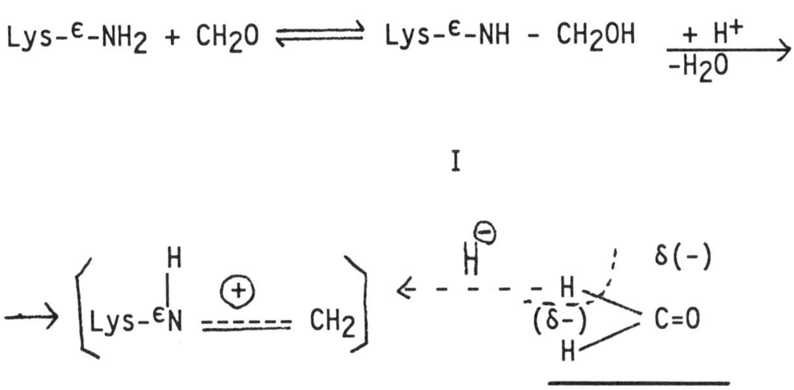

Scheme 1

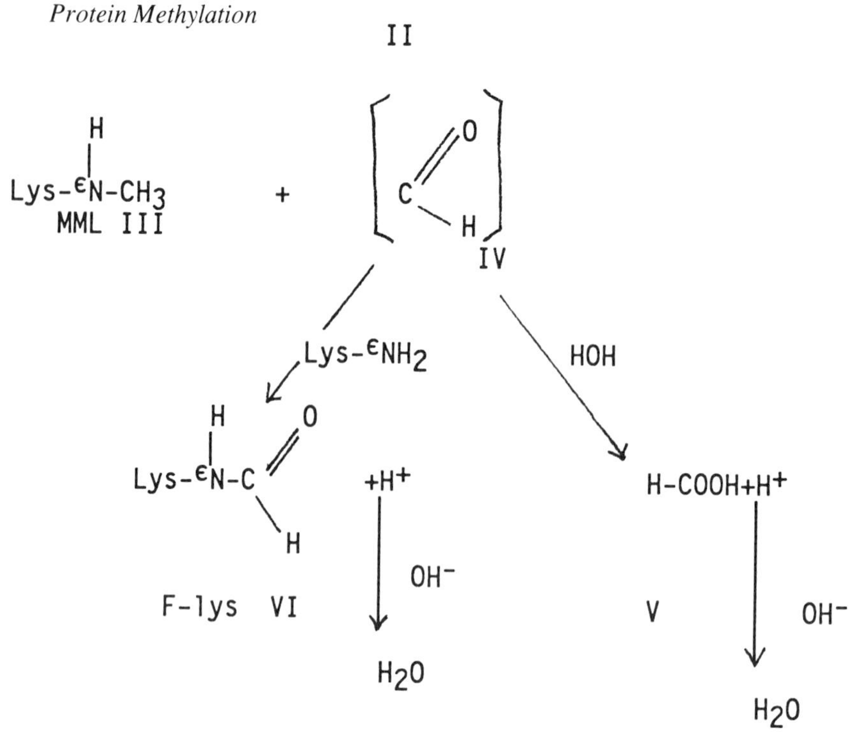

This reaction mechanism differs basically from that of the classical Eschweiler-Clarke reaction.[37] In that reaction, the aliphatic primary amines or amino acids are methylated by formaldehyde in the presence of an excess of formic acid (1 mol of amine or amino acid, 2.2 mol of formaldehyde and 5 mol of 90% formic acid m/v), the reaction mixture being heated under reflux at about 100°C for 2 to 4 h.

The reaction is as follows:

$$R\text{-}NH_2 + 2\ CH_2O + 2\ HCOOH \longrightarrow R\text{-}N\text{-}(CH_3)_2 + 2\ H_2O + 2\ CO_2$$

In the Eschweiler-Clarke reaction, the hydride anion H^{θ} donor is formic acid. Thus, the liberation of CO_2 is stoichiometric and F-Lys cannot be formed. By contrast, in the direct spontaneous methylation reaction, formaldehyde is itself the hydride anion H^- donor, and formyl cation could be formed and the formaldehyde reaction can take place spontaneously at room temperature without any formic acid.

Formic acid could be replaced by $NaBH_4$ in the reductive methylation processes of proteins by formaldehyde. In that case, $NaBH_4$ could serve as a hydride anion donor and ε-*N*-formyl-L-lysine could not be formed.

$$NaBH_4 \xrightarrow{HOH} 4\ H^{\theta}$$

This fact can also confirm the previously described mechanism of the spontaneous methylation reaction by formaldehyde. Furthermore, the mechanism of the reaction is supported by the fact that NADH plus H^+ coenzyme[54] could also serve as a hydride anion donor. In that case, F-Lys could also not be formed in the model reaction.

The hydride anion formation from formaldehyde itself at pH 7 is a key step in our postulated reaction mechanism. This is a new concept in organic chemistry as well. It is well known that, in the case of Cannizaro reaction, hydride anion could be formed in a strong alkaline medium, but not at pH 7.

Recently, Narasimhan et al.[72] reinvestigated the reductive methylation of amine salts by aqueous formaldehyde under acidic conditions (pH ~1.5; 25°C). They studied the reaction mechanism with [^{13}C]NMR investigations and assumed that there is an intramolecular hydride anion transfer through a hydroxymethylamine intermediate (Scheme 2).

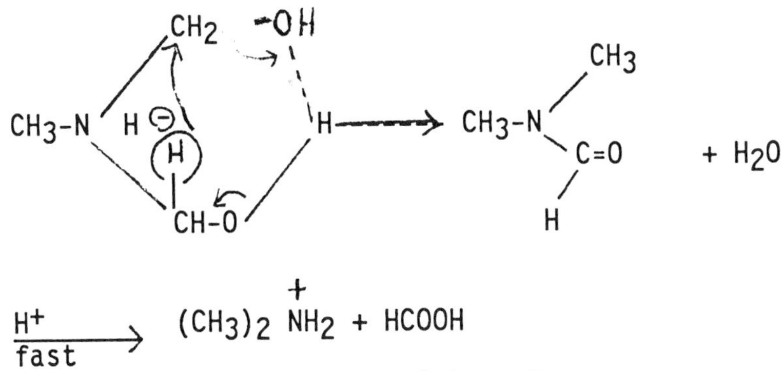

Scheme 2.

Scheme 2 indicates a 1,3-hydride anion shift producing *N*-dimethylformamide (formyl group) as an intermediate which undergoes hydrolysis to give rise to the dimethylamine. There is another possibility of the direct methylation reaction of formaldehyde in acidic medium, but this mechanism also assumes the existence of a hydride anion.

1. The Significance of the Spontaneous Methylation and Formylation Reactions

Spontaneous methylation and formylation reactions may have importance if these processes take place in proteins and in free amino acids in biological systems. In 1985, Tome et al.[69,70] reinvestigated and verified our results. They have shown the direct simultaneous formation of ε-*N*-methyl and ε-*N*-formyl groups on lysine and on lysine residues of bovine serum albumin (BSA) protein. These investigators used a very sensitive method to detect the formation of methyl and formyl groups. BSA was treated with [^{13}C]-enriched formaldehyde at 37°C, and in 0.1 *M* sodium phosphate buffer at pH 7.4. After the dialysis of the treated samples, [^{13}C]NMR spectra were recorded on a Bruker WM 250 spectrometer operating at 62.896 Mhz. Some BSA samples were treated with ^{13}C-enriched formaldehyde and $NaBH_4$ together (reductive methylation), some treated BSA samples were hydrolyzed with pronase E enzyme, and the [^{13}C]-NMR spectra were recorded after these manipulations.

The results of the French research group are as follows:

1. Simultaneous methylation and formylation reactions can take place only on lysine or lysine residues in BSA.
2. The [^{13}C]-chemical shifts were in good agreement with our previous results (N^{ε}-CH_3 = 31.1 ppm, N^{ε}-CHO = 164.5 ppm).
3. The ε-*N*-formyl groups were not resistant to hydrolysis to 6 *N* HCl. In the acid hydrolysate, the formyl groups were not detected, but the formyl groups were detected in the hydrolysate of BSA treated with pronase E enzyme.
4. If the BSA was methylated with the method of reductive methylation together with $^{13}CH_2O$ plus $NaBH_4$, only ε-*N*-methyl or ε-*N*-dimethyl groups but not ε-*N*-formyl groups could be detected.

The results of the French research team are in good agreement with our previous data thus supporting the mechanism of the spontaneous methylation and formylation reactions. We strongly believe that the formations of MML and F-Lys during the reaction between formaldehyde and L-lysine are very important in biological systems, since endogenously formed formaldehyde is known to occur at low concentrations in different tissues and fluids. Thus, 0.4 to 0.6 μg of formaldehyde per milliliter of normal human blood and 2.8 to 4.0 μg of formaldehyde per milliliter of urine were detected with [^{14}C]dimedone reagent.[73]

What is the importance of the *N*-formyl groups? We have accumulated sufficient knowledge of the role of ε-*N*-methyl groups in biological systems (see Chatper 20). However, our knowledge on the role of ε-*N*-formyl groups in biological systems is very limited. Up to now, ε-*N*-formyl-lysine containing peptides were detected among the peptides of bee venom of European honey bee *Apis mellifera*.[74,75] A mixture of variant forms of peptide has been isolated from bee venom. They have the same amino acid sequence and arrangement of disulfide bridges as the parent peptide but differ from it in that the side chains of lysine residues at positions 2, 17, and possibly 21 are chemically modified. The modifying group has been identified by mass spectrometry as formyl group. Each variant probably contains only one modified lysine residue:

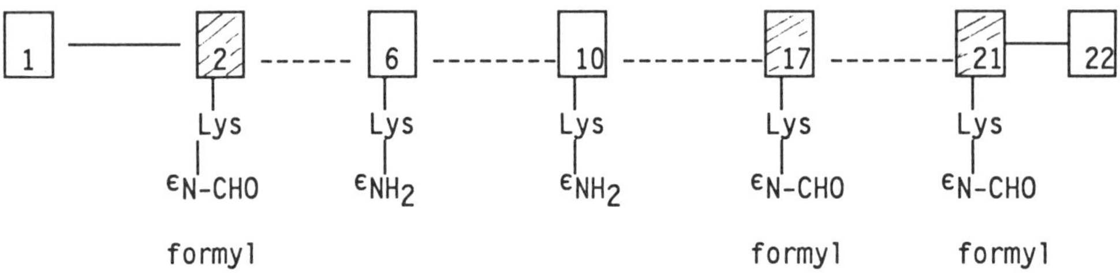

The peptide, known either as MCD peptide or peptide 401, is the most potent among the basic peptides and synthetic polymers which cause the degranulation and release of histamine from rat peritoneal mast cells, an effect resulting in inflammatory responses on injection of peptide-401 into rats.[75]

We have to assume that ε-*N*-formyl-lysine residues may also occur in other peptides and proteins, and may have a great biological significance such as *N*-formylmethionyloligopeptides which have chemotactic activity. During the hydrolytic processes of the naturally occurring peptides, only the ε-*N*-methyl groups have been detected in the hydrolysate of proteins because these ε-*N*-formyl groups disappear due to sensitivity to hydrolysis.

It appears that the formylation reaction of formaldehyde with free and bound L-lysine may be important mainly in the protective reactions of the biological systems.

C. ANTAGONISM BETWEEN L-LYSINE AND L-ARGININE AGAINST FORMALDEHYDE

During the potentiometric titration of L-arginine with formaldehyde, the pH of the system decreases in accordance with what may be expected on the basis of Sörensen formol titration similar to L-lysine. However, after the neutralization point N^G-monomethyl-L-arginine (MMA) and N^G,N^G-dimethyl-L-arginine (DMA) were not detected in the mixture of L-arginine and formaldehyde by the TLC method. Only N^G-hydroxymethyl-L-arginine derivatives could be detected on the TLC plate instantaneously after titration. The N^G-hydroxymethylation reaction of L-arginine is very fast and the products have been identified by TLC and NMR analysis to be N^G-hydroxymethyl derivatives of L-arginine. Solid phase NMR-investigations showed very similar chemical shifts to those seen with NMR-investigations in D_2O solvent (see Table 1).

Table 1 shows that the N^G-hydroxymethyl-L-arginines are stable compounds[61] and in D_2O will retain the chemical shifts of $-CH_2OH$ groups on N-atoms. Because of this chemical stability

TABLE 1
^{13}C-NMR-Investigations of L-Arginine Derivatives

	^{13}C-chemical shifts ppm		
N^G-CH_2OH groups (in D_2O)	68.9	65.6	63.2
N^G-CH_2OH groups (solid phase)	68.5	65.3	62.3
N^G-CH_3 group (solid phase)		28.3	

these compounds could be isolated from the reaction mixture by chromatographic columns.[55,56] In the case of L-lysine however, the ε-*N*-hydroxymethyl derivatives could not be detected by TLC and NMR methods and could not be isolated from the reaction mixtures because these derivatives are very labile products. These are the main antagonistic behaviors between the two basic amino acids.

The reaction mechanism as depicted for L-lysine in Scheme 1 cannot be adapted for L-arginine because the reaction ends at the hydroxymethylated product (see Scheme 3) and formation of the methylene iminium cation group II is unfavored.

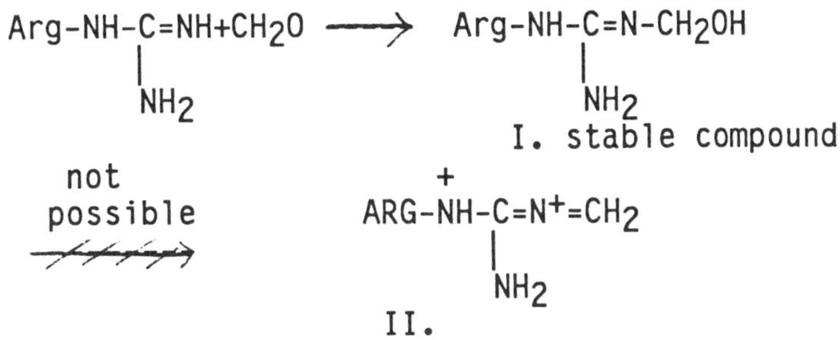

Scheme 3

The methylated compound could be formed only through the iminium cation group, thus L-arginine could not be methylated directly by formaldehyde. The results indicate very clearly that the N–C– in the N–CH_2OH group is relatively stable in arginine, but is very weak in lysine.

Why is there such a great difference between the two N–C bonds? Is it possible to interpret this difference with quantum chemical calculations? The exact quantum chemical calculations with orbital steering carried out by the Nobel laureate Lipscomb and co-workers[76-78] elucidated the real nature of N–C bond forming in nucleophilic attack on carbonyl system between ammonia, amine, and formaldehyde. They concluded that only a very weak bond is formed between ammonia and formaldehyde, the length of the bond being 0.26 nm and the strength of the bond being –4.4 kcal/mol. It is very weak and is similar to a hydrogen bond (compared with a covalent bond: 0.15 nm, 80 to 100 kcal/mol).

We adopted[60] the concept of Lipscomb's school to arginine. The quantum chemical calculations verified that the N–C bond in the –N–CH2OH group is much stronger in arginine. In arginine, there is a guanidine group and its nature (electron distribution) is not similar to the ε-amino group of L-lysine. Scheme 3 indicates the N^G-hydroxymethylation reaction only on the imino N-atom, but the hydroxymethylation can take place on the other N-atoms too, in addition to the imino N-atom.

1. Influence on Cell Proliferation

The methylated lysines, especially TML, increase cell proliferation and increase the crop yield by 10 to 20% when sprayed in small concentrations (10 to 50 ppm) on crops such as potato, red beet, sugar beet, maize, and wheat leaves.[61,63] However, N^G-hydroxymethylated L-arginines have antagonistic effects such as inhibiting the cell proliferation of healthy[57] and tumorous cells.[58,59] Significant inhibition of cell proliferation was observed *in vitro* with K-562 and P-388 cells and *in vivo* on tumor bearing mice, Ehrlich ascites, and Lewis lung[58] (see more detail in Chapter 20).

It is important to note that since N^G-hydroxymethyl-L-arginine derivatives were found in human blood and urine[55] as new components, it appears that L-arginine is an endogenous formaldehyde carrier in human blood and urine.

2. Summary

The previous results have demonstrated that there is a real antagonistic behavior between the two basic amino acids against formaldehyde. From their different chemical natures which were also demonstrated by quantum chemical calculations,[60] it follows that these two amino acids yield different reactions with formaldehyde and the compounds which are formed in the reaction cause antagonistic effects on cell proliferation. These effects may also have important practical applications.

D. METHYLATING COMPOUNDS ACTING ONLY THROUGH METABOLIC ACTIVATION IN BIOLOGICAL SYSTEMS

The original discovery of the production of malignant hepatic tumors in the rat by feeding DMN was made by Magee and Barnes[8] in 1956. It was subsequently discovered by Magee et al.[9,10] that DMN when injected into rats produced both methylated proteins and nucleic acids in the liver. There has been sufficient evidence of correlation between carcinogenicity of nitrosamines and their transformation into a methylating agent.

Since 1956, more than 300 nitroso compounds have been examined to elucidate their organ and species specificity in chemical carcinogenesis and nearly 90% of them have been found to be carcinogenic. Nitroso compounds are found in the environment and occupational exposure to *N*-nitroso compounds has been linked with the production of pesticides and detergents. The formation of nitrosamines in the body by nitrosation of secondary amines presents a threat which may be even more serious than the environmental pollution.

Some *N*-methylated-hydrazines which are formaldehyde and CH_3^+ donors (see Scheme 4) are extensively used in agricultural, pharmaceutical, and industrial applications. Compared to other chemical carcinogens such as nitrosamines, hydrazines have received little attention; however, about 37 hydrazines have been reported to induce tumors in experimental animals. Some hydrazines are found only in research laboratories, but others are used as rocket propellants and chemical intermediates and hence can be considered as environmental pollutants. Hydrazines also occur in tobacco. Edible mushrooms are natural sources of methylhydrazine.

Formaldehyde and an active methylating species CH_3^+ could be liberated from the following chemical compounds by metabolic activation in biological systems:[24,25]

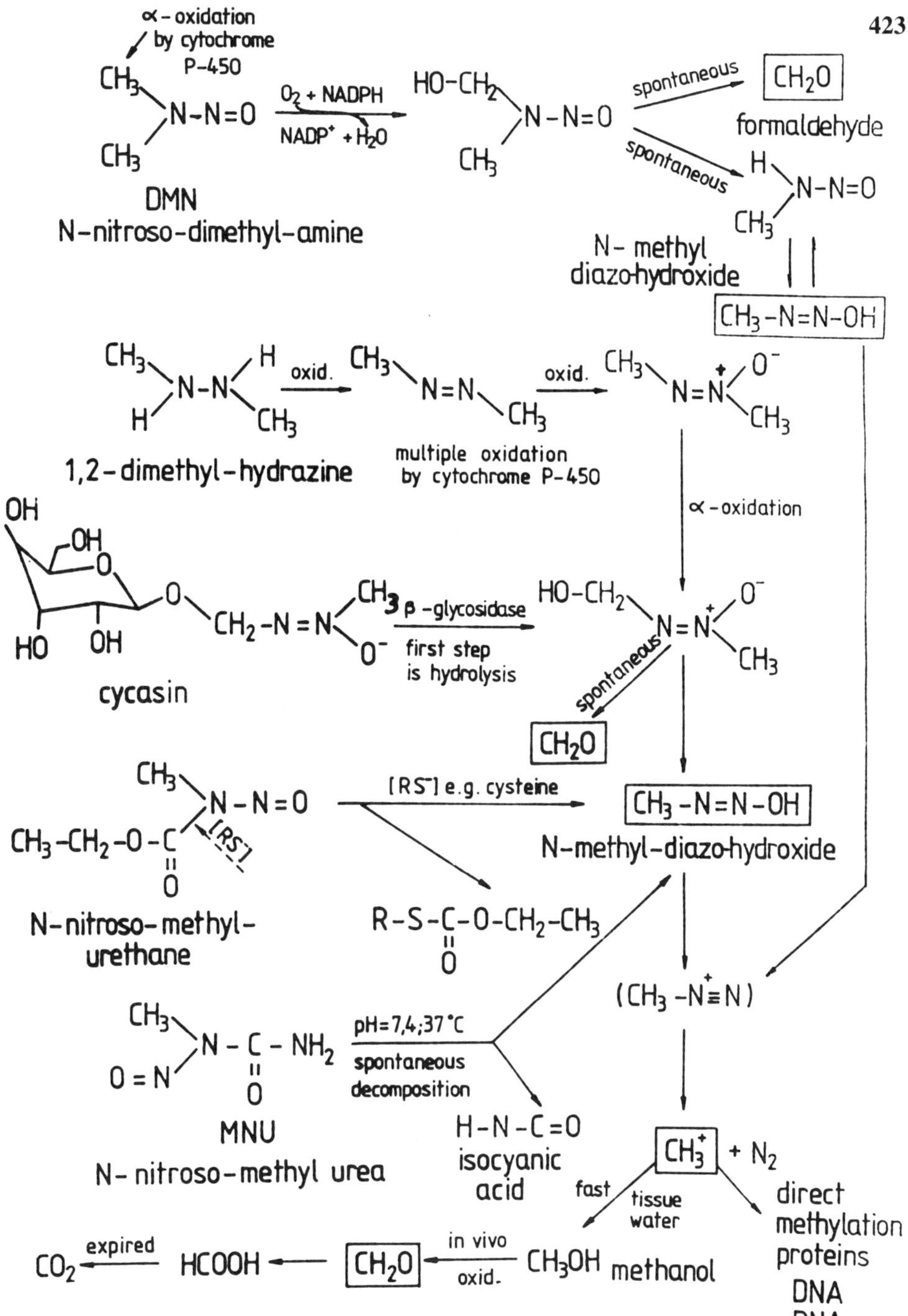

Scheme 4.

Scheme 4 shows the common metabolic pathways of different alkylating agents and the common origin of formaldehyde and CH_3^+ through *N*-methyl-diazohydroxide (CH_3–N=N–OH). The liberated CH_3^+ is assumed to be the ultimate carcinogenic metabolite methylating the macromolecular cellular components such as proteins, DNA, RNA, etc. Up to now, this scheme has been generally accepted.

The fact that the liberated formaldehyde in the metabolic pathways can also methylate the target macromolecules has never previously been suspected. The role of formaldehyde in the methylation processes was left out of consideration. Recently, Reynolds and Thomson[79] reinvestigated the metabolic pathways of DMN and the short half-life of the reactive intermediates CH_3^+, CH_3–N≡N, CH_3–N=N–OH with *ab initio* quantum chemical calculations. They stated that due to the short half-life of these species *in vivo,* there are possibly two important questions to be answered: one is how such reactive intermediates can travel far enough across the cell from the endoplasmic reticulum where the enzymatic α-oxidation occurs to the nucleus, and where the ultimate carcinogen is believed to methylate histone proteins, DNA, RNA, etc.

The other relates to the nature of the ultimate carcinogens, because the reactive intermediates react very rapidly with tissue water forming methanol which then can be oxidized to formaldehyde *in vivo* systems.

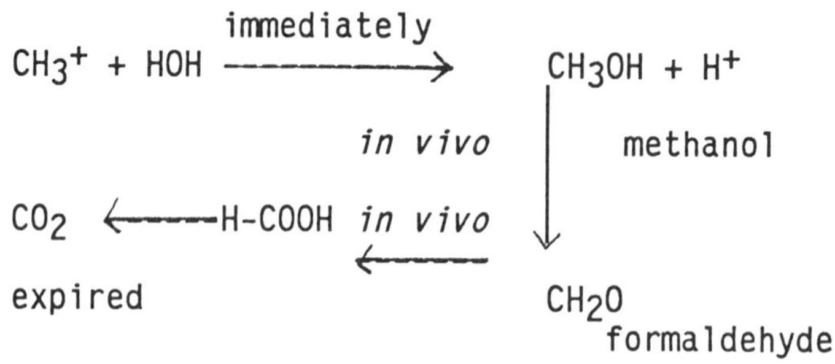

These authors,[79] however, have not considered the possibility that the liberated formaldehyde may play an important role in methylation processes. Their exact quantum chemical calculations suggested that the half-life of the methyl cation CH_3^+ is very short and is unable to travel far enough across the cell. The diazohydroxide CH_3–N=N–OH will be stable, unless its decomposition is catalyzed and may have sufficient half-life to diffuse across the cell and this reactive intermediate may methylate proteins, DNA, RNA, etc. But, according to the authors' calculations, CH_3–N=N–OH can also react with water very rapidly forming methanol as well.

From methanol, formaldehyde could be formed in *in vivo* systems. Hence, from 1 mol of

$$CH_3\text{-}N{=}N\text{-}OH + 4H_2O \xrightarrow{\text{fast}} H_2O\cdot H_3O^+\cdot CH_3OH + N_2 + HOH\cdot OH^-$$

DMN, 2 mol of formaldehyde could be liberated in biological systems (see Scheme 4). DMN is therefore a very powerful formaldehyde precursor.

Why is this concept then not generally accepted in the literature? Lijinsky et al.[80] published a concept in 1968 that only methyl cation CH_3^+ can directly methylate proteins and DNA because CH_3^+ group was transferred intact from DMN to DNA in *in vivo* systems.

However, Lijinsky et al. overlooked the following:

1. They left out of consideration formaldehyde which is liberated by enzymatic α-oxidation.
2. They left out of consideration the reaction of CH_3^+ with tissue water.

Thus, in addition to CH_3^+ group from DMN being transferred intact into macromolecules as suggested by Lijinsky et al.,[80] the liberated formaldehyde from DMN can also methylate proteins, DNA or RNA in two ways:

1. Methylation through C_1-one carbon-pool, formaldehyde being incorporated enzymatically into *S*-adenosyl-L-methionine.

2. Direct, spontaneous methylation
   a. Through *S*-adenosyl-L-methionine directly[3-5]
   b. Through direct spontaneous methylation reaction of formaldehyde with proteins and bases in nucleic acids, etc.[45,48,49,52]

It was proven that since formaldehyde is itself a hydride anion donor H^{θ} in the spontaneous direct methylation process,[52] all three H atoms of the CH_3^+ group can be derived from formaldehyde (see Scheme 1).

Our results[52] strongly suggest that it is difficult to differentiate between the direct methylation with CH_3^+ and direct methylation with formaldehyde in *in vivo* experiments, because both compounds (CH_3^+; CH_2O) could be liberated simultaneously from DMN in *in vivo* metabolism.

For example, in the case of L-lysine:

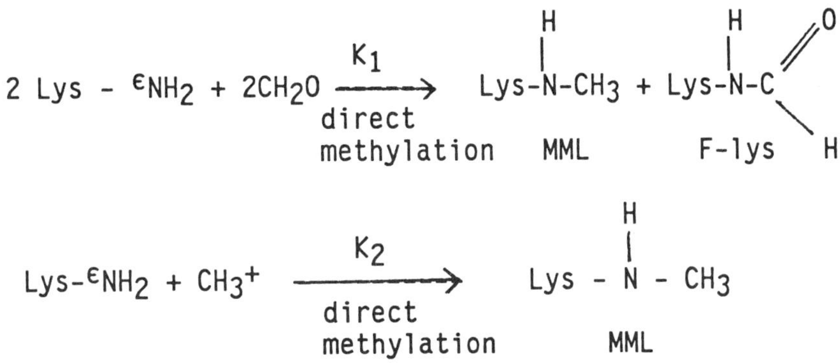

The ratio of the two direct methylation reactions (K_1/K_2) could only be determined correctly with *in vitro* model experiments.

The possibilities of the direct, nonenzymatic, and enzymatic methylation are summarized in Scheme 5.

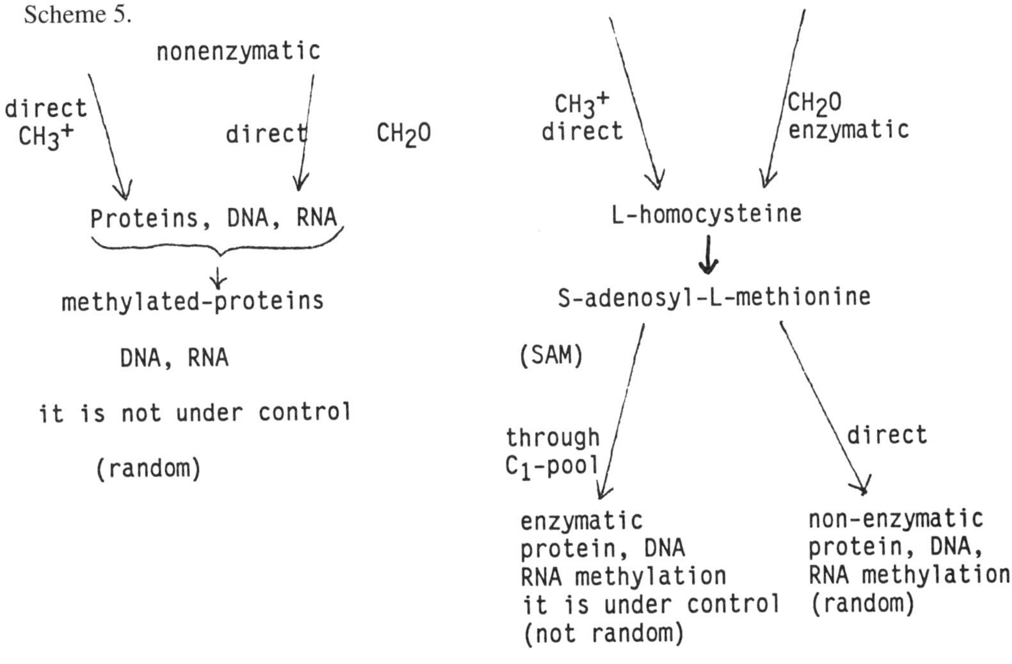

It is possible that the direct and random methylation processes are the most deleterious in biological systems. It is not yet known whether the methyl cation, CH_3^+ or formaldehyde is the more deleterious one. However, one thing that has to be taken into account is that there are several hundreds of carcinogenic, mutagenic compounds (precursors)[81] from which only formaldehyde is liberated in biological systems without any liberation of methyl cation CH_3^+. It is not possible to elucidate their carcinogenic effect with the mechanism of Lijinsky et al.[80] with the intact transfer of CH_3^+ group to proteins, DNA, or RNA if we have not known anything about the direct, nonenzymatic methylating capacity of formaldehyde.

1. The Results of Turberville and Craddock[11]

Turberville and Craddock published fundamental and very important *in vivo* results in 1971 about the possibilities of the simultaneous methylation processes (direct, nonenzymatic and enzymatic) which took place in nuclear histone proteins when [$^{14}CH_3$]DMN was injected into the liver of rats. Other *in vivo* experiments were also performed under similar conditions with the administration of either [$^{14}CH_3$]methionine or sodium-[^{14}C] formate to distinguish between nonenzymatic and enzymatic methylation processes, respectively.

Figure 4 shows the radioactivity incorporation of the three compounds into amino acids of liver histones of female rats 2 h after the injections. From the figure, the following conclusions could be drawn:

1. The amount of MML is much higher than DML in the course of direct, nonenzymatic methylation with [$^{14}CH_3$]DMN, compared with enzymatic methylation with [$^{14}CH_3$]methionine and [^{14}C]HCOONa, respectively, which indicates that a large part of MML was formed with direct, nonenzymatic methylation with CH_3^+ and formaldehyde after the treatment of [$^{14}CH_3$]DMN.
2. *S*-methylcysteine, 1-*N*-methylhistidine, and 3-*N*-methylhistidine could only be formed during the direct methylation due to the effect of CH_3^+ cation, because these compounds could not be formed with [$^{14}CH_3$]methionine and H-^{14}COONa due to the effect of enzymatic methylation.
3. Labeled methionine could be formed with the injection of either [$^{14}CH_3$]DMN or H-^{14}COONa.

There are two possibilities for the formation of labeled methionine from [$^{14}CH_3$]DMN: the liberated reactive methyl cation CH_3^+ will directly methylate L-homocysteine, or the liberated formaldehyde will enzymatically methylate L-homocysteine to form methionine.

In the case of sodium formate however, there is only one possibility: H-^{14}COONa will be reduced by NADPH coenzyme and [^{14}C]formaldehyde could be formed.

$$H\text{-}^{14}COONa \underset{NAD^+}{\overset{NADPH}{\rightleftharpoons}} H\text{-}^{14}CHO$$

This [^{14}C]formaldehyde could then be built up enzymatically into L-homocysteine forming methionine. As Figure 4 indicates, the rate of formation of methionine is very high.

Finally, large amounts of labeled serine are formed from [$^{14}CH_3$]DMN and sodium-[^{14}C]formate, but some labeled serine could also be formed from [$^{14}CH_3$]methionine.

It is well known that serine could be formed enzymatically by formaldehyde through the glycine-serine interconversion pathway.

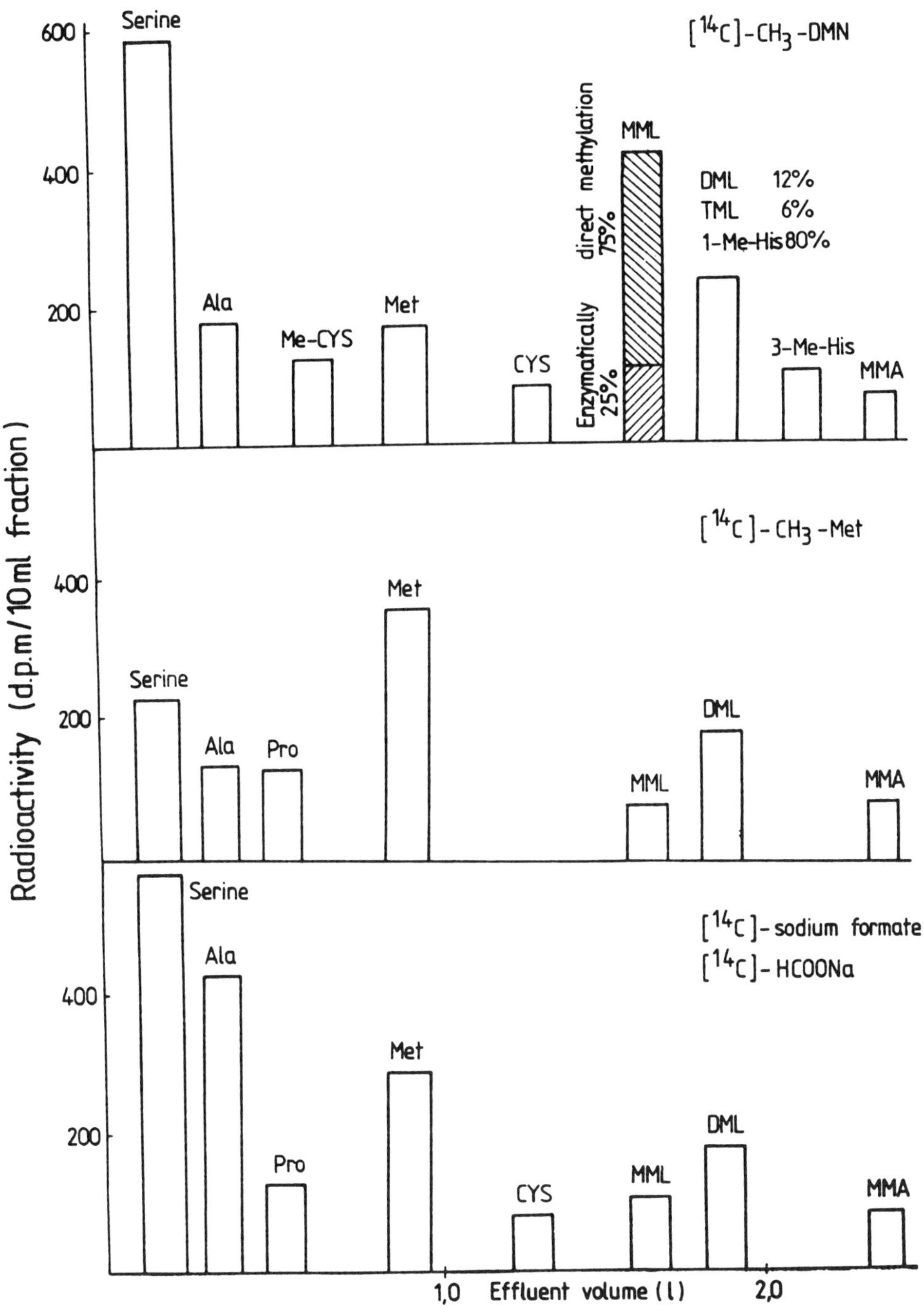

FIGURE 4. Radioactivity incorporation into amino acids of liver histones of female rats 2 h after injection of [^{14}C]-CH_3-DMN, ^{14}C-CH_3-methionine, and [^{14}C]-sodium formate, respectively.

The liberated formaldehyde is the common intermediate of the three injected compounds and serine could only be formed with this liberated formaldehyde. Thus, the *in vivo* results of Turberville and Craddock[11] provided confirmatory evidence to the existence of two simultaneous methylating processes in biological systems; direct and enzymatically catalyzed. Their data are also in agreement with our *in vitro* results that the lysine residues could be methylated

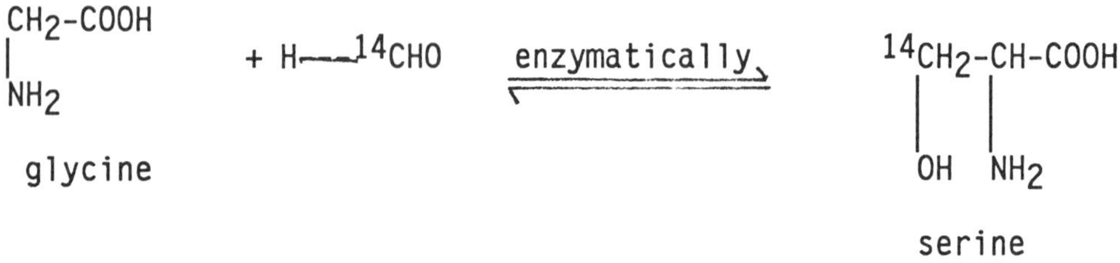

TABLE 2
Total Radioactivity Incorporation into Methylated Amino Acids of Nuclear Acid Soluble Histones in V79 Chinese Hamster Cultured Cell Lines with Alkylating Agents

Alkylating agents	Methyl cysteine (%)	MML (%)	DML (%)	3-Methyl histidine (%)	N,N′ DMA (%)
$[^{14}C]$–CH_3-MMS	10.2 ± 0.9	1.3 ± 1.3	1.2 ± 1.2	32.6 ± 0.9	1.1 ± 1.1
$[^{3}H]$–CH_3-DMS	11.1 ± 0.9	2.1 ± 1.0	1.7 ± 0.5	37.2 ± 1.0	2.1 ± 0.4
$[^{3}H]$–CH_3-MNU	0.9 ± 0.9	20.3 ± 0.6	15.4 ± 0.5	1.3 ± 1.3	6.6 ± 0.4
$[^{14}C]$–CH_3-MNNG	1.2 ± 1.2	30.1 ± 1.1	19.3 ± 1.2	Not detected	9.1 ± 1.1
$[^{3}H]$–CH_3-methionine	Not detected	6.7 ± 0.6	6.3 ± 0.2	Not detected	0.6 ± 0.5

directly by formaldehyde in *in vivo* systems as well. They reinforced the results of Mackenzie et al.[15,16] that L-methionine may serve as a formaldehyde precursor in biological systems.

Other authors also verified the high methylating capacity of the direct alkylating agents in *in vitro* and *in vivo* experiments. The –SH groups of cysteine in hemoglobin[82] and erythrocyte[83] proteins is very sensitive to the attack by these alkylating agents. It is in good agreement with the results of Trézl et al.[30]

Boffa and Bolognesi[84] investigated the effect of many direct alkylating agents such as MMS, DMS, MNU, and MNNG on *in vitro* cultured cell lines comparing the radioactivity incorporation of these compounds into acid soluble histones with the incorporation of tritiated L-methionine. Some of their data are presented in Table 2. These results are in good agreement with the data of Turberville nd Craddock.[11] Due to the effect of L-methionine, *S*-methylcysteine and 3-*N*-methylhistidine could not be formed.

E. THE PROMOTION OF THE SPONTANEOUS REACTIONS OF FORMALDEHYDE BY HYDROGEN PEROXIDE AND ITS SIGNIFICANCE

More than 10 years ago, Lichszteld and Kruk[85] studied the oxidation of formaldehyde by H_2O_2 in a highly basic solution (pH ≥12) and obtained very important effects: chemiluminescence emission, singlet molecular oxygen ($^1O_2$), and excited formaldehyde formation. These results suggested that we should investigate how H_2O_2 can influence the spontaneous reactions of formaldehyde under physiological conditions. We have recognized the biological importance of these reactions since excited formaldehyde could be liberated in biological systems from many carcinogenic precursors, e.g., from DMN mediated by cytochrome P-450 isoenzymes. These enzymes consume H_2O_2 during the oxidation, therefore, CH_2O could be liberated in excited state. The NADPH coenzyme produces H_2O_2 from O_2 present in the tissues; these enzymes are inactive under anaerobic conditions.

TABLE 3
Chemiluminescence Emission of Different Systems for 1 min in 10 ml 0.1 *M* Sodium Phosphate Buffer, pH 7.4 at 25°C

Reaction mixtures	Counts/min (c.p.m.)
1 m*M* L-lysine + 1 m*M* CH_2O	2.0×10^2
1 m*M* L-lysine + 1 m*M* CH_2O + 1 m*M* H_2O_2	1.05×10^5
1 m*M* L-arginine + 1 m*M* CH_2O	1.5×10^2
1 m*M* L-arginine + 1 m*M* CH_2O + 1 m*M* H_2O_2	8×10^2
1 m*M* L-lysine + 1 m*M* CH_2O + 1 m*M* H_2O_2 + 5 mg horseradish peroxidase	1.25×10^5
1 m*M* *N*-α-acetyl-L-lysine methylamide + 1 m*M* CH_2O + 1 m*M* H_2O_2	1.55×10^5
0.05 m*M* cytochrome c (horse heart) + 1 m*M* CH_2O + 1 m*M* H_2O_2	55.5×10^5
1 m*M* L-lysine + 1 m*M* CH_2O + 1 m*M* H_2O_2 + 3 m*M* L-ascorbic acid	6.5×10^3
1 m*M* L-lysine + 1 m*M* CH_2O + 1 m*M* H_2O_2 + 2 m*M* glutathione	1.8×10^3
1 m*M* L-lysine + 1 m*M* CH_2O + 1 m*M* H_2O_2 + 1 m*M* Na-azide	8.5×10^3

It was found[44,46,67,68] that the spontaneous methylation and formylation reactions between L-lysine and formaldehyde could be increased significantly by H_2O_2 (see Table 3). Singlet oxygen ($^1O_2$) formation could be detected instantaneously with chemiluminescence spectrometer[67,68] at 530, 633, and 705 nm with simultaneous burst of chemiluminescence emission.

$$2\ {}^1O_2 \longrightarrow 2\ {}^3O_2 + h\nu \quad (633\ \text{nm})$$

The blue light emission at 430 nm indicates the formation of an excited, biradical very reactive form of formaldehyde.

$$H_2C=\overset{*}{O} \longrightarrow H_2C=O + h\nu \ (430\ \text{nm})$$

Because of its biradical character and extremely long half-life[67,68] (0.025 s), this formaldehyde shows an extreme reactivity and can hypermethylate and formylate proteins, nucleic acids, etc.; this character may additionally trigger the initial step in chemical carcinogenesis.

Table 3 indicates very clearly the antagonism between the two amino acids: L-lysine promotes whereas L-arginine inhibits the chemiluminescence emission. Cytochrome *c* has a tremendous promoting effect whereas L-ascorbic acid and glutathione both have a significant inhibitory effect on the chemiluminescence emission.

F. INHIBITION OF THE SPONTANEOUS REACTION OF FORMALDEHYDE WITH L-ASCORBIC ACID AND ITS BIOLOGICAL IMPORTANCE

A new chemical reaction of L-ascorbic acid with its enediol group readily entering into addition reaction with formaldehyde was observed.[86] This reaction is significant since it can occur in biological systems as well and in this way L-ascorbic acid is able to eliminate the toxic, carcinogenic formaldehyde released from exogenous formaldehyde precursors. The nucleophilic addition reaction of formaldehyde takes place on the C-2 atom of L-ascorbic acid because the partial charge on the C-2 atom is negative. In this reaction, first an adduct (I) is formed which then undergoes a cyclization to generate the compound (II).

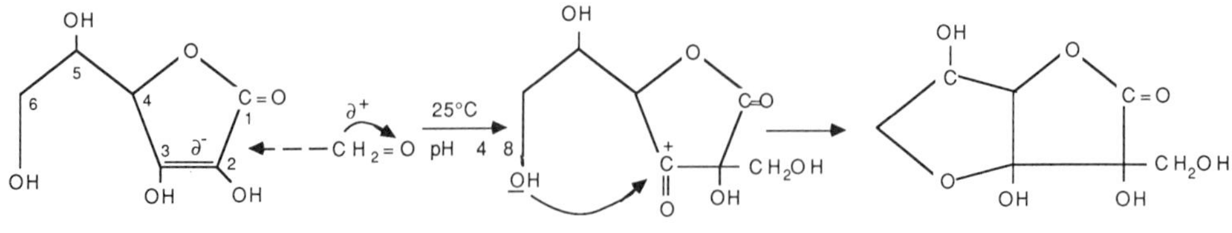

The [^{13}C]NMR investigations[86] confirmed these reaction steps.

Because of this addition reaction of L-ascorbic acid with formaldehyde, the spontaneous methylation reactions of formaldehyde with L-lysine could be inhibited.[43,49,62] This may be important in biological systems because the inhibition takes place with a biological compound and its endogenous concentration may be increased to a large extent. Indeed, it was proven by *in vitro* experiments that the formaldehyde released from DMN during enzymatic oxidation by microsomal fraction is trapped by L-ascorbic acid.[86]

There is a great practical application of this reaction as well. A tobacco smoke filter containing L-ascorbic acid and other ingredients can trap a large amount of the toxic formaldehyde from the smoke. This process has already been patented in Hungary[87] and is under consideration in several other countries.[87]

III. SUMMARY: THE SIGNIFICANCE OF NONENZYMATIC PROTEIN METHYLATION

The review described in this chapter has clearly shown that there are many different pathways for nonenzymatic protein methylation. It can be seen that there are two main direct methylation pathways: one with reactive methyl cation CH_3^+ and another with formaldehyde. Both compounds could also be incorporated into AdoMet and this may serve either as a direct methylating compound or may enzymatically methylate proteins or DNA through the C_1-pool.

Which process is more important or more deleterious? Presently, it is not possible to exactly elucidate these problems because these two reactive methylating compounds are liberated simultaneously from the carcinogenic compounds in *in vivo* systems (see Scheme 4). However, these two methylating species have one thing is common; they methylate peptides, proteins, DNA, or RNA randomly and it is generally accepted that the random methylation processes are the most deleterious in biological systems.

Paik et al.[29] demonstrated that cytochrome *c* could be trimethylated enzymatically by AdoMet only at the 72 position of lysine residue, whereas MMS and MNU methylated cytochrome *c* randomly at several lysine residues.

The results of Turberville and Craddock[11] brought confirmatory evidence to these facts that the two reactive species can methylate the protein residues simultaneously in the liver. ε-NH_2 groups of lysine residues could be methylated readily by formaldehyde, but –SH groups of cysteine residues could rapidly be methylated only by reactive methyl cation CH_3^+ species (see Figure 4).

It is well known that some enzymes will lose their activity by methylation producing an inactive enzyme, e.g., reductive methylation of ribonuclease with formaldehyde produced an enzymatically inactive protein with less than a single lysine residue remaining unmodified.[40]

If those enzymes which take part in the repair mechanisms were inactivated by random methylation then such methylation processes may also be very deleterious in chemical carcinogenesis.

The great emphasis of this chapter is that we draw attention to the fact that since formaldehyde is itself a direct methylating and formylating agent of proteins, DNA, etc., it may be potentially mutagenic especially if formaldehyde was liberated from carcinogenic compounds such as DMN. The simultaneous formylation process may also be very important in *in vivo* systems but so far no information is available in this area.

Before concluding our review, we would like to draw the readers' attention to a *Science* citation[88] stating that "Formaldehyde unlike most carcinogens identified in animal cancer tests, is not a synthetic industrial chemical but a normal metabolite in human biochemistry with an elaborate enzymatic system already in place for handling it. This suggests that our bodies are capable of handling it safely as long as exposure is not much higher than the amounts the body itself manufactures."[88]

The last conclusion is that: formaldehyde has a "double face", it is indispensable in biological systems, but under certain conditions (e.g., in excited state, in random methylation reaction or when its local concentration is increased to a high level, etc.) it may be deleterious, toxic, mutagenic, or a carcinogenic compound.

REFERENCES

1. **Houben-Weyl,** *Methoden der Organischen Chemie,* Georg Thieme Verlag, Stuttgart, 1957, Band XI/1, XI2.
2. **Novak, J.,** Alkylierung der Aminosäuren mit Dialkylsulfaten, *Berichte,* 45, 834, 1912.
3. **Paik, W. K., Lee, H. W., and Kim, S.,** Non-enzymatic methylation of proteins with S-adenosyl-L-methionine, *FEBS Lett.,* 58, 628, 1975.
4. **Rydberg, B. and Lindhal, T.,** Nonenzymatic methylation of DNA by the intracellular methyl group donor S-adenosyl-L-methionine is a potentially, mutagenic reaction, *EMBO J.,* 1, 211, 1982.
5. **Barrows, L. R. and Magee, P. N.,** Nonenzymatic methylation of DNA by S-adenosyl-L-methionine *in vitro, Carcinogenesis,* 3, 349, 1982.
6. **Borchardt, R. T., Creveling, C. R., and Ueland, P. M.,** *Biological Methylation and Drug Design,* Humana Press, Clifton, NJ, 1986.
7. **Paik, W. K. and Kim, S.,** *Protein Methylation,* John Wiley & Sons, New York, 1980.
8. **Magee, P. N. and Barnes, J. M.,** The production of malignant primer hepatic tumors in the rat by feeding dimethylnitrosamine, *Br. J. Cancer,* 10, 114, 1956.
9. **Magee, P. N. and Hultin, T.,** Toxic liver injury and carcinogenesis: methylation of proteins of rat-liver slices by dimethylnitrosamine, *Biochem. J.,* 83, 106, 1962.
10. **Magee, P. N. and Farber, E.,** Toxic liver injury and carcinogenesis: methylation of rat-liver nucleic acids by dimethylnitrosamine *in vivo, Biochem. J.,* 83, 114, 1962.
11. **Turberville, C. and Craddock, V. M.,** Methylation of nuclear proteins by dimethylnitrosamine and by methionine in the rat *in vivo, Biochem. J.,* 124, 725, 1971.
12. **Reis, P. J. and Tunks, D. A.,** Influence of formaldehyde-treated casein supplements on the concentration of ε-*N*-methyl-lysine in sheep plasma, *Aust. J. Biol. Sci.,* 26, 1127, 1973.
13. **Johansson, E. B. and Tjälve, H.,** Autoradiographic studies in mice with inhibited and non-inhibited dimethylnitrosamine metabolism and comparison with the distribution of [^{14}C]-formaldehyde, *Toxicol. Appl. Pharmacol.,* 45, 565, 1978.
14. **Tyihák, E., Rusznák, I., and Trézl, L.,** Antagonistic behaviour of L-lysine and L-arginine with formaldehyde in homogeneous and heterogeneous phase, *Proc. 15th Hung. Annu. Meet. Biochem.,* Miskolc, Hungary, 1975, Hungarian Chemical Society, Rosdy, B., Ed., Budapest; *Chem. Abstr.,* 85, 118, 182u, 1976.

15. **Mackenzie, C. G., Chandler, J. P., Keller, E. B., Rachele, J. R., Cross, N., Melville, D. B., and DuVigneaud, V.,** The demonstration of the oxidation *in vivo* of the methyl group of methionine, *J. Biol. Chem.*, 169, 757, 1947.
16. **Mackenzie, C. G.,** Formation of formaldehyde and formate in the biooxidation of the methyl group, *J. Biol. Chem.*, 186, 351, 1950.
17. **Szarvas, T., Simon, L. P., Trézl, L., and Tyihák, E.,** S-adenosyl-L-methionine (SAM) as formaldehyde precursor in animal liver tissue and its interdependence with biological methylation, *Izotoptechnika*, 29, 247, 1986 (in Hungarian).
18. **Kim, S., Benoiton, L., and Paik, W. K.,** ε-Alkyllysinase, purification and properties of the enzyme, *J. Biol. Chem.*, 239, 148, 1964.
19. **Tyihák, E., Balla, J., Báborjányi, R., and Balázs, A.,** Increased free formaldehyde level in crude extract of virus infected hypersensitive tobaccos, *Acta Phytopathol. Acad. Sci. Hung.*, 13, 29, 1978.
20. **Tyihák, E., Gáborjányi, R., János, E., Trézl, L., Rusznák, I., and Balla, J.,** Increased N-demethylase activity in TMV infected hypersensitive tobaccos and the supposed reactions of formed formaldehyde, *18th Hung. Annu. Meet. Biochem.*, Salgotarján, Hungary, Hungarian Chemical Society, Rosdy, B., Ed., Budapest, 1978.
21. **Szarvas, T., János, E., Gáborjányi, R., and Tyihák, E.,** Increased formaldehyde formation an early event of TMV infections in hypersensitive host, *Acta Phytopathol. Acad. Sci. Hung.*, 17, 7, 1982.
22. **Burgyán, J., Szarvas, T., and Tyihák, E.,** Increased formaldehyde production from L-methionine-(S-$^{14}CH_3$) by crude enzyme of TMV-infected tobacco leaves, *Acta Phytopathol. Acad. Sci. Hung.*, 17, 11, 1982.
23. **Tyihák, E., Szarvas, T., Trézl, L., and Kiraly, Z.,** Possibilities of spontaneous formylation reaction by endogenous formaldehyde after heat shock in plants, *Acta Biochim. Biophys. Acad. Sci. Hung.*, 19, 136, 1984.
24. **Grover, P. L.,** *Chemical Carcinogens and DNA*, Vol. I, CRC Press, Boca Raton, FL, 1979.
25. **Blackburn, G. M. and Kellerd, B.,** Chemical carcinogens, *Chem. Ind.*, 18, 607, 1986.
26. **Hofe, E., Kleihues, P., and Keefer, L. K.,** Extent of DNA 2-hydroxymethylation by *N*-nitrosomethylethylamine and *N*-nitroso-*N*-dimethylamine *in vivo, Carcinogenesis*, 7, 1335, 1986.
27. **Bridson, P. K., Markiewicz, W., and Reese, C. B.,** Conversion of guanosine into its N^2-methyl derivative, *J. C. S. Chem. Commun.*, 12, 447, 1977.
28. **Trézl, L., Rusznák, I., Szarvas, T., Török, G., and Börzsönyi, M.,** Spontaneous formation of L-homocitrullin in neutral medium from L-lysine and *N*-methyl-*N*-nitrosourea, *in vivo* and *in vitro* inhibition of its formation by L-ascorbic acid, *Acta Biochim. Biophys. Acad. Sci. Hung.*, 17, 181, 1982.
29. **Paik, W. K., DiMaria, P., Kim, S., Magee, P. N., and Lotlikar, P. D.,** Alkylation of protein by methyl methanesulfonate and 1-methyl-1-nitrosourea *in vitro, Cancer Lett.*, 23, 9, 1984.
30. **Trézl, L., Park, K. S., Kim, S., and Paik, W. K.,** Studies on *in vitro* S-methylation of naturally occurring thiol compounds with *N*-methyl-*N*-nitrosourea and methyl methanesulfonate, *Environ. Res.*, 43, 417, 1987.
31. **Walker, J. F.,** *Formaldehyde*, Reinhold, New York, 1964.
32. **Leuckart, R.,** Ueber eine neue Bildungsweise von Tribenzylamine, *Berichte*, 18, 2341, 1885.
33. **Wallach, O.,** Ueber die Überführung von ketonen und aldehyden, *Basen. Ann.*, 343, 54, 1905.
34. **Plöchl, J.,** Ueber eine reaction des formaldehyds, *Berichte*, 21, 2117, 1888.
35. **Eschweiler, W.,** Ersatz an Stickstoff gebundenen Wasserstoffatomen durch die Methyl Gruppe Mitochondria hülfe von Formaldehyd, *Berichte*, 38, 880, 1905.
36. **Hess, K.,** Zur Synthese des Hygrins, *Berichte*, 46, 4104, 1913.
37. **Clarke, H. T., Gillespie, H. B., and Weisshaus, S. Z.,** The action of formaldehyde on amines and amino acids, *J. Am. Chem. Soc.*, 55, 4571, 1933.
38. **Staple, E. and Wagner, E. C.,** A study of the Wallach reaction for alkylation of amines by action of aldehydes or ketones and formic acid, *J. Org. Chem.*, 14, 559, 1949.
39. **Lindeke, B., Anderson, E., and Jenden, D. J.,** Specific deuteromethylation by Eschweiler-Clarke reaction. Synthesis of differently labelled variants of trimethylamine and their use for the preparation of labelled choline and acetylcholine, *Biomed. Mass. Spectr.*, 3, 257, 1976.
40. **Means, G. E. and Feneey, E.,** Reductive alkylation of amino groups in proteins, *Biochemistry*, 7, 2192, 1968.
41. **Sörensen, S. P. L.,** Enzymstúdien über die quantitative Messung proteolytischer Spaltungen. Die Formoltitrierung, *Biochem. Z.*, 2, 45, 1908.
42. **Kitamoto, Y. and Maeda, H.,** Reevaluation of the reaction of formaldehyde at low concentration with amino acids, *J. Biol. Chem. (Japan)*, 87, 1519, 1980.
43. **Trézl, L., Rusznák, I., Tyihák, E., Balla, J., and Müller, T.,** The retarding effect of L-ascorbic acid on the spontaneous methylation reaction of L-lysine with formaldehyde, *Proc. 18th Hung. Annu. Meet. Biochem.*, Salgotarján, Hungary, 1978, Hungarian Chemical Society, Rosdy, B., Ed., Budapest, 1978; *Chem. Abstr.*, 90, 16482c, 1979.
44. **Trézl, L., Rusznák, I., Tyihák, E., Szarvas, T., Márkus, I., and Müller, T.,** Promotion and inhibition of the spontaneous methylation and formylation reactions between L-lysine and formaldehyde, *Proc. 19th Hung. Annu. Meet. Biochem.*, Budapest, 1979, Hungarian Chemical Society, Rosdy, B., Ed., Budapest, 1979; *Chem. Abstr.*, 92, 36263, 1980.

45. **Tyihák, E., Szarvas, T., and Trézl, L.,** Analogies and differences between spontaneous methylation of L-lysine and some nucleic bases with formaldehyde, *Proc. 19th Hung. Annu. Meet. Biochem.*, Budapest, 1979, Hungarian Chemical Society, Rosdy, B., Ed., Budapest, 1979; *Chem. Abstr.*, 92, 17390z, 1980.
46. **Trézl, L., Rusznák, I., Tyihák, E., Szarvas, T., and Kolonits, P.,** Singlet oxygen formation in the reaction between L-lysine and formaldehyde and hydrogen peroxide enhancing the formation with horseradish peroxidase, *Proc. 19th Hung. Annu. Meet. Biochem.*, Budapest, 1979, Hungarian Chemical Society, Rosdy, B., Ed., Budapest, 1980.
47. **Tyihák, E., Trézl, L., and Rusznák, I.,** Spontaneous N^{ε}-methylation of L-lysine by formaldehyde, *Pharmazie*, 35, 18, 1980.
48. **Tyihák, E., Tétényi, P., Trézl, L., Szarvas, T., Héthelyi, E., and Mincsovics, E.,** Spontane Methylierung von nor-Morphin durch Formaldehyd, *Pharmazie*, 36, 214, 1981.
49. **Trézl, L., Rusznák, I., Tyihák, E., Szarvas, T., and Szende, B.,** Spontaneous N^{ε}-methylation and N^{ε}-formylation reactions between L-lysine and formaldehyde inhibited by L-ascorbic acid, *Biochem. J.*, 214, 289, 1983.
50. **Trézl, L., Bako, P., Fenichel, L., and Rusznák, I.,** Detection of crown ethers by means of Dragendorff's reagent, *J. Chromatogr.*, 269, 40, 1983.
51. **Trézl, L., Rusznák, I., Tyihák, E., Szarvas, T., and Müller, T.,** Influence of crown ethers of the spontaneous N^{ε}-methylation reactions between L-lysine and formaldehyde, Collected Abstract, *Steric Effects in Biomolecules. Int. Symp. Eger*, Hungary, 1981, Hungarian Chemical Society, Náray-Szabó, G., Ed., Budapest, 1981.
52. **Tyihák, E., Trézl, L., and Kolonits, P.,** The isolation of N^{ε}-formyl-L-lysine from the reaction between formaldehyde and L-lysine and its identification by OPLC and NMR spectroscopy, *J. Pharmacol. Biomech. Anal.*, 3, 343, 1985.
53. **Ludányi, A., Trézl, I., Rusznák, I., and Tyihák, E.,** Reevaluation of the Eschweiler-Clarke and Leuckart-Wallach reactions for the N-methylation (abstr.), *1st Int. Conf. on "The Role of Formaldehyde in Biological Systems"*, Balatonfüred, Hungary, 1985, Hungarian Biochemical Society, Tyihák, E., Ed., Budapest, 1985.
54. **Csiba, A., Trézl, L., Rusznák, I., Tyihák, E., and Szarvas, T.,** Regulation of spontaneous methylation of L-lysine by formaldehyde with $NADH+H^{+}$-NAD^{+} coenzymes, *Proc. 21st Hung. Annu. Meet. Biochem.*, Veszprém, Hungary, 1981, Hungarian Chemical Society, Rosdy, B., Ed., Budapest, 1981.
55. **Trézl, L., Csiba, A., Rusznák, I., Tyihák, E., and Szarvas, T.,** L-Arginine as an endogenous formaldehyde carrier in human blood and urine, *Proc. 21st Hung. Annu. Meet. Biochem*, Veszprém, Hungary, 1981, Hungarian Chemical Society, Rosdy, B., Ed., Budapest, 1981; *Chem. Abstr.*, 96, 159 825b, 1982.
56. **Csiba, A., Trézl, L., Tyihák, E., Graber, H., Vari, E., Téglás, G., and Rusznák, I.,** Assumes role of L-arginine in mobilization of endogenous formaldehyde, *Acta. Physiol. Acad. Sci. Hung.*, 59, 35, 1982; *Chem. Abstr.*, 98, 49113b, 1983.
57. **Csiba, A., Trézl, L., and Rusznák, I.,** Biotransformation of N^{G}-hydroxymethyl-L-arginines in Baker's yeast, *Acta. Physiol. Acad. Sci. Hung.*, 68, 96, 1986.
58. **Trézl, L., Szende, B., Csiba, A., Rusznák, I., and Lapis, K.,** *In vivo* and *in vitro* inhibition of tumor cell proliferation by N^{G}-hydroxymethyl-L-arginines as a natural cell component, *Abstr. 14th Int. Cancer Congr.*, Budapest, 1986, Hungarian Academic Publishing, Lapis, K. and Echardt, S., Eds., Budapest, 1986.
59. **Csiba, A., Trézl, L., and Rusznák, I.,** A hypothesis for the mechanisms of tumor killing by N^{G}-hydroxymethyl-L-arginines, *Med. Hypothesis*, 19, 75, 1986.
60. **Trézl, L. and Náray-Szabo, G.,** Quantumchemical interpretation of spontaneous N^{G}-hydroxymethylation of L-arginine by formaldehyde, Book of Abstr. WATOC'87, *World Congress of Theoretical Organic Chemists*, Budapest, Hungary, 1987. Hungarian Chemical Society, Náray-Szabó, G. and Surjan, P., Eds., Budapest, 1987.
61. **Trézl, L., Rusznák, I., Tyihák, E., Szarvas, T., Csiba, A., and Szende, B.,** Antagonistic behaviour in spontaneous reaction of L-lysine and L-arginine with formaldehyde and their biological significance, *Proc. 2nd Int. Conf. on "The Role of Formaldehyde in Biological Systems"*, Keszthely, Hungary, 1987, Hungarian Biochemical Society, Tyihák, E. and Gullner, G., Eds., Budapest, 1987.
62. **Trézl, L., Rusznák, I., Tyihák, E., and Szarvas, T.,** Spontaneous N-methylation and N-formylation reactions between L-lysine and formaldehyde inhibited by L-ascorbic acid and their biochemical aspects, *Biologia*, 30, 55, 1982 (in Hungarian); *Chem. Abstr.*, 99, 49253t, 1983.
63. **Rusznák, I., Trézl, L., Tyihák, E., Kovacs, G., Földesi, D., and Szabó, B.,** Plant growth and increase in crop brought about by formaldehyde treatment of lysine rich soils and manures, *Proc. 1st Int. Conf. on "The Role of Formaldehyde in Biological Systems"*, Balatonfüred, Hungary, 1985, Hungarian Biochemical Society, Tyihák, E., Hungary, Hung. Pat. 182.677, 1984, European Pat. E.P. 49.536, 1984, Canadian Pat. 1.184.399, 1985, U.S. Pat. 4.532.214, 1985, Japanese Pat. 134.792, 1982.
64. **Tyihák, E., Szende, B., and Lapis, K.,** Biological significance of methylated derivatives of lysine and arginine, *Life Sci.*, 20, 385, 1977.
65. **Szende, B., Tyihák, E., Kopper, L., and Lapis, K.,** The tumor growth promoting effect of ε-trimethyl-lysine, *Neoplasma*, 17, 433, 1970.

66. **Suba, Zs., Szende, B., Lapis, K., Takacs, J., and Elek, G.,** ε-N-trimethyllysine a natural cell component with mitogenic activity, *Neoplasma,* 27, 11, 1980.
67. **Trézl, L. and Pipek, J.,** Formation of excited formaldehyde in a model biological system, Book of Abstracts WATOC'87, *World Congress of Theoretical Organic Chemists,* Budapest, 1987, Hungarian Chemical Society, Náray-Szabo, G. and Surjan, P., Eds., Budapest, 1987.
68. **Trézl, L. and Pipek, J.,** Formation of excited formaldehyde in model reactions simulating real biological systems, *J. Mol. Struct. Theochem,* 170, 213, 1988.
69. **Tome, D., Naulet, N., and Kozlowski, A.,** [^{13}C]NMR characterization of formaldehyde bonds in model mixtures and proteins containing lysine, *Int. J. Pept. Protein Res.,* 25, 258, 1985.
70. **Tome, D., Kozlowski, A., and Mabon, F.,** Carbon-13 NMR study on the combination of formaldehyde with bovine serum albumin, *J. Agric. Food Chem.,* 33, 449, 1985.
71. **Hemminki, K.,** Reactions of formaldehyde with guanosine, *Toxicol. Lett.,* 9, 161, 1981.
72. **Narasimhan, S., Balakrishnan, Kumar, A. S., and Venkatasubramanian, N.,** Carbon-13 NMR study of reaction of aqueous formaldehyde with amine salts; evidence for reductive methylation by intramolecular hydride transfer, *Indian J. Chem.,* 243, 568, 1985.
73. **Szarvas, T., Szatloczky, E., Volford, J., Trézl, L., Tyihák, E., and Rusznák, I.,** Determination of endogenous formaldehyde level in human blood and urine by dimedone-^{14}C-radiometric method, *J. Radioanal. Nucl. Chem. Lett.,* 106, 357, 1986.
74. **Doonan, S., Garman, A. J., Hanson, J. M., London, A. G., and Vernon, C. A.,** Identification by mass spectrometry of N^{ε}-formyl-lysine residues in peptide from bee venom, *J. Chem. Soc. Perkin Trans. I.,* 1157, 1978.
75. **Dempsey, C. E.,** Selective formylation of α- or ε-amino groups of peptides, *J. Chem. Soc. Perkin Trans. I.,* 2625, 1982.
76. **Scheiner, S. and Lipscomb, W. N.,** Comments on orbital steering, *Int. J. Quantum Chem. Quantum Biol. Symp.,* 3, 161, 1976.
77. **Scheiner, S., Lipscomb, W. N., and Kleier, D. A.,** Molecular orbital studies of enzyme activity. 2. Nucleophilic attack on carbonyl systems with comments on orbital steering, *J. Am. Chem. Soc.,* 98, 4770, 1976.
78. **Lipscomb, W. N.,** personal communication and discussion, Budapest, 1987.
79. **Reynolds, C. A. and Thomson, C.,** Ab. initio calculations relevant to the mechanism of chemical carcinogens by N-nitrosamines, *J. Mol. Struct. Theochem.,* 149, 345, 1987.
80. **Lijinsky, W., Loo, J., and Ross, A. E.,** Mechanism of alkylation of nucleic acids by nitroso-dimethylamine, *Nature (London),* 218, 1174, 1968.
81. **Sawicki, E. and Sawicki, C. R.,** Aldehydes-photometric analysis, in *Formaldehyde Precursors,* Vol. 5, Academic Press, New York, 1978.
82. **Bailey, E., Connors, T. A., Farmer, P. B., Gorf, S. M., and Rickard, J.,** Methylation of cysteine in hemoglobin following exposure to methylating agents, *Cancer Res.,* 41, 2514, 1981.
83. **Kim, S., Paik, W. K., Choi, J., Lotlikar, P., and Magee, P. N.,** Microsome-dependent methylation of erythrocyte proteins by dimethylnitrosamine, *Carcinogenesis,* 2, 179, 1981.
84. **Boffa, L. C. and Bolognesi, C.,** Methylating agents: their target amino acids in nuclear proteins, *Carcinogenesis,* 6, 199, 1985.
85. **Lichszteld, K. and Kruk, I.,** Singlet molecular oxygen in formaldehyde oxidation, *Z. Phys. Chem. Neue Folge.,* 108, 167, 1977.
86. **Trézl, L., Bitter, I., Szarvas, T., Töke, L., and Rusznák, I.,** The biological importance of the addition reaction between L-ascorbic acid and formaldehyde, *Izotoptechnika,* 30, 85, 1987 (in Hungarian).
87. **Trézl, L., Bitter, I., Gabor, J., Hernadi, S., Horváth, V., Irimi, S., Molnár, and Rusznák, I.,** Hungarian Pat. 192.213, 1987, Deutsches pat. offenlegungsschrift DE 35 32 618 41, 1986, Swiss, 3807, U.S.A., 788815, Japan, 48.191, 1986, Great Britain, 85-22-895, Austria, A. 2598.
88. **Wildawsky, A., Todhunter, J. A., Hevender, W. R., and Ashford, N. A.,** Formaldehyde regulation, *Science,* 224, 550, 1984.

INDEX

A

B

C

D

E

F

G

H

I

Q

R

S